MEDICAL INTELLIGENCE UNIT

Insulin-Like Growth Factors

Derek LeRoith, M.D., Ph.D.
Clinical Endocrinology Branch
National Institutes of Health, MSC
Bethesda, Maryland, U.S.A.

Walter Zumkeller, M.D.
Department of Pediatrics
Martin-Luther-University, Halle-Wittenberg
Children's University Hospital
Halle/Saale, Germany

Robert C. Baxter, Ph.D., DSc
Kolling Institute of Medical Research
Royal North Shore Hospital
St. Leonards, Australia

LANDES BIOSCIENCE / EUREKAH.COM
GEORGETOWN, TEXAS
U.S.A.

KLUWER ACADEMIC / PLENUM PUBLISHERS
NEW YORK, NEW YORK
U.S.A.

INSULIN-LIKE GROWTH FACTORS

Medical Intelligence Unit

Eurekah.com / Landes Bioscience
Kluwer Academic / Plenum Publishers

Copyright ©2003 Eurekah.com and Kluwer Academic / Plenum Publishers

All rights reserved.
No part of this book may be reproduced or transmitted in any form or by any means, electronic or mechanical, including photocopy, recording, or any information storage and retrieval system, without permission in writing from the publisher, with the exception of any material supplied specifically for the purpose of being entered and executed on a computer system; for exclusive use by the Purchaser of the work.

Printed in the U.S.A.

Kluwer Academic / Plenum Publishers, 233 Spring Street, New York, New York, U.S.A. 10013
http://www.wkap.nl/

Please address all inquiries to the Publishers:
Eurekah.com / Landes Bioscience, 810 South Church Street
Georgetown, Texas, U.S.A. 78626
Phone: 512/ 863 7762; FAX: 512/ 863 0081
www.Eurekah.com
www.landesbioscience.com

Insulin-Like Growth Factors edited by Derek LeRoith, Walter Zumkeller and Robert C. Baxter, Landes / Kluwer dual imprint / Landes series: Medical Intelligence Unit

ISBN: 0-306-47846-3

While the authors, editors and publisher believe that drug selection and dosage and the specifications and usage of equipment and devices, as set forth in this book, are in accord with current recommendations and practice at the time of publication, they make no warranty, expressed or implied, with respect to material described in this book. In view of the ongoing research, equipment development, changes in governmental regulations and the rapid accumulation of information relating to the biomedical sciences, the reader is urged to carefully review and evaluate the information provided herein.

Library of Congress Cataloging-in-Publication Data

Insulin-like growth factors / [edited by] Derek LeRoith, Walter
Zumkeller, Robert C. Baxter.
   p. ; cm. -- (Medical intelligence unit)
Includes bibliographical references and index.
 ISBN 0-306-47846-3
 1. Somatomedin. 2. Insulin-like growth factor-binding proteins.
 [DNLM: 1. Receptors, Somatomedin. 2. Somatomedins. WK 515 I576
2003] I. LeRoith, Derek, 1945- II. Zumkeller, Walter. III. Baxter, R.
C. (Robert C.) IV. Title. V. Series.
 QP552.S65I567 2003
 612'.015756--dc21

2003012143

CONTENTS

Preface xvii

1. **The Structure of the Type 1 Insulin-Like Growth Factor Receptor** 1
   *Colin W. Ward, Thomas P. J. Garrett, Mei Lou, Neil M. McKern,
   Timothy E. Adams, Thomas C. Elleman, Peter A. Hoyne,
   Maurice J. Frenkel, Leah J. Cosgrove, George O. Lovrecz,
   Lindsay G. Sparrow, Lynne Lawrence and V. Chandana Epa*
   Introduction 1
   IGF-I Receptor: Discovery and Sequence 2
   Domain Organization and Evolutionary Relationships
      of IGF-IR and Related Receptors 2
   Secondary Structure 4
   3D Structure of the L1/cys-rich/L2 Domains of the IGF-IR 5
   3D Structure of the Tyrosine Kinase Domain 8
   The IGF-IR and IR Ectodomain Dimers 10
   Ligand Binding by Receptor Chimeras 11
   Effect of Point Mutations on Ligand Binding 13
   IGF Mutations 14
   Concluding Remarks 18

2. **Structure and Function of the IGF-1 and Mannose
   6-Phosphate/IGF-2 Receptors** 22
   Susan L. Spence and Peter Nissley
   The IGF System 22
   The IGF-1 Receptor 23
   The M6P/IGF-2 Receptor 32
   Conclusion 37

3. **IGFBPs—Gene and Protein Structure** 48
   Leon A. Bach and Nigel J. Parker
   Genes 48
   mRNA stability 51
   Proteins 51
   Tertiary Structure 53
   IGF Binding 54
   Sequence Determinants of IGF Binding 54
   IGF Binding Preference 56
   Post-Translational Modification 56
   Glycosylation 56
   Phosphorylation 57
   The Basic Regions of IGFBP-3 and -5: An Interaction 'Hot Spot' 57
   Proteolytic Cleavage 59
   Conclusions 60

4. **Regulation of Insulin-Like Growth Factor-I Gene Expression** 64
   Xia Wang and Martin L. Adamo
   Introduction and Background 64
   Summary of Current Research 65
   Summary and Prospectives 83

5. **The Many Levels of Control of IGF-II Expression** 91
   P. Elly Holthuizen
   Introduction 91
   The Role of IGF-II in Growth and Development 91
   Biological Effects of IGF-II 92
   IGF-II Protein Structure 92
   IGF-II Gene Structure 93
   Site-Specific Endonucleolytic Processing of IGF-II mRNAs 98
   Translational Regulation of IGF-II Expression 99

6. **IGF-I Receptor Signaling in Health and Disease** 104
   Renato Baserga, Marco Prisco and Tina Yuan
   Introduction 104
   The IGF Axis and Aging 105
   Transformation and Cell Size 106
   Differentiation: Importance and Contradictions
      of the IRS Proteins 109
   The IGF-I Receptor and Cell Death 110
   The Nuclear Connection: The Id Proteins 112
   Conclusions (and Some Random Thoughts) 115

7. **Insulin-Like Growth Factor-I Stimulation of Growth:
   Autocrine, Paracrine and/or Endocrine Mechanisms of Action?** 121
   A. Joseph D'Ercole
   Introduction 121
   The Somatomedin Hypothesis 122
   Initial Evidence for Endocrine IGF-I Actions 122
   Initial Evidence for IGF-I Local Actions 124
   Evidence for Paracrine/Autocrine IGF-I Actions from Studies
      of Transgenic and Null Mutant Mice 124
   Other Evidence for IGF-I Autocrine/Paracrine Actions 125
   Further Evidence for IGF-I Endocrine Actions 127
   Studies of Conditional Null Mutant Mice 127
   Relative Roles of GH and IGFs in Somatic Growth 131
   Conclusions 131

8. **Insulin-Like Growth Factor 1 (IGF1) and Brain Development** 137
   Carolyn A. Bondy, Wei-Hua Lee and Clara M. Cheng
   Introduction 137
   IGF System Expression in the Brain 138
   IGF1 Promotes Neuronal Glucose Utilization and Growth 140
   Signaling Pathways Involved in IGF1's Neurotrophic Effects 145
   IGF1 Promotes Neuronal Survival 147
   IGF1 and Brain Myelination 150
   Discussion 152

9. **IGFs and the Nervous System** 158
   Gina M. Leinninger, Gary E. Meyer and Eva L. Feldman
   Introduction 158
   Developmental Expression of the IGFs and IGFRs
     in the Nervous System 158
   Neurotrophic Roles of the IGFs in the Nervous System 161
   IGFs Promote Myelination in the CNS and PNS 173
   IGFs in the Treatment of Neurological Diseases 176
   Summary 179

10. **The Insulin-Like Growth Factors in Mammary Development
    and Breast Cancer** 188
    *Teresa L. Wood, Malinda A. Stull, Dawn Kardash-Richardson,
    Michael A. Allar and Aimee V. Loladze*
    Introduction 188
    Postnatal Mammary Gland Development 189
    The IGFs in GH-Mediated Mammary Development 189
    Function of the IGFs in Postnatal Mammary Development:
      Transgenic Mice 190
    Endogenous Expression of the IGFs and IGF-IR in Developing
      Mammary Tissue 190
    Interactions of IGF and EGF-Related Ligands in Normal
      Mammary Epithelial Cell Cycle Progression 192
    The IGFBPs and Regulation of IGF Actions in the Developing
      Mammary Gland 194
    The IGFs in Mammary Tumorigenesis and Breast Cancer 196
    The IGF-IR in Breast Cancer 197
    Relationship between the IGF-IR and the Estrogen Receptor ... 197
    The Insulin Receptor and IGF-II Receptor in Breast Cancer 198
    The IGFBPs and Breast Cancer 198
    Conclusions 199

11. **The Insulin-Like Growth Factor System and Bone** 206
    Thomas L. Clemens and Clifford J. Rosen
    Introduction 206
    Origins of Skeletal IGFs and Their Activity within Bone 206
    IGF-I Gene Expression and Regulation in Bone 208
    Systemic and Local Regulation of IGF-I in Bone Cells 208
    IGF Binding Proteins (IGFBPs) and Bone 209
    IGFBP Proteases 210
    Effects of IGF-I on Osteoblasts in Vitro 211
    Effects of IGF-I on Osteoclasts 211
    Studies in vivo and Genetically Altered Mice 211
    Clinical Aspects of IGF-I and Bone: Implications
      from Recent Findings 215

12. **IGFBP-5, a Multifunctional Protein, Is an Important Bone Formation Regulator** 219
   Subburaman Mohan, Yousef Amaar and David J. Baylink
   Introduction 219
   Discovery of IGFBP-5 219
   Regulation of IGFBP-5 Levels by Proteases 220
   Regulation of IGFBP-5 Production in Osteoblasts in Vitro 221
   Regulation of IGFBP-5 Levels in Vivo 223
   Actions of IGFBP-5 227
   Models of IGFBP-5 Action in Bone 228
   Conclusions 231

13. **IGF Action and Skeletal Muscle** 235
    David T. Kuninger and Peter S. Rotwein
    Introduction 235
    Basics of Skeletal Muscle Development 235
    Regulation and Expression of the IGF System in Skeletal Muscle 236
    IGF Action in Muscle—Signaling Pathways
       and Molecular Mechanisms 237
    Perspectives 241

14. **Insulin-Like Growth Factor-I and the Kidney** 244
    Franz Schaefer and Ralph Rabkin
    Introduction 244
    Expression and Regulation of GH-IGF System Components
       in the Kidney 245
    Renal Handling of GH, IGFs and IGFBPs 246
    IGFBPs Have Direct and Indirect Actions on Kidney Function 247
    GH and IGF-I Increase Renal Blood Flow and Glomerular
       Filtration Rate 248
    GH and IGF-1 Modulate Renal Tubular Function 249
    GH and IGF-I Increase Renal Mass 250
    GH, IGF-I and Progression of Renal Disease 251
    The GH-IGF System in Chronic Renal Failure 252
    Therapeutic Manipulation of the GH/IGF-I System
       in Chronic Renal Failure 253
    The GH-IGF System and rhIGF Treatment in Acute Renal Failure 254
    Summary and Conclusion 255

15. **IGF-Independent Effects of the IGFBP Superfamily** 262
    Gillian E. Walker, Ho-Seong Kim, Yong-Feng Yang and Youngman Oh
    Introduction 262
    IGF-Dependent Effects for the IGFBP Superfamily 263
    IGF-Independent Effects for the IGFBP Superfamily 265
    Summary 274

16. **Functional Relationships between Transforming Growth Factor-β and the Insulin-Like Growth Factor Binding Proteins** 281
    Susan Fanayan and Robert C. Baxter
    Introduction 281

The Transforming Growth Factor-β (TGF-β) Superfamily 281
TGF-β Receptors 282
Smads 283
Alterations in TGF-β Signaling during Carcinogenesis 283
TGF-β Effects on IGFBP Production 284
IGFBP-3 Induction in TGF-β Growth Regulation 284
Involvement of TGF-β Signaling Pathways in IGFBP-3 Action 286
Concluding Comments 287

17. **Insulin-Like Growth Factor Binding Proteins (IGFBPs) and Apoptosis** 294
C.M. Perks and J.M.P. Holly
What Is Apoptosis? 294
Apoptosis during Development and Maturation 294
Activation of Apoptotic Signaling Pathways 295
The Sphingomyelin Pathway 295
Extracellular Control of Apoptosis 296
Intracellular Control 297
IGF-Dependent Modulation of Apoptosis by IGFBPs 297
IGF-Independent Modulation of Apoptosis by IGFBPs 297
Mechanism of IGFBP Intrinsic Actions 299
Therapeutic Interventions and Potential Modulation by IGFBPs 300

18. **The Role of Insulin-Like Growth Factor-1 and Extracellular Matrix Protein Interaction in Controlling Cellular Responses to This Growth Factor** 304
David R. Clemmons
Introduction 304
IGFs and Extracellular Matrix Component Composition 305
Interaction between IGF-I and ECM Components and Subsequent Cellular Responses 306
Stimulation of Cell Migration 306
Role of IGF Binding Proteins in ECM Localization 307
Role of Integrin Receptor Activation in Modulating IGF-I Biologic Actions 310
Molecular Mechanism Mediating Signaling between the αVβ3 Integrin and the IGF-I Receptor 311
IGF-I Stimulation of Integrin Activation 312

19. **Epidemiologic Approaches to Evaluating Insulin-Like Growth Factor and Cancer Risk** 317
Eva S. Schernhammer and Susan E. Hankinson
Epidemiologic Methods in Studying Insulin-Like Growth Factor 317
Introduction to Insulin-Like Growth Factor and Cancer 318
Demographic, Lifestyle, and Dietary Predictors of Levels of Insulin-Like Growth Factor 319
Etiologic Studies of Insulin-Like Growth Factor and Cancer 320
Summary 333

20. **IGF-I, Insulin and Cancer Risk: Linking the Biology and the Epidemiology** 338
    Michael Pollak
    Introduction 338
    Variation of Cancer Risk according to Circulating Level of IGF-I 338
    What about Insulin, IGF-II, and IGF Binding Proteins? 340
    Hypotheses to Explain the Epidemiologic Observations 340
    Challenges 342
    Potential Medical Relevance 342
    Conclusion 343

21. **The Molecular Basis of IGF-I Receptor Gene Expression in Human Cancer** 346
    Haim Werner
    Introduction 346
    Overexpression of the IGF-IR Gene as a Common Theme in Malignancy 346
    The Role of the IGF-IR in the Transformation Process 347
    Mapping of Receptor Domains Involved in Transformation 348
    Phosphorylation of the IGF-IR in Malignancy 348
    Transcriptional Regulation of the IGF-IR Gene 349
    Regulation of the IGF-IR Gene by Oncogenes 349
    Regulation of the IGF-IR Gene by Tumor Suppressor p53 350
    Regulation of the IGF-IR Gene by Tumor Suppressor WT1 351
    Regulation of the IGF-IR Gene by Disrupted Transcription Factors 352
    Regulation of the IGF-IR Gene by Growth Factors, Cytokines and Steroid Hormones 352
    Conclusions 353

22. **Antisense and Triple Helix Strategies in Basic and Clinical Research: Challenge for Gene Therapy of Tumors Expressing IGF-I** 357
    L. C. Upegui-Gonzalez, J. C. Francois, L. A. Trojan, A. Ly, R. Przewlocki, C. Malvy and Jerry Trojan
    Introduction 357
    General View of Antisense Strategy 357
    Antisense Strategy in Protein Function Studies 359
    Antisense Strategy in Tumor Gene Therapy 360
    Triple Helix Strategy in Tumor Gene Therapy 362
    Gene Therapy of Tumors Expressing IGF-I 363

23. **The IGF System in Breast Cancer** 367
    Janet L. Martin
    Introduction 367
    The IGF System in Breast Cancer: Clinical Studies 367
    Mechanisms of IGF Action in Breast Cancer 369
    The IGFBPs in Breast Cancer 373
    Concluding Remarks 378

24. **The Role of the IGF System in Prostate Cancer** 385
    Charles T. Roberts, Jr.
    Introduction 385
    IGF Action in Prostate Growth and Development 385
    The IGF-IR and IGF-II in Tumorigenesis-Molecular Studies 385
    IGF-I and Prostate Cancer-Epidemiological Studies 386
    IGF-II and Prostate Cancer-Epidemiological Studies 387
    Summary 387

25. **IGFs and Sarcomas** 390
    Fariba Navid and Lee J. Helman
    Introduction 390
    Rhabdomyosarcoma 391
    Osteosarcoma 392
    Ewing's Family of Tumors 394
    Other Sarcomas 395
    Conclusion 395

26. **IGFs and Epithelial Cancer** 399
    Walter Zumkeller
    Introduction 399
    Prostate Cancer 399
    Wilms' Tumour 400
    Colorectal Adenoma 401
    Breast Cancer 402
    Ovarian Cancer 403
    Cervical Cancer 404

27. **Insulin-Like Growth Factors and Hematological Malignancies** 410
    Anne J. Novak and Diane F. Jelinek
    Introduction 410
    IGFs and IGFBPs in the Immune System 410
    The Insulin-Like Growth Factor Receptor-I in the Immune System 412
    IGFs and IGF-IR: A Pathway to Malignancy 413
    IGFs and the IGF-IR in B Cell Malignancies 415
    Receptor Cross Talk 417
    Concluding Remarks 418

28. **Metabolic Effects of Insulin-Like Growth Factor I and Growth Hormone in vivo: A Comparison** 423
    Nelly Mauras
    Introduction 423
    GH/IGF-I: Effects on Protein Metabolism 424
    GH/IGF-I: Effects on Carbohydrate Metabolism 425
    GH/IGF-I Effects on Lipid Metabolism and Body Composition 425
    GH/IGF-I Effects on Bone 427
    GH/IGF-I: Use in Adult Replacement 427
    GH/IGF-I: Comparison of Anabolic Effects in Man 428
    In Summary 431

29. Insulin-Like Growth Factor II (IGF-II) and Non-Islet Cell
    Tumor Hypoglycemia (NICTH) 434
    Naomi Hizuka, Izumi Fukuda, Yukiko Ishikawa and Kazue Takano
    Introduction 434
    Clinical Features of IGF-II Producing NICTH 434
    Serum IGF-II and IGF-I Levels 435
    Characterization of Big IGF-II 436
    Circulating Form of IGF-II and Serum IGFBPs 437
    Mechanism of Hypoglycemia 438
    Treatment of NICTH 439

30. Diabetes 441
    Tero Saukkonen and David B. Dunger
    Introduction 441
    IGF-I and Glucose Metabolism 441
    Interaction between Insulin and IGF-I and Its Binding Proteins 442
    IGF-I and Diabetes 443
    The Use of rhIGF-I in Diabetes 446
    GH/IGF-I Axis and Microvascular Complications of Diabetes 448
    Conclusions and the Future Role of rhIGF-I Therapy 452

31. Insulin-Like Growth Factors in Critical Illness 457
    Greet Van den Berghe
    Introduction 457
    Changes within the IGF-System in the Acute Phase
      of Critical Illness 458
    Changes within the IGF-System in the Chronic Phase
      of Critical Illness 460
    Conclusion 464

32. Laron Syndrome: Primary GH Insensitivity or Resistance 467
    Zvi Laron
    History 467
    Nomenclature 467
    Geographical Distribution and Genetic Aspects 468
    Clinical Aspects 468
    Ageing, Longevity and Mortality 477
    Laboratory Findings 477
    The Pygmies 484

Index 493

EDITORS

Derek LeRoith, M.D., Ph.D.
Clinical Endocrinology Branch
National Institutes of Health, MSC
Bethesda, Maryland, U.S.A.
email: derek@helix.nih.gov

Walter Zumkeller, M.D.
Department of Pediatrics
Martin-Luther-University, Halle-Wittenberg
Children's University Hospital
Halle/Saale, Germany
Chapter 26

Robert C. Baxter, Ph.D., DSc
Kolling Institute of Medical Research
Royal North Shore Hospital
St. Leonards, Australia
email: robaxter@med.usyd.edu.au
Chapter 16

CONTRIBUTORS

Martin L. Adamo
Department of Biochemistry, MSC
The University of Texas Health Science
  Center at San Antonio
San Antonio, Texas, U.S.A.
Chapter 4

Timothy E. Adams
CSIRO Health Sciences and Nutrition
Parkville, Victoria, Australia
Chapter 1

Michael A. Allar
Department of Neuroscience
  and Anatomy
Penn State College of Medicine
Hershey, Pennsylvania, U.S.A.
Chapter 10

Yousef Amaar
Musculoskeletal Diseases Center
Jerry L. Pettis VA Medical Center, and
Departments of Medicine, Biochemistry
  and Physiology
Loma Linda University
Loma Linda, California, U.S.A.
Chapter 12

Leon A. Bach
Department of Medicine
University of Melbourne
Austin and Repatriation Medical Centre
Heidelberg, Victoria, Australia
Chapter 3

Renato Baserga
Jefferson Cancer Institute
Thomas Jefferson University
Philadelphia, Pennsylvania, U.S.A.
Chapter 6

David J. Baylink
Musculoskeletal Diseases Center
Jerry L. Pettis VA Medical Center, and
Departments of Medicine, Biochemistry
  and Physiology
Loma Linda University
Loma Linda, California, U.S.A.
Chapter 12

Carolyn A. Bondy
Developmental Endocrinology Branch
NICHD, National Institutes of Health
Bethesda, Maryland, U.S.A.
Chapter 8

Clara M. Cheng
Developmental Endocrinology Branch
NICHD, National Institutes of Health
Bethesda, Maryland, U.S.A.
Chapter 8

Thomas L. Clemens
Departments of Medicine and Molecular
  and Cellular Physiology
University of Cincinnati
Cincinnati, Ohio, U.S.A.
Chapter 11

David R. Clemmons
Department of Medicine
University of North Carolina
Chapel Hill, North Carolina, U.S.A.
Chapter 18

Leah J. Cosgrove
CSIRO Health Sciences and Nutrition
Parkville, Victoria, Australia
Chapter 1

A. Joseph D'Ercole
Department of Pediatrics
University of North Carolina
Chapel Hill, North Carolina, U.S.A.
Chapter 7

David B. Dunger
Department of Pediatrics
University of Cambridge
Addenbrooke's Hospital
Cambridge, U.K.
Chapter 30

Thomas C. Elleman
CSIRO Health Sciences and Nutrition
Parkville, Victoria, Australia
Chapter 1

V. Chandana Epa
CSIRO Health Sciences and Nutrition
Parkville, Victoria, Australia
Chapter 1

Susan Fanayan
Kolling Institute of Medical Research
Royal North Shore Hospital
St. Leonards, Australia
Chapter 16

Eva L. Feldman
Department of Neurology
University of Michigan
Ann Arbor, Michigan, U.S.A.
Chapter 9

J. C. Francois
Laboratory of Biophysics
INSERM, Museum d'Historie Naturelle
Paris, France
Chapter 22

Maurice J. Frenkel
CSIRO Health Sciences and Nutrition
Parkville, Victoria, Australia
Chapter 1

Izumi Fukuda
Department of Medicine II
Tokyo Women's Medical University
Tokyo, Japan
Chapter 29

Thomas P. J. Garrett
Walter and Elisa Hall Institute
Royal Melbourne Hospital
Parkville, Victoria, Australia
Chapter 1

Susan E. Hankinson
Department of Medicine
Brigham and Women's Hospital
  and Harvard Medical School, and
Department of Epidemiology
Harvard School of Public Health
Boston, Massachusetts, U.S.A.
Chapter 19

Lee J. Helman
Pediatric Oncology Branch
National Institutes of Health
Bethesda, Maryland, U.S.A
Chapter 25

Naomi Hizuka
Department of Medicine II
Tokyo Women's Medical University
Tokyo, Japan
Chapter 29

J. M. P. Holly
University Division of Surgery
Bristol Royal Infirmary
Bristol, U.K.
Chapter 17

P. Elly Holthuizen
Department of Physiological Chemistry
University Medical Center Utrecht
Utrecht, The Netherlands
Chapter 5

Peter A. Hoyne
CSIRO Health Sciences and Nutrition
Parkville, Victoria, Australia
Chapter 1

Yukiko Ishikawa
Department of Medicine II
Tokyo Women's Medical University
Tokyo, Japan
Chapter 29

Diane F. Jelinek
Department of Immunology
Mayo Graduate and Medical Schools
Mayo Clinic
Rochester, Minnesota, U.S.A.
Chapter 27

Dawn Kardash-Richardson
Department of Neuroscience
  and Anatomy
Penn State College of Medicine
Hershey, Pennsylvania, U.S.A.
Chapter 10

Ho-Seong Kim
Department of Pediatrics
Oregon Health Sciences University
Portland, Oregon, U.S.A.
Chapter 15

David T. Kuninger
Department of Medicine
Oregon Health Sciences University
Portland, Oregon, U.S.A.
Chapter 13

Zvi Laron
Endocrinology and Diabetes Research Unit
Schneider Children's Medical Center
Tel Aviv University
Tel Aviv, Israel
Chapter 32

Lynne Lawrence
CSIRO Health Sciences and Nutrition
Parkville, Victoria, Australia
Chapter 1

Wei-Hua Lee
Developmental Endocrinology Branch
NICHD, National Institutes of Health
Bethesda, Maryland, U.S.A.
Chapter 8

Gina M. Leinninger
Department of Neurology
University of Michigan
Ann Arbor, Michigan, U.S.A.
Chapter 9

Aimee V. Loladze
Department of Neuroscience
and Anatomy
Penn State College of Medicine
Hershey, Pennsylvania, U.S.A.
Chapter 10

Mei Lou
Walter and Elisa Hall Institute
Royal Melbourne Hosptial
Parkville, Victoria, Australia
Chapter 1

George O. Lovrecz
CSIRO Health Sciences and Nutrition
Parkville, Victoria, Australia
Chapter 1

A. Ly
Laboratory of Developmental Neurology
INSERM, University of Paris VII
Paris, France
Chapter 22

C. Malvy
Institute of Pharmacology
Polish Academy of Sciences
and Collegium Medicum
Cracow, Poland
Chapter 22

Janet L. Martin
Department of Molecular Medicine
University of Sydney
Royal North Shore Hospital
St. Leonards, Australia
Chapter 23

Nelly Mauras
Mayo Medical School, and
Division of Endocrinology
Nemours Children's Clinic
Jacksonville, Florida, U.S.A.
Chapter 28

Neil M. McKern
CSIRO Health Sciences and Nutrition
Parkville, Victoria, Australia
Chapter 1

Gary E. Meyer
Department of Neurology
University of Michigan
Ann Arbor, Michigan, U.S.A.
Chapter 9

Subburaman Mohan
Musculoskeletal Diseases Center
Jerry L. Pettis VA Medical Center, and
Departments of Medicine, Biochemistry
and Physiology
Loma Linda University
Loma Linda, California, U.S.A.
Chapter 12

Fariba Navid
Pediatric Oncology Branch
National Institutes of Health
Bethesda, Maryland, U.S.A
Chapter 25

Peter Nissley
Metabolism Branch
Center for Cancer Research
National Cancer Institute
National Institutes of Health
Bethesda, Maryland, U.S.A.
Chapter 2

Anne J. Novak
Department of Immunology
Mayo Graduate and Medical Schools
Mayo Clinic
Rochester, Minnesota, U.S.A.
Chapter 27

Youngman Oh
Department of Pediatrics
Oregon Health Sciences University
Portland, Oregon, U.S.A.
Chapter 15

Nigel J. Parker
Department of Medicine
University of Melbourne
Austin and Repatriation Medical Centre
Heidelberg, Victoria, Australia
Chapter 3

Claire M. Perks
University Division of Surgery
Bristol Royal Infirmary
Bristol, U.K.
Chapter 17

Michael Pollak
McGill University
Montreal, Canada
Chapter 20

Marco Prisco
Jefferson Cancer Institute
Thomas Jefferson University
Philadelphia, Pennsylvania, U.S.A.
Chapter 6

R. Przewlocki
Institute of Pharmacology
Polish Academy of Sciences
  and Collegium Medicum
Cracow, Poland
Chapter 22

Ralph Rabkin
Veterans Affairs Palo Alto Health
  Care System, and
Department of Medicine
Stanford University
Palo Alto, California, U.S.A.
Chapter 14

Charles T. Roberts, Jr.
Department of Pediatrics
Oregon Health Sciences University
Portland, Oregon, U.S.A.
Chapter 24

Clifford J. Rosen
Maine Center for Osteoporosis Research
  and Education
St. Joseph Hospital, and
The Jackson Laboratory
Bangor, Maine, U.S.A.
Chapter 11

Peter S. Rotwein
Department of Medicine
Oregon Health Sciences University
Portland, Oregon, U.S.A.
Chapter 13

Tero Saukkonen
Department of Pediatrics
University of Cambridge
Addenbrooke's Hospital
Cambridge, U.K.
Chapter 30

Franz Schaefer
Division of Pediatric Nephrology
University Children's Hospital
Heidelberg, Germany
Chapter 14

Eva S. Schernhammer
Department of Medicine
Brigham and Women's Hospital
  and Harvard Medical School, and
Department of Epidemiology
Harvard School of Public Health
Boston, Massachusetts, U.S.A.
Chapter 19

Lindsay G. Sparrow
CSIRO Health Sciences and Nutrition
Parkville, Victoria, Australia
Chapter 1

Susan L. Spence
Metabolism Branch
Center for Cancer Research
National Cancer Institute
National Institutes of Health
Bethesda, Maryland, U.S.A.
Chapter 2

Malinda A. Stull
Department of Neuroscience
  and Anatomy
Penn State College of Medicine
Hershey, Pennsylvania, U.S.A.
Chapter 10

Kazue Takano
Department of Medicine II
Tokyo Women's Medical University
Tokyo, Japan
Chapter 29

Jerry Trojan
Laboratory of Developmental Neurology
University of Paris VII
Paris, France, and
Institute of Pharmacology
Polish Academy of Sciences
   and Collegium Medicum
Cracow, Poland
Chapter 22

L. A. Trojan
Laboratory of Developmental Neurology
University of Paris VII
Paris, France, and
Department of Pharmacology
CWRU
Cleveland, Ohio, U.S.A.
Chapter 22

L. C. Upegui-Gonzalez
Laboratory of Developmental Neurology
INSERM, University of Paris VII
Paris, France
Chapter 22

Greet Van den Berghe
Department of Intensive Care Medicine
University Hospital Gasthuisberg
University of Leuven
Leuven, Belgium
Chapter 31

Gillian E. Walker
Department of Pediatrics
Oregon Health Sciences University
Portland, Oregon, U.S.A.
Chapter 15

Xia Wang
Department of Biochemistry, MSC
The University of Texas Health Science
   Center at San Antonio
San Antonio, Texas, U.S.A.
Chapter 4

Colin W. Ward
CSIRO Health Sciences and Nutrition
Parkville, Victoria, Australia
Chapter 1

Haim Werner
Department of Clinical Biochemistry
Sackler School of Medicine
Tel Aviv University
Tel Aviv, Israel
Chapter 21

Teresa L. Wood
Department of Neuroscience
   and Anatomy
Penn State College of Medicine
Hershey, Pennsylvania, U.S.A.
Chapter 10

Yong-Feng Yang
Department of Pediatrics
Oregon Health Sciences University
Portland, Oregon, U.S.A.
Chapter 15

Tina Yuan
Jefferson Cancer Institute
Thomas Jefferson University
Philadelphia, Pennsylvania, U.S.A.
Chapter 6

PREFACE

The insulin-like growth factors are ubiquitously expressed and are crucial to the normal growth and function of virtually all cells and tissues. Together with their binding proteins and receptors, they form a widely studied biological system characterized by complex interactions among its members, and involving many other proteins. In addition to its significance in growth and development, the insulin-like growth factor system also has important roles in a wide variety of pathological states. This has led to growing interest in the therapeutic potential of insulin-like growth factors and their binding proteins for numerous disease states, as well as the possibility that these proteins and their receptors might be candidate drug targets in cancer.

Because of their increasing clinical and therapeutic significance, we believe it was timely to compile another book on IGFs; the last one was published in 1999, and since then IGF-related research has continued to grow rapidly.

The current book contains chapters on a broad range of topics covering both the basic science and clinical aspects of IGFs and their regulatory proteins, with some emphasis on their relevance in cancer. Each chapter is written by an expert in the field and is current. There is liberal usage of references to enable the reader to keep updated on the latest significant research in each area.

We, the editors, are indebted to the authors for their outstanding contributions, without which this book would not be possible.

Derek LeRoith
Walter Zumkeller
Robert C. Baxter

CHAPTER 1

The Structure of the Type 1 Insulin-Like Growth Factor Receptor

Colin W. Ward, Thomas P. J. Garrett, Mei Lou, Neil M. McKern,
Timothy E. Adams, Thomas C. Elleman, Peter A. Hoyne, Maurice J. Frenkel,
Leah J. Cosgrove, George O. Lovrecz, Lindsay G. Sparrow, Lynne Lawrence
and V. Chandana Epa

Abstract

The type 1 insulin-like growth factor receptor (IGF-1R) is widely expressed across many cell types in fetal and postnatal tissues. Signalling through the IGF-1R is the principal pathway responsible for somatic growth in fetal mammals, while somatic growth in postnatal animals is achieved through the synergistic interaction of growth hormone (GH) and the IGFs. Forced overexpression of the IGF-1R results in the malignant transformation of cultured cells and elevated levels of IGF-1R are observed in a variety of human tumor types. Downregulation of IGF-1R levels can reverse the transformed phenotype of tumor cells, and may render them sensitive to apoptosis in vivo. These discoveries have led to the emergence of IGF-1R as a therapeutic target for the development of anti-tumor agents. This Chapter will review the key developments in our understanding of the structure of the type I insulin-like growth factor receptor drawing on parallel studies with the closely related insulin receptor. These tyrosine kinase receptors are large, transmembrane proteins consisting of several structural domains. Their ectodomains have a similar arrangement of two homologous domains (L1 and L2) separated by a cys-rich region. The L domains consist of five and a half leucine-rich repeats. The C-terminal half of their ectodomains consists of three fibronectin type 3 repeats, and an insert domain which contains the α-β cleavage site. The cytoplasmic portion of the receptor consists of a catalytic kinase domain flanked by a juxtamembrane and C-tail region, the sites of binding of various signalling molecules. Our current knowledge on the structure of these two receptors has come from a combination of multiple sequence analyses, site specific mutagenesis and chimera studies, single-molecule electron microscope images of receptor and receptor/ligand complexes, and the 3D structures of the first three domains of the IGF-1R and the tyrosine kinase domain of the insulin receptor, determined by X-ray crystallography.

Introduction

The insulin-like growth factors (IGFs) are essential for normal fetal and postnatal growth and development. The type 1 IGF receptor (IGF-IR) binds IGF-I with high affinity and initiates the physiological response to this ligand in vivo.[1] The IGF-IR also binds IGF-II, albeit with lower affinity, and is in part responsible for the mitogenic effects of this polypeptide during fetal development.[2] The alternately spliced form of the insulin receptor (IR) that lacks exon 11 and is expressed in many fetal tissues has recently been identified as binding IGF-II with high affinity,[3] confirming an earlier genetic study implicating the insulin receptor in the growth-promoting effects of IGF-II.[4]

The ligands, insulin, IGF-I and IGF-II share a common three-dimensional (3D) architecture[5,6,7] and can bind to both IR and IGF-IR in a competitive manner. The hIRR ligand is unknown and IRR knockouts lack a distinguishing phenotype.[8]

Insulin-Like Growth Factors, edited by Derek LeRoith, Walter Zumkeller
and Robert Baxter. ©2003 Eurekah.com and Kluwer Academic / Plenum Publishers.

Germline deletion of both *Igf-1r* alleles results in severe growth retardation during the second half of gestation.[9] Fibroblast cell lines established from *Igf-1r* knockout mice are impaired in their progression through the cell cycle in serum-rich conditions and are resistant to oncogenic transformation by a variety of viral and cellular oncogenes.[9] Conversely, overexpression of IGF-IRs promoted the neoplastic transformation of cell lines in a ligand-dependent manner.[10] Thus the IGF axis has an important role to play not only in normal cellular development, but also in malignant transformation.[11] As a result, the IGF-IR has emerged as a candidate therapeutic target for the treatment of human cancer.

IGF-I Receptor: Discovery and Sequence

The first evidence for the presence of an IGF receptor distinct from IR came in 1974 when ^{125}I-labelled insulin and ^{125}I-labelled NSILA-s (soluble fraction of non-suppressible insulin-like activity) were used to label distinct proteins in purified rat liver plasma membranes[12] and detergent solubilized fractions.[13] The IGF-IR was subsequently shown, by SDS gel electrophoresis, to be a homodimer composed of two α- and two β- chains held together by disulfide bonds.[14,15] The IGF-IR is synthesized as a 180 kDa precursor which is glycosylated, dimerized and proteolytically processed to yield the mature $\alpha_2\beta_2$ receptor.[16] The next key discovery was the demonstration that IGF-IR is a tyrosine kinase which is activated and autophosphorylated following IGF-I binding.[17,18]

The cDNA for human (h) IGF-IR was cloned and sequenced in 1986.[19] It consists of 4,989 nucleotides and codes for a 1,367 amino acid precursor (Fig. 1). The pre-proreceptor monomer includes a 30 residue signal peptide (residues −30 to −1) and an Arg.Lys.Arg.Arg furin protease cleavage site at residues 708-711, which on cleavage yields one α-chain and one β-chain. The α-chain (residues 1-707) and 195 residues of β-chain comprise the extracellular portion of the IGF-IR and contain eleven and five potential N-linked glycosylation sites, respectively.[19] There is a single transmembrane sequence (residues 906-929) and a 408 residue cytoplasmic domain containing the tyrosine kinase (residues 930-1337). The cDNA for the hIR and the third member of the IR family, hIRR, have been cloned and sequenced and are similarly organized.[20,21,22]

The human IGF-IR gene is greater than 100 kilobasepairs in size and contains 21 exons, ten in the α-chain and eleven in the β chain.[23] An alternate human IGF-IR mRNA transcript has been reported, in which a three-basepair (CAG) deletion results in the substitution of Arg for Thr898Gly899 (Fig. 1), eight residues upstream from the start of the transmembrane region of hIGF-IR.[24] The CAG- isoform shows reduced internalization and enhanced signalling properties compared to the CAG+ isoform.[25]

The hIGF-IR, like the hIR, is heavily glycosylated with 16 potential N-linked glycosylation sites.[19,20,21] The hIR has 18 N-linked sites while hIRR, the third member of this receptor sub-family, has 11 sites.[22] Most analyses have been conducted with hIR. Analytical ultracentgrifugation showed that hIR expressed in CHO-K1 cells contained 58-64 kDa of carbohydrate.[26] Oligosaccharides of both the high mannose and complex type are present, the latter containing additional fucose, N-acetylglucosamine, galactose and sialic acid residues.[27,28,29] O-linked glycosylation has been demonstrated only in the β subunit of the hIR.[27,30,31] Studies on the effects of removing N-linked glycosylation sites indicate that there are many redundancies in hIR glycosylation. Every site, with one exception, can be mutated individually without detriment to cell-surface expression, receptor processing and ligand binding (see refs 30 to 36). When combinations of sites are examined, it appears that the major domains of the receptor, particularly those closer to the N-terminus (i.e., L1, cys-rich, L2), require at least one intact glycosylation site to ensure correct folding and processing.[34]

Domain Organization and Evolutionary Relationships of IGF-IR and Related Receptors

Comparative sequence analyses have revealed that many proteins, particularly eukaryotic extracellular proteins, are composed of a number of different, sometimes repeated, structural units. In the case of the IR subfamily, 11 distinct regions have been identified in each monomer (Fig. 2). The N-terminal half of the IGF-IR ectodomain contains two homologous domains (L1 and L2), separated by a cys-rich region (Cys152 to Cys298) containing 22 cysteine residues.[37,38,39] The C-terminal half of the IGF-IR ectodomain consists of three fibronectin type III (FnIII) domains, the second of which contains a large insert domain of ~120–130 residues.[40,41,42,43] Intracellularly, each IGF-IR

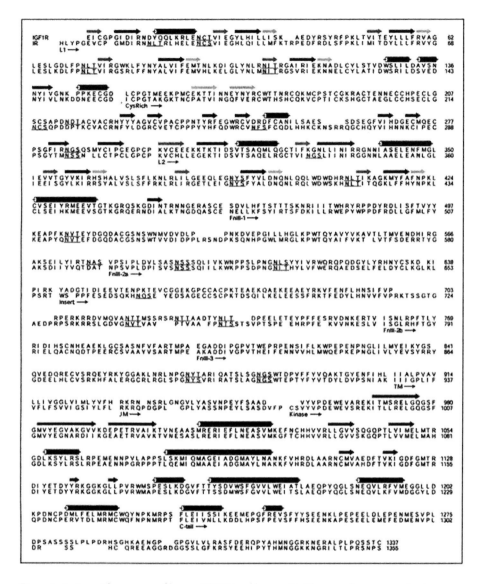

Figure 1. Amino acid sequences of human IGF-IR and human IR. The secondary structural assignments for L1/cys-rich/L2 domains and the tyrosine kinase domain are depicted above the sequences as cylinders for α-helices and arrows for β-strands. The demarcation boundaries for the various domains and modules are shown underneath the sequences. Reprinted with permission from: Adams TE, Epa VC, Garrett TPJ and Ward CW. Cell Mol Life Sci 2000; 57:1050-1093. © 2000 Birkhäuser Verlag, Basel.

monomer contains a tyrosine kinase catalytic domain (residues 973-1229) flanked by two regulatory regions—a juxtamembrane region, residues 930-972, and an 108 residue C-tail, residues 1230-1337- that contain the phosphotyrosine binding sites for signalling molecules (Fig. 1). The C-terminal boundary of the kinase domain of IGF-IR is Phe1229 or shorter, not Ile1236[44] since IGF-IRs truncated at 1229[45] and 1243[46] are still catalytically active as judged by receptor autophosphorylation, the phosphorylation/activation of cellular substrates and mitogenic responses to IGF-I. The 3D structure of the hIGF-1R kinase domain has been deposited in the PDB as

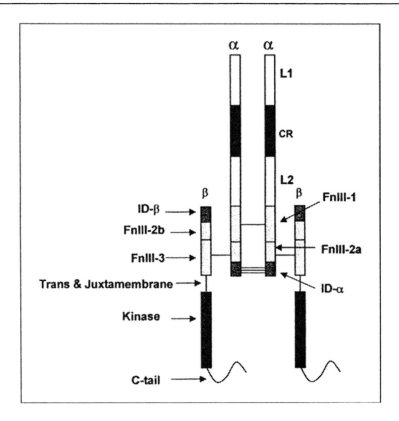

Figure 2. Cartoon of the IGF-IR dimer showing the distribution of domains across the α- and β- chains and the approximate location of the α-β disulfides and the α-α dimer disulfide bonds. Reprinted with permission from: Adams TE, Epa VC, Garrett TPJ and Ward CW. Cell Mol Life Sci 2000; 57:1050-1093. © 2000 Birkhäuser Verlag, Basel.

1JQH.[47] The 3D structure of the hIR kinase domain has been described for both the inactive[48] and active[49] states. Phe1229, which is 10 residues downstream from the conserved CysTrp sequence located at 1218-1219 (Fig. 1), appears to be very close to the catalytic domain boundary when compared with the 3D structure of the cAMP-dependent protein kinase.[48,50]

Representatives of the IGF-IR/IR receptor family have been characterized in some of the simplest multicellular animals, including cnidarians (polyps and jellyfish), nematodes, gastropods and insects (see 43). The IGF-IR/IRs from primitive organisms such as *Caenorhabditis elegans* (nematode) and *Drosophila melanogaster* (insect) have additional sequences at their N- and C-termini (see refs 51 and 52). The emergence of distinct IR and IGF-IR genes appears to coincide with the evolutionary transition from protochordates to vertebrates. The protochordate amphioxus contains only a single IR-like receptor cDNA in contrast to the hagfish, considered to be the most primitive extant vertebrate, which appears to contain two IR/IGF-IR-like cDNAs.[53]

Secondary Structure

The α-chain of IGF-IR has a total of 38 cysteine residues, while the β-chain has three extracellular and five intracellular cysteine residues (Fig. 1). There are disulphide bonds at each end of the L1 domain (Cys3-Cys22 and Cys120-Cys148) and L2 domain (Cys302-Cys323 and Cys425-Cys458) based on chemical analysis of hIR,[54,55] sequence alignments[38] and the 3D structure of the first three domains of the IGF-IR.[39] The cys-rich region consists of eight disulfide-linked modules,[38,39] similar to those found in the tumor necrosis factor (TNF) receptor[56] and subsequently seen in laminin.[57]

There are at least two α-α disulphide bonds involved in the IGF-I dimer. One involves Cys514, in the first FnIII domain, linked to Cys514 in the second monomer based on chemical analysis[55] and site specific mutagenesis[58,59,60] of the corresponding bond in hIR. The second involves the triplet of Cys residues at positions 669, 670 and 672 in the insert domain of IGF-IR, based on chemical analyses of the corresponding region of hIR.[54] It was not possible to determine which one of the three, or whether all three Cys residues were involved in dimer disulfides. The sequence around this triplet resembles those found in the hinge region of antibodies[61] where multiple disulfide bonds occur. The IGF-I receptor has an additional Cys residue (Cys662), seven residues upstream of the Cys triplet in the insert domain, and lacks the Cys residue equivalent to Cys884 in hIR (Fig. 1). *Drosophila* IR[62,63] also lacks the cysteine residue equivalent to Cys884 in hIR as well as the equivalent of one of Cys669 or Cys670. The two reports differ with regard to the presence or absence of the cysteine residue equivalent to the Cys514, known to form one of the α-α dimer bonds.[55]

There is only a single α-β disulfide link in the hIGF-IR, between Cys633 in the first FnIII repeat and Cys849 in the second FnIII repeat,[54] which is consistent with the mutagenesis data for hIR[60,64] and the predictions of Ward et al,[38] but not those of Schaefer et al.[65] The structural implications of the disulfide bond between Cys633 and Cys849 are that the two FnIII domains are aligned side by side,[38] not end to end, as is the more common configuration.[66] Finally, there is a single, intra-chain disulfide linkage between the β-chain residues Cys776 and Cys785 in the predicted F and G strands of the third FnIII domain based on chemical analysis of the corresponding disulphide bond in hIR.[54] Interestingly, the reported sequence for rat, but not mouse, IR[67] shows a Ser residue at the position equivalent to Cys785 in human IR.

3D Structure of the L1/cys-rich/L2 Domains of the IGF-IR

The 3D structure of the L1/cys-rich/L2 domain fragment of the IGF-IR has been solved.[39] As shown in Figure 3, the molecule adopts an extended bilobal structure (approximately 40 x 48 x 105 Å) with the L domains at either end. The cys-rich region runs two-thirds the length of the molecule, making contact along the length of the L1 domain but having very little contact with the L2 domain. This leaves a space at the centre of the molecule of approximately 24 Å diameter and of sufficient size to accommodate the ligands, IGF-I or IGF-II as shown (Fig. 3). The space is bounded on three sides by the regions of IGF-IR which are known to contribute to ligand binding, based on studies of chemical cross-linking, receptor chimeras, and natural or site-specific mutants.[39]

The L Domains

Each L domain (residues 1-150 and 300-460) adopts a compact shape (approx. 24 x 32 x 37 Å) being formed from a single-stranded, right-handed β-helix, capped at each end by short α-helices and a disulfide bond. The body of each L domain looks like a loaf of bread with three flat sides (β sheets) and an irregular top (Fig. 3). The two domains are superimposable.[39] The repetitive nature of the β-helix is reflected in the sequence where a five-fold repeat, centered on a conserved glycine, had been identified by sequence analyses.[38] The structure, however, revealed that the L domains comprise six helical turns and a fold that was quite unexpected.[39]

A notable difference between the L1 and L2 domains is found at their C-terminal ends. The indole ring of Trp176, from the first module in the cys-rich region, is inserted into a pocket in the hydrophobic core of L1 formed by residues Ile98, Gly99, Leu100, and Leu103 from the fourth turn and Val125, Trp127 and Ile130 from the fifth turn of L1.[39] The sequence motif of residues which form the Trp pocket in L1 does not occur in L2 of the IR/IGF-IR family. However, in EGFR, which has a second cys-rich region after the L2 domain, the motif can be found in both the L1 and L2 domains and the Trp is conserved in the first module of both cys-rich regions.[39] It appears to contribute to the stability of the L2 domain of the EGFR since a construct that extends to the end of the first module of the second cys-rich region (sEGFR501) binds ligand with high affinity while the shorter fragment (sEGFR476), which lacks this cys-rich module, fails to bind ligand.[68]

Recently, evidence has been presented to show that the L domains are members of the leucine-rich repeat superfamily.[69] Multiple sequence alignments, coupled with the 3 D structure of the L1 and L2 domains of the IGF-IR,[39] enabled the residues equivalent to the eight conserved positions in the known LRR motif, LxxLxLxxNx-Lxx-Lxx-Lxx-Lxx-, to be identified (Fig. 4). Isoleucine (or valine) is

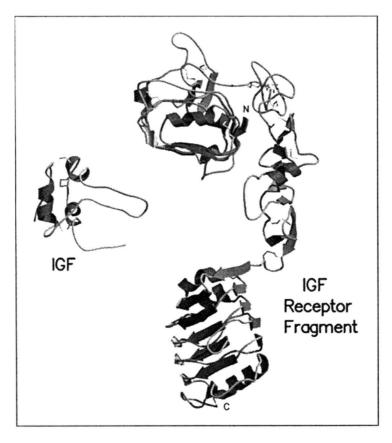

Figure 3. Polypeptide fold for residues 1-459 of the human IGF-I receptor and 1-70 of IGF-1. The L1 domain is at the top viewed from the N-terminal end. Helices are depicted as curled ribbons and β-strands as broad arrows. Based on refs 7 and 39.

preferred over leucine at some positions of the repeat and the L domains of the IR and EGFR families contain five full repeats with the 6th partially truncated.[69]

The Cys-rich Domain

As anticipated,[38] the cys-rich domain is composed of modules with disulfide bond connectivities resembling parts of the TNF receptor[56] and laminin[57] repeats. The first module sits at the end of L1 domain while the remaining seven form a curved rod running diagonally across L1 and reaching to L2 (Fig. 3). The strands in modules 2-7 run roughly perpendicular to the axis of the rod[39] in a manner more akin to laminin than to the TNF receptor, where the strands run parallel to the axis (Fig. 5). The modular arrangement of the IGF-IR cys-rich domain is different to other cys-rich proteins for which structures are known.[39] The first three modules of IGF-IR have a common core, containing a pair of disulfide bonds, but show considerable variation in the loops. These modules have been referred to as C2 (two disulfide bonds).[51,52] The connectivity of the cysteines is the same as the first part of an EGF motif (Cys 1-3 and 2-4). Modules 4 to 7 have a single disulfide bond and have been referred to as C1 modules.[51,52] Each C1 module is composed of three polypeptide strands, the first and third being disulfide bonded and the second and third forming a β-ribbon. The β-ribbon of each β-finger (or C1) module lines up antiparallel to form a tightly twisted eight-stranded β-sheet.[39] Module 6 deviates from the common pattern with the first segment being replaced by an α-helix followed by a large loop that is implicated in ligand binding (see below). As modules 4-7 are similar,

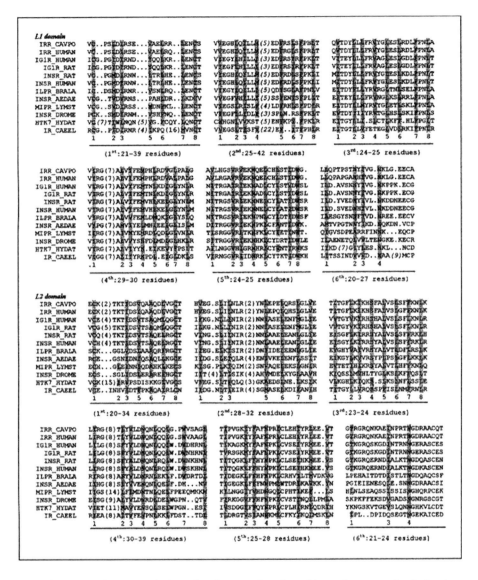

Figure 4. Leucine-rich repeats in the L1 and L2 domains of members of the insulin receptor family. Residues equivalent to conserved positions 1 to 8 in the sequence motif LxxLxLxxNxLxxLxxLxxLxx are shaded. The INSR_MOUSE is not shown as it differs from INSR_RAT at only Ser139Tyr in the 6th repeat of the L1 domain and Ser415Asn in the 4th repeat of the L2 domain. These two positions are not conserved in the LRR motif. . Reprinted with permission from: Ward CW and Garrett TPJ. BMC Bioinformatics. 2001; 2:4. BioMed Central.

it is possible that they arose from a series of gene duplications. The final (eighth) module is a disulfide-linked bend of five residues.[39]

The Fibronectin Type III Domains

The FnIII domain is one of the most common structural modules found in many proteins, including membrane-anchored receptors. 3D structures have been reported for several such

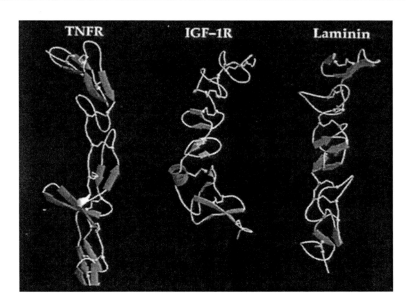

Figure 5. Comparisons of the folds and arrangement of modules 2-8 in the cys-rich region of the IGF-1R[39] with the four repeats (eight modules) in the extracellular domain of the tumor necrosis factor receptor (TNFR)[56] and the first three repeats (nine modules) of laminin.[57] The module arrangements are: C2-C2-C2-C1-C1-C1-C1-C1' in IGF-1R; (C1-C2)$_4$ in the TNFR and (C2-C1-C1)$_{12-16}$ in laminin. Reprinted with permission from: Ward CW, Garrett TPJ, McKern NM, Lou M, Cosgrove LJ, Sparrow LG, Frenkel MJ, Hoyne PA, Elleman TC, Adams TE, Lovrecz GO, Lawrence LJ. and Tulloch PA. Molecular Pathology, 2001; 54:125-132.© 2001 BMJ Publishing Group, London.

domains and have the fold and topology shown in Figure 6. The FnIII domain is relatively small (~100 residues) and has a fold similar to that of immunoglobulins but with a distinctive sequence motif. The domain consists of a seven-stranded β sandwich in a three-on-four (EBA:GFCC') topology. Its main functions appear to be to mediate protein-protein interactions including ligand binding and to act as spacers to correctly position functionally important regions of extracellular proteins.

O'Bryan et al[40] were the first to describe the existence of FnIII domains in members of the IR subfamily following their cloning and characterization of the tyrosine kinase axl. Their sequence alignments and descriptions covered the two C-terminal FnIII domains in the IR subfamily and these were given structural assignments by Schaffer et al[65] following comparisons with the FnIII modules present in the growth hormone receptor. Recently it has been shown that members of the IR subfamily contain an additional FnIII domain in the region previously referred to as the connecting domain.[41,42,43] This first FnIII domain (equivalent to residues 461-579 in IGF-IR) is 118-122 residues long, while the second FnIII domain (equivalent to 580-798 in IGF-IR) has a major insert of 120 to 130 amino acids. The third FnIII domain (equivalent to residues 799-901 in IGF-IR) is of normal size. The different authors differ in the residues assigned to the seven β-strands in each of these FnIII domains, with the regions of greatest disagreement being the locations of the C' and E strands in all three modules (Table 1).

3D Structure of the Tyrosine Kinase Domain

The crystal structure of the unphosphorylated (basal state) tyrosine kinse domain from the closely related hIR was solved by Hubbard et al[48] and is shown in Figure 7. Like the protein serine kinases,[50] the insulin receptor kinase is composed of two lobes with a single connection between them. The N-terminal lobe comprises a twisted β-sheet of five antiparallel β-strands (β1-β5) and one α-helix (αC). The larger C-terminal lobe comprises eight α-helices (αD, αE, αEF, αF, αG, αH, αI, αJ) and four β-strands (β7, β8, β10, β11).[48] The hIR kinase lacks β-strands 6 and 9 present

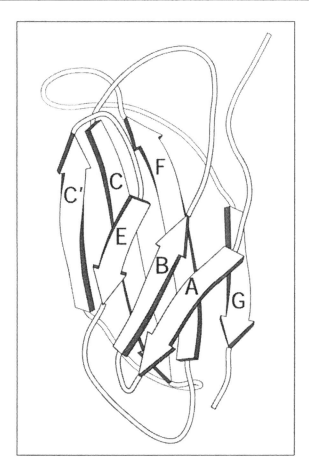

Figure 6. Polypeptide fold of the 10th type III repeat of fibronectin.[107] The seven β-strands that comprise the two sheet sandwich are labelled A, B, C, C', E, F and G. Reprinted with permission from: Adams TE, Epa VC, Garrett TPJ and Ward CW. Cell Mol Life Sci 2000; 57:1050-1093. © 2000 Birkhäuser Verlag, Basel.

in cAMP protein kinase.[50] In the unactivated kinase, one of the three tyrosines in the activation loop, Tyr1162, is bound in the active site but cannot be phosphorylated (in cis) because part of the A-loop interferes with the ATP binding site and the catalytic Asp1150 is improperly positioned to co-ordinate MgATP.[48] On activation, autophosphorylation of Tyr1162, Tyr1158 and Tyr1163 occurs in trans by the kinase domain of the second monomer. Thus in the basal state, Tyr1162 competes with the neighboring β-chain and other protein substrates, for binding to the active site, but is not cis-phosphorylated because of steric constraints that prevent simultaneous binding of Tyr1162 and MgATP.[48] The structure of the activated phosphorylated IR kinase reveals that autophosphorylation of the three tyrosines in the A-loop, leads to a dramatic change in its configuration.[49] In the phosphorylated state the A-loop is displaced by approximately 30 Å resulting in unrestricted access to the binding sites for ATP and protein substrates.[49] This movement facilitates the proper spatial arrangement of Lys1030 and Glu1047, the residues involved in MgATP coordination and Asp1150 of the highly conserved Asp-Phe-Gly triad.[49] The loop A rearrangement also leads to closure of the N and C-terminal lobes, which is necessary for productive ATP binding.[49] This closure involves significant rotation of the N-terminal lobe as shown in Figure 7.

Table 1. Comparison of alignments of the three fibronectin type-III domains in the hIGF-IR

A	B	C	C'	E	F	G	Ref.
FnIII-1							
462-466	475-480	492-499	**520-524**	**535-538**	547-554	572-574	[41]
461-467	476-482	492-498	**519-523**	**526-531**	545-553	565-573	[42]
463-469*	474-480	492-500	505-510	518-522	546-554	563-569	[43]
nd	nd	nd	nd	nd	nd	nd	[65]
FnIII-2							
nd	nd	nd	nd	nd	nd	nd	[41]
585-591	598-604	613-619	627-631	**754-759**	767-775	784-793	[42]
587-593	596-602	613-621	624-629	636-639	768-776	782-788	[43]
586-593	595-602	609-616	619-627	**632-636**	772-780	781-788	[65]
FnIII-3							
nd	nd	nd	nd	nd	nd	nd	[41]
806-812	819-825	**834-840**	848-854	857-862	869-877	885-895	[42]
808-814	817-823	834-841	843-848	855-859	870-878	885-891	[43]
807-814	816-823	830-837	845-853	859-865	869-881	885-897	[65]

The seven β-strands in each FnIII module are A, B, C, C', E. F. G. Regions of greatest diveregence between the four reports are shown in bold. Reprinted with permission from: Adams TE, Epa VC, Garrett TPJ and Ward CW. Cell Mol Life Sci 2000; 57:1050-1093. © 2000 Birkhäuser Verlag, Basel.

The IGF-IR and IR Ectodomain Dimers

The major feature which separates the IGF-IR and other members of the IR family from most other receptor families is that they exist on the cell surface as disulfide-linked dimers and require domain re-arrangements rather than receptor oligomerization for cell signalling. There is currently no high resolution 3D structure to reveal how the various domains are organized in the dimeric, native receptor. The first clues have come from electron microscopy (EM) of single-molecule images of the hIGF-IR ectodomain and, more particularly, the hIR ectodomain and its complexes with three different monoclonal antibody-derived Fab fragments.[70] These images show that the hIGF-IR and hIR ectodomains resemble a U-shaped prism of approximate dimensions 90 x 80 x 120 Å. The images show clearly the dimeric structure of these ectodomains, with the length of the images ~80 Å along each bar, and the width of ~90 Å across the two bars. The width of the cleft (assumed membrane-distal) between the two side arms is ~30 Å, sufficient to accommodate ligand.[70]

Fab molecules from the monoclonal antibody 83-7 bound hIR half-way up one end of each side arm in a diametrically opposite manner, indicating a two-fold axis of symmetry normal to the membrane surface.[70] Mab 83-7 recognizes an epitope between residues 191 and 297 in the cys-rich region of hIR.[71,72] Examination of the location of the sequence differences between mouse and human IRs and the EM images suggests residues 210 in the third cys-rich module and 236 in the fourth cys-rich module form part of the 83-7 epitope. They are located at one corner of the fragment in line with the centre of the L1 domain (Fig. 3). This suggests that the L1/cys-rich/L2 fragment spans the cleft between the parallel bars rather than lying within each parallel bar. Fabs 83-14 and 18-44, which have been mapped respectively to the first FnIII repeat (residues 469-592) and residues 765-770 in the insert domain,[73] bound near the base of the prism at opposite corners.[70]

The single molecule images, together with the 3D structure of the first three domains of hIGF-IR,[39] suggest that the ectodomain dimer is organized into two layers. The L1/cys-rich/L2 domains are suggested to occupy the upper (membrane-distal) region of the U-shaped prism with the Fn III domains, the insert domains (and the disulfide bonds involved in dimer formation)

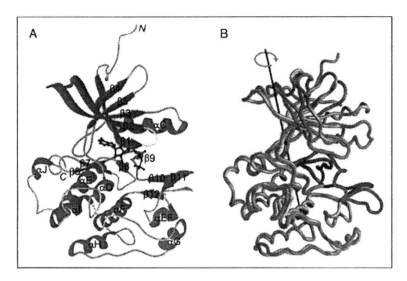

Figure 7. Polypeptide fold for the tyrosine kinase domain of the insulin receptor. (A) Ribbon diagram for the active (phosphorylated) form of the tyrosine kinase (see ref. 49 for colors). (B) Superposition of the C-terminal lobes of the phosphorylated form[49] (green and blue) and non-phosphorylated form[48] (orange and red) of the tyrosine kinase. Reprinted with permission from: Hubbard SR. EMBO J 1997; 16:5572-5581© 1997 Oxford University Press.

located predominantly in the membrane-proximal region.[70] The nature of the interaction between the two L1/cys-rich/L2 fragments in the ectodomain dimer is not clear from these EM studies. However, some clues to these domain associations may be gained by examining the surface properties of the L1/cys-rich/L2 fragment. Surfaces that are involved in the dimer interface would be expected to have complementary properties. In addition, the surfaces involved in either ligand binding or the interface with the bottom layer of fibronectin type III and insert modules, would be expected to be more conserved than those regions that are exposed on the surface of the dimer. The distribution of amino acid sequence conservation and the electrostatic potential of the IGF-IR L1/cys-rich/L2 domains have been mapped on to the molecular surface of the IGF-1R L1-cys rich-L2 domain structure allowing some predictions of domain association to be made.[51]

High-resolution data is required to establish the precise arrangement of the 14 modules that make up the ectodomain dimer and the way they interact with the respective ligands to generate signal transduction. Recently whole receptors solubilized from human placental membranes have been examined by electron cryomicroscopy and 3D reconstruction performed using a library of 700 images.[74,75,76] Gold-labeled insulin was used to locate the insulin-binding domain. The images seen were compact and globular, measuring 150 Å in diameter. Some domain-like features became evident at intermediate-density thresholds, which indicated a strong two-fold vertical rotational symmetry. When this symmetry was applied to the reconstruction, some structural features became evident. The overall model showed the L1/cys-rich/L2 domains arranged in an antiparallel manner and at an angle to each other when viewed from the side. The six FnIII domains and the two L2 domains are placed in a central band with the two tyrosine kinase domains at the base of the model.[74,75,76] The images described in these two studies[70,74,75,76] are substantially different from the "T"-, "X"- or "Y"- shaped objects reported for recombinant ectodomain,[65] detergent-solubilized whole receptors or vesicle-reconstituted whole receptors.[77,78,79]

Ligand Binding by Receptor Chimeras

The results of binding studies with IR/IGF-IR chimeras are summarized in Figure 8 and indicate that the determinants of specificity for insulin and IGF-I binding reside in different regions of the two receptors. Data for whole receptor chimeras showed that residues 1-137 in the L1 domain of

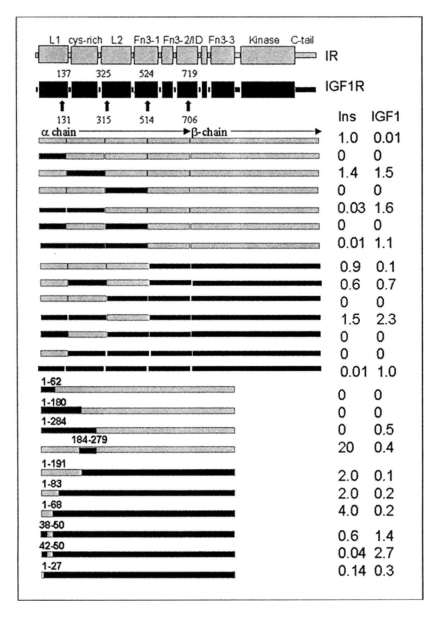

Figure 8. Schematic summary of IR/IGF-IR Chimeras. The domain organisation of the receptors is shown at the top. The approximate locations of the fragment boundaries exchanged in the whole receptor chimeras are shown, as are the α-β cleavage sites for hIR and hIGF-IR. The fragment composition of each chimera is shown by solid and open boxes. The residues substituted in the ectodomain IR-based or IGF-IR-based chimeras are indicated above each construct. The relative binding affinities of the parent receptors and the various chimeras are shown opposite each construct on the right hand side. Reprinted with permission from: Adams TE, Epa VC, Garrett TPJ and Ward CW. Cell Mol Life Sci 2000; 57:1050-1093. © 2000 Birkhäuser Verlag, Basel.

hIR and residues 325-524, comprising most of the L2 domain and part of the first fibronectin III domain of hIR, were important determinants of insulin binding while residues 131-315 in the IGF-IR (cys-rich plus flanking regions from L1 and L2), were prime determinants of IGF-I binding.[80,81] Further support to the importance of the cys-rich region in IGF-I binding is seen in the differential binding specificities of the hIR-based chimeric receptors containing residues 1-217 of IGF-IR and 1-274 of IGF-IR.[82] Finally, an hIR-based chimera where residues 450-601 were replaced with the corresponding residues from hIRR showed decreased insulin binding.[71]

The studies with ectodomain chimeras provide similar findings to those obtained with whole-receptor chimeras (Fig. 8). The hIR-based chimera containing IGF-IR residues 1-284 resembled the IGF-IR ectodomain and showed high affinity for IGF-I and poor binding for insulin.[83] The chimeras with smaller IGF-IR fragments, either 1-180 or 1-62, lack the appropriate specificity determinants for either ligand and showed poor binding of both ligands as expected.[84] The ectodomain chimera with IGF-IR residues 184-279 showed high affinity for both ligands, particularly insulin. The region controlling IGF-I specificity has been further narrowed to the 14 residues (amino acids 253-266) in the variable loop of module 6 of the cys-rich region.[85] The chimeric receptor where residues 260 to 277 of hIR were replaced with residues 253 to 266 from hIGF-IR was not significantly different from wild-type hIR in terms of insulin binding affinity, but was more amenable to displacement of radiolabelled insulin by IGF-I than the parent hIR.[85]

The importance of the N-terminal region in insulin binding was confirmed by examining a series of IGF-IR-based chimeric ectodomains[84] where the N-terminus contained decreasingly smaller proportions (191, 83 and 68 residues, respectively) of hIR-derived sequences. All showed similar binding affinities, binding insulin with comparable affinity to wild-type hIR while retaining relatively high (10-20%) binding affinity for IGF-I.[84,86] The 1-68/63-1337 IR/IGF-IR whole receptor chimera displayed relative ligand affinities similar to the corresponding ectodomain construct, validating the use of ectodomain constructs in such studies[86] and the IGF-IR chimera with only residues 38-50 from IR still bound insulin almost as well as the IR ectodomain.[87]

Residues 38-43 are predicted to lie in the second rung of the L1 β-helix domain[39] at the edge of the putative binding pocket (Fig. 3). The region 223-274 in IGF-IR, implicated in IGF-I specificity, contains major sequence differences when compared to hIR (Fig. 1). It corresponds to modules 4-6 in the cys-rich region and includes a large and somewhat mobile loop [residues 255-263, mean B(Cα atoms) = 57 Å^2] which extends into the central space (see Fig. 3). In hIR, this loop is four residues bigger, differs totally in sequence (Fig. 1) and is stabilized by an additional disulfide bond.[38,88] The improvement in IGF-I binding by the hIR cys-loop exchange chimera, hIR_CLX, suggests that the larger loop of hIR may serve to exclude IGF-I from the hormone binding site but allows the smaller insulin molecule to bind.[85] The third region implicated in insulin binding, residues 326-524, starts in the middle of the first rung of the L2 β-helix domain[39] and extends to Cys514 in the middle of the first Fn III domain (Fig. 1).

Effect of Point Mutations on Ligand Binding

As chimeras only address residues which differ between the two receptors, a more precise analysis of binding determinants can be obtained from single-site mutants. Alanine scanning mutagenesis has been carried out on three distinct regions of the receptor that have been implicated in ligand binding. The first region examined was the L1 domain where 29 residues in the second and third β-sheets of this domain were mutated.[89,90] Two mutants Tyr54 and Thr93 failed to yield detectable protein. Of the other 27 Ala mutants, 10 caused a significant impairment of IGF-1 binding as summarized in Table 2. The greatest effect was seen with the Phe90 mutant, which showed a 23-fold reduction in affinity while mutations of Asp8, Asn11, Tyr28, His30, Leu33, Leu56, Phe58 and Arg59 showed reductions in affinity of 3 to 9 fold.[89,90] The second region examined was the cys-rich region where 25 residues, predicted to be accessible to ligand on the basis of the IGF-1R fragment 3D structure,[39] were mutated to alanine.[90] The region scanned was residues 240-284 and the results are summarized in Table 2. Only 4 of the mutants produced significant decreases in affinity for IGF-I.[90] Three of these (residues 240-242) are located in cys-rich module 5 while Phe251 is at the start of module 6.[39]

The residues in the L1 domain (10 residues) and the cys-rich region (four residues) implicated in IGF-I binding are located in two discontinuous regions.[90] The first site includes Asp8, Asn11,

Table 2. Effect of Ala mutations on ligand binding by hIGF-IR

Residue Mutated	K_dMut/K_dWT	Residue Mutated	K_dMut/K_dWT
L1 domain		**Cys-rich region**	
wild type receptor	1	Arg240	2
Asp8	9	Phe241	6
Asn11	7	Glu242	4
Tyr28	4.5	Phe251	2.2
His30	4.5		
Leu33	6	**α-chain C-terminus**	
Leu56	5	Phe692	30
Phe58	3	Glu693	10
Arg59	5	His697	10
Trp79	3	Asn698	10
Phe 90	23	Asn694	12
		Leu696	25
		Ile700	27
		Phe701	>120

Based on ref 90

Tyr28, His30, Leu33, Leu56, Phe58, Arg59 and Phe90 which are distributed across the first four repeats of the L1 domain (Fig. 9) and form a footprint on the second β-sheet which faces the central cavity of the 3D structure of IGF-1R fragment (Fig. 10). The residues implicated in insulin binding to the insulin receptor by alanine scanning mutagenesis[89] occur in a similar but not identical location over the first four repeats of the hIR L1 domain (Fig. 9). The second site consists of Trp79 from the L1 domain loop and Arg240, Phe241, Glu242 and Phe251 from the cys-rich region. These residues form a small patch on the cys-rich domain[90] adjacent to and in the same plane as the L1 footprint, providing an extended flat face as shown in Figure 10. Ala mutations of residues in the negatively charged region 255-284 (Fig. 1) had negligible effects on binding.[90] This is in contrast to the effects of exchanging loop residues 260-277 of hIR with the corresponding loop residues 253-266 from hIGF-1R, which increased the capacity of IGF-1 to diplace bound insulin by more than 10 fold.[85]

The third region subjected to alanine mutations was the eleven residue sequence at the C-terminal end of the α-chain of hIGF-IR.[90] The results are summarized in Table 2 and reveal that this region of the receptor appears to provide the majority of the free energy of the interaction between the receptor and IGF-1.[90] Three mutations (Phe695, Ser699 and Val702) had no effect on IGF-I binding. Mutation at Phe701 produced a receptor with no detectable IGF-I binding, while mutation at Phe692, Glu693, Asn694, Leu696, His697, Asn698 and Ile700 resulted in decreases in affinity for IGF-I ranging from 10 to 29 fold (Table 2).[90]

The importance of Phe701 was reinforced in the studies of hIR- or hIGF-IR-based chimeric minireceptors.[91] Swapping the carboxy terminal domains (16 amino acids) with that from human hIRR completely abolished insulin and IGF-I binding by either hIR- or hIGF-IR-based minireceptor chimeras, while chimeras involving the carboxy terminal domains of hIR or hIGF-IR were little affected.[91] Sequence comparisons of hIRR, hIR and hIGF-IR suggest the substitution of Thr for Phe at the position equivalent to 701 in IGF-IR is responsible for this loss of ligand binding by the chimeras containing the carboxy-terminal peptide from the hIRR α-chain.

IGF Mutations

IGF-I and -II contain two extra regions compared to insulin; the C region between the B and A domains and the D region at the C-terminus (Fig. 11). Replacing residues 1-16 of IGF-I with the

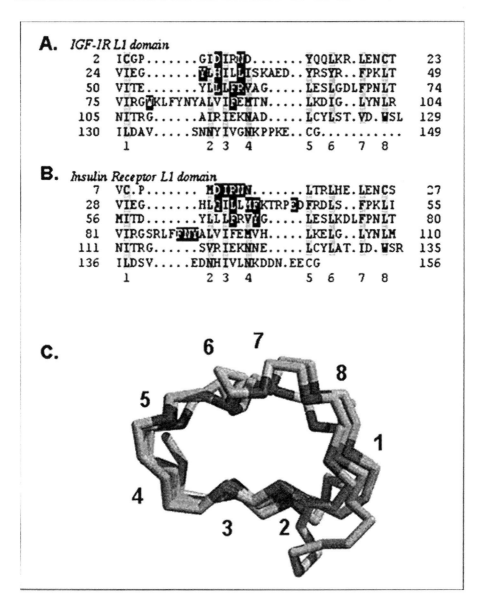

Figure 9. Location of the functional epitopes for IGF-1 and insulin binding on the leucine-rich repeats of the L1 domains of the IGF-1R and IR. The residues corresponding to the eight conserved positions in the leucine rich repeat motif[69] are shaded gray. The residues involved in ligand binding are shaded black. (A) Location of the 10 residues from the L1 domain that are involved in the binding of IGF-I to the IGF-1R[90] (see Table 2). (B) Location of the 14 residues involved in insulin binding to IR.[89] (C) Location of the eight conserved residues (numbered 1 to 8) in the LRR motif (shaded gray in panel A) on the polypeptide fold of the IGF-1R L1 domain.[69]

first 17 residues of the insulin B-chain, which involves 10 sequence differences (Fig. 11), gave only a two-fold reduction in IGF-IR binding.[92] Changing just two of these residues in IGF-I, to the corresponding sequence in insulin (Gln15Tyr/Phe16Leu), increased affinity for IR by 10-fold but

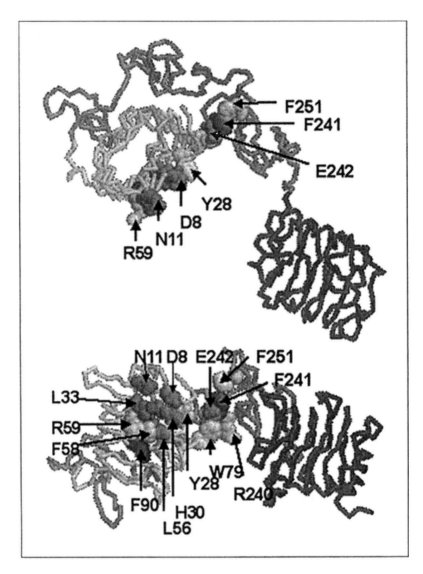

Figure 10. Location of the functional epitopes for IGF-1[90] on the polypeptide fold of the hIGF-IR L1/cys-rich/L2 fragment[39]. The L1 domain is coloured white, the cys-rich region is coloured cyan and the L2 domain is coloured magenta. The residues in the L1 domain and the cys-rich region that have been implicated in IGF-1 binding[90] (see Table 2) are shown in spacefill and are coloured red (residues 90 and 241), orange (residues 8, 11, 30, 33, 56 and 242) and yellow for ease of identification. The top panel is the view down the axis of the L1 domain. The bottom panel is a front on view where the image has been rotated 90 degrees around the x-axis.

had no change in affinity for IGF-IR or IGF-IIR.[92] A different double mutant, Glu3Gln/Thr4Ala, showed normal IGF1 binding to IR, IGF-IR and IGF-IIR.[92]

Residues in the IGF-I A-region are important for binding to the type 2 receptor and to the IGFBPs, but not the type 1 receptor. Replacing the A-domain residues 42-56 of IGF-I with A1-15 of insulin, together with a Thr41Ile mutation (eight differences, Fig. 11) had no effect on IGF-IR

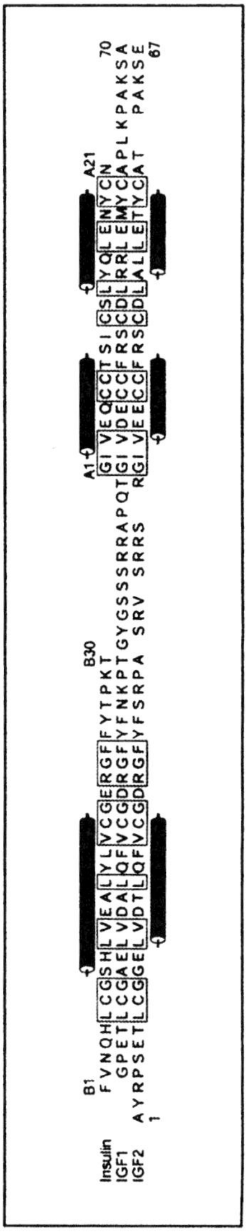

Figure 11. Sequence alignments of human insulin and human IGF-I and IGF-II. The secondary structural assignments are depicted as cylinders for α-helices. The assignments for insulin are shown above the sequences while the assignments for IGF-II are below. The secondary structure has been assigned using the 3D structures of insulin[57] and IGF-II.[5] Reprinted with permission from: Adams TE, Epa VC, Garrett TPJ and Ward CW. Cell Mol Life Sci 2000; 57:1050-1093. © 2000 Birkhäuser Verlag, Basel.

binding, reduced IGF-IIR binding 20-fold and increased IR binding by seven-fold.[93] The areas of greatest divergence between the A domains of insulin and the IGFs are residues 49-51 and 55-56 (Fig. 11). Replacement of residues 49-51 with the corresponding sequence from insulin had no effect on IGF-IR or IR binding, but reduced IGF-IIR binding by 20-fold.[93] The double mutant Arg55Tyr/Arg56Gln, where the residues in IGF-I are replaced by the corresponding residues in the insulin sequence, had no effect on IGF1-R or IR binding but resulted in a seven-fold increase in affinity for the type 2 receptor.[93] The triple exchange mutant Arg50Ser/Arg55Tyr/Arg56Gln in a recombinant construct gave a similar result, with no effects on IGF-IR or IR binding. [94]

The structure of IGF-I is very sensitive to amino acid substitutions, with the loss of binding due to either direct effects on receptor interactions or indirect effects on ligand structure.[95] Residues implicated in IGF-IR binding are Phe23 and Tyr24 in the B-region, Tyr31 in the C-domain and Tyr60 in the A-region of IGF-I[96,97] and the corresponding residues Phe26 and Tyr27 in IGF-II.[98,99] Tyr24 and Phe25 can be changed to the Phe.Tyr sequence found in insulin without affecting IGF-IR, IGF-IIR and IR binding.[96] Tyr24, Tyr31 and Tyr60 are protected from iodination when bound to IGF-IR.[100] Residues Phe23, Tyr24 and Tyr31 form part of a hydrophobic patch on the surface of IGF-I while Tyr60 is largely buried and unlikely to interact directly with the IGF-IR. Similarly, the reduction in IGF-IR binding by a Phe16Ala mutant is concluded to be due to the loss of structural integrity rather than direct binding interactions.[95] The mutations Ala8Leu and Met59Phe, on opposite sides of the IGF-I molecule, decreased binding to IGF-IR and IR ectodomains five- to six-fold and 17- to 28-fold, respectively.[101] Asp12, which is adjacent to Ala8 on the IGF-I surface, appears to contribute to IGF-IR binding, as the Asp12Ala mutant showed a four-fold reduction in affinity.[95] In addition, an Ala62Leu mutant showed eight-fold and two-fold reductions in binding to the IGF-IR and IR ectodomains, respectively.[101] The B-domain helix mutants, Val11Ala, Val11Thr, Gln15Ala and Gln15Glu, showed smaller (1.5 to three-fold) reductions in affinity, which have been correlated with structural defects such as the loss of α-helix content in the mutant ligand.[95,102] Finally, the A-chain helix mutant of IGF-II, Val43Leu, showed a 16-fold reduction in IGF-IR binding, and a 220-fold reduction in IR binding, but normal binding to the IGF-IIR.[99] The corresponding mutant in IGF-I has not been examined.

One notable feature of IGF-I and -II is the large number of charged residues and their uneven distribution over the surface. Replacement of the C-region of IGF-I by a four Gly linker reduced affinity for IGF-IR by 40-100-fold, with a 10-fold or no reduction in binding to IR.[103,104] An IGF-I analogue in which residues 29-41 of the C-region have been deleted (mini IGF-I) showed no affinity for either receptor, indicating that the C-region of IGF-I contributes directly to the free energy of binding to the IGF-IR.[104] These authors suggested that binding of IGF-I to the IGF-IR resembles insulin/IR binding, and involves a conformational change in which the C-terminal B-region residues are displaced from the body of the molecule to expose the underlying A-region residues. It has been shown that deletion of the C-region of IGF-I results in a substantial tertiary structural rearrangement that can account for the loss of receptor affinity.[105] Truncation of the nearby D peptide in IGF-II reduced IGF-IR binding six-fold,[98] while the corresponding truncation of IGF-I had no effect.[99] Ala mutants have implicated Arg21, Arg36 and Arg37 in IGF-IR binding while the Arg50, Arg55 and Arg56 mutations had smaller effects,[106] in agreement with the IGF-I/insulin sequence exchanges involving Arg50, Arg55 and Arg56.[93,94] The region of the receptor responsible for recognizing Arg36Arg37 was shown to be the cys-rich region 217-284.[94] The putative binding site of the receptor, which incorporates these residues, has a sizeable patch of acidic residues in the corner where the cys-rich domain departs from L1. Other acidic residues which are specific to this receptor are found along the inside face of the cys-rich domain and the loop (residues 255-263) extending from module 6. Thus it is possible that electrostatics play an important part in IGF-I binding, with the C-region binding to the acidic patch of the cys-rich region near L1, and the acidic patch on the other side of the hormone directed towards a small patch of basic residues (residues 307-310) on the N-terminal end of L2.[39,51]

Concluding Remarks

The weight of experimental and, more recently, epidemiological evidence points to deregulated signalling through the IGF-IR as a contributing factor in the pathogenesis of some cancers. With respect to the IGF-IR, a number of experimental strategies, particularly those based on antisense technology, offer clinical promise. The recent determination of the 3D structures of the IGF-IR kinase domain[47] and the L1/cys rich/L2 domain fragment of the IGF-IR[39] provide high resolution templates for the development of selective tyrosine kinase inhibitors or small molecule antagonists of receptor function. Clearly the goal is to obtain atomic resolution data for the whole IGF-IR ectodomain in complex with ligand to elucidate the precise nature of ligand/receptor interactions. Ultimately, the goal is to obtain 3D structural information of the high affinity complex, to reveal the conformational changes associated with ligand binding and signal transduction.

References

1. LeRoith D, Werner H, Beitner-Johnson D et al. Molecular and cellular aspects of the insulin-like growth factor I receptor. Endocr Rev 1995; 16:143-163.
2. Baker J, Liu JP, Robertson EJ et al. Role of insulin-like growth factors in embryonic and postnatal growth. Cell 1993; 75:73-82.
3. Frasc F, Pandini G, Scalia P et al. Insulin receptor isoform A, a newly recognized, high-affinity insulin-like growth factor II receptor in fetal and cancer cells. Mol Cell Biol 1999; 19:3278-3288.
4. Louvi A, Accili D, Efstratiadis A. Growth-promoting interaction of IGF-II with the insulin receptor during mouse embryonic development. Devel Biol 1997; 189:33-48.
5. Torres AM, Forbes BE, Aplin SE et al. Solution structure of human insulin-like growth factor II—relationship to receptor and binding protein interactions. J Mol Biol 1995; 248:385-401.
6. Vajdos FF, Ultsch M, Schaffer M et al. Crystal structure of human insulin-like growth factor-1: determinant binding inhibits binding protein interactions. Biochemistry 2001; 40:11022-11029.
7. Ciszak E, Smith GD. Crystallographic evidence for dual coordination around zinc in the T_3R_3 human insulin hexamer. Biochemistry 1994; 33:1512-1517.
8. Kitimura T, Kido Y, Nef S et al. Preserved pancreatic β-cell development and function in mice lacking the insulin receptor-related receptor. Molec Cell Biol 2001; 21:5624-5630.
9. Sell C, Dumenil G, Deveaud C et al. Effect of a null mutation of the insulin-like growth factor I receptor gene on growth and transformation of mouse embryo fibroblasts. Mol Cell Biol 1994; 14:3604-3612.
10. Kaleko M, Rutter WJ, Miller AD. Overexpression of the human insulin-like growth factor I receptor promotes ligand-dependent neoplastic transformation. Mol Cell Biol 1990; 10:464-473.
11. Baserga R, Hongo A, Rubini M et al. The IGF-I receptor in cell growth, transformation and apoptosis. Biochim Biophys Acta 1997; 1332:F105-F126.
12. Megyesi K, Kahn CR, Roth J et al. The NSILA-s receptor in liver plasma membranes. Characterization and comparison with the insulin receptor. J Biol Chem 1975; 250:8990-8996.
13. Marshall RN, Underwood LE, Voina SJ et al. Characterization of the insulin and somatomedin-C receptors in human placental cell membranes. J Clin Endocrinol Metab 1974; 39:283-292.
14. Bhaumick B, Bala RM, Hollenberg MD. Somatomedin receptor of human placenta: solubilization, photolabeling, partial purification, and comparison with insulin receptor. Proc Natl Acad Sci USA 1981; 78:4279-4283.
15. Chernausek SD, Jacobs S, Van Wyk JJ. Structural similarities between human receptors for somatomedin C and insulin: analysis by affinity labeling. Biochemistry 1981; 20:7345-7350.
16. Jacobs S, Kull Jr FC, Cuatrecasas P. Monensin blocks the maturation of receptors for insulin and somatomedin C: identification of receptor precursors. Proc Natl Acad Sci USA 1983; 80:1228-1231.
17. Jacobs S, Kull Jr FC, Earp HS et al. Somatomedin-C stimulates the phosphorylation of the beta-subunit of its own receptor. J Biol Chem 1983; 258:9581-9584.

18. Rubin JB, Shia MA, Pilch PF. Stimulation of tyrosine-specific phosphorylation in vitro by insulin-like growth factor. Nature 1983; 305:438-440.
19. Ullrich A, Gray A, Tam AW et al. Insulin-like growth factor 1 receptor primary structure: comparison with insulin receptor suggests structural determinants that define functional specificity. EMBO J 1986; 5:2503-2512.
20. Ullrich A, Bell JR, Chen EY et al. Human insulin receptor and its relationship to the tyrosine kinase family of oncogenes. Nature 1985; 313:756-761.
21. Ebina Y, Ellis L, Jarnagin K et al. The human insulin receptor cDNA: the structural basis for hormone-activated transmembrane signalling. Cell 1985; 40:747-758.
22. Shier P, Watt VM. Primary structure of a putative receptor for a ligand of the insulin family. J Biol Chem 1989; 264:14605-14608.
23. Abbott AM, Bueno R, Pedrini MT et al Insulin-like growth factor I receptor gene structure. J Biol Chem 1992; 267:10759-10763.
24. Yee D, Lebovic GS, Marcus RR et al. Identification of an alternate type I insulin-like growth factor receptor beta subunit mRNA transcript. J Biol Chem 1989; 264:21439-21441.
25. Condorelli G, Bueno R, Smith RJ. Two alternatively spliced forms of the human insulin-like growth factor I receptor have distinct biological activities and internalization kinetics. J Biol Chem 1994; 269:8510-8516.
26. Cosgrove L, Lovrecz GO, Verkuylen A et al. Purification and properties of insulin receptor ectodomain from large-scale mammalian cell culture. Protein Express Purif 1995; 6:789-798.
27. Hedo JA, Kasuga M, Van Obberghen E et al. Direct demonstration of glycosylation of insulin receptor subunits by biosynthetic and external labeling: evidence for heterogeneity. Proc Natl Acad Sci USA 1981; 78:4791-4795.
28. Hedo JA, Gorden P. Biosynthesis of the insulin receptor. Horm Metab Res 1985; 17:487-490.
29. Herzberg VL, Grigorescu F, Edge AS et al. Characterization of insulin receptor carbohydrate by comparison of chemical and enzymatic deglycosylation. Biochem Biophys Res Commun 1985; 129:789-796.
30. Collier E, Gorden P. O-linked oligosaccharides on insulin receptor. Diabetes 1991; 40:197-203.
31. Collier E, Carpentier JL, Beitz L et al. Specific glycosylation site mutations of the insulin receptor alpha-subunit impair intracellular transport. Biochemistry 1993; 32:7818-7823.
32. Leconte I, Carpentier JL, Clauser E. The functions of the human insulin receptor are affected in different ways by mutation of each of the four N-glycosylation sites in the beta subunit. J Biol Chem 1994; 269:18062-18071.
33. Wiese RJ, Herrera R,d Lockwood DH. Glycosylation sites encoded by exon 2 of the human insulin receptor gene are not required for the oligomerization, ligand binding, or kinase activity of the insulin receptor. Receptor 1995; 5:71-80.
34. Elleman TC, Frenkel MJ, Hoyne PA et al. Mutational analysis of the N-linked glycosylation sites of the human insulin receptor. Biochem J 2000; 347:771-779.
35. Caro LHP, Ohali A, Gorden P et al. Mutational analysis of the NH2-terminal gycosylation sites of the insulin receptor alpha-subunit. Diabetes 1994; 43:240-246.
36. Bastian W, Zhu J, Way B et al. Glycosylation of Asn397 or Asn418 is required for normal insulin receptor biosynthesis and processing. Diabetes 1993; 42:966-974.
37. Bajaj M, Waterfield MD, Schlessinger J et al. On the tertiary structure of the extracellular domains of the epidermal growth factor and insulin receptors. Biochim Biophys Acta 1987; 916:220-226.
38. Ward CW, Hoyne PA, Flegg RH. Insulin and epidermal growth factor receptors contain the cysteine repeat motif found in the tumor necrosis factor receptor. Proteins Struct Funct Genet 1995; 22:141-153.
39. Garrett TPJ, Mckern NM, Lou MZ et al. Crystal structure of the first three domains of the type-1 insulin-like growth factor receptor . Nature 1998; 394:395-399.
40. O'Bryan JP, Frye RA, Cogswell PC et al. axl, a transforming gene isolated from primary human myeloid leukemia cells, encodes a novel receptor tyrosine kinase. Mol Cell Biol 1991; 11:5016-5031.
41. Marino-Buslje C, Mizuguchi K, Siddle K et al. A third fibronectin type III domain in the extracellular region of the insulin receptor family. FEBS Lett 1998; 441:331-336.
42. Mulhern TD, Booker GW, Cosgrove L. A third fibronectin type-III domain in the insulin-family receptors. Trends Biochem Sci 1998; 23:465-466.
43. Ward CW. Members of the insulin receptor family contain three fibronectin type III domains. Growth Factors 1999; 16:315-322.
44. Hanks SK. Eukaryotic protein kinases. Curr Opin Struct Biol 1991; 1:369-383.
45. Surmacz E, Sell C, Swantek J et al. Dissociation of mitogenesis and transforming activity by C-terminal truncation of the insulin-like growth factor-I receptor. Exp Cell Res 1995; 218:370-380.
46. Hongo A, Dambrosio C, Miura M et al. Mutational analysis of the mitogenic and transforming activities of the insulin-like growth factor I receptor . Oncogene 1996; 12:1231-1238.
47. Pautsh A, Zoephel A, Ahorn H et al. IGF-1 receptor kinase domain PDB. 2001; ID:1JQH.
48. Hubbard SR, Wei L, Elis L et al. Crystal structure of the tyrosine kinase domain of the human insulin receptor. Nature 1994; 372:746-754.
49. Hubbard SR. Crystal structure of the activated insulin receptor tyrosine kinase with peptide substrate and ATP analog. EMBO J 1997; 16:5572-5581.
50. Knighton DR, Zeng J, Eyck LFT et al. Crystal structure of the catalytic subunit of cyclic adenosine monophosphate-dependent protein kinase. Science 1991; 253:407-414.

51. Adams TE, Epa VC, Garrett TPJ et al. Structure and function of the type-I insulin-like growth factor receptor. Cell Mol Life Sci 2000; 57:1050-1093.
52. Ward CW, Garrett TPJ, McKern NM et al. Structure of the insulin receptor family: unexpected relationships with other proteins. Today's Life Sciences 1999; 11:26-32.
53. Pashmforoush M, Chan SJ, Steiner DF. Structure and expression of the insulin-like peptide receptor from amphioxus. Mol Endocrinol 1996; 10:857-866.
54. Sparrow LG, Mckern NM, Gorman JJ et al. The disulfide bonds in the C-terminal domains of the human insulin receptor ectodomain. J Biol Chem 1997; 272:29460-29467.
55. Schaffer L, Ljungqvist L. Identification of a disulfide bridge connecting the alpha-subunits of the extracellular domain of the insulin receptor. Biochem Biophys Res Commun 1992; 189:650-653.
56. Banner DW, D'Arcy A, Janes W et al. Crystal structure of the soluble human 55 kd TNF receptor-human TNF-β complex: implications for TNF receptor activation. Cell 1993; 73:431-445.
57. Stetefeld J, Mayer U, Timpl R et al. Crystal structure of three consecutive laminin-type epidermal growth factor-like (LE) modules of laminin gamma-1 chain harboring the nidogen binding site. J Mol Biol 1996; 257:644-657.
58. Macaulay SL, Polites M, Hewish DR et al. Cysteine-524 is not the only residue involved in the formation of disulfide-bonded dimers of the insulin receptor. Biochem J 1994; 303:575-581.
59. Bilan PJ, Yip CCQ. Unusual insulin binding to cells expressing an insulin receptor mutated at cysteine 524. Biochem Biophys Res Commun 1994; 205:1891-1898.
60. Lu K, Guidotti G. Identification of the cysteine residues involved in the class I disulfide bonds of the human insulin receptor: properties of insulin receptor monomers. Mol Biol Cell 1996; 7:679-691.
61. Kabat EA, Wu TT, Perry HM et al. Sequences of Proteins of Immunological Interest. 5th ed. Bethesda: US Department of Health and Human Services, 1991.
62. Fernandez R, Tabarini D, Azpiazu N et al. The *Drosophila* insulin receptor homolog—a gene essential for embryonic development encodes two receptor isoforms with different sigaling potential . EMBO J 1995; 14:3373-3384.
63. Ruan YM, Chen C, Cao YX et al. The *Drosophila* insulin receptor contains a novel carboxy-terminal extension likelt to play an important role in signal transduction . J Biol Chem 1995; 270:4236-4243.
64. Cheatham B, Kahn CR. Cysteine 647 in the insulin receptor is required for normal covalent interaction between α- and β-subunits and signal transduction. J Biol Chem 1992; 267:7108-7115.
65. Schaefer EM. Erickson HP, Federwisch M et al. Structural organization of the human insulin receptor ectodomain. J Biol Chem 1992; 267:23393-23402.
66. Campbell ID, Spitzfaden C. Building proteins with fibronectin type III modules. Structure 1994; 2:333-337.
67. Goldstein BJ, Dudley AL. The rat insulin receptor: primary structure and conservation of tissue-specific alternative mRNA splicing. Mol Endocrinol 1990; 4:235-244.
68. Elleman TC, Domagala T, Mckern NM et al. Identification of a determinant of epidermal growth factor receptor ligand-binding specificity using a truncated, high-affinity form of the ectodomain. Biochemistry 2001; 40:8930-8939.
69. Ward CW, Garrett TPJ. The relationship between the L1 and L2 domains of the insulin and epidermal growth factor receptors and leucine-rich repeat modules. BMC Bioinformatics 2001; 2:4 [http://w.w.w.biomedcentral.com/1471-2105/2/4].
70. Tulloch PA, Lawrence LJ, McKern NM et al. Single-molecule imaging of human insulin receptor ectodomain and its Fab complexes. J Struct Biol 1999; 125:11-18.
71. Zhang B, Roth RA. A region of the insulin receptor important for ligand bimnding (residues 450-601) is recognized by patient's autoimmune antibodies and inhibitory monoclonal antibodies. Proc Natl Acad Sci USA 1991; 88:9858-9862.
72. Schaefer EM, Siddle K, Ellis L. Deletion analysis of the human insulin receptor ectodomain reveals independently folded soluble subdomains and insulin binding by a monomeric α-subunit. J Biol Chem 1990; 265:13248-13253.
73. Prigent SA, Stanley KK, Siddle K. Identification of epitopes on the human insulin receptor reacting with rabbit polyclonal antisera and mouse monoclonal antibodies. J Biol Chem 1990; 265:9970-9977.
74. Luo RZT, Beniac DR, Fernandes A et al. Quaternary structure of the insulin-insulin receptor complex. Science 1999; 285:1077-1080.
75. Ottensmeyer FP, Beniac DR, Luo RZT et al. Mechanism of transmembrane signaling: Insulin binding and the insulin receptor. Biochemistry 2000; 39:12103-12112.
76. Ottensmeyer FP, Beniac DR, Luo RZT et al. Mechanism of transmembrane signaling: Insulin binding and the insulin receptor—Correction. Biochemistry 2001; 40:6988-6988.
77. Christiansen K, Tranum-Jensen J, Carlsen J et al. A model for the quaternary structure of human placental insulin receptor deduced from electron microscopy. Proc Natl Acad Sci USA 1991; 88:249-252.
78. Tranum-Jensen J, Christiansen K, Carlsen J et al. Membrane topology of insulin receptors reconstituted into lipid vesicles. J Membrane Biol 1994; 140:215-223.
79. Woldin CN, Hing FS, Lee J et al. Structural studies of the detergent-solubilized and vesicle-reconstituted insulin receptor. J Biol Chem 1999; 274:34981-34992.
80. Schumacher R, Mosthaf L, Schlessinger J et al. Insulin and insulin-like growth factor-1 binding specificity is determined by distinct regions of their cognate receptors. J Biol Chem 1991; 266:19288-19295.
81. Schumacher R, Soos MA, Schlessinger J et al. Signaling-competent receptor chimeras allow mapping of major insulin receptor binding domain determinants. J Biol Chem 1993; 268: 1087-1094.

82. Gustafson TA, Rutter WJ. The cysteine-rich domains of the insulin and insulin-like growth factor I receptors are primary determinants of hormone binding specificity. J Biol Chem 1990; 265:18663-18667.
83. Andersen AS, Kjeldsen T, Wiberg FC et al. Changing the insulin receptor to possess insulin-like growth factor 1 ligand specificity. Biochemistry 1990; 29:7363-7366.
84. Kjeldsen T, Andersen AS, Wiberg FC et al. The ligand specificities of the insulin receptor and the insulin-like growth factor I receptor reside in different regions of a common binding site. Proc Natl Acad Sci USA 1991; 88:4404-4408.
85. Hoyne PA, Elleman TC, Adams TE et al. Properties of an insulin receptor with an IGF-1 receptor loop exchange in the cysteine-rich region. FEBS Lett 2000; 469:57-60.
86. Andersen AS, Kjeldsen T, Wiberg FC et al. Identification of determinants that confer ligand specificity on the insulin receptor. J BiolChem 1992; 267:13681-13686.
87. Kjeldsen T, Wiberg FC, Andersen AS. Chimeric receptors indicate that phenylalanine 29 is a major contributor to insulin specificity of the insulin receptor. J Biol Chem 1994; 269:32942-32946.
88. Schäffer L, Hansen PH. Partial characterization of the disulfide bridges of the soluble insulin receptor. Exp Clin Endocrinol Diabetes 1996; 104:89.
89. Mynarcik DC, Williams PF, Schaffer L et al. Identification of common ligand binding determinants of the insulin and insulin-like growth factor receptors—insights into mechanisms of ligand binding. J Biol Chem 1997; 272:18650-18655.
90. Whittaker J, Groth AV, Mynarcik D et al. Alanine scanning mutagenesis of a type-I insulin-like growth factor receptor ligand binding site. J Biol Chem 2001; 276:43980-43986.
91. Kristensen C, Wiberg FC, Andersen AS. Specificity of insulin and insulin-like growth factor receptors investigated using chimeric minireceptors. Role of carboxy terminal of receptor a-subunit. J Biol Chem 1999; 274:37351-37356.
92. Bayne ML, Applebaum J, Chicchi GG et al. Structural analogs of human insulin-like growth factor 1 with reduced affinity for serum binding proteins and the type 2 insulin-like growth factor receptor. J Biol Chem 1988; 263:6233-6239.
93. Cascieri MA, Chicci GG, Applebaum J et al. Structural analogues of human insulin-like growth factor (IGF) I with altered affinity for type 2 IGF receptors. J Biol Chem 1989; 264:2190-2202.
94. Zhang WG, Gustafson TA, Rutter WJ et al. Positively charged side chains in the insulin-like growth factor-1 C- and D-regions determine receptor binding specificity. J Biol Chem 1994; 269:10609-10613.
95. Jansson M, Uhlen M, Nilsson B. Structural changes in insulin-like growth factor (IGF) I mutant proteins affecting binding kinetics to IGF binding protein 1 and IGF-I receptor. Biochemistry 1997; 36:4108-4117.
96. Cascieri MA, Chicci GG, Applebaum J et al. Mutants of human insulin-like growth factor I with reduced affinity for the type I insulin-like growth factor receptor. Biochemistry 1988; 27:3229-3233.
97. Bayne ML, Applebaum J, Chicchi GG et al. Role of tyrosines 24, 31, and 60 in the high affinity binding of insulin-like growth factor-1 to the type 1 insulin-like growth factor receptor. J Biol Chem 1990; 265:15648-15652.
98. Roth BV, Burgisser DM, Luthi C et al. Mutants of human IGF2: expression and characterization of analogues with a substitution of Tyr27 and/or a deletion of residues 62-67. Biochem Biophys Res Commun 1991; 181:907-914.
99. Sakano K, Enjoh T, Numata F et al. The design, expression, and characterization of human insulin-like growth factor II (IGF-II) mutants specific for either the IGF-II/cation-independent mannose 6-phosphate receptor or IGF-I receptor. J Biol Chem 1991; 266:20626-20635.
100. Maly P, Luthi C. The binding sites of insulin-like growth factor I (IGF I) to type I IGF receptor and to a monoclonal antibody. Mapping by chemical modification of tyrosine residues. J Biol Chem 1988; 263:7068-7072.
101. Shooter GK, Magee B, Soos MA et al. Insulin-like growth factor (IGF)-I A- and B-domain analogues with altered type 1 IGF and insulin receptor binding specificities. J Mol Endocrinol 1996; 17:237-246.
102. Hodgson DR, May FEB, Westley BR. Mutations at positions 11 and 60 of insulin-like growth factor 1 reveal differences between its interactions with the type I insulin-like-growth-factor receptor and the insulin receptor . Eur J Biochem 1995; 233:299-309.
103. Bayne ML, Applebaum J, Underwood D et al. The C-region of human insulin-like growth factor (IGF) I is required for high affinity binding to the type I IGF receptor. J Biol Chem 1988; 264:11004-11008.
104. Gill R, Wallach B, Verma C, et al. Engineering the C-region of human insulin-like growth-factor-1—implications for receptor binding. Protein Engin 1996; 9:1011-1019.
105. Dewolf E, Gill R, Geddes S et al. Solution structure of a mini IGF-1. Protein Sci 1996; 5:2193-2202.
106. Jansson M, Andersson G, Uhlen M et al. The insulin-like growth factor (IGF) binding protein 1 binding epitope on IGF-1 probed by heteronuclear NMB spectroscopy and mutational analysis. J Biol Chem 1998; 273:24701-24707.
107. Dickinson CD, Veerapandian B, Dai X-P et al. Crystal structure of the tenth type III cell adhesion module of human fibronectin. J Mol Biol 1992; 236:1079-1092.

CHAPTER 2

Structure and Function of the IGF-1 and Mannose 6-Phosphate/IGF-2 Receptors

Susan L. Spence and Peter Nissley

Abstract

The IGF-1 receptor (IGF1R) gene is transcriptionally regulated by growth factors and tumor suppressors and encodes a 210 kDa $\alpha\beta$ chain which is proteolytically cleaved and disulfide bonded to produce an $\alpha_2\beta_2$ structure. IGF-1 binding to a cysteine-rich domain in the extracellular α subunit causes autophosphorylation of three tyrosines in the activation loop of the tyrosine kinase domain in the cytoplasmic portion of the β subunit, which results in amplification of tyrosine kinase activity and further autophosphorylation of additional tyrosine residues. These phosphotyrosine containing motifs are binding sites for adaptor and effector molecules in receptor signaling pathways. Binding of the adaptor protein Shc and the large docking protein IRS-1 to receptor phosphotyrosine 950 results in activation of MAP kinase and PI3 kinase/Akt, major signaling pathways of the IGF1R. Other IGF1R binding partners have been described which include Grb10, Crk-II, CSK, 14-3-3 proteins, p85 and p55γ regulatory subunits of PI3-K, SHP-2, SH2-B, $G\alpha_i$, JAKs, and SOCS proteins. Most aspects of the roles of these binding partners in IGF1R stimulated proliferation, differentiation, transformation, and inhibition of apoptosis still remain to be defined.

The mammalian mannose 6-phosphate/IGF-2 receptor (M6P/IGF2R), whose main function is to deliver a large number of acid hydrolases to lysosomes, also provides a degradative pathway for IGF-2 via receptor-mediated endocytosis. Separate receptor binding sites for lysosomal enzymes and IGF-2 have been identified among the 15 repeats of the large extracellular portion of the receptor. There is genetic evidence that the M6P/IGF2R is a tumor suppressor, consistent with its role in IGF-2 degradation.

The IGF System

The IGFs, IGF-1 and IGF-2, and the IGF-1 receptor (IGF1R) are important mediators of cell growth and differentiation. Both IGF-1, and at reduced affinity, IGF-2, activate the IGF1R, which has been shown to signal mitogenic, antiapoptotic, and transforming activities within the cell and to have important roles in many tissues. IGF-2 can also signal via the A isoform of the insulin receptor. These activities are opposed, in part, by the activity of the mannose 6-phosphate/IGF-2 receptor (M6P/IGF2R), which can bind extracellular IGF-2 and target it for degradation in lysosomes (Fig. 1). The M6P/IGF2R, as a result, behaves as a tumor suppressor, and is found to be deleted or mutated in various types of cancer, whereas the IGF1R is often found to be overexpressed. In this Chapter, we briefly outline the structure of the genes which encode the IGF-1 and M6P/IGF-2 receptors and structure/function relationships in the proteins that they encode.

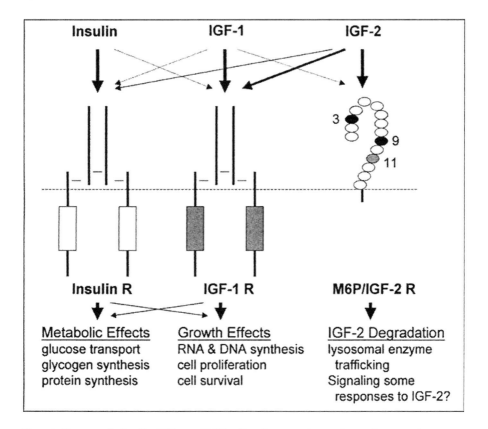

Figure 1. Receptors for insulin, IGF-1, and IGF-2: ligand-receptor interactions and receptor functions.

The IGF-1 Receptor

The IGF1R Gene

The human IGF1R gene maps to chromosome 15q25-26.[1] It spans more than 100 kb of genomic DNA and contains 21 exons.[2] The IGF1R is very similar in its organization to the human insulin receptor gene, which maps to chromosome 19p13.2, spans 120 kb and is encoded by 22 exons[3] and to the human insulin-related receptor, which maps to 1q21-23, spans more than 120 kb and is encoded by 22 exons.[4] In the mouse, the IGF1R, IR and IRR map to chromosomes 7, 8, and 3, respectively. The exon/intron structure of these three sets of genes is virtually identical except that the vertebrate IGF1Rs lack expression of an equivalent to exon 11 of the insulin and insulin-related receptors.

The IGF1R promoter has been cloned from both rat[5] and from human.[6,7] Although the IGF1R promoter lacks TATA and CAAT motifs, transcription initiates from a single "initiator" motif 940-1000 bp upstream from the coding region. The IGF1R promoter and 5' UTR are very GC rich and contain multiple potential binding sites for Sp1 and for Egr family transcription factors.[8,9] The long 5' UTR also has potential to form hairpins which may be used in transcriptional and/or translational control.[10]

The IGF1R promoter appears to be activated by transcription factors activated by growth factor receptor signaling and repressed by a variety of tumor suppressors including WT1, p53 and BRCA1. Expression of the IGF1R can be upregulated by PDGF signaling by means of a 100 bp promoter element immediately upstream from the transcription start site[11] or by bFGF signaling by means of a promoter element -476 to -100 from the start site.[12] Stat1, Stat3, and c-jun may mediate transcriptional activation by bFGF, while NFκB and c-jun may mediate angiotensin II-upregulation of IGF1R expression.[13] In an apparent negative feedback loop, IGF-1 downregulates transcription of the IGF1R using a response element which appears to lie outside the -2358 to +640 region.[12]

Expression of the IGF1R can be repressed by binding of WT1 to 11 sites in its promoter and 5' UTR[14,15] but transcription is greatly stimulated by the oncogenic fusion protein EWS/WT1.[16] Expression of the IGF1R is also regulated by the tumor suppressor p53. Wild type p53 was found to inhibit IGF1R transcription by direct binding to the promoter, whereas mutant p53 was found to stimulate IGF1R expression.[17,18] p73α, in contrast, transactivates the IGF1R promoter.[19] In another study, BRCA1 was found to suppress Sp1-mediated transactivation of the IGF1R promoter, suggesting another possible mechanism for IGF1R overexpression in cancer patients carrying mutations in BRCA1.[20] Unlike the mouse IGF-2, M6P/IGF2R and insulin loci and the human IGF-2 gene, the IGF1R locus is not imprinted in either humans or rodents.[21-23]

The IGF1R gene is expressed nearly ubiquitously, with highest levels in brain, significant levels in most other tissues and lowest levels in liver. In the rat, high levels of expression are found in the embryo and lower levels of expression in the adult.[24] While IGF-2 is the predominant ligand expressed in embryonic and fetal stages, IGF-1 predominates postnatally in rodents. In the absence of IGF1R expression, mice are born at 45% of normal birthweight, with defects in bone and muscle development and perinatal death due to respiratory failure.[25] Some growth may be attributed to IGF-2 signaling through the insulin receptor in such animals, in that double knockout of both the IGF1R and insulin receptor results in more severely growth retarded animals, 30% of normal size.[26] Double knockout of the IGF1R and IGF2R, furthermore, results in normal sized embryos. Here, in the absence of IGF2R-mediated IGF-2 degradation, IGF-2 signals through the insulin receptor to a sufficient extent to sustain normal prenatal growth.[27] In the absence of IGF-2, IGF1R, and IGF2R, this effect is eliminated, and mice are born at 30% of normal size.[27]

From mRNA to Protein

The predominant human IGF1R transcript is approximately 11 kb, which includes ~5.1 kb of open reading frame flanked by ~1 kb of 5' UTR and ~5 kb of 3' UTR. A minor ~7 kb transcript also appears in Northern blots of human mRNA from some tissues,[28] although only the 11 kb transcript was found in the rat,[29] and in several other species.

Full length IGF1R cDNAs have now been sequenced from several vertebrate species ranging from mammals to bird to frog and fish.[28,30-33] Each of these is similar to the human cDNA which encodes a 1367 amino acid precursor protein including a 30 aa signal sequence and the alpha and beta subunits of the receptor. During translocation of the nascent polypeptide in the ER, the 30 aa signal sequence is clipped off. As with all insulin-family receptors, the nascent precursor is glycosylated and cleaved at a consensus Arg-Lys-Arg-Arg sequence (aa 707-710) into α and β subunits by furin, a Ca^{++}-dependent serine endoprotease in the trans-Golgi compartment.[34] The mature receptor is assembled as an α2/β2 heterodimer, in which two disulfide-linked αβ half-receptors are linked by secondary disulfide bonds. While the α subunits are entirely extracellular, the β subunits include a short extracellular domain followed by a hydrophobic transmembrane domain (aa 906-929). The majority of the β subunit lies intracellularly, and is composed of juxtamembrane (aa 930-972), tyrosine kinase (aa 973-1229) and C-terminal domains (aa 1230-1337) (Fig. 2).

The predicted sizes for the α and β subunits are 80 kDa and 71 kDa, respectively, but due to glycosylation, they are observed at approximately 135 kDa and 95 kDa on SDS-PAGE (28). There are 11 potential N-glycosylation sites in the α subunit and 5 in the extracellular domain of the β subunit. The maturation of the proreceptor has been studied[35,36] and has a half-life of about one hour. In the insulin receptor, mutation of glycosylation sites can result in defective proreceptor processing or affect kinase activity.[37,38]

Multiple molecular weight forms of the IGF1R α and β subunits have been reported to occur in some tissues. Such differences might be attributed to altered glycosylation or other unknown posttranslational modifications. In one case they have been associated with alternative splicing. Use of a different splice site at the 5' end of exon 14 can result in replacement of the Thr-Gly found at aa 898-899 in the extracellular domain of the β subunit by a single Arg.[39] The Arg form of the receptor exhibits 50% decreased ligand-mediated endocytosis and 2-fold higher autophosphorylation and activity.[40] However, no such alternative splice site occurs in the rat.[30] Hybrid receptors are known to

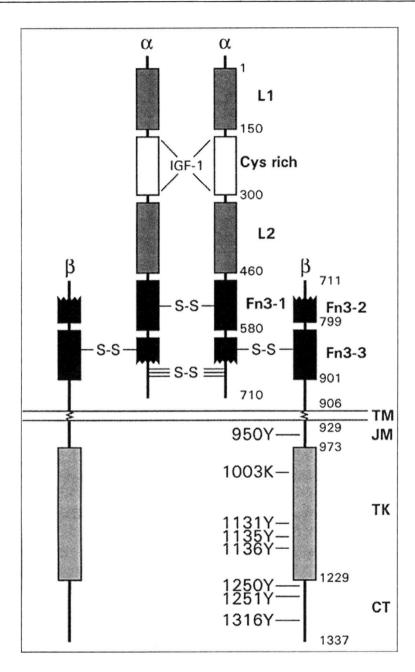

Figure 2. Domains of the IGF-1 receptor.

form between IGF1R and insulin receptor half-receptors and to be preferentially activated by IGF-1 compared to insulin.[41] The number of hybrid receptors can vary between tissues and developmental stages from a small minority to the vast majority of IGF-1 binding sites present.[42,43] Atypical IGF1Rs are characterized by atypical ligand binding characteristics marked by increased affinity for insulin.[41] The structure and significance of most of the reported receptor variants remains unclear.

The α Subunit

Further examination of the sequence for the α subunit reveals a correspondence between its intron/exon and domain structure. Exon 1 (aa -30-1) encodes the signal peptide. Exon 2 (aa 2-183) encodes the first L domain (1-150). Exon 3 (aa 184-288) encodes most of a furin-like cysteine rich region (151-299). Exons 4, 5, and 6 (aa 289-337, 338-386, 387-457) encode a second L domain (300-460). Exons 7 and 8 (aa 458-500, 501-579) encode a fibronectin type III domain (461-579) and exon 9 (aa 580-635) encodes part of a second, interrupted, fibronectin type III domain (580-798). Exon 10 (636-704) encodes the C-terminus of the α subunit, which lacks the extra amino acids encoded by exon 11 in the insulin receptor. There are 4 conserved cysteines in the first L domain, 20 conserved cysteines in the cys-rich domain, and 7-8 conserved cysteines scattered along the rest of the α subunit. These are used not only to stabilize intrasubunit folding but also to mediate α-α dimerization as well as binding of the α subunit C-termini to β subunit extracellular domains (Fig. 2).

The crystal structure of the N-terminus of the α subunit from the IGF1R (aa 1-462) has been reported.[44] It consists of two globular L domains attached by a curved cysteine-rich arm which includes a series of modules connected by 10 intramolecular disulfide bonds. An approximately 24 Å cleft lies between these three domains which appears to serve as the ligand binding site. Studies with chimeric receptors have shown that while insulin binds to the two L domains at 1-137 and 325-524 of the insulin receptor[45,46] IGF-1 binds to the cysteine-rich region of the IGF1R. Replacement of the insulin receptor cysteine-rich region with IGF1R aa 223-274 confers IGF-1 binding.[47,48] Replacement of insulin receptor aa 260-277 with IGF1R aa 253-266 also confers increased IGF-1 binding.[49] These amino acids form a large loop in the Cys-rich region in which much sequence divergence exists between the insulin and IGF-1 receptors.

The first three domains of the α subunit are followed by two fibronectin type III repeats.[50] Such motifs are commonly found in extracellular proteins and appear to mediate protein-protein interactions and/or to serve as spacers between important domains. Experiments indicate that some areas both N-terminal and C-terminal to the cysteine-rich region can influence IGF-1 binding. While most IGF1R blocking antibodies such as αIR3 bind to the cysteine-rich domain, some antibodies raised against L-1 and Fn3-1 domain epitopes (aa 38-44 and 440-586) can also block IGF-1 binding, reflecting an influence of adjacent sequences on binding at the cysteine-rich domain.[47,51,52] Mutation of Phe 701 at the C-terminus of the α subunit completely inhibits IGF-1 binding while mutation of Asp8, Asn11, Phe 58, Phe 692, Glu 693, His 697, and Asn 698 reduce IGF-1 binding 3-6 fold.[53] Swapping of the insulin receptor and IGF1R α subunit C-termini has been reported to have little effect on ligand affinities.[54] However, absence of exon 11 at the C-terminus of the insulin receptor α subunit increases IGF-2's affinity for "isoform A" relative to "isoform B", which contains exon 11 and is largely insulin-specific.[55]

A number of disulfide bonds link the α and β subunits of the IGF1R. Studies have shown that the insulin receptor α subunits are linked in part by disulfide bonds between Cys 524 and Cys 524 of two monomers.[56] Some or all of cysteines 682, 683, and 685 are also involved in α-α dimer formation in the insulin receptor.[57,58] These correspond to cysteines 514, 669, 670, and 672 of the IGF1R. Other evidence for dimerization through the IGF1R α chain's first Fn3 domain and C-terminus has been presented.[59] A single disulfide bond between Cys 647 and Cys 872 (exon 11+ numbering) links the α and β subunits of the insulin receptor.[58] These cysteines correspond to Cys 633 in the second Fn3 domain of the α subunit and Cys 849 in the Fn3 domain of the β subunit of the IGF1R. An intramolecular disulfide bond has been reported to exist in the IR β chain extracellular domain at Cys 798-Cys 807 (58), which corresponds to IGF1R Cys 776-Cys 785. Others have found that mutation of IGF1R cysteine 776 to serine prevents proreceptor dimerization and proreceptor cleavage.[60]

The β Subunit

The extracellular domain of the IGF1R β subunit encodes 3 conserved cysteine residues with potential to form disulfide bridges with the α subunit and a fibronectin type III repeat adjacent to the plasma membrane. The intracellular portion of the β subunit includes a juxtamembrane region, the tyrosine kinase domain, and a C-terminal tail (Fig. 2). Based on its genomic sequence, IGF1R exon 11 (aa 705-798) encodes the Arg-Lys-Arg-Arg proteolytic cleavage site (707-710) which directs separation of aa 711-1337 as the β subunit. Exons 12 and 13 (aa 799-844; aa 845-897) encode another extracellular fibronectin type III domain (Ward, 1999). Exon 14 (aa 898-932) encodes the transmembrane region (906-929). Exon 15 (aa 933-955) encodes the juxtamembrane region. Exons 16-20 (aa 956-1032, 1033-1069, 1070-1122, 1123-1166, 1167-1211) encode the tyrosine kinase domain of the receptor (973-1229), including the catalytic loop (1104-1110), the activation loop (1121-1144) and the P+1 loop that confers specificity for tyrosine phosphorylation (1145-1153). The tyrosine kinase domain ends at approximately aa 1229, hence exon 21 (aa 1212-1337) encodes mostly the C-terminal tail domain.

Activation of receptor tyrosine kinase activity occurs in several steps: ligand binding results in a conformational change which allows receptor autophosphorylation at tyrosines 1131, 1135 and 1136 in the kinase activation loop, which greatly amplifies kinase activity.

In the unoccupied receptor, the extracellular domains of the β subunits appear to maintain an inhibitory conformation on the receptor tyrosine kinase. Ligand binding or deletion of the extracellular domain releases this inhibition and allows the kinase to be more active. For example, tryptic digestion of the α subunit of the insulin receptor appeared to release its β subunit from inhibitory control.[61] Retroviral gag-IGF1R β fusion constructs were found to be more actively tumorigenic if 36 amino acids (870-905) from the immediate extracellular domain of the IGF1R β subunit were deleted.[62] Deletion of these same amino acids from the intact IGF1R results in a mutant receptor which is autophosphorylated in the absence of ligand stimulation and, oddly, constitutively activates the PI3 kinase, but not ras/MAPK downstream pathways.[63] Expression of an IGF1R consisting of only the transmembrane and cytoplasmic domain also resulted in preferential activation of PI3 kinase. In addition, anchorage-independent growth, but not proliferation, was increased.[64]

The transmembrane domain is also important in transmitting the kinase-activating conformation. In ErbB2 (Her2/Neu), a Val 664 to Glu mutation in the transmembrane helix was found to cause ligand-independent dimerization and constitutive activation of that receptor.[65] IGF-1 receptors with an analogous Val922Glu mutation in the transmembrane domain were also constitutively activated.[66]

Autophosphorylation of the α2β2 IGF1R occurs intramolecularly, as the rate of the reaction was shown to be independent of receptor concentration.[67] The question of whether autophosphorylation occurs through a *cis* mechanism (within a single β subunit) or a *trans* mechanism (between two β subunits in the heterodimer) was studied by producing hybrid receptors in intact cells, then carrying out autophosphorylation reactions in vitro in the presence of ligand. In hybrid insulin or IGF-1 receptors composed of kinase-negative αβ and carboxy-truncated, kinase-positive αβ half-receptors, the kinase-negative half receptor was shown to be autophosphorylated, indicating a *trans* mechanism.[68] Thus, IGF1 receptors deleted of the entire cytoplasmic domain can act as dominant negative inhibitors of endogenous IGF1R function and of transformation by overexpressed receptors.[69]

The crystal structures of the unphosphorylated and tris-phosphorylated insulin receptor tyrosine kinase domains reveal how autophosphorylation of tyrosines in the activation loop results in amplified kinase activity and provide a model for the closely related IGF1R. The kinase is maintained in a low activity state prior to ligand binding because Tyr 1162 and Asp 1150 of the activation loop block substrate binding to the catalytic site and Mg-ATP binding, respectively.[70] Autophosphorylation of the insulin receptor's Tyr 1158, 1162, and 1163 causes repositioning of the activation loop resulting in open access of the kinase active site to Mg-ATP and substrates (exon 11+ numbering).[71]

In the IGF1R kinase domain, Tyr 1131, 1135, and 1136 are autophosphorylated first upon ligand stimulation. Mutation of all three tyrosines to phenylalanines results in loss of autophosphorylation and all receptor kinase activity.[72-74] Mutation of individual tyrosines in this region or of Trp 1173 results in reduced autophosphorylation but retention of kinase activity and some downstream effects.[74-77] Mutation of Lys 1003 to Ala or Arg also completely blocks tyrosine kinase activity and receptor function purportedly by blocking ATP binding.[78] Conversely, mutation of insulin receptor Asp 1161 to Ala (= IGF1R Asp 1134) relieves autoinhibition and activates ATP binding.[79]

Following autophosphorylation of the kinase domain, subsequent autophosphorylation of other tyrosines on the receptor β subunit creates binding motifs for downstream signaling molecules with

SH2 and PTB domains. The cytoplasmic domain of the IGF1R is phosphorylated at at least four tyrosines in addition to the three in the kinase domain. Tyrosine 950 is phosphorylated in the juxtamembrane domain. Tyrosines 1250, 1251, and 1316 are phosphorylated in the C-terminal tail. Although there has been one report of intrinsic serine kinase activity in purified IGF1R preparations,[80] this has been disputed by others.[81] The insulin receptor has been reported to be serine phosphorylated both in the juxtamembrane domain and in the C-terminal region (Sers 1287, 1321 and 1305/6, exon 11+ numbering).[82-84] While some have attributed this to intrinsic serine kinase activity of the insulin receptor,[85] others have shown a serine kinase to copurify with the insulin receptor.[86]

IGF1R Binding Partners: Juxtamembrane Region

IRS-1

The phosphorylation of tyrosine 950 in the juxtamembrane domain is important for the activation of both major pathways which lie downstream from the IGF1R, the PI3 kinase pathway and the MAP kinase pathway. One major substrate of the receptor kinase is the 180 kDa downstream docking molecule IRS-1 or its relatives.[87,88] The IRS proteins are composed of N-terminal PH and PTB domains followed by a large number of potential phosphotyrosine docking sites. Activation of IRS-1 creates binding sites for a large number of signaling proteins, which have been reported to include the p85 subunit of PI3K, Grb2, SHP-2, Nck, 14-3-3, crk, fyn, FAK, and possibly JAK kinases and others. Most importantly, it activates the PI3K/Akt pathway through p85 binding and the ras/MAPK pathway through Grb2/Sos. Several groups using the yeast two-hybrid system have shown that IRS-1 binds to $NPEY_{950}$ of the IGF1R by means of its PTB domain within aa 160-516.[89-91] When the insulin receptor system was studied using peptide competition and chimeric constructs in mammalian cells, however, the N-terminal IRS-1 PH domain was found to be necessary for optimal insulin receptor-IRS-1 PTB binding.[92,93] A novel protein, PHIP, has been reported which interacts with the IRS-1 PH domain and may facilitate coupling of IRS-1 to activated insulin receptors.[94] While interaction with the phosphorylated NPEY950 motif remains speculative for IRS-4, both IRS-2 and IRS-3 have also been shown to bind to the IGF1R at this motif.[95,96] In the case of IRS-2, a third region between amino acids 591 and 786 was found to engage in phosphotyrosine-dependent interactions with the activation loop (Tyr 1158, 1162, 1163; exon 11+ numbering) of the insulin receptor kinase domain.[95,97,98] Since the IRS-1 does not have a similar domain, this interaction may contribute to IRS-2 specific signaling.

Shc

Another major IGF1R substrate which binds at tyrosine 950 is Shc, which also activates the ras/MAPK pathway through Grb2 and Sos. The Shc family includes the p46, p52, and p66 isoforms of ShcA. All encode an N-terminal PTB domain followed by a central Pro/Gly rich region carrying three phosphotyrosines and a C-terminal SH2 domain, while p66 Shc encodes an extra Pro/Gly rich region at its extreme N-terminus.[99,100] Several groups using the yeast two-hybrid system have shown that Shc binds to Tyr 950 of the activated IGF1R by means of its PTB domain within aa 1-232.[89-91] IGF1R mutants at Tyr 950 bind neither IRS-1 nor Shc. In the insulin receptor system, mutations of the Asn and Pro from its $NPEY_{960}$ motif prevent IRS-1 but only partially prevent Shc binding.[101,102] In addition, several other mutations in the surrounding sequence affect IRS-1, but not Shc binding, or vice versa.[102,103] For example, although Shc binds with lower affinity than IRS-1 for the insulin receptor NPEY motif, mutation of insulin receptor Ser 955 to Ile at the pY-5 position results in stable association of Shc with insulin receptor in insulin-stimulated cells. A Pro951Gln, Leu952Gly, Ser955Ile triple mutation in the insulin receptor pY-9, -8, -5 resulted in even stronger Shc binding.[104] Shc appears to be activated over a longer time course than IRS-1[105] and in contrast to IRS-1, receptor internalization may be required.[106]

GAP

Although the ras GTPase-activating protein, GAP, was reported to bind to the IGF1R at tyrosine 950,[107] a later study found this interaction to be weak, at best.[108] GAP contains an SH2 domain and acts as a negative regulator of ras-dependent signaling. It is most often described in an insulin-stimulated complex of GAP, PI3K, SHP-2, Grb2, Nck and Abl with an IRS-like docking protein now called p62 Dok.[109] Such a complex may also be induced by IGF-1 stimulation.[110]

Crk-II

A peptide fragment from the IGF1R juxtamembrane domain was found to bind to the SH2 domain of GST-Crk-II, and mutation of Tyr 943 or Tyr 950 was found to eliminate binding.[111] Crk-II is the largest among four members of the Crk family of adaptor proteins and encodes an N-terminal SH2 domain followed by two SH3 domains separated by a tyrosine phosphorylation site.[112] Whereas the SH2 domain of Crk-II binds to growth factor receptors or the focal adhesion proteins paxillin or p130cas, its SH3 domains bind to the kinase Abl or to the nucleotide exchange factors C3G and Sos. IGF-1 or insulin stimulation induces tyrosine phosphorylation of Crk-II, its dissociation from IRS-1 and presumably also from the IGF1R.[113,114] Upon phosphorylation, Crk-II appears to fold as to mask its SH3 domain, thereby dissociating the Crk-C3G complex, inactivating Rap1/Raf1 interactions and promoting Raf1/Ras association.[115,116] Crk phosphorylation, therefore, appears to enhance IGF-1 stimulated mitogenesis via the Raf/MAPK pathway. It appears to also regulate cytoskeletal rearrangement[117,118] and may activate JNK and/or PI3K/Akt signaling.[119,120]

CSK

C-terminal Src kinase (CSK) was found to bind by means of its SH2 domain to phosphotyrosines 943 and 1316 of the IGF1R[121] and is known to inhibit Src activity by phosphorylation of its C-terminal Tyr 527.[122] Since CSK is not stably associated with Src,[123] its binding to the IGF1R or to paxillin may be necessary for localized inhibition of Src activity. It has been suggested that v-crk may transform by displacing CSK from paxillin and relieving Src from its inhibition there.[124] Since presence of the IGF1R is required for transformation by Src,[125] it may be possible that displacement of CSK from the IGF1R may be necessary for Src-mediated transformation.

IGF1R Binding Partners: C-Terminus

Deletion analysis has indicated that the C-terminal domain of the IGF1R is dispensable for its antiapoptotic and mitogenic properties but essential for cellular transformation as assayed by growth of cells in soft agar (Fig. 3). Receptors deleted of aa 1293-1337 have reduced transforming capacity, while receptors deleted of aa 1245-1337 are negative for transformation.[126,127] Chimeric receptors composed of an IGF1R carrying a cytoplasmic domain from the insulin receptor are much less effective in stimulating DNA synthesis[128] and those carrying just the insulin receptor C-terminal domain are non-transforming[126] suggesting that IGF1R-specific downsteam pathways are responsible for these properties. One laboratory has indicated receptors with double mutation at tyrosines 1250 and 1251 (Y1250F, Y1251F) are negative for both the transforming and antiapoptotic, but not mitogenic activities.[126,129] Others reported that Y1250F, Y1251H mutant receptors were impaired for both transformation and mitogenesis.[130,131] Mutation of the serine quartet at Ser 1280-1283 to alanines reduces transforming ability, but neither mitogenicity nor inhibition of apoptosis[127,132] suggesting interaction of this region with signals promoting anchorage-independent growth, although a role in inhibition of apoptosis has been suggested by others.[133,134] Mutation of His 1293 to Phe and Lys 1294 to Arg affects mostly the antiapoptotic activity of the receptor, suggesting interaction with survival pathways.[126,127] The extreme C-terminus appears nonessential to most activities of the IGF1R, since a receptor deleted of aa 1310-1337 retained normal proliferative and transforming activities.[126,135] However, mutation of tyrosine 1316 has been variously reported to partially impair mitogenesis and transformation[130] or to have no effect on these parameters.[126]

14-3-3

Although most of the potential C-terminal binding sites may be known, much of what binds there and how it functions remains elusive. Among the most interesting C-terminal binding proteins to be identified so far is 14-3-3. 14-3-3 β and ζ were found to bind to phosphoserine 1283 from the Ser 1280-1283 cluster in a yeast two hybrid screen[136] and 14-3-3 ε was also found to be capable of binding to serines 1272 and 1283.[137] 14-3-3 proteins are members of a family of dimeric phosphoserine binding proteins. They have been shown to interact with a large number of cell

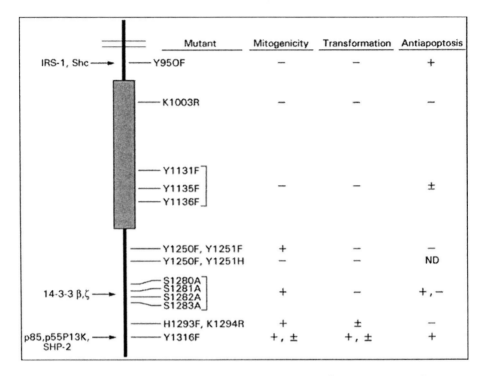

Figure 3. Mutations of the β subunit of the IGF-1 receptor: consequences for mitogenicity, transformation and antiapoptosis.

proliferation-associated signaling proteins including Raf-1, cdc25, and IRS-1 and to help inhibit the activity of other proteins which promote apoptosis such as ASK1, FKHRL1, and Bad.[138] 14-3-3 β, ζ and ε do not bind to the insulin receptor[136,137] consistent with the absence of the 1280-1283 serine cluster. Although the precise role of the 14-3-3-IGF1R interaction remains unknown, it has been reported to influence an antiapoptotic pathway in 32Dcells[133,134] and mutation of the serine quartet at 1280-1283 to alanine reduces IGF1R transforming ablility in fibroblasts.[127,132]

p85, p55γ

PI3-kinase, which has been thought to mostly be activated by binding of its regulatory subunit p85α to IRS-1, also binds independently to the receptor at phosphotyrosine 1316.[107,139,140] p85 is composed of an N-terminal SH3 domain followed by proline-rich and BCR homology regions, then by two SH2 domains separated by an inter-SH2 linker which binds to the p110 catalytic subunit of the kinase.[141] Phosphotyrosine binding to either of p85's SH2 domains induces a conformational change which activates p85/p110 dimers.[142] While both of p85's SH2 domains have been found to bind to the insulin receptor, only the C-terminal SH2 domain appears to bind to the IGF1R, which binds more weakly.[139,140] Mutation of Y1316 diminishes, but does not eliminate binding of full length p85 to the IGF1R in a yeast two-hybrid assay, suggesting that other regions of the receptor may contribute to p85 binding.[140] In addition to p85α, the p55γ isoform of the regulatory subunit of PI3K has been demonstrated to bind to the IGF1R.[143,144] In contrast to p85, p55γ can bind using either of its SH2 domains.[143]

SHP-2

The phosphatase SHP-2 has also been reported to bind to the IGF1R at tyrosine 1316.[107,145] Further analysis has indicated that while IGF1R Tyr 1316 does appear to bind to SHP-2's SH2 domains, this binding is weak compared to that between SHP-2 and IRS-1, and the phosphotyrosines from the IGF1R kinase activation loop influence both SHP-2 and IRS-1 binding.[108,145] SHP-2 (also called Syp, PTP1D, and SHPTP2) is a cytoplasmic protein tyrosine phosphatase which contains two N-terminal SH2 domains followed by a C-terminal catalytic region. It is unusual among phosphatases in that it can promote signaling from growth factor receptors as opposed to playing an exclusively negative role.[146] Microinjection of anti-SHP-2 antibodies inhibits mitogenesis by IGF-1 and insulin[147] and fibroblasts from SHP-2 exon 3 -/- mice show impaired growth factor and integrin signaling.[148,149] In the insulin receptor system, SHP-2 has been reported to enhance ERK activation and JNK activation[150,151] and to bind to a transmembrane glycoprotein SHPS-1/SIRPα1 upon insulin stimulation or integrin engagement.[152-154]

Additional IGF1R Binding Partners

Grb10

Other proteins have been reported to bind to the IGF1R, but their binding sites are less clear. Grb10, also called Grb-IR, is an adaptor protein with a proline-rich N-terminus followed by a central PH and a C-terminal SH2 domain which comes in a number of splice variants. Grb10 has been shown to bind to Raf-1 and MEK-1 and to colocalize with Raf-1 to mitochondria where it may function in apoptosis.[155,156] It also binds to the E3 ubiquitin protein ligase Nedd4[157] and to Bcr-Abl,[158] and is phosphorylated by Src.[159] There is disagreement about the IGF1R and insulin receptor binding sites for Grb10. Based on differential binding of IGF1R deletion mutants in a two hybrid assay, Morrione et al[160] mapped the binding site between amino acids 1229 and 1245 in the carboxy tail of the receptor. Binding of the Grb10 SH2 domain to the insulin receptor has been variously localized to the carboxy tail,[161] kinase activation loop[162] and juxtamembrane region.[162] More recently, He et al[163] identified a second receptor-binding site in Grb10 termed the BPS domain (between the PH and SH2 domains). Binding of the BPS domain to the insulin receptor was shown to require tyrosines 1162/1163 in the activation loop (exon 11+ numbering)[163] but peptide competition experiments showed that the BPS does not bind directly to phosphotyrosines in the activation loop.[164] The BPS domain bound to a tris-phosphorylated core insulin receptor kinase which was devoid of juxtamembrane and carboxy tail domains. Thus, autophosphorylation of the core kinase appears to induce a conformational change that enables the BPS domain to bind. In the case of binding of Grb10 to the IGF1R, the BPS domain appears to be more important than the SH2 domain.[163] Preference for binding of Grb10 to the insulin receptor as compared to the IGR1R in intact cells[165] may be explained by the relative importance of BPS and SH2 domains for binding to the two receptors.[163] There is controversy about whether Grb10 enhances or inhibits insulin receptor and IGF1R signaling. Wang et al[166] reported evidence for a stimulatory role for Grb10 in IGF-1 and insulin action, whereas Morrione et al[160] reported that overexpression of Grb10 resulted in inhibition of IGF-1 stimulated proliferation of mouse fibroblasts overexpressing the IGF1R. Recently, Stein et al[164] reported that the BPS domain of Grb10 directly inhibited substrate phosphorylation by the activated tyrosine kinase domain of the IGF-1 receptor.

SH2-B

SH2-B, also known as PSM, is a large adaptor protein which exists in at least three differentially spliced isoforms(α,β,γ) each with PH and SH2 domains interspersed with several proline-rich regions and phosphorylation sites. Although it can promote IGF-1 stimulated mitogenesis[167] and it has been reported to bind Grb2 or to bind and help activate JAK2 in other systems, the precise role of SH2-B in insulin and IGF-1 receptor I signaling remains unknown. A related molecule, APS, binds cbl and can inhibit erythropoietin-induced STAT activation.[168] SH2-B binds to both the activated IGF-1 and insulin receptors by means of its SH2 domain. One group, studying mouse SH2-Bβ, mapped its interaction to IGF1R tyrosines 950 and 1316 or their insulin receptor equivalents.[169] Two other groups, working with the human α or mouse γ isoforms, found it to interact with the activation loop of the insulin receptor kinase domain.[170,171]

Gα_i

A heterotrimeric Gi protein has been reported to bind the IGF1R at an undefined site. Upon IGF-1 stimulation, G$\beta\gamma$, but not Gα_i, was found to dissociate from the IGF1R.[172] In keeping with earlier findings in Rat-1 fibroblasts,[173] pertussis toxin, which inactivates Gi proteins, was found to inhibit IGF-1/Gβγ-mediated activation of MAP kinase in neuronal cells[172] and in intestinal smooth muscle cells.[174] β-arrestins, which bind to IGF1Rs and appear to promote their endocytosis by a G protein-coupled receptor-related mechanism, were found to enhance IGF-1 dependent MAPK activation.[175] Microinjection of anti-β-arrestin 1 antibodies inhibits IGF-1 stimulated mitogenesis, but does not affect mitogenic activity through the insulin or EGF receptors, which do not bind $G\alpha_i$.[176] Instead, the insulin receptor associates with the $G\alpha_{q/11}$ protein, which plays an important role in insulin-stimulated, PI3-kinase mediated, GLUT4 translocation in 3T3-L1 adipocytes.[177]

JAKs

JAK-1 kinase has been reported to bind to undefined portions of the activated IGF1R using its JH1 and JH6/7 domains[178] and JAKs have also been reported to bind to the insulin receptor, IRS-1, and Grb2.[179,180] JAK-1 and -2 are indirectly activated by IGF-1 stimulation[178] and may or may not be required for the activation of STAT transcription factors by the usual paradigm.[181] IGF-1 receptor signaling has been reported to activate Stat3 by a JAK-dependent mechanism.[182,183] While direct binding of Stat3 to the IGF1R has not yet been reported, in the insulin receptor system Stat5B has been reported to bind to the $NPEY_{960}$ motif through its SH2 domain[184,185] where it may be activated by JAK-dependent and JAK-independent mechanisms.

SOCS 1-3

Lastly, the suppressor of cytokine signaling SOCS-1 (also called JAB, SSI-1), SOCS-2 (CIS2, SSI-2) and SOCS-3 (CIS3, SSI-3) proteins bind to the IGF1R at undefined sites.[186,187] The suppressor of cytokine signaling (SOCS) 1-7 and cytokine-inducible SH2-containing protein (CIS) comprise a family of related proteins which encode a variable N-terminal domain followed by an SH2 domain and a domain now called the SOCS box. SOCS proteins appear to negatively regulate JAK/STAT signaling either by binding to activated JAKs, as in the case of SOCS 1 and 3, or by competing with STAT proteins for binding to the receptor, as in the case of CIS.[188,189] The SOCS proteins have also been found to interact with elongin BC complex and a putative E3 ubiquitin ligase, suggesting that binding of SOCS proteins to signaling molecules may prevent or promote their degradation in proteasomes.[190,191] The accelerated growth phenotype of SOCS-2 knockout mice most strongly implicates SOCS-2 in growth hormone and/or IGF1R function.[192] SOCS-2 was found to interact with the activated wild type IGF1R, but also to interact with receptors mutated at tyrosines 950, 1250, 1251, and 1316.[186] SOCS-1 also was found to interact in a kinase-dependent manner,[186] but SOCS-3 was found to bind to the IGF1R in both the presence and absence of IGF-1 in mammalian cells.[187] In contrast, others reported SOCS-3 to bind to phosphotyrosine 960 of the insulin receptor.[193] Although the role of SOCS-IGF1R interaction has not been clearly defined, SOCS-1 and -3, and to a lesser degree, SOCS-2 and CIS, have been reported to inhibit IGF-1 induced STAT3 activation, and overexpression of JAK1 or JAK2, to override SOCS-1 mediated STAT3 inhibition.[183]

The M6P/IGF-2 Receptor

In the insulin-like growth factor field the IGF-2 receptor was characterized by having a much higher binding affinity for IGF-2 than for IGF-1 and not binding insulin (Fig. 1).[194] The affinity crosslinked IGF-2:receptor complex behaved as a 250 kDa species on SDS/PAGE. Meanwhile in the lysosomal enzyme field a similar sized receptor was identified which recognized mannose 6-phosphate (M6P) residues on lysosomal hydrolases.[195] After a decade and a half, molecular cloning of the cation-independent M6P receptor and the IGF-2 receptor revealed a common bifunctional receptor (M6P/IGF2R).[196] Thus, during evolution the cation-independent M6P receptor which targets newly synthesized lysosomal enzymes from the trans Golgi network to lysosomes and delivers extracellular lysosomal enzymes to the same organelle, acquired an IGF-2 binding site, providing a degradative pathway for a growth factor.[197]

The M6P/IGF2R Gene

The gene that codes for the M6P/IGF2R maps to the centromeric third of chromosome 17 in the mouse, where it spans 130 kb and 48 exons,[198] and to the long arm of human chromosome 6, region 6q25-q27, where it also spans 48 exons.[199] The M6P/IGF2R is imprinted in the mouse, being expressed from the maternal chromosome,[200] but for most humans expression is biallelic.[201,202] Like IGF-2, the M6P/IGF2R is expressed at highest levels during fetal development, and declines to somewhat lower levels in the adult.[203]

M6P/IGF-2 Receptor Structure

The 270 kDa M6P/IGF2R consists of a large extracellular domain (2269 residues), a 23 amino acid transmembrane domain, and a small cytoplasmic domain (163 residues) (Fig. 1).[204] The striking feature of the extracellular domain is that it consists of 15 contiguous repeats with an average size of 147 amino acids. The percent of identical residues within repeats ranges from 16-38% and there is a clear pattern in the arrangement of 8 cysteines in each repeat. Repeat 13 is different from the others because it contains a 43 aa insertion that is similar to the type II repeat of fibronectin. The extracellular portion of the M6P/IGF2R is involved in ligand binding whereas the short cytoplasmic domain contains motifs that are important for the binding of factors required for receptor trafficking.[205,206] It has been inferred that at least 2 of the 19 glycosylation sites within the extracellular domain are glycosylated,[207] and phosphorylation has been reported for its cytoplasmic domain.[208]

There is a second M6P receptor, the cation-dependent M6P receptor.[209] This second receptor is only 46 kDa in size and does not bind IGF-2.[210] The extracellular domain of the 47 kDa M6P receptor is similar to each of the repeats in the M6P/IGF2R, the sequence identity ranging from 14-28%. The 47 kDa M6P exists as a dimer in the membrane.[211]

Soluble Form of the M6P/IGF-2 Receptor

A binding component for IGF-2 that was considerably larger than the known IGF binding proteins was first observed in fetal rat serum.[212] This large binding protein was identified as a truncated M6P/IGF2R. Using antibodies raised against synthetic peptides from the cytoplasmic or extracellular domain, MacDonald et al[213] demonstrated that the cytoplasmic domain was missing from the serum receptor. Presumably, the serum receptor arises by proteolytic cleavage of the extracellular domain at the cell surface, and may represent a major degradative pathway for the receptor.[214] The circulating M6P/IGF2R has been shown to be associated with IGF-2;[215] however, measurement of the soluble M6P/IGF2R by enzyme-linked immunosorbent assay indicates that the receptor could not be a significant carrier for IGF-2.[216] The soluble M6P/IGF2R has also been shown to be capable of inhibiting biological responses to IGF-2 in cell culture[217] and when overexpressed in vivo,[218] but it is not clear whether or not the soluble receptor plays a regulatory role in normal physiology.

Receptor Ligands

Lysosomal Enzymes

The M6P/IGF2R recognizes terminal M6P residues in N-linked high mannose oligosaccharides present in a large family of lysosomal hydrolases.[197] The protein portion of the lysosomal enzymes does not directly contribute to binding to the receptor. The stoichiometry values for binding are 2.17 for M6P and 0.9 for the lysosomal enzyme β-galactosidase.[219] Thus, there are two receptor binding sites for M6P residues and lysosomal enzymes utilize multivalent binding which results in higher binding affinity compared to monovalent M6P (K_d 2 nM versus 7 μM for M6P).[219] Binding of proteolytic fragments of the receptor to pentamannosyl phosphate-agarose together with site-directed mutagenesis studies have localized the two M6P binding sites to repeats 1-3 and 7-9 of the extracellular domain of the receptor.[220,221] Mutation of critical arginine residues in repeats 3 and 9 abolishes binding to M6P residues (Fig. 1).[221] Evidence has been presented that the two M6P binding sites are not functionally equivalent.[222] Recently, evidence for dimerization of the M6P/IGF2R has been presented.[223-225] Therefore, high affinity, multivalent binding of lysosomal enzyme to the M6P/IGF2R could occur between two receptor molecules in a dimer rather than being intramolecular between the two M6P binding sites. Indeed, truncated receptors containing only one M6P binding domain or mutant receptor constructs containing an Arg1325 to Ala mutation that eliminates binding to the repeats 7-9 binding domain, were capable of high affinity binding of pentamannose phosphate-BSA.[223]

IGF-2

In evolutionary terms, the M6P/IGF2R bound lysosomal enzymes first; IGF-2 binding appeared later. Thus, the chicken and frog 270 kDa M6P receptors do not bind IGF-2[226,227] and the oppossum and kangaroo receptors exhibit only low affinity binding.[228,229] Measurements of binding affinity of IGF-2 for highly purified M6P/IGF2R have ranged from 0.7 nM to 0.017 nM (K_d).[230-233] A binding stoichiometry value of 0.95 was obtained from experiments in which receptor concentration was accurately measured and all of the radioligand was shown to be capable of binding to the receptor.[233] IGF-1 binds to the M6P/IGF2R with very low affinity such that physiologic concentrations of IGF-1 would not interact with the receptor.[233] Insulin does not bind to the M6P/IGF2R.

To localize the IGF-2 binding site within the 15 repeats of the M6P/IGF2R extracellular domain, proteolytic fragments of the receptor have been tested for IGF-2 binding.[234] A second approach has been to express fusion proteins containing truncated M6P/IGF2R (minireceptors) and test for IGF-2 binding.[235,236] These studies pointed to repeat 11 as being important for IGF-2 binding (Fig. 1). Furthermore, a point mutation substituting threonine for isoleucine at residue 1572 located in the N-terminal half of repeat 11, abolished IGF-2 binding.[236] Repeat 11 was also sufficient to mediate internalization of IGF-2 as shown by experiments in which repeat 11 was fused to the transmembrane and cytoplasmic domain of the M6P/IGF2R and transfected into mouse embryonic fibroblasts.[237] In experiments with minireceptors containing repeat 11, it was observed that IGF-2 binding affinity was approximately 10 fold lower than the binding affinity for the holoreceptor.[237] Examination of IGF-2 binding to minireceptors containing repeats 11-12, 11-13, and 11-15 suggested that an affinity enhancing domain was present in repeat 15, although repeat 15 itself does not bind IGF-2.[238] Interestingly, repeat 15 contains the 43 residue fibronectin homology domain which when deleted from the holoreceptor resulted in decreased IGF-2 binding affinity. Real time kinetic studies of the binding of IGF-2 to soluble, truncated forms of the M6P/IGF2R confirmed the importance of repeat 15 as a binding enhancer, demonstrating that repeat 15 decreased the rate of dissociation of IGF-2 from the receptor.[239]

Although the receptor binding sites for lysosomal enzymes and IGF-2 are different there is evidence that a lysosomal enzyme (β-galactosidase) inhibits binding of IGF-2[240] and conversely, IGF-2 inhibits the binding of β-galactosidase.[241] The earlier observation that M6P actually increased the binding of lysosomal enzymes to the M6P/IGF2R is probably explained by M6P stripping inhibitory lysosomal enzymes from the receptor. The reciprocal inhibition of binding of these two classes of ligands is probably explained by steric inhibition of binding. These observations would predict that extracellular lysosomal enzymes could inhibit the receptor mediated degradation of IGF-2 resulting in increased activation of the IGF-1 receptor and that IGF-2 could increase the concentration of extracellular lysosomal enzymes. Indeed, overexpression of IGF-2 by MCF-7 breast cancer cells[242] or 293 embryonic kidney cells[243] resulted in increased extracellular cathepsin D and β-hexosaminadase.

TGF-β1

Latent TGF-β1 binds to the M6P/IGF2R at M6P recognition sites,[244] leading to activation of TGF-β1.[245] Latency-associated peptide (LAP) and TGF-β1 are synthesized as a preprotein which is cleaved intracellularly producing mature TGF-β1 and LAP. LAP remains associated with TGF-β1 through noncovalent interactions. LAP is glycosylated at three N-linked sites and M6P residues are present on at least two of these carbohydrate side chains. After binding of latent TGF-β1 to the M6P/IGF2R, activation of TGF-β1 involves proteolytic cleavage of LAP by extracellular plasmin. Presumably the cleavage of LAP leads to a conformational change in LAP and release of TGF-β1. There is evidence that the M6P/IGF2R complexes with urokinase (plasminogen activator) receptor and directly binds plasminogen, leading to the generation of plasmin.[246] While this scheme for TGF-β1 activation has been shown to operate in a cell culture model it is not known to what extent this pathway involving M6P/IGF2R operates in vivo.

Proliferin

A cDNA for a mRNA that appeared in mouse 3T3 cells within a few hours after stimulation with serum or PDGF was cloned and the encoded protein was named proliferin. Proliferin binds to the M6P/IGF2R at M6P recognition sites.[247] A second, high affinity receptor has also been identified for proliferin.[248]

Thyroglobulin

M6P residues are found in thyroglobulin and radiolabeled thyroblobulin binds to the M6P/IGF2R.[249] Although thyroblobulin can be endocytosed by human skin fibroblasts via the M6P/IGF2R, the pathway for the delivery of thyroglobulin to lysosomes in the thyroid gland does not involve the M6P/IGF2R.[250]

Leukemia Inhibitory Factor (LIF)

M6P-sensitive LIF binding to the M6P/IGF2R results in rapid internalization and degradation of the cytokine in numerous cell lines, suggesting that the M6P/IGF2R may regulate the availability of LIF in vivo.[251]

Granzyme B

The serine proteinase granzyme B is important for the rapid induction of cell apoptosis by cytotoxic T cells. Motyka et al[252] provided evidence that granzyme B enters cells via binding to the M6P/IGF2R through binding to the receptor M6P binding site and proposed that tumors carrying mutated non-functional M6P/IGF2R would have an inherent resistance to immune surveillance.

Retinoic Acid

Kang et al[253] provided evidence that retinoic acid binds to the M6P/IGF2R and that binding was not inhibited by either IGF-2 or M6P. Correlation of growth inhibition and apoptosis in response to retinoic acid with the presence of the M6P/IGF2R led to the proposal that the receptor may play an important role in mediating retinoid-induced apoptosis/growth inhibition.[254]

M6P/IGF-2 Receptor Functions

Delivery of Acid Hydrolases to Lysosomes

Targeting of lysosomal enzymes to lysosomes can be divided into biosynthetic and endocytic pathways.[255] In the major, biosynthetic arm, newly synthesized lysosomal enzymes are directed to lysosomes by first binding to M6P/IGF2R in the trans Golgi network, then traveling in coated vesicles to early and late endosomes where in the acidic environment of late endosomes, acid hydrolases dissociate from the receptor and are incorporated into lysosomes (Fig. 4). The unoccupied receptor then cycles back to the trans Golgi network to bind newly synthesized lysosomal enzymes. The second minor pathway utilizes approximately 10% of the total cellular M6P/IGF2R population which is present on the plasma membrane to bind extracellular lysosomal enzymes that have escaped the intracellular lysosomal targeting pathway. These extracellular lysosomal enzymes are delivered to lysosomes via clathrin coated pits and vesicles, and early and late endosomes. The M6P/IGF2R cycles continuously among the membrane compartments of both the biosynthetic and endocytic pathways. The rate of receptor internalization has been shown to be increased by binding lysosomal enzyme.[223]

The 46 kDa M6P receptor also targets acid hydrolases to lysosomes. Studies with M6P/IGF2R and 46 kDa M6P receptor knockout mice and fibroblasts derived from these mice indicate that the complementary activities of these two closely related M6P receptors are required for targeting a diverse array of lysosomal enzymes to lysosomes, although lack of the M6P/IGF2R appears to perturb lysosomal enzyme trafficking to a greater degree than lack of the 46 kDa M6P receptor.[256-260]

Degradation of IGF-II

A role for the M6P/IGF2R in receptor mediated internalization and degradation of IGF-2 has been documented in cultured cells including rat adipocytes,[261] L6 rat myoblasts,[262] rat C6 glial cells[263] and mouse L cells (Fig. 4).[264] For example, during an 8 hr incubation of IGF-2 (45ng/ml) with L6 cells, 34% of the IGF-2 was degraded and addition of M6P/IGF2R blocking antibody decreased degradation to only 4%.[262]

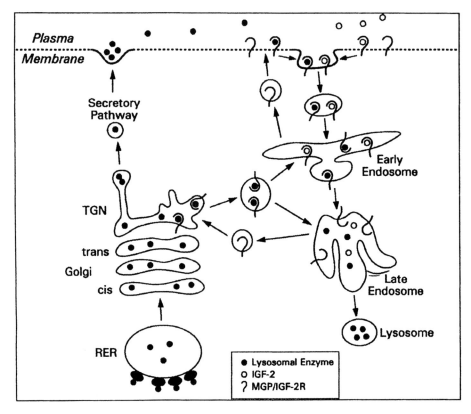

Figure 4. Trafficking of lysosomal enzymes and IGF-2 by the M6P/IGF-2 receptor.

The most compelling evidence for an important role of the M6P/IGF2R in the degradation of IGF-2 comes from gene knockout experiments in mice.[27,265,266] Disruption of the M6P/IGF2R gene resulted in death in the perinatal period and fetuses were 30% larger than normal.[265] Lau et al[266] found that serum IGF-2 in the M6P/IGF2R deficient mice was 2 to 2.7 fold increased compared to wild type littermates. IGF-2 mRNA expression in the M6P/IGF2R knockout embryos was not increased. Ludwig et al[27] found that serum IGF-2 was 4.4 fold higher in the receptor knockout embryos and tissue levels of IGF-2 were also elevated. These studies provide direct evidence for a role of the M6P/IGF2R in IGF-2 degradation.

The overgrowth phenotype of these M6P/IGF2R knockout mice also provide in vivo evidence that the M6P/IGF2R indirectly controls signaling by the IGF-1 receptor in response to IGF-2 by providing a degradative pathway for IGF-2. The hearts of the M6P/IGF2R knockout mice were nearly three times larger than normal and perinatal lethality was attributed to this abnormality.[266] Histological analysis revealed hyperplasia of the myocardium which was supported by increase in total DNA content. During normal development, the heart has the highest expression of M6P/IGF2R of any fetal tissue and IGF-2 is also abundant in the heart. It is plausible that the heart pathology is explained by increased local IGF-2 in the absence of the M6P/IGF2R leading to a proliferative response to IGF-2 that is mediated by the IGF-1 receptor (Fig. 1). Perinatal lethality in the M6P/IGF2R knockout mouse is reversed by generating M6P/IGF2R knockout mice in an IGF-2 deficient background or by generating mice that were deficient in both M6P/IGF2R and IGF-1 receptors.[27]

Signaling of Biological Responses to IGF-2

The overgrowth phenotype of the M6P/IGF2R knockout mouse suggests that the M6P/IGF2R does not signal major growth stimulatory pathways. Nonetheless, there are reports that are consistent with a signaling function for the M6P/IGF2R. The evidence is based on relative dose response curves for IGF-2 versus IGF-1, stimulation by IGF-2 analogs which recognize the M6P/IGF2R but not the IGF-1 receptor, and the use of various M6P/IGF2R antibodies to block or mimic responses to IGF-2. Evidence based solely on relative potency of IGF-2 versus IGF-1 needs to be re-examined because of recent reports that IGF-2 potently activates the A isoform of the insulin receptor.[55]

Biological responses to IGF-2 that are purported to be signaled by the M6P/IGF2R include calcium influx in mouse embryo fibroblasts,[267] phosphorylation of kidney proximal tubule membranes,[268] alkalinization of proximal tubule cells,[269] generation of inositol trisphosphate and diacylglycerol in basolateral kidney membranes,[270] thymidine incorporation into DNA in undifferentiated mouse embryonic limb buds,[271] proteoglycan synthesis in human chondrosarcoma cells,[272] calcium mobilization in rabbit articular chondrocytes,[273] cell motility in rhabdomyosarcoma cells,[274] aromatase activity in placental cytotrophoblasts,[275] extravillous trophoblast migration,[276] insulin exocytosis in pancreatic beta cells,[277] and Ca^{++} transients in rat calvarial osteoblasts.[278] For the most part, in these reports the pathways that connect the M6P/IGF2R to the particular biologic response have not been elucidated. In a series of papers, Nishimoto and his colleagues have studied the molecular basis of their original observation that pertussis toxin, which causes ADP ribosylation of the α subunit of heterotrimeric G proteins, inhibited IGF-2 stimulated Ca^{++} influx in 3T3 mouse embryo fibroblasts.[267,279-282] Ultimately, they provided evidence that a segment of the cytoplasmic domain of the activated M6P/IGF2R (peptide 14) interacts directly with the $G\alpha_i$ resulting in exchange of GTP for GDP on the α subunit. However, Korner et al (283) attempted to confirm these later observations and were unable to demonstrate stimulation of GTPγS binding to $G\alpha_i$ in reconstituted phospholipid vesicles that was dependent upon the presence of peptide 14 in the receptor.

Evidence that the M6P/IGF-2 Receptor is a Tumor Suppressor

IGF-2 overexpression has been reported in a number of human cancers.[284] TGF-β1 is a growth inhibitor for most epithelial cells.[285] Because the M6P/IGF2R mediates the degradation of IGF-2 and may play a role in the activation of TGF-β1, the receptor is a candidate for being a tumor suppressor. An inactivating mutation in one allele of a tumor suppressor is recessive but becomes manifest when the other allele is lost. Loss of the other allele is detected by loss of heterozygosity (LOH) at polymorphic DNA marker loci in or near the gene. LOH and mutation in the remaining allele of the M6P/IGF2R has been demonstrated in hepatocellular carcinoma[286-288] and breast cancer,[289] providing genetic evidence that the M6P/IGF2R is a tumor suppressor. Microsatellites are oligonucleotide repeat sequences present throughout the human genome. Disruption of the DNA mismatch repair system plays a major role in human carcinogenesis and is characterized by microsatellite instability. Most microsatellite instability has been described in noncoding DNA but microsatellite instability has been described for the TGF-β1 type II receptor[290] and more recently for the M6P/IGF2R.[291] Microsatellite instability within the M6P/IGF2R coding region has been reported in gastric, colorectal and endometrial cancers.[291,292]

Conclusion

Since the identification of two different receptors for IGFs by affinity crosslinking almost three decades ago, much has been learned about the IGF-1 receptor and the M6P/IGF-2 receptor. Future work will be directed toward understanding the signaling pathways which account for the diverse roles of the IGF-1 receptor in stimulating proliferation, transformation, and differentiation, and inhibiting apoptosis. These studies may provide a more complete explanation for why insulin's in vivo effects are largely metabolic while the IGFs promote growth. The M6P/IGF-2R has clearly been shown to regulate IGF-2 levels by receptor mediated internalization and degradation and thereby to influence activation of the IGF-1 receptor by IGF-2. It remains to be seen whether or not the M6P/IGF-2 receptor directly signals biological responses to IGF-2.

References

1. Francke U, Yang-Feng TL, Brissenden JE et al. Chromosomal mapping of genes involved in growth control. Cold Spring Harbor Symp Quant Biol 1986; 51:855-866.
2. Abbott AM, Bueno R, Pedrini MT et al. Insulin-like growth factor I receptor gene structure. J Biol Chem 1992; 267:10759-10763.
3. Seino S, Seino M, Nishi S et al. Structure of the human insulin receptor gene and characterization of its promoter. Proc Nat Acad Sci USA 1989; 86:114-118.
4. Shier P, Williard HF, Watt VM. Localization of the insulin receptor-related receptor gene to human chromosome 1. Cytogenet Cell Genet 1990; 54:80-81.
5. Werner H, Stannard B, Bach MA et al. Cloning and characterization of the proximal promoter region of the rat insulin-like growth factor I (IGF-1) receptor gene. Biochem Biophys Res Commun 1990; 169:1021-1027.
6. Mamula PW, Goldfine ID Cloning and characterization of the human insulin-like growth factor-I receptor gene 5'-flanking regions. DNA Cell Biol 1992; 11:43-50.
7. Cooke DW, Bankert LA, Roberts CT et al. Analysis of the human type I insulin-like growth factor receptor promoter region. Biochem Biophys Res Commun 1991; 177:1113-1120.
8. Werner H, Bach MA, Stannard B et al. Structural and functional analysis of the insulin-like growth factor I receptor gene promoter. Mol Endocrinol 1992; 6:1545-1558.
9. Beitner-Johnson D, Werner H, Roberts CT et al. Regulation of insulin-like growth factor I receptor gene expression by Sp1: physical and functional interactions of Sp1 at GC boxes and at a CT element. Mol Endocrinol 1995; 9:1147-1156.
10. Cooke DW, Casella SJ. The 5' untranslated region of the IGF-1 receptor gene modulates gene expression by both pre- and post-transcriptional mechanisms. Mol Cell Endocrinol 1994; 101:77-84.
11. Rubini M, Werner H, Gandini W et al. Platelet derived growth factor increases the activity of the promoter of the insulin-like growth factor-I (IGF-I) receptor gene. Exp Cell Res 1994; 211:374-379.
12. Hernandez-Sanchez C, Werner H, Roberts C. et al. Differential regulation of insulin-like growth factor-1 (IGF-1) receptor gene expression by IGF-1 and basic fibroblastic growth factor. J Biol Chem 1997; 272:4663-4670.
13. Scheidegger KJ, Du J, Delafontaine P. Distinct and common pathways in the regulation of insulin-like growth factor-1 receptor gene expression by angiotensin II and basic fibroblast growth factor. J Biol Chem 1999; 274:3522-3530.
14. Werner J, Rauscher FJ, Sukhatme VP et al. Transcriptional repression of the insulin-like growth factor 1 receptor (IGF-1 R) gene by the tumor suppressor WT1 involves binding to sequences both upstream and downstream of the IGF-1-R gene transcription start site. J Biol Chem 1994; 269:12577-12582.
15. Werner H, Roberts CT, Rauscher FJ et al. Regulation of insulin-like growth factor I receptor gene expression by the Wilms' tumor suppressor WT1. J Mol Neurosci 1996; 7:111-123.
16. Karnieli E, Werner H, Rauscher FJ et al. The IGF-1 receptor gene promoter is a molecular target for the Ewing's sarcoma-Wilm's tumor 1 fusion protein. J. Biol Chem. 1996; 271:19304-19309.
17. Werner H, Karnieli E, Rauscher FJ et al. Wild type and mutant p53 differentially regulate transcription of the insulin-like growth factor 1 receptor gene. Proc Nat Acad Sci USA 1996; 93:8318-8323.
18. Ohlsson C, Kley N, Werner J et al. p53 regulates insulin-like growth factor-1 (IGF-1) receptor expression and IGF-1-induced tyrosine phosphorylation in an osteosarcoma cell line: interaction between p53 and Sp1. Endocrinology 1998; 139:1101-1107.
19. Deb D, Lanyi A, Scian M et al. Differential modulation of cellular and viral promoters by p73 and p53. Int J Oncol 2001; 18:401-409.
20. Maor SB, Abramovitch S, Erdos MR et al. BRCA1 suppresses insulin-like growth factor-1 receptor promoter activity: potential interaction between BRCA1 and Sp1. Molec Genet Metab 2000; 69:130-136.
21. Howard TK, Algar EM, Glatz JA et al. The insulin-like growth factor 1 receptor gene is normally biallelically expressed in human juvenile tissue and tumors. Hum Molec Genet 1993; 2:2089-2092.
22. Ogawa O, McNoe LA, Eccles MR et al. Human insulin-like growth factor type I and II receptors are not imprinted. Hum Molec Genet 1993; 1:2163-2165.
23. Bartolomei MS, Tilghman S M (1997) Genomic imprinting in mammals. Ann Rev Genetics 1997; 3:493-515.
24. Werner H, Woloschak M, Adamo M et al. Developmental regulation of the rat insulin-like growth factor I receptor gene. Proc Nat Acad Sci USA 1989; 86:7451-7455.
25. Liu JP, Baker J, Perkins AS et al. Mice carrying null mutations of the genes encoding insulin-like growth factor I (Igf-1) and type I IGF receptor (Igf1r). Cell 1993; 75:59-72.
26. Louvi A, Accili D, Efstradiatis A. Growth-promoting interaction of IGF-II with the insulin receptor during mouse embryonic development. Dev Biol 1991; 189: 33-48.
27. Ludwig T, Eggenschwiler J, Fisher P et al. Mouse mutants lacking the type 2 IGF receptor (IGF2R) are rescued from perinatal lethality in the Igf2 and Igfr null backgrounds. Dev Biol 1996; 177:517-535.
28. Ullrich A, Gray A, Tam AW et al. Insulin-like growth factor I receptor primary structure: comparison with insulin receptor suggests structural determinants that define functional specificity. EMBO J 1986; 5: 2503-2512.
29. Lowe W.L, Adamo M, Werner H et al. Regulation by fasting of rat insulin-like growth factor I and its receptor. Effects on gene expression and binding. J Clin Invest 1989; 84: 619-626.
30. Pedrini MT, Giorgino F, Smith RJ. cDNA cloning of the rat IGF-1 receptor: structural analysis of rat and human IGF-1 and insulin receptors reveals differences in alternative splicing and receptor-specific domain conservation. Biochem Biophys Res Commun 1994; 202:1038-1046.

31. Holtzenberger M, Lapointe F, Leibovici M et al. The avian IGF type I receptor: cDNA analysis and in situ hybridization reveal conserved sequence elements and expression patterns relevant for the development of the nervous system. Dev Brain Res 1996; 97:76-87.
32. Zhu L, Ohan N, Agazie Y et al. Molecular cloning and characterization of Xenopus insulin-like growth factor I receptor: its role in mediating insulin-induced Xenopus oocyte maturation and expression during embryogenesis. Endocrinology 1998; 139: 949-954.
33. Elies G, Duval H, Bonnec G et al. Insulin- and insulin-like growth factor I receptors in an evoluted fish, the turbot: cDNA cloning and mRNAs expression. Mol Cell Endocrinol 1999; 158:173-185.
34. Bass J, Turck C, Rounar M et al. Furin-mediated processing in the early secretory pathway: Sequential cleavage and degradation of misfolded insulin receptors. Proc Nat Acad Sci 2000; 97:11905-11909.
35. Jacobs S, Kull FSV, Cuatracasas P. Monensin blocks the maturation of receptors for insulin and somatomedin C: identification of receptor precursors. Proc Nat Acad Sci 1993; 80: 1228-1231.
36. Duronio V, Jacobs S, Romero PA et al. Effects of inhibitors of N-linked oligosaccharide processing on the biosynthesis and function of insulin and insulin-like growth factor-I receptors. J Biol Chem 1998; 263:5436-5445.
37. Collier E, Carpentier JL, Beitz L et al. Specific glycosylation sites of the insulin receptor alpha subunit impair intracellular import. Biochemistry 1993; 32:7818-7832.
38. Elleman TC, Frenkel MJ, Hoyne PA et al. Mutational analysis of the N-linked glycosylation sites of the human insulin receptor. Biochem J 2000; 347:771-779.
39. Yee D, Lebovic GS, Marcus RR et al. Identification of an alternate type I insulin-like growth factor receptor beta subunit mRNA transcript. J Biol Chem 1989; 264: 21439-21441.
40. Condorelli G, Bueno R, Smith R. Two alternatively spliced forms of the human insulin-like growth factor I receptor have distinct biological activities and internalization kinetics. J Biol Chem 1994; 269:8510-8516.
41. Siddle K, Soos MA, Field CE et al. Hybrid and atypical insulin/insulin-like growth factor I receptors. Horm Res 1994; 41:56-65.
42. Federici M, Porzio O, Zucaro L et al. Distribution of insulin/insulin-like growth factor-I hybrid receptors in human tissues. Mol Cell Endocrinol 1997; 129:121-126.
43. Bailyes EM, Naveacute BT, Soos MA et al. Insulin receptor/IGF-1 receptor hybrids are widely distributed in mammalian tissues: quantification of individual receptor species by selective immunoprecipitation and immunoblotting. Biochem J 1997; 327:209-215.
44. Garrett TP, McKern NM, Lou M et al. Crystal structure of the first three domains of the type-1 insulin-like growth factor receptor. Nature 1998; 394:395-399.
45. Kjeldsen T, Andersen AS, Wiberg FC et al. Ligand specificities of insulin receptor and the insulin like growth factor I receptor reside in different regions of a common binding site. Proc Nat Acad Sci USA 1991; 88: 4404-4408.
46. Schumacher R, Soos MA, Schlessinger J et al. Signaling-competent receptor chimeras allow mapping of major insulin receptor binding domain determinants. J Biol Chem 1993; 268:1087-1094.
47. Gustafson TA, Rutter WJ. The cysteine-rich domains of the insulin-like growth factor I receptors are primary determinants of hormone binding specificity. Evidence from receptor chimeras. J Biol Chem 1990; 265:18663-18667.
48. Andersen AS, Kjeldsen T, Wiberg FC et al. Changing the insulin receptor to possess insulin-like growth factor I ligand specificity. Biochemistry 1990; 29:7363-7366.
49. Hoyne PA, Ellemann TC, Adams TE et al. Properties of an insulin receptor with an IGF-1 receptor loop exchange in the cysteine-rich region. FEBS Lett. 2000; 469:57-60.
50. Ward CW. Members of the insulin receptor family contain three fibronectin type III domains. Growth Factors 1999; 16:315-322.
51. Soos MA, Field CE, Lammers R et al. A panel of monoclonal antibodies for the type 1 insulin-like growth factor receptor. J Biol Chem 1992; 267:12955-12963.
52. Delafontaine P, Ku L, Ververis JJ et al. Epitope mapping of the alpha-chain of the insulin-like growth factor I receptor using antipeptide antibodies. J Mol Cell Cardiol 1994; 26:1659-1673.
53. Mynarcik DC, Williams PF, Schaffer L et al. Identification of common ligand binding determinants of the insulin and insulin-like growth factor I receptors. Insights into mechanisms of ligand binding. J Biol Chem. 1997; 272:18650-18655.
54. Kristensen C, Wiberg FC, Andersen AS. Specificity of insulin and insulin-like growth factor I receptors investigated using chimeric mini-receptors. Role of C-terminal of receptor alpha subunit. J Biol Chem. 1999; 274:37351-37356.
55. Frasca F, Pandini G, Scalia P et al. Insulin receptor isoform A., a newly recognized high-affinity insulin-like growth factor II receptor in fetal and cancer cells. Mol Cell Biol 1999; 19: 3278-3288.
56. Schaffer L, Ljungqvist L. Identification of a disulphide bridge connecting the alpha-subunits of the extracellular domain of the insulin receptor. Biochim Biophys Res Commun 1992; 189: 650-653.
57. Lu K, Guidotti C. Identification of the cysteine residues involved in the class I disulfide bonds of the human insulin receptor: Properties of insulin receptor monomers. Mol Biol Cell 1996; 7:679-691.
58. Sparrow LG, McKern NM, Gorman JJ et al. The disulphide bonds in the C-terminal domains of the human insulin receptor ectodomain. J Biol Chem 1997; 272:29460-29467.
59. Molina L, Marino-Buslje C, Quinn DR et al. Structural domains of the insulin receptor and IGF receptor required for dimerisation and ligand binding. FEBS Lett 2000; 467:226-230.
60. Maggi D, Cordera R. Cys 786 and Cys 776 in the posttranslational processing of the insulin and IGF-1 receptors. Biochem Biophys Res Commun 2001; 280:836-841.

61. Shoelson SE, White MF, Kahn CR. Tryptic activation of the insulin receptor. Proteolytic truncation of the α subunit releases the β subunit from inhibitory control. J Biol Chem 1988; 263:4852-4860.
62. Liu D, Rutter WJ, Wang LH Modulating effects of the extracellular sequence of the human insulin-like growth factor I receptor on its transforming and tumorigenic potential. J Virol 1993; 67:9-18.
63. Li S, Zhang H, Hoff H et al. Activation of the insulin-like growth factor type I receptor by deletion of amino acids 870-905. Exp Cell Res 1998; 243:326-333.
64. Himmelmann B, Terry C, Dey BR et al. Anchorage-independent growth of fibroblasts that express a truncated IGF-1 receptor. Biochem Biophys Res Commun 2001; 286:472-477.
65. Bargmann CI, Hung, MC, Weinberg RA. Multiple independent activations of the neu oncogene by a point mutation altering the transmembrane domain of p185. Cell 1986; 45:649-657.
66. Takahashi K, Yonezawa K, Nishimoto I. Insulin-like growth factor I receptor activated by a transmembrane mutation. J. Biol. Chem 1995; 270:19041-19045.
67. Sasaki N, Rees-Jones RW, Zick Yet al. Characterization of insulin-like growth factor stimulated tyrosine kinase activity associated with the β subunit of type I insulin-like growth factor receptors of rat liver cells. J Biol Chem 1985; 260:9793-9804.
68. Treadway JL, Morrison BD, Soos MA et al. Trans-dominant inhibition of tyrosine kinase activity in mutant insulin/insulin-like growth factor I hybrid receptors. Proc Natl Acad Sci USA 1991; 88:214-218.
69. Prager D, Li HL, Asa S et al. Dominant negative inhibition of tumorigenesis by human insulin-like growth factor I receptor mutant. Proc Natl Acad Sci USA 1994; 91:2181-2185.
70. Hubbard SR, Wei L, Ellis L et al. Crystal structure of the tyrosine kinase domain of the human insulin receptor. Nature 1994; 372:746-754.
71. Hubbard SR. Crystal structure of the activated insulin receptor tyrosine kinase in complex with peptide substrate and ATP analog. EMBO J 1997; 16:5572-5581.
72. Gronborg M, Wulff BS, Rasmussen JS et al. Structure-function relationship of the insulin-like growth factor-I receptor tyrosine kinase. J Biol Chem 1993; 258:23435-23440.
73. Kato H, Faria TM, Stannard B et al. Essential role of tyrosine residues 1131, 1135, and 1136 of the insulin-like growth factor-I (IGF-I) receptor. J Biol Chem 1994; 265:2655-2661.
74. Li S, Ferber A, Miura M et al. Mitogenicity and transforming activity of the insulin-like growth factor-I receptor with mutations in the tyrosine kinase domain. J Biol Chem 1994; 269:32558-32564.
75. Hernandez-Sanchez C, Blakesley V, Kalebic T et al. The role of the tyrosine kinase domain of the insulin-like growth factor-I receptor in intracellular signaling, cellular proliferation, and tumorigenesis. J Biol Chem 1995; 270:20953-20958.
76. Stannard B, Blakesley V, Kato H et al. Single tyrosine substitution in the insulin-like growth factor I receptor inhibits ligand-induced receptor autophosphorylation and internalization, but not mitogenesis. Endocrinology 1995; 136:4918-4924.
77. Blakesley VA, Kato H, Roberts CT et al. Mutation of a conserved amino acid residue (tryptophan 1173) in the tyrosine kinase domain of the IGF-1 receptor abolishes autophosphorylation but does not eliminate biological function. J Biol Chem 1995; 270:2764-2769.
78. Kato H, Faria TN, Stannard B et al. Role of tyrosine kinase activity in signal transduction by the insulin-like growth factor receptor. Characterization of kinase-deficient IGF-1 receptors and the action of an IGF-1 mimetic antibody (alpha IR-3). J Biol Chem 1993; 268:2655-2661.
79. Till JH, Ablooglu AJ, Frankel M et al. Crystallographic and solution studies of an activation loop mutant of the insulin receptor tyrosine kinase: insights into kinase mechanism. J Biol Chem 2001; 276:10049-10055.
80. Tennegels N, Hube-Magg C, Wirth A et al. Expression, purification and characterization of the cytoplasmic domain of the human IGF-1 receptor using a baculovirus expression system. Biochem Biophys Res Commun 1999; 260:724-728.
81. Lopaczynski W, Terry C, Nissley P Autophosphorylation of the insulin-like growth factor I receptor cytoplasmic domain. Biochem Biophys Res Commun 2000; 279:955-960.
82. Lewis, RE, Wu, P, MacDonald, RG et al. Insulin-sensitive phosphorylation of serine 1293/1294 on the human insulin receptor by a tightly associated serine kinase. J Biol Chem 1990; 265:947-954.
83. Feener EP, Backer JM, King GL et al. Insulin stimulates serine and tyrosine phosphorylation in the juxtamembrane region of the insulin receptor. J Biol Chem 1993; 268:11256-11264.
84. Noelle V, Tennagels N, Klein W. A single substitution of the insulin receptor kinase inhibits serine autophosphorylation in vitro: evidence for an interaction between the C-terminus and the activation loop. Biochemistry 2000; 39:7170-7177.
85. Tauer, TJ, Volle, DJ, Rhode, SL et al. Expression of the insulin receptor with a recombinant vaccinia virus. Biochemical evidence that the insulin receptor has intrinsic serine kinase activity. J Biol Chem 1996; 271:331-336.
86. Carter WG, Sullivan AC, Asamoah KA et al. Purification and characterization of an insulin-stimulated insulin receptor serine kinase. Biochemistry 1996; 35:14340-14351.
87. White MF, Yenush L. The IRS-signaling system: A network of docking proteins that mediate insulin and cytokine action. Curr Top Micro Immunol. 1998; 228:179-208.
88. Giovannone B, Scaldaferri ML, Federici M et al. Insulin receptor substrate (IRS) transduction system: distinct and overlapping signaling potential. Diabetes Metab Rev 2000; 16: 434-441.
89. Craparo A, O'Neill TJ, Gustafson TA. Non-SH2 domains within insulin receptor substrate-1 and SHC mediate their phosphotyrosine-dependent interaction with the NPEY motif of the insulin-like growth factor I receptor. J Biol Chem 1995; 270:23456-23460.

90. Tartare-Deckert S, Sawka-Verhelle D, Murdaca J et al. Evidence for a differential interaction of SHC and the insulin receptor substrate-1 (IRS-1) with the insulin-like growth factor-1 (IGF-1) receptor in the yeast two-hybrid system. J Biol Chem 1995; 270:23456-23460.
91. Dey BR, Frick K, Lopaczynski W et al. Evidence for the direct interaction of the insulin-like growth factor I receptor with IRS-1, Shc, and Grb-10. Mol Endocrinol 1996; 10:631-641.
92. Yenush L, Makati KJ, Smith-Hall J et al. The pleckstrin homology domain is the principle link between the insulin receptor and IRS-1. J Biol Chem 1996; 271:24300-243006.
93. Burks DJ, Pons S, Towery H et al. Heterologous pleckstrin homology domains do not couple IRS-1 to the insulin receptor. J Biol Chem 1997; 272:27716-27721.
94. Farhang-Fallah J, Yin X, Trentin G et al. Cloning and characterization of PHIP, a novel insulin receptor substrate-1 pleckstrin homology domain interacting protein. J Biol Chem. 2000; 275:40492-40497.
95. He W, Craparo A, Zhu Y et al. Interaction of insulin receptor substrate-2 (IRS-2) with the insulin and insulin-like growth factor I receptors. Evidence for two distinct phosphotyrosine-dependent interaction domains within IRS-2. J Biol Chem 1996; 271:11641-11645.
96. Xu P, Jacobs AR, Taylor SI. Interaction of insulin receptor substrate 3 with insulin receptor, insulin receptor-related receptor, insulin-like growth factor-1 receptor and downstream signaling proteins. J Biol Chem 1999; 274:15262-15270.
97. Sawka-Verhelle D, Tartare-Deckert S, White MF et al. Insulin receptor substrate-2 binds to the insulin receptor through its phosphotyrosine-binding domain and through a newly identified domain comprising amino acids 591-786. J Biol Chem 1996; 271:5980-5983.
98. Sawka-Verhelle D, Baron V, Mothe I et al. Tyr624 and Tyr628 in insulin receptor substrate-2 mediate its association with the insulin receptor. J Biol Chem 1997; 272:16414-16420.
99. Bonfini L, Migliaccio E, Pelicci G et al. Not all Shc's roads lead to ras. Trends Biochem Sci 1996; 21:257-261.
100. Sasaoka T, Kobayashi M. The functional significance of Shc in insulin signaling as a substrate of the insulin receptor. Endocr J 2000; 47: 373-381.
101. Gustafson TA, He W, Craparo A et al. Phosphotyrosine-dependent interaction of SHC and insulin receptor substrate I with the NPEY motif of the insulin receptor via a novel non-SH2 domain. Mol Cell Biol 1995; 15:2500-2508.
102. He W, O'Neill TJ, Gustafson TA. Distinct modes of interaction of SHC and insulin receptor substrate-1 with the insulin receptor NPEY region via non-SH2 domains. J Biol Chem 1995; 270:23258-23262.
103. Farooq A, Plotnikova O, Zeng L et al. Phosphotyrosine binding domains of Shc and insulin receptor substrate I recognize the NPXpY motif in a thermodynamically distinct manner. J Biol Chem 1999; 274:6114-6121.
104. van der Geer P, Wiley S, Pawson T. Re-engineering the target specificity of the insulin receptor by modification of a PTB domain binding site. Oncogene 1999; 18:3071-3075.
105. Sasaoka T, Ishiki M, Sawa T et al. Comparison of the insulin and insulin-like growth factor I mitogenic intracellular signaling pathways. Endocrinology 1996; 137:4427-4434.
106. Chow JC, Condorelli G, Smith RJ. Insulin-like growth factor I receptor internalization regulates signaling via the Shc/mitogen-activated protein kinase pathway, but not the insulin receptor substrate-1 pathway. J Biol Chem 1998; 273:4672-4680.
107. Seely BL, Reichart DR, Staubs PA et al. Localization of the insulin-like growth factor I receptor binding sites for the SH2 domain proteins p85, Syp, and GTPase activating protein. J Biol Chem 1995; 270:19151-19157.
108. Lamothe B, Bucchini D, Jami J et al. Reexamining interaction of the SH2 domains of SYP and GAP with insulin and IGF-1 receptors in the yeast two-hybrid system. Gene 1996; 182:77-80.
109. Yamanashi Y, Baltimore D. Identification of the Abl- and rasGAP-associated 62 kDa protein as a docking protein, Dok. Cell 1997; 88:205-211.
110. Sanchez-Margalet V, Zoratti R, Sung CK. Insulin-like growth factor-1 stimulation of cells induces formation of complexes containing phosphatidylinositol-3-kinase, guanosine triphosphatase-activating protein (GAP) and p62 GAP-associated protein. Endocrinology 1995; 136:316-321.
111. Koval AP, Blakesley VA, Roberts CT, et al. Interaction in vitro of the product of the c-Crk-II proto-oncogene with the insulin-like growth factor I receptor. Biochem J 1998; 330:923-932.
112. Feller SM, Posern G, Voss J, et al. Physiological signals and oncogenesis mediated through Crk family adapter proteins. J. Cell Physiol. 1998; 177:535-552.
113. Beitner-Johnson D, LeRoith D. Insulin-like growth factor-1 stimulates tyrosine phosphorylation of endogenous c-Crk. J Biol Chem 1995; 270:5187-5190.
114. Beitner-Johnson D, Blakesley,VA, Shen-Orr Z et al. The proto-oncogene product c-crk associates with Insulin receptor substrate-1 and 4PS. Modulation by insulin growth factor-1 (IGF) and enhanced IGF1-signaling. J Biol Chem 1996; 271:9287-9290.
115. Okada, S, Pessin JE. Insulin and epidermal growth factor stimulate a conformational change in Rap1 and dissociation of the CrkII-C3G complex. J Biol Chem 1997; 272:28179-28182.
116. Okada S, Matsuda M, Anafi M et al. Insulin regulates the dynamic balance between ras and rap1 signaling by coordinating the assembly states of the Grb2-SOS and CrkII-C3G complexes. EMBO J 1998; 17:2554-2565.
117. Ishiki M, Sasaoka T, Ishihara H et al. Evidence for functional roles of Crk-II in insulin and epidermal growth factor signaling in rat-1 fibroblasts overexpressing insulin receptors. Endocrinology 1997; 138:4950-4958.

118. Nakashima N, Rose DW, Xiao S et al. The functional role of CrkII in actin cytoskeletal organization and mitogenesis. J Biol Chem 1997; 274:3001-3008.
119. Tanaka S, Ouchi T, Hanafusa H. Downstream of Crk adaptor signaling pathway: activation of Jun kinase by v-crk through the guanine nucleotide exchange protein C3G. Proc Nat Acad Sci USA 1997; 94:2356-2361.
120. Akagi T, Shishido T, Murata K et al. v-crk activates the phosphoinositide 3-kinase/AKT pathway in transformation. Proc Nat Acad Sci USA 2000; 97:7290-7295.
121. Arbet-Engels C, Tartare-Deckart S, Eckhart W. C-terminal Src kinase associates with ligand-stimulated insulin-like growth factor-1 receptor. J Biol Chem. 1999; 274:5422-5428.
122. Neet K, Hunter T. Vertebrate nonreceptor protein tyrosine kinase families. Genes Cells 1996; 1:147-169.
123. Sabe H, Knudsen B, Okada M et al. Molecular cloning and expression of chicken C-terminal Src kinase: lack of stable association with c-Src protein. Proc Nat Acad Sci USA 1992; 89:2190-2194.
124. Sabe H, Shoelson SE, Hanafusa H. Possible v-Crk-induced transformation through activation of Src kinases. J Biol Chem 1995; 270:31219-31224.
125. Valentinis B, Morrione A, Taylor SJ et al. Insulin-like growth factor I receptor signaling in transformation by src oncogenes. Mol Cell Biol 1997; 17:3744-3754.
126. Hongo A, D'Ambrosio C, Miura M et al. Mutational analysis of the mitogenic and transforming activities of the insulin-like growth factor I receptor. Oncogene 1996; 12:1231-1238.
127. O'Connor R, Kauffmann-Zeh A, Liu Y et al. Identification of domains of the insulin-like growth factor I receptor that are required for protection from apoptosis. Mol Cell Biol 1997; 17: 427-435.
128. Lammers R, Gray A, Schlessinger J et al. Differential signalling potential of insulin- and IGF-1 receptor cytoplasmic domains. EMBO J 1989; 8:1369-1375.
129. Miura M, Surmacz E, Burgaud JL et al. Different effects on mitogenesis and transformation of a mutation at tyrosine 1251 of the insulin-like growth factor-I receptor. J Biol Chem 1995; 270:22639-22644.
130. Blakesley VA, Kalebic T, Helman LJ et al. Tumorigenic and mitogenic capacities are reduced in transfected fibroblasts expressing mutant insulin-like growth factor (IGF)-I receptors. The role of tyrosine residues 1250, 1251, and 1316 in the carboxy-terminus of the IGF-1 receptor. Endocrinology 1996; 137:410-417.
131. Blakesley VA, Koval. AP, Stannard BS et al. Replacement of tyrosine 1251 in the carboxyl terminus of the insulin-like growth factor-I receptor disrupts the actin cytoskeleton and inhibits proliferation and anchorage-independent growth. J Biol Chem 1998; 273:18411-18422.
132. Li S, Resnicoff M, Baserga R. Effect of mutations at serines 1280-1283 on the mitogenic and transforming activities of the insulin-like growth factor I receptor. J Biol Chem 1996; 271: 12254-12260.
133. Peruzzi F, Prisco M, Dews M et al. Multiple signaling pathways of the insulin-like growth factor I receptor in protection from apoptosis. Mol Cell Biol 1999; 19:7203-7215.
134. Peruzzi F, Prisco M, Morrione A et al. Anti-apoptotic signaling of the insulin-like growth factor-I receptor through mitochondrial translocation of c-Raf and Nedd4. J Biol Chem 2001; 276:25990-25996.
135. Jiang Y, Chan JL, Zong CS et al. Effect of tyrosine mutations on the kinase activity and transforming potential of an oncogenic human insulin-like growth factor I receptor. J Biol Chem 1996; 271:160-167.
136. Furlanetto RW, Dey BR, Lopaczynski W et al. 14-3-3 proteins interact with the insulin-like growth factor receptor but not the insulin receptor. Biochem J 1997; 327:765-771.
137. Craparo A, Freund R, Gustafson TA. 14-3-3 (ε) interacts with the insulin-like growth factor I receptor and insulin receptor substrate I in a phosphoserine-dependent manner. J Biol Chem 1997; 272:11663-11669.
138. Fu H, Subramanian RR, Masters SC. 14-3-3 proteins: Structure, function, and regulation. Ann Rev Pharmacol Toxicol 2000; 40:617-647.
139. Lamothe B, Bucchini D, Jami J et al. Interaction of p85 subunit of PI 3-kinase with insulin and IGF-1 receptors analysed by using the two-hybrid system. FEBS Lett 1995; 373:51-55.
140. Tartare-Deckert S, Murdaca J, Sawka-Verhelle D et al. Interaction of the molecular weight 85K regulatory subunit of the phosphatidylinositol 3-kinase with the insulin receptor and the insulin-like growth factor-1 (IGF-I) receptor: comparative study using the yeast two-hybrid system. Endocrinology 1996; 137:1019-1024.
141. Vanhaesebroeck B, Leevers S, Panayotou G et al. Phosphoinositide 3-kinases: a conserved family of signal transducers. Trends Biochem Sci 1977; 22: 267-272.
142. Yu J, Wjasow, C, Backer JM. Regulation of the p85/p110α phosphatidylinositol 3'-kinase. Distinct roles for the N-terminal and C-terminal SH2 domains. J Biol Chem 1998; 273:30199-30203.
143. Mothe I, Delahaye L, Filloux C. et al. Interaction of wild type and dominant-negative p55PIK regulatory subunit of phosphatidylinositol 3-kinase with insulin-like growth factor-1 signaling proteins. Mol Endocrinol 1997; 11:1911-1923.
144. Dey BR, Furlanetto RW, Nissley SP. Cloning of human p55 gamma, a regulatory subunit of phosphatidylinositol 3-kinase, by a yeast two-hybrid library screen with the insulin-like growth factor-I receptor. Gene 1998; 209:175-183.
145. Rocchi, S, Tartare-Deckert S, Sawka-Verhelle D et al. Interaction of SH2-containing protein tyrosine phosphatase 2 with the insulin receptor and the insulin-like growth factor receptor: studies of the domains involved using the yeast two-hybrid system. Endocrinology 1996; 137:4944-4952.
146. Tonks NK, Neel BG. Combinatorial control of the specificity of protein tyrosine phosphatases. Curr Opin Cell Biol 2001; 13:182-195.

147. Xiao S, Rose DW, Sasaoka T et al. Syp (SH-PTP2) is a positive mediator of growth factor-stimulated mitogenic signal transduction. J Biol Chem 1994; 269:21244-21248.
148. Shi Z, Lu W, Feng G. The Shp-2 tyrosine phosphatase has opposite effects in mediating the activation of extracellular signal-regulated and c-jun NH2-terminal mitogen-activated protein kinases. J Biol Chem 1998; 273:4904-4908.
149. Yu DH, Qu,CK, Henegariu O et al. Protein-tyrosine phosphatase SHP-2 regulates cell spreading, migration, and focal adhesion. J Biol Chem 1998; 273:21125-21131
150. Sawada T, Milarski, KL, Saltiel AR. Expression of a catalytically inert Syp blocks activation of the MAP kinase pathway downstream of p21 ras. Biochem Biophys Res Commun 1995; 214:737-743.
151. Fukunaga K, Noguchi T, Takeda H et al. Requirement for protein-tyrosine phosphatase SHP-2 in insulin-induced activation of c-jun NH2-terminal kinase. J Biol Chem 2000; 275: 5208-5213.
152. Kharitonenkov A, Chen Z, Sures I et al. A family of proteins that inhibit signalling through tyrosine kinase receptors. Nature 1997; 386: 181-186.
153. Takada T, Matozaki T, Takeda H et al. Roles of the complex formation of SHPS-1 with SHP-2 in insulin-stimulated mitogen-activated protein kinase activation. J Biol Chem 1998; 273:9234-9242.
154. Tsuda M, Matozaki T, Fukunaga K et al. Integrin-mediated tyrosine phosphorylation of SHPS-1 and its association with SHP-2. Roles of Fak and Src family kinases. J Biol Chem 1998; 273:13223-13229.
155. Nantel A, Mohammed-Ali K, Sherk J et al. Interaction of the Grb10 adapter protein with the Raf1 and MEK1 kinases. J Biol Chem 1998; 273:10475-10484.
156. Nantel A, Huber M, Thomas DY. Localization of endogenous Grb10 to the mitochondria and its interaction with the mitochondrial-associated Raf-1 pool. J Biol Chem 1999; 274:35719-35724.
157. Morrione A, Plant P, Valentinis B et al. mGrb10 interacts with Nedd4. J Biol Chem 1999; 274:24094-24099.
158. Bai RY, Jahn T, Schrem S et al. The SH2-containing adapter protein GRB10 interacts wtih BCR-ABL. Oncogene 1998; 27:941-948.
159. Langlais P, Dong LQ, Hu D et al. Identification of Grb10 as a direct substrate for members of the Src tyrosine kinase family. Oncogene 2000; 19:2895-2903.
160. Morrione A, Valentinis B, Li S et al. Grb10: a new substrate of the insulin-like growth factor I receptor. Cancer Res 1996; 56:3165-3167.
161. Hansen H, Svensson U, Zhu J et al. Interaction between the Grb10 SH2 domain and the insulin receptor carboxyl terminus. J Biol Chem 1996; 271:8882-8886.
162. Frantz JD, Giorgetti-Peraldi S, Ottinger EA et al. Human Grb-IRβ/Grb10: Splice variants of an insulin and growth factor receptor-binding protein with PH and SH2 domains. J Biol Chem 1997; 272:2659-2667.
163. He W, Rose DW, Olefsky JM et al. Grb10 interacts differentially with the insulin receptor, insulin-like growth factor I receptor, and epidermal growth factor receptor via the Grb10 Src homology 2 (SH2) domain and a second novel domain located between the pleckstrin homology and SH2 domains. J Biol Chem 1998; 273:6860-6867.
164. Stein EG, Gustafson TA, Hubbard SR. The BPS domain of Grb10 inhibits the catalytic activity of the insulin and IGF1 receptors. FEBS Lett. 2001; 493:106-111.
165. Laviola L, Giorgino F, Chow JC et al. The adapter protein Grb10 associates preferentially with the insulin receptor as compared with the IGF-1 receptor in mouse fibroblasts. J Clin. Invest 1997; 99:830-837.
166. Wang J, Dai H, Yousaf N et al. Grb10, a positive, stimulatory signaling adapter in platelet-derived growth factor BB-, insulin-like growth factor I- and insulin-mediated mitogenesis Mol Cell Biol 1999; 19:6217-6228.
167. Reidel H, Yousaf N, Zhao Y et al. PSM, a mediator of PDGF-BB-, IGF-1, and insulin-stimulated mitogenesis. Oncogene 2000; 19:39-50.
168. Wakioka T, Sasaki A, Mitsui K et al. APS, an adaptor protein containing PH and SH2 domains inhibits the JAK/STAT pathway in collaboration with c-cbl. Leukemia 1999; 13:760-767.
169. Wang J, Riedel H. Insulin-like growth factor-I receptor and insulin receptor association with a Src homology-2 domain-containing putative adapter. J Biol Chem 1998; 273:3136-3139.
170. Kotani K, Wilden P, Pillay TS. SH2Bα is an insulin-receptor adapter protein and substrate that interacts with the activation loop of the insulin-receptor kinase. Biochem. J 1998; 335:103-109.
171. Nelms K, O'Neill TJ, Li S et al. Alternative splicing, gene localization, and binding of SH2-B to the insulin receptor kinase domain. Mamm Genome 1999; 10:1160-1167.
172. Hallak H, Seiler AEM, Green JS et al. Association of heterotrimeric Gi with the insulin-like growth factor I receptor. Release of Gβγ subunits upon receptor activation. J Biol Chem 2000; 275:2255-2258.
173. Luttrell LM, van Biesen T, Hawes BE et al. Gβγ subunits mediate mitogen-activated protein kinase activation by the tyrosine kinase insulin-like growth factor I receptor. J Biol Chem 1995; 270:16495-16498.
174. Kuemmerle JF, Murthy KS Coupling of insulin-like growth factor I receptor tyrosine kinase to Gi2 in human intestinal smooth muscle. Gβγ-dependent mitogen-activated protein kinase activation and growth. J Biol Chem 2001; 276:7187-7194.
175. Lin FT, Daaka Y, Lefkowitz RJ. β-arrestins regulate mitogenic signaling and clathrin-mediated endocytosis of the insulin-like growth factor I receptor. J Biol Chem 1998; 273:31640-31643.
176. Dalle S, Ricketts W, Imamura T et al. Insulin and insulin-like growth factor I receptors utilize different G protein signaling components. J Biol Chem. 2001; 276:15688-15695.
177. Imamura T, Vollenweider P, Egawa K et al. G-alpha-q/11 protein plays a key role in insulin-induced glucose transport in 3T3-L1 adipocytes. Mol Cell Biol 1999; 19:6765-6774.

178. Gual P, Baron V, Lequoy V et al. Interaction of Janus kinases JAK-1 and JAK-2 with the insulin receptor and the insulin-like growth factor-1 receptor. Endocrinology 1998; 139:884-893.
179. Giorgetti-Peraldi S, Peyrade F, Baron V et al. Involvement of Janus kinases in the insulin signaling pathway. Eur J Biochem 1995: 234:656-660.
180. Saad MJA, Carvalho CRO, Thirone ACP et al. Insulin induces tyrosine phosphorylation of JAK2 in insulin-sensitive tissues of the intact rat. J Biol Chem 1996; 271:22100-22104.
181. Bromberg J. Activation of STAT proteins and growth control. BioEssays 2001; 23:161-169.
182. Zong CS, Zeng L, Jiang Y et al. Stat3 plays an important role in oncogenic ros- and insulin-like growth factor I receptor-induced anchorage-independent growth. J Biol Chem 1998; 273: 28065-28072.
183. Zong CS, Chan J, Levy D, et al. Mechanism of STAT3 activation by insulin-like growth factor I receptor. J Biol Chem 2000; 275:15099-15105.
184. Sawka-Verhelle D, Filloux C, Tartare-Deckert S et al. Identification of Stat 5B as a substrate of the insulin receptor. Eur J Biochem 1997; 250:411-417.
185. Chen J, Sadowski HB, Kohanski RA et al. Stat5 is a physiological substrate of the insulin receptor. Proc Nat Acad Sci USA 1997; 94:2295-2300.
186. Dey BR, Spence SL, Nissley P et al. Interaction of human suppressor of cytokine signaling (SOCS)-2 with the insulin-like growth factor-I receptor. J Biol Chem 1998; 273:24095-24101.
187. Dey BR, Furlanetto RW, Nissley P. Suppressor of cytokine signaling (SOCS)-3 protein interacts wtih the insulin-like growth factor-1 receptor. Biochem Biophys Res Commun 2000; 278:38-43.
188. Hilton DJ. Negative regulators of cytokine signal transduction. Cell Mol Life Sci 1999; 55: 1568-1577.
189. Yasukawa H, Sasaki A, Yoshimura A. Negative regulation of cytokine signaling pathways. Ann Rev Immunol 2000; 18:143-164.
190. Kamura T, Sato S, Haque D et al. The elongin BC complex interacts with the conserved SOCS box motif present in members of the SOCS, ras, WD-40 repeat, and ankyrin repeat families. Genes Dev 1998; 12:3872-3881.
191. Zhong JG, Farley, A, Nicholson SE et al. The conserved SOCS box motif in supressors of cytokine signalling binds to elongins B and C and may couple proteins to proteosomal degradation. Proc Nat Acad Sci USA 1999; 96:2071-2076.
192. Metcalf D, Greenleigh C, Viney E et al. Gigantism in mice lacking suppressor of cytokine-signaling 2. Nature 2000; 405:1069-1073.
193. Emmanuelli V, Peraldi P, Filloux C et al. SOCS-3 is an insulin-induced negative regulator of insulin-signaling. J Biol Chem 2000; 275:15985-15994.
194. Rechler MM, Nissley SP. The nature and regulation of the receptors for insulin-like growth factors. Ann Rev Physiol 1985; 47:425-442.
195. Kaplan A, Achord DT, Sly WS. Phosphomannosyl components of a lysosomal enzyme are recognized by pinocytosis receptors on human fibroblasts. Proc Natl Acad Sci USA 1977; 74:2026-2030.
196. Morgan DO, Edman JC, Standring DN et al. Insulin-like growth factor II receptor as a multifunctional binding protein. Nature 1987; 329:301-307.
197. Kornfeld S. Structure and function of the mannose 6-phosphate/insulin-like growth factor II receptor. Ann Rev Biochem 1992; 61:307-330.
198. Szebenyi G, Rotwein P. The mouse IGF-II/cation-independent Man6P receptor gene: molecular cloning and genomic organization. Genomics 1994; 19:120-129.
199. Laureys G, Barton DE, Ullrich A et al. Chromosomal mapping of the gene for the type II IGF receptor/ cation-independent M6P receptor in man and mouse. Genomics 1988; 3:224-229.
200. Barlow DP, Stöger R, Hermann BG et al. The mouse IGF-type-2 receptor is imprinted and closely linked to the Tme locus. Nature 1991; 349:84-87.
201. Xu YQ, Goodyer CG, Deal C et al. Functional polymorphism in the parental imprinting of the human *IGF2R* gene. Biochem Biophys Res Commun 1993; 197:747-754.
202. Killian JK, Byrd JC, Jirtle JV et al., M6P/IGF2R imprinting evolution in mammals. Mol Cell 2000; 5:707-716.
203. Sklar MM, Thomas CL, Municchi G et al. Developmental expression of rat insulin-like growth factor II/ mannose 6-phosphate receptor messenger ribonucleic acid. Endocrinology 1992; 130:3484-3491.
204. Lobel P, Dahms NM, Kornfeld S. Cloning and sequence analysis of the cation-independent mannose 6-phosphate receptor. J Biol Chem 1988; 263:2563-2570.
205. Hille-Rehfeld A. Mannose 6-phosphate receptors in sorting and transport of lysosomal enzymes. Biochim Biophys Acta 1995; 1241:177-194.
206. Dell'Angelica EC, Payne GS. Intracellular cycling of lysosomal enzyme receptors: cytoplasmic tails' tales. Cell 2001; 106:395-398.
207. Lobel P, Dahms NM, Breitmeyer J et al. Cloning of the bovine 215-kDa cation-independent mannose 6-phosphate receptor. Proc Natl Acad Sci USA 1987; 84:2233-2237.
208. Rosorius, O, Issinger, OG, Braulke, T. Phosphorylation of the cytoplasmic tail of the 300 kDa mannose 6-phosphate recetpor is required for interaction with a cytosolic protein. J Biol Chem 1993; 268: 21470-21473.
209. Dahms NM, Lobel P, Breitmeyer J et al. 46 kd mannose 6-phosphate receptor: cloning, expression, and homology to the 215 kd mannose 6-phosphate receptor. Cell 1987; 50:181-192.
210. Tong PY, Tollefsen SE, Kornfeld S. The cation-independent mannose 6-phosphate receptor binds insulin-like growth factor II. J Biol Chem 1988; 263:2585-2588.

211. Tong PY, Kornfeld S. Ligand interactions of the cation-dependent mannose 6-phosphate receptor. Comparison with the cation-independent mannose 6-phosphate receptor. J Biol Chem 1989; 264:7970-7975.
212. Kiess W, Greenstein LA, White RM et al. Type II IGF receptor is present in rat serum. Proc Natl Acad Sci USA 1987; 84:7720-7724.
213. MacDonald RG, Trepper MA, Clairmont KB et al. Serum form of the rat IGF-II/M6P receptor is truncated in the carboxyl-terminal domain. J Biol Chem 1989; 264:3256-3261.
214. Clairmont KB, Czech MP. Extracellular release as the major degradative pathway of the IGF-II/M6P receptor. J Biol Chem 1991; 266:12131-12134.
215. Valenzano KJ, Remmler J, Lobel P. Soluble insulin-like growth factor-II mannose 6-phosphate receptor carries multiple high molecular weight forms of insulin-like growth factor-II in fetal bovine serum. J Biol Chem 1995; 270:16441-16448.
216. Costello M, Baxter RC, Scott CD. Regulation of soluble insulin-like growth factor II/mannose 6-phosphate receptor in human serum: measurement by enzyme-linked immunosorbent assay. J Clin Endocrinol Metab 1999; 84:2611-2617.
217. Scott CD, Ballesteros M, Madrid J et al. Soluble insulin-like growth factor-II/mannose 6-phosphate receptor inhibits deoxyribonucleic acid synthesis in cultured rat hepatocytes. Endocrinology 1996; 137:873-878.
218. Zaina S, Squire S. The soluble type 2 insulin-like growth factor (IGF-II) receptor reduces organ size by IGF-II-mediated and IGF-II-independent mechanisms. J Biol Chem 1998; 273:28610-28616.
219. Tong PY, Gregory W, Kornfeld S. Ligand interactions of the cation-independent mannose 6-phosphate receptor. The stoichiometry of mannose 6-phosphate binding. J Biol Chem 1989; 264:7962-7969.
220. Westlund B, Dahms NM, Kornfeld S. The bovine Man6P/IGF-II receptor. Localization of Man6P binding sites to domains 1-3 and 7-11 of the extracytoplasmic region. J Biol Chem 1991; 266:23233-23239.
221. Dahms NM, Rose PA, Molkentin JD et al. The bovine Man6P/IGF-II receptor. The role of arginine residues in Man6P binding. J Biol Chem 1993; 268:5457-5463.
222. Marron-Terada PG, Brzycki-Wessell MA, Dahms NM. The two mannose 6-phosphate binding sites of the insulin-like growth factor-II/mannose 6-phosphate receptor display different ligand binding properties. J Biol Chem 1998; 273:22358-22366.
223. York SJ, Arneson LS, Gregory WT et al. The rate of internalization of the mannose 6-phosphate/insulin-like growth factor II receptor is enhanced by multivalent ligand binding. J Biol Chem 1999; 274:1164-1171.
224. Byrd JC, Park JHY, Schaffer BS et al. Dimerization of the insulin-like growth factor II/mannose 6-phosphate receptor. J Biol Chem 2000; 18647-18656.
225. Byrd JC, MacDonald RG. Mechanisms for high affinity mannose 6-phosphate ligand binding to the insulin-like growth factorII/mannose 6-phosphate receptor. Negative cooperativity and receptor oligomerization. J Biol Chem 2000; 275:18638-18646.
226. Canfield WM, Kornfeld S. The chicken liver cation-independent mannose 6-phosphate receptor lacks the high affinity binding site for insulin-like growth factor II. J Biol Chem 1989; 264:7100-7103.
227. Clairmont KB, Czech MP. Chicken and Xenopus mannose 6-phosphate receptors fail to bind insulin-like growth factor II. J Biol Chem 1989; 264:16390-16392.
228. Dahms NM, Brzycki-Wessell MA, Ramanujam KS et al. Characterization of mannose 6-phosphate receptors (MPRs) from opossum liver: opossum cation-independent MPR binds insulin-like growth factor II. Endocrinology 1993; 133:440-446.
229. Yandell CA, Dunbar AJ, Wheldrake JF et al. The kangaroo cation-independent mannose 6-phosphate receptor binds insulin-like growth factor II with low affinity. J Biol Chem 1999; 274:27076-27082.
230. Oppenheimer CL, Czech MP. Purification of the type II insulin-like growth factor from rat placenta. J Biol Chem 1983; 258:8539-8542.
231. August GP, Nissley SP, Kasuga M et al. Purification of an insulin-like growth factor II receptor from rat chondrosarcoma cells. J Biol Chem 1983; 258:9033-9036.
232. Scott CD, Baxter RC. Purification and immunological characterization of the rat liver insulin-like growth factor-II receptor. Endocrinology 1987; 120:1-9.
233. Tong PYK, Tollefsen SE, Kornfeld S. The cation-independent mannose 6-phosphate receptor binds insulin-like growth factor II. J. Biol Chem 1988; 263:2585-2588.
234. Schmidt B, Kiecke-Siemsen C, Waheed A et al. Localization of the IGF-II binding site to amino acids 1508-1566 in repeat 11 of the Man6P/IGF-II receptor. J Biol Chem 1995; 270:14975-14982.
235. Dahms NM, Rose PA, Molkentin JD et al. The bovine Man6p/IGF-II receptor. Localization of the IGF-II binding site to domains 5-11. J Biol Chem 1994; 269:3802-3809.
236. Garmroudi F, Devi G, Slentz DH et al. Truncated forms of the insulin-like growth factor II (IGF-II)/mannose 6-phosphate receptor encompassing the IGF-II binding site: characterization of a point mutation that abolishes IGF-II binding. Mol Endocrinol 1996; 10:642-651.
237. Grimme S, Höning S, von Figura K et al. Endocytosis of insulin-like growth factor II by a mini-receptor based on repeat 11 of the mannose 6-phosphate/insulin-like growth factor II receptor J Biol Chem 2000; 275: 33697-33703.
238. Devi GR, Byrd JC, Slentz DH et al. An insulin-like growth factor II (IGF-II) affinity-enhancing domain localized within extracytoplasmic repeat 13 of the IGF-II/mannose 6-phosphate receptor. Mol endocrinol 1998; 12:1661-1672.
239. Linnell J, Groeger G, Hassan AB. Real time kinetics of insulin-like growth factor II (IGF-II) interaction with the IGF-II/mannose 6-phosphate receptor. The effects of domain 13 and pH. J Biol Chem 2001; 276:23986-23991.

240. Kiess W, Blickenstaff GD, Sklar MM et al. Biochemical evidence that the type II insulin-like growth factor receptor is identical to the cation-independent mannose 6-phosphate receptor. J Biol Chem 1988; 263:9339-9344.
241. Kiess W, Thomas CL, Greenstein et al. Insulin-like growth factor-II (IGF-II) inhibits both the cellular uptake of β-galactosidase and the binding of β-galactosidase to purified IGF-II/mannose 6-phosphate receptor. J Biol Chem 1989; 264: 4710-4714.
242. DeLeon DD, Terry C, Asmerom Y et al. Insulin-like growth factor II modulates the routing of cathepsin D in MCF-7 breast cancer cells. Endocrinology 1996; 137: 1851-1859.
243. Hoeflich A, Wolf E, Braulke T et al. Does the overexpression of pro-insulin-like growth factor-II in transfected human embryonic kidney fibroblasts increase the secretion of lysosomal enzymes. Eur J Biochem 1995; 232:172-178.
244. Kovacina KS, Steele-Perkins G, Purchio AF et al. Interactions of recombinant and platelet transforming growth factor-β1 precursor with the insulin-like growth factor II/mannose 6-phosphate receptor. Biochem Biophys Res Commun 1989: 160:393-403.
245. Flaumenhaft R, Kojima S, Abe M et al. Activation of latent transforming growth factor β Adv Pharmacology 1993; 24:51-76.
246. Godár S, Horejsi V, Weidle UH et al. M6P/IGFII-receptor complexes urokinase receptor and plasminogen for acivation of transforming growth factor-β1. Eur J Immunol 1999; 29:1004-1013.
247. Lee S-J, Nathans D. Proliferin secreted by cultured cells binds to mannose 6-phosphate receptors. J Biol Chem 1988; 263:3521-3527.
248. Nelson JT, Rosenzweig N, Nilsen-Hamilton M. Characterization of the miitogen-regulated protein (proliferin) receptor. Endocrinology 1995; 136:283-288.
249. Herzog V, Neumuller W, Holzmann B. Thyroglobulin, the major and obligatory exportable protein of thyroid follicle cells, carries the lysosomal recognition marker mannose 6-phosphate. EMBO J 1987; 6:555-560.
250. Lemansky P, Herzog V. Endocytosis of thyroglobulin is not mediated by mannose-6-phosphate receptors in thyrocytes. Evidence for low-affinity-binding sites operating in the uptake of thyroglobulin. Eur J Biochem 1992; 209:111-119.
251. Blanchard F, Duplomb L, Raher S et al. Mannose 6-phosphate/insulin-like growth factor II receptor mediates internalization and degradation of leukemia inhibitory factor but not signal transduction. J Biol Chem 1999; 274:24685-24693.
252. Motyka B, Korbutt G, Pinkoski MJ et al. Mannose 6-phosphate/insulin-like growth factor II receptor is a death receptor for granzyme B during cytotoxic T cell-induced apoptosis. Cell 2000; 103:491-500.
253. Kang JX, Li Y, Leaf A. Mannose-6-phosphate/insulin-like growth factor-II receptor is a receptor for retinoic acid. Proc Natl Acad Sci USA 1997; 95:13671-13676.
254. Kang JX, Bell J, Beard RL et al. Mannose 6-phosphate/insulin-like growth factor II receptor mediates the growth-inhibitory effects of retinoids. Cell Growth & Differentiation 1999; 10:591-600.
255. Dahms NM, Lobel P, Kornfeld SA. Mannose 6-phosphate receptors and lysosomal enzyme targeting. J Biol Chem 1989; 12115-12118.
256. Munier-Lehmann H, Mauxion F, Bauer U et al. Re-expression of the mannose 6-phosphate receptors in receptor-deficient fibroblasts. Complementary function of the two mannose 6-phosphate receptors in lysosomal enzyme targeting. J Biol Chem 1996; 271:15166-15174.
257. Kasper D, Dittmer F, von Figura K et al. Neither type of mannose 6-phosphate receptor is sufficient for targeting of lysosomal enzymes along intracellular routes. J Cell Biol 1996; 34:615-623.
258. Pohlmann R, Boeker MWC, von Figura K. The two mannose 6-phosphate receptors transport distinct complements of lysosomal proteins. J Biol Chem 1995; 270:27311-27318.
259. Ludwig T, Munier-Lehmann H, Bauer U et al. Differential sorting of lysosomal enzymes in mannose 6-phosphate receptor-deficient fibroblasts. EMBO J 1994; 13:3430-3437.
260. Sohar I, Sleat D, Liu C-G et al. Mouse mutants lacking the cation-independent mannose6-phosphate/ insulin-like growth factor II receptor are impaired in lysosomal enzyme transport: comparison of cation-independent and cation-dependent mannose 6-phosphate receptor-deficient mice. Biochem J 1998; 330:903-908.
261. Oka Y, Roze LM, Czech MP Direct demonstration of rapid insulin-like growth factor II receptor internalization and recycling in rat adipocytes. Insulin stimulates ^{125}I-insulin-like growth factor II degradation by modulating the IGF-II receptor recycling process. J Biol Chem 1985;260: 9435-9442.
262. Kiess W, Haskell JF, Lee L et al. An antibody that blocks insulin-like growth factor (IGF) binding to the type II IGF receptor is neither an agonist nor an inhibitor of IGF-stimulated biologic responses in L6 myoblasts. J Biol Chem 1987; 262: 12745-12751.
263. Kiess W, Lee L, Graham DE et al. Rat C6 glial cells synthesize insulin-like growth factor-I (IGF-I) and express IGF-I receptors and IGF-II/mannose 6-phosphate receptors. Endocrinology 1989; 124: 1727-1736.
264. Nolan CM, Kyle JW, Watanabe H et al. Binding of insulin-like growth factor II (IGF-II) by human cation-independent mannose 6-phosphate/IGF-II receptor expressed in receptor-deficient mouse L cells. Cell Regulation 1990; 1:197-213.
265. Wang Z-Q, Fung MR, Barlow DP et al. Regulation of embryonic growth and lysosomal enzyme targeting by the imprinted igf2/Mpr gene. Nature 1994; 372:464-467.
266. Lau MMH, Stewart CEH, Liu Z et al. Loss of the imprinted IGF2/cation-independent mannose 6-phosphate receptor results in fetal overgrowth and perinatal lethality. Genes & Development 1994; 8: 2953-2963.

267. Nishimoto I, Hata Y, Ogata E et al. Insulin-like growth factor II stimulates calcium influx in competent Balb/c 3T3 cells primed with epidermal growth factor. Characteristics of calcium influx and involvement of GTP-binding protein. J Biol Chem 1987: 262:12120-12126.
268. Hammerman MR, Gavin JR Binding of insulin-like growth factor II and multiplication-stimulating activity-stimulated phosphorylation in basolateral membranes from dog kidney. J Biol Chem 1984: 259:13511-13517.
269. Mellas J, Gavin JR, Hammerman MR. Multiplication-stimulating activity-induced alkalinization of canine renal proximal tubular cells. J Biol Chem 1986; 261:14437-14442.
270. Rogers SA, Hammerman MR. Insulin-like growth factor II stimulates production of inositol trisphosphate in proximal tubular basolateral membranes from canine kidney. Proc Natl Acad Sci USA 1988; 85:4037-4041.
271. Bhaumick B, Bala RM. Receptors for insulin-like growth factors I and II in developing embryonic mouse limb bud. Biochim Biophys Acta 1987; 927:117-128.
272. Takigawa M, Okawa T, Pan H-O et al. Insulin-like growth factors I and II are autocrine factors in stimulating proteoglycan synthesis, a marker of differentiated chondrocytes, acting through their respective receptors on a clonal human chondrosarcom-derived chondrocyte cell line, HCS-2/8. Endocrinology 1997; 138:4390-4400.
273. Poiraudeau S, Lieberherr MA Kergosie N et al. Different mechanisms are involved in intracellular calcium increase by insulin-like growth factors 1 and 2 in articular chondrocytes: voltage-gated calcium channels, and/or phospholipase C coupled to a pertussis-sensitive G-protein. J Cell Biochem 1997; 64:414-422.
274. El-Badry OM, Minniti C, Kohn EC et al. Insulin-like growth factor II acts as an autocrine growth and motility factor in hu;man rhabdomyosarcoma tumors. Cell Growth & Differentiation 1990; 1:325-331.
275. Nestler JE. Insulin-like growth factor II is a potent inhibitor of the aromatase activity of human placental cytotrophoblasts. Endocrinology 1990; 127:2064-2070.
276. McKinnon T, Chakraborty C, Gleeson LM et al. Stimulation of human extravillous trophoblast migration by IGF-II is mediated by IGF type 2 receptor involving inhibitory G protein(s) and phosphorylation of MAPK. J Clin Endocrinol Metab 2001; 3665-3674.
277. Zhang Q, Tally M, Larsson O et al. Insulin-like growth factor II signaling through the insulin-like growth factor II/mannose-6-phosphate receptor promotes exocytosis in insulin-secreting cells. Proc Natl Acad Sci USA 1997; 94:6232-6237.
278. Martinez DA, Zuscik MJ Ishibe M et al. Identification of functional insulin-like growth factor-II/mannose-6-phosphate receptors in isolated bone cells. J Cell Biochem 1995; 59:246-257.
279. Nishimoto I. The IGF-II receptor system: a G protein-linked mechanism. Mol Reprod Dev 1993;35:398-407.
280. Okamoto T, Nishimoto I. Analysis of stimulation-G protein subunit coupling by using active insulin-like growth factor II receptor peptide. Proc Natl Acad Sci USA 1991; 88: 8020-8023.
281. Takahashi K, Murayama Y, Okamoto T et al. Conversion of G-protein specificity of insulin-like growth factor II/mannose 6-phosphate receptor by exchanging of a short region with β-adrenergic receptor. Proc Natl Acad Sci USA 1993; 90:11772-11776.
282. Ikezu T, Okamoto T, Giambarella U et al. Yokota, T. In vivo coupling of insulin-like growth factor II/mannose 6-phosphate receptor to heteromeric G proteins. Distinct roles of cytoplasmic domains and signal sequestration by the receptor. J Biol Chem 1995; 270:29224-29228.
283. Korner C, Nurnberg B, Uhde M et al. Mannose 6-phosphate/insulin-like growth factor II receptor fails to interact with G-proteins. Analysis of mutant cytoplasmic receptor domains. J Biol Chem 1995; 270:287-295.
284. Toretsky JA, Helman LJ. Involvement of IGF-II in human cancer. J Endocrinology 1996; 149:367-372.
285. Massague J, Wotton D. Transcriptional control by the TGF-β/Smad signaling system. EMBO J 2000; 19:1745-1754.
286. De Souza AT, Hankins GR, Washington MK et al. Frequent loss of heterozygosity on 6q at the mannose 6-phosphate/insulin-like growth factor II receptor locus in human hepatocellular tumors. Oncogene 1995; 10: 1725-1729.
287. De Souza AT, Hankins GR, Washington MK et al. *M6P/IGF2R* gene is mutated in human hepatocellular carcinomas with loss of heterozygosity. Nature Genetics 1995;11:447-449.
288. Yamada T, De Souza AT, Finkelstein S et al. Loss of the gene encoding mannose 6-phosphate/insulin-like growth factor II receptor is an early event in liver carcinogenesis. Proc Natl Acad Sci USA 1997; 94: 10351-10355.
289. Hankins GR, De Souza AT, Bentley RC et al. M6P/IGF2 receptor: a candidate breast tumor suppressor gene. Oncogene 1996; 12:2003-2009.
290. Markowitz S, Wang J, Myeroff L et al. Inactivation of the type II TGF-b receptor in colon cancer cells with microsatellite instability. Science 1995; 268:1336-1338.
291. Souza RF, Appel R, Yin J et al. Microsatellite instability in the insulin-like growth factor II receptor gene in gastrointestinal tumors. Nature Genetics 1996; 14:255-257.
292. Ouyang H, Shiwaku HO, Hagiwara H et al. The insulin-like growth factor II receptor gene is mutated in genetically unstable cancers of the endometrium, stomach, and colorectum. Cancer Res 1997; 57:1851-1854.

CHAPTER 3

IGFBPs—Gene and Protein Structure

Leon A. Bach and Nigel J. Parker

Abstract

IGF actions are regulated by a family of six high affinity binding proteins (IGFBPs), some of which also have IGF-independent actions. The IGFBPs have highly conserved N- and C-domains, each of which contain internal disulfide links. The middle, 'linker' L-domains of the IGFBPs are not conserved. The N- and C-domains are important for high affinity IGF binding. The L-domains are sites of post-translational modification such as glycosylation and phosphorylation as well as being sites of IGFBP proteolysis. A basic region in the C-domains of IGFBP-3 and IGFBP-5 is involved in binding of these IGFBPs to a range of biomolecules, including the acid labile subunit of the circulating 150 kDa ternary complex, heparin, extracellular matrix proteins and proteases. This region also underlies properties such as cell association and nuclear localization. This region appears to be important for IGF-independent as well as IGF-dependent actions of the IGFBPs.

Insulin-like growth factor (IGF) actions are regulated by a family of six high-affinity IGF binding proteins (IGFBP 1-6).[1] The N- and C-terminal domains of the IGFBPs have highly conserved amino acid sequences which underlie common properties such as high affinity IGF binding. However, each of the IGFBPs also has unique sequences that underlie differentiating biochemical and functional properties such as post-translational modifications, binding to glycosaminoglycans and integrins, and nuclear localization. Unravelling the structural basis of the properties of the IGFBPs is therefore important for fully understanding their actions.

Major advances in understanding the structural basis of IGFBP functions have arisen because of the use of powerful technologies in recent years. Among these is the use of recombinant DNA technology for large-scale IGFBP expression, site-directed mutagenesis, and construction of IGFBP chimeras containing regions of different IGFBPs. Mass spectrometry has allowed more detailed studies of disulfide linkages, proteolytic fragments and post-translational modifications. Very recently, nuclear magnetic resonance and X-ray crystallography have allowed the first glimpses of the three-dimensional structure of IGFBPs. Finally, kinetics of IGF:IGFBP interactions are now being analyzed using surface plasmon resonance with the BiaCore instrument.

Genes

Genomic Structure

The genes encoding each of the human IGFBPs have been cloned, sequenced and characterized (Table 1).[2-7] The human genome project has provided the complete sequence of the introns and flanking sequences for each of the genes as well as definitively locating and orienting the genes. Some interesting information regarding neighbouring genes has also been obtained. Two pairs of IGFBP genes are arranged as chromosomal neighbours in a "tail-to-tail" orientation; *IGFBP1* and *IGFBP3* are separated by 18.7 kb while *IGFBP2* and *IGFBP5* are separated by 11.6 kb. For the tandemly arranged genes, each is the other's nearest neighbour and they are transcribed convergently from opposite strands. The nearest gene to *IGFBP4* is topoisomerase II α (*TOP2A*) which is transcribed divergently from the opposite strand. *IGFBP6* lies between eukaryotic translation initiation

---

Insulin-Like Growth Factors, edited by Derek LeRoith, Walter Zumkeller and Robert Baxter. ©2003 Eurekah.com and Kluwer Academic / Plenum Publishers.

Table 1. Features of the IGFBP genes

Gene	Chromosomal Location	Gene Size (kb)	mRNA Size (kb)	Intron 1 Size (kb)	Prepeptide (a.a.'s)	Mature Protein (a.a.'s)
IGFBP1	7p12-13	5.2	1.46	1.5	25	234
IGFBP2	2q33-34	31	1.43	26.5	39	289
IGFBP3	7p12-13	8.9	2.5	3.3	27	264
IGFBP4	17q	14.2	2.1	8.9	21	237
IGFBP5	2q33	19.5	1.72	15.4	20	252
IGFBP6	12q13	4.7	1.3	2.7	24	216

factor 4B (*EIF4B*) and sterol O-acyltransferase 2 (*SOAT2*), These three genes are oriented in the same direction.

Conservation of Genomic Organization

The IGFBP genes have a similar genomic structure, consisting of four exons, except for *IGFBP3* which has an additional 3' non-coding exon. This additional exon appears to have arisen from disruption of the fourth exon just 3' of the termination codon. The considerable variation in gene size is primarily due to the size of the first intron, which in the case of *IGFBP2*, for example, is larger than 25 kb (Table 1). The sizes of mature mRNAs also vary, in part due to 5' and 3' untranslated regions but there are also some differences in exon size.

The exon structure of the genes corresponds to the domain structure of the proteins. The N-domain and a few amino acids of the L-domain are encoded by exon 1. Most of the L-domain is encoded by exon 2, whereas the C-domain is encoded by exons 3 and 4. Interestingly, the first eight amino acids of the highly basic amino acid regions in the C-domains of IGFBP-3 and -5 are encoded by exon 3 whereas the remaining nine amino acids are encoded by exon 4. IGFBP-3 and -5 have five basic amino acids in the nine residues encoded by exon 4, whereas the other IGFBPs have 0-3 basic amino acids. In contrast, all six IGFBPs have 3-5 basic amino acids in the eight residues encoded by exon 3.

cDNAs

The cDNAs for the human IGFBPs have been isolated from various sources.[1] *IGFBP1* was isolated from liver, placenta, HepG2 liver cells and decidua libraries and found to encode a 234 amino acid mature protein with a 25 amino acid leader peptide (Table 1).[8,9] *IGFBP2* was cloned from fetal liver and encodes a 289 amino acid mature protein with a 39 amino acid leader peptide.[10] *IGFBP3* was isolated from liver and encodes a 264 amino acid mature protein with a 27 amino acid leader peptide.[11] *IGFBP4* was cloned from placenta and TE89 osteosarcoma cells and encodes a 237 amino acid mature protein with a 21 amino acid leader peptide.[12,13] *IGFBP5* was cloned from placenta and osteosarcoma cells and encodes a 252 amino acid mature protein, including a 20 amino acid leader peptide.[14,15] *IGFBP6* was also cloned from placenta and osteosarcoma cells and encodes a 216 amino acid mature protein with a 24 amino acid leader peptide.[16,17]

Promoters

Regulation of transcription of the IGFBP genes is an important aspect of their biology. Interestingly, some of the genes lack obvious TATA boxes (Table 2). While TATA-less promoters are generally associated with housekeeping genes and genes that are required at specific times in all cells, it would appear that the IGFBPs use a variety of promoter elements to regulate expression. The following discussion is not exhaustive but covers recent evidence of functional *cis*-elements within IGFBP promoters.

Table 2. Features of IGFBP promoters

Gene	TATA Sequence	CCAAT Sequence	Other Elements
IGFBP1	+	+	HNF-1, AP-1, AP-2, IRE, hypoxia response element (HRE)
IGFBP2	-	-	Sp1, HRE, NF-κB
IGFBP3	+	-	Sp1, AP-2, HRE, IRE
IGFBP4	+	+	Sp1, AP-1
IGFBP5	+	+	AP-2, CACCC box, E-box, C/EBP, NF-1
IGFBP6	-	+	RARE

IGFBP1

The *IGFBP1* promoter contains an insulin response element (IRE). How these elements function is not fully understood, nor are the proteins that bind them fully characterized, but the forkhead/winged helix (FKH or Fox) class of transcription factors has been implicated. In the case of *IGFBP1*, both insulin and glucocorticoids function through this single site to inhibit and increase transcription, respectively. Initial studies on the *IGFBP1* IRE binding proteins identified the FKH hepatic nuclear factor (HNF)3β. However, HNF3β itself is unlikely to be the central factor in the response. Transcription factors related to the Caenorhabditis elegans DAF-16 have been found to mediate the insulin and glucocorticoid responses. These proteins are FKHR[18] and FKHRL1.[19] Nasrin et al (2000) went on to show that FKHR recruited the p300/CREB-binding protein/steroid receptor coactivator complex to the promoter. [18] *IGFBP1* transcription also increases with hypoxia and the promoter contains 3 potential hypoxia response elements (HREs), at least one of which is functional.[20]

IGFBP2

Basal *IGFBP2* expression is regulated by Sp1.[21] *IGFBP2* expression was increased by nuclear factor-κB and the promoter contains 4 putative binding sites.[22] The *IGFBP2* promoter contains HREs, but the response to hypoxia is variable. [20]

IGFBP3

The *IGFBP3* promoter contains an insulin response element (IRE) which mediates insulin's stimulatory effect on *IGFBP3* expression.[23] There is an HRE element in *IGFBP3* and hypoxia modestly decreases expression. [20] IGF-I and cAMP effects on *IGFBP3* transcription are mediated by a promoter region that contains multiple AP-2 and Sp1 sites.[24] Sp1 and AP-2 have also been implicated in the effect of histone deacetylase inhibitors on *IGFBP3* transcription.[25] Single nucleotide polymorphisms within the *IGFBP-3* promoter modulate its activity and correlate with circulating IGFBP-3 levels.[26]

IGFBP4

Deletion studies on the *IGFBP4* promoter revealed that a cAMP-response element lay between nt -869 and -6 relative to the transcription initiation site, although treatment with a cAMP analogue resulted in only a two-fold increase in promoter activity.[5] Estrogen regulation of *IGFBP4* transcription appears to be dependent on Sp1.[27]

IGFBP5

AP-2 regulates basal and cAMP-dependent *IGFBP-5* transcription.[28] The effects of prostaglandin E2 are mediated by a region containing E box, C/EBP, nuclear factor 1 and AP-2 binding sites.[29] E box binding proteins may also be involved in inhibition of *IGFBP5* transcription by cortisol.[30] Progesterone increases *IGFBP5* transcription through interaction of transcription factors with CACCC sequences rather than via progesterone response element half-sites.[31]

IGFBP6

The *IGFBP6* gene does not have a TATA-box and, like many TATA-less promoters, has multiple transcription start sites clustered downstream of a GC-rich region.[7,32] Interestingly, the two groups to have investigated the transcription start sites have identified different bases, although they are physically close. Retinoic acid increases IGFBP6 mRNA levels in osteosarcoma cells.[32] The promoter region contains three retinoic acid response elements (RAREs); disruption of the most proximal RARE abolishes responsiveness to retinoic acid. Additional upstream elements are required for high level expression and potentially include HSF, NF-1, AP-2, Sp1 and CACCC binding sites, although involvement of these factors has not been shown.

mRNA Stability

Instability of the IGFBP mRNAs is at least in part due to AUUUA sequences in their 3' untranslated regions, which are bound by proteins and result in mRNA degradation. *IGFBP1* contains six of these motifs and its mRNA has a half life of approximately 3 h. Gay and Babajko (2000) removed each of the AUUUA motifs and found that the half life of the message was increased as each was removed.[33] Removal of all six motifs extended the half life of the mRNA to 26 h. In contrast, the cAMP-mediated increase in half-life of *IGFBP3* mRNA was associated with a reduction in binding of a 42 kDa protein to a uridine-rich region in the 3' untranslated region which did not contain the typical AUUUA sequence.[34]

Proteins

Primary Structure

Mature human IGFBPs vary in length from 216-289 amino acids (Table 1). They have three domains of approximately equal size, the N- and C-domains being joined by a 'linker' L-domain (Fig. 1). The N- domains of different IGFBPs share significant homology, as do the C-domains (Fig. 2). In contrast, the L-domains show no obvious homology. A general concept of IGFBP structure suggests that they consist of two compact domains (the N- and C-domains), both of which are involved in IGF binding, joined by a flexible 'hinge' region (the L-domain, Fig. 1).

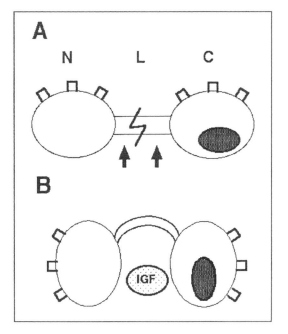

Figure 1. IGFBP structure. (A) IGFBPs consist of 3 domains. The N- and C-domains contain internal disulfide linkages and are involved in IGF binding. The L-domain is a site of post-translational modification (arrows) and proteolysis (zig-zag). The C-domain of some IGFBPs contains a region (dark shading) involved in binding to the acid labile subunit, glycosaminoglycans, importin β and other proteins. (B) The N- and C-domains are required for high affinity IGF binding. The L-domain is thought to facilitate binding by acting as a flexible 'linker' region.

```
IGFBP-1   --------APWQCAPCSAEKLALCPPVSAS----------------CSEVTRS----AGCG  33
IGFBP-2   --------EVLFRCPPCTPERLAACGPPPVAPPAAVAAVAGGARMPCAELVRE----PGCG  49
IGFBP-3   GASSGGLGPVVRCEPCDARALAQCAPPPA-----------------VCAELVRE----PGCG  41
IGFBP-4   --------DEAIHCPPCSEEKLARCRPP----------VG------CEELVRE----PGCG  33
IGFBP-5   --------LGSFVHCEPCDEKALSMCPPSPL---------------GC-ELVKE----PGCG  34
IGFBP-6   --------ALARCPGCGQGVQAGCPG--------------------GCVEEEDGGSPAEGCA  34
Consensus         :*    *    :  *                            * *       **.

IGFBP-1   CCPMCALPLGAACGVATARCARGLSCRALPGEQQPLHALTRGQGAC»VQES----------  83
IGFBP-2   CCSVCARLEGEACGVYTPRCGQGLRCYPHPGSELPLQALVMGDGTC»EKRR----------  99
IGFBP-3   CCLTCALSEGQPCGIYTERCGSGLRCQPSPDEARPLQALLDGRGLC»VMASAVSRLRAYLL 101
IGFBP-4   CCATCALGLGMPCGVYTPRCGSGLRCYPPRGVEKPLHTLMHGQGVC»MEL----------  82
IGFBP-5   CCMITCALAEGQSCGVYTERCAQGLRCLPRQDEEKPLHALLHGRGVC»LNE------KSYRE  88
IGFBP-6   EAEGCLRREGSQECGVYTPNCAPGLQCHPPKDDEAPLRALLIGRGRC»LPA----------  83
Consensus .   *       *   **:  *  .*.  **  *  .   .      **::* *.* *

IGFBP-1   --DASAPHAAEAGSPESPE.STEIT-----EEELLDNFHLMAPSE--------EDHSILWD 128
IGFBP-2   --DAEYGASPEQVADNGDDHSEGGLVENHVDSTMNMLGGGGSAGRKPLKSGMKELAVFRE  157
IGFBP-3   PAPPAPGMA.SE.SEEDRSAGSVESPSVSSTHRVSDPKFHPLHSKIIIIKKGHAKDSQRYKV 161
IGFBP-4   ---AEIEAIQESLQPS--DKDEG-------DHPNNSFSPCSAHDRRCLQ---KHFAKIRD 127
IGFBP-5   QVKIER-DSREHEEPITSEMAEETYSPKIFRPKHTRISELKAEAVKKDRRKKLTQSKFVG 147
IGFBP-6   ---------------RAPAVAEE----------NPKESKPQAGTARPQDVNRRDQQRNPG 118
Consensus                          *                    .

IGFBP-1   AIST------YDGSKALHVTNIKKWK»EP-----CRIELYRVVESLAKA----QETSGEEI 173
IGFBP-2   KVTEQHRQMGKGGKHHLGLEEPKKLRPPPAR»TPCQQELDQVLERISTMRLPDERGPLEHL 217
IGFBP-3   DYES--------QSTDTQMFSSESKRETEY»GPCRREMEDTLNHLKFL--------NVLSP 205
IGFBP-4   RSTS-------GGKMKVNGAPREDARPVPQ»GS-CQSELHRAIERLAAS----QSRTHEDL 175
IGFBP-5   GAEN---------TAHPRIISAPEMRQESEQ»GPCRRHMEASLQELKAS-------PRMVP 191
IGFBP-6   TSTT---------PSQP----NSAGVQDTEM»GPCRRHLDSVLQQLQTE-------VYRGA 158
Consensus                                 *:.:    ::  :

IGFBP-1   SKFYLPNCNKNGFYHSRQCETSMDGEAGLCWCVYPWNGKRIPGSPEIRG-DPNCQIYFNV 232
IGFBP-2   YSLHIPNCDKHGLYNLKQCKMSLNGQRGECWCVNPNTGKLIQGAPTIRG-DPECHLFYNE 276
IGFBP-3   RGVHIPNCDKKGFYKKKQCRPSKGRKRGFCWCVD-KYGQPLPGYTTKGKEDVHCYSMQSK 264
IGFBP-4   YIIPIPNCDRNGNFHPKQCHPALDGQRGKCWCVDRKTGVKLPGGLEPKG-ELDCHQLADS 234
IGFBP-5   RAVYLPNCDRKGFYKRKQCKPSRGRKRGICWCVD-KYGMKLPGMEYVDG-DFQCHTFDSS 249
IGFBP-6   QTLYVPNCDHRGFYRKRQCRSSQGQRRGPCWCVD-RMGKSLPGSPDGNG-SSSCPTGSSG 216
Consensus .  :***::.* :. :**.  :  . * ****    *  : *       .  *

IGFBP-1   QN---------- 234
IGFBP-2   QQEARGVHTQRMQ 289
IGFBP-3   -------------
IGFBP-4   FRE---------- 237
IGFBP-5   NVE---------- 252
IGFBP-6   -------------
```

Figure 2. Sequence alignment of human IGFBPs 1-6. Alignment was performed using CLUSTALW version 1.81. Cysteines are shown in bold. Glycosylated or phosphorylated amino acids are in italics. » indicates the start of L- and C-domains. In the 'Consensus' line,* indicates conserved amino acids in all sequences, : indicates conservative substitutions, and . indicates semi-conservative substitutions. The first amino acid of IGFBP-6 is shown as Ala-1 based on the cDNA sequence by Shimisaki et al.[16] In most studies, the predominant form of IGFBP-6 commences at Arg-4.

Disulfide Linkages

Disulfide linkages between cysteines are important in maintaining the folded structures of proteins. IGFBPs 1-5 have 18 conserved cysteines whereas IGFBP-6 has 16. The N-domains of IGFBPs 1-5 contain 12 cysteines and share a conserved GCGCC motif. Human IGFBP-6 lacks the two adjacent cysteines in this motif. Rat IGFBP-6 lacks two further N-domain cysteines. The C-domains of IGFBP-1-6 contain six conserved cysteines and all share a conserved CWCV sequence. IGFBP-4 has two additional cysteines in the L-domain.

Early studies showed that the N- and C-domains of IGFBPs were not disulfide-linked. More recently, using proteolytic digestion and mass spectrometry, the disulfide linkages have been completely defined for IGFBP-6[35] and partially defined for IGFBP-1.[35] The disulfide linkages of IGFBP-2,[36] IGFBP-3,[37] IGFBP-4[38,39] and IGFBP-5[40] have also been partially characterized.

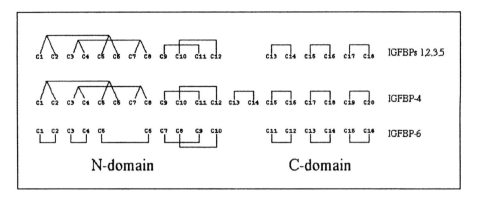

Figure 3. Disulfide linkages in the IGFBPs. Cysteine residues are numbered sequentially from the N-terminus.

N-Domain

As mentioned above, human IGFBP-6 lacks the two adjacent cysteines that are conserved in the N-domain of IGFBPs 1-5. Numbering the cysteines sequentially from the N-terminus, the three N-terminal disulfide linkages of IGFBP-6 join C1-C2, C3-C4 and C5-C6. In contrast, in IGFBP-1, C1 is not linked to C2 and C3 is not linked to C4,[35] showing that the disulfide linkages of IGFBP-6 differ from those of IGFBP-1 and, by implication, the other IGFBPs (Fig. 3).

More recently, the linkages in the N-terminus of IGFBP-4 have been further characterized.[39] This study confirmed that, as in IGFBP-1, C1 is not linked to C2 and C3 is not linked to C4. Further, C1 and C2 were found to be linked to C5 and C6, whereas C3 and C4 were found to be linked to C7 and C8. The authors concluded on the basis of very limited mass spectrometric fragmentation data that C3 was linked to C8 and C4 to C7, but this requires confirmation. A summary of the most likely disulfide linkage patterns is shown in Figure 3.

The major difficulty in completely solving the N-terminal disulfide linkages of IGFBPs 1-5 is the particularly close spacing of 3 of the 4 cysteines in the conserved GCGCCXXC sequence and the lack of suitable sites for proteolytic cleavage between them.

The recently described high affinity IGF binding region[40] in the N-domain of the IGFBPs contains two disulfide linkages that are conserved in IGFBP-1,[35] IGFBP-3,[37] IGFBP-4,[39] IGFBP-5[40] and IGFBP-6;[35] they have not been determined for IGFBP-2. The conserved disulfide linkages are C9 linked to C11 and C10 linked to C12 (C7-C9 and C8-C10 in IGFBP-6).

L-Domain

The two additional cysteine residues that are unique to the L-domain of IGFBP-4 are disulfide linked to each other.[39]

C-Domain

The three C-domain disulfide linkages are the same in IGFBP-2,[36] IGFBP-4[38,39] and IGFBP-6.[35] In each of these IGFBPs, sequential cysteines in the primary sequences are disulfide-linked (Fig. 3).

The highly conserved disulfide binding patterns exhibited by the IGFBPs implies that they contribute to maintaining the appropriate three-dimensional structure for IGF binding, and it would therefore be predicted that their disruption might adversely affect IGF binding. Consistent with this notion, disruption experiments performed with IGFBP-4[41] showed reduced binding when cysteine residues were mutated or deleted (see below).

Tertiary Structure

The only three-dimensional structure available for any region of the IGFBPs is confined to a region corresponding to amino acids 40-92 of the N-domain of IGFBP-5.[40] The corresponding polypeptide, named mini-IGFBP-5, was shown by nuclear magnetic resonance (NMR) to adopt a relatively rigid, globular structure consisting of a three-stranded, anti-parallel β-sheet stabilized by two disulfide bonds.[40] Mini-IGFBP-5 bound IGFs with 100-fold lower affinity than full-length

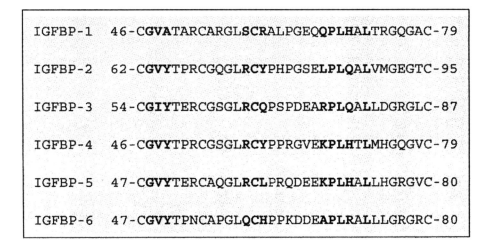

Figure 4. High affinity IGF binding domain of IGFBPs. Structural studies indicate that the high affinity IGF binding domain of IGFBP-5 is localised to the N-domain.[40,42] Amino acids in mini-IGFBP-5 that make contact with IGF-I within 4 Å (and corresponding amino acids in other IGFBPs) are shown in bold.

IGFBP-5. Very recently, the same group published a crystal structure of this domain of IGFBP-5 bound to IGF-I.[42] Both the NMR and crystal structures indicated that the principal interactions in IGF binding are hydrophobic (Fig. 4). Additionally, IGF binding made an otherwise mobile solvent-exposed loop of IGFBP-5 less flexible.

The CWCV motif in the C-domain of IGFBPs is characteristic of thyroglobulin type-1 domains, which are found in a range of other proteins including thyroglobulin and cysteine protease inhibitors. The crystal structure of the MHC class II-associated p41 fragment, which inhibits cathepsin L, has been solved.[43] It consists of two subdomains, one containing an α-helix-β-strand arrangement, whereas the second has a β-strand arrangement. Given the sequence homology of this protein with IGFBPs, it is likely that they share this structure, which has been implicated in modulating proteolytic activity.

IGF Binding

Although all IGFBPs bind IGFs with high affinity, they differ in their absolute affinities and relative specificities for IGF-I and IGF-II.[44] Over many years, a number of structural principles have emerged from numerous studies of IGF binding to IGFBPs. (1) Disulfide linkages are important since IGFBPs in the reduced form do not bind IGFs. (2) Both the N- and C-domains are required for maximal high affinity IGF binding. (3) Although the L-domain is not directly involved in IGF binding, it nevertheless has an important role in high affinity binding.

Sequence Determinants of IGF Binding

N-Domain

Two sources of IGFBP fragments containing the N-domain have been studied. Some fragments were produced by proteolytic cleavage whereas others were specifically expressed using recombinant technology. In general, these fragments retain IGF binding but with markedly lower affinity than full-length IGFBPs. Specifically, an IGFBP-2 fragment containing the N- and L-domains bind IGFs with ~40-fold lower affinity than full-length IGFBP-2.[45] Isolated N-domains of IGFBP-3 bind IGFs with 20-1000-fold lower affinity than full-length IGFBP-3.[37,46,47] Similarly, N-terminal fragments of IGFBP-4 bind IGFs with 4-20 fold-lower affinity than full-length IGFBP-4,[38,48,49] and N-terminal fragments of IGFBP-5 bind IGFs with 10-100 fold-lower affinity than full-length IGFBP-5.[40,50,51]

Kinetic binding studies show that the N-domains of IGFBP-2,[45] IGFBP-3[47] and IGFBP-5[40] associate with IGFs at least as rapidly as the full-length IGFBPs. The lower affinities of the N-domains correlate with the much more rapid dissociation that has been observed in these studies.

Further detailed studies of the N-domain shows that the high affinity IGF binding domain of the IGFBPs encompasses the C9-C11 and C10-C12 disulfide linkages (C7-C9 and C8-C10 of IGFBP-6). Chemical modification of Tyr-60 within this region of bovine IGFBP-2 reduced IGF binding 2-5-fold.[52] Progressive deletion analysis of IGFBP-4 indicated that Leu-72 to Ser-91 are critical for IGF-I binding.[49] The three-dimensional NMR studies of mini-IGFBP-5, which is homologous to this region of IGFBP-4, indicated that a series of hydrophobic residues is implicated in high affinity IGF binding.[40] More recently, an X-ray crystal structure of IGF-I bound to this region of IGFBP-5 confirmed that these residues formed the IGF binding site (Fig. 4).[42] Substitution of five of these hydrophobic residues in IGFBP-3 and IGFBP-5 decreased their IGF-I binding affinities more than 1000-fold and greatly reduced the inhibitory effects of these IGFBPs on IGF actions.[53]

The importance of the disulfide linkages in this region was demonstrated by the ~100-fold reduction in IGF-II binding by IGFBP-4 following the substitution of C9 by Arg, thereby disrupting the C9-C11 disulfide linkage.[41] Substitution of C12 by Arg, resulting in disruption of the C10-C12 linkage, had a more modest effect.

The unique disulfide linkages at the N-terminus of IGFBP-6 might be important in the IGF-II specificity of this binding protein, although this remains unproven. It is interesting to note that rat IGFBP-6, which lacks the additional pair of cysteine residues, C3 and C4, found in human IGFBP-6 nevertheless has similar IGF binding characteristics.[54]

L-Domain

There is no homology between the L-domains of the IGFBPs and they are not thought to contribute directly to high-affinity IGF binding. Deletion of ~30% of the L-domain of IGFBP-4 had no effect on IGF binding.[49] However, deletion of the entire L-domain of IGFBP-3 reduced IGF binding ~30-fold.[46] Further, coincubation of an IGFBP-2 fragment containing the N-domain and half of the L-domain with another fragment containing the C-domain and the other half of the L-domain failed to restore full high-affinity IGF binding, suggesting that N- and C-domains must be physically linked for high affinity binding.[45]

The L-domains of the IGFBPs are sites of post-translational modification, including glycosylation and phosphorylation (see below). O-glycosylation of IGFBP-6, which occurs on five Ser/Thr residues within the L-domain,[55] has no effect on IGF binding.[56,57] Similarly, phosphorylation of IGFBP-3 on residues within the mid-region has no effect on IGF binding,[58] although phosphorylation increases the IGF-I binding affinity of human[59] but not rat IGFBP-1.[60]

C-Domain

Since full-length IGFBPs have higher IGF binding affinities than isolated N-terminal or N+L domains,[40,46,49,50] the C-domain must also play a role in high affinity IGF binding. Although some studies suggest that isolated C-terminal IGFBP fragments do not bind IGFs,[49,61] a number of studies have demonstrated IGF binding of the C-domains of IGFBPs. Specifically, C-terminal fragments of IGFBP-2 bind IGFs with ~10-fold lower affinity than full-length IGFBP-2.[45,62,63] A C-domain fragment of IGFBP-3 binds IGF-I with ~15-fold lower affinity than full-length IGFBP-3,[47] and a proteolytic fragment of IGFBP-4 containing the C-domain bound IGF-I and IGF-II with 50- and 500-fold reduced affinity respectively.[38]

The C-domain of IGFBP-2 had slightly slower association kinetics than full length IGFBP-2, but its overall binding affinity was also greater than that of the N-domain due to substantially slower dissociation.[45] The C-domain of IGFBP-3 also bound IGFs with higher affinity than the N-domain, and it had slower dissociation kinetics.[47] These findings, together with the rapid dissociation of IGFs from isolated N-domains, support the notion that the C-domain is necessary for the maintenance of IGF:IGFBP complexes.

A study using photoaffinity labelled IGF-I suggested that the primary binding site on IGFBP-2 resides in the C-domain.[64] More specifically, IGF-I was reported to contact amino acids 212-227 and 266-287 of human IGFBP-2. However, deletion of the C-terminal 48 amino acids of bovine IGFBP-2, which includes the homologous residues to 266-287 in human IGFBP-2, had no substantive effect on IGF binding.[36] Further deletion analysis suggested that amino acids 222-236 were very important for IGF binding; the first three amino acids of this sequence correspond to the last three amino acids in the fragment identified by photoaffinity labelling. Further studies are clearly necessary to confirm and extend these observations.

Mutation of two conserved amino acids (Gly-203 and Gln-209) within the heparin binding domain of IGFBP-5 (see below) resulted in a 3-6-fold decrease in IGF binding affinity.[65] These residues are found in the region homologous to amino acids 222-236 of bovine IGFBP-2 described above. Site-specific substitution of Lys-251-Glu-Asp of IGFBP-3 with Arg-Gly-Asp from IGFBP-1 and -2 resulted in a 4-6-fold decrease in IGF binding affinity.[46]

Mutation of individual cysteines in the C-domain of IGFBP-4, which would disrupt disulfide linkages, resulted in a 4-5-fold decrease in IGF binding affinity, apart from mutation of C18 which resulted in a far more dramatic effect on IGF-I binding.[41] The latter is somewhat surprising given that mutation of C17, to which C18 is linked, did not have a similar effect, suggesting that the C17 mutation may have had other effects on IGFBP-4 structure.

IGF Binding Preference

IGFBPs-1, -3, -4 and -5 bind IGF-I and IGF-II with approximately equal affinity, whereas IGFBP-2 has a 4-20-fold and IGFBP-6 has a 20-100-fold binding preference for IGF-II.[44] Both IGF-I and IGF-II associate rapidly with IGFBP-6, but dissociation of IGF-I is far more rapid than that of IGF-II, thereby accounting for the IGF-II binding preference.[57]

Studies of IGFBP-5/IGFBP-6 chimeras in which domains are interchanged do not clearly localize the IGF-II binding preference of IGFBP-6 to any single domain.[51,66] It may be that determinants in both the N- and C-domains are necessary. In contrast, C-terminal deletion analysis of IGFBP-2 suggests that its IGF-II binding preference may be determined by amino acids 222-236.[36]

Post-Translational Modification

IGFBPs are post-translationally modified by processes including glycosylation and phosphorylation (Table 3). Most post-translational modifications occur within the variable L-domains of the IGFBPs.

Glycosylation

IGFBP-3 and IGFBP-4 are N-glycosylated. The L-domain of IGFBP-3 has 3 N-glycosylation sites (Asn-89, Asn-109, Asn-172) and variability in glycosylation accounts for the 40-45 kDa forms of IGFBP-3 seen on Western ligand blotting and immunoblotting.[67] IGFBP-4 is N-glycosylated in its L-domain on Asn-104; the N-glycosylated and non-glycosylated forms migrate with apparent molecular masses of 28 and 24 kDa respectively.

IGFBP-5 and IGFBP-6 are O-glycosylated. In contrast to N-glycosylation, no amino acid consensus sequence exists for O-glycosylation so sites must be determined empirically. An exhaustive study of human IGFBP-5 glycosylation sites has not been performed, but it has been shown that Thr-152 is O-glycosylated.[68] An extensive study of O-glycosylation sites of human IGFBP-6 showed five sites in the L-domain (Thr-102, Ser-120, Thr-121, Thr-122 and Ser-128) which are variably glycosylated.[55] Although human IGFBP-6 has a potential N-glycosylation site in its C-domain, it is not N-glycosylated.[55]

Glycosylation has no effect on high affinity IGF binding by IGFBP-3,[67] IGFBP-4,[39] or IGFBP-6.[56,57] N-glycosylation of IGFBP-3 also has no effect on ternary complex formation.[67] However, glycosylation of IGFBP-3[67] and IGFBP-6[69] inhibit cell surface binding, the latter probably by inhibiting binding to glycosaminoglycans. O-glycosylation of IGFBP-6 also inhibits proteolysis of IGFBP-6.[55,69] Since proteolysis and glycosaminoglycan binding decrease the IGF-II binding affinity of IGFBP-6, O-glycosylation therefore indirectly maintains the high IGF-II binding affinity of IGFBP-6.

O-glycosylation prolongs the circulating half-life of IGFBP-6 via a mechanism that is independent of protection from proteolysis.[70] In contrast, N-glycosylation of IGFBP-3 has no effect on its pharmacokinetics,[71] presumably because ternary complex formation is the prime determinant of the circulating half-life of IGFBP-3.

Table 3. Post-translational modification of IGFBPs

	Glycosylation	Phosphorylation
IGFBP-1	-	+
IGFBP-2	-	-
IGFBP-3	N-glycosylated	+
IGFBP-4	N-glycosylated	-
IGFBP-5	O-glycosylated	+
IGFBP-6	O-glycosylated	-

Phosphorylation

IGFBPs -1, -3 and -5 may be phosphorylated on Ser residues. Human IGFBP-1 may be phosphorylated on Ser-101, Ser-119 (both in the L-domain) and Ser-169 (in the C-domain). Ser-101 is the predominant phosphorylation site.[59] Rat IGFBP-1 may be phosphorylated on Ser-107 and Ser-132 (both in the L-domain).[60] Human IGFBP-3 may be phosphorylated on Ser-111 and Ser-113 (both in the L-domain).[58] Phosphorylation of IGFBP-5 has been reported in one abstract (Jones JI et al, 74th Annual Meeting of the Endocrine Society, 1992, p. 372).

Phosphorylation of human IGFBP-1 increases its affinity for IGF-I 6-fold[72] but has no effect on its affinity for IGF-II.[73] Phosphorylation of rat IGFBP-1[60] and human IGFBP-3[58] has no effect on IGF binding. The higher IGF-I binding affinity of phosphorylated IGFBP-1 may contribute to its largely inhibitory effect on IGF-I actions whereas non-phosphorylated IGFBP-1 may potentiate IGF-I actions. Phosphorylation of IGFBP-3 by casein kinase 2 inhibited ternary complex formation, susceptibility to proteolysis and cell surface binding.[74] However, phosphorylation had no effect on modulation of IGF-I action by IGFBP-3.

The Basic Regions of IGFBP-3 and -5: An Interaction 'Hot Spot'

Most IGFBPs contain putative heparin binding domains (HBD: XBBXBX or XBBBXXBX, where B=basic amino acid and X=other amino acids). In particular, IGFBPs -3, -5 and -6 have HBDs in their C-domains, at Lys-220 to Arg-225, Lys-206 to Arg-211 and Arg-197 to Arg-202 respectively.[75] These HBDs are found in the context of more extensive highly basic regions (Fig. 4). These regions of IGFBP-3 and IGFBP-5 are involved in a remarkable range of interactions with other biologically important molecules and appear to underlie a number of interesting properties of these IGFBPs as outlined in the following sections.

Binding to the Acid-Labile Subunit (ALS)

Most of the IGFs circulating in human serum are bound in ~150 kDa ternary complexes with an IGFBP and the acid labile subunit (ALS).[76] The predominant IGFBP in ternary complexes is IGFBP-3. IGFBP-5, but not other IGFBPs, can also form ternary complexes. The ternary complex is unable to leave the circulation and therefore prolongs the half-life of circulating IGFs.

Domain swapping experiments show that the C-domains of IGFBP-3[37] and -5[66] are involved in ALS binding. The binding site was further mapped to the basic region of the C-domain, encompassing amino acids 228-232 of IGFBP-3[46] and 215-232 of IGFBP-5.[66] An additional ALS binding site in the L-domain of IGFBP-5 has been identified.[51]

ALS is N- glycosylated and complete deglycosylation abolishes binding of ALS to binary complexes.[77] Removal of negatively charged sialic acids from the N-linked carbohydrates also substantially decreases ALS binding, indicating that ionic interactions may be involved in ALS binding to binary IGF:IGFBP complexes.[77]

Transferrin is the major circulating iron transport protein. Similarly to ALS, it binds to the basic C-domain region of IGFBP-3.[78] Interestingly, IGFBP-5 does not bind to transferrin.

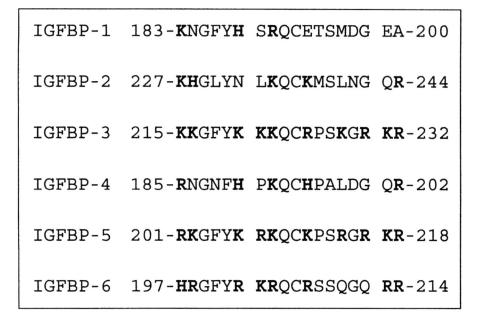

Figure 5. C-domain basic regions of IGFBP-3 and -5. Putative heparin binding regions in the C-domains of human IGFBP-3, IGFBP-5 and IGFBP-6. In contrast, there are relatively few basic amino acids in the homologous regions of the other IGFBPs.

Heparin Binding

Cell association is an important modulator of IGFBP actions; one mechanism of cell association is binding of IGFBPs to glycosaminoglycans within proteoglycans. Binding of IGFBP-2,[79] IGFBP-3,[80] IGFBP-5,[80] and IGFBP-6[69] to glycosaminoglycans reduces their IGF binding affinities.

As well as the C-domain HBDs described above, IGFBP-5 has two further putative HBDs in its L-domain, whereas the L-domains of IGFBP-1 and -2 each have one HBD. However, IGFBP-1 has not been shown to bind to heparin.

The basic amino acid region of the C-domain is a major determinant of heparin binding by IGFBP-3.[81] In particular, substitution of amino acids 228-232 decreases the heparin binding affinity of IGFBP-3.[46] Basic amino acids within the C-domain are also important for heparin binding by IGFBP-5.[82] Although the C-domain basic amino acid regions of IGFBP-3 and IGFBP-5 contain putative HBDs, the latter are not strict determinants of heparin binding, extracellular matrix binding or cell association. Rather, basic amino acids within this region, including some within the HBDs, are important for each of these properties; the set of basic amino acids responsible for each of these properties overlap but are distinct. For example, an eight amino acid peptide containing the HBD of IGFBP-3 alone bound heparin poorly whereas a peptide with the same amino acid composition as Lys-215 to Arg-232 of IGFBP-3 but with the HBD scrambled bound heparin similarly to the intact peptide.[81]

Although a peptide consisting of the basic region of IGFBP-6 binds cell surfaces[75] and heparin,[81] full-length IGFBP-6 binds poorly; this may be due to inhibition by glycosylation (see above).

Extracellular Matrix

IGFBP-5 is the predominant IGFBP in extracellular matrix. The basic amino acid region and the L-domain HBD are important for binding of IGFBP-5 to extracellular matrix,[83] although the L-domain site may be active only in the absence of the C-domain.[84]

Within extracellular matrix, the basic region of the C-domain of IGFBP-5 is important for binding to a number of proteins, including plasminogen activator inhibitor-1,[85] thrombospondin and osteopontin.[86] Interestingly, binding of these proteins to IGFBP-5 in solution had no effect on IGF binding affinity, whereas the IGF binding affinity of IGFBP-5 bound to extracellular matrix is decreased.

Cell Association

The basic amino acid region of the C-domain is a major determinant of binding of IGFBP-3 and -5 to endothelial cell monolayers.[75] Amino acids 228-232 are essential for association of IGFBP-3 with Chinese hamster ovary cell surfaces.[46]

IGFBP-5 binds to a 420 kDa membrane protein with resulting serine autophosphorylation.[87] A truncated form of IGFBP-5 incorporating amino acids 1-169 binds to this receptor, as does a C-domain peptide containing amino acids 201-218, suggesting that there are multiple binding sites.

Following perfusion into rat hearts, IGFBP-4 and IGFBP-3 preferentially localize to connective tissue and cardiac muscle respectively.[88] Chimeras in which the basic C-domain regions of IGFBP-3 and IGFBP-4 were swapped had the localization characteristics of the IGFBP from which the basic C-domain region derived. In addition, endothelial cell binding of these IGFBPs was dependent on the basic C-domain region.

Integrin Binding

IGFBP-1 and IGFBP-2 have Arg-Gly-Asp sequences in their C-domains. The Arg-Gly-Asp motif in extracellular matrix proteins is recognized by cell surface integrins which modulate many cellular processes including adhesion and migration. IGFBP-1 binds to the $\alpha 5\beta 1$ integrin which is an important fibronectin receptor. Binding to this integrin promotes cell migration independently of the presence of IGFs.[89] It is unclear whether the Arg-Gly-Asp sequence of IGFBP-2 mediates binding to integrins or has a role in the function of this IGFBP.

Nuclear Localization

IGFBP-3 and IGFBP-5 contain bipartite nuclear localization sequences (NLSs) in their C-domains; these consist of 2 highly basic sequences separated by a spacer region. In IGFBP-3, the NLS spans Lys-215 to Arg-232, whereas Arg-201 to Arg-218 forms the NLS of IGFBP-5. Both IGFBP-3 and IGFBP-5 may localize to the nucleus[90,91] although the role of nuclear localization of these IGFBPs is unknown. Nuclear localization of these IGFBPs is dependent on their NLSs and is mediated by the importin β subunit.[91] Amino acids 228-232 of IGFBP-3 appear to be most important for nuclear localization.[91]

Proteolytic Cleavage

IGFBPs are proteolysed by proteases from at least 3 classes: serine proteases, metalloproteases and aspartic proteases. Most cleavage sites for IGFBPs are in their L-domains, possibly because this region is more surface-exposed.[92] Following proteolysis, the IGF binding affinities of N- and C-terminal fragments are markedly reduced as described above. This may be a mechanism for modulation of IGFBP activity.

Trypsin-like proteases and chymotrypsin-like proteases may also cleave the N- and C-domains of IGFBPs. For example, chymotrypsin cleaves IGFBP-6 at a number of sites in the high affinity IGF binding domain, thus abrogating IGF binding.[69] The extent to which this occurs physiologically requires further exploration.

IGFBP-3 and -5 are cleaved by plasmin. Plasminogen, a precursor of active plasmin, binds to the highly basic C-domain regions of IGFBP-3[93] and IGFBP-5.[94] Similarly, the same region of IGFBP-3 is involved in binding prekallikrein,[95] fibrin and fibrinogen.[96] Binding of protease precursors to IGFBPs may be a mechanism whereby proteolysis of specific IGFBPs is achieved. Further, glycosaminoglycans inhibit proteolysis of IGFBP-5, and a possible mechanism is via inhibition of binding of protease precursors.[94] Similarly, a potential mechanism whereby ALS prolongs the circulating half-life of IGFBP-3 and -5 may be inhibition of binding of protease precursors, since ALS binds to the same regions of IGFBP-3 and -5.

Conclusions

It is obvious from this brief survey that considerable progress has been made in recent years in understanding the structural basis of the properties of IGFBPs. New and powerful technologies have recently accelerated the pace of discovery. However, many questions still remain. A complete three-dimensional structure of the IGFBPs will be an important advance in understanding the interactions of the IGFBPs not only with IGFs but also with the myriad other proteins and biomolecules with which they interact. Understanding the ways in which binding of IGFBPs with these molecules affects IGF-dependent and IGF-independent actions is another important issue. The possibility of modifying IGFBPs to confer particular properties such as cell association, resistance to proteolysis and targeting to specific tissues will help to answer questions about the physiological roles of the IGFBPs. Finally, given that dysregulation of the IGF system is implicated in many disease processes, understanding the structural basis of IGFBP actions may help in the design of new therapies for these diseases.

Acknowledgements

The authors would like to thank Dr. Greg Newmann (La Trobe University) for his thoughtful comments. Work reported here was supported by grants from National Health and Medical Research Council of Australia and the Australian Research Council.

References

1. Rechler MM. Insulin-like growth factor binding proteins. Vitam Horm 1993; 47:1-114.
2. Brinkman A, Groffen CAH, Kortleve DJ et al. Organization of the gene encoding the insulin-like growth factor binding protein IBP-1. Biochem Biophys Res Commun 1988; 157:898-907.
3. Ehrenborg E, Vilhelmsdotter S, Bajalica S et al. Structure and localization of the human insulin-like growth factor-binding protein 2 gene. Biochem Biophys Res Commun 1991; 176:1250-1255.
4. Cubbage ML, Suwanichkul A, Powell DR. Insulin-like growth factor binding protein-3. Organization of the human chromosomal gene and demonstration of promoter activity. J Biol Chem 1990; 265:12642-12649.
5. Zazzi H, Nikoshkov A, Hall K et al. Structure and transcription regulation of the human insulin-like growth factor binding protein 4 gene (IGFBP4). Genomics 1998; 49:401-410.
6. Allander SV, Larsson C, Ehrenborg E et al. Characterization of the chromosomal gene and promoter for human insulin- like growth factor binding protein-5. J Biol Chem 1994; 269:10891-8.
7. Ehrenborg E, Zazzi H, Lagercrantz S et al. Characterization and chromosomal localization of the human insulin-like growth factor binding protein 6 gene. Mammalian Genome 1999; 10:376-380.
8. Brewer MT, Stetler GL, Squires CH et al. Cloning, characterization, and expression of a human insulin-like growth factor binding protein. Biochem Biophys Res Commun 1988; 152:1289-1297.
9. Lee Y-L, Hintz RL, James PM et al. Insulin-like growth factor (IGF) binding protein complementary deoxyribonucleic acid from human HEP G2 hepatoma cells: predicted protein sequence suggests an IGF binding domain different from those of the IGF-I and IGF-II receptors. Mol Endocrinol 1988; 2:404-411.
10. Binkert C, Landwehr J, Mary J-L et al. Cloning, sequence analysis and expression of a cDNA encoding a novel insulin-like growth factor binding protein (IGFBP-2). EMBO J 1989; 8:2497-2502.
11. Wood WI, Cachianes G, Henzel WJ et al. Cloning and expression of the growth hormone-dependent insulin-like growth factor-binding protein. Mol Endocrinol 1988; 2:1176-1185.
12. Shimasaki S, Uchiyama F, Shimonaka M et al. Molecular cloning of the cDNAs encoding a novel insulin-like growth factor-binding protein from rat and human. Mol Endocrinol 1990; 4:1451-1458.
13. La Tour D, Mohan S, Linkhart TA et al. Inhibitory insulin-like growth factor binding protein: cloning, complete sequence, and physiological regulation. Mol Endocrinol 1990; 4:1806-1814.
14. Shimasaki S, Shimonaka M, Zhang H-P et al. Identification of five different insulin-like growth factor binding proteins (IGFBPs) from adult rat serum and molecular cloning of a novel IGFBP-5 in rat and human. J Biol Chem 1991; 266:10646-10653.
15. Kiefer MC, Ioh RS, Bauer DM et al. Molecular cloning of a new human insulin-like growth factor binding protein. Biochem Biophys Res Commun 1991; 176:219-225.
16. Shimasaki S, Gao L, Shimonaka M et al. Isolation and molecular cloning of insulin-like growth factor-binding protein-6. Mol Endocrinol 1991; 5:938-948.
17. Kiefer MC, Masiarz FR, Bauer DM et al. Identification and molecular cloning of two new 30-kDa insulin-like growth factor binding proteins isolated from adult human serum. J Biol Chem 1991; 266:9043-9049.
18. Nasrin N, Ogg S, Cahill CM et al. DAF-16 recruits the CREB-binding protein coactivator complex to the insulin-like growth factor binding protein 1 promoter in HepG2 cells. Proc Natl Acad Sci USA 2000; 97:10412-10417.
19. Hall RK, Yamasaki T, Kucera T et al. Regulation of phosphoenolpyruvate carboxykinase and insulin-like growth factor-binding protein-1 gene expression by insulin—The role of winged helix/forkhead proteins. J Biol Chem 2000; 275:30169-30175.

20. Tazuke SI, Mazure NM, Sugawara J et al. Hypoxia stimulates insulin-like growth factor binding protein 1 (IGFBP-1) gene expression in hepg2 cells—a possible model for IGFBP-1 expression in fetal hypoxia. Proc Natl Acad Sci USA 1998; 95:10188-10193.
21. Boisclair YR, Brown AL, Casola S et al. Three clustered Sp1 sites are required for efficient transcription of the TATA-less promoter of the gene for insulin-like growth factor-binding protein-2 from the rat. J Biol Chem 1993; 268:24892-24901.
22. Cazals V, Nabeyrat E, Corroyer S et al. Role for NF-kappa B in mediating the effects of hyperoxia on IGF binding protein 2 promoter activity in lung alveolar epithelial cells. Biochim Biophys Acta 1999; 11:349-362.
23. Villafuerte BC, Zhao WD, Herington AC et al. Identification of an insulin-responsive element in the rat insulin-like growth factor-binding protein-3 gene. J Biol Chem 1997; 272:5024-5030.
24. Cohick WS, Wang BJ, Verma P et al. Insulin-like growth factor I (IGF-I) and cyclic adenosine 3 ',5 '-monophosphate regulate IGF-binding protein-3 gene expression by transcriptional and posttranscriptional mechanisms in mammary epithelial cells. Endocrinology 2000; 141:4583-4591.
25. Walker GE, Wilson EM, Powell D et al. Butyrate, a histone deacetylase inhibitor, activates the human IGF binding protein-3 promoter in breast cancer cells: molecular mechanism involves an Sp1/Sp3 multiprotein complex. Endocrinology 2001; 142:3817-27.
26. Deal C, Ma J, Wilkin F et al. Novel promoter polymorphism in insulin-like growth factor-binding protein-3: correlation with serum levels and interaction with known regulators. J Clin Endocrinol Metab 2001; 86:1274-80.
27. Qin C, Singh P, Safe S. Transcriptional activation of insulin-like growth factor-binding protein-4 by 17beta-estradiol in MCF-7 cells: role of estrogen receptor- Sp1 complexes. Endocrinology 1999; 140:2501-8.
28. Duan C, Clemmons DR. Transcription factor AP-2 regulates human insulin-like growth factor binding protein-5 gene expression. J Biol Chem 1995; 270:24844-51.
29. Ji CH, Chen Y, Centrella M et al. Activation of the insulin-like growth factor-binding protein-5 promoter in osteoblasts by cooperative E box, CCAAT enhancer-binding protein, and nuclear factor-1 deoxyribonucleic acid-binding sequences. Endocrinology 1999; 140:4564-4572.
30. Gabbitas B, Pash JM, Delany AM et al. Cortisol inhibits the synthesis of insulin-like growth factor binding protein-5 in bone cell cultures by transcriptional mechanisms. J Biol Chem 1996; 271:9033-8.
31. Boonyaratanakornkit V, Strong DD, Mohan S et al. Progesterone stimulation of human insulin-like growth factor-binding protein-5 gene transcription in human osteoblasts is mediated by a CACCC sequence in the proximal promoter. J Biol Chem 1999; 274:26431-26438.
32. Dailly YP, Zhou Y, Linkhart TA et al. Structure and characterization of the human insulin-like growth factor binding protein (IGFBP)-6 promoter: identification of a functional retinoid response element. Biochem Biophys Res Commun 2001; 1518:145-151.
33. Gay E, Babajko S. AUUUA sequences compromise human insulin-like growth factor binding protein-1 mRNA stability. Biochem Biophys Res Commun 2000; 267:509-515.
34. Erondu NE, Nwankwo J, Zhong Y et al. Transcriptional and posttranscriptional regulation of insulin-like growth factor binding protein-3 by cyclic adenosine 3',5'-monophosphate: messenger RNA stabilization is accompanied by decreased binding of a 42-kDa protein to a uridine-rich domain in the 3'-untranslated region. Mol Endocrinol 1999; 13:495-504.
35. Neumann GM, Bach LA. The N-terminal disulfide linkages of human insulin-like growth factor binding protein-6 (hIGFBP-6) and hIGFBP-1 are different as determined by mass spectrometry. J Biol Chem 1999; 274:14587-14594.
36. Forbes BE, Turner D, Hodge SJ et al. Localization of an insulin-like growth factor (IGF) binding site of bovine IGF binding protein-2 using disulfide mapping and deletion mutation analysis of the C-terminal domain. J Biol Chem 1998; 273:4647-4652.
37. Hashimoto R, Ono M, Fujiwara H et al. Binding sites and binding properties of binary and ternary complexes of insulin-like growth factor-II (IGF-II), IGF-binding protein-3, and acid-labile subunit. J Biol Chem 1997; 272:27936-27942.
38. Standker L, Braulke T, Mark S et al. Partial IGF affinity of circulating N- and C-terminal fragments of human insulin-like growth factor binding protein-4 (IGFBP-4) and the disulfide bonding pattern of the C-terminal IGFBP-4 domain. Biochemistry 2000; 39:5082-5088.
39. Chelius D, Baldwin MA, Lu X et al. Expression, purification and characterization of the structure and disulfide linkages of insulin-like growth factor binding protein-4. J Endocrinol 2001; 168:283-296.
40. Kalus W, Zweckstetter M, Renner C et al. Structure of the IGF-binding domain of the insulin-like growth factor-binding protein-5 (IGFBP-5): implications for IGF and IGF-I receptor interactions. EMBO J 1998; 17:6558-6572.
41. Byun D, Mohan S, Baylink DJ et al. Localization of the IGF binding domain and evaluation of the role of cysteine residues in IGF binding in IGF binding protein-4. J Endocrinol 2001; 169:135-143.
42. Zeslawski W, Beisel HG, Kamionka M et al. The interaction of insulin-like growth factor-I with the N-terminal domain of IGFBP-5. EMBO J 2001; 20:3638-3644.
43. Guncar G, Pungercic G, Klemencic I et al. Crystal structure of MHC class II-associated p41 Ii fragment bound to cathepsin L reveals the structural basis for differentiation between cathepsins L and S. EMBO J 1999; 18:793-803.
44. Bach LA, Hsieh S, Sakano K et al. Binding of mutants of human insulin-like growth factor II to insulin-like growth factor binding proteins 1-6. J Biol Chem 1993; 268:9246-9254.

45. Carrick FE, Forbes BE, Wallace JC. BIAcore analysis of bovine insulin-like growth factor (IGF)-binding protein-2 identifies major IGF binding site determinants in both the amino- and carboxyl-terminal domains. J Biol Chem 2001; 276:27120-27128.
46. Firth SM, Ganeshprasad U, Baxter RC. Structural determinants of ligand and cell surface binding of insulin-like growth factor-binding protein-3. J Biol Chem 1998; 273:2631-2638.
47. Galanis M, Firth SM, Bond J et al. Ligand-binding characteristics of recombinant amino- and carboxyl-terminal fragments of human insulin-like growth factor-binding protein-3. J Endocrinol 2001; 169:123-133.
48. Cheung PT, Wu J, Banach W et al. Glucocorticoid regulation of an insulin-like growth factor-binding protein-4 protease produced by a rat neuronal cell line. Endocrinology 1994; 135:1328-35.
49. Qin XZ, Strong DD, Baylink DJ et al. Structure-function analysis of the human insulin-like growth factor binding protein-4. J Biol Chem 1998; 273:23509-23516.
50. Andress DL, Loop SM, Zapf J et al. Carboxy-truncated insulin-like growth factor binding protein-5 stimulates mitogenesis in osteoblast-like cells. Biochem Biophys Res Commun 1993; 195:25-30.
51. Twigg SM, Kiefer MC, Zapf J et al. A central domain binding site in insulin-like growth factor binding protein-5 for the acid-labile subunit. Endocrinology 2000; 141:454-457.
52. Hobba GD, Forbes BE, Parkinson EJ et al. The insulin-like growth factor (IGF) binding site of bovine insulin- like growth factor binding protein-2 (bIGFBP-2) probed by iodination. J Biol Chem 1996; 271:30529-36.
53. Imai Y, Moralez A, Andag U et al. Substitutions for hydrophobic amino acids in the N-terminal domains of IGFBP-3 and -5 markedly reduce IGF-I binding and alter their biologic actions. J Biol Chem 2000; 275:18188-18194.
54. Bach LA, Tseng LY-H, Swartz JE et al. Rat PC12 pheochromocytoma cells synthesize insulin-like growth factor-binding protein-6. Endocrinology 1993; 133:990-995.
55. Neumann GM, Marinaro JA, Bach LA. Identification of O-glycosylation sites and partial characterization of carbohydrate structure and disulfide linkages of human insulin-like growth factor binding protein 6. Biochemistry 1998; 37:6572-6585.
56. Bach LA, Thotakura NR, Rechler MM. Human insulin-like growth factor binding protein-6 is O-glycosylated. Biochem Biophys Res Commun 1992; 186:301-307.
57. Marinaro JA, Jamieson GP, Hogarth PM et al. Differential dissociation kinetics explain the binding preference of insulin-like growth factor binding protein-6 for insulin-like growth factor-II over insulin-like growth factor-I. FEBS Lett 1999; 450:240-244.
58. Hoeck WG, Mukku VR. Identification of the major sites of phosphorylation in IGF binding protein 3. J Cell Biochem 1994; 56:262-273.
59. Jones JI, Busby WH Jr, Wright G et al. Identification of the sites of phosphorylation in insulin-like growth factor binding protein-1. Regulation of its affinity by phosphorylation of serine 101. J Biol Chem 1993; 268:1125-1131.
60. Peterkofsky B, Gosiewska A, Wilson S et al. Phosphorylation of rat insulin-like growth factor binding protein-1 does not affect its biological properties. Arch Biochem Biophys 1998; 357:101-110.
61. Lalou C, Lassarre C, Binoux M. A proteolytic fragment of insulin-like growth factor (IGF) binding protein-3 that fails to bind IGFs inhibits the mitogenic effects of IGF-I and insulin. Endocrinology 1996; 137:3206-3212.
62. Wang J-F, Hampton B, Mehlman T et al. Isolation of a biologically active fragment from the carboxy-terminus of the fetal rat binding protein for insulin-like growth factors. Biochem Biophys Res Commun 1988; 157:718-726.
63. Ho PJ, Baxter RC. Characterization of truncated insulin-like growth factor-binding protein-2 in human milk. Endocrinology 1997; 138:3811-3818.
64. Horney MJ, Evangelista CA, Rosenzweig SA. Synthesis and characterization of insulin-like growth factor (IGF)-1 photoprobes selective for the IGF-binding proteins (IGFBPs)—Photoaffinity labeling of the IGF-binding domain on IGFBP-2. J Biol Chem 2001; 276:2880-2889.
65. Song H, Beattie J, Campbell IW et al. Overlap of IGF- and heparin-binding sites in rat IGF-binding protein-5. J Mol Endocrinol 2000; 24:43-51.
66. Twigg SM, Kiefer MC, Zapf J et al. Insulin-like growth factor-binding protein 5 complexes with the acid-labile subunit—role of the carboxyl-terminal domain. J Biol Chem 1998; 273:28791-28798.
67. Firth SM, Baxter RC. Characterisation of recombinant glycosylation variants of insulin-like growth factor binding protein-3. J Endocrinol 1999; 160:379-87.
68. Standker L, Wobst P, Mark S et al. Isolation and characterization of circulating 13-kDa C-terminal fragments of human insulin-like growth factor binding protein-5. FEBS Lett 1998; 441:281-286.
69. Marinaro JA, Neumann GM, Russo VC et al. O-glycosylation of insulin-like growth factor (IGF) binding protein-6 maintains high IGF-II binding affinity by decreasing binding to glycosaminoglycans and susceptibility to proteolysis. Eur J Biochem 2000; 267:5378-5386.
70. Marinaro JA, Casley DJ, Bach LA. O-glycosylation delays the clearance of human IGF-binding protein-6 from the circulation. Eur J Endocrinol 2000; 142:512-516.
71. Sommer A, Spratt SK, Tatsuno GP et al. Properties of glycosylated and non-glycosylated human recombinant IGF binding protein-3 (IGFBP-3). Growth Regul 1993; 3:46-49.
72. Jones JI, D'Ercole AJ, Camacho Hubner C et al. Phosphorylation of insulin-like growth factor (IGF)-binding protein 1 in cell culture and in vivo: effects on affinity for IGF-I. Proc Natl Acad Sci USA 1991; 88:7481-5.

73. Westwood M, Gibson JM, White A. Purification and characterization of the insulin-like growth factor-binding protein-1 phosphoform found in normal plasma. Endocrinology 1997; 138:1130-1136.
74. Coverley JA, Martin JL, Baxter RC. The effect of phosphorylation by casein kinase 2 on the activity of insulin-like growth factor-binding protein-3. Endocrinology 2000; 141:564-570.
75. Booth BA, Boes M, Andress DL et al. IGFBP-3 and IGFBP-5 association with endothelial cells; role of C-terminal heparin binding domain. Growth Regul 1995; 5:1-17.
76. Boisclair YR, Rhoads RP, Ueki I et al. The acid-labile subunit (ALS) of the 150 kDa IGF-binding protein complex: an important but forgotten component of the circulating IGF system. J Endocrinol 2001; 170:63-70.
77. Janosi JBM, Firth SM, Bond JJ et al. N-linked glycosylation and sialylation of the acid-labile subunit— Role in complex formation with insulin-like growth factor (IGF)-binding protein-3 and the IGFs. J Biol Chem 1999; 274:5292-5298.
78. Weinzimer SA, Gibson TB, Collett-Solberg PF et al. Transferrin is an insulin-like growth factor-binding protein-3 binding protein. J Clin Endocrinol Metab 2001; 86:1806-1813.
79. Russo VC, Bach LA, Fosang AJ et al. Insulin-like growth factor binding protein-2 binds to cell surface proteoglycans in the rat brain olfactory bulb. Endocrinology 1997; 138:4858-4867.
80. Arai T, Parker A, Busby W Jr et al. Heparin, heparan sulfate, and dermatan sulfate regulate formation of the insulin-like growth factor-1 and insulin-like growth factor-binding protein complexes. J Biol Chem 1994; 269:20388-93.
81. Booth BA, Boes M, Dake BL et al. Structure-function relationships in the heparin-binding c-terminal region of insulin-like growth factor binding protein-3. Growth Regul 1996; 6:206-213.
82. Arai T, Clarke J, Parker A et al. Substitution of specific amino acids in insulin-like growth factor (IGF) binding protein 5 alters heparin binding and its change in affinity for IGF-I in response to heparin. J Biol Chem 1996; 271:6099-6106.
83. Parker A, Clarke JB, Busby WH Jr et al. Identification of the extracellular matrix binding sites for insulin- like growth factor-binding protein 5. J Biol Chem 1996; 271:13523-9.
84. Song H, Shand JH, Beattie J et al. The carboxy-terminal domain of IGF-binding protein-5 inhibits heparin binding to a site in the central domain. J Mol Endocrinol 2001; 26:229-239.
85. Nam TJ, Busby W Jr, Clemmons DR. Insulin-like growth factor binding protein-5 binds to plasminogen activator inhibitor-I. Endocrinology 1997; 138:2972-8.
86. Nam TJ, Busby WH, Rees C et al. Thrombospondin and osteopontin bind to insulin-like growth factor (IGF)-binding protein-5 leading to an alteration in IGF-I-stimulated cell growth. Endocrinology 2000; 141:1100-1106.
87. Andress DL. Insulin-like growth factor-binding protein-5 (IGFBP-5) stimulates phosphorylation of the IGFBP-5 receptor. Am J Physiol 1998; 37:744-750.
88. Knudtson KL, Boes M, Sandra A et al. Distribution of chimeric IGF binding protein (IGFBP)-3 and IGFBP-4 in the rat heart: importance of C-terminal basic region. Endocrinology 2001; 142:3749-55.
89. Jones JI, Gockerman A, Busby WH Jr et al. Insulin-like growth factor binding protein 1 stimulates cell migration and binds to the alpha 5 beta 1 integrin by means of its Arg-Gly-Asp sequence. Proc Natl Acad Sci USA 1993; 90:10553-7.
90. Schedlich LJ, Young TF, Firth SM et al. Insulin-like growth factor-binding protein (IGFBP)-3 and IGFBP-5 share a common nuclear transport pathway in T47D human breast carcinoma cells. J Biol Chem 1998; 273:18347-18352.
91. Schedlich LJ, Le Page SL, Firth SM et al. Nuclear import of insulin-like growth factor-binding protein-3 and-5 is mediated by the importin beta subunit. J Biol Chem 2000; 275:23462-23470.
92. Fowlkes JL. Insulinlike growth factor-binding protein proteolysis—an emerging paradigm in insulinlike growth factor physiology. Trends Endocrinol Metab 1997; 8:299-306.
93. Campbell PG, Durham SK, Suwanichkul A et al. Plasminogen binds the heparin-binding domain of insulin-like growth factor-binding protein-3. Am J Physiol 1998; 38:331.
94. Campbell PG, Andress DL. Plasmin degradation of insulin-like growth factor-binding protein-5 (IGFBP-5)—regulation by IGFBP-5-(201-218). Am J Physiol 1997; 36:E996-E1004.
95. Durham SK, Suwanichkul A, Hayes JD et al. The heparin binding domain of insulin-like growth factor binding protein (IGFBP)-3 increases susceptibility of IGFBP-3 to proteolysis. Horm Metab Res 1999; 31:216-25.
96. Campbell PG, Durham SK, Hayes JD et al. Insulin-like growth factor-binding protein-3 binds fibrinogen and fibrin. J Biol Chem 1999; 274:30215-30221.

CHAPTER 4

Regulation of Insulin-Like Growth Factor-I Gene Expression

Xia Wang and Martin L. Adamo

Abstract

Insulin-like growth factor-I (IGF-I) mRNAs is expressed in adults principally in liver, which secretes IGF-I to act as an endocrine growth factor. Liver IGF-I mRNAs are transcribed from multiple start sites within two promoters, giving rise to mRNAs containing either exon 1 or exon 2 as alternative first exons. Extra-hepatic tissues express lower levels of IGF-I mRNA, most of which is transcribed from the exon 1 promoter. Exons 1 and 2 encode alternative 5'untranslated regions (UTRs), which may affect translational efficiency, and encode alternative amino-terminal ends of the IGF-I signal peptide. The 380 bp exon 1 sequence itself contains cis-acting elements essential for stimulation of IGF-I exon 1 transcription in tumor cells and by insulin in liver hepatocytes, which acts via a novel insulin response element binding protein and Sp1, and cAMP-elevating agents in osteoblasts, which act via C/EBP-delta. STAT-3 and-5 and HNF-1alpha transcription factors are strongly implicated in GH stimulation of IGF-I transcription in liver, although the mechanisms by which GH regulates IGF-I transcription have not been elucidated, nor has the mechanism of the GH-independent initial post-natal induction of exon 1 IGF-I transcription. Moreover, mechanisms responsible for the more limited tissue expression of exon 2 containing transcripts and the delayed expression of these transcripts in liver during post-natal development have not yet been elucidated. Studies from tumor cells suggest that silencers in the 5'flanking region may inhibit exon 2 transcription in non or low exon 2 producing cells. GH stimulates IGF-I expression in at least some extra-hepatic tissues, with evidence in muscle cells and glioma cells for JAK/STAT-dependent pathways. However, there are clear examples of tissue-specific regulation of IGF-I gene expression in response to tropic hormones, wounding, and inflammatory and repair scenarios, which are in some cases GH-independent. An example is the stimulation of uterine IGF-I expression and uterine growth by estradiol. In addition, several hormonal factors in addition to GH and insulin regulate hepatic IGF-I gene expression. For example, thyroid hormone appears to facilitate GH action in liver, whereas estradiol and glucocorticoids may inhibit GH stimulation of hepatic IGF-I expression. Alternative splicing of IGF-I mRNAs involving exons 4, 5, and 6 alters the E-peptide coding sequence, although no major functional significance has yet been ascribed to these differences. However, exon 6, which is present in all rodent IGF-I mRNAs, and which makes up a large proportion of human IGF-I mRNAs, contains a number of AU-sequences in the 3'-UTR that may de-stabilize IGF-I mRNA. IGF-I pre mRNA splicing is delayed, and IGF-I mRNA stability may be decreased in livers of fasted and dietary restricted animals; however, the role if any of the exon 6 3'-UTR in mediating reduced mRNA stability has not been elucidated.

Introduction and Background

Manipulation of IGF-I gene expression in mice has provided clear evidence that IGF-I plays an important role in normal growth and development, maintenance of tissue function and neoplastic growth. Mice nullizygous for both IGF-I alleles exhibit embryonic growth retardation, which

Insulin-Like Growth Factors, edited by Derek LeRoith, Walter Zumkeller and Robert Baxter. ©2003 Eurekah.com and Kluwer Academic / Plenum Publishers.

becomes more pronounced in those mice that survive into adulthood.[1-5] Inhibition of musculoskeletal[1-5] and reproductive[6] tissue growth and development, and reversible inhibition of brain myelination[7,8] are observed. Moreover, the IGF-I null mice do not respond to GH with increased growth,[4] thus confirming the somatomedin hypothesis that GH promotes growth by stimulating IGF-I production. The role of liver-derived versus extra-hepatic IGF-I expression in growth and development has come into focus with studies showing that disruption of both liver IGF-I alleles using an albumin promoter drive cre-lox system (LID mouse) has very little effect on longitudinal growth.[9,10] Since this disruption decreased serum IGF-I by about 75%, it can be concluded that either the residual 25% serum IGF-I (presumably of extra-hepatic origin) sustains normal growth, that IGF-I produced by extra-hepatic tissues which acts at or near its site of production sustains growth, or both. These hypotheses have been addressed by crossing LID mice with mice in which the acid-labile subunit (ALS) of the serum IGF carrier is deleted.[10] These mice display only 10 to 15% of wild-type serum IGF-I and show significant inhibition of skeletal growth despite normal bone IGF-I mRNA levels. Moreover, the 75% reduction in serum IGF-I in LID mice is sufficient to inhibit growth of transplanted tumors[11] and to lead to increased serum GH, presumably as the result of reduced serum IGF-I and loss of feedback inhibition of GH.[12,13] LID mice also display reduced spleen weight,[9,14] and skeletal muscle insulin resistance, which is either secondary to increased GH or due to loss of a direct effect of serum IGF-I to sensitize muscle to insulin action, or both.[12] These results strongly implicate serum IGF-I in normal and neoplastic growth. Although more definitive results await the development of tissue-specific knockouts, extra-hepatic IGF-I also appears to be important for growth. For example, PTH is unable to stimulate bone formation in vivo in IGF-I null mice.[15] Moreover, targeted expression of IGF-I in bone osteoblasts,[16] thyroid gland[17] and skeletal muscle[18] increases growth of these tissues without altering systemic IGF-I or growth. And, targeted expression of IGF-I in basal keratinocytes caused epidermal hyperplasia and increased skin tumors.[19]

Given the clear role of both serum and locally produced IGF-I, the purpose of this review is to present a critical analysis of recent developments in our understanding of the regulation of liver and extra-hepatic IGF-I biosynthesis. In order to place these new data in proper context, a systematic presentation of the major features of the IGF-I gene and gene products is presented first, followed by more specific information on mechanisms of regulation.

Summary of Current Research

IGF-I Gene and Transcripts

Overview

Alternative transcription initiation, alternative splicing, and alternative polyadenylation of IGF-I premRNA transcripts result in the production of multiple IGF-I mRNAs. These transcripts differ in 5'-and 3'-untranslated regions and in the coding sequence of the IGF-I precursor.[20-48] Exon 1 and exon 2 are alternative first exons in IGF-I mRNAs. Each of these first exons is transcribed from its own start sites and each contains a unique translation start codon that is in-frame with the IGF-I coding sequence. Thus, the 5' ends of exon 1 and exon 2 encode alternative 5'-UTRs and the 3' ends of exon 1 and exon 2 encode alternative amino terminal sequences of the IGF-I signal peptide (Fig. 1). In IGF-I premRNAs, either exon 1 or exon 2 is spliced to exon 3. Exon 3 is always spliced to exon 4 in all IGF-I mRNAs. The 5' end of exon 3 encodes the carboxyl terminal amino acids of the IGF-I signal peptide, and the remainder of exon 3 and the initial 5' portion of exon 4 encode the B, C, A, and D domains of the mature 7.5 kilo Dalton (KDa) IGF-I peptide found in serum. However, the IGF-I coding sequence does not stop at this point. Rather, the 3' end of exon 4 encodes the common amino-terminal portion of each of two different E peptides, and exon 5 and exon 6 encode divergent carboxyl terminal sequences of the two E peptides. The E peptide is not, however, present in the mature 7.5 KDa IGF-I peptide purified from serum. Exons 4, 5, and 6 are alternatively spliced. In rodents, the splicing pattern is exon 4-exon 5-exon 6 or exon 4-exon 6, whereas in humans the splicing pattern is exon 4-exon 5 or exon 4-exon 6. Exon 6 contains alternative polyadenylation sites whose use result in 3'-UTRs of varying length.

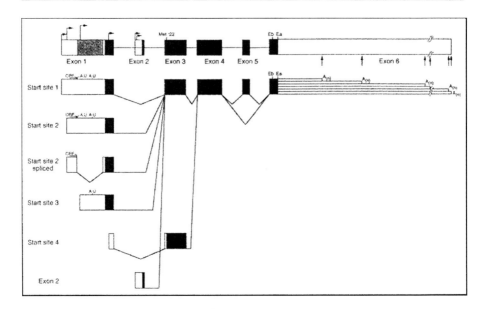

Figure 1. The rat IGF-I gene and transcripts. The gene is shown schematically at the top, and below are shown the various transcripts resulting from alternative transcription initiation, splicing and polyadenylation. Arrows above exon 1 and exon 2 indicate transcription start sites. ORF = the putative open reading frame in the exon 1 5'-UTR. A:U = the AUGUGA start stop codons in the exon 1 5'-UTR. Black areas are the preproIGF-I coding sequences. Met −22 indicates the exon 3 translation start site; Met −48 and Met −32 are not labeled, but begin the preproIGF-I coding sequence in exons 1 and 2, respectively. Eb and Ea indicate the stop codons for the Ea and Eb forms of preproIGF-I. The arrows below exon 6 indicate the polyadenylation sites, and the small rectangles below show the resulting lengths of 3'-UTR and their terminating polyA tails (An). Modified from Adamo ML. Diabetes Reviews 1995; 3:2-27.

IGF-I Gene Transcription Initiation Sites: Exon 1 and Exon 2 Are Transcribed from Separate Promoters

Exon 1 transcription initiation occurs from multiple sites or clusters of sites[38-40,42,44,47,48] (Fig. 1) that are located about 380 nucleotides (nt) (start site 1), 345 nt (start site 2), 245 nt (start site 3), and about 50 nt (start site 4) upstream of the 3' end of exon 1, respectively. IGF-I mRNAs initiated from start sites 1 and 4 are of extremely low abundance, but the location of start site 1 is used to designate the +1 position in this review for purposes of orienting exon 1 and numbering its nucleotides. The transcripts initiated from start sites 2 and 3 account for over 90% of total rat liver exon 1 mRNAs and are expressed in similar abundance in rat liver. A minor transcription start site cluster located about 1150 nt upstream of the 3' end of exon 1 has been mapped in human liver,[39] but has not been reported in rat liver. A 186 nt segment within exon 1 is spliced out of 10-20% of rat liver IGF-I mRNAs, virtually all of which are initiated from start site 2.[41,42] This alternatively spliced transcript has also been detected in a mouse adipocyte cell line.[49]

IGF-I mRNAs can also contain exon 2 as their first exon. Initiation of these transcripts occurs at two clusters of sites located at about 70 nt (accounting for 90% of exon 2 transcripts) and at about 52 nt (accounting for about 10% of exon 2 transcripts) upstream of the 3' end of exon 2.[38,41,42,44,47] In human liver, exon 2 start sites are similarly located about 65 to 75 nt upstream of the 3' end of exon 2.[35,39] A site or sites of transcription initiation appears to occur further upstream giving rise to longer versions of exon 2 in both rat and human liver IGF-I mRNAs.[35,38,39,42,44] The abundance of these transcripts appears to be quite low, with a possible exception apparent in a human liver sample based on results of RNase protection assays.[35]

Exon 1 transcripts account for about 70% of adult liver IGF-I mRNA and exon 2 transcripts account for about 30%.[42,44] In extra-hepatic tissues, exon 1 transcripts account for 90 to 100% of

IGF-I mRNAs.[44,46] In a few extra-hepatic tissues start site 2 and start site 3 transcripts are expressed in similar abundance, whereas in some extra-hepatic tissues, most of the exon 1 transcripts are initiated from start site 3.[42,44,46] In contrast, in extra-hepatic tissues the locations of the exon 2 start sites are the same as in liver.[44,46] Early studies[40,44,50-53] used plasmids containing DNA sequences either flanking and including exon 1 or flanking and including exon 2 fused to a luciferase reporter gene in transfection assays to generate data strongly suggesting that exon 1 and exon 2 transcription initiation are each driven by their own proximal promoters. Specific elements and factors important for transcriptional control will be discussed in pertinent sections below.

Alternative Splicing Generates mRNAs Encoding Different E-Peptides

Alternative splicing also results in the two forms of mRNA encoding two isoforms of the E-peptide. In humans, mice, and rats, exon 4 encodes the common amino terminal portion of both E-peptides. In humans, exon 4 is spliced either to exon 5 or to exon 6. Both exon 5 and exon 6 in humans contain stop codons, and thus encode the divergent carboxyl terminal proteins of Eb and Ea E peptides, respectively.[21,23] In mice and rats (Fig. 1), IGF-I mRNAs contain either exons 4, 5, and 6, spliced in that order, to encode the Eb peptide, or exon 4 spliced directly to exon 6 to encode the Ea peptide.[24,27,31] Exon 5 is 52 nt long, and its inclusion in mRNA changes the translational reading frame. Thus, in rodents, exon 6 uses different stop codons in the two different reading frames to terminate translation of preproIGF-IEa and preproIGF-I Eb, respectively (Fig. 1). The Ea form of mRNA makes up the majority of liver (88% of the total) and extra-hepatic tissue (~ 95% of the total) IGF-I mRNA.[31] The evidence for translation of the E-peptide isoforms will be presented in a later section.

Alternative Polyadenylation Results in 3'-UTRs of Different Lengths and is Responsible for Most of the Size Variation in IGF-I mRNAs

IGF-I mRNAs in humans and rats exist in three sizes of about 0.8 to 1.2 KB, about 2.2 KB and about 7 to 7.5 Kb.[21,27] Four proximal polyadenylation sites are located in exon 6 whose use would result in IGF-I mRNAs with 3'-UTRs with lengths of between ~ 347 and ~ 1140 nt.[24,25,27] (Fig. 1). The 7.5 kb transcript results from use of a more downstream polyadenylation signal in exon 6, resulting in a 3'-UTR that is about 6.5 Kb long (Fig. 1)[34,44,45] In humans, exon 5 and exon 6 each encode unique 3'-UTR sequences.[21,23] Two small human mRNAs of 1.3 KB and 1.1 KB result from splicing of the exon 1-exon 3-exon 4 unit to exon 5 and exon 6, respectively.[37] The 7.6 KB mRNA contains the common exon 1-exon 3-exon 4 block spliced to exon 6 and has 6.5 KB of 3'-UTR resulting from use of the downstream polyadenylation site in exon 6.[37] The long exon 6-derived 3'-UTR in both humans[37] and rats[45] contains many AU-rich destabilizing sequences, and the 7.5 Kb mRNA is indeed unstable when incubated with a cellular extract.[54]

Translation and Stability of IGF-I mRNAs

Signal Peptides and Effect of Alternative 5'-Untranslated Regions

Potential translation initiation AUG codons in-frame with the preproIGF-I coding sequence are located at Met −48 in exon 1, Met −32 in exon 2 and Met −22 in exon 3. In human IGF-I mRNAs there is also a start codon at Met −25 in exon 3. In vitro studies show that in rat and human exon 1-containing transcripts, preproIGF-I translation is initiated at Met −48 (Fig. 1).[30,47,48] In rat transcripts with exon 2 as their first exon, preproIGF-I translation is initiated in vitro at Met −32 in exon 2 (Fig. 1).[36,47,48] Met −22 in rat exon 3 can also be used for translation initiation[36,47,48] and is the only available start codon in the start site 4 transcript in rats (Fig. 1).[48] The resulting signal sequences all appear to be cleaved co-translationally by pancreatic microsomes,[30,36,47,48] strongly suggesting that they function as signal sequences in vivo. An in-frame CUG in exon 2[47] is not used as a start site in vitro.[48] Transfection of human preproIGF-I expression plasmids, which contained only the Met −25 and Met −22 initiation codons, into 293 cells resulted in the appearance of metabolically labeled mature IGF-I in the conditioned medium, indicating that the signal peptide(s) initiated at one or both of these codons properly targeted the nascent peptide and were cleaved.[55]

In vitro, exon 1 start site 4 and exon 2 mRNAs were translated about 10-fold more efficiently than was the exon 1 start site 2 RNA. The spliced start site 2 RNA was translated about 6-fold more efficiently than was the full-length RNA.[48] Studies with rat liver polyribosomes,[41,56] suggested that IGF-I mRNAs with alternative 5'-UTRs could be translated, although the start site 2-spliced mRNA was present as a higher percentage of translatable IGF-I mRNA than it was in the total IGF-I mRNA population. Three AUG codons are located upstream of the Met –48 codon in rat exon 1 that could affect initiation at Met –48 (Fig. 1). The first of these AUGs initiates translation of a putative 13 amino acid open reading frame (ORF; Fig. 1), which was first noted in the xenopus exon 1 sequence.[57] The next two of the AUGs are each immediately followed by a stop codon[25] (shown as A:U in Fig. 1). The greater in vitro[48] and in vivo[41,57] translation of the start site 2-spliced mRNA was postulated to result from the fact that this mRNA lacked the two A:U motifs (Fig. 1). Mutation of the AUG initiating the upstream ORF did not affect translation of either the start site 2 or the start site 2-spliced mRNA. Mutation of the AUG in the first A:U motif slightly increased translation of the start site 2 mRNA, whereas mutation of the AUG in the second A:U motif stimulated translation of both the start site 2 and the start site 3 mRNAs.[48] These results are consistent with re-initiation theory, which states that the closer the termination of a short upstream ORF (in this case the second AUGUGA sequence) is to the major ORF (in this case the preproIGF-I coding sequence initiated at the Met –48 codon), the more likely that translation of that major ORF will be impeded.[58] An initiation/termination event may occur at the first A:U as well; in this case it would be interesting to know whether the array of the two A:U motifs could result in translational control of IGF-I gene expression in a manner analogous to what occurs with the GCN-4 mRNA.[59] The efficient translation of the start site 2 spliced transcript and the lack of effect of mutation of the AUG initiating the putative upstream 14 amino acid ORF suggests that this ORF is not translated in vitro, which is consistent with its poor sequence context.[48,58]

Transient transfection of IGF-I 5'-UTR-luciferase plasmids under viral promoter control into HUH-7 cells produced results partly confirming the in vitro data, in that the start site 2 spliced fusion mRNA was translated the greatest and the start site 2 full-length fusion mRNA was translated least.[48] However, mutation of either of the A:U motifs in the start site 2 full-length construct did not increase its translation.[48] This result suggested that there were differences between intact HUH-7 cells and the rabbit reticulocyte lysate in vitro translation system or alternatively, the additional 50 nt of 5'-UTR between the start site of transcription and the first nt of the IGF-I encoded 5'-UTR in the transfected plasmids could have influenced the results. The effect of the exon 2 and start site 4-encoded 5'-UTRs on translation in intact cells was not evaluated.

A subsequently published study[49] yielded different results regarding translation of mouse IGF-I mRNAs with different 5'-UTRs. Two synthetic exon 1 transcripts, one initiated from a site intermediate between start site 2 and start site 3, and another initiated from around start site 3, were translated much more efficiently than a synthetic exon 1 transcript from which the 186 nt spliced region was removed. Moreover, the appearance of IGF-I peptide in conditioned medium and inside cells in a differentiating Ob 1771 cell line was associated with the appearance of the non-spliced exon 1 mRNA, but not with the appearance of spliced exon 1 or exon 2 mRNA suggesting that the translationally competent IGF-I mRNA was that containing the longer exon 1-encoded 5'-UTR. The reason(s) for these different results is not clear. Obviously, there could be differences in trans-acting factors influencing translation between HUH-7 cells, Ob1771 cells, rat liver in vivo, and the two different commercial preparations of rabbit reticulocyte lysate used in the two in vitro studies.

Effect of Alternative 3'-Untranslated Regions on mRNA Stability and Translation

The 7.5 KB mRNA that contains 6.5 Kb of 3'-UTR is variably found on rat liver polysomes and in cytoplasmic extract.[41,45,60] The 7.5 Kb transcript was unstable in cell-free extracts and decayed faster in vivo than IGF-I transcripts with shorter 3'-UTRs,[54] which was postulated[45] to explain the lack of cytoplasmic 7.5 Kb transcript observed in ref 41. Overall, results showing differences in the translatability and stability of IGF-I mRNAs with different 5'-UTRs and 3'-UTRs, respectively, suggest that these regions may play important roles in the regulation of IGF-I gene expression.

Translation of Alternative E Peptides

In vitro translation studies indicate that the Ea and Eb domains are indeed translated as part of preproIGF-I from the appropriate synthetic human and rat RNAs.[30,36] Co-translational processing using canine pancreatic microsomes results in cleavage to proIGF-I forms retaining the respective

E-peptide.[36] In vitro, the Ea peptide in rat preproIGF-I undergoes N-linked glycosylation by canine pancreatic microsomes.[36,47] However, this glycosylation only occurred when translation of the preproIGF-I was initiated from Met –32 or Met –22.[47] The significance of these observations is unknown. The Eb peptide is not glycosylated in vitro, presumably because it's sequence does not contain the N-linked glycosylation sites.[36]

The form of IGF-I initially isolated from the conditioned medium of cultured human fibroblasts had a molecular weight and amino acid composition indicating that it was larger than serum IGF-I[61] and was hypothesized to be secreted preproIGF-I Eb.[62] Using an antibody to the Ea peptide, this work[62] also reported that the serum of patients with chronic renal failure contained an immunoreactive IGF-I with a molecular weight from 13,000 to 19,000 Daltons, i.e., possibly prepro or proIGF-I Ea. A subsequent study reported that cultured human fibroblasts secreted a peptide which reacted with this antiserum, and whose molecular weight was larger than mature IGF-I.[63] Moreover, these cells did not secrete any detectable 7.5 KDa IGF-I. These data collectively suggest that the E-peptides are indeed synthesized in vivo, and may be retained in a stable, secreted form of proIGF-I. In studies in which human preproIGF-IEa expression vectors were transfected into human 293 cells, followed by metabolic labeling and immunoprecipitation,[55] small amounts of proIGF-I Ea were detected in the CM, along with larger amounts of mature IGF-I. ProIGF-I Ea forms were detected in cellular lysates, and mutational analysis indicated that intracellular removal of the E-peptide occurred as the result of proteolytic processing to generate the mature 70-amino acid peptide. Of interest, glycosylation of the Ea peptide was also observed in this intact cell system. Mutational analysis suggested the importance of several amino acids around the E peptide cleavage site in order for processing to occur. There was also evidence that cleavage, probably by furin, occurred to generate a secreted 1-76 amino acid form of IGF-I. The relationship of this cleavage to the cleavage(s) that generates the 1-70 form was not clear. Moreover, the significance of secreted IGF-I peptides that retain part or all of the E-peptide with respect to endocrine, autocrine, or paracrine control of cell function is unclear.

Control of Liver IGF-I Gene Expression

Liver expresses the highest amount of IGF-I mRNA in adults, and disruption of the IGF-I alleles in liver decreases serum IGF-I by 75%.[10,28] Four variables, as discussed in the sections below, exert the principal control of liver IGF-I gene expression: growth hormone, development, and insulin, which regulate liver IGF-I gene transcription, and nutritional status (fasting/refeeding and dietary protein restriction), which regulate IGF-I mRNA at a post-transcriptional level.

Growth Hormone and Developmental Regulation of Hepatic IGF-I Transcription

GH is a major stimulator of liver IGF-I gene expression.[5,44,64-66] In a model of acute single GH injection to hypophysectomized rats, the levels of all the variant IGF-I mRNAs were increased by a similar magnitude.[44] Moreover, nuclear run-on assays using target DNAs corresponding to various regions of the IGF-I gene, measures of nuclear pre mRNA transcripts, and measures of either total or nuclear mRNAs resulting from use of the various exon 1 and exon 2 transcription start sites, all support a conclusion that GH stimulates liver IGF-I mRNA by increasing transcription initiation coordinately from all start sites within both the exon 1 and exon 2 promoters.[44,66-69]

Injection of GH into hypox rats caused the rapid appearance of a DNase I hypersensitive site (HS7; Fig. 2) within intron 2 of the rat IGF-I gene.[66] Subsequent dimethylsulfate (DMS) in vivo footprinting studies revealed that the GH-induced chromatin re-arrangement occurred over an approximately 350 bp region of intron 2, located between about 800 and 1100 bp downstream of the 5' end of intron 2.[70] Within this region, assays of in vivo and in vitro DNA-protein interaction using intact liver nuclei or hepatic nuclear extracts revealed two distinct sites of DNA-protein binding, that, nevertheless were not different between hypox and GH-injected hypox rats.[70]

Liver IGF-I mRNAs containing exon 1 are induced during post-natal development. Increased liver IGF-I gene transcription from all exon 1 start sites was shown to be responsible for the initial induction of liver IGF-I mRNA and increased serum IGF-I observed in the peri-natal, (i.e., within the first two weeks) period. This initial induction of hepatic IGF-I exon 1 transcription was accompanied by appearance of two DNase 1 hypersensitive sites located upstream of, and within, exon 1, termed HS2 (Fig. 3) and HS3.[71] In vitro and in some cases in vivo footprints and protein-DNA interactions measured by gel mobility shift assay were characterized within these HS sites in adult

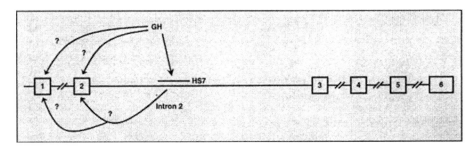

Figure 2. Location of GH-inducible HS7 in the rat IGF-I gene. GH causes a chromatin rearrangement that reveals a DNaseI hypersensitive site in intron 2. This is not associated with any change in protein binding to this region, nor is it known how HS7 influences exon 1 or exon 2 transcription initiation.

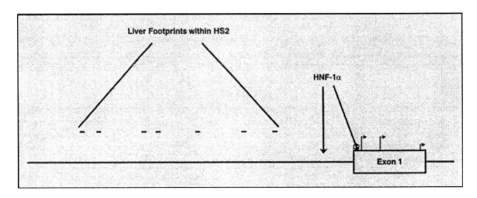

Figure 3. Exon 1 promoter 5'flanking region. The sites of DNA-protein binding within the first 1400 bp of 5' exon 1 flanking sequence are shown, along with the HNF1-α binding sites at –145 and +19.

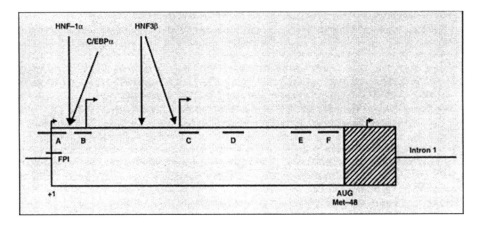

Figure 4. Locations of potential liver transcription factor binding sites in the exon 1 promoter. The entire 382 bp exon 1 is shown schematically. The arrows indicate the transcription start sites, and Met –48 indicates the translation start site. A, B, C, D, E and F indicate the location of binding sites for liver proteins (unidentified) within HS3. Also shown is the location of IGFIFP1 (FP1), which binds unidentified liver protein(s). Sites which show in vitro binding of, and can be transactivated by, HNF1-α and C/EBP-α (+19) and HNF3-β are indicated.

rats, specifically in the region between ~ -1400 to +322 (where +1 is the first exon 1 transcription start site). HS3 was found to include six distinct sites of DNA-protein interaction occurring at intervals throughout the untranslated exon 1 sequence from ~ +1 to ~ +322, that were termed HS3A, B, C, D, E, and F, respectively (Fig. 4).[68] However, the proteins from liver that bind to these sites have not been identified. Moreover, GH did not alter DNA-protein interactions on any of these sites in rat liver nor in sites in the 5'-flanking region upstream of exon 1 to –1400.[68] Furthermore, during the post-natal induction of hepatic exon 1 transcription, there was no change in the amount of liver nuclear protein binding to a region of the exon 1 promoter termed IGFI-FP1 (Fig. 4), which corresponds to HS3A, located around +1 of exon 1.[72]

With respect to regulation of exon 2 transcription, six sites of DNA-protein interaction were also characterized in the 5' region flanking exon 2 from ~ -950 to ~ -50 (where +1 is the first exon 2 start site in the start site cluster) using adult rat liver nuclear extracts.[69] However, as with the proximal exon 1 promoter, there were no differences in DNA-protein binding using nuclear extracts from hypoxed compared to GH-injected hypox rats. Thus, in these studies using rat liver extracts from an in vivo model of GH-induction of exon 1 and exon 2 transcription, cis-acting elements or trans-acting factor responsible for GH regulation of IGF-I transcription have not yet been identified.

In a more chronic model of GH replacement, in which GH was supplied to hypox rats daily for three days, GH stimulated exon 2 transcripts to a greater extent than exon 1 transcripts.[73] However, this result was not observed in a chronic model of GH administration (i.e., daily GH injections over 14 days) to intact rats.[44] Moreover, it was never formally shown that the greater sensitivity of exon 2 transcripts to GH[73] was due to increased transcription. In addition, there is some evidence that within exon 1, IGF-I mRNAs initiated from start site 3 may be more sensitive to chronic GH than start site 2 transcripts,[44] although this has also not yet been shown to be due to increased transcription. These results are mentioned in view of consistent observations that during post-natal development, hepatic IGF-I transcripts initiated from the exon 2 start sites appear in total RNA preparations later than do exon 1 transcripts.[71,74,75] Specifically, exon 1 transcription is induced during the first one to two post-natal weeks,[42,71] whereas exon 2 transcripts do not appear until 2 to 3 weeks post-natal,[42,71,74,75] which correlates with the onset of GH-regulated growth and the beginning of the time period in which large increases in serum IGF-I and continued increases in exon 1 transcripts are observed.[76-78] Although it has not yet been formally shown whether this delayed exon 2 induction is indeed transcriptional, it is of interest that the DNase 1 HS7 (Fig. 2) site that is induced by GH injection to hypox adult rats, does not appear before post-natal day 14,[71] which is concomitant with the induction of exon 2 transcripts and the approximate onset of hepatic GH sensitivity. Collectively, these results suggest that the increase in hepatic exon 1 transcripts during the first one to two post-natal weeks of life are GH-independent, whereas the delayed increase in exon 2 transcripts and the continued increase in exon 1 transcripts beginning at 2 weeks post-natal is GH-dependent.

To determine whether a more intact IGF-I gene structure could indeed be used to detect cis-acting elements responsible for tissue-specific and/or GH regulation of transcription an 11.3 Kb IGF-I mini-gene consisting of 5.3 Kb of 5' flanking region, the 382 bp exon 1, the 1.8 Kb intron 1. (i.e., exon 2 5'-flanking region), the 72 bp exon 2, the entire 3 Kb intron 2 and the initial 48 bp of exon 3, was fused to luciferase.[79] This mini-gene-reporter construct was then used to generate transgenic mice, and luciferase expression was measured in a number of tissues as an index of transcription of the mini-gene. Liver expression of the trans-gene was unfortunately not detected, although as described in a later section, this mini-gene could prove useful for characterizing mechanisms of regulation of extra-hepatic IGF-I gene transcription.

Role of Liver-Enriched Transcription Factors and Signaling Mechanisms Involved in GH Regulation of Liver IGF-I Transcription

HNF-1α knockout mice show reduced liver IGF-I mRNA levels and evidence of GH resistance.[80] HNF-1α can bind to and trans-activate the human IGF-I exon 1 promoter through conserved sites at -145 and +19 (Fig. 3 and 4) when expressed in Hep 3B cells.[81] Moreover, GH stimulated the activity of a salmon IGF-I exon 1 promoter-luciferase construct containing this site in Hep3B cells co-transfected with HNF1-α and STAT 5 expression vectors. The salmon IGF-I promoter construct contained a potential STAT 5 binding site located at –32.[82] In adult rat hepatocytes,

GH stimulated IGF-I expression concomitant with stimulation of tyrosine phosphorylation of STAT 5.[83] Most compelling were results showing that STAT 5B knockout mice had reduced liver IGF-I mRNA compared to wild-type mice. Hypophysectomy decreased liver IGF-I mRNA to a greater degree in the STAT 5B knockout mice and GH was unable to restore liver IGF-I mRNA in these mice.[84] Although these results argue for important and interactive roles of HNF-1-alpha and STAT 5B in stimulating liver IGF-I transcription through the exon 1 promoter, there is no evidence available that would explain the mechanism. First, there is no evidence that the HNF1-α sites at either – 145 or at +19 are bound by endogenous liver HNF-1α either in vitro or in vivo.[68] Moreover, binding of liver proteins to the nearby HS3A and HS3B sites, which flank the HNF1-α site at +19 is not altered by GH.[68] This is perhaps not surprising in view of data indicating that these binding sites are part of the HS3 induced during the first post-natal week before the onset of hepatic GH responsiveness.[71] Binding of nuclear proteins from liver to IGFIFP1, which corresponds to HS3A, is not altered during the developmental induction and increase of exon 1 transcription.[72] Secondly, there are no functionally characterized STAT 5 binding sites in the mammalian exon 1 promoter. Third, a conserved C/EBP site overlaps with the HNF1-α site at +19 (Fig. 4),[85] and recently it was shown that phorbol esters increased C/EBP levels and binding to this site, as well as exon 1 promoter activity in HepG2 cells;[86] whether this is of significance in normal liver is not known. Moreover, HNF3β activates the exon 1 promoter in co-transfection experiments in Hep3B cells through sites at +105 and +142 (Fig. 4), and also by augmenting HNF-1alpha activity.[87] Again, it is uncertain whether HNF3-beta acts through these or other sites to regulate transcription of the endogenous IGF-I gene in hepatocytes or elsewhere in vivo. Fourth, non-STAT 5 factors are implicated in regulation of liver IGF-I transcription by GH. For example, it was recently reported that although GH increases STAT 5 tyrosine phosphorylation in rat hepatocytes, it also increases ERK and Akt phosphorylation. Moreover, inhibition of the PI3 kinase/Akt pathway, while it did not block GH from stimulating STAT 5b tyrosine phosphorylation, did prevent GH from stimulating IGF-I mRNA.[83] GH increases liver STAT 3 and STAT 5 tyrosine phosphorylation concomitant with increased IGF-I transcription in vivo.[88,89] Thus, GH may signal through both JAK2/STAT 5 and PI3 kinase-dependent/non-STAT 5 factors to stimulate liver IGF-I transcription (Fig. 5). However, the cis-acting elements and trans-acting factors that mediate developmental and GH-stimulated exon 1 and exon 2 transcription in liver have not been unequivocally identified. If STATs and liver-enriched transcription factors HNF1-α, HNF3β or C/EBP are responsible for these regulations, they 1) act at sites in the IGF-I gene not yet identified; and/or 2) act indirectly through sites and factors not yet identified.

Insulin Regulation of Hepatic IGF-I Transcription

Insulin is a key regulator of hepatic IGF-I gene expression and serum IGF-I levels, as evidenced by the markedly reduced liver IGF-I mRNA and serum levels in the streptozotocin (STZ) rat model of insulinopenic diabetes.[90] Moreover, this effect of insulin is physiologically important because type 1 diabetes mellitus in rats and humans is associated with growth retardation, which in rats can be reversed by IGF-I treatment.[91] In STZ-diabetic rats, levels of both exon 1 and exon 2 mRNAs are coordinately reduced,[42] and there is a reduction of liver IGF-I gene transcription rate, as measured by nuclear run-on assay.[90] These effects can be reversed by provision of insulin to the diabetic animals.[42,90] Moreover, in isolated rat hepatocytes cultured in vitro, insulin stimulates IGF-I transcription,[92] including initiation of both exon 1 and exon 2 transcripts.[65] An IGF-I exon 1-promoter fragment extending from -326 to +364 of the proximal exon 1 promoter was fused to a G-free cassette and programmed into an in vitro transcription system driven by nuclear liver nuclear extracts from control and STZ-diabetic rats. An in vitro transcript corresponding to initiation at start site 3 was obtained using control liver extracts. The amount of this in vitro transcript product was reduced by about 10-fold when nuclear extracts from diabetic rat liver were used.[93] Moreover, deletion of the 3' end of the exon 1 promoter construct between +148 and +364 resulted in loss of inhibition by diabetic extracts, leading to the conclusion that this downstream exon 1 region contained important sequence elements for down-regulation of liver IGF-I transcription in STZ-diabetes. The liver extracts bound specifically to the +90 to +364 region of the exon 1 promoter, and the binding was diminished with extracts from diabetic rats.

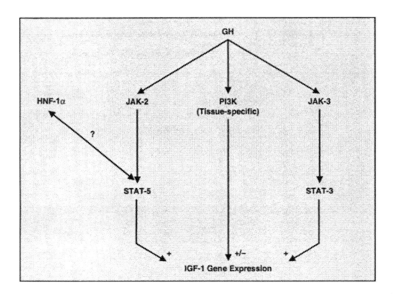

Figure 5. Signaling pathways for GH regulation of IGF-I gene expression. In vivo and cell culture data implicate activation of JAK2, JAK3, STAT3, and STAT5 by GH in stimulating IGF-I mRNA in liver, muscle cells and glioma cells. Transfection results suggest that GH requires the functional interaction of STAT 5 and HNF1-α. GH activation of PI3K stimulates liver cell IGF-I (+), but activation of PI3K inhibits IGF-I mRNA in muscle cells.

Six in vitro footprints, located from about +145 to about +310 were mapped in this region,[93] and were termed footprints I (~ +145), II (~ +160), III (~+190), IV (~ +235), V (~ +295), and VI (+305). Deletion of regions III or V caused loss of inhibition of in vitro transcription by diabetic liver extracts. Moreover, these footprints bound less nuclear protein specifically from diabetic than from control liver, suggesting that insulinopenic diabetes inhibited IGF-I promoter activity by reducing the levels of critical transcriptional activating factors that bound to this downstream exon 1 region. In subsequent studies, site mutants of specific nucleotide residues within footprint III suggested that it bound C/EBP transcription factor, whereas footprint V bound a homeodomain-containing transcription factor.[94]

More recently published studies have focused on the region V footprint (Fig. 6). Within this 25-bp footprint an AT-rich and GC-rich region have been characterized. The GC-rich region appeared to bind Sp like transcription factors, whereas the AT-rich region bound a factor(s) in nuclear extracts termed B1.[95] Further characterization was done in insulin-responsive cultured cells, in which insulin stimulated the amount of factor in nuclear extracts binding to the AT-rich region and corresponding to B1 observed in rat liver nuclear extracts.[95] This factor has been termed the insulin-responsive binding protein (IRBP), and does not appear capable of binding to insulin response elements found in other insulin regulated genes. The stimulation of IRBP by insulin in cultured cells was partially inhibited by a PI3-kinase inhibitor, but not by an ERK inhibitor or a p70S6K inhibitor. These workers have further shown that Sp1 interacts with the GC-region to support IGF-I promoter activity, and insulin activates IGF-I transcription by stimulating appearance of IRBP in a PI3-kinase dependent mechanism. IRBP appears to promote Sp1 binding as part of its stimulatory effect on exon 1 promoter activity.[96] The identity of IRBP remains unknown, including whether it is related to the 65 Kd proteins from rat liver nuclear extracts identified by these workers earlier as interacting with footprints III and V, and which was reduced in STZ-diabetic extracts.

Altered Nutritional Status Regulates Liver IGF-I mRNA Processing and Stability

Fasting potently reduces serum IGF-I and liver IGF-I mRNA levels.[97,98] Moreover, restriction of dietary calories or protein content reduces serum IGF-I.[99,100] In the case of protein restriction, liver

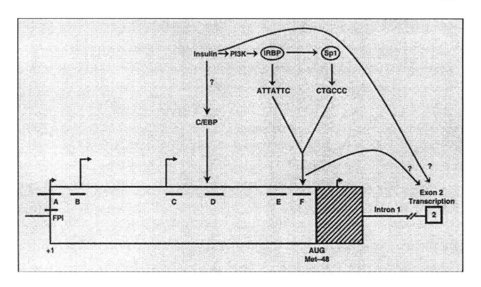

Figure 6. Insulin stimulates exon 1 promoter in liver. In a PI3K-dependent mechanism, insulin increases binding of a (novel) insulin response element binding protein (IRBP) to a sequence in footprint V (also corresponding to HS3F). This binding event facilitates Sp1 binding to an adjacent sequence and stimulates exon 1 transcription. It is not known if this mechanism is involved in stimulation of exon 2 transcription initiation by insulin.

IGF-I mRNA levels are also decreased.[101,102] The effects of fasting and dietary restriction on IGF-I production and action in animals and humans[103] are significant in at least two respects. First, reduction in IGF-I levels as a result of mal-nutrition is associated with impaired growth.[100-105] Interestingly, there is evidence that dietary protein restriction also induces resistance to musculoskeletal growth promoting effects of IGF-I,[100,105] whereas exogenous IGF-I appears able to attenuate weight loss in fasted animals.[103] Secondly, reductions in serum IGF-I associated with short term fasting or dietary restriction may inhibit tumor growth.[99,106,107]

The mechanism(s) by which fasting and protein restriction reduce liver IGF-I mRNA levels is not completely elucidated. Initial studies showed a non-significant decrease in nuclear run-on IGF-I gene transcription rates in fasted rats,[108] although a subsequent study showed that fasting caused a decrease in nuclear premRNA abundance, suggesting an effect on transcription.[109] However, results of more recent nuclear run-on assays indicated that fasting did not reduce liver IGF-I gene transcription, but rather inhibited the maturation of IGF-I premRNA.[110] The major effect of restriction of dietary protein in rats on liver IGF-I mRNA abundance in vivo also appears to be post-transcriptional, and not a reduced gene transcription rate.[101,111]

Rat hepatocytes in primary culture have been used in an attempt to characterize the mechanisms by which nutrients regulate IGF-I gene expression. Lowering the amino acid concentration in the conditioned medium (CM) reduced the IGF-I gene transcription rate despite the presence of a high concentration of insulin.[92] In another study, increasing the amino acid concentration in CM of rat hepatocyte cultures from 0.2 times arterial plasma level up to 5 times arterial plasma level caused a dose-dependent increase in IGF-I mRNA levels. The dose-response effect of amino acids was observed in the presence or absence of GH in the CM, although the absolute level of IGF-I mRNA was higher in the presence of GH at all amino acid concentrations.[112] Subsequently, it was shown that lowering the amino acid concentration in the CM from 1 times arterial plasma level to 0.2 times arterial plasma level reduced the stability of IGF-I mRNAs.[110] Thus, the effects of dietary protein restriction on IGF-I gene expression may result from the direct ability of amino acids to alter mRNA stability in liver hepatocytes. Consistent with these results, glucose regulates IGF-I gene expression by altering IGF-I mRNA stability in cultured fetal hepatocytes.[113] Since no effect of glucose on IGF-I mRNA was observed in adult hepatocytes, the significance of this observation is unclear.

Overall, however, data on the effects of amino acids and glucose in cultured hepatocytes suggest that changes in IGF-I gene expression in liver during fasting and dietary restriction in vivo may be due to direct effects of altered nutrients rather than due to changes in GH or insulin levels or action. Thus, neither the resistance of IGF-I and growth to GH observed in fasted[103,108] and protein restricted animals[114,115] nor decreased insulin levels primarily cause the decreased liver IGF-I, since GH and insulin regulate liver IGF-I transcription. In contrast, altered nutrition principally alters IGF-I mRNA processing and stability without having major effects on transcription or mRNA translation.[116] However, it can be postulated that the resistance of liver IGF-I mRNA to GH in protein-restricted rats[115] is due to decreased IGF-I mRNA processing/stability offsetting increases in IGF-I gene transcription.

Other Modulators of Hepatic IGF-I Gene Expression

Other hormones and stimuli modulate IGF-I gene expression in liver, including glucagon, glucocorticoids, thyroid hormone, tumor necrosis factor α, and estrogen. Glucagon, acting in a cAMP-dependent fashion, stimulates IGF-I mRNA in cultured hepatocytes,[117,118] although neither the mechanism nor physiological significance of this effect are known. Activation of protein kinase C (PKC) by phorbol ester did not increase hepatocyte IGF-I mRNA, although inhibition or down-regulation of PKC lowered GH stimulation of IGF-I mRNA. Interestingly, GH increased DAG levels, suggesting that activation of PKC is part of the GH signaling mechanism to increase IGF-I mRNA levels.[117] As described earlier, activation of PKC using phorbol ester increased IGF-I exon 1 promoter activity in HepG2 cells by increasing levels and binding of C/EBP β to the C/EBP site at +19 in the exon 1 promoter. Whether this mechanism is related to activation of IGF-I transcription by GH or any other agonists that activate PKC in hepatocytes is not known.

Glucocorticoid administration to hypox rats inhibited GH-stimulation of hepatic IGF-I mRNA and serum IGF-I levels. Moreover, glucocorticoid injection of pituitary-intact rats caused decreased growth rate and decreased liver IGF-I mRNA, although serum IGF-I levels were not reduced.[119] In a model of fetal growth retardation induced by injecting glucocorticoid into pregnant dams, liver IGF-I levels in the fetus and dams were not reduced (fetal liver IGF-I gene expression was undetectable in control animals).[120] However, adrenalectomy partially prevents the reduction in liver IGF-I mRNA and serum IGF-I levels in streptozotocin diabetic rats, while having no effect in non-diabetic rats.[121] In contrast, a glucocorticoid receptor antagonist prevented decreased serum IGF-I seen in a burn model without preventing decreased liver IGF-I mRNA.[122]

Hypothyroidism prevents the post-natal increase in liver IGF-I mRNA and serum IGF-I that occur during the GH-dependent stage,[123,124] consistent with a GH-augmenting action of thyroid hormone on liver IGF-I expression.[125,126]

The decreased liver IGF-I mRNA and decreased serum IGF-I observed in septic rats was blocked by a tumor necrosis factor α (TNF) antagonist.[127] Moreover, TNF-α inhibited GH stimulation of IGF-I mRNA in hepatocytes in association with decreased STAT 5 tyrosine phosphorylation and DNA binding activity.[127] Overall, therefore, glucocorticoids, thyroid hormone and TNF-α modulate hepatic IGF-I responses to GH (and insulin), although translational and post-translational control of serum IGF-I by glucocorticoids may also play a role in regulating serum IGF-I.

Estrogens may also modulate GH regulation of liver IGF-I mRNA, serum IGF-I levels, and growth. Daily administration of estradiol to ovariectomized-hypophysectomized rats for 10 days blocked GH stimulation of liver IGF-I mRNA and serum IGF-I levels, and body weight gain.[128] Neither chronic nor acute estradiol injection in this model altered liver IGF-I mRNA or serum levels, although acute estradiol slightly potentiated GH stimulation of liver IGF-I mRNA. Consistent with the lack of effect of a direct effect of estradiol on hepatic IGF-I mRNA in mammals, estradiol did not stimulate a human exon 1 promoter construct in transiently transfected Hep3B cells.[129] In contrast, a specific anti-estrogen stimulated the IGF-I promoter construct, apparently through C/EBP sites located at −12 and at +19 in the exon 1 promoter. This effect was blocked by estradiol. The site at −12 has not been previously characterized as a C/EBP site in the IGF-I promoter, although it corresponds to HS3A and IGFIFP1. Moreover, it is not known if this anti-estrogen increases liver IGF-I transcription or serum levels in vivo, although the authors speculate that such an action could be therapeutically useful in raising serum IGF-I levels in women with osteoporosis. However, the relevance of these observations to regulation of liver IGF-I by estradiol itself is unclear.

Regulation of Extra-Hepatic IGF-I Gene Expression

Adipose Tissue

In pubertal rats, epididymal fat expresses more IGF-I mRNA than other soft extra-hepatic tissues.[130] Moreover, both stromal cells and adipocytes from the epididymal fat pad express IGF-I mRNA, which is reduced by hypox. GH restores epididymal IGF-I gene expression, and GH increases IGF-I mRNA in fat pads cultured in vitro.[130] The role of IGF-I expression and its regulation by GH in adipose tissue is not known. Fat tissue has been postulated to be an extra-hepatic source of serum IGF-I.[14] Moreover, GH-regulated IGF-I may play an important autocrine/paracrine role in pre-adiopcyte differentiation. In the mouse Ob1771 pre-adipocyte cell line, GH stimulated IGF-I transcription and adipose differentiation.[131] However, a subsequent study showed that in the presence of GH alone, Ob1771 cells expressed exon 1 transcripts deriving from start sites 1, 2 or 3, as well as exon 2 transcripts transiently, while expressing start site 1/2 spliced transcripts throughout the 4-day treatment period. However, combined GH and IGF-I treatment resulted in a greater abundance of non-spliced exon 1 transcripts, without transient exon 2 expression, and also resulted in maximal adipose differentiation. It can be hypothesized that GH stimulated transcription initiation from start sites 1 and/or 2, and most of this mRNA was subjected to removal of the 186 nt segment of exon 1 by post-transcriptional splicing. In the presence of IGF-I, much less of the exon 1 mRNA was spliced, i.e., there was a greater abundance of non-spliced exon 1 mRNA resulting from the use of start sites 1, 2, or 3. And, as described earlier, the longer IGF-I mRNA was translated, resulting in production of autocrine IGF-I mRNA and maximum adipose differentiation.[49] Because the primer used to assess non-spliced exon 1 mRNA was complementary to a sequence downstream of start site 3, it cannot be ruled out that IGF-I increased transcription initiation from start site 3, as well as decreasing splicing of start site 1/2 transcripts. Moreover, no mechanisms to explain these stimulus-specific differences in transcription and splicing have been reported. In contrast to the stimulatory effect of GH-stimulated IGF-I on Ob 1771 pre-adipocyte differentiation, GH inhibited porcine pre-adipocyte differentiation concomitant with stimulation of IGF-I mRNA and peptide secretion.[132] In this case, it was proposed that GH might increase local IGF-I production to insure that adequate numbers of mitoses occur prior to terminal differentiation.

Bone

IGF-I mRNA is expressed in growth plate chondrocytes and in osteoblasts.[5,133] Physiologically, bone tropic hormones that increase intracellular cAMP levels such as parathyroid hormone (PTH) and prostaglandin E2 (PGE2) stimulate local IGF-I production, which then mediates anabolic effects of these hormones.[15,134] Initial nuclear run-on assays and assays of IGF-I mRNA abundance clearly showed that PGE2 increased transcription of the IGF-I gene, including coordinate stimulation of mRNAs resulting from the use of the exon 1 start sites 2 and 3, in rat calvarial osteoblast cultures.[135] There was no effect of PGE2 to increase IGF-I mRNA stability. Subsequently it was shown that PGE2 stimulated the activity of an IGF-I exon 1 promoter-luciferase fusion gene in transiently transfected calvarial cells in a PKA-dependent manner through the HS3D site located within the transcribed exon 1 region. Binding of factors from nuclear extracts to this site was stimulated within hours by PGE2 treatment in a mechanism that did not depend on new protein synthesis.[136,137] Subsequently the nuclear protein whose binding to this site was increased by PGE2 was identified as the C/EBPδ transcription factor.[138] Ectopic expression of C/EBPδ stimulated both rat and human IGF-I promoter activity through this site.[138,139]

Recent studies have elucidated the mechanisms by which PKA regulates nuclear levels of C/EBP-δ and have revealed an interesting species difference between rat and human (Fig. 7). In rat calvarial cells, PKA activation causes pre-existing C/EBPδ to move from the cytoplasm into the nucleus, and also maintains C/EBPδ in the nucleus.[140] Once in the nucleus, C/EBPδ is then available to trans-activate the IGF-I promoter. C/EBPδ has not been found to be phosphorylated by PKA, and indeed, the specific phosphorylation event responsible for cytoplasmic nuclear transport and nuclear maintenance of C/EBPδ in rat cells has not yet been identified. The basic domain containing the nuclear localization signal and the leucine zipper dimerization face are required for PKA-stimulated transport into the nucleus. Perhaps, PKA-mediated phosphorylation events block a cytoplasmic inhibitor of C/EBPδ translocation or stimulate a component or components that activate

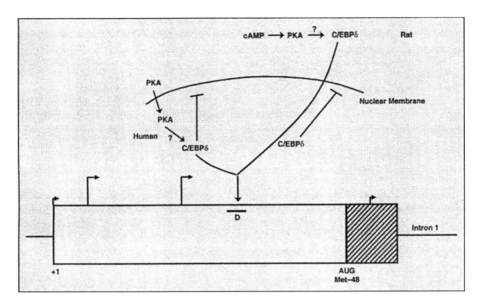

Figure 7. Stimulation of exon 1 promoter by protein kinase A activation in osteoblasts. PTH or PGE2 increase cAMP and PKA activity. In rats, PKA causes C/EBP-δ to translocate into the nucleus, where it binds to a response element in HS3D and activates the exon 1 promoter. In humans, activated PKA translocates into the nucleus and stimulates binding of nuclear-resident C/EBP-δ to HS3D. PKA promotes nuclear retention of C/EBP-δ in rats and human osteoblasts.

nuclear transport.[140] Inhibition of the nuclear export receptor CRM1 prevented loss of nuclear C/EBPδ even when cAMP was removed from the cells, suggesting that the nuclear retention signal generated by PKA also acts to block export.[140]

In human osteoblasts, unlike in rat calvarial cells, C/EBPδ was found in the nucleus in the absence of added cAMP (Fig. 7).[141] Treatment with the cAMP-elevating agent forskolin caused the catalytic subunit of PKA to translocate to the nucleus and stimulate transcriptional trans-activation of the IGF-I promoter by C/EBPδ, although C/EBPδ was not phosphorylated. In addition to a nuclear action of PKA, inhibition of cytoplasmic PKA caused C/EBPδ to leave the nucleus, reminiscent of a similar mechanism in rat cells.[141]

Estrogen is believed to be anabolic for bone and was reported to enhance osteoblast IGF-I transcription.[142] However no stimulatory effect of estrogen on IGF-I transcription or promoter activity in osteoblasts was observed in a subsequent study.[143] Rather, estrogen suppressed cAMP stimulated IGF-I promoter activity in rat osteoblasts by reducing cAMP-stimulated nuclear protein binding (presumably C/EBPδ) to the cAMP-responsive HS3D site. The significance of this effect is not clear, but it is possible that estradiol suppresses both serum and skeletal IGF-I. Measurements of the effect of estradiol on endogenous IGF-I expression in osteoblasts using conditions described in ref. 143 would be necessary to determine whether this possibility is plausible. Estradiol was reported to increase chicken IGF-I exon 1 promoter activity in HepG2 cells through an AP-1 site.[144] However, this particular site, which also conveyed phorbol ester responsiveness to the chicken exon 1 promoter[145] has not been identifed in mammalian exon 1 promoters, and furthermore, it is unclear whether results from liver cells would be pertinent to control of osteoblast cell IGF-I transcription.

Chronic injection of dexamethsone in rats reduces tibial IGF-I mRNA levels.[119] Glucocorticoids reduced IGF-I release from osteoblasts cultured in vitro by reducing levels of both exon 1 and exon 2 mRNAs. A region of the IGF-I exon 1 promoter from +34 to +192 contained sequences that conferred inhibition of promoter activity by glucocortiocids.[146] In contrast, pre-treatment of cultured rat osteoblasts with glucocorticoid followed by its removal potentiated subsequent cAMP stimulation of local IGF-I production and bone protein synthesis.[147] This effect was associated with glucocorticoid-stimulated increase in expression of C/EBPδ and C/EBPβ, which were subsequently activated by cAMP to stimulate IGF-I promoter activity through the HS3D site.[147] These results were interpreted as reflecting the effect of chronically high levels of glucocorticoid to inhibit bone formation as opposed to the effect of transiently increased glucocorticoid to participate in increased bone formation.

Bone morphogenetic protein –7 (BMP-7), also known as osteogenic protein-1 (OP-1), stimulates DNA synthesis and alkaline phosphatase as well as mineralization, in osteoblast cultures. This effect was associated with increased IGF-I mRNA, including increase in the abundance of all exon 1 transcripts and exon 2 transcripts. Start site 2-transcript abundance was significantly higher as a percentage of total IGF-I mRNA after OP-1 treatment compared to other species.[148] However, a subsequent study indicated that OP-1 did not alter IGF-I gene transcription, suggesting that the effect on leader exon abundance was secondary to a change in IGF-I mRNA stability.[149] Of interest, one other study also showed that fetal rat calvarial cells express very low levels of exon 2 transcripts, and that PGE2 increased the abundance of these transcripts as well as the exon 1 transcripts initiated from the various exon 1 start sites.[150] Exon 1 and exon 2 promoter activities were increased by PGE2, but no mechanism has been reported to explain the stimulatory effect of PGE2 on the exon 2 promoter. In contrast to the effects of bone tropic hormones, GH may have only a very small effect on chondrocyte and osteoblast IGF-I mRNA.[5,135]

Muscle

In rats, skeletal muscle expresses exon 1 transcripts initiated principally from start site 3, and little, if any, exon 2 transcripts.[46,73] Also, 95% of muscle IGF-I transcripts contain exon 4 spliced to exon 6, i.e., Ea transcripts.[31] During post-natal development, muscle transcripts are highest during the first one to 2 weeks of life, and then begin to decline.[72,75] GH signaling may be a key determinant of these changes since levels of GH receptor remain high in muscle throughout post-natal development, and levels of STAT 1, 3 and 5 as well as JAK2 correlate well with muscle IGF-I expression during development.[72] Moreover, hypophysectomy dramatically lowers skeletal muscle IGF-I expression in rats, and GH restores levels.[28] The effect of the lit or GHR null mutations in mice on muscle IGF-I mRNA was not reported.[5,66] However, in mouse C2C12 muscle cells, GH stimulates IGF-I mRNA in association with increased tyrosine phosphorylation of Jak 2, STAT 5a, 5b, and 3.[151,152] Moreover, a JAK 3 inhibitor blocked GH activation of STAT 3 and stimulation of IGF-I mRNA, but did not inhibit STAT 5 activation.[152] These results implicate JAK3 and STAT3 as well as JAK2 and STAT 5 signaling as important for IGF-I expression in skeletal muscle (Fig. 5). In these cells, inhibition of ERK and PI3 kinase/Akt pathways actually potentiated the response of IGF-I mRNA to GH.[151] Moreover, activation of PI3K by IGF-I down-regulated IGF-I mRNA.[152] Thus, whereas in liver, the PI3K pathway was required for GH stimulation of IGF-I mRNA, in the C2C12 muscle cells, mitogen-activated pathways may actually be inhibitory. Whether GH actually increases muscle IGF-I transcription in this model remains to be determined.

In addition to GH, other variables regulate muscle IGF-I expression. Catabolic conditions such as burn injury,[122] increased TNF-α (e.g., during sepsis),[153,154] and fasting[155] decrease muscle IGF-I mRNA. Whether these effects occur at the level of gene transcription or mRNA stability is not yet known. The decrease seen in burn injury is apparently not caused by glucocorticoid,[122] and TNF-α may act directly on muscle since it inhibits IGF-I mRNA in cultured C2C12 cells.[154] In contrast, muscle injury,[156] or mechanical overload[157] induce muscle IGF-I, apparently as part of the compensatory growth response. In the former case, the effect is GH-independent. Interestingly, the form of IGF-I mRNA stimulated by mechanical overload is the Eb form, in which exon 4 is spliced to exon 5 and exon 5 is spliced to exon 6,[157] suggesting post-transcriptional regulation. Moreover, the ability of mechanical overload to increase the Eb mRNA declines with age.[157] This result is consistent with the observation that a muscle-specific transgenic expression of IGF-I is associated with prevention of muscle atrophy and promotion of compensatory muscle growth after injury in aged mice.[18] IGF-I mRNA in cardiovascular muscle is increased by angiotensin-II, which may modulate increased cardiac muscle growth occurring with increased blood pressure.[158]

Brain

In adult male rats, exon 1 mRNAs initiated from transcription start sites 2 and 3 are expressed in brain, with no exon 2 expression.[46] Also, brain expresses primarily IGF-I Ea transcripts.[31] GH may be an important regulator of brain IGF-I transcription. Brain IGF-I mRNA is reduced in the GH deficient lit/lit mice, and GH treatment increases brain IGF-I mRNA levels.[66] Moreover, hypophysectomy decreases brain IGF-I mRNA which can be restored by exogenous GH,[159] although the GHR null mouse does not have reduced brain IGF-I mRNA.[5] Also, the developmental peak of brain IGF-I mRNA during the first post-natal week correlates with GHR gene expression in brain.[72] In mice carrying the IGF-I mini-gene-luciferase transgene described previously, luciferase expression was detected in brain and testis, but not in liver.[79] GH stimulated expression of the transgene when injected into the transgenic mice or when added to brain cells cultured from the transgenic mice. Moreover, the brain-region-specifc, developmental regulation of the transgene expression was similar to that of the endogenous IGF-I gene expression.[79] A conceptually similar IGF-I mini-gene fused to luciferase was transiently transfected into rat C6 glioma cells along with plasmids encoding the GH receptor and JAK-2.[160] The C6 cells already contained endogenous STATS 1, 3, 5a, and 5b. Luciferase activity, presumably reflective of promoter activity, was stimulated slightly by GH, although GH did not stimulate the activities of the isolated proximal exon 1 and exon 2 promoters fused to luciferase in similar types of transient transfection assays. Thus, either a putative GH response element is not in the proximal exon 1 and exon 2 promoters, or the contiguous structure of the 5' portion of the IGF-I gene, which includes the proximal promoters and intron 2, with its GH-induced HS site, properly spaced, is required for transcriptional induction by GH. It should be emphasized that this latter possibility does not allow one to conclude that HS 7 does contain the GH-response element. Moreover, these results are not necessarily pertinent to liver, but do suggest that GH may stimulate brain IGF-I transcription in a JAK-2-STAT (1, 3, and/or 5) dependent mechanism through elements present in the first 15 KB of the IGF-I gene. Moreover, they suggest that sequences mediating alterations in brain IGF-I transcription during development are also located in the 11.3 KB mini-gene,[79] although there were no consistent changes in brain nuclear protein binding to IGFIFP1 during development.[72] Glucocorticoids inhibited expression of the mini-gene,[79] consistent with their effect to decrease IGF-I mRNA levels in primary cultures of neurons and astrocytes.[161] This IGF-I inhibitory effect of glucocorticoids could be related to their growth inhibitory action on brain, and further suggests that glucocorticoid responsive elements mediating effects on brain IGF-I transcription could be located in the first 11.3 kb of the IGF-I gene. Although transgene expression was noted in astrocytes and neurons, oligodendrocytes transiently express IGF-I mRNA during post-natal development.[162]

Fasting lowers brain IGF-I mRNA,[155] whereas central hypoxic-ischemic injury transiently up-regulates IGF-I mRNA in brain astrocytes.[163] The up-regulation of brain IGF-I expression could contribute to compensatory growth/neuroprotective responses. However, the mechanisms responsible for these changes in brain IGF-I expression have not yet been characterized. Indeed, despite the promising results with the regulation of expression of the mini-gene in brain, there is no evidence that GH, developmental or glucocorticoid-induced changes in IGF-I gene expression in brain are transcriptional.

Reproductive System

Uterine and ovarian IGF-I mRNA levels are somewhat reduced in GHR null mice,[5] and GH slightly increases uterine IGF-I mRNA when given acutely to hypox-ovx rats.[128,164] However, chronic GH has no effect on uterine IGF-I mRNA levels. Estradiol strongly increases uterine IGF-I mRNA without increasing serum levels in hypox-ovx rats,[128,164] whereas GH inhibits estradiol-stimulated levels. And, estradiol, but not GH stimulates uterine growth. Moreover, estradiol and not GH increases ovarian IGF-I mRNA.[165] The mechanisms by which estradiol increases IGF-I mRNA in the female reproductive tract are unknown.

FSH increases IGF-I mRNA levels in cultured granulosa cells in a cAMP-dependent process in vitro.[166] However, no mechanism for this increase has yet been reported and ovarian IGF-I mRNA was unaltered in both hypox and FSH-null mice, suggesting that neither gonadotrophins nor GH regulate ovarian IGF-1 mRNA in vivo.[167] In males, IGF-I mRNA was unaltered in GHR null mice,[5] although GH increases IGF-I mRNA levels in Leydig cells of hypox rats.[168] In vitro cultures of Leydig cells release more IGF-I when stimulated by LH or cAMP.[169] However, hCG and cAMP inhibited IGF-I mRNA potently and gene transcription rate slightly in cultured Leydig cells.[170] Therefore, the significance of regulation of testicular IGF-I by gonadotrophins is unclear.

Kidney

Kidney IGF-I mRNA is stimulated by GH as evidenced in both mouse genetic models and hypox rats.[5,28,66] During post-natal development, kidney IGF-I and GHR mRNA levels increase mildly,[72] with evidence for delayed expression of exon 1 start site 2 and exon 2 transcripts.[46] Fasting[155] and dietary protein restriction[102] decrease kidney IGF-I mRNA, and glucocorticoids have a slight inhibitory effect on GH stimulation of kidney IGF-I mRNA in hypox rats and on kidney IGF-I mRNA in intact rats.[119] However, it is not known whether any of these changes in kidney IGF-I mRNA are regulated at a transcriptional or post-transcriptional level. Two models of increased kidney growth, specifically unilateral nephrectomy and insulinopenic diabetes have been examined for changes in kidney IGF-I gene expression. Immature uninephrectomized rats exhibit non-GH-dependent increases in IGF-I mRNA in the remaining hypertrophying kidney. In contrast, in older rats, there are no increases in IGF-I mRNA, but the compensatory growth in the remaining kidney is GH-dependent.[33,171-173] In insulin deficient diabetic rodents, kidney size is increased; however, recent studies indicate that while kidney IGF-I peptide levels increase, there is no increase in IGF-I mRNA.[174 178] Thus, the increased kidney IGF-I peptide observed in diabetic and older uninephrectomized animals is either due to increased translation[33,174] or increased trapping of serum-derived IGF-I.[178] It is possible that increased IGFBP-1 expression, which occurs in kidney possibly due to reduced insulin in the diabetes models, traps and enhances the action of IGF-I. Interestingly, octreotide transiently prevents increased kidney IGFBP-1 mRNA and IGF-I peptide in diabetic rats without altering insulin levels.[178] In vitro cultures of mesangial cells show increased IGF-I mRNA in response to the binding of advanced glucosylation end products (AGE) to their receptor,[179] but neither the mechanism nor in vivo significance of these observations is clear.

Fibroblasts

In vitro cultures of fibroblasts have been used to characterize regulation of IGF-I gene expression, and are thought to represent local sources of IGF-I for tissue repair or remodeling scenarios.[180, 181] Protein kinase C and protein kinase A pathway activation have been examined for their impact on IGF-I mRNA. PKC activation or increased intracellular calcium inhibited fibroblast IGF-I gene expression, although some ligands that activated PKC inhibited IGF-I in a PKC-independent manner.[180,182] In contrast, parathyroid hormone-related protein (PTH-rp) increased fibroblast IGF-I initiated from the exon 1 promoter in a cAMP-dependent manner.[181] This effect, which was postulated to explain the growth promoting effect of keratinocytes-derived PTH-rp, was associated with increased IGF-I mRNA stability. It was not reported whether transcription was altered by PTH-rp, but neither PTH-rp nor cAMP itself altered the activity of the IGF-I exon 1 promoter-luciferase construct containing the element that is cAMP responsive in osteoblasts. Thus, the mechanisms of these IGF-I regulatory events in fibroblasts are not known.

Macrophages

IGF-I gene expression is regulated in macrophages during inflammatory responses and in wound healing, and may thereby contribute to fibrosis. In macrophages, certain tissue damaging inflammatory stimuli increase TNF-α release, which in contrast to its inhibitory effects in other tissues, stimulated IGF-I biosynthesis in macrophages. The IGF-I stimulatory effect on macrophages was associated with a small increase in IGF-I mRNA.[183, 184] Moreover, interferon-gamma, which may function to limit fibrosis, inhibits IGF-I gene expression at the transcriptional level.[185] Even more interesting, PGE2, which is involved in blocking certain actions of IFN-gamma in macrophage inflammatory responses, stimulates IGF-I biosynthesis while actually inhibiting IGF-I mRNA, suggesting that PGE2 increases translation of IGF-I mRNA.[184]

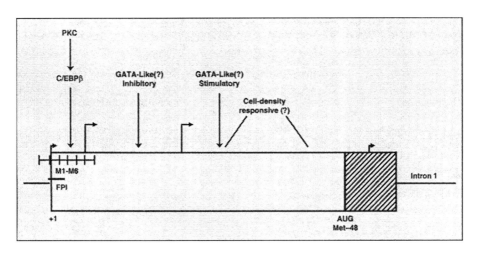

Figure 8. Regulation of IGF-I promoter activity in tumor cells. The locations of elements within exon 1 required for maximal promoter activity in C6 (FP1) and SK-N-MC (M1-M6) cells are shown. Also, the PKC-dependent C/EBP site essential for SK-N-MC promoter activity is shown. The locations of stimulatory and inhibitory sites that bind GATA-like factors from SK-N-MC cells are shown along with the location of a region that may mediate stimulation of exon 1 promoter activity by cell density in C6 cells.

Regulation of IGF-I in Tumor Cells

Considerable evidence implicates both serum and locally (i.e., by tumors) produced IGF-I in tumor growth.[186] Indeed, tumor cell lines in culture because of their ease of growth and transfection have been used to characterize the IGF-I promoters. The first 50 or so bp of exon 1 contains elements essential for promoter activity in SK-N-MC neuroblastoma cells and C6 glioma cells (Fig. 8).[187, 188] Binding sites for nuclear proteins are present in this region, and mutations in this region (FP1 and M1-M6) variably lower promoter activity. Recently, C/EBP-β was identified as a transcription factor endogenous in SK-N-MC cells that bound to the C/EBP site at +19.[86] Mutations of this site lowered promoter activity. More importantly, inhibition of protein kinase C activity decreased binding of C/EBP, decreased exon 1 promoter activity, and decreased IGF-I mRNA. While mutation of this site was also shown to reduce promoter activity by about 30% in another study in SK-N-MC cells, mutations of other nearby sites for which cognate transcription factors have not yet been identified produced even greater decreases in promoter activity.[188] Promoter constructs containing only 18 bp of 5'flanking sequence and the first 75 bp of exon 1 had significant activity in C6 cells.[187] Interestingly, this upstream exon 1 region includes HS3A and HS3B. However, it is not known whether liver proteins binding to these sites are related in any way to tumor cell proteins binding to sites within this region.

Sequences downstream of +75 region are also important for IGF-I promoter activity in tumor cells. In C6 cells and in SK-N-MC cells, promoter constructs that begin at +75 and extend to +192 or +319 stimulate significant luciferase activity.[189-191] Conversely, deletion of the downstream sequence between +107 and +194 had very little effect on the activity of a promoter construct that retained the first 107 bp of exon 1.[188] These results suggest that transcription from start site 2 (located at +40) and start site 3 (located at +140) can occur independently of each other. Within the +75 to +192 region are two perfect matches to the consensus binding site for GATA transcription factors, located at +108 and +183, which are, respectively, inhibitory and stimulatory to promoter activity in SK-N-MC cells and which bind nuclear proteins that appear to be distinct from one another and distinct from GATA-1, -2, -3, and-4 (Fig. 8).[190] Whether these proteins are other members of the GATA family is not yet known. Moreover, these putative GATA elements do not regulate promoter activity in C6 cells. Deletion of exon 1-promoter sequences between +192 and +280 prevents the cell-density associated increase in exon 1 promoter activity in C6 cells (Fig. 8).[191]

It is of interest that rat liver sites HS3C, D, E, and F, as well as sites important for binding insulin-regulated factors from liver are located within the +75 to +319 sequence of the exon 1 promoter. Whether these sites correspond functionally to elements important for nuclear protein binding or promoter activity in tumor cells is not known.

Although sequences within the transcribed portion of exon 1 are necessary for promoter activity in tumor cells, it is unclear if they are sufficient for exon 1 promoter activity. Initial studies suggested that sequences between −1700 and about −800 were required for maximum promoter activity in SK-N-MC cells, and that deletion to −238 or −132 led to loss of promoter activity despite the presence of exon 1 in the promoter constructs.[40,44] Studies from a different lab also using SK-N-MC cells differed slightly in showing that a construct with 553 bp of 5′ flanking sequence gave maximal promoter activity and that a construct with 212 bp of 5′ flank retained activity.[51] However, as described above, other studies in C6 and SK-N-MC showed that constructs beginning at −18 or even at +75 retained maximal promoter activity provided most of exon 1 was present.[187, 189-191] Nuclear protein binding to regions of the exon 1 promoter located upstream of IGFIFP1 have not been reported in tumor cells, although rat liver nuclear proteins bind to the upstream regions between −1400 and +1 (HS2).[68] And, ectopically expressed HNF-1α binds to a site −145.[81] The role of these sites in regulating IGF-I exon 1 promoter activity in tumor cells or normal tissues is not known.

Anti-proliferative agents including glucocorticoids,[192] retinoic acid,[192] PGA2,[193] cAMP,[194] double-stranded RNA,[195-197] and type 1 interferon,[196,197] all markedly inhibit IGF-I production at the mRNA level in C6 glioma cells, which results in inhibition of autocrine cell growth. Although the effects of PGA2, cAMP and dsRNA were transcriptional, transient transfection assays failed to reveal response elements in the exon 1 promoter from about −1500 to +319. Moreover, the effect of cAMP on exon 1 expression is cell-type-specific in that cAMP had no effect on IGF-I mRNA in SK-N-MC cells, and an extremely small, if any, stimulatory effect on GH3 cells, which express exon 1 and exon 2 mRNAs or on OVCAR-3 cells, which express mostly exon 2 mRNAs, with small amounts of exon 1 mRNAs.[194] Of interest, cAMP also de-stabilized IGF-I mRNA in C6 cells.[194] Moreover, the inhibitory effect of cAMP on C6 cell IGF-I mRNA was independent of PKA activation, but rather occurred as a result of inhibition of PI3 kinase and MAP kinase (ERK 1,2) pathways.[198] In the case of dsRNA, inhibition of IGF-I mRNA occurred in a mechanism dependent on protein kinase R activity, but independent of interferon induction.[196] Also, it has been shown that glucose regulates IGF-I mRNA in C6 cells,[199] with glucose starvation decreasing IGF-I mRNA stability in both C6 and GH3 pituitary tumor cells.[200] These effects could relate to nutrient control of cell growth and survival.[200] In GH3 cells, in the absence of added agents, IGF-I mRNAs containing either exon 1 or exon 2 were shown to be equally stable.[201] Regulation of IGF-I mRNA expression resulting from use of the exon 2 promoter has also been characterized to some degree in tumor cells. OVCAR-3 cells express mostly exon 2-initiated mRNA, although exon 1 mRNA is detectable. Exon 2 promoter activity was compared in OVCAR-3 cells to that in SK-N-MC cells, which express much more exon 1 mRNA than exon 2 mRNA and to that in HepG2 cells, which did not express IGF-I mRNA under non-stimulated conditions.[51] Exon 2 promoter-luciferase constructs containing 1.8 or 1.2 Kb of 5′flanking sequence upstream of exon 2 and the first 58 nt of exon 2 (i.e., most of exon 2) had 4-5-fold higher activity in OVCAR-3 cells than in SK-N-MC cells and 14-fold higher activity than in HepG2 cells, suggesting that elements important for high level exon 2 transcription are contained in the 1250 bp of proximal exon 2 promoter.[51] Moreover, sequences between −1200 and −1800 appeared to inhibit exon 2 promoter activity. In subsequent studies, 5′ flanking deletion mutants suggested that silencers located between −1500 and about −400 inhibit exon 2 promoter activity in non-exon 2 expressing C6 cells, without having major effects on exon 2 promoter activity in OVCAR-3 cells.[189] However, this hypothesis obviously requires much additional testing. Silencer elements and pertinent trans-acting factors would need to be identified in C6 cells and more importantly in non-exon 2 expressing tissues in vivo. Moreover, the relationship of putative silencers to the sites of DNA-protein binding observed in rat liver would have to be determined. For example, is binding of proteins to the upstream sequences different using nuclear extracts from liver versus extra-hepatic tissues? In respect to basal promoter activity, deletion and mutation analysis have identified a CACCC-box located at −52 relative to the first exon 2 start site, which is essential for promoter activity. This CACCC box binds a complex of factors, including at least one immunologically related to Sp1.[202] The precise role of the CACCC box in exon 2 transcription is unclear, but since it is in a region of the proximal promoter that shows activity in all cell types, and because of its proximity to the transcription initiation sites, it may perhaps be a basal positioning factor.

Summary and Prospectives

The literature reviewed herein clearly points to major gains and major deficits in our understanding of how IGF-I gene expression is regulated. Major questions remain regarding GH regulation of IGF-I transcription and the mechanisms responsible for high levels of liver IGF-I expression. Although STAT 5 expression is important for GH regulation of IGF-I gene expression, no STAT 5 (or STAT 3) binding sites have as yet been demonstrated in the mammalian IGF-I gene, including the GH-regulated HS7 site in intron 2. And, although HNF1-α expression has been shown to be important for GH regulation of liver IGF-I, and HNF1-lα sites have been identified in the proximal exon 1 promoter, no evidence is yet available to indicate that HNF-1α binding to these sites or any other sites in the IGF-I gene plays a role in GH regulation of liver IGF-I transcription in vivo. Indeed, the roles of the proximal exon 1 promoter C/EBP-α/β, HNF1-α, or HNF3-β sites in regulation of IGF-I transcription in vivo is unknown, except that protein kinase C may activate IGF-I transcription in SK-N-MC cells by increasing binding of C/EBP to the proximal C/EBP-α site at +19. Thus, the detailed molecular mechanisms that cause high-level exon 1 transcription in liver hepatocytes are not known. Moreover, factors and sites responsible for regulating exon 2 transcription in liver or other tissues in which exon 2 mRNAs are expressed have not yet been identified. The only mechanism clearly elucidated for regulation of IGF-I transcription is that of cAMP stimulation of IGF-I in osteoblasts, which involves binding of C/EBP-δ to a site in the downstream portion of exon 1. Future studies will be important for determining how PKA promotes nuclear localization and retention of C/EBP-δ in osteoblasts. However, current evidence suggests that this particular transcriptional mechanism does not explain regulation of IGF-I by cAMP in other cell types or tissues such as macrophages, fibroblasts and tumor cells; moreover, physiological relevance and/or mechanisms of cAMP stimulation of IGF-I gene expression in other cell types such as granulosa and thyroid [203] await clarification. An area that also is promising is that of insulin regulation of IGF-I transcription, which awaits definitive identification of the insulin-response element binding protein(s). With respect to many extra-hepatic tissues, the role of GH versus tissue-specific factors in regulating IGF-I gene expression is unclear. Some extra-hepatic tissues show some dependence on GH and in cultured muscle cells, GH stimulates IGF-I gene expression in a mechanism that may involve STATs. However, there is evidence that extra-hepatic IGF-I expression may be GH-independent in certain repair scenarios. Mechanisms responsible have not yet been elucidated, nor has, for example, the mechanism by which estradiol regulates IGF-I gene expression been elucidated. The hormone and brain development-responsive IGF-I mini-gene may prove useful in characterizing mechanisms of GH, glucocorticoid, and developmental stage-regulated IGF-I transcription. Finally, it is very important to continue work into post-transcriptional mechanisms of regulation of IGF-I gene expression, including possible regulation of mRNA stability and translation. Altered nutritional states such as fasting and dietary protein restriction appear to alter IGF-I pre mRNA processing and reduce IGF-I mRNA stability and these in vivo models, as well as cell culture models would appear to offer excellent experimental models in which to determine whether, for example, the long 3'-UTR of IGF-I mRNAs contains sequences that mediate regulation of IGF-I mRNA degradation by stimulus- and tissue-specific trans-acting factors. And, alternative 5'-UTRs may indeed affect translational efficiency of IGF-I mRNAs in physiologically important ways, the possibility of which should be addressed in future studies.

References

1. Liu J-P, Baker J, Perkins AS et al. Mice carrying null mutation of the genes encoding insulin-like growth factors and type I IGF receptor (IGF Ir). Cell 1993; 75:59-72.
2. Baker J, Liu J-P, Perkins AS et al. Role of insulin-like growth factors in embryonic and postnatal growth. Cell 1993; 75:73-90.
3. Wang J, Zhou, J, Powell-Braxton L et al. Effects of Igf1 gene deletion on postnatal growth patterns. Endocrinology 1999; 140:3391-3394.
4. Liu, J-L, LeRoith D. Insulin-like growth factor I is essential for postnatal growth in response to growth hormone. Endocrinology 1999;140:5178-5184.

5. Lupu F, Terwilliger JD, Lee K et al. Roles of growth hormone and insulin-like growth factor I in mouse postnatal growth. Develop Biol 2001; 229:141-162.
6. Baker J, Hardy MP, Zhou J et al. Effect of an igf1 gene null mutation on mouse reproduction. Mol Endocrinol 1996; 10:903-918.
7. Beck KD, Powell-Braxton L, Widmer H-R et al. Igf1 gene disruption results in reduced brain size, CNS hypomyelination, and loss of hippocampal granule and striatal parvalbumin-containing neurons. Neuron 1995; 14:717-730.
8. Ye P, Li L, Richards RG, DiAugustine RP et al. Myelination is altered in insulin-like growth factor-I null mutant mice. J Neuroscience 2002; 22:6041-6051.
9. Yakar S, Liu J-L, Stannard B et al. Normal growth and development in the absence of hepatic insulin-like growth factor I. Proc Natl Acad Sci USA 1999; 96:7324-7329.
10. Yakar S, Rosen CJ, Beamer WG et al. Circulating levels of IGF-I directly regulate bone growth and density. J Clin Invest 2002; 110:771-781.
11. Wu Y, Yakar S, Zhao L et al. Circulating insulin-like growth factor-I levels regulate colon cancer growth and metastasis. Cancer Res 2002; 62:1030-1035.
12. Yakar S, Liu J-L, Fernandez AM et al. Liver-specific igf-1 gene deletion leads to muscle insulin insensitivity. Diabetes 2001; 50:1110-1118.
13. Wallenius K, Sjogren K, Peng X-D et al. Liver-derived IGF-I regulates GH secretion at the pituitary level in mice. Endocrinology 2001; 142:4762-4770.
14. Naranjo WM, Yakar S, Sanchez-Gomez M et al. Protein calorie restriction affects non-hepatic IGF-I production and the lymphoid system: studies using the liver-specific IGF-I gene-deleted mouse model. Endocrinology 2002;143:2233-2241.
15. Mikakoshi N, Kasukawa Y, Linkhart T et al. Evidence that anabolic effects of PTH on bone require IGF-I in growing mice. Endocrinology 2001;142:4349-4356.
16. Zhao G, Monier-Faugere M-C, Langub MC et al. Targeted overexpression of insulin-like growth factor I to osteoblasts of transgenic mice:increased trabecular bone volume without increased osteoblast proliferation. Endocrinology 2000; 141:2674-2682.
17. Clement S, Refetoff S, Robaye B et al. Low TSH requirement and goiter in transgenic mice overexpressing IGF-I and IGF-I receptor in the thyroid gland. Endocrinology 2001;142:5131-5139.
18. Musaro A, McCullagh K, Paul A et al. Localized Igf-1 transgene expression sustains hypertrophy and regeneration in senescent muscle. Nature Genetics 2002; 27:195-200.
19. DiGiovanni J, Bol DK, Wilker E et al. Constitutive expression of insulin-like growth factor-I in epidermal basal cells of transgenic mice leads to spontaneous tumor promotion. Cancer Res 2000; 60:1561-1570.
20. Jansen M, van Schaik FMA, Ricker AT et al. Sequence of cDNA encoding human insulin-like growth factor I precursor. Nature 1983; 306:609-611.
21. Rotwein P. Two insulin-like growth factor I messenger RNAs are expressed in human liver. Proc Natl Acad Sci USA 1986; 83:77-81.
22. De Pagter-Holthuizen P, Vanschaik FMA, Verduijn GM et al. Organization of the human genes for insulin like growth factors I and II. FEBS Lett 1986; 195:179-184.
23. Rotwein P, Pollock KM, Didier DK et al. Organization and sequence of the human insulin-like growth factor I gene. J Biol Chem 1986; 261:4828-4832.
24. Bell GI, Stempien MM, Fong NM et al. Sequences of liver cDNAs encoding two different mouse insulin-like growth factor I precursors. Nuc Acids Res 1986; 14:7873-7882.
25. Shimatsu A, Rotwein P. Mosaic evolution of insulin-like growth factors. Organization, sequence and expression of the rat insulin-like growth factor I gene. J Biol Chem 1987; 262:7894-7900.
26. Casella SJ, Smith EP, Van Wyk JJ et al. Isolation of rat testis cDNA encoding an insulin-like growth factor I precursor. DNA 1987; 6:325-330.
27. Roberts, CT Jr, Lasky SR, Lowe WL Jr et al. Molecular cloning of rat insulin-like growth factor I complementary deoxyribonucleic acids:differential messenger ribonucleic acid processing and regulation by growth hormone in extrahepatic tissues. Mol Endocrinol 1987; 1:243-248.
28. Murphy LJ, Bell GI, Duckworth ML et al. Identification, characterization, and regulation of a rat complementary deoxyribonucleic acid which encodes insulin-like growth factor-I. Endocrinology 1987;121:684-691.
29. Shimatsu A, Rotwein P. Sequence of two rat insulin-like growth factor I mRNA differing within the 5'-untranslated region. Nuc Acids Res 1987; 15:7196.
30. Rotwein P, Folz RJ, Gordon JI. Biosynthesis of human insulin-like growth factor I (IGF-I). J Biol Chem 1987; 262:11807-11812.
31. Lowe WL Jr, Lasky SR, LeRoith D et al. Distribution and regulation of rat insulin-like growth factor I messenger ribonucleic acids encoding alternative carboxyterminal E-peptides:evidence for differential processing and regulation of liver. Mol Endocrinol 1988; 2:528-535.
32. Bucci C, Mallucci P, Roberts CT et al. Nucleotide sequence of a genomic fragment of the rat IGF-I gene spanning an alternate 5'-non coding exon. Nucleic Acid Res 1989; 17:3596.
33. Lajara R, Rotwein P, Bortz JD et al. Dual regulation of insulin-like growth factor I expression during renal hypertrophy. Am J Physiol 1989; 257:F252-F261.
34. Lund PK, Hoyt EC, Van Wyk JJ. The size heterogeneity of rat IGF-I mRNA is due to primarily to differences in the length of the 3'-untranslated region. Mol Endocrinol 1989; 3:2054-2061.
35. Tobin G, Yee D, Brunner N et al. A novel human insulin-like growth factor I messenger RNA is expressed in normal and tumor cells. Mol Endocrinol 1990; 4:1914-1920.

36. Bach MA, Roberts Jr CT, Smith EP et al. Alternative splicing produces messenger RNAs encoding insulin-like growth factor-I prohormones that are differentially glycosylated in vitro. Mol Endocrinol 1990; 4:899-904.
37. Steenbergh PH, Koonen-Reemst AMCB, Cleutjens CBJM et al. Complete nucleotide sequence of the high molecular weight human IGF-I mRNA. Biochem Biophys Res Comm 1991;175:507-514.
38. Adamo M, Ben-Hur H, LeRoith D et al. Transcription initiation in the two leader exons of the rat IGF-I gene occurs from disperse versus localized sites. Biochem Biophys Res Comm 1991; 176:887-893.
39. Jansen E, Steenbergh PH, LeRoith D et al. Identification of multiple transcription start sites in the human insulin-like growth factor I gene. Mol Cell Endocrinol 1991; 78:115-125.
40. Kim S, Lajara R, Rotwein P. Structure and function of a human insulin-like growth factor I gene promoter. Mol Endocrinol 1991; 9:1964-1972.
41. Foyt HL, LeRoith D, Roberts Jr CT. Differential association of insulin-like growth factor I mRNA variants with polysomes in vivo. J Biol Chem 1991; 266:7300-7305.
42. Adamo M, Ben-Hur H, Roberts CT Jr et al. Regulation of start site usage in the leader exons of the rat insulin-like growth factor I gene by development, fasting, and diabetes. Mol Endocrinol 1991; 5:1677-1686.
43. Rotwein P. Structure, evolution, expression, and regulation of insulin-like growth factors I and II. Growth Factors 1991; 5:3-18.
44. Hall LJ, Kajimoto Y, Bichell D et al. Functional analysis of the insulin-like growth factor I gene and identification of an IGF-I gene promoter. DNA Cell Biol 1992;11:301-313.
45. Hoyt EC, Hepler JE, Van Wyk JJ et al. Structural characterization of exon 6 of the rat IGF-I gene. DNA Cell Biol 1992; 11:433-441.
46. Shemer J, Adamo M, Roberts CT Jr et al. Tissue-specific transcription start site usage in the leader exons of the rat insulin-like growth factor-I gene: evidence for differential regulation in the developing kidney. Endocrinology 1992; 131:2793-2799.
47. Simmons JG, Van Wyk JJ, Hoyt EC et al. Multiple transcription start sites in the rat insulin-like growth factor I gene give rise to IGF-I mRNAs that encode different IGF-I precursors and are processed differently in vitro. Growth Factors 1993; 9:205-221.
48. Yang H, Adamo ML, Koval AP et al. Alternative leader sequences insulin-like growth factor I mRNAs modulate translational efficiency and encode multiple signal peptides. Mol Endocrinol 1995; 9:1380-1395.
49. Kamai Y, Mikawa S, Endo K et al. Regulation of insulin-like growth factor-I expression in mouse preadipocyte Ob 1771 cells. J Biol Chem 1996; 271:9883-9886.
50. Kajimoto Y, Rotwein P. Structure of the chicken insulin-like growth factor I gene reveals conserved promoter elements. J Biol Chem 1991; 266:9724-9731.
51. Jansen E, Steenbergh PH, van Schaik FMA et al. The human IGF-I gene contains two cell type-specifically regulated promoters. Biochem Biophys Res Comm 1992;187:1219-1226.
52. Lowe WL Jr, Teasdale RM. Characterization of a rat insulin-like growth factor I gene promoter. Biochem Biophys Res Comm 1992; 189:972-978.
53. Adamo M, Lanau F, Neuenschwander S et al. Distinct promoters in the rat insulin-like growth factor (IGF-I) gene are active in CHO cells. Endocrinology 1993; 132:935-937.
54. Hepler JE, Van Wyk JJ, Lund P K. Different half-lives of insulin-like growth factor I mRNAs that differ in length of 3' untranslated sequence. Endocrinology1990; 127:1550-1552.
55. Duguay SJ, Lai-Zhang J, Steiner DF. Mutational analysis of the insulin-like growth factor I prohormone processing site. J Biol Chem 1995; 270:17566-17574.
56. Foyt HL, Lanau F, Woloschak M et al. Effect of growth hormone on levels of differentially processed insulin-like growth factor I mRNAs in total and polysomal mRNA populations. Mol Endocrinol 1992; 6:1881-1888.
57. Kajimoto Y, Rotwein P. Evolution of insulin-like growth factor I (IGF-I): structure and expression of an IGF-I precursor from Xenopus laevis. Mol Endocrinol 1990; 4:217-226.
58. Kozak M. Regulation of translation in eukaryotic systems. Ann Rev Cell Biol 1992; 8:197-225.
59. Dever TE, Feng L, Wek RC et al. Phosphorylation of initiation factor 2a by protein kinase GCN2 mediates gene-specific translational control of GCN4 in yeast. Cell 1992;68:585-596.
60. Thissen J-P, Underwood LE. Translational status of the insulin-like growth factor I mRNAs in liver of protein restricted rats. J Endocrinol 1992; 132:141-147.
61. Clemmons DR, Shaw DS. Purification and biological properties of fibroblast somatomedin. J Biol Chem 1986; 261:10293-10298.
62. Powell DR, Lee PDK, Chang D et al. Antiserum developed for the E peptide region of insulin-like growth factor IA prohormone recognizes a serum protein by both immunoblot and radioimmunoassay. J Clin Endocrinol Metab 1987; 65:868-875.
63. Conover CA, Baker BK, Hintz RL. Cultured human fibroblasts secrete insulin-like growth factor IA prohormone. J Clin Endocrinol Metab 1989; 69:25-30.
64. Thissen J-P, Pucilowska JB, Underwood LE. Differential regulation of insulin-like growth factor I (IGF-I) and IGF binding protein messenger ribonucleic acids by amino acid availability and growth hormone in rat hepatocyte primary culture. Endocrinology 1994; 134:1570-1576.
65. Krishna AY, Pao C-I, Thule PM et al. Transcription initiation of the rat insulin-like growth factor-I gene in hepatocyte primary culture. J Endocrinol 1996; 151:215-223.
66. Mathews LS, Norstedt G, Palmiter RD. Regulation of insulin-like growth factor I gene expression by growth hormone. Proc Natl Acad Sci 1986; USA 83:9343-9347.

67. Bichell D, Kikuchi K, Rotwein P. Growth hormone rapidly activates insulin-like growth factor I gene transcription in vivo. Mol Endocrinol 1992; 6:1899-1908.
68. Thomas MJ, Kikuchi K, Bichell DP et al. Rapid activation of rat insulin-like growth factor-I gene transcription by growth hormone reveals no alterations in deoxyribonucleic acid-protein interactions within the major promoter. Endocrinology 1994; 135:1584-1592.
69. LeStunff CE, Bichell DP, Thomas MJ et al. Rapid activation of rat insulin-like growth factor-I gene transcription by growth hormone reveals no change in deoxyribonucleic acid-protein interactions within the second promoter. Endocrinology 1995; 136:2230-2237.
70. Thomas MJ, Kikuchi K, Bichell DP et al. Characterization of deoxyribonucleic acid-protein interactions at a growth hormone-induced nuclease hypersensitive site in the rat insulin-like growth factor-I gene. Endocrinology 1995;136:562-569.
71. Kikuchi K, Bichell DP, Rotwein P. Chromatin changes accompany the developmental activation of insulin-like growth factor I gene transcription. J Biol Chem 1992; 267:21505-21511.
72. Shoba L, An MR, Frank SJ et al. Developmental regulation of insulin-like growth factor-I and growth hormone receptor gene expression. Mol Cell Endocrinol 1999; 152:125-136.
73. Lowe WL Jr, Roberts CT Jr, Lasky SR et al. Differential expression of alternative 5'-untranslated regions in mRNA encoding rat insulin-like growth factor I. Proc Natl Acad Sci USA 1987; 84:8946-8950.
74. Hoyt EC, Van Wyk JJ, Lund PK. Tissue and development-specific regulation of a complex family of rat insulin-like growth factor I messenger ribonucleic acids. Mol Endocrinol 1988; 2:1077-1086.
75. Adamo M, Lowe WL Jr, LeRoith D et al. Insulin-like growth factor I messenger ribonucleic acids with alternative 5'-untranslated regions are differentially expressed during development of the rat. Endocrinology1989; 124:2737-2744.
76. Eicher EM, Beamer WG. Inherited ateliotic dwarfism in mice. Characteristics of the mutation, little, on chromosome 6. J Hered 1976; 67:87-91.
77. Zhou Y, Xu BC, Maheshwari HG et al. A mammalian model for Laron syndrome produced by targeted disruption of the mouse growth hormone receptor/binding protein gene (the Laron mouse). Proc Natl Acad Sci USA 1997; 94:13215-13220.
78. Sara VR, Hall K, Lins P-E et al. Serum levels of immunoreactive somatomedin A in the rat:some developmental aspects. Endocrinology 1980;107:622-625.
79. Ye P, Umayahara Y, Ritter D et al. Regulation of insulin-like growth factor I (IGF-I) gene expression in brain of transgenic mice expressing an IGF-I-luciferase fusion gene. Endocrinology 1997; 138:5466-5475.
80. Lee Y-H, Sauer B, Gonzales FJ. Laron dwarfism and non-insulin-dependent diabetes mellitus in the $Hnf-1\alpha$ knockout mouse. Mol Cell Biol 1998; 18:3059-3068.
81. Nolten LA, Steenbergh PH, Sussenbach JS. Hepatocyte nuclear factor 1 alpha activates promoter 1 of the human insulin-like growth factor I gene via two distinct binding sites. Mol Endocrinol 1995; 9:1488-1499.
82. Meton I, Boot EPJ, Sussenbach JS et al. Growth hormone induces insulin-like growth factor-I gene transcription by a synergistic action of STAT 5 and HNF-1alpha. FEBS Lett 1999; 444:155-159.
83. Shoba LNN, Newman M, Liu et al. LY 294002, an inhibitor of phosphatidylinositol 3-kinase, inhibits GH-mediated expression of the IGF-I gene in rat hepatocytes. Endocrinology 2001; 142:3980-3986.
84. Davey HW, Xie T, McLachlan MJ et al. STAT5b is required for GH-induced liver$Igf-1$ gene expression. Endocrinology 2001;142:3836-3841.
85. Nolten LA, van Schaik FMA, Steenbergh PH et al. Expression of the insulin-like growth factor I gene is stimulated by the liver-enriched transcription factors C/EBPa and LAP. Mol Endocrinol 1994; 8:1636-1645.
86. Umayahara Y, Kajimoto Y, Fujitani Y et al. Protein kinase C-dependent, CCAAT/enhancer binding protein beta-mediated expression of insulin-like growth factor I gene. J Biol Chem 2002; 277:15261-15270.
87. Nolten LA, Steenbergh PH, Sussenbach JS. The hepatocyte nuclear factor 3β stimulates the transcription of the human insulin-like growth factor I gene in a direct and indirect manner. J Biol Chem 1996; 271:31846-31854.
88. Gronowski AM, Zhong Z, Wen Z et al. In vivo growth hormone treatment rapidly stimulates the tyrosine phosphorylation and activation of Stat3. Mol Endocrinol 1995; 9:171-177.
89. Gronowski AM, LeStunff C, Rotwein P. Acute nuclear actions of growth hormone (GH): cycloheximide inhibits inducible activator protein-1 activity, but does not block GH-regulated signal transducer and activator of transcription activation or gene expression. Endocrinology 1996; 137:55-64.
90. Pao C-I, Farmer PK, Begovic S et al. Expression of hepatic insulin-like growth factor and insulin-like growth factor-binding protein-1 gene is transcriptionally regulated in streptozotocin-diabetic rats. Mol Endocrinol 1992; 6:967-977.
91. Scheiwiller E, Guler H-P, Merryweather J et al. Growth restoration of insulin-deficient diabetic rats by recombinant human insulin-like growth factor I. Nature 1986; 323:169-171.
92. Pao C-I, Farmer PK, Begovic S et al. Regulation of insulin-like growth factor I (IGF-I) and IGF-I binding protein 1 gene transcription by hormones and amino acids in rat hepatocytes. Mol Endocrinology 1993; 7:1561-1568.
93. Pao C-I, Zhu J-L, Robertson DG et al. Transcriptional regulation of the rat insulin-like growth factor-I gene involves metabolism-dependent binding of nuclear proteins to a downstream region. J Biol Chem 1995; 270:24917-24923.
94. Zhu J-L, Pao C-I, Hunter Jr et al. Identification of core sequences involved in metabolism-dependent nuclear protein binding to the rat insulin-like growth factor I gene. Endocrinology 1999; 140:4761-4771.
95. Kaytor EN, Zhu JL, Pao C-I et al. Physiological concentrations of insulin promote binding of nuclear proteins to the insulin-like growth factor I gene. Endocrinology 2001;142:1041-1049.

96. Kaytor EN, Zhu JL, Pao C-I et al. Insulin-responsive nuclear proteins facilitate Sp1 interactions with the insulin-like growth factor-I gene. J Biol Chem 2001; 276:36896-36901.
97. Emler CA, Schalch DS. Nutritionally-induced changes in insulin-like growth factor I (IGF-I) gene expression in rats. Endocrinology 1986;120:832-834.
98. Goldstein S, Harp JB, Phillips LS. Nutrition and somatomedin XXII: Molecular regulation of insulin-like growth factor-I during fasting and refeeding in rats. J Endocrinol 1991; 6:33-43.
99. Dunn SE, Kari FW, French J et al. Dietary restriction reduces insulin-like growth factor I levels, which modulates apoptosis, cell proliferation, and tumor progression in p53-deficient mice. Cancer Res 1997; 57:4667-4672.
100. Bourrin S, Ammann P, Bonjour JP et al. Dietary protein restriction lowers plasma insulin-like growth factor I (IGF-I), impairs cortical bone formation, and induces osteoblastic resistance to IGF-I in adult female rats. Endocrinology 2000; 141:3149-3155.
101. Straus DS, Takemoto CD. Effect of dietary protein deprivation on insulin-like growth factor (IGF)-I and -II, IGF binding protein-2 and serum albumin gene expression in rat. Endocrinology 1990; 127:1849-1860.
102. Lemozy S, Pucilowska JB, Underwood LE. Reduction of insulin-like growth factor-I (IGF-I) in protein-restricted rats is associated with differential regulation of IGF-binding protein messenger ribonucleic acids in liver and kidney, and peptides in liver and serum. Endocrinology 1994; 135:617-623.
103. Thissen JP, Ketelslegers JM, Underwood LE. Nutritional regulation of the insulin-like growth factors. Endocrine Rev 1994; 15:80-101.
104. O'Sullivan U, Gluckman PD, Breier BH et al. Insulin-like growth factor-I (IGF-I) in mice reduces weight loss during starvation. Endocrinology 1989; 125:2793-2794.
105. Thissen J-P, Underwood LE, Maiter D et al. Failure of insulin-like growth factor-I (IGF-I) infusion to promote growth in protein-restricted rats despite normalization of serum IGF-I concentrations. Endocrinology 1991; 128:885-890.
106. Hursting SD, Perkins SN, Brown CC et al. Calorie restriction induces a p53-independent delay of spontaneous carcinogenesis in p53-deficient and wild-type mice. Cancer Res 1997; 57:2843-2846.
107. Berrigan D, Perkins SN, Haines DC et al. Adult-onset calorie-restriction and fasting delay spontaneous tumorigenesis in p53-deficient mice. Carcinogenesis 2002; 23:817-822.
108. Straus DS, Takemoto CD. Effect of fasting on insulin-like growth factor-I (IGF-I) and growth hormone receptor mRNA levels and IGF-I gene transcription in rat liver. Mol Endocrinol 1990; 4:91-100.
109. Hayden JM, Marten NW, Burke EJ et al. The effect of fasting on insulin-like growth factor-I nuclear transcript abundance in rat liver. Endocrinology 1994; 134:760-768.
110. Zhang J, Chrysis D, Underwood LE. Reduction of hepatic insulin-like growth factor I (IGF-I) messenger ribonucleic acid (mRNA) during fasting is associated with diminished splicing of IGF-I pre-mRNA and decreased stability of cytoplasmic IGF-I mRNA. Endocrinology 1998; 139:4523-4530.
111. Hayden JM, Straus D. IGF-I and serine protease inhibitor 2.1 nuclear transcript abundance in rat liver during protein restriction. J Endocrinol 1995; 145:397-407.
112. Thissen J-P, Pucilowski JB, Underwood LE. Differential regulation of insulin-like growth factor I (IGF-I) and IGF binding protein-1 messenger ribonucleic acids by amino acid availability and growth hormone in rat hepatocyte primary culture. Endocrinology 1994; 134:1570-1576.
113. Goya L, de la Puente A, Ramos S et al. Regulation of insulin-like growth factor-I and -II by glucose in primary cultures of fetal rat hepatocytes. J Biol Chem 1999; 274:24633-24640.
114. Thissen J-P, Triest S, Underwood LE et al. Divergent responses of serum insulin-like growth factor-I and liver growth hormone (GH) receptors to exogenous GH in protein-restricted rats. Endocrinology 1990; 126:908-913.
115. Thissen J-P, Triest S, Moats-Staats BM et al. Evidence that pretranslational and translational defects decrease serum insulin-like growth factor-I concentrations during dietary protein restriction. Endocrinology 1991;129:429-435.
116. Thissen J-P, Underwood LE. Translational status of the insulin-like growth factor-I mRNAs in liver of protein restricted rats. J Endocrinol 1992; 132:141-147.
117. Tollet P, Legraverend C, Gustafsson J-A et al. A role for protein kinases in the growth hormone regulation of cytochrome P450C12 and insulin-like growth factor-I messenger RNA expression in primary adult hepatocytes. Mol Endocrinol 1991; 5:1351-1358.
118. Kachra Z, Barash I, Yannopoulos C et al. The differential regulation by glucagon and growth hormone of insulin-like growth factor (IGF)-I and IGF binding proteins in cultured rat hepatocytes. Endocrinology1991; 128:1723-1730.
119. Luo J, Murphy LJ. Dexamethasone inhibits growth hormone induction of insulin-like growth factor-I (IGF-I) messenger ribonucleic acid (mRNA) in hypophysectomized rats and reduces IGF-I mRNA abundance in the intact rat. Endocrinology 1989; 125:165-171.
120. Price WA, Stiles AD, Moats-Staats BM et al. Gene expression of insulin-like growth factors (IGFs) the type 1 IGF receptor, and IGF-binding proteins in dexamethasone-induced fetal growth retardation. Endocrinology 1992; 130:1424-1432.
121. Unterman TG, Jentel JJ, Oehler DT et al. Effects of circulating levels and hepatic expression of insulin-like growth factor (IGF)-binding proteins and IGF-I in the adrenalectomized streptozotocin-diabetic rat. Endocrinology 1993; 133:2531-2539.
122. Lang CH, Nystrom GJ, Frost RA. Burn-induced changes in IGF-I and IGF-binding proteins are partially glucocorticoid dependent. Am J Physiol: Regul Integr Comp Physiol 2002; 282:R207-R215.

123. Gallo G, de Marchis M, Voci A et al. Expression of hepatic mRNAs for insulin-like growth factors-I and –II during the development of hypothyroid rats. J Endocrinol 1991; 131:367-372.
124. Nanto-Salonen K, Glasscock GF, Rosenfeld RG. The effects of thyroid hormone on insulin-like growth factor (IGF-I) and IGF-binding protein (IGFBP) expression in the neonatal rat:prolonged high expression of IGFBP-2 in methimazole-induced congenital hypothyroidism. Endocrinology 1991; 129:2563-2570.
125. Tollet P, Enberg B, Mode A. Growth hormone (GH) regulation of cytochrome P-450IIC12, insulin-like growth factor-I (IGF-I), and GH receptor messenger RNA expression in primary rat hepatocytes: a hormonal interplay with insulin, IGF-I, and thyroid hormone. Mol Endocrinol 1990; 4:1934-1942.
126. Wolf M Ingbar SH, Moses AC. Thyroid hormone and growth hormone interact to regulate insulin-like growth factor-I messenger ribonucleic acid and circulating levels in the rat. Endocrinology 1989; 125:2905-2914.
127. Yumet G, Shumate ML, Bryant P et al. Tumor necrosis factor mediates hepatic growth hormone resistance during sepsis. Am J Physiol Endocrinol Metab 2002; 283:E472-E481.
128. Murphy LJ, Friesen HG. Differential effects of estrogen and growth hormone on uterine and hepatic insulin-like growth factor I gene expression in the ovariectomized hypophysectomized rat. Endocrinology 1988;122:325-332.
129. Fournier B, Gutzwiller S, Dittmar T et al. Estrogen receptor (ER)-α, but not ER-β, mediates regulation of the insulin-like growth factor I gene by antiestrogens. J Biol Chem 2001; 276:35444-35449.
130. Peter MA, Winterhalter KH, Boni-Schnetzler M et al. Regulation of insulin-like growth factor-I (IGF-I) and IGF-binding proteins by growth hormone in rat white adipose tissue. Endocrinology 1993; 133:2624-2631.
131. Doglio A, Dani C, Fredrikson,G et al. Acute regulation of insulin-like growth factor-I gene expression by growth hormone during adipose cell differentiation. EMBO J 1987; 6:4011-4020.
132. Gaskins HR, Kim J-W, Wright T et al. Regulation of insulin-like growth factor-I ribonucleic acid expression, polypeptide secretion, and binding protein activity by growth hormone in porcine preadipocyte cultures. Endocrinology 1990;126:622-630.
133. Shinar DM, Endo N, Halperin D et al. Differential expression of insulin-like growth factor-I (IGF-I) and IGF-II messenger ribonucleic acid in growing rat bone. Endocrinology 1993; 132:1158-1167.
134. Edwall D, Prisell PT, Levinovitz A et al. Expression of insulin-like growth factor I messenger ribonucleic acid in regenerating bone after fracture:influence of indomethacin. J Bone Mineral Res 1992; 7:207-213.
135. Bichell DP, Rotwein P, McCarthy TL. Prostaglandin E$_2$ rapidly stimulates insulin-like growth factor-I gene expression in primary rat osteoblast cultures: evidence for transcriptional control. Endocrinology 1993; 133:1020-1028.
136. McCarthy TL, Thomas M, Centrella M et al. Regulation of insulin-like growth factor I transcription by cyclic adenosine 3', 5'-monophosphate (cAMP) in fetal rat bone cells through a element within exon 1: protein kinase A-dependent control without a consensus AMP response element. Endocrinology 1995; 136:3901-3908.
137. Thomas MJ, Umayahara Y, Shu H et al. Identification of the cAMP response element that control transcriptional activation of the insulin-like growth factor-I gene by prostaglandin E$_2$ in osteoblasts. J Biol Chem 1996; 271:21835-21841.
138. Umayahara Y , Ji, C, Centrella M et al. CCAAT/enhancer-binding protein δ activates insulin-like growth factor-I gene transcription in osteoblasts. Identification of a novel cyclic AMP signalling pathway in bone. J Biol Chem 1997; 272:31793-31800.
139. Umayahara Y, Billiard J, Ji C et al. CCAAT/enhancer-binding protein δ is a critical regulator of insulin-like growth factor-I gene transcription in osteoblasts. J Biol Chem 1999; 274:10609-10617.
140. Billiard J, Umayahara Y, Wiren K et al. Regulated nuclear-cytoplasmic localization of CCAAT/enhancer –binding protein δ in osteoblasts. J Biol Chem 2001; 276:15354-15361.
141. Billiard J, Grewal SS, Lukaesko L et al. Hormonal control of insulin-like growth factor I gene transcription in human osteoblasts. Dual actions of cAMP-dependent protein kinase on CCAAT/enhancer –binding protein δ. J Biol Chem 2001; 276:31238-31246.
142. Ernst M, Rodan GA. Estradiol regulation of insulin-like growth factor I gene expression in osteoblastic cells: evidence for transcriptional control. Mol Endocrinol 1991; 5:1081-1089.
143. McCarthy TL, Ji C, Shu H et al. 17β-estradiol potently suppresses cAMP-induced insulin-like growth factor-I gene activation in primary rat osteoblast cultures. J Biol Chem 1997; 272:18132-18139.
144. Umayahara Y, Kawamori R, Watada H et al. Estrogen regulation of the insulin-like growth factor I gene transcription involves an AP-1 enhancer. J Biol Chem 1994; 269:16433-16442.
145. Kajimoto Y, Kawamori R, Umayahara Y et al. An AP-1 enhancer mediates TPA-induced transcriptional activation of the chicken insulin-like growth factor I gene. Biochem Biophys Res Comm 1993; 190:767-773.
146. Delany AM, Canalis E. Transcriptional regulation of insulin-like growth factor I by glucocorticoids in rat bone cells. Endocrinology 1995; 136:4776-4781.
147. McCarthy TL, Ji C, Chen Y et al. Time- and dose-related interactions between glucocorticoid and cyclic adenosine 3',5'-monophosphate on CCAAT/enhancer-binding protein-dependent insulin-like growth factor I expression by osteoblasts. Endocrinology 2000; 141:127-137.
148. Yeh LCC, Adamo ML, Kitten AM et al. Osteogenic protein-1 mediated insulin-like growth factor gene expression in primary cultures of rat osteoblastic cells. Endocrinology 1996; 137:1921-1931.
149. Yeh L-CC, Adamo ML, Duan C et al. Osteogenic protein-1 regulates insulin-like growth factor-I (IGF-I), IGF-II, and IGF-binding protein-5 (IGFBP-5) gene expression in fetal rat calvarial cells by different mechanisms. J Cell Physiol 1998; 175:78-88.

150. Pash JM, Delany AM, Adamo ML et al. Regulation of insulin-like growth factor I transcription by prostaglandin E_2 in osteoblast cells. Endocrinology 1995; 136:33-38.
151. Sadowski CL, Wheeler TT, Wang L-H et al. GH regulation of IGF-I and suppressor of cytokine signaling gene expression in C2C12 skeletal muscle cells. Endocrinology 2001; 142:3890-3900.
152. Frost RA, Nystrom GJ, Lang CH. Regulation of IGF-I mRNA and signal transducers and activators of transcription-3 and −5 (Stat-3 and-5) by GH in C2C12 myoblasts. Endocrinology 2002; 143:492-503.
153. Lang CH, Nystrom GJ, Frost RA. Tissue-specific regulation of IGF-I and IGF binding proteins in responses to TNF alpha. Growth Horm IGF Res 2001; 11:250-260.
154. Fernandez-Celemin L, Pasko N, Blomart V et al. Inhibition of muscle insulin-like growth factor I expression by tumor necrosis factor-alpha. Am J Physiol Endocrinol Metab 2002; 283:E1279-E1290.
155. Lowe Jr WL, Adamo M, Werner H et al. Regulation by fasting of rat insulin-like growth factor I and its receptor. Effects on gene expression and binding. J Clin Invest 1989; 84:619-626.
156. Edwall D, Schalling M, Jennische E et al. Induction of insulin-like growth factor I messenger ribonucleic acid during regeneration of rat skeletal muscle. Endocrinology 1989; 124:820-830.
157. Owino V, Yang SY, Goldspink G. Age-related loss of skeletal muscle function and inability to express the autocrine form of insulin-like growth factor-I (MGF) in response to mechanical overload. FEBS Lett 2001; 505:259-263.
158. Brink M, Chrast J, Price SR et al. Angiotensin II stimulates gene expression of cardiac insulin-like growth factor I and its receptor through effects on blood pressure and food intake. Hypertension 1999; 34:1053-1059.
159. Hynes MA, VanWyk JJ, Brooks PJ et al. Growth hormone dependence of somatomedin-C/insulin-like growth factor-I and insulin-like growth factor-II messenger ribonucleic acids. Mol Endocrinol 1987; 1:233-242.
160. Benbassat C, Shoba LNN, Newman M et al. Growth hormone-mediated regulation of insulin-like growth factor I promoter activity in C6 glioma cells. Endocrinology 1999; 140:3073-3081.
161. Adamo M, Werner H, Farnsworth W et al. Dexamethasone reduces steady state insulin-like growth factor I messenger ribonucleic acid levels in rat neuronal and glial cells in primary culture. Endocrinology 1988; 123:2565-2570.
162. Shinar Y, McMorris FA. Developing oligodendroglia express mRNA for insulin-like growth factor-I, a regulator of oligodendrocyte development. J Neurosci Res 1995; 42:516-527.
163. Gluckman PD, Klempt N, Guan J et al. A role for IGF-I in the rescue of CNS neurons following hypoxic-ischemic injury. Biochem Biophys Res Comm 1992; 182:593-599.
164. Murphy LJ, Murphy LC, Friesen HG. Estrogen induces insulin-like growth factor-I expression in the rat uterus. Mol Endocrinol 1987; 1:445-450.
165. Hernandez ER, Roberts Jr CT, LeRoith D et al. Rat ovarian insulin-like growth factor I (IGF-I) gene expression is granulose-cell-selective: 5'-untranslated mRNA variant representation and hormonal regulation. Endocrinology 1989; 125:572-574.
166. Hatey F, Langlois I, Mulsant P et al. Gonadotrophins induce accumulation of insulin-like growth factor I mRNA in pig granulosa cells in vitro. Mol Cell Endocrinol 1992; 86:205-211.
167. Zhou J, Kumar R, Matzuk MM et al. Insulin-like growth factor I regulates gonadotrophins responsiveness in the murine ovary. Mol Endocrinol 1997; 11:1924-1933.
168. Lin T, Wang G, Calkins JH et al. Regulation of insulin-like growth factor-I-messenger ribonucleic acid expression in Leydig cells. Mol Cell Endocrinol 1990; 73:147-152.
169. Callieau J, Vermeire S, Verhoeven G. Independent control of the production of insulin-like growth factor I and itss binding protein by cultured rat testicular cells. Mol Cell Endocrinol 1990; 69:79-89.
170. Lin T, Wang D, Nagpal ML et al. Human chorionic gonadotrophins decreases insulin-like growth factor-I gene transcription in rat Leydig cells. Endocrinology 1994; 134:2142-2149.
171. Mulroney SE, Haramati A, Roberts Jr CT et al. Renal IGF-I mRNA levels are enhanced following unilateral nephrectomy in immature but not adult rats. Endocrinology 1991; 128:2660-2662.
172. Mulroney SE, Haramati A, Werner H et al. Altered expression of insulin-like growth factor-I (IGF-I) andIGF receptor genes after unilateral nephrectomy in immature rats. Endocrinology 1992; 130:249-256.
173. Mulroney SE, Lumpkin MD, Roberts Jr CT et al. Effect of a growth hormone-releasing factor antagonist on compensatory renal growth, insulin-like growth factor-I (IGF-I), and IGF-I receptor gene expression after unilateral nephrectomy in immature rats. Endocrinology 1992; 130:2697-2702.
174. Flyvberg A, Bornfeldt KE, Marshall SM et al. Kidney IGF-I mRNA in initial renal hypertrophy in experimental diabetes in rats. Diabetologia 1990; 33:334-338.
175. Landau D, Chin E, Bondy C et al. Expression of insulin-like growth factor binding proteins in the rat kidney:effects of long-term diabetes. Endocrinology 1995; 136:1835-1842.
176. Segev Y, Landau D, Marbach M et al. Renal hypertrophy in hyperglycemic non-obese diabetic mice is associated with persistent renal accumulation of insulin-like growth factor-I. J Am Soc Nephrol 1997; 8:436-444.
177. Miyatake N, Shikata K, Wada J et al. Differential distribution of insulin-like growth factor-I and insulin-like growth factor binding proteins in experimental diabetic rat kidney. Nephron 1999; 81:317-323.
178. Raz I, Rubinger D, Popovtzer M et al. Octreotide prevents the early increase in renal insulin-like growth factor binding protein 1 in streptozotocin diabetic rats. Diabetes 1998; 47:924-930.
179. Pugliese G, Pricci F, Romeo G et al. Upregulation of mesangial growth factor and extracellular matrix synthesis by advanced glycation end products via a receptor-mediated mechanism. Diabetes 1997; 46:1881-1887.

180. Lowe Jr WL, Yorek MA, Teasdale RM. Ligands that activate protein kinase –C differ in their ability to regulate basic fibroblast growth factor and insulin-like growth factor-I messenger ribonucleic acid levels. Endocrinology 1993; 132:1593-1602.
181. Shin JH, Ji C, Casinghino S et al. Parathyroid hormone-related protein enhances insulin-like growth factor-I expression by fetal rat dermal fibroblasts. J Biol Chem 1997; 272:23489-23502.
182. Hovis JG, Meyer T, Teasdale RM et al. Intracellular calcium regulates insulin-like growth factor-I messenger ribonucleic acid levels. Endocrinology 1993; 132:1931-1938.
183. Noble PW, Lake FR, Henson PM et al. Hyaluronate activation of CD44 induces insulin-like growth factor-I expression by a tumor necrosis factor-α-dependent mechanism in murine macrophages. J Clin Invest 1993; 91:2368-2378.
184. Fournier T, Riches DWH, Winston BW et al. Divergence in macrophage insulin-like growth factor-I (IGF-I) synthesis induced by TNF-α and prostaglandin E2. J Immunol 1995; 155:2123-2133.
185. Arkins S, Rebeiz N, Brunke-Reese DL et al. Interferon-γ inhibits macrophage insulin-like growth factor-I synthesis at the transcriptional level. Mol Endocrinol 1995; 9:350-360.
186. Khandwala HM, McCutcheon IE, Flyvberg A et al. The effects of insulin-like growth factors on tumorigenesis and neoplastic growth. Endocrine Rev 2000; 21:215-244.
187. An MR, Lowe WL Jr. The major promoter of the rat insulin-like growth factor-I gene binds to protein complex that is required for basal expression. Mol Cell Endocrinol 1995; 114:77-89.
188. Mittanck DW, Kim SW, Rotwein P. Essential promoter elements are located within the 5'-untranslated region of human insulin-like growth factor-I exon 1. Mol Cell Endocrinol 1997; 126:153-163.
189. Wang X, Yang Y, Adamo ML. Characterization of the rat insulin-like growth factor I gene promoter and identification of a minimal exon 2 promoter. Endocrinology 1997; 138:1528-1536.
190. Wang L, Wang X, Adamo ML. Two putative GATA motifs in the proximal exon 1 promoter of the rat insulin-like growth factor I gene regulate basal promoter activity. Endocrinology 2000; 141:1118-1126.
191. Wang L, Adamo ML. Cell density influences insulin-like growth factor I gene expression in a cell type-specific manner. Endocrinology 2000; 141:2481-2489.
192. Lowe WL Jr, Meyer T, Karpen CW et al. Regulation of insulin-like growth factor I production in rat C6 glioma cells:possible role as an autocrine/paracrine factor. Endocrinology 1992; 130:2683-2691.
193. Bui T, Kuo C, Rotwein P et al. Prostaglandin A2 specifically represses insulin-like growth factor-I gene expression in C6 rat glioma cells. Endocrinology 1997; 138:985-993.
194. Wang L, Adamo ML. Cyclic adenosine 3',5'-monophosphate inhibits insulin-like growth factor I gene expression in rat glioma cell lines: evidence for regulation of transcription and messenger ribonucleic acid stability. Endocrinology 2001; 142:3041-3050.
195. Chacko MS, Adamo ML. Double-stranded ribonucleic acid decreases C6 rat glioma cell numbers: effects on insulin-like growth factor I gene expression and action. Endocrinology 2000; 141:3546-3555.
196. Chacko MS, Adamo ML. Double-stranded RNA decreases IGF-I gene expression in a protein kinase R-dependent, but type I interferon-independent mechanism in C6 rat glioma cells. Endocrinology 2002; 143:525-534.
197. Chacko MS, Ma X, Adamo ML. Double-stranded ribonucleic acid decreases C6 rat glioma cell proliferation in part by activating protein kinase R and decreasing insulin-like growth factor I levels. Endocrinology 2002; 143:2144-2154.
198. Wang L, Liu F, Adamo ML. Cyclic AMP inhibits extracellular signal-regulated kinase and phosphatidylinositol 3-kinase/Akt pathways by inhibiting Rap1. J Biol Chem 2001; 276:37242-37249.
199. Straus DS, Burke EJ. Glucose stimulates IGF-I gene expression in C6 glioma cells. Endocrinology 1995; 136:365-368.
200. Wang L, Yang H, Adamo ML. Glucose starvation reduces IGF-I mRNA in tumor cells: evidence for an effect on mRNA stability. Biochem Biophys Res Comm 2000; 269:336-346.
201. Lowe WL, Jr, Adamo M, LeRoith D et al. Expression and stability of insulin-like growth factor-I (IGF-I) mRNA splicing variants in the GH3 pituitary cell line. Biochem Biophys Res Comm 1989; 162:1174-1179.
202. Wang X, Talamantez JL, Adamo ML. A CACCC box in the proximal exon 2 promoter of the rat IGF-I gene in required for basal promoter activity. Endocrinology 1998; 139:1054-1066.
203. Hofbauer LC, Rafferzeder M, Jannsen OE et al. Insulin-like growth factor I messenger ribonucleic acid expression in porcine thyroid follicles is regulated by thyrotropin and iodine. Eur J Endocrinol 1995; 132:605-610.

CHAPTER 5

The Many Levels of Control of IGF-II Expression

P. Elly Holthuizen

Abstract

The functions of IGF-II are very diverse. IGF-II plays a role in the development of the embryo, it is an important mitogenic factor for growing cells, it is involved in differentiation processes and it is highly expressed in many tumor tissues. Because of its pivotal role in these diverse processes it is not surprising that the bioavailability of IGF-II must be regulated at many levels. The IGF-II gene is a complex transcription unit with several interesting regulatory aspects. Expression of IGF-II is controlled at multiple levels from tissue-specific and developmental-dependent transcription initiation, alternative splicing, polyadenylation at multiple sites, to post-transcriptional processing through endonucleolytic cleavage of IGF-II mRNAs and post-translational modification of the protein.

Introduction

The insulin-like growth factors play a central role in growth and development. They are potent mitogens for many different cell types. IGFs are produced primarily in hepatocytes and they direct their endocrine functions in general growth synthesis in a wide variety of tissues from early embryonic stages on to adulthood. In addition, IGFs are synthesized in a number of cell types where they exhibit local paracrine and autocrine functions resulting in both proliferative and differentiating effects that influence processes such as cell cycle control, tissue regeneration, tumor growth and apoptosis. All of these processes involve mitogenic stimulation and cell division and need to be controlled accurately by an intricate mechanism of regulating the levels of protein required in a particular tissue and at a distinct developmental stage.

Various regulatory mechanisms can modulate the biological effects of IGFs. Serum concentrations of IGFs vary widely as a result of their individual expression profiles during prenatal and postnatal growth and development. First, the number of membrane associated IGF receptors expressed on target cells determines the extent of the IGF signal. Both IGF-I and IGF-II act predominantly through the type I IGF receptor resulting in a mitogenic response. The type II IGF receptor, also known as the mannose-6-phosphate receptor, is involved in degradation of IGF-II. Second, the biological activities of the IGFs are controlled by their interaction with high affinity IGF binding proteins (IGF-BPs) in serum. The IGF-BPs differ in tissue distribution and binding characteristics but all six IGF-BPs protect the IGFs against degradation and they facilitate transport to distinct body compartments. Third, the bioavailability of IGFs is regulated by the amounts of these growth factors produced and secreted by the cells capable of their synthesis.

In this Chapter the IGF-II gene structure, the bioavailability of the IGF-II protein and the regulation of expression of IGF-II at the level of transcription and translation will be discussed. In addition, a unique mechanism will be described that regulates IGF-II protein synthesis through post-transcriptional processing of IGF-II mRNAs. Although the main focus will be on human IGF-II, the corresponding genes of other species will also be discussed.

The Role of IGF-II in Growth and Development

IGF-II expression is observed in many different tissues and at various stages of development. In contrast to IGF-I, expression of IGF-II is not growth hormone dependent. The significance of IGF-II

Insulin-Like Growth Factors, edited by Derek LeRoith, Walter Zumkeller and Robert Baxter. ©2003 Eurekah.com and Kluwer Academic / Plenum Publishers.

in various growth and developmental processes was clearly demonstrated in mice by the dramatic effects of lack of IGF-II gene expression. Targeted disruption of the IGF-II gene leads to proportionate growth retardation from embryonic day E11 onwards. Mice with an IGF-II null-mutation are born as viable, fertile, proportionate dwarfs with a body weight that is 60% that of wild-type littermates.[1] Furthermore, it was found that homozygous IGF-II mutants and heterozygous progeny carrying a paternally derived mutated IGF-II gene are phenotypically indistinguishable as a consequence of parental imprinting. It was demonstrated that the paternal allele is transcriptionally active during mouse embryonic development, while the maternal allele is transcriptionally silent.[2] This will be further discussed in Chapter 3.

During early development IGF-II expression is high in liver, both in human and in rodents. Many other tissues express IGF-II during embryonic development. For example, in second trimester human fetal tissue IGF-II is expressed in cerebral cortex, costal cartilage, skeletal muscle, adrenal, kidney, lung and placental tissue.[3,4] Recently an extensive IGF-II expression study in fetal rhesus monkey tissues during second and third trimester was performed. It was shown that IGF-II is expressed in a similar manner in human and monkey and that expression of IGF-II and IGF-BPs in specific cell types indicates also a paracrine and autocrine role for IGF-II during development.[5]

After birth and throughout adult life, IGF-II expression remains high in humans, while it is markedly decreased in all rodent tissues with the exception of the leptomeninges and choroid plexus of the brain, where IGF-II expression persists throughout adult life.[6] In human, IGF-II gene expression not only persists in these neuronal tissues after birth, but IGF-II is expressed also in the liver and at a lower level in several adult tissues such as heart, brain, kidney, muscle, skin, the reproductive organs and the nervous system.[7] Thus although the hypothesis that IGF-II is a fetal growth factor applies to rodents, it is not true for humans. For several other species i.e., human, monkey, horse, sheep, pig, guinea pig, mink and salmonids it was shown that IGF-II can be expressed in selected adult tissues, suggesting that there must be a distinct function for IGF-II in adult tissues as well.

Biological Effects of IGF-II

The biological effects of IGF-II have been studied extensively using both in vitro and in vivo experimental systems. The biologically relevant effects for IGF-II at the usual nanomolar concentrations are the stimulation of cell proliferation, and at least in certain tissues, cell differentiation, an effect that has been characterized in detail in myoblasts.[8] Furthermore, IGF-II may act as a survival factor that counteracts apoptosis in various cell systems.[9] For example, in vitro cytokine induced programmed β-cell destruction, which normally takes place in the rat pancreas 2-3 weeks after birth, can be inhibited by IGF-II.[10] Several hormones have been described that can activate or inhibit IGF-II expression. In cultured human ovarian granulosa cells IGF-II mRNA synthesis is stimulated by FSH, chorionic gonadotropin, and dibutyryl cyclic AMP, while in cultured human fetal adrenal cells IGF-II synthesis is stimulated by ACTH and dibutyryl cyclic AMP.[11] In contrast, ACTH and cortisol decrease IGF-II gene expression in ovine fetal adrenal cells.[12] IGF-II mRNA expression is inhibited by glucocorticoid and thyroid hormone in rat hepatic cells in 8 days old pups.[13] Glucose can increase IGF-II mRNA expression 3-fold in the highly differentiated rat insulin-producing beta-cell INS-1 cell line.[14] The mechanisms by which these hormones act or the identification of specific responsive elements that may be involved in regulation of IGF-II are in most cases still elusive.

IGF-II Protein Structure

Human IGF-II is a single-chain polypeptide of only 67 amino acids, structurally related to IGF-I and proinsulin. It contains the structural domains B, C, A, and D. In contrast to insulin, the C-domain is not removed by proteolytic processing and the D-domain does not exist in insulin. IGF-II is synthesized as a large 180 amino acids precursor protein with a 24 amino acids signal peptide at the N-terminal end and a C-terminal extension of 89 amino acids called the E-peptide.[15] The biologically most active form of IGF-II is the fully processed 7.5 kDa protein, but larger forms of up to 20 kDa have also been isolated from serum.[16-20] In addition, several variant human IGF-II proteins have been characterized. A 70-amino acid form of IGF-II arises through alternative use of a 5' acceptor splice site in exon 9 of the human IGF-II gene, which results in the substitution of serine residue 29 for the RLPG tetrapeptide of IGF-II.[21] This variant molecule comprises approximately 25% of serum IGF-II and has about one third of the potency of the 67-residue IGF-II for the

IGF-1R.[22] A 69-residue variant IGF-II molecule was identified in which serine residue 33 was substituted by the CGD tripeptide.[23] This IGF-II molecule is probably encoded by an allelic variant of the IGF-II gene. Due to incomplete processing of the IGF-II precursor, so-called "big IGF-II" can be formed that still contains 21 residues of the E domain.[24] These large forms of IGF-II generally constitute less than 10% of the total IGF-II in human serum, but patients with certain tumors may have up to 75% of their serum IGF-II in the form of these larger IGF-II variants resulting in extrapancreatic tumor hypoglycemia.

The structure and composition of other vertebrate IGF-II proteins is highly conserved, and at the most 4 out of 67 amino acids are different, which results in an overall homology of 94%. Of the non-vertebrate species several bony fish IGF-II peptides were characterized and all known fish IGF-II peptides are 70 amino acids in length. The overall homology between the mature IGF-II peptides of fish and mammalian species is approximately 80%. Despite the fact that the mature fish IGF-II peptide is similar to that of its mammalian counterpart, sequences of other domains of the fish IGF-II precursor i.e., the signal peptide and the E domain deviate strongly from that of their mammalian counterparts.[25]

IGF-II Gene Structure

Since the isolation of the first IGF-II cDNA from human liver in 1985, many other IGF-II cDNAs have been characterized from which the amino acid sequence of IGF-II was deduced. The protein sequence of IGF-II in all species is highly conserved. IGF-II cDNAs were isolated for a number of species, and based on nucleotide sequences the IGF-II protein sequences were determined for horse, deer, kangaroo, cow, sheep, pig, guinea pig, mink, rat, mouse, chicken, shark, and various fish such as salmon, tilapia and barramundi.[9,25,26] The complete genomic structure has only been determined in detail for the human, ovine, rat and mouse IGF-II genes, and recently for the salmon IGF-II gene.[9,27]

The IGF-II gene is a complex transcription unit with many interesting regulatory aspects. Expression of IGF-II is controlled at multiple levels from tissue-specific and developmental-dependent transcription initiation, alternative splicing, and multiple polyadenylation sites to endonucleolytic cleavage of IGF-II mRNAs. Since the functions of IGF-II are so diverse, it is not surprising that the regulation at the transcriptional level is complex.

The human IGF-II gene is located on chromosome 11, it consists of 9 exons, of which exons 7, 8 and the first 237 nt of exon 9 encode the 180 aa long pre-pro-IGF-II, that is further processed into the 67 aa long mature IGF-II polypeptide (Fig. 1). The gene further contains six 5'-untranslated exons (exons 1-6) of which exons 1, 4, 5, and 6 are each preceded by a separate promoter (P1-P4), resulting in the transcription of a family of mRNAs that contain four different leader sequences.[28] Recently an additional exon of 164 nt, named exon 4b, was identified in several tumor tissues and in liver and placenta tissues. The exon is located between exons 4 and 5 and its expression is under the control of promoter P2. In most instances a P2 transcript will consist of exons 4, 7, 8, and 9 yielding a mRNA of 5.0 kb. In addition a P2 transcript is co-expressed consisting of exons 4, 4b, 7, 8, and 9, yielding a mRNA of 5.1 kb.[29] The four IGF-II promoters P1 to P4, are differentially active in a tissue- and development-specific manner and give rise to mRNAs ranging from 6.0 kb to 4.8 kb. Further diversity is created at the 3' end where two functional polyadenylation signals are present in exon 9 that are separated by 3.7 kb, resulting in an additional P3 transcript of only 2.2 kb.[30] Summarizing, all IGF-II transcripts share the IGF-II coding region, but have different 5'- and 3'-untranslated regions (UTRs).

The overall structure of the IGF-II genes is extremely well conserved (Fig. 2). All vertebrate species examined contain at least 6 common exons: three exons that encode the IGF-II prepropeptide and at least three 5'-non-coding exons that are each preceded by a promoter. A subset of vertebrate IGF-II genes among which the human, sheep, and pig IGF-II genes contain additional 5'-non-coding exons preceded by a fourth promoter. The presence and activation of this additional promoter is directly correlated with the presence of circulating IGF-II after birth, when IGF-II expression is maintained in the liver throughout life. In rat and mouse, which lack the adult liver promoter, expression rapidly declines after birth.[31-34]

Recently the first complete non-vertebrate IGF-II gene structure was determined (Fig. 2). The chum salmon IGF-II gene is much smaller and simpler organized than its known mammalian

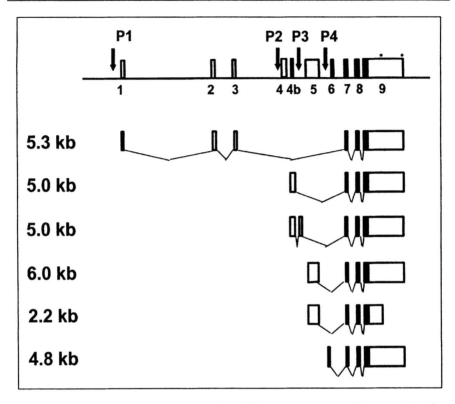

Figure 1. Structure of the human IGF-II gene with its different mRNAs generated by transcription from four promoters, P1 to P4, which are indicated by arrows. The composition of the various transcripts are indicated as well as the sizes of the mRNAs in kb. The prepro-IGF-II encoding region is located in exons 7, 8, and the first part of exon 9, indicated with filled black boxes. Open boxes represent 5'- and 3'-untranslated sequences. The asterisks in exon 9 represent the two alternative poly(A) sites in the 3'-UTR.

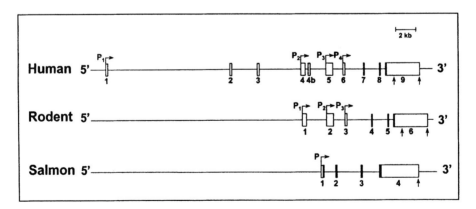

Figure 2. Structural comparison of the IGF-II genes from human, rodents, and salmon. Open boxes represent 5'- and 3'-untranslated sequences; filled black boxes represent the prepro-IGF-II encoding exons. Promoters are represented by the letter "P" in front of a bent arrow. Polyadenylation sites are indicated by vertical arrows below the last exon. The rodent IGF-II gene structure has been determined for rat and mouse.[32,33] The salmon IGF-II gene structure was determined for the chum salmon.[27]

counterparts and consists of four exons and three introns, spanning 8.1 kb of chromosomal DNA. IGF-II is expressed throughout its life span from a single promoter, giving rise to single 4 kb long transcript.[27]

After the elucidation of the IGF-II gene structure, attention shifted towards interpreting the complex regulation of transcription and the characterization of the transcription factors involved in activation of the multiple IGF-II promoters. A variety of approaches have been applied to identify cis-acting elements and their cognate binding proteins involved in IGF-II gene regulation. In the next paragraph the factors regulating the four human promoters will be described in more detail. In addition some regulatory aspects of the rodent and fish genes will be discussed.

Regulation of the Human Promoter P1

Expression of IGF-II in human adult liver tissue is derived from promoter P1, giving rise to a 5.3 kb mRNA, containing the transcript of leader exons 1, 2 and 3 (Fig. 1).[35] This promoter, located only 1.4 kb downstream of the insulin gene, comprises approximately 1 kb. It is the main activator of IGF-II transcription in the adult liver, which functions as the source of circulating endocrine IGF-II.

Regulation of the human P1 promoter has been studied in detail and both activating and repressing elements have been identified.[26] The promoter is very GC-rich and lacks elements resembling a TATA-box or a CAAT-box. Based on transient transfection assays the promoter can be subdivided into two regions. The distal promoter region (positions -900 to -175) contains two homologous 67 nucleotides (nt) long inverted repeat elements (IR1 and IR2) that act as cell type-dependent suppressors of P1 activity. Similar to previously identified silencer elements, the IR elements affect promoter activity in an orientation- and position-independent manner.[36]

The first 175 nt upstream of the transcription start site contain several regions to which nuclear proteins bind, resulting in activation of transcription. A single binding site for the ubiquitous transcription factor Sp1 is located around position -48 that is essential for basal and activated transcription of P1. Mutations in the Sp1 binding site result in an 85% decrease in P1 promoter activity and without a functional Sp1 site the promoter can no longer be activated by other transcription factors. This suggests that Sp1 may facilitate the interaction between transcription factors and recruitment of the transcriptional machinery to this TATA-less promoter.[37] Several liver-specific transcription factors activate P1 expression. Around position -100 the most important activator element, a functional C/EBP binding site, was identified. Two of the C/EBP family members are able to activate P1, C/EBPα (6-fold) and C/EBPβ (10-fold).[38] Interestingly, C/EBP itself is expressed in a tightly controlled manner, the concentration being minimal in fetal liver, increasing around birth and reaching maximum levels in adult liver tissue.[39] The matching expression patterns of IGF-II and C/EBP suggest that the C/EBPs are the major contributors to the postnatal liver-specific activation of the human IGF-II promoter P1. The role of the hepatocyte nuclear factors (HNFs) in IGF-II P1 expression was also examined. P1 activity is stimulated 10-fold by the presence of HNF-3β, whereas HNF-3α and HNF-3γ and HNF-1 hardly affect P1 promoter activity. Interestingly, HNF-4, a factor belonging to the family of steroid/thyroid hormone nuclear receptors, strongly suppresses IGF-II P1 activity. This suppression was observed in liver-derived Hep3B cells but not in kidney-derived 293 cells. The negative effect of HNF-4 on P1 activity was completely abolished by coexpression of C/EBP transcription factors, but not by HNF-3β. The IGF-II promoter P1 contains at least two tandem binding sites for nuclear hormone receptor responsive elements that can be activated up to 15-fold when cotransfected with the retinoic acid receptor (RAR) and the retinoic X receptor (RXR).[26] The observed effects of liver-specific transcription factors influencing P1 in combination with what is known about the expression of the HNFs during liver development have led to the following model for P1 expression.[40] During the early phase of fetal liver development, HNF-4 is the dominant transcription factor and its presence suppresses P1 activation via a still unknown mechanism. Although the stimulatory factor HNF-3β is also expressed, this does not counteract the suppressing effect of HNF-4. Around the time of birth, the expression of C/EBPα and C/EBPβ is initiated and these factors do have the ability to counteract the suppression of P1 by HNF-4. After birth, expression of HNF-4 decreases, while the expression of HNF-3β and the C/EBPs increases, resulting in maximal expression of P1 in adult liver tissue.

Regulation of the Human Promoter P2

Transcription from the human IGF-II promoter P2 results in a 5.0 kb mRNA, containing the transcript of exon 4, or the combination of exon 4 and 4b, as leader exons (Fig. 1).[40-42] P2 transcripts have been detected at low levels in fetal and adult tissues and may be elevated in some human tumor tissues.[43] The P2 promoter is a very weak promoter that lacks TATA- and CAAT-boxes and transcription starting at multiple initiation sites is observed. No Sp1 recognition sequences were found, and no other enhancer elements have been described to date. Promoters homologous to the human P2 were detected and characterized in sheep and in rodents, where this promoter is designated P1, since rodents lack a homologue to the human promoter P1. Based on Northern blotting experiments with rat and mouse tissues, it can be concluded that also in these species the homologous mRNA represents a minor transcript.[44]

Regulation of the Human Promoter P3

The human IGF-II promoter P3 is active in many fetal and non-hepatic adult tissues and in most IGF-II expressing cell lines. The human P3 transcripts, containing exon 5 derived leader sequences, give rise to an abundant 6.0 kb mRNA and a minor 2.2 kb transcript that contains a shorter 3' untranslated region (UTR) due to usage of an internal polyadenylation signal (Fig. 1).[35] P3 contains a functional TATA-box and CAAT-box and 70-80% of the promoter consists of G/C-basepairs. Transient transfection and EMSA experiments using truncated promoter P3 constructs have revealed that the distal promoter region (-1300/-1050) is responsible for cell-type specific IGF-II expression and multiple cell-type specific DNA-protein complexes are formed although the exact nature of these binding proteins has not yet been identified.[45,46] Additional mutation/deletion analysis revealed that the distal segment from -1091 to -1048 contains a cis element named P3-D that plays a critical role in regulating IGF-II P3 promoter activity in a cell density/differentiation-dependent manner. The promoter elements between -1048 and -515 appear to lack any significant regulatory elements.

The proximal region (-289/+140) exhibits basal levels of expression that are similar in several different cell lines, including HeLa and Hep3B.[47] In the proximal region a number of elements that are recognized by nuclear proteins were identified by DNaseI footprint analysis, EMSA and in vitro transcription. TBP (TATA-binding protein) binds to the TATA-motif around position –25 and CTF (CAAT transcription factor), also known as Nuclear Factor 1, binds to the CAAT-box. Several members of the zinc-finger containing transcription factors can bind to the GC-rich proximal promoter region. There are three functional Sp1 sites and at least 5 binding sites for the early growth response proteins Egr-1 and Egr-2 in promoter P3. The activation by Egr has received much attention, since the specific DNA motif that can be bound by the zinc-fingers of Egr-1 and Egr-2 is also recognized by the Wilms' tumor WT1 gene product. Binding studies have revealed that indeed both Egr and WT1 proteins are able to recognize and bind to multiple sites in P3, albeit with different affinities.[48,49] Of the four isotypes of WT1, the WT1(-KTS) form which is a minor component in vivo, was shown to bind to all Egr/WT1 sites of P3, while WT1(+KTS) could only bind with high affinity to the site located within exon 5 at position +63/+71. Using transient transfection experiments it was shown that Egr-1 has a strong stimulatory effect on P3 activity and a high level of WT1 expression of the -KTS type, represses IGF-II P3 activity.[50] Based on these and other results it has been postulated that the Egr proteins may play a role in stimulating expression of the IGF-II gene resulting in autocrine growth stimulation of specific tumors, whereas WT1 may act as a repressor of P3 activity.[51] In fact, enhanced EGR-1 expression in cells exposed to hypoxic stress activates P3 expression.[52] As shown by footprinting analysis there are at least five binding sites for the G/C-box transcription factor AP-2 within the P3 region form -290 to -7.[53] AP-2 was shown to activate P3 expression up to 3-fold and all binding sites contribute to this enhanced expression.[54] It was demonstrated that AP-2 has a dual role in the regulation of P3 activity. At high endogenous amounts, overexpression of AP-2 causes inhibition of P3 activity, whereas expression of AP-2 in cells with low endogenous AP-2 levels leads to P3 activation.[53] Two as yet unknown proteins, tentatively named IPBP3 and IPBP4/5 bind to a key regulatory element P3-4, located between positions -192 and -172 relative to the transcription start site. They each bind to a separate domain of this element simultaneously without steric hindrance; IPBP4/5 binds Box A (-193/-188), whereas IPBP3 binds Box B (-183/-172).[53]

Summarizing the P3 results, it is clear that promoter P3 is a very complex promoter and activation involves binding of multiple trans-acting factors that act in concert in an intricate mechanism of regulation. It is not known whether some of the above mentioned factors compete for the same binding sites or that they cooperate in transactivation. Moreover, not all of these factors are present in each cell type.

Regulation of the Human Promoter P4

Transcription of the human promoter P4, preceding leader exon 6, results in an mRNA species of 4.8 kb (Fig. 1). P4 contains a TATA-box and expression is moderate in most fetal and non-hepatic adult human tissues examined. The major regulator of P4 is the ubiquitously expressed transcription factor Sp1 of which four binding sites are located within the first 125 nt upstream of the transcription start site.[55,56] Two of the sites fit the consensus sequence of Sp1 perfectly, the remaining two sites contain a few mismatches, suggesting that the contribution of the latter to transcriptional activation is less significant.[57] In the absence of Sp1 no P4 activity can be detected by transient transfection assays.[58] Also, enhanced Sp1 binding results in enhancement of P4 activation.[59] Flanking the Sp1 site between positions -67 to -58, two retinoblastoma control elements (RCE) were identified, and it was shown that the retinoblastoma gene product could act as a positive regulator on Sp1-mediated transcription of P4.[58] In addition, the human P4 promoter can also be activated by Egr-1, and inhibited by WT1.[60,61] Recently it was shown that activation of Sp1 and EGR-1 by hepatitis C virus core gene product increases the IGF-II expression from P4 up to 7-fold.[62]

Regulation of Expression of the Rodent IGF-II Promoters

The rat and mouse promoters P1 and P2, homologous to the human promoters P2 and P3, respectively, are not very active in most tissues examined (Fig. 2). Very little is known about the regulation of these promoters by specific transcription factors.

The rat and mouse promoter P3, homologous to the human promoter P4, was shown to be the most active promoter in these species.[63] Transcripts of 3.6 kb derived from this promoter are in general 10-fold more abundant than the 4.6 kb transcripts from rodent promoter P2.

Interestingly, in the rat and the mouse P3 promoters also four Sp1 binding sites were identified, and all four sites were shown to bind Sp1.[32,33,57] Furthermore, it was shown that the mouse IGF-II promoter P3 contains two non-canonical AP-1 binding sites, to which recombinant c-jun protein can bind. In a transient expression system using CP-1 embryonic stem cells it was demonstrated that the mouse P3 is activated by c-jun and c-fos through the AP-1 binding sites.[64]

Regulation of Expression of the Fish IGF-II Promoter

The salmon IGF-II gene has only one promoter that is active in all stages of development.[27] Because fish have the ability to continue growing, the IGF-II promoter is active throughout the entire life span of the animal. Liver is the main producer of IGF-II in the circulation. Interestingly, several homologues of the mammalian liver-specific transcription factors have now been identified in fish. Of five different liver-enriched transcription factors tested, two are activators of the sIGF-II promoter, HNF-3β and C/EBPβ. In addition, the ubiquitously expressed transcription factor Sp1 is involved in the activation of the salmon IGF-II promoter. Interestingly, these are the same factors that can activate the human P1 promoter, which is also active in the adult stage.

The promoter of the salmon IGF-II gene comprises approximately 500 nucleotides upstream of the translation initiation codon (Fig. 3). Three HNF-3β binding sites with similar binding affinity for the HNF-3β protein are present in the sIGF-II promoter and stimulation of transcription by HNF-3β is 6-fold.[65] The fish homolog of mammalian HNF-3b, *Axial*, can bind to HNF-3β binding sites and is functionally interchangeable with rat HNF-3β.[66] It is thus conceivable that *Axial*, that plays an important role in fish embryonic development, is the real stimulator of salmon IGF-II expression mainly at the embryonic stage of the fish.

The transcription factor C/EBPβ, also highly expressed in liver, is another important regulator of the salmon IGF-II promoter. Four functional C/EBPβ binding sites were identified that each contribute to activate transcription from the sIGF-II promoter up to 20-fold.[67] In addition, two binding sites for the ubiquitous transcription factor Sp1 were identified. Although mutation of these Sp1 sites only reduces basal promoter activity two-fold, the integrity of the two Sp1 binding

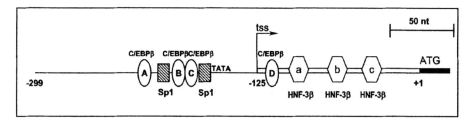

Figure 3. Structure of the chum salmon IGF-II promoter region. The position of the transcription start site (tss) and the translation initiation codon (ATG) is indicated. Binding sites identified for the various transcription factors are shown by different symbols.[65,67]

sites was shown to be critical for activation of the sIGF-II promoter by C/EBPβ. Mutation of both Sp1 binding sites abolished almost all C/EBPβ-dependent activation of the sIGF-II promoter. The presence of transcription factors Sp1 and C/EBPβ that work in synergy to activate the sIGF-II promoter, results most likely in the continued high IGF-II expression in the adult salmon.

Site-Specific Endonucleolytic Processing of IGF-II mRNAs

As illustrated in the previous paragraphs, IGF-II gene expression is regulated extensively at the level of transcription. In addition, regulation occurs at the level of post-transcriptional processing of IGF-II mRNAs by a mechanism known as site-specific endonucleolytic cleavage. In human, rat, and mouse a non-IGF-II encoding RNA of 1.8 kb was identified as a product generated by cleavage of the 4 kb long 3'-UTR of full length IGF-II mRNAs.[68] Endonucleolytic cleavage of IGF-II mRNAs yields an unstable 5'-cleavage product that contains a cap structure, but no polyA tail and a 3'-cleavage product of 1.8 kb RNA that is polyadenylated but lacks a cap structure and in spite of this is very stable (Fig. 4). Endonucleolytic cleavage of IGF-II mRNAs was shown to occur in the cytoplasm for all types of full-length IGF-II transcripts, irrespective of the IGF-II promoter from which the transcripts are derived.[69] The cleavage site in IGF-II mRNA has been mapped to the single nucleotide resolution. Cleavage destabilizes the full-length IGF-II mRNAs and thus removes it from the pool of protein-producing mRNAs, thereby providing the cell with an additional way to control IGF-II protein synthesis. Thus, as cleavage abolishes the protein-coding potential of IGF-II mRNA, it provides the cell with a putative additional mechanism to control IGF-II protein production, acting at the level of mRNA stability. The 3'-terminal 1.8 kb cleavage product lacks a cap structure and it is not actively engaged in protein synthesis, but in spite of that it is an extremely stable RNA that can be detected in IGF-II expressing fetal and post-natal tissues and in cell lines expressing IGF-II.[70] This has led to the interesting possibility that the 3'-cleavage product itself, rich in structural motifs, performs an intrinsic cellular function.

Studies so far have addressed two questions: 1. What are the (structural) determinants for recognition of the IGF-II mRNA cleavage site by an endoribonuclease? 2. Is cleavage of full-length IGF-II mRNA a regulated process and, if so, what are the determinants for regulated degradation of IGF-II mRNA?

In search for structural determinants, an in vivo IGF-II minigene expression assay was developed. Employing many different mutant IGF-II genes two widely separated elements were identified, located in the 3'-UTR of IGF-II mRNAs, that are essential for the site-specific endonucleolytic cleavage reaction to occur. It was shown that a 323 nt region (element II, -173 to +150) surrounding the cleavage site is necessary but not sufficient for cleavage. A second domain comprising an additional 103 nt long element (element I, -2116 to -2013) located 2 kb upstream of element II is also required to confer cleavage. When the two elements were transferred into the 3'-UTR of a heterologous gene, β-globin, it was shown that the presence of both elements I and II in the 3'-UTR of β-globin mRNA was sufficient for cleavage.[71] Based on these results a model was tested suggesting that elements I and II, that are 2 kb apart in the primary sequence, are brought into proximity of each other by RNA folding, thereby forming the recognition determinant for endonucleolytic cleavage. Using the in vivo IGF-II minigene expression system, it was confirmed that formation of a stable stem-structure between element I and a G-rich part of element II is necessary for cleavage of IGF-II

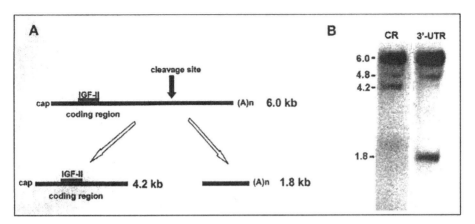

Figure 4. Schematic representation of the endonucleolytic cleavage at a distinct nucleotide in the 3'-UTR of the IGF-II mRNAs. IGF-II cleavage generates a capped but unstable 5' cleavage product containing the coding region and a stable 3' cleavage product that consists of 3'-UTR sequences and a poly(A) tail. Shown is the 6.0 kb IGF-II mRNA as a target for cleavage, but the other IGF-II mRNAs are also cleaved. The cleavage site (CS) is indicated by an arrow, and the preproIGF-II coding region is indicated by a box. Northern blot of total RNA isolated from SHSY-5Y cells that endogenously express IGF-II. When an IGF-II coding region-specific probe is used (CR), the full-length IGF-II mRNAs of 6.0 kb (P3) and 4.8 kb (P4) are detected. The 4.2 kb RNA that contains the coding region is the 5' cleavage product derived from the 6.0 kb mRNA. With an IGF-II 3'-UTR-specific probe (3'-UTR), the full-length IGF-II mRNAs of 6.0 kb (P3) and 4.8 kb (P4) are detected as well as the 1.8 kb 3' cleavage product. The sizes of the different RNA species are indicated on the left.

mRNAs (Fig. 5). In addition, the cleavage site itself must be located in an open conformation, flanked by two stem-loop structures.[72,73] After the full length IGF-II mRNA has been cleaved, the 3'-product remains very stable, although it lacks a cap-structure. The most likely explanation for this is that the G-rich stretch directly downstream of the cleavage site region, when no longer forming the long double stranded RNA structure, can now adopt another highly folded structure, the so-called guanosine-quadruplex structure, that protects the 1.8 kb RNA from rapid 5'-exonucleolytic degradation.[74] This would account for the unusual stability of the 3'-terminal cleavage product.

To determine if IGF-II mRNA cleavage a regulated process, the kinetics and the efficiency of endonucleolytic cleavage of IGF-II mRNAs under various growth conditions were examined.[75] Under standard cell culture conditions cleavage is a slow process that only plays a limited role in destabilisation of IGF-II mRNAs. However, for a limited set of cell types that endogenously express IGF-II, a regulated endonucleolytic cleavage occurs, leading to 5-fold enhanced cleavage efficiency. Closer examination of the cell types revealed that this phenomenon is restricted to cells for which altered IGF-II expression is related to specific cell processes such as growth arrest or cell differentiation. The regulated IGF-II mRNA endonucleolytic cleavage is thus one of the few known examples of functional RNA processing, involving both structural RNA determinants as well as specific physiological conditions of the cells.

Translational Regulation of IGF-II Expression

Regulation of translation of mRNAs is another important regulatory step in gene expression. Although translation takes place in the cytoplasm, compartmentalization can occur by interaction of the translational machinery with the cytoskeleton or by selective mobilization of specific mRNAs.

As indicated above, transcription from the four different human IGF-II promoters generates a population of mRNAs that differ in their 5'-UTRs. These leaders vary significantly in length from 139 nt to 1171 nt long and some have a high GC content potentially leading to higher ordered structures in the leader RNAs. On the basis of its minor length of 109 nt and lack of secondary structure, leader 4 can be expected to allow efficient translation, and that is in fact what has been

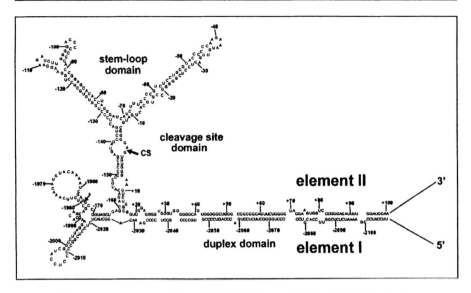

Figure 5. Schematic representation of the 3'-UTR of the human IGF-II mRNAs showing the computer folding prediction of the RNA structural domains involved in secondary structure formation. For clarity, a folding is shown where the region from −1955 to −174, which is dispensable for cleavage, is omitted. Elements I and II, although 2 kb apart in the 3'-UTR of IGF-II mRNA, form an 80 nt long ds RNA duplex domain. The cleavage site (CS) must be located in an open loop structure in order to create access for the endonuclease.

observed; mRNAs with leader 4 sequences are associated with the membrane-bound polysomes and are actively translated. Likewise, leader 2-containing mRNAs were found to be completely in the polysomal fraction in proliferating tissue culture cells and in fetal liver, indicating that mRNAs containing leader 2 will not be inhibited in translation.[76,77] The high GC content of leader 1 (32%G, 35% C) and the presence of an upstream open reading frame as well as an internal ribosome entry site, make it likely that mRNAs containing leader 1 will be poorly translated.[78] However, in adult liver this mRNA species is responsible for almost all IGF-II mRNA and thus most of the IGF-II protein in the circulation. Indeed, although contrary to the expectation, mRNAs with leader 1 can be translated efficiently. In contrast to mRNAs with leaders 1, 2, and 4, the mRNAs carrying leader 3 are hardly found in the polyribosome fraction, indicating that this mRNA species is in general not efficiently translated.[77,79] However, also the leader 3-containing IGF-II transcripts can be efficiently translated depending on the growth conditions; in rhabdomyosarcoma cells, the leader 3 transcripts are translationally silent when the cells are quiescent but become selectively mobilized and actively translated in exponentially growing cells.[80] Activation of translation is inhibited by rapamycin and mimicked by anisomycin, suggesting that the translation of leader 3 mRNAs is regulated by the $p70^{s6k}$ signaling pathway. Thus, IGF-II mRNAs containing leader 3 mRNA are translated efficiently during active cell growth and this is regulated by the $p70^{s6k}$ signaling pathway.

References

1. DeChiara TM, Efstratiadis A, Robertson EJ. A growth-deficiency phenotype in heterozygous mice carrying an insulin-like growth factor II gene disrupted by targeting. Nature 1990; 345:78-80.
2. DeChiara TM, Robertson EJ, Efstratiadis A. Parental imprinting of the mouse insulin-like growth factor II gene. Cell 1991; 64:849-859.
3. Brice AL, Cheetham JE, Bolton VN et al. Temporal changes in the expression of the insulin-like growth factor II gene associated with tissue maturation in the human fetus. Development 1989; 106:543-554.
4. Birnbacher R, Amann G, Breitschopf H et al. Cellular localization of insulin-like growth factor II mRNA in the human fetus and the placenta: Detection with a digoxigenin-labeled cRNA probe and immunocytochemistry. Pediatr Res 1998; 43:614-620.
5. Lee CI, Goldstein O, Han VK et al. IGF-II and IGF binding protein (IGFBP-1, IGFBP-3) gene expression in fetal rhesus monkey tissues during the second and third trimesters. Pediatr Res 2001; 49(3):379-387.

6. Stylianopoulou F, Herbert J, Soares MB et al. Expression of the insulin-like growth factor II gene in the choroid plexus and the leptomeninges of the adult rat central nervous system. Proc Natl Acad Sci USA 1988; 85:141-145.
7. Gray A, Tam AW, Dull TJ et al. Tissue-specific and developmentally regulated transcription of the insulin-like growth factor 2 gene. DNA 1987; 6:283-295.
8. Florini JR, Magri KA, Ewton DZ et al. "Spontaneous" differentiation of skeletal myoblasts is dependent upon autocrine secretion of insulin-like growth factor-II. J Biol Chem 1991; 266:15917-15923.
9. Engström W, Shokrai A, Otte K et al. Transcriptional regulation and biological significance of the insulin like growth factor II gene. Cell Prolif 1998; 31:173-189.
10. Petrik J, Arany E, McDonald TJ et al. Apoptosis in the pancreatic islet cells of the neonatal rat is associated with a reduced expression of insulin-like growth factor II that may act as a survival factor. Endocrinology 1998; 139:2994-3004.
11. Voutilainen R. Adrenocortical cells are the site of secretion and action of insulin-like growth factors and TNF-alpha. Horm Metab Res 1998; 30:432-435.
12. Lu F, Han VKM, Milne WK et al. Regulation of insulin-like growth factor-II gene expression in the ovine fetal adrenal gland by adrenocorticotropic hormone and cortisol. Endocrinology 1994; 134(6):2628-2635.
13. Kitraki E, Philippidis H, Stylianopoulou F. Hormonal control of insulin-like growth factor-II gene expression in the rat liver. J Mol Endocrinol 1992; 9:131-136.
14. Asfari M, De W, Nöel M et al. Insulin-like growth factor-II gene expression in a rat insulin- producing beta-cell line (INS-1) is regulated by glucose. Diabetologia 1995; 38:927-935.
15. Rinderknecht E, Humbel RE. Primary structure of human insulin-like growth factor II. FEBS Lett 1978; 89:283-286.
16. Schmitt S, Ren-Qiu Q, Torresani T et al. High molecular weight forms of IGF-II ('big-IGF-II') released by Wilms' tumor cells. Eur J Endocrinol 1997; 137:396-401.
17. Hunter SJ, Daughaday WH, Callender ME et al. A case of hepatoma associated with hypoglycaemia and overproduction of IGF-II (E-21): Beneficial effects of treatment with growth hormone and intrahepatic adriamycin. Clin Endocrinol (Oxf) 1994; 41:397-401.
18. Zapf J, Futo E, Peter M et al. Can "big" insulin-like growth factors II in serum of tumor patients account for the development of extrapancreatic tumor hypoglycemia? J Clin Invest 1992; 90:2574-2584.
19. Gowan LK, Hampton B, Hill DJ et al. Purification and characterization of a unique high molecular weight form of insulin-like growth factor II. Endocrinology 1987; 121:449-458.
20. Kotani K, Tsuji M, Oki A et al. IGF-II producing hepatic fibrosarcoma associated with hypoglycemia. Intern Med 1993; 32:897-901.
21. Jansen M, Van Schaik FMA, Van Tol H et al. Nucleotide sequences of cDNAs encoding precursors of human insulin-like growth factor II (IGF-II) and an IGF-II variant. FEBS Lett 1985; 179:243-246.
22. Hampton B, Burgess WH, Marshak DR et al. Purification and characterization of an insulin-like growth factor II variant from human plasma. J Biol Chem 1989; 264:19155-19160.
23. Zumstein P, Luthi C, Humbel RE. Amino acid sequence of a variant pro-form of insulin-like growth factor II. Proc Natl Acad Sci USA 1985; 82:3169-3172.
24. Daughaday WH, Emanuele MA, Brooks MH et al. Synthesis and secretion of insulin-like growth factor II by a leiomyosarcoma with associated hypoglycemia. N Engl J Med 1988; 319:1434-1440.
25. Reinecke M, Collet C. The phylogeny of the insulin-like growth factors. Int Rev Cytol 1998; 183:1-94.
26. Holthuizen PE, Steenbergh PH, Sussenbach JS. Regulation of IGF gene expression. In: Rosenfeld RG, Roberts CT Jr, eds. The IGF System. Molecular Biology, Physiology and Clinical Applications. Totowa: Humana Press, 1999:37-61.
27. Palamarchuk AY, Holthuizen PE, Müller WEG et al. Organization and expression of the chum salmon insulin-like growth factor II gene. FEBS Lett 1997; 416:344-348.
28. Steenbergh PH, Holthuizen PE, Sussenbach JS. Molecular aspects of the insulin-like growth factor (IGF) genes. Adv Mol Cell Endo 1997; 1:83-121.
29. Mineo R, Fichera E, Liang SJ et al. Promoter usage for insulin-like growth factor-II in cancerous and benign human breast, prostate, and bladder tissues, and confirmation of a 10th exon. Biochem Biophys Res Commun 2000; 268:886-892.
30. De Pagter-Holthuizen P, Jansen M, van der Kammen RA et al. Differential expression of the human insulin-like growth factor II gene. Characterization of the IGF-II mRNAs and an mRNA encoding a putative IGF-II-associated protein. Biochim Biophys Acta 1988; 950:282-295.
31. Ueno T, Takahashi K, Matsuguchi T et al. Reactivation of rat insulin-like growth factor II gene during hepatocarcinogenesis. Carcinogenesis 1988; 9:1779-1783.
32. Matsuguchi T, Takahashi K, Ikejiri K et al. Functional analysis of multiple promoters of the rat insulin-like growth factor II gene. Biochim Biophys Acta 1990; 1048:165-170.
33. Rotwein P, Hall LJ. Evolution of insulin-like growth factor II: Characterization of the mouse IGF-II gene and identification of two pseudo-exons. DNA Cell Biol 1990; 9:725-735.
34. Ohlsen SM, Lugenbeel KA, Wong EA. Characterization of the linked ovine insulin and insulin-like growth factor-II genes. DNA Cell Biol 1994; 13:377-388.
35. De Pagter-Holthuizen P, Jansen M, Van Schaik FMA et al. The human insulin-like growth factor II contains two development- specific promoters. FEBS Lett 1987; 214:259-264.
36. Rodenburg RJT, Krijger JJT, Holthuizen PE et al. The liver-specific promoter of the human insulin-like growth factor-II gene contains two negative regulatory elements. FEBS Lett 1996; 394:25-30.

37. Rodenburg RJT, Holthuizen PE, Sussenbach JS. A functional Sp1 binding site is essential for the activity of the adult liver-specific human insulin-like growth factor II promoter. Mol Endocrinol 1997; 11:237-250.
38. van Dijk MA, Rodenburg RJT, Holthuizen P et al. The liver-specific promoter of the human insulin-like growth factor II gene is activated by CCAAT/enhancer binding protein (C/EBP). Nucleic Acids Res 1992; 20:3099-3104.
39. Birkenmeier EH, Gwynn B, Howard S et al. Tissue-specific expression, developmental regulation, and genetic mapping of the gene encoding CCAAT/enhancer binding protein. Genes Dev 1989; 3:1146-1156.
40. Rodenburg RJT. Transcriptional regulation of the liver-specific promoter of the human IGF-II gene. Thesis; Utrecht University, The Netherlands; 1996.
41. Holthuizen P, van der Lee FM, Ikejiri K et al. Identification and initial characterization of a fourth leader exon and promoter of the human IGF-II gene. Biochim Biophys Acta 1990; 1087:341-343.
42. Mineo R, Fichera E, Liang SJ et al. Promoter usage for insulin-like growth factor-II in cancerous and benign human breast, prostate, and bladder tissues, and confirmation of a 10th exon. Biochem Biophys Res Commun 2000; 268:886-892.
43. Ikejiri K, Furuichi M, Ueno T et al. The presence and active transcription of three independent leader exons in the mouse insulin-like growth factor II gene. Biochim Biophys Acta 1991; 1089:77-82.
44. Kou K, Rotwein P. Transcriptional activation of the insulin-like growth factor-II gene during myoblast differentiation. Mol Endocrinol 1993; 7:291-302.
45. Schneid H, Holthuizen PE, Sussenbach JS. Differential promoter activation in two human insulin-like growth factor-II-producing tumor cell lines. Endocrinology 1993; 132:1145-1150.
46. Dai B, Wu H, Holthuizen E et al. Identification of a novel cis element required for cell density-dependent down-regulation of insulin-like growth factor-2 promoter activity in CaCo2 cells. J Biol Chem 2001; 276(10):6937-6944.
47. van Dijk MA, Holthuizen P, Sussenbach JS. Elements required for activation of the major promoter of the human insulin-like growth factor II gene. Mol Cell Endocrinol 1992; 88:175-185.
48. Rauscher FJ, III. Tumor suppressor genes which encode transcriptional repressors: Studies on the EGR and Wilms' tumor (WT1) gene products. Adv Exp Med Biol 1993; 348:23-29.
49. Drummond IA, Rupprecht HD, Rohwer-Nutter P et al. DNA recognition by splicing variants of the Wilms' tumor suppressor, WT1. Mol Cell Biol 1994; 14:3800-3809.
50. Drummond IA, Madden SL, Rohwer-Nutter P et al. Repression of the insulin-like growth factor II gene by the Wilms tumor suppressor WT1. Science 1992; 257:674-678.
51. Madden SL, Rauscher FJ, III. Positive and negative regulation of transcription and cell growth mediated by the EGR family of zinc-finger gene products. Ann NY Acad Sci 1993; 684:75-84.
52. Bae MH, Lee MJ, Bae SK et al. Insulin-like growth factor II (IGF-II) secreted from HepG2 human hepatocellular carcinoma cells shows angiogenic activity. Cancer Lett 1998; 128:41-46.
53. Rietveld LEG, Koonen-Reemst AMCB, Sussenbach JS et al. Dual role for transcription factor AP-2 in the regulation of the major fetal promoter P3 of the gene for human insulin-like growth factor II. Biochem J 1999; 338:799-806.
54. Zhang LJ, Zhan SL, Navid F et al. AP-2 may contribute to IGF-II overexpression in rhabdomyosarcoma. Oncogene 1998; 17:1261-1270.
55. van Dijk MA, Van Schaik FMA, Bootsma HJ et al. Initial characterization of the four promoters of the human IGF-II gene. Mol Cell Endocrinol 1991; 81:81-94.
56. Hyun SW, Kim S-J, Park K et al. Characterization of the P4 promoter region of the human insulin-like growth factor II gene. FEBS Lett 1993; 332:153-158.
57. Holthuizen PE, Cleutjens CBJM, Veenstra GJC et al. Differential expression of the human, mouse and rat IGF- II genes. Regul Pept 1993; 48:77-89.
58. Kim SJ, Wagner S, Liu F et al. Retinoblastoma gene product activates expression of the human TGF-β2 gene through transcription factor ATF-2. Nature 1992; 358:331-334.
59. Lee YI, Lee S, Lee Y et al. The human hepatitis B virus transactivator X gene product regulates Sp1 mediated transcription of an insulin-like growth factor II promoter 4. Oncogene 1998; 16:2367-2380.
60. Hyun SW, Park K, Lee YS et al. Inhibition of protein phosphatases activates P4 promoter of the human insulin-like growth factor II gene through the specific promoter element. J Biol Chem 1994; 269:364-368.
61. Lee YI, Kim SJ. Transcriptional repression of human insulin-like growth factor-II P4 promoter by Wilms' tumor suppressor WT1. DNA Cell Biol 1996; 15:99-104.
62. Lee S, Park U, Lee YI. Hepatitis C virus core protein transactivates insulin-like growth factor II gene transcription through acting concurrently on Egr1 and Sp1 sites. Virology 2001; 283(2):167-177.
63. Gangji V, Rydziel S, Gabbitas B et al. Insulin-like growth factor II promoter expression in cultures rodent osteoblasts and adult rat bone. Endocrinology 1998; 139:2287-2292.
64. Caricasole A, Ward A. Transactivation of mouse insulin-like growth factor II (IGF- II) gene promoters by the AP-1 complex. Nucleic Acids Res 1993; 21:1873-1879.
65. Palamarchuk AY, Kavsan VM, Sussenbach JS et al. The chum salmon IGF-II gene promoter is activated by hepatocyte nuclear factor 3b. FEBS Lett 1999; 446:251-255.
66. Chang BE, Blader P, Fischer N et al. Axial (HNF3b) and retinoic acid receptors are regulators of the zebrafish *sonic hedgehog* promoter. EMBO J 1997; 16:3955-3964.
67. Palamarchuk AY, Kavsan VM, Sussenbach JS et al. The chum salmon insulin-like growth factor II promoter requires Sp1 for its activation by C/EBPb. Mol Cell Endocrinol 2001; 172(1-2):57-67.
68. Meinsma D, Holthuizen P, Van den Brande JL et al. Specific endonucleolytic cleavage of IGF-II mRNAs. Biochem Biophys Res Commun 1991; 179:1509-1516.

69. Scheper W, Holthuizen PE, Sussenbach JS. The *cis*-acting elements involved in endonucleolytic cleavage of the 3' UTR of human IGF-II mRNAs bind a 50 kDa protein. Nucleic Acids Res 1996; 24:1000-1007.
70. Nielsen FC, Christiansen J. Endonucleolysis in the turnover of insulin-like growth factor II mRNA. J Biol Chem 1992; 267:19404-19411.
71. Meinsma D, Scheper W, Holthuizen P et al. Site-specific cleavage of IGF-II mRNAs requires sequence elements from two distinct regions of the IGF-II gene. Nucleic Acids Res 1992; 20:5003-5009.
72. Van Dijk EL, Sussenbach JS, Holthuizen PE. Identification of RNA sequences and structures involved in site-specific cleavage of IGF-II mRNAs. RNA 1998; 4:1623-1635.
73. Van Dijk EL, Sussenbach JS, Holthuizen PE. Distinct RNA structural domains cooperate to maintain a specific cleavage site in the 3'-UTR of IGF-II mRNAs. J Mol Biol 2000; 300:449-467.
74. Christiansen J, Kofod M, Nielsen FC. A guanosine quadruplex and two stable hairpins flank a major cleavage site in insulin-like growth factor II mRNA. Nucleic Acids Res 1994; 22:5709-5716.
75. Van Dijk EL, Sussenbach JS, Holthuizen PE. Kinetics and regulation of site-specific endonucleolytic cleavage of human IGF-II mRNAs. Nucleic Acids Res 2001; 29(17):3477-3486.
76. de Moor CH, Jansen M, Sussenbach JS et al. Differential polysomal localization of human insulin-like growth factor 2 mRNAs, both in cell lines and foetal liver. Eur J Biochem 1994; 222:1017-1022.
77. de Moor CH, Jansen M, Bonte E et al. Influence of the four leaders of the human insulin-like growth factor mRNAs on the expression of reporter genes. Eur J Biochem 1994; 226:1039-1047.
78. Teerink H, Voorma HO, Thomas AAM. The human insulin-like growth factor II leader 1 contains an internal ribosomal entry site. Biochim Biophys Acta Gene Struct Expression 1995; 1264:403-408.
79. Nielsen FC, Gammeltoft S, Christiansen J. Translational discrimination of mRNAs coding for human insulin-like growth factor II. J Biol Chem 1990; 265:13431-13434.
80. Nielsen FC, Ostergaard L, Nielsen J et al. Growth-dependent translation of IGF-II mRNA by a rapamycin-sensitive pathway. Nature 1995; 377:358-362.

CHAPTER 6

IGF-I Receptor Signaling in Health and Disease

Renato Baserga, Marco Prisco and Tina Yuan

Abstract

The type 1 insulin-like growth factor (IGF-I) receptor is a tyrosine kinase receptor, conserved through evolution in metazoans, from *C. elegans* to mammals. It displays several functions, some of which are seemingly contradictory. This Chapter examines only selected functions of the IGF-I receptor, with emphasis on some of the most recent developments.

Introduction

The type 1 insulin-like growth factor I receptor (IGF-IR) has emerged in recent years as a receptor that plays an important role in the physiology of the cell. The known functions of the IGF-IR are summarized in Figure 1. No claims are made here that the IGF-IR is unique in these functions, which are shared with other growth factor receptors. Put in proper perspective, the IGF-IR is one of several growth factor receptors that contribute to the physiology of the cell. If one wished to find some unique features of the IGF-IR, there are two characteristics that, if not unique, are more pronounced in the IGF-IR. The first is that the IGF-IR, unlike other growth factor receptors, is ubiquitous. The distribution of growth factor receptors among different cell types depends on their preference for specific ligands. Whereas epithelial cells prefer the epidermal growth factor, fibroblasts the platelet-derived growth factor, and neuronal and hemopoietic cells entire constellations of growth factors, all cells (with two exceptions) use IGF-I or IGF-II as a second growth factor (the exceptions are hepatocytes and B lymphocytes). Both IGF-I and IGF-II interact efficiently with the IGF-IR. The second quasi-unique characteristic is that the IGF-IR plays a special role in anchorage-independent growth (see below). Among the functions of the IGF-IR, the first to be established was its mitogenicity. The IGF-IR, activated by its ligands, sends a strong mitogenic signal and is a crucial component in the process of malignant transformation. It also increases cell size, which is obligatory for mitosis, and is a powerful inhibitor of apoptosis. These four functions are all growth promoting functions, and it is therefore mildly surprising that the IGF-IR can also induce differentiation of cells. As a general rule, proliferation and differentiation (especially terminal differentiation) are mutually exclusive. The contradictions of the IGF-IR have been discussed in a previous review, see ref. 1, but we will consider them again here in the light of the new information that has become available.

The IGF-IR has two other known functions (Fig. 1), identified only more recently. The receptor itself and one of its major substrates, the insulin receptor substrate-1 (IRS-1) regulate cell adhesion and motility, reviewed in ref. 1. Finally, the IGF axis plays an important role in longevity, a negative role, in fact, as inhibition of IGF-I signaling lengthens the life span. This is, conceptually, another contradiction. The IGF-IR promotes malignant transformation of mammalian cells, and transformation has often been made the equivalent of "immortalization" of cells, an unfortunate term which is in open defiance of possible mass extinctions not unknown to planet Earth. It seems that, while "immortalizing" cells, the IGF-IR shortens the life of the organism.

In this Chapter, I shall touch on some new aspects of IGF-IR functions, especially transformation, differentiation, apoptosis and aging. For more detailed information on the basic aspects and the various functions of the IGF-IR, the reader is referred to recent reviews by Baserga[1], Blakesley et al[2], Baserga,[3] and especially to the book by Rosenfeld and Roberts.[4]

Insulin-Like Growth Factors, edited by Derek LeRoith, Walter Zumkeller and Robert Baxter. ©2003 Eurekah.com and Kluwer Academic / Plenum Publishers.

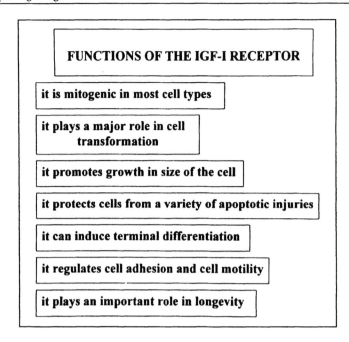

Figure 1. Functions of the Type 1 Insulin-like Growth Factor Receptor. These are the well-established functions of the IGF-IR receptor, and do not preclude the possibility of other functions.

The IGF Axis and Aging

The insulin/IGF-I receptor pathway in C. elegans and Drosophila is a pro-aging pathway (Fig. 2). Mutations at daf-2 (which encodes the receptor homologue) or age-1 prolong the life span of C. elegans,[5] while mutations in daf-18 (the PTEN homologue) suppress the life-span extensions of daf-2 and age-1 mutants.[6,7] Age-1 is the C. elegans homologue of PI3K, and, like PI3K, it signals through the PDK/Akt pathway. Similarly, the life span of Drosophila is substantially increased by mutations of the insulin/IGF-I receptor[8] and by the loss of CHICO, the Drosophila equivalent of the IRS proteins.[9] The IRS proteins are among the major substrates of the insulin and IGF-I receptors.[10] The fact that the life-span of C. elegans can be regulated by modulating the daf-2 pathway in neurons only[11] invites the reflection that old age may be a state of mind. A mutation in the gene encoding the p66 isoform of Shc (another major substrate of the insulin and IGF-I receptors) extends the life span of mice by about 30%.[12] Similarly, mice with a disruption of the GHBP gene, which causes a marked decrease in IGF-I levels, are smaller and have a significantly longer life span than their wild type littermates.[13] In addition, caloric restriction, which increases the life span of rodents,[14] is associated with a decrease in insulin and IGF-I levels.[15]

In dog breeds, there is a clear correlation between body size, IGF-I plasma levels and longevity.[16,17] The relationship between body size and longevity in dogs is illustrated in Figure 3, which is taken from the paper by Deeb and Wolf.[17] In turn, and not surprisingly, there is a strong correlation between body size and IGF-I plasma levels. Giant breeds (like Newfoundlands, Great Danes) have an average life-span of less than 7 years, and IGF-I plasma levels close to and sometimes even exceeding 400 ng/mL. Small breeds (like Toy poodles, Cocker spaniels, Chihuahua) often reach 15 years of age (or longer) and have as low as 40 ng/mL of plasma IGF-I. The correlation between body size and the IGF axis has been known for a long time, and it has been elegantly confirmed in Drosophila, where CHICO,[18] the Drosophila equivalents of Akt[19] and of the p70^{S6K} kinase[20] determine cell and body size. Both Akt and p70^{S6K} are important components of IGF-IR (and insulin receptor) signaling.[20] IGF-I is especially important in determining body size in the early stages of development, and growth at this time in mice correlate negatively with longevity.[21] Samaras and Elrick[22] have marshaled the evidence that, in humans, height is negatively correlated with longevity. This is true

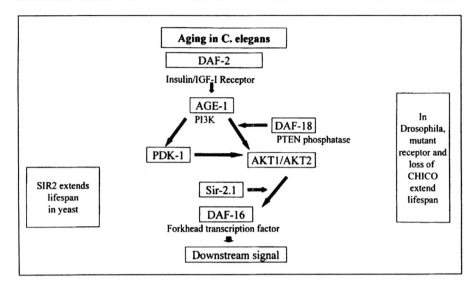

Figure 2. Aging in Lower Animals. The IGF axis in *C. elegans* is given in more detail, but similar pathways have been described in yeast and in Drosophila. Explanation in the text.

whether one looks at ethnic groups, baseball players, football players or US presidents. Their conclusion is firm: "The health and longevity of shorter, smaller people appears to be better than that of taller, bigger people based on extensive and consistent data on humans and animals".

However, there is a price to pay for an increased life span. Some daf-2 mutants have low fertility, although others have normal fertility.[5] Some mice mutant for the growth hormone axis have increased longevity, but also dwarfism and obesity. Both sexes of adult mice homozygous for a targeted mutation of the IGF-I gene (null mutants) are infertile dwarfs,[23] and the males have drastically reduced levels of serum testosterone (18% of normal). Igf1 nullizygous males do not exhibit aggressive behavior when caged with other males, a trait that can be reasonably attributed to the reduced testosterone levels. It is common knowledge that 95% of human violence is closely associated with the phenotype of the Y chromosome. The correlation between longevity and inhibition of IGF-IR signaling, however, is not a simple one. In *C. elegans*, the ligand of daf-2, insulin-1 actually has a negative effect on daf-2 signaling, inducing longevity, rather than shortening the life span.[24]

At any rate, it seems that height (tallness) and IGF-I levels are negatively correlated to longevity in humans. IGF-I plasma levels vary with age and are indeed known to decrease substantially with age.[25] In fact, a few years ago, a number of clinicians, arguing that IGF-I plasma levels are lower in old people, championed the administration of IGF-I as a possible adjuvant against the ravages of old age. Not surprisingly, the results were far from being encouraging, but the point I would like to make is another one. We have always assumed that decreased IGF-I plasma levels in elderly people are a consequence of old age. Perhaps we should ask a different question: is it possible that it is the lower levels of IGF-I that have allowed these individuals to reach old age? In other words, longevity may be the consequence, not the cause, of decreased IGF-I plasma levels. Such a prospective would open up fascinating possibilities, as for instance, the genetic background that allows certain people to decrease IGF-I production with age.

Transformation and Cell Size

The evidence for a role of the IGF system in transformation has been discussed in recent reviews.[1-3] Briefly, the importance of the IGF-IR in transformation rests primarily on the demonstration that R-cells[26,27] are refractory to transformation (colony formation in soft agar). R-cells are 3T3-like fibroblasts derived from mouse embryos with a targeted disruption of the IGF-IR genes.[28,29] R-cells are resistant to transformation by a variety of viral and cellular oncogenes that readily transform 3T3 cells with endogenous IGF-IR. In fact, 3T3 cells, especially NIH 3T3 cells, have a

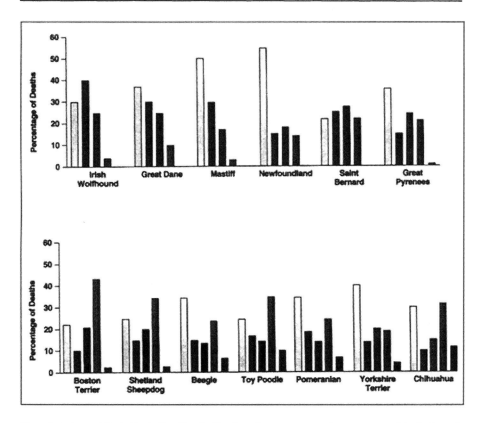

Figure 3. Aging and Body Size in Dog Breeds. The five columns represent the percentage of deaths (in each breed) at <4 years, 4-7 years, 7-10 years, 10-15 years and > 15 years. The upper half has only 4 columns, because most of the dogs in those breeds are dead before reaching 15 years of age (modified after figure in ref. 17). The body size of the dog breeds correlated with IGF-I plasma levels (see text).

tendency to transform spontaneously, which makes even more remarkable the resistance of R-cells to transformation. The agents that fail to transform R-cells are listed in Table 1. As of this date, there are only two oncogenes that can transform R-cells, v-src[30] and a mutant G_{a13} protein.[31] The resistance of R-cells to transformation is abrogated when the IGF-IR is re-introduced into these cells, indicating that the defect in transformation of these cells is exclusively dependent on the absence of the IGF-IR. A non-functional IGF-IR fails to restore the transformed phenotype in R-cells.[27] Interestingly, there are some mutants of the IGF-IR that are still mitogenic (i.e., they respond to IGF-I stimulation) when expressed in R- cells, but are incapable of transforming them.[32] In other words, there are domains of the IGF-IR that are dispensable for mitogenesis, but not for transformation.

The second important piece of evidence for a role of the IGF-IR in transformation is that down-regulation of IGF-IR function causes apoptosis of tumor cells, while having only a moderate effect on normal cells, reviewed in ref. 3. It suggests that the IGF-IR may be crucial, not for normal growth, but for anchorage-independent growth. Targeting of the IGF-IR would then be a good candidate for chemotherapeutic interventions against cancer, as it would discriminate, at least to a certain extent, between normal and tumor cells. Targeting of the ligand would probably be less effective in humans. There are several reports in the literature that a reduction in IGF-I plasma levels reduces tumor growth in rodents, see for instance ref. 15. This is undoubtedly correct, because adult mice and rats produce only IGF-I, the production of IGF-II having been turned off at birth.[28] Thus, reduction in IGF-I levels would take away a favorite growth factor from tumor cells. However, humans do not turn off IGF-II production after birth, and both ligands are produced throughout

Table 1. R-cells are refractory to transformation by a variety of cellular and viral oncogenes

Oncogenes That Do Not Transform R-Cells	Oncogenes That Transform R-Cells
SV40 large T antigen	v-src
Bovine papilloma virus	G_{a13}
Ha-Ras	
Activated c-src	
Human papilloma virus	
Ewing sarcoma fusion protein	
Over-expression of:	
EGF receptor	
PDGF β receptor	
Insulin receptor substrate-1	
Insulin receptor	

All the agents in the left column readily transform mouse embryo fibroblasts with a physiological number of IGF-I receptors. Re-introduction of the IGF-I receptor in R-cells restores the transformed phenotype. References to this table are given in ref. 3.

the life span.[33] Although IGF-II does not bind the IGF-IR as efficiently as IGF-I, it does bind and can activate the receptor, thus substituting for IGF-I. This observation brings out an important consideration. There is a substantial difference between inhibition of growth and induction of apoptosis. Tumor cells seem to possess an uncanny ability to circumvent growth inhibition, and the tumors eventually grow even in the presence of an inhibitor. The latest and most dramatic example of how tumor cells can become resistant to chemotherapy is the clinical resistance to STI-571, which develops in patients treated with this drug for chronic myeloid leukemia.[34] Apoptosis is irrevocable or, to put it in simple words, the only good cancer cell is a dead cancer cell.

Tumor growth obviously requires sustained cell division and cell division, in turn, requires growth in size of the cell. Without a doubling in size between G1 and G2, the cell, after mitosis, would be a smaller cell and, after a few such divisions, it would eventually vanish. The IGF-IR sees to it that its mitogenic signal is accompanied by an increase in cell size. We have mentioned above that body size in Drosophila is dependent on the IRS-1/Akt/p70^{S6K} pathway, which is activated by the IGF-I and insulin receptors. This is also true in mice, since p70^{S6K} knock-out mice are somewhat smaller than their wild type littermates.[35] Body size is a reflection of cell size, as several recent reports have shown. As already mentioned, CHICO, the Drosophila homologue of IRS proteins, determines both cell size and body size.[18] The Drosophila equivalents of Akt[19] and of the p70^{S6K} kinase[20] have the same effects, increasing cell size as well as body size. It is also true, in general, in mammalian cells, where increase in size of the cell is correlated with sustained activation of p70^{S6K}.[36] Increase in cell size and activity of p70^{S6K} correlate with malignant transformation, which brings us logically to the mechanism of IGF-IR signaling in malignant transformation of cells.

As a model, we can take the one by Valentinis et al[36,37] in murine hemopoietic 32D cells. 32D cells are IL-3-dependent, and they quickly undergo apoptosis if deprived of IL-3.[37,38] 32D cells have very low levels of IGF-IR,[39] and do not express IRS-1 or IRS-2.[37,40] When expressing a human IGF-IR, 32D cells (now designated as 32D IGF-IR cells) survive and grow vigorously for 48 hrs after shifting from IL-3 to IGF-I. However, after 48 hrs, 32D IGF-IR cells stop growing and begin to differentiate along the granulocytic pathway.[37] Ectopic expression of IRS-1 in 32D IGF-IR cells completely inhibits IGF-I-mediated differentiation.[37] In fact 32D IGF-IR/IRS1 cells become permanently IL-3-independent, and even form tumors in syngeneic and nude animals.[36] Cell size is increased in 32D IGF-IR/IRS-1 cells, when compared to 32D IGF-IR cells, as one would expect in cells with a very active IRS-1/Akt/p70^{S6K} pathway. Interestingly, the size of 32D IGF-IR/IRS1 cells decreases when the cells are treated with rapamycin. Rapamycin is a specific inhibitor of mTOR,[41] which is required for the activation of p70^{S6K}.[42] Rapamycin inhibits p70^{S6K} and causes the differentiation of 32D IGF-IR/IRS1 cells.[36] This result is especially remarkable, since it indicates that cell size and malignant transformation are both dependent on the activity of the p70^{S6K} pathway, at least in this model.

This may not be true of other models of malignant transformation. Vogt and co-workers[43] have reported that PI3K-transformed chicken embryo fibroblasts (CEF) are very sensitive to rapamycin, but v-src or Ha-Ras-transformed cells are not. Rapamycin abrogates transformation of CEF by constitutively activated PI3K or Akt, but has no effect on foci formation of CEF transformed by v-src. CEF transformed by Ras actually make more foci in the presence of rapamycin. One would like to formulate the hypothesis that src and ras transform cells by mechanisms that are largely if not completely independent of the PI3K/p70^{S6K} pathway, which is instead very important in transformation caused by the IGF-IR and, by extrapolation, other growth factor receptors. It is a hypothesis that can be easily tested. The intervention of Rho family GTPases in transformation by the IGF-IR[44] may also be an alternative not operating in 32D-derived cells, which grow in suspension.

Differentiation: Importance and Contradictions of the IRS Proteins

In certain hemopoietic cells in culture and in vivo, the Granulocytic-Colony Stimulating Factor (G-CSF) sends at the same time a proliferative and a differentiating signal.[45] The cells grow in number, but also begin to differentiate, and eventually (as terminally differentiated cells begin to die) they decrease in number. Differentiation occurs only if the cells can undergo one or two rounds of replication.[46] The IGF-IR, in most cell types (mouse embryo fibroblasts, like 3T3 cells, human diploid fibroblasts, some epithelial cells, etc.), sends a mitogenic, anti-apoptotic signal.[2,47] But in other cell types, as already mentioned, for instance myoblasts, osteoblasts, neurons and hemopoietic cells IGF-I and IGF-II can stimulate either proliferation or differentiation, or both,[48] and reviewed in refs. 1 and 32. Terminal differentiation is usually followed by cell death.

A good model in which to study IGF-IR-directed differentiation or transformation is again that of 32D cells, a model already discussed above. We have mentioned above that ectopic expression of IRS-1 in 32D IGF-IR cells inhibits differentiation. Conversely, a dominant negative mutant of Shc induces partial differentiation of 32D IGF-IR/IRS-1 cells.[37] Thus, in 32D cells stimulated by IGF-I, the cells are programmed for proliferation when IRS-1 is the predominant substrate. If Shc proteins predominate, the cells, after a brief burst of proliferation, differentiate. If 32D cells over-express only IRS-1, they do not even survive in the absence of IL-3.[37,49,50] Thus, IRS-1 or the IGF-IR, singly, cannot transform 32D cells, or even induce their prolonged survival. In combination, they cause malignant transformation of 32D cells.

In our model, we had to increase the levels of IGF-IR, because parental 32D cells have very low levels. However, in other cell lines with normal levels of IGF-IR, IGF-I-induced differentiation does not require an exogenous receptor.[51,52] An intriguing aspect of this system (32D IGF-IR cells versus 32D IGF-IR/IRS-1 cells) is that 32D IGF-IR cells are stimulated to proliferate for the first 48 hrs, before they differentiate, as they do with G-CSF. Yet, markers of differentiation are already apparent in 32D IGF-IR cells in the first 24 hrs, when these cells are growing exponentially. For instance, myeloperoxidase (MPO) RNA increases in 32D IGF-IR cells in the first 24 hrs after the cells are shifted from IL-3 to IGF-I.[36] At the same time, Id2 RNA and proteins remain low (see below), while they increase dramatically in 32D IGF-IR/IRS1 cells, which will not differentiate.[53] Thus, it seems that IGF-I in 32D IGF-IR cells induces two programs, one for differentiation and a second one for proliferation, both activated in the first hours after shifting the cells from IL-3 to IGF-I. Ectopic expression of IRS-1 suppresses the differentiation signal, while leaving intact the proliferative signal.

This duality of mitogenesis versus differentiation can also be observed in 32D IGF-IR cells expressing a dominant negative mutant of Stat3 (DNStat3). In these cells, DNStat3 inhibits IGF-I-mediated differentiation.[54] Again, the presence of DNStat3 abrogates the increase in MPO RNA that is usually observed in 32D IGF-IR cells, indicating that both IRS-1 and DNStat3 extinguish the differentiation program without affecting the proliferative program. The abrogation of the differentiation program entails the up-regulation of Id2 gene expression,[53,54] to be discussed further below.

While the absence of IRS-1 seems to predispose cells to differentiation,[37,55] its complete absence is not a requirement. There are a number of cell types that have low levels of IRS-1, that further decrease when they are induced to differentiate.[56,57] A good example is that of rat hippocampal cells, H19-7 that can be induced to differentiate by IGF-I. When these cells differentiate, there is a decrease in IRS-1 levels. Again, over-expression of IRS-1 inhibits IGF-I-mediated differentiation of these neuronal cells.[58]

It is therefore surprising that IRS-1 is not expressed in LNCaP cells, a human prostatic cancer cell line from a metastatic tumor.[59] It is the conventional wisdom that cancer cells tend to utilize all possible ways to increase their growth potential, including production of growth factors, over-expression of growth factor receptors or activation of mitogenic pathways. IRS-1, as repeatedly mentioned, sends a powerful mitogenic signal, one can almost say a transforming signal. Why would a prostatic cancer cell turn off IRS-1? One possible reason for turning off the expression of IRS-1 in metastatic prostatic cancer cells may be the fact that IRS-1 promotes cell adhesion and decreases cell motility,[60] which would tend to inhibit metastatic spread. IGF-I is known to promote cell adhesion, especially to laminin and collagen I,[61,62] and also cell-to-cell adhesion.[63,64] It also increases cell motility and promotes invasion through collagen IV.[62] Interestingly, ectopic expression of IRS-1 in LNCaP cells increases cell adhesion but decreases cell motility.[60] By turning off IRS-1, the cancer cell could favor its metastasizing ability. A consequence of IRS-1 down-regulation is a decrease in PI3-kinase signaling to Akt/PKB, which is one of the major pathways for the mitogenic action of several growth factor receptors. LNCaP cells may not need to express IRS-1, because of a frame-shift mutation in the tumor suppressor gene PTEN.[65,66] This mutation causes a constitutive activation of Akt/PKB in the absence of stimulation by growth factors.[67] However, since, in the absence of IRS-1, the IGF-IR can induce differentiation,[37] the metastatic cancer cell, to avoid differentiation, must also reduce the IGF-IR levels, which is indeed what happens in LNCaP cells.[68] If the levels of IGF-IR are increased, then the prostatic cancer cells tend to undergo growth arrest.[69]

In fact, one can speculate that the three events observed in LNCaP cells (extinction of IRS-1 expression, PTEN mutation and down-regulation of the IGF-I R) may not be unrelated. One could hypothesize a succession of events as follows:
1. The prostate cancer cell down-regulates IRS-1, which increases its ability to metastasize.
2. The absence of IRS-1 expression, though, leads to a decrease in PI3-kinase signaling, which causes a decrease in the ability of the cell to respond to growth factors. This is compensated by a mutation in PTEN, resulting in a constitutive activation of the Akt/PKB pathway.
3. However, the lack of IRS-1 can also lead to IGF-I-mediated differentiation,[37] which can be obviated by down-regulation of the IGF-IR.

This hypothesis has received strong support from two recent papers by Reiss and co-workers.[59,60] IRS-1 does indeed increase cell adhesion. Over-expression of the IGF-IR in LNCaP cells leads to growth arrest in the absence of IRS-1, while the combined expression of the IGF-IR and IRS-1 increases the transformed phenotype of these cells. Interestingly, re-introduction of IRS-1 in LNCaP cells results in its serine phosphorylation.[60]

It seems therefore that IRS-1, like the IGF-IR, is capable of sending contradictory signals. In fact, an attractive possibility is that IRS-1 (as well as other transducing molecules) may be sending contradictory signals, depending on whether they are tyrosyl- or seryl phosphorylated. It is a suggestion that can be easily tested.

The IGF-I Receptor and Cell Death

The credit for discovering the anti-apoptotic function of the IGF-IR must go to Evan and co-workers,[70] who showed that IGF-I, added to serum-free medium, protected fibroblasts from apoptosis induced by over-expression of c-myc. However, other investigators had previously demonstrated a protective effect of IGF-I against cell death, the type of cell death being unspecified.[71,72] Since then, evidence has accumulated on the anti-apoptotic effect of IGF-I and the IGF-IR in a variety of cell types. A non-comprehensive list of the reports in the literature is given in Figure 4. The variety of the procedures used to induce apoptosis suggests that the wild type IGF-IR has a widespread anti-apoptotic effect against many death signals. IGF-I is not the only growth factor and the IGF-IR is not the only growth factor receptor to have an anti-apoptotic function. For instance,

The IGF-IR activated by its ligands protects cells from apoptosis induced by a variety of agents:	
1) etoposide 2) interleukin-3 withdrawal 3) osmotic shock 4) tumor necrosis factor alpha 5) c-myc over-expression 6) ICE proteases expression 7) anti-cancer drugs 8) TGFbeta-1 9) growth factor withdrawal 10) ionizing and non-ionizing radiations 11) p53 12) okadaic acid 13) anoikis (anchorage-independence) 14) high concentrations of serum 15) inhibitors of PI3-kinase 16) high glucose 17) N-myc and many others	A) As a survival factor, the IGF-IR has a peculiar characteristic: its anti-apoptotic effect is much more dramatic when the cells are in anchorage independence than when they are growing in monolayer. It suggests a possible discrimination between normal cells and tumor cells. B) The IGF-IR uses at least three different pathways for its anti-apoptotic signaling.

Figure 4. The IGF-I Receptor and Cell Death. The IGF-IR protects different cell types from a variety of apoptotic injuries (partial list only). In general, its protective effect is demonstrated by the addition of IGF-I or by an increase in the levels of IGF-IR. Alternatively, it can be shown that down-regulation of the IGF-IR induces apoptosis. Down-regulation or impairment of IGF-IR function can also sensitize cells to apoptosis induced by other agents.

interleukin-3 (IL-3) prevents apoptosis in a variety of cell lines (like 32D cells, see above). There is, however, one peculiarity that seems to distinguish the IGF-IR from other growth factor receptors. When the function of the IGF-IR is impaired by different strategies (see below), the effect on cells in monolayers is detectable, but very modest. It is only when the cells are in anchorage independence that the IGF-IR becomes crucial for their survival, reviewed in ref. 3. One can say that the IGF-IR is not absolutely necessary for normal growth but is required for anchorage-independent growth. This conclusion is also supported by genetic experiments in knock-out mice.[73]

The protective action of the IGF-IR against cell death can be demonstrated either by addition of IGF-1,[71,72,74,75,76] or over-expression of the IGF-IR.[39,77] As a corollary, down-regulation of the IGF-IR function, either by antisense strategies or by the use of dominant negative mutants causes apoptosis of tumor cells in vitro. This effect results in abrogation of tumorigenesis,[78] and reviewed in ref. 1 and 2, and metastases in experimental animals.[62,79,80] Apart from its direct effect, targeting of the IGF-IR can also impair the cells' resistance to apoptotic injuries, and thus sensitize cancer cells to chemotherapeutic agents[76] or ionizing radiation.[81,82] Thus, manipulation of the IGF-IR could be of practical interest in the treatment of human cancer, review in ref. 83. Indeed, targeting of the IGF-IR has recently yielded promising results in a clinical trial involving patients with recurrent malignant astrocytomas.[84]

As to the anti-apoptotic signaling, the data can be summarized very briefly (Fig. 5). The main anti-apoptotic pathway of the IGF-IR goes through IRS-1, PI3K and the p70^{S6K} pathway, but there are at least two other pathways that are operative, and become important in cells that do not express IRS-1. One is the MAPK pathway, and the third one requires the integrity of the serine 1280-1283 of the receptor.[85] This third pathway induces the mitochondrial translocation of both Raf-1 and of Nedd4, a target of caspases.[86] One interesting aspect of these three pathways is their limited redundancy. At least two pathways must be operative for protection from apoptosis, but any two pathways are sufficient.[87]

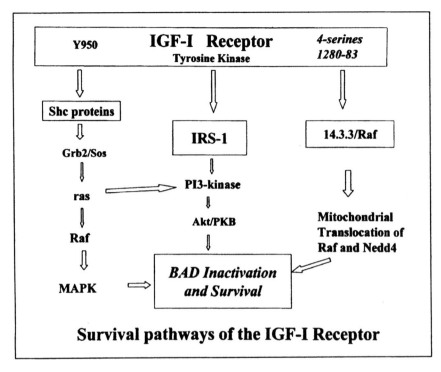

Figure 5. The Anti-apoptotic Pathways of the IGF-I Receptor. This is a very simplified scheme, representing the most important and direct signals. The relevant message is that the IGF-I receptor has multiple pathways for protection of cells from apoptosis, as described in the papers by Parrizas et al,[74] Peruzzi et al,[85] Blakesley et al,[2] Peruzzi et al[86] and Navarro and Baserga.[87]

In the case of 32D cells and other IL-3-dependent cell lines, a recent development has opened new insights on the mechanism by which the IGF-IR protects cells from apoptosis. Devireddy et al[88] have shown that withdrawal of IL-3 from IL-3-dependent cell lines, induces the activation of 24p3, a gene encoding a secreted lipocalin. The lipocalin protein family is a large group of small extracellular proteins. Although there is low conservation at the sequence level, analysis of the lipocalins' crystal structures shows that their overall folding patterns are highly conserved. The lipocalin fold is a symmetrical all-b protein dominated by a single eight-stranded antiparallel b-sheet closed back on itself, reviewed in ref. 89. 24p3 is also transcribed by 32D cells after IL-3 withdrawal, and the protein is secreted into the medium, where it causes apoptosis. Indeed, the 24p3 protein, added to the medium, can induce apoptosis even when IL-3-dependent cells are not IL-3-deprived.[88] The transcription of the 24p3 gene is inhibited by addition of IGF-I.[88] Interestingly, 24p3 is also induced by SV40 T antigen.[90] In agreement with these findings, expression of SV40 T antigen in 32D cells accelerates apoptosis,[50] as shown in Figure 6. Parental 32D cells and 32D cells expressing IRS-1 also undergo apoptosis after IL-3 withdrawal, but more slowly than 32D cells expressing the T antigen (Fig. 6). Interestingly, co-expression of T antigen and IRS-1 renders the cells resistant to apoptosis (Fig. 6) and totally IL-3-independent.[50] Since the SV40 T antigen is known to interact and co-precipitate with IRS-1,[50,91] one is left with two interesting hypotheses. The first is that IRS-1 activation inhibits the transcription of 24p3, the second hypothesis is that, somehow, the SV40 T antigen activates IRS-1 in the absence of IGF-IR activation.

The Nuclear Connection: The Id Proteins

The pathway described in Figure 2 ends in a rather non-committal "downstream signal". It is an ending not unique to this figure. Discussions of the many and complicated signaling pathways of

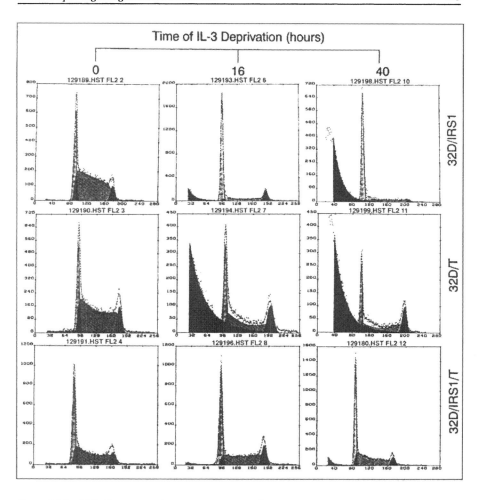

Figure 6. The SV40 T antigen and IRS-1 Co-operate in the Protection of 32D Cells from apoptosis. The determination of apoptosis was carried out by FACS analysis. The peak at the extreme left of each panel gives the percentage of apoptotic cells in the population. Parental 32D cells behave exactly as 32D/IRS-1 cells. Notice that 32D cells expressing the SV40 T antigen (32D/T) die very quickly, several hrs before 32D/IRS-1 cells (and parental 32D cells). Note also that while neither the T antigen nor IRS-1 can protect 32D cells from apoptosis, their combined expression results in IL-3-independence, modified from ref. 50. See text for an explanation.

growth factor receptors often end this way or, alternatively, in a statement that signaling activates a proliferative (or a differentiation) program. This is obvious also for the IGF-IR, as we discussed above. "Downstream signaling" or "cell cycle program", in this respect, join the "cell context" as answers which are really not answers, but questions. What is the downstream signaling? Clearly, a nuclear connection is necessary for the signaling pathway to initiate the proliferative or differentiation programs. We already know that both the PI3K/p70^{S6K} and MAPK pathways connect to the nucleus, by translocation of their products into the nuclei.[92,93] And then? No one has a definitive answer to this question, but some connections between the IGF-IR and transcription in the nucleus have been established.

One connection is with the Id proteins.[53,54] The Id proteins are a family of helix-loop-helix proteins that form heterodimers with another family of helix-loop-helix proteins, called basic helix-loop-helix (bHLH) proteins. These bHLH proteins[94] are generally transcription factors that play a role in the differentiation of a variety of cell types. The Id proteins usually function as negative regulators of bHLH proteins through the formation of inactive heterodimers.[94,95] MyoD is the best known transcription factor inhibited by Id proteins, but other genes important in neurogenic and hemopoietic differentiation are also inhibited, reviewed in ref. 96. There are at least 4 Id proteins, but there is evidence in the literature the Id1 and Id3 are overlapping, while expression of Id4 is limited to specific tissues. Id gene expression is markedly increased in proliferating cells, in cycling cells and tumor cell lines.[95,96,97,98] High levels of Id gene expression inhibit differentiation.[54,95,99,100] The Id1 protein promotes mammary epithelial cell invasion,[101] and increases the aggressive phenotype of human breast cancer cells.[102] The Id2 protein, instead, has been reported to enhance cellular proliferation by associating with the retinoblastoma protein.[103] More recently, it has been reported that the Id2 promoter is the target of the proto-oncogene N-Myc in neuroblastoma cells.[104] The IGF-IR is a powerful inducer of Id2 gene expression. It can increase Id2 gene expression in 32D cells, through a signal originating from tyrosine residue 950.[54] However, signaling from IRS-1 dramatically increases IGF-I-mediated Id2 gene expression,[53] and, as expected, IRS-1 signals through the PI3K/p70^{S6K} pathway. The induction of Id2 gene expression in 32D cells is also increased by a dominant negative mutant of Stat3, which, like IRS-1 inhibits the differentiation of 32D IGF-IR cells (see above). Although other growth factors (and serum) may up-regulate Id gene expression,[97] the IGF-IR is definitely a powerful inducer. Over-expression of Id2 in 32D cells, however, is not sufficient to confer IL-3 independence, although it does inhibit differentiation.[54,105] It seems that, for transformation (IL-3 independence), 32D IGF-IR cells require some other signal, besides the inhibition of differentiation caused by Id2 up-regulation. This was a particularly significant observation, since it clearly and unequivocally separated inhibition of differentiation from transformation.

In mouse embryo fibroblasts (MEF), instead, it is the Id1 gene that responds to IGF-I stimulation. The increase in Id1 protein in MEF is regulated, at least in part, at the transcriptional level. An Id1 promoter driving a reporter gene (luciferase) is stimulated by IGF-1 (Fig. 7). All of the cells described in Figure 7 are derived from R- cells (first bar). R12 cells have $7x10^3$ IGF-I receptors/cell, R508 have $15x10^3$, R503 have $22x10^3$ and R600 have $30x10^3$ receptors/cell.[106] R-, R12 and R508 do not grow when stimulated with IGF-I, while R503 and R600 cells do.[106] Activation of Id1 promoter by IGF-I therefore correlates with mitogenic stimulus. Interestingly, Id1 promoter activity is also increased by a constitutively activated Stat3 (Stat3C), and by v-src, even in serum-free medium and in R- cells (Fig. 7). This is seemingly in contradiction with the observation mentioned above that a DNStat3 also increases Id2 gene expression. However, it is not surprising, as Stat3 signaling is radically different in 32D cells and in MEF.[54,107,108] Finally, the Id1 promoter is also activated by 10% fetal bovine serum (FBS), which cannot activate Id2 gene expression in 32D cells.[53] Related to the induction of Id proteins by IGF-I is the recent report by Dupont et al[109] that Twist expression is also involved in the anti-apoptotic effects of the IGF-IR. Twist belongs to the bHLH family of transcription factors, which, as mentioned above, interact with the Id proteins.

A second connection is the phosphorylation of nuclear phospholipase C (PLC) b1, which is tied to the nuclear translocation of ERK1 and ERK2.[110] PLC b1 is a nuclear enzyme responsible for initiation of the nuclear phosphoinositide cycle. Inhibition of this pathway by a dominant negative mutant of PCL b1 decreases IGF-I-mediated DNA synthesis. It will be interesting to determine how this connection relates to the other nuclear connections of IGF-IR signaling.

Another nuclear connection is with the Forkhead transcription factors. As obvious from the previous discussion, Akt is part of the pathway originating from the IGF-IR that protects cells from apoptotic injuries. Brunet et al[111] have shown that Akt promotes cell survival by phosphorylating and inhibiting a forkhead transcription factor, FKHRL1. Forkhead transcription factors (at least three are known in humans) have been identified in chromosomal breakpoints in human tumors, and, interestingly, they have a homolog in *C. elegans*, DAF16, for references, see ref. 111. It seems, therefore, that the connections to the nucleus from IGF-IR signaling are slowly emerging, presenting possibilities for studying how the receptor activates the cell cycle and differentiation programs.

Cell size is largely determined by the protein content of the cells, which, in turn, depends on the amount of ribosomal RNA, for a discussion see ref. 112. Since differentiated cells are smaller than transformed cells, they should have decreased ribosomal RNA synthesis. This prediction was confirmed by Comai et al,[113] who found that TPA-induced differentiation of human promyelocytic leukemic cells results in inhibition of RNA polymerase I transcription. This inhibition is apparently due to the nucleolar translocation of the retinoblastoma (Rb) gene product.[114,115,116,117,118] It is

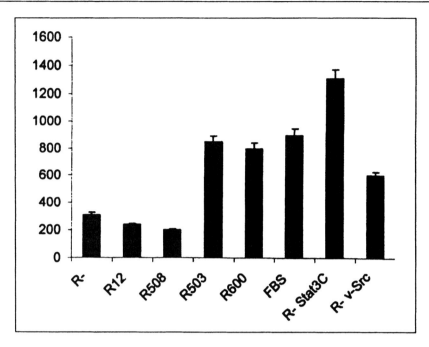

Figure 7. Activation of the Id1 promoter in Mouse Embryo Fibroblasts. The plasmid with the Id1 promoter driving luciferase,[125] was transfected transiently into mouse embryo fibroblasts. The luciferase activity was determined after 24 hrs, and is shown here (in arbitrary units on the ordinate) after normalization with b-galactosidase (to monitor for transfection efficiency). The different cell types are indicated under each bar. The cell lines are described in the text.

interesting that IGF-I is known to activate the rDNA promoter.[119] We have indicated above that the IGF-IR /IRS-1 pathway increases cell size and inhibits differentiation. It would be interesting to determine the relationship between IGF-IR signaling, ribosomal RNA synthesis and the retinoblastoma protein. An exploration of this relationship is made even more attractive by the fact that the Id2 proteins (regulated by the IGF-IR/IRS-1 pathway) interact with the retinoblastoma protein.[103]

Conclusions (and Some Random Thoughts)

The IGF-IR activated by its ligands plays an important role in cell growth and differentiation. How important it is, is an irrelevant question. There are many growth factors and many growth factor receptors, and most of them play a significant role in the physiology of different types of cells. The IGF-IR is one of these receptors and its ligands are among the many growth factors that regulate growth in vivo and in vitro. Like other growth factor receptors, the IGF-IR could be a good target for chemotherapeutic interventions against cancer. Conversely, the IGF-IR and its ligands could be used to increase cell survival, for instance of stem cells injected into animals in order to repair damaged tissue. Its possible applications have not passed unnoticed, and many pharmaceutical and biotechnology companies have active programs involving the IGF-IR. We have mentioned in the Introduction two quasi-unique characteristics of the IGF-IR: its role in anchorage-independent growth and the fact that it is almost ubiquitous. I would like to add here a third property of the IGF-IR, which has intrigued us for some years. Down-regulation of the IGF-IR in experimental animals elicits a strong host response, with all the characteristics of an immune response.[120] This observation has been confirmed in at least two other laboratories[121,122,123] and in a clinical trial.[84] The mechanism of this host response is obscure, and the hypothesis advanced two years ago by Baserga and Resnicoff[120] has yet to receive adequate testing. But the effect of the host response in experimental animals is to markedly increase the anti-tumor effect of IGF-IR targeting.[122,124]

From a basic point of view, the IGF-IR and its signaling pathway offer a very interesting challenge in their contradictions. We have given examples above. The IGF-IR can induce transformation or differentiation depending on the availability of its immediate substrates, IRS-1 and Shc proteins. IRS-1 itself is a powerful mitogen, but its presence may inhibit the metastatic spread of cancer cells. These contradictions beg to be resolved. I would like to offer here a speculation, which can be easily tested. My speculation is that tyrosyl phosphorylation (for instance of IRS-1 and p66 Shc) induces a mitogenic response, while seryl phosphorylation produces an inhibitory response. If this hypothesis turns out to be correct, then the next wave of investigation will be to determine why and when this differential phosphorylation occurs.

However, another intriguing possibility is the multiplicity of signals originating from the IGF-IR that must converge at some point to give the full impact of its functions. As we have seen in our discussion, the IGF-IR inhibits the transcription of 24p3, which induces apoptosis of 32D cells after IL-3 withdrawal.[88] The IGF-IR, through IRS-1 and other domains, induces Id2 gene expression, which results in the inhibition of cell differentiation.[54] All this is not sufficient, yet, for IL-3-independent growth. The IGF-IR also requires the activation, through IRS-1, of the PI3K pathway, for IL-3 independent growth.[36] We would like to propose the following hypothesis (Fig. 8): there are three separate signals from the IGF-IR that must converge to transform 32D cells to IL-3 independence (and therefore to ability to form tumors in animals). The first is an anti-apoptotic signal, presumably the inhibition of 24p3 transcription.[88] The second is an inhibition of the differentiation program through the Id proteins,[53] and the third signal is the actual stimulus to cell proliferation, proceeding from IRS-1, through the PI3K pathway.[36] Whether these three pathways converge at some point or remain independent is unclear. But this hypothesis, if confirmed, could serve as a model for other transforming agents and other cell types.

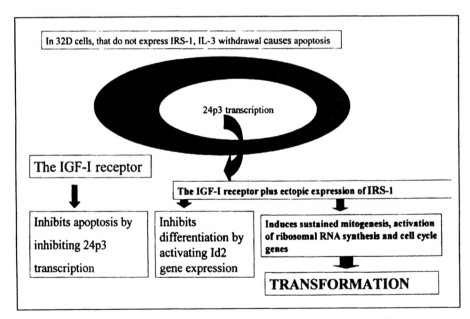

Figure 8. The Pathway to Transformation in 32D Cells. Transformation is defined here as ability to grow indefinitely in the absence of Interleukin-3 (IL-3). Removal of IL-3 induces transcription of 24p3, which is secreted and induces apoptosis. The IGF-IR, activated by its ligands, inhibits transcription of 24p3, thus preventing apoptosis. Because 32D cells do not express IRS-1, the IGF-IR tends to cause differentiation, which is inhibited when IRS-1 is ectopically expressed. Inhibition of differentiation occurs through the activation of Id2 gene expression, which, by itself, is not sufficient for IL-3 independence. IRS-1 must also activate ribosomal RNA synthesis and the cell cycle program to render 32D cells IL-3-independent.

Acknowledgements
This work was supported by Grant CA56309 from the National Institutes of Health.

References
1. Baserga R. The contradictions of the insulin-like growth factor 1 receptor. Oncogene 2000; 19:5574-5581.
2. Blakesley VA, Butler AA, Koval AP et al. IGF-I receptor function: Transducing the IGF-I signal into intracellular events. In: Rosenfeld, RG, Roberts CT Jr, eds. The IGF System. Totowa: Humana Press, 1999; 143-163.
3. Baserga R. The IGF-I receptor in cancer research. Exp Cell Res 1999; 253:1-6.
4. Rosenfeld RG, Roberts CT Jr., eds. The IGF System. Totowa: Humana Press, 1999.
5. Guarente L, Kenyon C. Genetic pathways that regulate ageing in model organisms. Nature 2000; 408:255-262.
6. Ogg S, Ruvkun G. The C.elegans PTEN homolog, DAF-18, acts in the insulin receptor-like netabolic signaling pathway. Molecular Cell 1998; 2:887-893.
7. Mihaylova VT, Borland CZ, Manjarrez I et al. The PTEN tumor suppressor homolog in Caenorhabditis elegans regulates longevity and dauer formation in an insulin receptor-like signaling pathway. Proc Natl Acad Sci 1999; 96:7427-7432.
8. Tatar M, Kopelman A, Epstein D et al. A mutant Drosophila insulin receptor homolog that extends life-span and impairs neuroendocrine function. Science 2001; 292:107-110.
9. Clancy DJ, Gems D, Harshman LG et al. Extension of life-span by loss of CHICO, a Drosophila insulin receptor substrate protein. Science 2001; 292:104-106.
10. White MF. The IRS-signalling system: A network of docking proteins that mediate insulin action. Mol Cell Bioch 1998; 182:3-11.
11. Wolkow CA, Kimura K, Lee MS et al. Regulation of *C. elegans* life-span by insulinlike signaling in the nervous system. Science 2000; 290:147-150.
12. Migliaccio E, Giorgio M, Mele S et al. The p66^{shc} adaptor protein controls oxidative stress response and life span in mammals. Nature 2000; 402:309-313.
13. Coschigano KT, Clemmons D, Bellush LL et al. Assessment of growth parameters and life span of GHR/BP gene-disrupted mice. Endocrinology 2000; 41:2608-2613.
14. Tannenbaum A, Silverstone H. Nutrition in relation to cancer. Advan Cancer Res 1953; 1:452-501.
15. Dunn SE, Kari FW, French J et al. Dietary restriction reduces insulin-like growth factor I levels, which modulates apoptosis, cell proliferation and tumor progression in p53-deficient mice. Cancer Res 1997; 57:4667-4672.
16. Eigenman JE, Amador A, Patterson DF. Insulin-like growth factor I levels in proportionate dogs, chondrodystrophic dogs and in giant dogs. Acta Endocrinol 1988; 118:105-108.
17. Deeb B, Wolf N. Studying longevity and mortality in giant and small breeds of dogs. Veterinary Med 1994; 89:702-713.
18. Bohni R, Riesco-Escovar J, Oldham S et al. Autonomous control of cell and organ size by CHICO, a Drosophila homolog of vertebrate IRS-1. Cell 1999; 97:865-875.
19. Verdu J, Buratovich MA, Wilder EL et al. Cell-autonomous regulation of cell and organ growth in Drosophila by Akt/PKB. Nature Cell Biol 1999; 1:500-506.
20. Montagne J, Stewart MJ, Stocker H et al. Drosophila 6 kinase: A regulator of cell size. Science 1999; 285:2126-2129.
21. Miller RA, Chrisp C, Atchley W. Differential longevity in mouse stock selected for early life growth trajectory. J Gerontol 2000; 55A:B455-B461.
22. Samaras TT, Elrick H. Height, body size and longevity. Acta Med Okayama 1999; 53:149-169.
23. Baker J, Hardy MP, Zhou J et al. Effects of an Igf1 gene null mutation on mouse reproduction. Mol Endocr 1996; 10:903-918.
24. Pierce SB, Costa M, Wysotzkey R et al. Regulation of DAF-2 receptor signaling by human insulin and ins-1, a member of the unusually large and diverse *C. elegans* insulin gene family. Genes & Develop 2001; 15:672-686.
25. Goodman-Gruen D, Barrett-Connor E. Epidemiology of insulin-like growth factor 1 in elderly men and women. A J Epidemiol 1997; 145:970-976.
26. Sell C, Rubini M, Rubin R et al. Simian virus 40 large tumor antigen is unable to transform mouse embryonic fibroblasts lacking type-1 IGF receptor. Proc Natl Acad Sci USA 1993; 90:11217-11221.
27. Sell C, Dumenil G, Deveaud C et al. Effect of a null mutation of the type 1 IGF receptor gene on growth and transformation of mouse embryo fibroblasts. Mol Cell Biol 1994; 14:3604-3612.
28. Liu J-P, Baker J, Perkins, AS et al. Mice carrying null mutations of the genes encoding insulin-like growth factor I (igf-1) and type 1 IGF receptor (Igf1r). Cell 1993; 75:59-72.
29. Baker J, Liu J-P, Robertson EJ et al. Role of insulin-like growth factors in embryonic and postnatal growth. Cell 1993; 75:73-82.
30. Valentinis B, Morrione A, Taylor SJ et al. Insulin-like growth factor 1 receptor signaling in transformation by src oncogenes. Mol Cell Biol 1997; 17:3744-3754.
31. Liu JL, Blakesley VA, Gutkind JS et al. The constitutively active mutant G_{a13} transforms mouse fibroblast cells deficient in insulin-like growth factor-I receptor. J Biol Chem 1997; 272:29438-29441.
32. Baserga R, Morrione, A. Differentiation and malignant transformation: two roads diverged in a wood. J Cell Biochem 1999; 32/33:68-75.

33. Conlon MA, Tomas FM, Owens PC et al. Long R/3 insulin-like growth factor I (IGF-I) infusion stimulates organ growth but reduces plasma IGF-I, IGF-II and IGF binding protein concentrations in the guinea pig. J Endocrin 1995; 146:247-253.
34. Gorre ME, Mohammed M, Ellwood K et al. Clinical resistance to STI-571 cancer therapy caused by BCR-ABL gene mutation or amplification. Science 2001; 293:876-880.
35. Shima H, Pende M, ChenY et al. Disruption of the $p70^{S6K}/p85^{S6K}$ gene reveals a small mouse phenotype and a new functional S6 kinase. EMBO J 1998; 17:6649-6659.
36. Valentinis B, Navarro, M, Zanocco-MaraniT et al. Insulin receptor substrate-1, p70S6K and cell size in transformation and differentiation of hemopoietic cells. J Biol Chem 2000; 275:25451-25459.
37. Valentinis B, Romano G, Peruzzi F et al. Growth and differentiation signals by the insulin-like growth factor 1 receptor in hemopoietic cells are mediated through different pathways. J Biol Chem 1999; 274:12423-12430.
38. Askew DS, Ashmun RA, Simmons BC, Cleveland JL. Constitutive c-myc expression in an IL-3-dependent myeloid cell line suppresses cell cycle arrest and accelerate apoptosis. Oncogene 1991; 6:1915-1922.
39. Prisco M, Hongo A, Rizzo MG et al. The IGF-I receptor as a physiological relevant target of p53 in apoptosis caused by Interleukin-3 withdrawal. Mol Cell Biol 1997; 17:1084-1092.
40. Wang LM, Myers MG Jr, Sun XJ et al. IRS-1: Essential for insulin- and IL-4-stimulated mitogenesis in hemopoietic cells. Science 1993; 261:1591-1594.
41. Schmelzle T, Hall MN. TOR, a central controller of cell growth. Cell 2000; 103:253-262.
42. Dufner A, Thomas G. Ribosomal S6 kinase signaling and the control of translation. Exp Cell Res 1999; 253:100-109.
43. Aoki M, Schetter C, Himly M et al. The catalytic subunit of phosphatidylinositol 3-kinase: Requirements for oncogenicity. J Biol Chem 2000; 275:6267-6275.
44. Sachdev P, Jiang YX, Li W et al. Differential requirement for Rho family GTPases in an oncogenic insulin-like growth factor 1 receptor-induced cell transformation. J Biol Chem 2001; 276:26461-26471.
45. Ward AC, Smith L, de Koning JP et al. Multiple signals mediate proliferation, differentiation and survival from the granulocyte-colony stimulating factor receptor in myeloid 32D cells. J Biol Chem 1999; 274:14956-14962.
46. Valtieri M, Tweardy DJ, Caracciolo D et al. Cytokine dependent granulocytic differentiation. J Immunol 1987; 138:3829-3835.
47. Baserga R, Hongo A, Rubini M et al. The IGF-I receptor in cell growth, transformation and apoptosis. Biochim Biophys Acta 1997; 1332:105-126.
48. Stewart CEH, James PL, Fant ME et al. Overexpression of insulin-like growth factor II induces accelerated myoblast differentiation. J Cell Physiol 1996; 169:23-32.
49. Zamorano J, Wang HY, Wang L-M et al. IL-4 protects cells from apoptosis via the insulin receptor substrate pathway and a second independent signaling pathway. J Immunol 1996; 157:4926-4934.
50. Zhou-Li F, Xu S-Q, Dews M et al. Cooperation of simian virus 40 T antigen and insulin receptor substrate-1 in protection from apoptosis induced by interleukin-3 withdrawal. Oncogene 1997; 15:961-970.
51. Navarro M, Barenton B, Garandel V et al. Insulin-like growth factor 1 (IGF-I) receptor overexpression abolishes the IGF-I requirement for differentiation and induces a ligand-dependent transformed phenotype in C2 inducible myoblasts. Endocrinology 1997; 138:5210-5219.
52. Liu Q, Ning W, Dantzer R, Freund GG et al. Activation of protein kinase C-zeta and phosphatidylinositol 3'-kinase and promotion of macrophage differentiation by insulin-like growth factor 1. J Immunol 1998; 160:1393-1401.
53. Belletti B, Prisco M, Morrione A et al. Regulation of Id2 gene expression by the IGF-I receptor requires signaling by phosphatidylinositol-3 kinase. J Biol Chem 2001; 276:13867-13874.
54. Prisco M, Peruzzi F, Belletti B et al. Regulation of Id gene expression by the type 1 insulin-like growth factor: Roles of Stat3 and the tyrosine 950 residue of the receptor. Mol Cell Biol 2001; 21:5447-5458.
55. Kim B, Cheng H-L, Margolis B et al. Insulin receptor substrate 2 and Shc play different roles in insulin-like growth factor I signaling. J Biol Chem 1998; 273:34543-34550.
56. Sarbassov DD, Peterson CA. Insulin receptor substrate-1 and phosphatidylinositol3-kinase regulate extracellular signal- regulated kinase-dependent and –independent signaling pathways during differentiation. Mol Endocrinol 1998; 12:1870-1878.
57. Sadowski CL, Choi TS, Le M et al. Insulin induction of SOCS-2 and SOCS-3 mRNA expression in C2C12 skeletal muscle cells is mediated by Stat5. J Biol Chem 2001; 276:20703-20710.
58. Morrione A, Navarro M, Romano G et al. The role of the insulin receptor substrate-1 in the differentiation of rat hippocampal neuronal cells. Oncogene 2001; 20:4842-4852.
59. Reiss K, Wang JY, Romano G et al. IGF-I receptor signaling in a prostatic cancer cell line with a PTEN mutation. Oncogene 2000; 19:2687-2694.
60. Reiss K, Wang JY, Romano G et al. Mechanism of regulation of cell adhesion and motilityby insulin receptor substrate-1 in prostate cancer cells. Oncogene 2001; 20:490-500.
61. Doerr ME, Jones JL. The roles of integrins and extracellular matrix proteins in the insulin-like growth factor 1-stimulated chemotaxis of human breast cancer cells. J Biol Chem 1996; 271:2443-2447.
62. Dunn SE, Ehrlich M, Sharp NJH et al. A dominant negative mutant of the insulin-like growth factor 1 receptor inhibits the adhesion, invasion and metastasis of breast cancer. Cancer Res 1998; 58:3353-3361.
63. Guvakova MA, Surmacz E. Overexpressed IGF-I receptors reduce estrogen growth requirements, enhance survival and promote E-cadherin-mediated cell-cell adhesion in human breast cancer cells. Exp Cell Res 1997; 231:149-162.

64. Valentinis B, Reiss K, Baserga R. Insulin-like growth factor-1-mediated survival from anoikis: Role of cell aggregation and focal adhesion kinase. J Cell Physiol 1998; 176:648-657.
65. Li DM, Sun H. PTEN/MMAC1/TEP1 suppresses the tumorigenicity and induces G/1 cell cycle arrest in human glioblastoma cells. Proc Natl Acad Sci 1998; 95:15406-15411.
66. Li J, Simpson L, Takahashi M et al. The PTEN/MMAC1 tumor suppressor induces cell death that is rescued by the Akt/Protein Kinase B oncogene. Cancer Res 1998; 58:5667-5672.
67. Davies MA, Koul D, Dhesi H et al. Regulation of Akt/PKB activity, cellular growth and apoptosis in prostate carcinoma cells by MMAC/PTEN. Cancer Res 1999; 59,2551-2556.
68. Reiss K, D'Ambrosio C, Tu X et al. Inhibition of tumor growth by a dominant negative mutant of the insulin-like growth factor I receptor with the by-stander effect. Clinical Cancer Res 1998; 4: 2647-2655.
69. Plymate SR, Bae VL, Maddison L et al. Reexpression of the type 1 insulin-like growth factor receptor inhibits the malignant phenotype of simian virus 40 T antigen immortalized human prostate epithelial cells. Endocrinology 1997; 138:1728-1735.
70. Harrington EA, Bennett MR, Fanidi A et al. c-myc-induced apoptosis in fibroblasts is inhibited by specific cytokines. EMBO J 1994; 13:3286-3295.
71. McCubrey JA, Stillman LS, Mayhew MW et al. Growth promoting effects of insulin-like growth factor 1 (IGF-1) on hematopoietic cells. Overexpression of introduced IGF-1 receptor abrogates interleukin-3 dependency of murine factor dependent cells by ligand dependent mechanism. Blood 1991; 78:921-929.
72. Rodriguez-Tarduchy G, Collins MKL, Garcia I et al. Insulin-like growth factor I inhibits apoptosis in IL-3 dependent hemopoietic cells. J Immunol 1992; 149:535-540.
73. Ludwig T, Eggenschwiler J, Fisher P, D'Ercole JP, Davenport ML, Efstratiadis A. Mouse mutants lacking the type 2 IGF receptor (IGF2R) are rescued from perinatal lethality in Igf2 and Igf1r null backgrounds. Develop Biol 1996; 177:517-535.
74. Parrizas M, Saltiel AR, LeRoith D. Insulin-like growth factor I inhibits apoptosis using the phosphatidylinositol 3'-kinase and mitogen-activated protein kinase pathways. J Biol Chem 1997; 272:154-161.
75. Sell C, Baserga R, Rubin R. Insulin-like growth factor I (IGF-I) and the IGF-I receptor prevent etoposide-induced apoptosis. Cancer Res 1995; 55:303-306.
76. Dunn SE, Hardman RA, Kari FW et al. Insulin-like growth factor 1 (IGF-I) alters drug sensitivity of HBL100 human breast cancer cells by inhibition of apoptosis induced by diverse anticancer drugs. Cancer Res 1997;5 7:2687-2693.
77. O'Connor R, Kauffmann-Zeh A, Liu Y et al. The IGF-I receptor domains for protection from apoptosis are distinct from those required for proliferation and transformation. Mol Cell Biol 1997; 17:427-435.
78. Wang JY, DelValle L, Gordon J et al. Activation of the IGF-IR system contributes to malignant growth of human and mouse medulloblastomas. Oncogene 2001; 20:3857-3868.
79. Long L, Rubin R, Baserga R et al. Loss of the metastatic phenotype in murine carcinoma cells expressing an antisense RNA to the insulin-like growth factor I receptor. Cancer Res 1995; 55:1006-1009.
80. Burfeind P, Chernicky CL, Rininsland F et al. Antisense RNA to the type I insulin-like growth factor receptor suppresses tumor growth and prevents invasion by rat prostate cancer cells in vivo. Proc Natl Acad Sci 1996; 93:7263-7268.
81. Peretz S, Jensen R, Baserga R et al. ATM-dependent expression of the insulin-like growth factor 1 receptor in a pathway regulating radiation response. Proc Natl Acad Sci. 2001; 98:1676-1681.
82. Macaulay VM, Salisbury AJ, Bohula EA et al. Downregulation of the type 1 insulin-like growth factor receptor in mouse melanoma cells is associated with enhanced radiosensitivity and impaired activation of Atm kinase. Oncogene 2001; 20:4029-4040.
83. Macaulay V M. Insulin-like growth factors and cancer. Bri J Cancer 1992; 65:311-320.
84. Andrews DW, Resnicoff M, Flanders AE et al. Results of a pilot study involving the use of an antisense oligodeoxynucleotide directed against the insulin-like growth factor type 1 receptor in malignant astrocytomas. J Clin Oncol 2001; 19:2189-2200.
85. Peruzzi F, Prisco M, Dews M et al. Multiple signaling pathways of the IGF-I receptor in protection from apoptosis. Mol Cell Biol 1999; 19:7203-7215.
86. Peruzzi F, Prisco M, Morrione A et al. Anti-apoptotic signaling of the insulin-like growth factor-I receptor through mitochondrial translocation of c-Raf and Nedd4. J Biol Chem 2001; 276:25990-25996.
87. Navarro M, Baserga R. Limited redundancy of survival signals from the type 1 insulin-like growth factor receptor. Endocrinology 2001; 142:1073-1081.
88. Devireddy LR, Teodoro JG, Richard FA et al. Induction of apoptosis by a secreted lipocalin that is transcriptionally regulated by IL-3 deprivation. Science 2001; 293:829-834.
89. Flower DR. The lipocalin protein family: structure and function. Biochem J 1998; 318:1-14.
90. Hraba-Renevey S, Turler H, Kress M et al. SV40-induced expression of mouse gene 24p3 involves a post-transcriptional mechanism. Oncogene 1989; 4:601-608.
91. Zhou-Li F, D'Ambrosio C, Li S et al. Association of insulin receptor substrate 1 with Simian Virus 40 large T antigen. Mol Cell Biol 1995; 15:4232-4239.
92. Reinhard C, Fernandez A, Lamb NJC et al. Nuclear localization of p85^{S6K}: functional requirement for entry into S phase. EMBO J 1994; 13:1557-1565.
93. Lenormand P, Sardet C, Pages G et al. Growth factors induce nuclear translocation of MAP kinases (p42^{mpk} and p44^{mpk}) but not of their activator MAP kinase kinase (p45^{mpkk}) in fibroblasts. J Cell Biol 1993; 122:1079-1088.

94. Sun XH, Copeland NG, Jenkins NA et al. Id proteins Id1 and Id2 selectively inhibit DNA binding by one class of helix-loop-helix proteins. Mol Cell Biol 1991; 11:5603-5611.
95. Benezra R, Davis RL, Lockshon D et al. The protein Id: A negative regulator of helix-loop-helix DNA binding proteins. Cell 1990; 61:49-59.
96. Norton JD, Deed RW, Craggs G et al. Id helix-loop-helix proteins in cell growth and differentiation. Trends in Cell Biol 1998; 8:58-65.
97. Barone MV, Pepperkok R, Peverali FA et al. Id proteins control growth induction in mammalian cells. Proc Natl Acad Sci 1994; 91:4985-4988.
98. Hara E, Yamaguchi T, Nojima H et al. Id related genes encoding HLH proteins are required for G1 progression and are repressed in senescent human fibroblasts. J Biol Chem 1994; 269:2139-2145.
99. Kreider BL, Benezra R, Rovera G et al. Inhibition of myeloid differentiation by the helix-loop-helix protein Id. Science 1992; 255:1700-1702.
100. Desprez PY, Hara E, Bissell MJ et al. Suppression of mammary epithelial cell differentiation by the helix-loop-helix protein Id-1. Mol Cell Biol 1995; 15:3398-3404.
101. Desprez PY, Lin CQ, Thomasset N et al. A novel pathway for mammary epithelial cell invasion by the helix-loop-helix protein Id-1. Mol Cell Biol 1998; 18:4577-4588.
102. Lin CQ, Singh J, Murata K et al. A role for Id-1 in the aggressive phenotype and steroid hormone response of human breast cancer cells. Cancer Res 2000; 60:1332-1340.
103. Iavarone A, Garg P, Lasorella A et al. The helix-loop-helix protein Id-2 enhances cell proliferation and binds to the retinoblastoma protein. Genes Develop 1994; 8:1270-1284.
104. Lasorella A, Noseda M, Beyna M et al. Id2 is a retinoblasoma protein target and mediates signaling by Myc oncoproteins. Nature 2000; 407:592-598.
105. Florio M, Hernandez MC, Yang H et al. Id2 promotes apoptosis by a novel mechanism independent of dimerization to basic helix-loop-helix factors. Mol Cell Biol 1998; 18:5435-5444.
106. Rubini M, Hongo A, D'Ambrosio C et al. The IGF-I receptor in mitogenesis and transformation of mouse embryo cells: Role of receptor number. Exp Cell Res 1997; 230:284-292.
107. Zong CS, Zong L, Jiang Y et al. Stat3 plays an important role in oncogenic ros- and insulin-like growth factor 1 receptor-induced anchorage-independent growth. J Biol Chem 1998; 273:28065-28072.
108. Zong CS, Chan J, Levy DE et al. Mechanism of STAT3 activation by insulin-like growth factor I receptor. J Biol Chem 2000; 275:15099-15105.
109. Dupont J, Fernandez AM, Glackin CA et al. Insulin-like growth factor-1 (IGF-I)-induced Twist expression is involved in the anti-apoptotic effects of the IGF-I receptor. J Biol Chem 2001; 276:26699-26707.
110. Xu A, Suh P-G, Marmy-Conus N et al. Phosphorylation of nuclear phospholipase C b1 by extracellular signal-regulated kinase mediates the mitogenic action of insulin-like growth factor 1. Mol Cell Biol 2001; 21:2981-2990.
111. Brunet A, Bonni A, Zigmond MJ et al. Akt promotes cell survival by phosphorylating and inhibiting a forkhead transcription factor. Cell 1999; 96:857-868.
112. Baserga R. The Biology of Cell Reproduction. Cambridge: Harvard University Press, 1985.
113. Comai L, Song Y, Tan C et al. Inhibition of RNA polymerase I transcription in differentiated myeloid leukemia cells by inactivation of Selectivity Factor-1. Cell Growth Diff. 2000; 11:63-70.
114. Rogalsky V, Todorov G, Moran D. Translocation of retinoblastoma protein associated with tumor cell growth inhibition. Biochem Biophys Res Comm 1993; 192:1139-1146.
115. Cavanaugh AH, Hempel WM, Taylor LJ et al. Activity of RNA polymerase I transcription factor UBF blocked by Rb gene product. Nature 1995; 374:177-180.
116. Voit R, Schafer K, Grummt I. Mechanism of repression of RNA polymerase I transcription by the retinoblastoma protein. Mol Cell Biol 1997; 17:4230-4237.
117. Hannan KM, Hannan RD, Smith SD et al. Rb and p130 regulate RNA polymerase I transcription: Rb disrupts the interaction between UBF and SL-1. Oncogene 2000; 19:4988-4999.
118. Hannan KM, Kennedy BK, Cavanaugh AH et al. RNA polymerase I transcription in confluent cells: Rb downregulates rDNA transcription during confluence-induced cell cycle arrest. Oncogene 2000; 19:3487-3497.
119. Surmacz E, Kaczmarek L, Ronning O et al. Activation of the ribosomal DNA promoter in cells exposed to insulin-like growth factor 1. Mol Cell Biol 1987; 7:657-663.
120. Baserga R, Resnicoff M. Insulin-like growth factor 1 receptor as a target for anticancer therapy. In: Hickman JA, Dive C, eds. Apoptosis and Cancer. Totowa: Humana Press, 1999:189-203.
121. Lafarge-Frayssinet C, Duc HT, Frayssinet C et al. Antisense insulin-like growth factor I transferred into a rat hepatoma cell line inhibits tumorigenesis by modulating major histocompatibility complex 1 cell surface expression. Cancer Gene Therapy 1997; 4:276-285.
122. Liu X, Turbyville T, Fritz A et al. Inhibition of insulin-like growth factor 1 receptor expression in neuroblastoma cells induces the regression of established tumors in mice. Cancer Res 1998; 58:5432-5438.
123. Ly A, Duc HT, Kalamarides M et al. Human glioma cells transformed by IGF-I triple helix technology show immune and apoptotic characteristics determining cell selection for gene therapy of glioblastoma. J Clin Path: Mol Pathol 2001; 54:230-239.
124. Resnicoff M, Abraham D, Yutanawiboonchai W et al. The insulin-like growth factor I receptor protects tumor cells from apoptosis in vivo. Cancer Res 1995; 55:2463-2469.
125. Nehlin JO, Hara E, Kuo WL et al. Genomic organization, sequence, and chromosomal localization of the human helix-loop-helix Id1 gene. Biochem Biophys Res Comm 1997; 231:628-634.

CHAPTER 7

Insulin-Like Growth Factor-I Stimulation of Growth:
Autocrine, Paracrine and/or Endocrine Mechanisms of Action?

A. Joseph D'Ercole

Abstract

The relationship between the sites of insulin-like Growth factor-I (IGF-I) expression and the sites of its growth promoting actions have been debated for years. There is compelling evidence for each potential mode of IGF-I action. Arguments for IGF-I endocrine actions include the findings that: a) circulating IGF-I concentrations are relatively high and are regulated by GH and nutrition, as well as other agents, b) IGF-I administration stimulates somatic growth, making it certain that IGF-I transiting in the circulation can exert growth stimulating actions, and c) genetically manipulated mice made deficient in blood-borne IGF-I and the acid labile subunit of the major circulating IGF-I complex exhibit low blood IGF-I levels accompanied by postnatal somatic and bone growth retardation, as well as metabolic abnormalities. Findings supporting autocrine/paracrine IGF-I actions at or near sites of synthesis include: a) the expression of IGF-I, the type 1 IGF receptor (IGF1R) and IGF binding proteins in virtually all tissue/organs, b) growth enhancement in the organs/tissues of transgenic (Tg) mice engineered to overexpress IGF-I, while overexpression of inhibitory IGFBPs in specific organs/tissues results in growth retardation of these organs/tissues, c) IGF-I null mutant mice exhibit somatic growth retardation throughout developmental, notably during embryonic life when circulating IGF-I levels are low, and d) IGF-I expression often increases following injury and during recovery and/or regeneration. The increased expression of IGF-I and/or the IGF1R in a number of cancers also argues for IGF-I autocrine actions. It seems certain, therefore, that IGF-I acts in each of these ways. In this review it is argued that paracrine IGF-I actions predominate during embryogenesis through early postnatal life, and that postnatally endocrine actions progressively assume a greater role. Paracrine effects of IGF-I, however, likely serve a function in repair following some injuries and in organ regeneration. IGF-I also appears to advance the growth of some tumors by autocrine mechanisms.

Introduction

The requirement of insulin-like growth factor-I (IGF-I) expression for normal in utero and postnatal growth is now well established.[1,2] The relationship between IGF-I sites of synthesis and actions, however, remains a subject of debate and active experimentation. The questions of the ongoing debate are whether IGF-I expressed in specific tissues acts locally to promote growth, i.e., autocrine/paracrine actions, or whether it is the IGF-I circulating in the blood that is responsible for stimulating growth, i.e., endocrine actions (Fig. 1). Significant evidence supports each scenario. In this review it is argued that growth is promoted by both locally produced and blood-borne IGF-I, and that the contribution of each differs depending upon the stage of development. At the same time we attempt to provide a balanced summary of the arguments for and against local and endocrine IGF-I actions (Table 1). Much of this review is presented in a historical context because the chronological sequence of research findings explains how we have come to our current understanding

Insulin-Like Growth Factors, edited by Derek LeRoith, Walter Zumkeller and Robert Baxter. ©2003 Eurekah.com and Kluwer Academic / Plenum Publishers.

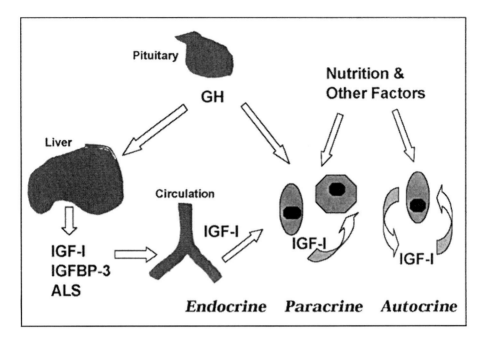

Figure 1. IGF-I Regulation and Modes of Action. GH stimulates the formation of the circulating 150 kDa, IGF-I complex—composed of IGF-I bound to IGFBP-3 which is in turn bound to the acid labile subunit (ALS). This complex, which accounts for most blood-borne IGF-I, traverses the circulation and appears to be responsible in large part for IGF-I endocrine actions. GH, as well as many other factors, stimulates IGF-I expression in many tissues where it also acts in a paracrine and/or autocrine fashion.

of IGF-I regulation of growth. The axiom of historians applies to science: "To know nothing of the past is to understand little of the present and to have no concept of the future" (anonymous).

The Somatomedin Hypothesis

The general perception of the nature of IGF-I actions has evolved over the years. The concept that growth hormone (GH) actions are mediated by an intermediate substance came from the 1957 report by Salmon and Daughaday.[3] In an attempt to develop an assay for GH they measured the capacity of cartilage to incorporate sulfate when exposed to GH or plasma with various GH concentrations. While plasma from intact rats was capable of stimulating much more sulfate uptake than that from hypophysectomized rats, purified GH had little to no apparent activity. They postulated, therefore, that GH stimulated another substance(s), termed sulfation factor, which was responsible for the activity in plasma. Further experiments were consistent with the concept that GH actions on cartilage were indirect, and that the substance(s) which mediate GH action act as hormones—that is substance(s) synthesized at a single site and secreted into the circulation where they travel to distant sites of action (see reviews in refs. 4, 5).

In 1972 this notion was stated in a consensus letter to *Nature* written by most of the active investigators in the field.[6] They proposed the name somatomedin, and formally put forth "the somatomedin hypothesis" which stated: "A GH dependent plasma factor stimulates in cartilage not only the incorporation of sulphate into chondroitin sulfate, but also the incorporation of thymidine into DNA, proline into the hydroxyproline of collagen and uridine into RNA."

Initial Evidence for Endocrine IGF-I Actions

Clearly the pioneers in IGF-I research had a hormone in mind, and did not envision a ubiquitously expressed growth factor. The fact that virtually all investigators studying GH-dependent factors

Table 1. Evidence for endocrine, paracrine and autocrine IGF-I growth promoting actions

Mode of Action	Finding
Endocrine	Sulfation factor identified in plasma.
	Somatomedin purified from plasma.
	Circulating IGF-I concentrations are comparable to those of hormones.
	Somatomedin-C/IGF-I serum levels are regulated, as are those of hormones.
	IGF-I administration stimulates somatic growth in rodents and man.
	Ablation of liver expression of IGF-I and ALS results in low circulating IGF-I levels and modest postnatal growth retardation.
Paracrine/Autocrine	IGF-I is expressed by multiple cell types in culture.
	IGF-I is expressed in virtually every organ and tissue in vivo.
	The IGF1R and IGFBPs are expressed in virtually all organs/tissues.
	Tg mice made to overexpress IGF-I in specific organs/tissues exhibit overgrowth in the organ/tissue of IGF-I transgene expression.
	Tg mice made to overexpress inhibitory IGFBPS in specific organs/tissues exhibit growth retardation of the organ/tissue of IGFBP expression.
	IGF-I KO mice exhibit marked growth retardation in utero and early postnatal life—developmental times when blood IGF-I levels are very low.
	IGF-I is often increased at sites of injury and/or repair following injury or during regeneration.
Autocrine	A variety of cultured cells express IGF-I, the IGF1R, and IGFBPs.
	Antibodies against IGF-I and/or the IGF1R often block or blunt proliferation of cells in culture.
	Some tumors overexpress IGF-I and/or the IGF1R

were Endocrinologists had a major influence on this thinking. This be as it may, there also was good reason to think that GH-stimulated the synthesis of an intermediary hormone. Those attempting to purify somatomedin found that plasma was the most abundant source. While not a traditional endocrine organ, the liver was considered the site of somatomedin origin. There was ample reason to accept the liver as the "gland" of somatomedin secretion. The liver was known to be rich in GH binding sites (for example ref. 7), suggesting that stimulation of somatomedin synthesis is a GH-dependent hepatic function. Additional support for this concept came from partial hepatectomy experiments.[8] These investigators found that serum somatomedin activity fell following partial hepatectomy, and that the magnitude of the decline correlated with the amount of liver resected. While the assumption that removal of liver tissue per se results in a decreased somatomedin synthesis was reasonable, later evidence showed that the fall in serum somatomedin was more likely due to the reduced food intake that accompanied these stressful procedures.[9] Regardless of the precise interpretation of these partial hepatectomy experiments, it is now clear that liver is the major source of circulating IGF-I. This evidence comes from liver-specific IGF-I gene ablation experiments[10,11] that are discussed below. Prior to gene ablation experiments, however, other evidence made a compelling argument for liver as the primary source of serum IGF-I. Using perfused rat liver, Schwander et al[12] demonstrated that liver could synthesize sufficient somatomedin to account for the activity in the circulation.

In the late 1970s, the availability of highly purified somatomedin-C allowed the development of a specific radioimmunoassay (RIA).[13] Clinical studies using this RIA provided convincing evidence that serum somatomedin-C levels were highly dependent on GH secretory status, i.e., low in hypopituitarism and high in acromegaly. Furthermore, the serum concentrations of somatomedin-C were found to be similar to those of hormones, as opposed to the very low plasma levels of the growth factors known at the time, e.g., EGF and NGF. The physiologic regulation of serum somatomedin-C, therefore, was entirely consistent with somatomedin-C acting in an endocrine fashion. Amino acid sequencing of somatomedins confirmed their suspected insulin-like structure, and led to the use of the term insulin-like growth factor to describe the two genes and their respective peptides, IGF-I and IGF-II (see review in ref. 14). Subsequent studies showed that while IGF-II has significant homology with IGF-I (and insulin), it's serum concentrations are not highly regulated by GH.

The availability of purified IGF-I also allowed direct testing of somatomedin-C/IGF-I actions in intact animals. The first such study showed that injection of somatomedin-C into hypophysectomized frogs stimulated mitosis in lens epithelial cells.[15] The importance of this study rested on the fact that it gave credence to the hypothesis that somatomedin can stimulate a biologic event related to growth. At the time many investigators in the area of GH physiology doubted that GH's growth-promoting actions were indirect and believed that somatomedin was either a reflection of growth or fictitious. Such thinking became more difficult to accept when Schoenle et al[16] demonstrated that injection of IGF-I stimulated growth in hypophysectomized rats. Numerous in vivo studies followed (see early reviews in refs. 17-19). Such studies re-enforced the concept of IGF-I endocrine actions because regardless of the site of IGF-I administration, travel through the circulation was necessary for stimulation of somatic growth.

Initial Evidence for IGF-I Local Actions

Following a series of experiments using explants of fetal mouse tissues, we postulated that IGF-I actions were exerted locally, i.e., exerted either on their cells of synthesis (autocrine actions) or on nearby cells (paracrine actions).[20] We observed that multiple fetal mouse tissues released immunoreactive IGF-I into their culture media, and that this material was much greater than could be due to serum remaining in the explants. Furthermore, it appeared that a large portion of the immunoreactive material was active in a membrane binding assays. This finding made it likely that the immunoreactive IGF-I was recognized by membrane associated receptors and was biologically active. Later we showed that multiple adult rat tissues had IGF-I concentrations much higher than could be explained by the blood they contained and that IGF-I levels in many tissues were regulated by GH.[21] Shortly after, Clemmons et al showed that IGF-I was synthesized by cultured fibroblasts,[22] that IGF-I synthesis in fibroblasts is regulated,[23-25] that fibroblasts express cell surface IGF receptors,[26] and that fibroblast IGF-I exerts actions on the cells that produced it.[27] Subsequently this group demonstrated that cultured smooth muscle cells expressed IGF-I[28] where IGF-I could also stimulate biologic responses.[29]

With the isolation of the IGF-I cDNA,[30] studies to identify IGF-I mRNA indisputably showed that IGF-I expression was widespread in rodent and human tissues and, in significant part, dependent on GH secretion. Such studies employed Northern analysis[31-32] and in situ hybridization histochemistry.[33] This ubiquitous IGF-I expression is accompanied by similarly widespread expression of the type 1 IGF receptor (IGF1R; the primary IGF-I receptor)[34] and IGFBPs.[35] It was, therefore, possible that IGF-I could be regulated and signal biologic effects in a local milieu. Despite the circumstantial nature of this evidence, autocrine and/or paracrine actions for IGF-I became generally accepted.

Evidence for Paracrine/Autocrine IGF-I Actions from Studies of Transgenic and Null Mutant Mice

The development of the ability to alter the genome of experimental animals provided powerful tools to study IGF-I actions. Creation of transgenic (Tg) mice that overexpress IGF-I provided strong support for local IGF-I actions. The first IGF-I overexpressing transgenic (Tg) mice utilized a fusion gene that linked the metallothionein-I promoter, which directs expression in many tissues, to an human IGF-IA cDNA.[36,37] In these mice organ overgrowth and the degree of transgene expression were generally concordant (see review in ref. 38), suggesting that IGF-I stimulated growth locally in the tissues of its overexpression. Such evidence, however, is circumstantial. It was equally possible that the relative overgrowth in some tissues represented heightened sensitivity of these tissues to the elevated circulating IGF-I levels. Later using the same transgene, we generated mice with brain IGF-I overexpression in which blood IGF-I concentrations were not elevated.[39] The marked increases in brain weight provided more convincing evidence of local IGF-I actions, at least in the central nervous system.

By exploiting the capacity of DNA to undergo homologous recombination, it also became possible to ablate the expression of native genes and thus create models of absent gene function. The development of null mutant mice, also called knockout (KO) mice, demonstrated the central role of IGF-I in control of somatic growth throughout development.[40,41] Mice carrying mutations that disrupt both IGF-I alleles exhibit growth retardation from day 13 of embryonic life such that their birth weights are reduced by about 40%. Postnatally, the IGF-I KO mice that survive continue to grow poorly and as adults are only 25% of normal size. Mice carrying homozygous disruptions of the IGF1R gene exhibit even more profound growth retardation at birth, being 45% of normal weight, and invariably die at or very near birth.[40,41] Because circulating IGF-I levels are very low in utero and during early postnatal life in the rodent [42,43] the marked growth retardation in IGF-I and IGF1R KO mice argues for local IGF-I actions. Such a paradigm appears to be the rule for growth factors during embryonic development (see reviews in refs. 44;45).

More convincing evidence of local IGF-I actions comes from lines of Tg mice made to overexpress IGF-I in specific tissues, e.g., brain, mammary gland, and muscle, etc. Each of these Tg mouse models exhibits specific overgrowth in the organ or tissue of IGF-I overexpression, and none has an alteration in circulating IGF-I levels (Table 2 reviews reports of such mouse models). In every model studied biologic actions in the organ of IGF-I transgene expression has been demonstrated. IGF-I, therefore, can exert local in vivo actions. On the other hand, because IGF-I is overexpressed, these models represent an aberrant milieu, and as such do not prove that IGF-I functions in the same way under conditions of normal growth or homeostasis.

Other experiments that address IGF-I local actions are the generation of Tg mice that overexpress IGFBPs in specific tissues. Here the expectation is that these IGFBPs will inhibit the actions of locally expressed IGF-I. Such studies have yielded results consistent with those obtained from studies of site-specific IGF-I overexpression. An example is the overexpression of rat IGFBP-4 in smooth muscle driven by the regulatory region of the α-actin gene.[77] Transgene IGFBP-4 expression results in smooth muscle hypoplasia. The lack of any change in circulating IGFBP-4 or IGF-I and the restriction of hypoplasia to smooth muscle argues for the inhibition of IGF-I growth promoting effects on smooth muscle. Alternative, but unlikely, interpretations are that IGFBP-4 inhibited the actions of IGF-I derived from the circulation and/or that IGFBP-4 inhibits growth by mechanisms independent of IGF-I. Other Tg mouse models have yielded consistent results. For example, a number of lines of IGFBP-1 Tg mice exhibit organ growth retardation that appears due to the capacity of IGFBP-1 to inhibit IGF activity in specific tissues, for example, in brain.[39,78-80] In each, however, the IGFBP-1 transgene is expressed in many tissues, and because circulating IGFBP-1 levels are elevated, blood borne IGFBP-1 could contribute to the growth retardation.

Other Evidence for IGF-I Autocrine/Paracrine Actions

Other lines of evidence also support the proposition that IGF-I exerts local actions. There is abundant experimental evidence for autocrine/paracrine roles of IGF-I during organ regeneration and in repair following injury. In 1985 we reported that IGF-I concentrations increase in the kidney during regeneration following unilateral nephrectomy, and that this occurs without a discernable change in blood IGF-I levels.[81] Since then a local role for IGF-I in kidney regeneration and repair following injury has been firmly established (see review in ref. 82). In addition to the evidence that IGF-I has a central role in brain development, IGF-I also is involved in recovery from a variety of injuries to the nervous system (see reviews in refs. 83-85). For example, IGF-I expression frequently is increased in or near brain lesions. Importantly, IGF-I administration also has been shown to ameliorate brain injury (references 86-89 provide examples of such IGF-I actions). When overexpressed

Table 2. IGF-I transgenic mice with tissue-specific IGF-I overexpression

Organ	Promoter	IGF-I Action	Ref.
Brain	m IGF-II 5' flanking region	Increased brain size, characterized by increased neuron number.	46-49
Bone	bovine osteocalcin	Increased trabecular bone.	50
Heart	r α myosin heavy chain	Increased myocyte proliferation.	51
		Protection against myocardial infarction.	52
	avian Skeletal α Actin	Induces cardiac hypertrophy.	53
Mammary gland	m Whey Acidic Protein	Inhibition of involution	54
	r Whey Acidic Protein	Des(1-3) IGF-I overexpression Inhibits apoptosis during natural, but not forced, involution.	55-56
		Ductal morphogenesis.	57
	m Mammary Tumor Virus-LTR (MMTV-LTR)	Enhanced rate of aveolar bud development.	58
Lens	αA-crystallin	Expanded lens transitional zone.	59
Muscle: Skeletal	avian Skeletal α Actin	Stimulates differentiation and myofibrile hypertrophy.	60
		Increased numbers of dihydropyridine receptors (DHPR) and increased DHPR to ryanodine receptor (RyR1) ratio without an age-related decline.	61, 62
		Increased DHPRα and RyR1mRNA expression; increased DHPRα transcription.	63
		No amelioration of hind-limb unloading atrophy.	64
		Extends the replicative life span of skeletal muscle satellite cells, but this activity is lost with aging.	65, 66
	m Myosin Light Chain (MLC)	Sustains hypertrophy and regeneration of Senescent muscle.	67
Muscle: Smooth	r Smooth Muscle α Actin flanking fragments (mSMA)	Smooth muscle hyperplasia in many organs/tissues.	68
		Increased vascular contractility.	69
		Enhanced neointimal formation after injury.	70
Ovary	m LH receptor	Increased testosterone and cyst formation in some mice.	71
Prostate	bovine keratin-5	Epithelial neoplasia.	72
Skin	h Keratin-1	Epidermal hyperplasia, hyperkeratosis, and tumor formation.	73
	m ultra-high sulfur keratin	Increased vibrissa growth.	74
Testes	CMV	Increased IGFBP expression.	75
Thyroid	bovine thyroglobulin	When the IGF1R is also overexpressed, there is a decreased TSH requirement and goiter.	76

in the brain of Tg mice IGF-I ameliorates demyelinating injury and hastens repair.[90] The heart provides yet another example of IGF-I exerting local actions to promote compensatory growth or in repair following injury (see reviews in refs.. 91,92).

While there is no evidence that IGF-I causes or is involved in the initiation of cancer, IGF-I may serve an autocrine role in enhancing tumor growth and proliferation (see reviews in refs. 93-94). Changes in the expression of IGF-I, the IGF1R, IGFBPs, and IGFBP proteases have been observed in a variety of cancers, including those of the prostate, breast, lung and colon, or in cell lines derived from such malignancies. Because antibodies to IGF-I or the IGF1R often have been observed to either decrease proliferation or increase apoptosis in cancer cell lines, a strong case has been made for autocrine IGF-I actions. In addition, a great deal of attention has been paid to measurement of serum IGF-I and IGFBP-3 as potential markers of cancer, especially of prostate cancer. While a comprehensive discussion of IGF-I actions in reparative and regenerative growth and in carcinogenesis is beyond the scope and topic of this review, evidence from these area of research provide further significant evidence of local IGF-I actions.

Further Evidence for IGF-I Endocrine Actions

Because systemic administration of IGF-I to laboratory animals and man stimulates somatic growth, there is no question that IGF-I has the capacity to act in an endocrine fashion (see reviews: 95). As with local actions, however, this demonstrates that IGF-I can act by endocrine action, but does not prove that endocrine mechanisms represent the physiologically relevant mode of IGF-I action. The fact that IGF-I blood levels are regulated by GH and nutritional status, as well as by other factors such as thyroxine, argue for endocrine actions (see reviews in refs. 96-98). Unless one postulates that this apparent regulation is simply a reflection IGF-I gene expression in multiple tissues, it is difficult to dismiss circulating IGF-I as physiologically unimportant. It is also empirically difficult to ignore the presence of IGFBPs in the circulation. Overwhelming evidence demonstrates that IGFBPs account for the relatively high IGF-I levels in the circulation (see reviews in refs. 99,100). There is also strong evidence that IGFBPs can control the amount of free IGF-I in the blood and the delivery of IGF-I to tissues. In addition, blood concentrations of many IGFBPs are hormonally regulated, e.g., IGFBP-1 by insulin, and IGFBP-3 and the acid labile subunit (ALS; a component of the 150 kda IGF binding complex in blood) by GH. Why is there such tight control of blood-borne IGF-I if not to regulate IGF-I bioactivity?

Studies of Conditional Null Mutant Mice

The direct studies of IGF-I endocrine actions have come from the generation and study of conditional null mutant mice. Use of homologous recombination in combination with traditional transgenic technology allows the generation of mice with null mutations in specific tissues. These experiments exploit the Cre recombinase-Lox system.[101] This strategy requires the generation of two lines of genetically altered mice that are subsequently bred to create the mice of interest (Fig. 2). Mice expressing Cre recombinase are created by traditional transgenic technology using a fusion gene in which the Cre coding sequence is under the control of a promoter that drives expression in a limited or conditional fashion, i.e., in a specific tissue, at a specific developmental time, in response to an administered agent, etc. Because Cre does not appear to have a normal function in mammals, Tg mice expressing Cre exhibit no apparent consequences. The other line of mice are made by homologous recombination such that sequences recognized by Cre, called lox sequences, flank sequences crucial to coding of the gene of interest. Typically, lox sequences are inserted in introns that flank a coding exon. After manipulation of a DNA fragment to insert the lox sequences, embryonic stem (ES) cells are transfected with the "lox'ed" DNA, selected for appropriate recombinations and amplified. Mutant mice are then generated by placing the mutant ES cells into blastocysts. The blastocysts are transferred to pseudo-pregnant females, and the pups carrying the mutation in their germ line are identified. Once the lox'd mouse line is generated, they are cross bred with mice carrying the Cre transgene. In the resultant double mutant mice, Cre excises the DNA flanked by the lox sequences in the cells where it is expressed, yielding a gene ablation in a specific fashion and in a limited number of cells.

Two groups of investigators have taken advantage of this approach to ablate liver IGF-I expression.[10,11] The aim of both groups was to assess IGF-I endocrine actions and each based their approach on the assumption that blood-borne IGF-I is predominantly liver-derived. Both groups were successful in creating mice with deficient liver IGF-I expression, termed LID mice (liver IGF-I deficiency). Each found that young adult mutant mice had marked reductions of circulating IGF-I

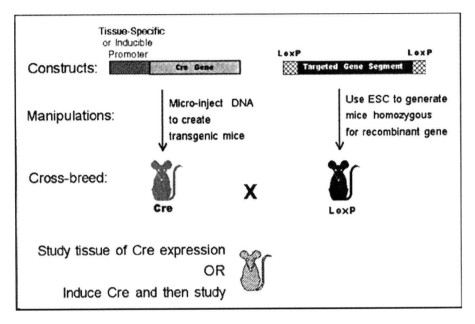

Figure 2. Strategy for Generation of Conditional Null mutant mice. Two different lines are mutant mice are generated. 1) A fusion gene is constructed linking a conditional promoter to the gene for Cre recombinase (left side of the figure). The promoter is selected because it has the capacity to drive expression in a fashion that suits the experimental design, e.g., it drives expression in defined cell types or at a specific developmental time, or it can be induced to express with the administration of exogenous agents, etc. Using traditional methods a Tg mouse line is generated with this fusion gene. 2) A genomic fragment is engineered that contains loxP sequences flanking a coding sequence in the gene of interest (right side of the figure). Taking advantage of homologous recombination, this fragment is manipulated to replace the native gene in embryonic stem cells (ESC). ESC with the mutant gene are injected into blastocysts and allowed to develop in psuedo-pregnant female mice. The resultant mice carrying the recombinant gene are selected and bred to homozygosity. These two lines of mice are than cross bred to obtain mice that carry both the Cre-recombinase and the alleles with lox sequences. If the experiment is successful the gene of interest will be nonfunctional in the cells that express Cre, and the consequences of an absence or deficiency of the gene of interest can be studied.

(66-75% reductions), but no apparent reduction in somatic growth. These results show that the liver is the major source of IGF-I in the circulation of the mature mouse. Given the lack of somatic growth retardation, it was tempting to also conclude that circulating IGF-I is not important in the stimulation of somatic growth. Alternatively one could conclude that blood IGF-I levels that are 25-33% of normal are sufficient to stimulate normal growth. This conclusion, however, also is not necessarily appropriate once the nature of these experimental mouse models is examined.

For a number of reason it is not clear when serum IGF-I levels become lower than in normals, and it is possible that a reduction in serum IGF-I was not achieved until at least several weeks of postnatal life. Yakar et al[11] utilized the albumin promoter to drive liver Cre expression. In these mice the Cre expression, and thus induction of the liver IGF-I null mutation, occurred by at least 10 days of postnatal life, and at 26 days of life serum IGF-I level were ~50% of normal. It is not clear, however, whether this reduction represents a total lack of liver derived IGF-I or whether the liver null mutation had yet to exert is maximum effect on serum IGF-I. The LID mice of Sjögren et al used an interferon-responsive promoter, which allows induction of Cre expression by administration of interferon.[10] Interferon induction was begun at 24 days of life. In these LID mice, therefore, IGF-I blood levels were not maximally reduced until sometime between the time of Cre expression (24 days) and 6-8 weeks of age—the time when circulating IGF-I was assessed. In both of these lines of mice, circulating IGF-I levels may not be dramatically reduced during periods of rapid postnatal growth. These mutant mice, therefore, provide excellent models to evaluate circulating IGF-I actions in more mature and adult mice (see below), but may not be ideal models of IGF-I endocrine actions in early post-natal growth. An obvious remedy for the limitations of the above LID mice is to develop a mutant mouse in which blood IGF-I concentrations are maintained at or below the normally low levels of fetal and early postnatal life.

IGF-I expression, however, is not the sole determinant of blood IGF-I levels. IGF-I circulates predominately in a large molecular weight form of ~150 kDa composed of IGF-I bound to IGFBP-3 which in turn is bound by another protein called, the acid labile subunit (ALS) (see review in ref. 100). ALS only binds IGFBP-3 when it is complexed to IGF-I. This ternary complex, that is IGF-I/IGFBP-3/ALS, is regulated by GH and is believed to maintain a reservoir of IGF-I in the circulation. The binary complex of IGF-I and IGFBP-3 can traverse the epithelium, and it, thus, appears to facilitate IGF-I delivery to tissue. Because IGFBP-3 and/or ALS control blood-borne IGF-I available to growing cells, it would seem possible that a lower then normal concentration of circulating IGF-I might suffice for normal growth if IGFBP-3 and ALS concentrations were adequate. Several recent studies suggest that this may be the case.

Ueki et al[102] have created mice carrying an ALS null mutation. Mice with a homozygous null mutation exhibit reduced circulating IGF-I and IGFBP-3 levels, 62% and 88% reductions, respectively, at 10 weeks of age. The low latter results are not surprising because ALS is responsible for the formation of 150 kDa complex which accounts for the stabilization of IGF-I bound to IGFBP-3 in the blood stream. These mice also exhibited somatic growth retardation that first achieved significance at 3 weeks of age and resulted in 13-20% reductions in adult body weight. The ALS null mutation thus causes growth retardation, while the LID mutation does not. If the growth retardation in ALS null mutants is posited to be due to the reduction in circulating IGF-I, the question becomes why are 60% reductions in blood-borne IGF-I sufficient to retard growth in ALS null mutants, while the 75% reductions in liver IGF-I null mutants are not. There are a number of plausible explanations for the growth failure that are consistent with endocrine IGF-I actions:

1. ALS is necessary to maintain a sufficient reservoir of blood IGF-I to optimally stimulate growth;
2. The total absence of ALS results in consistently low circulating IGF-I levels throughout development—unlike LID mice that may have exhibited a rise in blood IGF-I prior to the initiation of IGF-I gene ablation. Given that the post-weaning rise in blood IGF-I is temporally associated with the onset of ALS expression, it seems possible that this was the case, i.e., that circulating IGF-I levels in ALS KO mice never rise above those of the perinatal period. It is unfortunate that IGF-I and IGFBP-3 were only assayed at 10 weeks in ALS null mutant.
3. In the absence of ALS, IGFBP-3 is rapidly degraded and less available to facilitate IGF-I delivery to tissue; and
4. all or a combination of the above scenarios are correct. Alternately, it is possible that ALS deficiency has effects on growth that are independent of IGF-I, and these ALS actions are not related to circulating IGF-I and IGFBP-3 concentrations.

Recently, the cross breeding of LID and ALS null mutant mice has been reported.[103,104] Mice with both liver IGF-I (those expressing the albumin promoter driven transgene) and ALS deficiency exhibited significant postnatal growth retardation. The growth retardation is characterized by an ~30% reduction in body weight and length in adults, as well as 15% and 20% reductions in femoral and tibial lengths, respectively. Furthermore, the growth retardation is first observed at about two weeks of age. Detailed evaluations of bone growth were performed in the LID, ALS null and double mutant mice. While femoral total and cortical density is decreased in the double mutants (7 and 10%, respectively), perhaps the most striking findings are the decreases in femoral cortical and periosteal thickness and femoral cross-sectional areas, which were reduced by 35%, 33% and 55%, respectively, compared to normal mice. Smaller reductions in these parameters were also observed in LID and ALS null mutants. In double mutants tibial growth plates also were reduced in size.

As expected with double mutants, circulating IGF-I levels are lower (~10% of normal) than with either null mutation, as are serum IGFBP-3 levels. Free IGF-I is elevated and the half-life of IGF-I appears to be about 25% of normal (~18 min as opposed to ~70 min in normal and liver IGF-I KO mice). Not surprisingly, the absence of ALS appears to destabilize the 150 kDa circulating IGF-I complex because the IGF-I half-life of ALS KO mice is also shortened (~32 min). The straight forward interpretation of these data is that with a sufficient reduction in circulating IGF-I values (10% of normal in these double mutants versus 25-33% of normal in the liver null mutants), the endocrine contribution of IGF-I to postnatal growth is unmasked. Perhaps, a better interpretation is as follows: During postnatal life circulating IGF-I makes a significant contribution to somatic and bone growth. This contribution depends upon the adequacy of circulating ALS and IGFBP-3 concentrations that together serve to maintain a sufficient reservoir of IGF-I, and in the case of IGFBP-3, to facilitate IGF-I translocation to tissues.

Despite the appeal of the above hypothesis, other physiologic alterations occurring in these mice could account for the phenotype of the double mutant mice, and thus, make the above interpretation incorrect in whole or part. Both LID and the double null mutant mice have high circulating GH levels. The GH levels are extraordinarily high in the double null mutants (~15 fold increased over normal, as opposed to a 4.5 fold increase in LID mice). This is presumably secondary to the lower circulating IGF-I levels, and this may represent further evidence of endocrine IGF-I actions (see below). It is reasonable to ask, therefore, whether the elevated GH alters the growth phenotype of these mutant mice. Could the increased GH in LID and double mutants have altered IGF-I expression in multiple tissues sufficiently to influence the conclusions about IGF-I endocrine actions? Put another way, in LID mice could increased IGF-I expression secondary to increased GH signaling compensate for a loss of circulating IGF-I? In double mutants, could the extremely high GH secretion induce GH resistance, which in turn results in decreased IGF-I expression and growth retardation? The data of Yakar et al[105] argues against these possibilities because: a) no differences in IGF-I mRNA were observed among LID, ALS null mutant, double mutants and control mice in bone, kidney, muscle, spleen, lung, heart or fat, and b) no changes in the abundance of GH receptor mRNA were found. Other evidence against GH resistance in states of IGF-I deficient mutant mice comes from the findings of Liu and LeRoith.[105] They found that GH treatment of IGF-I null mutant mice, while not stimulating somatic growth in the absence of IGF-I, increased liver size despite greater then ten fold increases in serum GH. Despite these findings, it is difficult to dismiss the possibility that phenotypic changes result from the extra-ordinarily high GH levels in mutant mice with IGF-I deficiency, at least in some tissues.

Accumulating data shows that intracellular signaling pathways are shared by many receptors, and this likely holds true for the GH receptor and the IGF1R. For example, both can signal through IRS-1.[106] While speculative, it is possible to describe a pathway by which excessive GH exposure blunts IGF-I signaling. A class of signaling proteins, called suppressers of cytokine signaling or SOCS, have been described that are induced by cytokines and can suppress cytokine signaling, thus, providing negative regulation of cytokine action.[107] GH is known to induce SOCS-2 and SOCS-3.[108,109] SOCS-2 can bind to the IGF1R,[110] although the consequences of this interaction have yet to be described. While the evidence is lacking, it seems possible that the high GH secretion in LID and double mutant mice accounts in whole or in part for the growth retardation observed by inducing an IGF-I resistant state.

Other studies in LID mice provide evidence supporting IGF-I endocrine actions. In studies of LID mice carrying the albumin promoter driven Cre transgene, Yakar et al[111] found evidence of insulin insensitivity in muscle. While glucose tolerance was normal in LID mice, serum insulin was elevated after fasting and in response to a glucose load. Furthermore, glucose clearance in response to insulin was abnormal. Because insulin-induced phosphorylation of the insulin receptor and IRS-1 were decreased in muscle, but not in liver or fat, it appears that muscle insulin insensitivity is responsible for the decreased glucose clearance. Furthermore, the elevated GH levels appear to cause the muscle insulin resistance because reduction of circulating GH by administration of a GH-releasing hormone antagonist improves insulin sensitivity. Sjogren et al[112] also reported hyperinsulinemia with normoglycemia in their LID mice carrying the interferon induced Cre transgene. They also found elevations in serum leptin, despite a decrease in fat mass, and increased serum cholesterol.[113] These studies indicate that IGF-I has endocrine roles in carbohydrate and lipid metabolism. Such actions also likely exert an influence on postnatal somatic growth. Finally data from interferon-induced LID mice demonstrates that blood IGF-I exerts negative feedback control on pituitary GH secretion. While the elevated GH secretion observed in both LID mice models suggested an endocrine IGF-I negative feedback mechanism, the studies of Wallenius et al[114] provide more direct evidence of this. They showed that GH release is augmented in response to both GH releasing hormone

(GHRH) and GH secretagogue (GHS), and that this is likely the result of increased mRNA expression of the receptors for both GHRH and GHS.

Taken together, the data from these mutant mouse models make a compelling case for endocrine IGF-I action on growth and metabolism. Nonetheless, it also is clear that we do not yet understand the precise contribution of circulating IGF-I to postnatal somatic growth.

Relative Roles of GH and IGFs in Somatic Growth

While not the only factor that controls circulating IGF-I, GH regulation of liver IGF-I, ALS and IGFBP-3 expression clearly has a major influence on blood IGF-I levels. GH deficient mice usually have blood IGF-I levels that are ~10% of normal (see reviews in refs. 97,98,115). Similarly, patients with GH deficiency often have plasma IGF-I concentrations that are even lower, and acromegaly is often associated with markedly elevated IGF-I levels. The interaction of GH and IGF-I, therefore, may speak to IGF-I endocrine actions. The elegant genetic analyses of the growth-promoting effects of IGF-I and GH reported by Lupu et al[2] address this issue. These investigators analyzed the growth of mice with homozygous null mutations of the IGF-I and GH receptor (GHR) genes, and compared them to mice carrying both null mutations. IGF-I KO mice are ~60% of normal size at birth with the growth deficit beginning in the last third of fetal life (see review of IGF null mutant mice in ref. 1, and Fig. 3, panel A). Their growth deficit is magnified postnatally, such that as adults they are only ~25% of normal weight. In contrast, GHR KO mice are normal in weight at birth and exhibit no growth retardation until about 10 days after birth. Mice carrying both null mutations do not differ in size from IGF KO mice until after 10 days of age, indicating that GH does not regulate growth or IGF-I before early postnatal life, at least in the mouse. If GH is assumed to be the primary regulator of blood-borne IGF-I in well-nourished mice, it follows that virtually all IGF-I stimulated growth prior to the mid-weanling period is due to locally expressed IGF-I. This conclusion is consistent with the low circulating levels of IGF-I during fetal and early postnatal life, and the lack of convincing evidence that GH regulates IGF-I during this developmental time.[42,43]

Lupu et al[2] also calculated the contribution of each IGF-I and GH to adult weight by comparing the adult weights of single and double mutant mice with that of wild type mice (Fig. 3, panel B). IGF-I acting alone could account for 35% of adult weight, while GH alone could account for 14%. IGF-I and GH, however, exhibited an overlapping contribution of 34%. Only 17% of adult weight was not accounted for IGF-I and/or GH stimulation. Because IGF-II KO mice have a major in utero growth deficit[116] beginning at about embryonic day 10.5 and resulting in a 40% reduction in birth weight, it may be responsible for control of the remaining portion of adult weight. The overlapping GH and IGF-I contribution appears to represent the GH-stimulated growth that is mediated by IGF-I. If one accepts this analysis and assumes that IGF-I acts in an endocrine fashion, GH stimulation of circulating IGF-I could account for as much as 34% of adult size in the mouse. Because IGF-I expression is regulated by GH in many tissues[5], such a large contribution from circulating IGF-I seems unlikely. Generation of mouse models with ablation of IGF-I and/or the IGF1R in specific tissues, especially those whose growth is thought to be stimulated by both GH and IGF-I, such as kidneys and muscle, should help to resolve this issue.

Conclusions

The weight of the evidence indicates that IGF-I regulates somatic growth by both local and endocrine IGF-I mechanisms. Defining the contributions of each mode of IGF-I action, however, is problematic. In the mouse it seems likely that IGF-I predominately stimulates growth locally in an autocrine and/or paracrine fashion during in utero and early postnatal life. In the second week of postnatal life when GH begins to have somatic growth promoting action, circulating IGF-I acting as a hormone likely begins to act in an endocrine fashion. Because many trophic agents are capable of stimulating IGF-I expression in specific tissue fashion, e.g., thyroxine, FSH, etc. (see review in ref. 117), local IGF-I actions most likely continue through out life.

Development in man, however, differs significantly from that of rodents. In addition to being much more prolonged, the maturation of organ systems in man differs from that in developing rodents. Many maturational processes that occur in the early postnatal period in mice occur before birth in human fetuses. For these reasons it seems possible that endocrine IGF-I growth-promoting actions commence earlier in man. Regardless of the contribution of blood-borne IGF-I to somatic

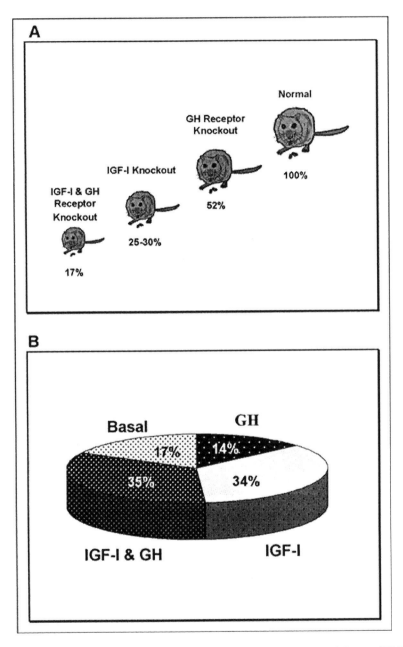

Figure 3. The Contribution of IGF-I and GH to Adult Size. The top panel depicts adult sizes of GHR and IGF-I null mutant mice, and mice carrying both null mutations, compared to a normal mouse. The bottom panel illustrates the percent of adult size that appears to be the result of the growth-stimulation accruing from GH independent of IGF-I (GH), IGF-I independent of GH (IGF-I) and IGF-I stimulated by GH (IGF-I & GH). The data depicted is from Lupu et al.[2]

growth, it is appealing to view IGF-I as a modulator of metabolism. Because circulating IGF-I levels are highly dependent upon nutritional status (see review in ref. 118), it seems possible that IGF-I provides an endocrine signal of the overall nutritional status.

Acknowledgements

The research reported in this review performed in the author's laboratory was supported by grants RO1 HD08299 and NS3891 from NICHD and NIMH, respectively, of the US National Institutes of Health.

References

1. Efstratiadis A. Genetics of mouse growth. Int J Dev Biol 1998; 42:955-976.
2. Lupu F, Terwilliger JD, Lee K et al. Roles of growth hormone and insulin-like growth factor 1 in mouse postnatal growth. Dev Biol 2001; 229:141-162.
3. Salmon WD Jr, Daughaday WH. A hormonally controlled serum factor which stimulates sulfate incorporation by cartilage in vitro. J Lab Clin Med 1957; 116:408-419.
4. Van Wyk JJ. The somatomedins: Biological actions and physiologic control mechanisms. In: Li CH, ed. Hormonal Proteins and peptides: Growth Factors. Orlando: Academic Press, Inc.. 1984:82-127.
5. Daughaday WH, Rotwein P. Insulin-like growth factors I and II. Peptide, messenger ribonucleic acid and gene structures, serum, and tissue concentrations. Endocr Rev 1989; 10:68-91.
6. Daughaday WH, Hall K, Raben MS et al. Somatomedin: proposed designation for sulphation factor. Nature 1972; 235:107.
7. Mayberry HE, Vandenabeele P, Van Wyk JJ et al. Early localization of 125I-labeled growth hormone in adrenals and other tissues of immature hypophysectomized rats. Endocrinology 1971; 88:1309-1317.
8. Uthne K, Uthne T. Influence of liver resection and regeneration on somatomedin (sulfation factor) activity in sera from normal and hypophysectomized rats den. Acta Endocrinol (Copenh) 1972; 71:255-264.
9. Russell WE, D'Ercole AJ, Underwood LE. Somatomedin C/insulinlike growth factor I during liver regeneration in the rat. Am J Physiol 1985; 248:E618-E623.
10. Sjögren K, Liu JL, Blad K et al. Liver-derived insulin-like growth factor I (IGF-I) is the principal source of IGF-I in blood but is not required for postnatal body growth in mice. Proc Natl Acad Sci USA 1999; 96:7088-7092.
11. Yakar S, Liu JL, Stannard B et al. Normal growth and development in the absence of hepatic insulin-like growth factor I. Proc Natl Acad Sci USA 1999; 96:7324-7329.
12. Schwander JC, Hauri C, Zapf J et al. Synthesis and secretion of insulin-like growth factor and its binding protein by the perfused rat liver: dependence on growth hormone status. Endocrinology 1983; 113:297-305.
13. Furlanetto RW, Underwood LE, Van Wyk JJ et al. Estimation of somatomedin-C levels in normals and patients with pituitary disease by radioimmunoassay. J Clin Invest 1977; 60:648-657.
14. Humbel RE. Insulin-like Growth Factors, somatomedins, and multiplication stimulating activity: Chemistry. In: Li CH, ed. Hormonal Proteins and Peptides. 12 ed. New York: Academic Press, 1984:57-79.
15. Rothstein H, Van Wyk JJ, Hayden JH et al. Somatomedin-C: restoration in vivo of cycle traverse in Go/G1 blocked cell of hypophysectomized animals. Science 1980; 208:410-412.
16. Schoenle E, Zapf J , Humbel RE et al. Insulin-like growth factor I stimulates growth in hypophysectomized rats. Nature 1982; 296:252-253.
17. Guler H-P, Wettstein K, Schurr W et al. Recombinant human insulin-like growth factor I: Effects in normal subjects and implications for use in patients. Adv Exper Med Biol 1991; 293:97-104.
18. Froesch ER, Guler H-P, Schmid C et al. Therapeutic potential of insulin-like growth factor I. TEM 1990; 1:254-260.
19. Gluckman PD, Ambler GR. Therapeutic use of insulin-like growth factor I: Lessons from in vivo animal studies. Acta Paediatr Scand 1992; 81(Suppl 383):134-136.
20. D'Ercole AJ, Applewhite GT, Underwood LE. Evidence that somatomedin is synthesized by multiple tissues in the fetus. Dev Biol 1980; 75:315-328.
21. D'Ercole AJ, Stiles AD, Underwood LE. Tissue concentrations of somatomedin C: further evidence for multiple sites of synthesis and paracrine or autocrine mechanisms of action. Proc Natl Acad Sci USA 1984; 81:935-939.
22. Clemmons DR, Underwood LE, Van Wyk JJ. Hormonal control of immunoreactive somatomedin production by cultured human fibroblasts. J Clin Invest 1981; 67:10-19.
23. Clemmons DR, Shaw DS. Variables controlling somatomedin production by cultured human fibroblasts. J Cell Physiol 1983; 115:137-142.
24. Clemmons DR. Multiple hormones stimulate the production of somatomedin by cultured human fibroblasts. J Clin Endocrinol Metab 1984; 58:850-856.
25. Clemmons DR, Shaw DS. Purification and biologic properties of fibroblast somatomedin. J Biol Chem 1986; 261:10293-10298.
26. Clemmons DR, Elgin RG, James PE. Somatomedin-C binding to cultured human fibroblasts is dependent on donor age and culture density. J Clin Endocrinol Metab 1986; 63:996-1001.

27. Clemmons DR, Van Wyk JJ. Evidence for a functional role of endogenously produced somatomedinlike peptides in the regulation of DNA synthesis in cultured human fibroblasts and porcine smooth muscle cells. J Clin Invest 1985; 75:1914-1918.
28. Clemmons DR. Variables controlling the secretion of a somatomedin-like peptide by cultured porcine smooth muscle cells. Circ Res 1985; 56:418-426.
29. Clemmons DR. Exposure to platelet-derived growth factor modulates the porcine aortic smooth muscle cell response to somatomedin- C. Endocrinology 1985; 117:77-83.
30. Jansen M, van Schaik FM, Ricker AT et al. Sequence of cDNA encoding human insulin-like growth factor I precursor. Nature 1983; 306:609-611.
31. Lund PK, Moats-Staats BM, Hynes MA et al. Somatomedin-C/insulin-like growth factor-I and insulin-like growth factor-II mRNAs in rat fetal and adult tissues. J Biol Chem 1986; 261:14539-14544.
32. Han VK, Lund PK, Lee DC et al. Expression of somatomedin/insulin-like growth factor messenger ribonucleic acids in the human fetus: identification, characterization, and tissue distribution. J Clin Endocrinol Metab 1988; 66:422-429.
33. Han VK, D'Ercole AJ, Lund PK. Cellular localization of somatomedin (insulin-like growth factor) messenger RNA in the human fetus. Science 1987; 236:193-197.
34. Lowe WL Jr, Adamo M, Werner H et al. Regulation by fasting of rat insulin-like growth factor I and its receptor. Effects on gene expression and binding. J Clin Invest 1989; 84:619-626.
35. Cerro JA, Grewal A , Wood TL et al. Tissue-specific expression of the insulin-like growth factor binding protein (IGFBP) mRNAs in mouse and rat development. Regul Pept 1993; 48:189-198.
36. Mathews LS, Hammer RE, Behringer RR et al. Growth enhancement of transgenic mice expressing human insulin-like growth factor I. Endocrinology 1988; 123:2827-2833.
37. Behringer RR, Lewin TM, Quaife CJ et al. Expression of insulin-like growth factor I stimulates normal somatic growth in growth hormone-deficient transgenic mice. Endocrinology 1990; 127:1033-1040.
38. D'Ercole AJ. Actions of IGF system proteins from studies of transgenic and gene knockout models. In: Rosenfeld R, Roberts CT Jr, eds. The IGF System: Molecular Biology, Physiology and Clinical Applications. Totowa: Humana Press, Inc., 1999:545-574.
39. Ye P, Carson J, D'Ercole AJ. In vivo actions of Insulin-like Growth Factor-I (IGF-I) on brain myelination: Studies of IGF-I and IGF binding protein-1 (IGFBP-1) transgenic mice. J Neurosci 1995; 15:7344-7356.
40. Liu J-P, Baker J, Perkins AS et al. Mice carrying null mutations of the genes encoding insulin-like growth factor I (Igf-1) and type 1 IGF receptor. Cell 1993; 75:59-72.
41. Baker J, Liu J-P, Robertson EJ et al. Role of insulin-like growth factors in embryonic and postnatal growth. Cell 1993; 75:73-82.
42. Chard T. Hormonal control of growth in the human fetus. J Endocrinol 1989; 123:3-9.
43. Hill DJ. What is the role of growth hormone and related peptides in implantation and the development of the embryo and fetus. Horm Res 1992; 38(Suppl 1):28-34.
44. Mercola M, Stiles CM. Growth Factor Superfamilies and mammalian embryogenesis. Development 1988; 102:451-459.
45. Adamson ED. Growth factors and their receptors in development. Dev Genet 1993; 14:159-164.
46. Ye P, Xing YZ, Dai ZH et al. In vivo actions of insulin-like growth factor-I (IGF-I) on cerebellum development in transgenic mice: Evidence that IGF-I increases proliferation of granule cell progenitors. Dev Brain Res 1996; 95:44-54.
47. Ye P, D'Ercole AJ. Insulin-like growth factor I (IGF-I) regulates IGF binding protein-5 gene expression in the brain. Endocrinology 1998; 139:65-71.
48. Chrysis D, Calikoglu AS, Ye P et al. Insulin-like growth factor-I overexpression attenuates cerebellar apoptosis by altering the expression of Bcl family proteins in a developmentally specific manner. J Neurosci 2001; 21:1481-1489.
49. Dentremont KD, Ye P, D'Ercole AJ et al. Increased insulin-like growth factor-I (IGF-I) expression during early postnatal development differentially increases neuron number and growth in medullary nuclei of the mouse. Dev Brain Res 1999; 114:135-141.
50. Zhao G, Monier-Faugere MC, Langub MC et al. Targeted overexpression of insulin-like growth factor I to osteoblasts of transgenic mice: Increased trabecular bone volume without increased osteoblast proliferation. Endocrinology 2000; 141:2674-2682.
51. Reiss K, Cheng W, Ferber A et al. Overexpression of insulin-like growth factor-1 in the heart is coupled with myocyte proliferation in transgenic mice. Proc Natl Acad Sci USA 1996; 93:8630-8635.
52. Li Q, Li B, Wang X et al. Overexpression of insulin-like growth factor-1 in mice protects from myocyte death after infarction, attenuating ventricular dilation, wall stress, and cardiac hypertrophy. J Clin Invest 1997; 100:1991-1999.
53. Delaughter MC, Taffet GE, Fiorotto ML et al. Local insulin-like growth factor I expression induces physiologic, then pathologic, cardiac hypertrophy in transgenic mice. FASEB J 1999; 13:1923-1929.
54. Neuenschwander S, Schwartz A, Wood TL et al. Involution of the lactating mammary gland is inhibited by the IGF system in a transgenic mouse model. J Clin Invest 1996; 97 :2225-2232.
55. Hadsell DL, Greenberg NM, Fligger JM et al. Targeted expression of des(1-3) human insulin-like growth factor I in transgenic mice influences mammary gland development and IGF- Binding protein expression. Endocrinology 1996; 137:321-330.
56. Hadsell DL, Alexeenko T, Klemintidis Y et al. Inability of overexpressed des(1-3)human insulin-like growth factor I (IGF-I) to inhibit forced mammary gland involution is associated with decreased expression of IGF signaling molecules. Endocrinology 2001; 142:1479-1488.

57. Ruan WF, Kleinberg DL. Insulin-like growth factor I is essential for terminal end bud formation and ductal morphogenesis during mammary development. Endocrinology 1999; 140:5075-5081.
58. Weber MS, Boyle PL , Corl BA et al. Expression of ovine insulin-like growth factor-1 (IGF-1) stimulates alveolar bud development in mammary glands of transgenic mice. Endocrine 1998; 8:251-259.
59. Shirke S, Faber SC , Hallem E et al. Misexpression of IGF-I in the mouse lens expands the transitional zone and perturbs lens polarization. Mech Dev 2001; 101:167-174.
60. Coleman ME, DeMayo F, Yin KC et al. Myogenic vector expression of insulin-like growth factor 1 stimulates muscle cell differentiation and myofiber hypertrophy in transgenic mice. J Biol Chem 1995; 270:12109-12116.
61. Renganathan M, Messi ML, Schwartz R et al. Overexpression of hIGF-1 exclusively in skeletal muscle increases the number of dihydropyridine receptors in adult transgenic mice. FEBS Letters 1997; 417:13-16.
62. Renganathan M, Messi ML, Delbono O. Overexpression of IGF-1 exclusively in skeletal muscle prevents age-related decline in the number of dihydropyridine receptors. J Biol Chem 1998; 273:28845-28851.
63. Zheng ZL, Messi ML , Delbono O. Age-dependent IGF-1 regulation of gene transcription of Ca^{2+} channels in skeletal muscle. Mech Ageing Dev 2001; 122:373-384.
64. Criswell DS, Booth FW, DeMayo F et al. Overexpression of IGF-I in skeletal muscle of transgenic mice does not prevent unloading-induced atrophy. Am J Physiol: Endocrinol Metab 1998; 275:E373-E379.
65. Chakravarthy MV, Abraha TW, Schwartz RJ et al. Insulin-like growth factor-I extends in vitro replicative life span of skeletal muscle satellite cells by enhancing G_1/S cell cycle progression via the activation of phosphatidylinositol 3'-kinase/Akt signaling pathway. J Biol Chem 2000; 275:35942-35952.
66. Chakravarthy MV, Fiorotto ML, Schwartz RJ et al. Long-term insulin-like growth factor-1 expression in skeletal muscles attenuates the enhanced in vitro proliferation ability of the resident satellite cells in transgenic mice. Mech Ageing Dev 2001; 122:1303-1320.
67. Musarò A, McCullagh K, Paul A et al. Localized Igf-1 transgene expression sustains hypertrophy and regeneration in senescent skeletal muscle. Nature Genet 2001; 27:195-200.
68. Wang JW, Niu W, Nikiiforov Y et al. Targeted overexpression of IGF-I evokes distinct patterns of organ remodeling in smooth muscle cell tissue beds in transgenic mice. J Clin Invest 1997; 100:1425-1439.
69. Zhao GS, Sutliff RL, Weber CS et al. Smooth muscle-targeted overexpression of insulin-like growth factor I results in enhanced vascular contractility. Endocrinology 2001; 142:623-632.
70. Zhu BH, Zhao GS, Witte DP et al. Targeted overexpression of IGF-I in smooth muscle cells of transgenic mice enhances neointimal formation through increased proliferation and cell migration after intraarterial injury. Endocrinology 2001; 142:3598-3606.
71. Dyck MK, Parlow AF , Sénéchal JF et al. Ovarian expression of human insulin-like growth factor-I in transgenic mice results in cyst formation. Mol Reprod Develop 2001; 59:178-185.
72. DiGiovanni J, Kiguchi K, Frijhoff A et al. Deregulated expression of insulin-like growth factor 1 in prostate epithelium leads to neoplasia in transgenic mice. Proc Natl Acad Sci USA 2000; 97:3455-3460.
73. Bol DK, Kiguchi K, Gimenez-Conti I et al. Overexpression of insulin-like growth factor-1 induces hyperplasia, dermal abnormalities, and spontaneous tumor formation in transgenic mice. Oncogene 1997; 14:1725-1734.
74. Su HY, Hickford JGH, The PHB et al. Increased vibrissa growth in transgenic mice expressing insulin-like growth factor 1. J Invest Dermatol 1999; 112:245-248.
75. Dyck MK, Ouellet M , Gagné M et al. Testes-specific transgene expression in insulin-like growth factor-I transgenic mice. Mol Reprod Develop 1999; 54:32-42.
76. Clément S, Refetoff S, Robaye B et al. Low TSH requirement and goiter in transgenic mice overexpressing IGF-I and IGF-I receptor in the thyroid gland. Endocrinology 2001; 142:5131-5139.
77. Wang J, Niu W, Witte DP et al. Overexpression of Insulin-like Growth Factor Binding protein-4 (IGFBP-4) in smooth muscle cells of transgenic mice through a smooth muscle Actin -IGFBP-4 fusion gene induces smooth muscle hypoplasia. Endocrinology 1998; 139:2605-2614.
78. D'Ercole AJ, Dai Z , Xing Y et al. Brain growth retardation due to the expression of human insulin-like growth factor binding protein-1 (IGFBP-1) in transgenic mice: An in vivo model for analysis of IGF function in the brain. Dev Brain Res 1994; 82:213-222.
79. Murphy LJ, Rajkumar K, Molnar P. Phenotypic manifestations of insulin-like growth factor binding protein-1 (IGFBP-1) and IGFBP-3 overexpression in transgenic mice. Prog Growth Factor Res 1996; 6:425-432.
80. Ni W, Rajkumar K, Nagy JI et al. Impaired brain development and reduced astrocyte response to injury in transgenic mice expressing IGF binding protein-1. Brain Res 1997; 769:97-107.
81. Stiles AD, Sosenko IR, D'Ercole AJ et al. Relation of kidney tissue somatomedin-C/insulin-like growth factor I to postnephrectomy renal growth in the rat. Endocrinology 1985; 117:2397-2401.
82. Hammerman MR, Miller SB. Effects of growth hormone and insulin-like growth factor I on renal growth and function. J Pediatr 1997; 131(Suppl):S17-S19.
83. D'Ercole AJ, Ye P, Calikoglu AS et al. The role of the insulin-like growth factors in the central nervous system. Mol Neurobiol 1996; 13:227-255.
84. Folli F, Ghidella S, Bonfanti L et al. The early intracellular signaling pathway for the insulin/insulin- like growth factor receptor family in the mammalian central nervous system. Mol Neurobiol 1996; 13:155-183.
85. Anlar B, Sullivan KA, Feldman EL. Insulin-like growth factor-I and central nervous system development. Horm and Metabol Res 1999; 31:120-125.
86. Johnston BM, Mallard EC, Williams CE et al. Insulin-like growth factor-1 is a potent neuronal rescue agent after hypoxic-ischemic injury in fetal lambs. J Clin Invest 1996; 97:300-308.

87. Guan J, Williams CE, Skinner SJM et al. The effects of insulin-like growth factor (IGF)-I, IGF-2, and des-IGF-I on neuronal loss after hypoxic-ischemic brain injury in adult rats: Evidence for a role for IGF binding proteins. Endocrinology 1996; 137:893-898.
88. Guan J, Bennet L, George S et al. Selective neuroprotective effects with insulin-like growth factor-1 in phenotypic striatal neurons following ischemic brain injury in fetal sheep. Neurosci 2000; 95:831-839.
89. Wang JM, Hayashi T, Zhang WR et al. Reduction of ischemic brain injury by topical application of insulin-like growth factor-I after transient middle cerebral artery occlusion in rats. Brain Res 2000; 859:381-385.
90. Mason JL, Ye P, Suzuki K et al. Insulin-like growth factor-1 inhibits mature oligodendrocyte apoptosis during primary demyelination. J Neurosci 2000; 20:5703-5708.
91. Delafontaine P. Insulin-like growth factor I and its binding proteins in the cardiovascular system. Cardiovasc Res 1995; 30:825-834.
92. Ren J, Samson WK, Sowers JR. Insulin-like growth factor I as a cardiac hormone: Physiological and pathophysiological implications in heart disease. J Mol Cell Cardiol 1999; 31:2049-2061.
93. Shim M, Cohen P. IGFs and human cancer: Implications regarding the risk of growth hormone therapy. Hormone Res 1999; 51:42-51.
94. Cohen P, Clemmons DR, Rosenfeld R. Does the GH-IGF axis play a role in cancer pathogenesis? GH & IGF Res 2000; 10:297-305.
95. Bondy CA, Underwood LE, Clemmons DR et al. Clinical uses of insulin-like growth factor I. Ann Intern Med 1994; 120:593-601.
96. Gluckman PD, Douglas RG, Ambler GR et al. The endocrine role of insulin-like growth factor I. Acta Paediatr Scand 1991; 80(Suppl 372):97-105.
97. LeRoith D, Clemmons D, Nissley P et al. Insulin-like growth factors in health and disease. Ann Intern Med 1992; 116:854-862.
98. Cohick WS, Clemmons DR. The insulin-like growth factors. Annu Rev Physiol 1993; 55:131-153.
99. Clemmons DR. Role of insulin-like growth factor binding proteins in controlling IGF actions. Mol Cell Endocrinol 1998; 140:19-24.
100. Baxter RC. Insulin-like growth factor (IGF)-binding proteins: interactions with IGFs and intrinsic bioactivities. Am J Physiol: Endocrinol Metabol 2000; 278:E967-E976.
101. Schwenk F, Baron U, Rajewsky K. A *cre*-transgenic mouse strain for the ubiquitous deletion of *loxP*-flanked gene segments including deletion in germ cells. Nucleic Acids Res 1995; 23:5080-5081.
102. Ueki I, Ooi GT, Tremblay ML et al. Inactivation of the acid labile subunit gene in mice results in mild retardation of postnatal growth despite profound disruptions in the circulating insulin-like growth factor system. Proc Natl Acad Sci USA 2000; 97:6868-6873.
103. Yakar S, Ooi GT, Boisclair YR et al. Inactivation of both liver IGF-1 and the acid labile subunit genes cause a marked reduction in serum IGF-1 and postnatal growth retardation. Programs & Abstracts of Annual Meeting of the Endocrine Society. 2001:75.
104. Yakar S, Rosen CJ, Beamer W et al. Circulating levels of insulin-like growth factor-I directly regulate bone growth and density. J Clin Invest 2002; 110:771-781.
105. Liu JL, LeRoith D. Insulin-like growth factor I is essential for postnatal growth in response to growth hormone. Endocrinology 1999; 140:5178-5184.
106. Smit LA, Meyer DJ et al. Molecular events in growth hormone-receptor interaction and signaling. In: Kostyo J, ed. Handbook of Physiology; Section 7: The Endocrine System 1999:445-480.
107. Krebs DL, Hilton DJ. SOCS proteins: negative regulators of cytokine signaling. Stem Cells 2001; 19:378-387.
108. Tollet-Egnell P, Flores-Morales A, Stavréus-Evers A et al. Growth hormone regulation of SOCS-2, SOCS-3, and CIS messenger ribonucleic acid expression in the rat. Endocrinology 1999; 140:3693-3704.
109. Davey HW, McLachlan MJ, Wilkins RJ et al. STAT5b mediates the GH-induced expression of SOCS-2 and SOCS-3 mRNA in the liver. Mol Cell Endocrinol 1999; 158:111-116.
110. Dey BR, Spence SL, Nissley P et al. Interaction of human suppressor of cytokine signaling (SOCS)-2 with the insulin-like growth factor-I receptor. J Biol Chem 1998; 273:24095-24101.
111. Yakar S, Liu JL, Fernandez AM et al. Liver-specific *igf-1* gene deletion leads to muscle insulin insensitivity. Diabetes 2001; 50:1110-1118.
112. Sjögren K, Wallenius K, Liu JL et al. Liver-derived IGF-I is of importance for normal carbohydrate and lipid metabolism. Diabetes 2001; 50:1539-1545.
113. Isaksson OGP, Jansson JO, Sjögren K et al. Metabolic functions of liver-derived (endocrine) insulin-like growth factor I. Hormone Res 2001; 55:18-21.
114. Wallenius K, Sjögren K, Peng XD et al. Liver-derived IGF-I regulates GH secretion at the pituitary level in mice. Endocrinology 2001; 142:4762-4770.
115. Rotwein P. Structure, evolution, expression and regulation of insulin- like growth factors I and II. Growth Factors 1991; 5:3-18.
116. DeChiara TM, Efstratiadis A, Robertson EJ. A growth-deficiency phenotype in heterozygous mice carrying an insulin-like growth factor II gene disrupted by targeting. Nature 1990; 345:78-80.
117. Rotwein P, Bichell DP, Kikuchi K. Multifactorial regulation of IGF-I gene expression. Mol Reprod Dev 1993; 35:358-364.
118. Thissen J-P, Ketelslegers J-M, Underwood LE. Nutritional regulation of the insulin-like growth factors. Endocr Rev 1994; 15:80-101

CHAPTER 8

Insulin-Like Growth Factor 1 (IGF1) and Brain Development

Carolyn A. Bondy, Wei-Hua Lee and Clara M. Cheng

Abstract

During the course of brain development, insulin-like growth factor (IGF1) and the IGF1 receptor are most highly expressed by maturing projection neurons during a time of rapid process growth and synaptogenesis. During this time, IGF1-expressing neurons grow the most extensive dendritic arbors found in the brain, forming the circuitry for complex information processing and higher level brain functions. IGF1 deletion results in reduced neuronal size, dendrite growth, branching and synapse formation and a selective decrease in dentate gyrus and olfactory neuron survival. Glucose utilization is reduced by ~30-60% in the developing IGF1 null brain, with the greatest decrease in structures where IGF1 expression is normally highest. IGF1 promotes neuronal glucose utilization and growth by activation of PI3K/Akt(PKB)/GSK3β signaling pathways. IGF1 null neurons demonstrate reduced GLUT4 expression, reduced hexokinase activity and reduced glycogen synthesis contributing to reduced growth. Since GSK3β, in addition to inhibiting glycogen and protein synthesis, promotes apoptosis, GSK3β over-activity may contribute to the loss of dentate and olfactory neurons as well as reduced neuronal growth in the IGF1 null brain. Oligodendrocyte differentiation is normal and oligodendrocyte numbers reflect neuronal numbers in the IGF1 null mouse brain. Myelin content is reduced in IGF1 null and increased in IGF1 over-expressing transgenic brains in proportion to changes in neuroaxonal mass. Taken together with the fact that IGF1 and IGF1 receptor expression are predominantly neuronal, these findings suggest that IGF1's effects on developmental myelination are secondary to its direct effects on neuronal growth and survival.

In summary, IGF1 promotes neuronal growth by 'insulin-like' anabolic effects. IGF1 promotes neuronal survival during normal brain development mainly in hippocampal and olfactory systems that depend on postnatal neurogenesis. IGF1's anabolic and neuroprotective roles may be coordinated by inhibition of GSK3β. IGF1's effects on developmental myelination are secondary to its effects on neuroaxonal mass. While these observations were derived from murine models, *IGF1* gene deletion in humans is also associated with reduced brain growth and mental retardation, suggesting IGF1 has a similar role in human brain development.

Introduction

Insulin-like growth factor 1 (IGF1) and the IGF1 receptor are highly expressed in the developing brain. The particular timing and cellular pattern of IGF1 expression in most brain structures suggests that it is involved in neuronal process growth, synaptogenesis and/or myelination.[1] IGF1 has pleiotrophic effects on cultured neural tissue, where, depending on the cell type and experimental conditions, it may promote proliferation, survival or differentiation (reviewed in refs. 2, 3). IGF1's actual effects in vivo are, however, determined by when and where the peptide and its receptor are expressed. To elucidate IGF1's role in brain development, this article reviews in vivo IGF system expression patterns and functional correlations together with observation on brain development in IGF1 null and over-expressing transgenic mice.

Insulin-Like Growth Factors, edited by Derek LeRoith, Walter Zumkeller and Robert Baxter. ©2003 Eurekah.com and Kluwer Academic / Plenum Publishers.

IGF System Expression in the Brain

Normal Brain Development

The IGF1 receptor is highly expressed in the developing nervous system from the time of neural tube formation.[4] As differentiated neurons emerge and mature brain structures form, IGF1 receptor mRNA is preferentially concentrated in neurons as opposed to glial components of the central nervous system (CNS).[5] The IGF1 receptor mRNA level per neuron increases in IGF1-expressing neurons postnatally.[5] IGF1 receptor mRNA decreases as a percent of total brain RNA in older animals due to the increasing abundance of glial cells, which express relatively little IGF1 receptor mRNA, during the course of development.[6] IGF1 receptor expression is very widespread, in general reflecting neuronal density throughout the brain. IGF1 binding sites generally coincide with IGF1 receptor mRNA concentrations in the brain[7-11] suggesting that the receptor is expressed on neuronal soma and dendrites.

In contrast to the IGF1 receptor, IGF1 expression in the brain is neuroanatomically focal and demonstrates strong developmental regulation. IGF1 mRNA is relatively scarce prenatally and increases rather dramatically in the early postnatal period, followed by a decline several weeks after birth.[1,12-15] IGF1 mRNA is most abundant in growing projection neurons belonging to the cerebellar and sensory relay systems.[1] The Purkinje cell is the prototypical IGF1-expressing neuron. In maturity, this primary neuron of the cerebellar cortex has a huge soma and an incredibly branched, highly complex dendritic arbor, resulting in the largest surface area of any cell type in mammals. IGF1 mRNA is first detected in Purkinje cells perinatally, after the postmitotic neurons have migrated to the cerebellar plate.[1] IGF1 mRNA levels increase until about postnatal day 10 (P10) and remain stable through approximately P20 after which there is a dramatic decline. IGF1 receptor expression increases in Purkinje cells in parallel with IGF1 expression (Fig. 1). Peak IGF1 expression in Purkinje cells and other cerebellar and sensory projection neurons coincides developmentally with somatic and dendritic growth and synaptogenesis.[1] IGF1 immunoreactivity is localized in Purkinje perikarya and processes and increases almost immediately after secretion is blocked with colchicine,[16] suggesting that the peptide is released from processes relatively close to the perikarya. The fact that IGF1 binding sites and IGF1 immunoreactivity are both concentrated in the immediate vicinity of neurons synthesizing the receptor and peptide strongly suggests a local, autocrine or paracrine mode of action for neuronal IGF1.

To elucidate IGF1's role in brain development, we have focused mainly on the IGF1-expressing, large projection neuron systems. IGF1 expression in these systems has been confirmed repeatedly by different workers using different methodologies, and thus is well-established. Moreover, these neurons have similar, well-defined morphological and functional phenotypes apt to provide important insights into IGF1's role in neuronal development. IGF1 mRNA is detected in a few other cell types that don't have such well-defined functions or connections. For example, IGF1 mRNA is localized in scattered cells in the developing hippocampal formation during postnatal development,[17] but this cell type remains unknown as is IGF1's function in these cells. IGF1 mRNA has also been detected in the anterolateral ventricle subventricular zone[18] during early postnatal development, suggesting a potential role in promoting proliferation of neurons or glial cells originating in this zone.

As shown in Figure 1, IGF2 is also synthesized within the brain, for the most part by adventitial structures such as the leptomeninges and choroid plexii.[5,12,19,20] Some investigators have detected IGF2 mRNA in cerebellar granule cells during early postnatal development.[21] These authors have also noted that IGF2 gene expression in the meninges and choroids originates from both parental alleles, in contrast to the situation for most tissues, where IGF2 expression is derived solely from the paternal allele. We have noted an abundance of IGF2 anti-sense transcripts in the brain (Wang J and Bondy C, unpublished data) and wonder whether much of brain IGF2 mRNA may be untranslated. In any case, IGF2's role in the brain remains unclear, and there is no apparent effect of IGF2 deletion on brain morphology, chemistry or neurological function (Reinhardt R and Bondy C, unpublished data).

IGFBPs

IGFs 1 and 2 bind with high affinity to a number of IGF binding proteins (IGFBPs), which may protect the IGFs from proteolysis and modulate their interaction with their receptor.[22] IGFBPs are

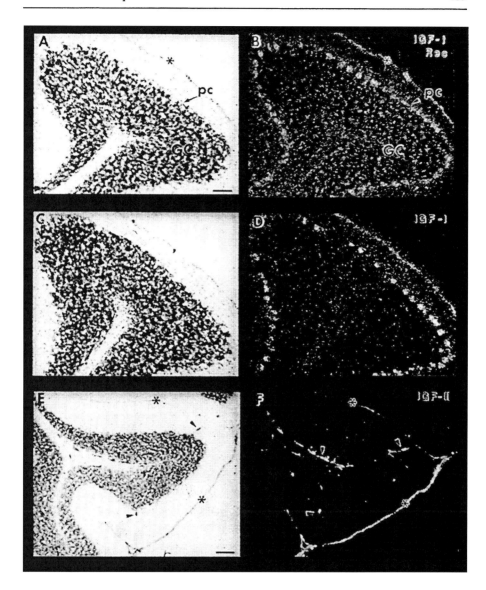

Figure 1. IGF system expression in the cerebellar cortex. Paired bright and dark field micrographs show that IGF1 receptor (A&B) and IGF1 (C&D) mRNAs are co-localized in Purkinje cells (pc). IGF1 receptor mRNA is also found in granule cells (gc). IGF2 mRNA in contrast is concentrated in non-neuronal elements including leptomeninges (asterisks) and blood vessels (arrowheads).

expressed in brain in addition to diverse peripheral tissues.[17,23-28] IGFBP2 and 5 are clearly the most abundant in brain, and are expressed in spatiotemporal co-ordination with IGF1 (Fig. 2). Early in development IGFBP5 mRNA is concentrated in germinal zones and is co-localized with IGF1 in developing sensory and cerebellar relay neurons.[27] As might be expected from this close association, IGF1 appears to induce IGFBP5 expression.[23] IGFBP2 mRNA is concentrated in astroglia adjacent to IGF1-expressing neurons[17] and co-localizes with IGF2 in the meninges and choroid plexei.[24] IGFBP2 is also highly abundant in capillary endothelium, median eminence and other circumventricular sites,[17] suggesting a potential role in carrier-mediated transcytosis of circulating

IGFs into the brain. Thus, each IGFBP may play a specific role in modulating IGF1's bioactivity in brain development.

IGF System Expression in Brain Injury

During normal brain development, IGF1 gene expression is primarily or exclusively neuronal and reduced after neuronal maturation, except in the olfactory system.[1] During the early postnatal period, hypoxic-ischemic insult decreases neuronal IGF-I expression abruptly.[29] If exogenous IGF-I is administered within two hours, neuronal injury can be alleviated,[30] suggesting IGF1 helps developing neurons to survive oxygen and/or glucose deprivation. This view is supported by in vitro studies.[31] Although IGF1 expression is reduced in mature neurons, it is induced in reactive astrocytes and microglia in response to a variety of insults, including ischemic, traumatic and chemotoxic injury.[26,32-38] The same types of insult also induce the expression of IGF2, IGFBP2 and other IGFBPs.[26,39] The specific timing and cellular patterns of IGF and IGFBP expression following brain insult suggest that these factors play specific roles in remodeling processes following acute brain injury.

IGF1, Insulin and the Blood Brain Barrier

Very little insulin is synthesized within the brain[40] and much of the material earlier thought to be 'brain' insulin, reviewed in[41] probably represented insulin derived from the circulation. In situ analysis has identified insulin mRNA in a few small periventricular cells in the anterior hypothalamus[42] where insulin may play a role in the neuroendocrine systems regulating appetite, but seems quite unlikely to be involved in regulation of substrate homeostasis in diverse brain regions. The insulin receptor, however, is widely expressed in brain in a pattern that overlaps that of the IGF1 receptor.[11] The significance of this widespread expression of the insulin receptor in brain is unclear. Targeted deletion of brain insulin receptor gene expression does not affect brain development or neuronal survival, but results in alterations in appetite and weight control,[43] supporting the view that insulin serves as a neuroendocrine signal to brain concerning nutritional status.

Both IGF1 and insulin receptors are abundant in choroid plexii, meninges and vascular elements of the brain, including the periventricular structures[5,44] thought to be important for transporting systemic molecules across the blood brain barrier (BBB). We have compared transport of IGF1 and insulin across the BBB using an in vivo carotid artery perfusion system and found that 3-fold more radio-labeled IGF1 accumulates in the cerebral cortex and almost 10 times more IGF1 accumulates in the hypothalamic paraventricular nuclei compared with insulin.[45] These findings are consistent with studies suggesting that augmentation of circulating IGF1 levels promotes brain growth and repair processes.[46-48] A number of factors may explain IGF1's relative facility in crossing the BBB. The co-expression of insulin and IGF1 receptors in brain capillary endothelium may result in formation of hybrid receptors, which bind IGF1 with substantially greater affinity than insulin.[49] In addition as noted above, IGFBP 2 is abundant in capillary endothelium, median eminence and other circumventricular sites[17] suggesting possible carrier-mediated IGF transport across the BBB.

IGF1 Promotes Neuronal Glucose Utilization and Growth

Regional glucose utilization as measured by 14C-2-deoxyglucose uptake (2DGU) parallels IGF1 and IGF1 receptor gene expression in the developing murine brain (Figs. 3&4). High level IGF1 expression is seen in concert with intense 2DGU in maturing cerebellar, somatosensory, auditory-vestibular, olfactory and visual system neurons. Generally speaking, IGF1 expression is most abundant after birth, except for early maturing structures such as the olfactory bulb, where IGF1 expression is abundant several days before birth, concordant with the precocious initiation of glucose utilization by this structure[1] (Fig. 3). As noted above, IGF1 and other elements of the IGF system are strongly induced in response to diverse types of CNS injury. Interestingly, this injury-invoked IGF1 expression is generally in astrocytes,[32-36] where it is also strongly correlated with local 2DGU.[50] In this setting the reactive astroglia are involved in scar formation with high level metabolic needs for synthesis and secretion of matrix proteins.

Targeted IGF1 gene deletion results in impaired glucose utilization by the developing brain.[50] Regional 2DGU patterns parallel IGF1 expression in the developing wildtype (WT) mouse brain (Fig. 4) and, as expected, 2DGU is significantly reduced in the IGF1 null brain, most profoundly in those structures where IGF1 is highly expressed in the WT. For example, 2DGU is reduced by more

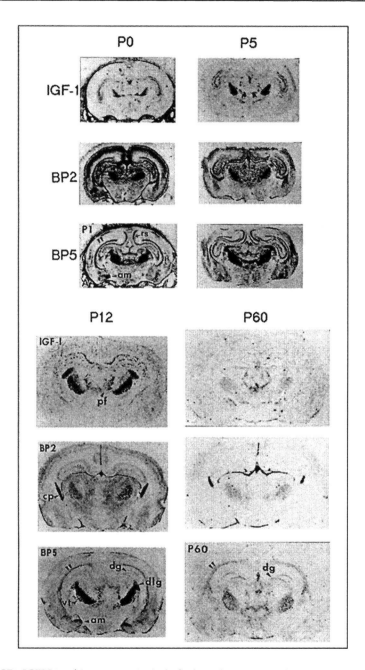

Figure 2. IGF1, IGFBP2 and 5 gene expression in the forebrain during postnatal brain development, shown by film autoradiography of anatomically matched forebrain sections from postnatal day 0, 5, 12 and 60 rats. IGF1 mRNA is concentrated in the ventrobasilar, geniculate, intralaminar and parafascilular nuclei of the anterior thalamus, with expression peaking between days 7 and 14 and receding by the beginning of the 4th week of life. IGFBP2 and 5 expression parallel IGF1 in these thalamic nuclei, but the binding proteins are more widely expressed in the developing cerebral cortex. IGFBP2 is also expressed in the choroid plexei (CP) and meninges. IGFBP5 is also expressed in white matter, presumably by oligodendrocytes (double arrowheards).

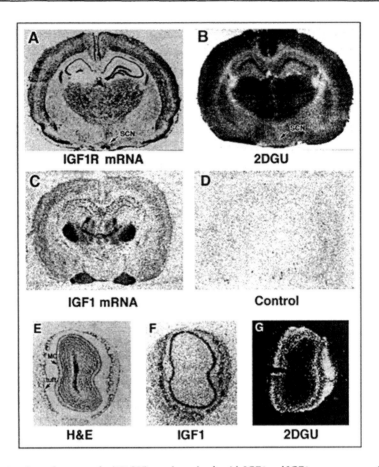

Figure 3. 2-deoxyglucose uptake (2DGU) correlates closely with IGF1 and IGF1 receptor expression in the normal developing brain. Panels A-D are film autoradiographs from anatomically matched sections from the anterior forebrain from P14 rats. Panels E-G show the rat olfactory bulb in hematoxylin/eosin-stained view (E), with F showing the film autoradiograph of IGF1 mRNA from the same section shown in E. Panel G shows a dark field micrograph of 14C-2DGU in an anatomically matched olfactory bulb section. IGF1 mRNA is concentrated in the olfactory projection neurons (mitral and tufted cells) and 2DGU is concentrated in their dendritic synaptic fields.

than 50% in thalamic and brain stem sensory nuclei (Fig. 4). Glucose uptake is also reduced in isolated nerve terminals, or synaptosomes, prepared from IGF1 null brains and addition of IGF1 to the incubation medium normalizes the defect.[50] Thus, the defective glucose utilization observed in the IGF1 null brain in vivo is associated with reduced 2DGU at the nerve terminal level in vitro, which is completely reversed by IGF1. These in vitro findings show that the defect in glucose utilization seen in the IGF1null brain in vivo is not due to reduced neural activity or reduced brain blood flow, neither of which affects the synaptosome preparation. Furthermore, the finding of reduced glucose uptake in isolated neuron terminals shows that IGF1 normally promotes glucose uptake by terminals independent of glial effects, since glia are not present in the synaptosome fraction.

Interestingly, brain 2DGU is globally increased in IGF1 over-expressing adult mice.[51] It is not certain which cell types are responsible for the ectopic IGF1 expression in these mice, but apparently IGF1 is in excess through most of the brain for much of development. Thus, the generalized increase in 2DGU likely reflects local field potentials originating from more highly ramified dendritic arbors with greater synaptic density in IGF1-over-expressing brains. The fact that pentobarbital anesthesia

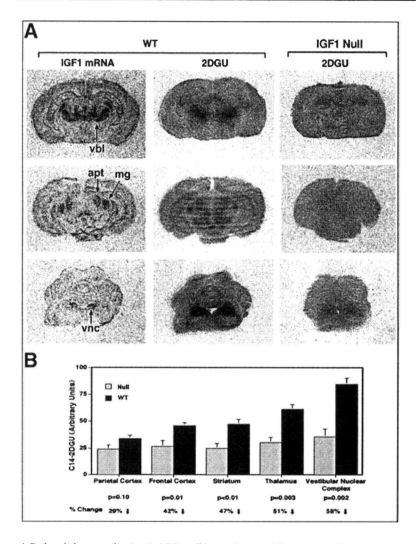

Figure 4. Reduced glucose utilization in IGF1 null brains (postnatal day 10). The 1st column in (A) shows IGF1 mRNA expression in WT fore-, mid- and hindbrain. The 2nd column shows 14C-2-deoxyglucose (2DGU) in these same brain regions in WT mice, and the 3rd column shows 2DGU in anatomically-matched brain sections from IGF1 null mice. Quantitative comparisons of 2DGU in select brain regions for the 2 groups are shown in (B). Adapted from reference 50.

abolished the differences in glucose uptake between transgenic and WT mice in that study was suggested by the authors to prove that IGF1 does not promote brain glucose utilization. However, pentobarbital interferes with glucose transporter function per se[52] and thus may obscure the study of the regulation of glucose transport. Furthermore, we have suggested that IGF1, over a time frame of days to weeks, promotes glucose uptake and use by developing neurons, allowing them to grow larger and develop a more complex dendritic architecture. In maturity, these neurons will clearly be capable of more complex information-processing activity, associated with increased glucose utilization. We have not suggested that the rapid, transient alterations in glucose uptake associated with neural activity in the mature brain are due to immediate effects by IGF1. These effects are related to changes in cerebral blood flow and neurotransmitter metabolism occurring in a matter of milliseconds,

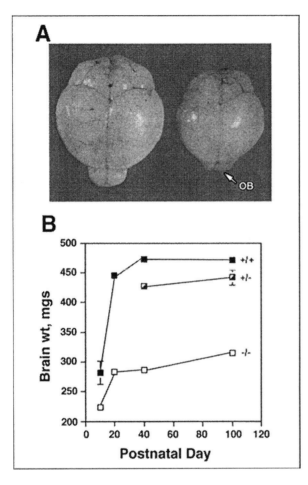

Figure 5. Brain growth in WT and IGF1 knockout mice mice. Panel A compares brains from representative WT and IGF1 null mice at P40. Note the hypoplastic olfactory bulb in the IGF1 null brain, but otherwise brain gross anatomy appears normal. Brain weights are compared graphically in panel B. The homozygous IGF1 deletion results in significantly reduced brain size from P10 onward. The heterozygous IGF1 deletion results in significant reduction in brain size at P40 and P100. Adapted from reference 75.

and involve catabolic rather than anabolic pathways associated with IGF1 activity (i.e., glycogenoloysis and glycolysis). Moreover, there is very little IGF1 expressed in the normal mature brain. We do speculate, however, that continued or repetitive activity in a specific brain region or structure may well induce a local increase in IGF1 expression, associated with anabolic activities in support of synaptic remodeling.

IGF1 deficiency results in decreased postnatal brain growth (Fig. 5). This effect is clearly more profound in the nullizygous state, but even partial IGF1 deficiency, as in IGF1 +/- mice, results in significantly diminished brain growth, even though these mice are not significantly smaller than wildtype.[53] Heterozygous brains are ~10% smaller than WT at P40 (P<0.0001). Despite the reduction in size, IGF1 null brain anatomy and cell numbers are for the most part normal, with the notable exceptions of the dentate gyrus and the olfactory bulb, which are discussed in a later section. Neuronal numbers in the cerebellum, thalamus, neocortex and brain stem, however, are normal in the IGF1 null brain. Most of the 30-40% reduction in brain size in adult IGF1 null mice is due to a reduction in neuropil, or neuronal processes. Cell density is significantly increased in the IGF1 null brain (Fig. 6 and refs. 54,55,56) suggesting decreased process growth, since the space between neurons is normally occupied by extensively branched neuronal processes. The cross sectional area of projection neurons in the IGF1 null brain is reduced by ~25%, and nerve process length and branching are reduced by a similar amount (Fig. 6).[56] Thus it appears that IGF1 deficiency impairs neuronal somatic growth and process formation, accounting in large part for the reduction in IGF1 null brain size. Supporting these in vivo findings, a recent study using IGF1 treatment of cortical

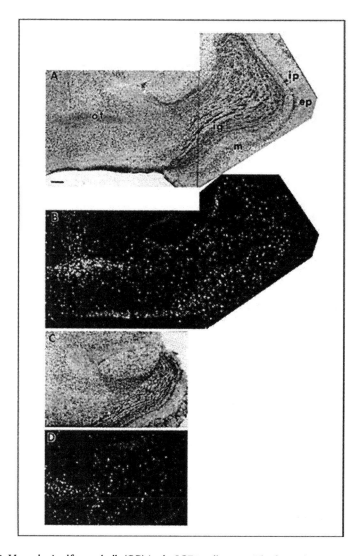

Figure 10. Hypoplastic olfactory bulb (OB) in the IGF1 null mouse. The figure shows paired bright- and dark field views of OB sections from representative WT (A&B) and IGF1 null (C&D) mice at P40. The IGF1 null OB is only about 25% the size of the WT, and has greatly reduced numbers of projection and local neurons. It also has reduced numbers of oligodendrocytes, which are seen in the dark-field expressing the oligo-specific marker PLP.

expressed by mitral and tufted neurons, but not by granule cells (Fig. 3 and ref. 1). IGF1 expression has been reported in the anterolateral subventricular germinal zone, however, where olfactory granule cells originate,[14] so it is possible that IGF1 promotes granule cell proliferation or survival at their point of origin. Alternatively, it is possible that the number of olfactory granule cells is dependent upon the number of olfactory projection neurons, since the small interneurons exist to modulate neurotransmission by the mitral and tufted projection neurons. Similarly, the number of oligodendrocytes, which serve to myelinate projection neuron processes, is also substantially reduced in the olfactory system (Fig. 10 and ref. 55).

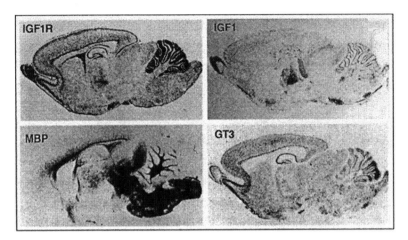

Figure 11. Dissociation between IGF system expression and myelination during postnatal brain development. These are film autoradiograph taken from serial sagittal P12 rat brain sections, comparing mRNA localization for the IGF1 receptor (IGF1R), IGF1, myelin basic protein (MBP), which is an oligodendrocyte-specific signal, and GLUT3, a neuron-specific glucose transporter. There is almost a negative or inverse relation between IGF1 receptor expression and myelination, as reflected by MBP expression. In contrast, there is a relatively close correlation between IGF1 receptor and GLUT3 expression.

IGF1 and Brain Myelination

IGF1 has been attributed a primary role in brain myelination.[81] This view originated with in vitro studies showing that IGF1 promotes oligodendrocyte proliferation and differentiation[82] and was bolstered by the finding of increased myelin in transgenic brains over-expressing IGF1 regulated by a metallothionine promoter.[83] The view that IGF1 has a direct role in oligodendrocyte development and myelination is not supported by in vivo IGF system expression patterns, where both IGF1 and its receptor are concentrated in neurons rather than oligodendrocytes, as illustrated in Figure 11. Moreover, there is no sign of central or peripheral myelinopathy in IGF1 null mice.[55,84] Myelin concentration, normalized to brain weight or protein, is equal in IGF1 null and wild type littermate mice. Likewise, concentrations of myelin-specific proteins (MBP, PLP, MAG and CNPase) and their corresponding mRNAs are equal in IGF1 and wild type littermate mice (Fig. 12). Oligodendrocyte numbers and myelin are reduced in the IGF1 null olfactory system (Figs. 10 and 12), which is profoundly reduced in size and depleted of neurons, with efferent tracts depleted of myelin. In brain structures where neurons are preserved, however, such as the cerebellum, myelination appears perfectly normal (Fig. 12). This observation suggests that if the system projection neurons survive despite the lack of IGF1, as in the cerebellum, oligodendrocytes prosper and appropriate myelination occurs. The peripheral nervous system of IGF1 null mice demonstrates reduced axonal diameter and proportionately reduced myelin sheath thickness, with no evidence of peripheral myelinopathy.[84] Finally, the mentally retarded man with IGF1 gene deletions shows no evidence of dysmyelination or myelinopathy.[58,59]

An early study reporting that IGF1 gene disruption resulted in brain "hypomyelination" was based primarily on the observation of the profound reduction in size of olfactory and hippocampal white matter tracts in the anterior forebrain.[54] However, that study did not assess myelination in other brain areas or the brain as a whole, and did not account for the attenuation of the neural structures giving rise to the selectively affected forebrain myelin tracts. Observations of increased myelin content in the brains of transgenic mice over-expressing IGF1 have been invoked to support a primary role for IGF1 in myelination.[83] That study reported that both brain size and myelin content are increased in the transgenic, but DNA content and oligodendrocyte numbers are not, suggesting that the increased brain mass is due primarily to increased cell size and/or process growth. Further investigation showed that myelin sheath thickness was increased in proportion to increased axonal diameter in this transgenic model.[85] These findings in IGF1 null and over-expressing brains

IGF1 and Brain Development

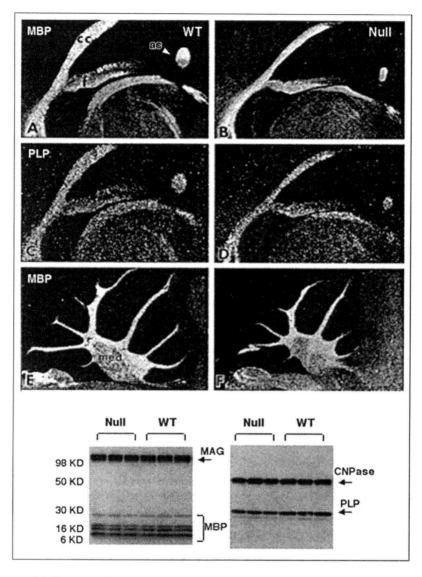

Figure 12. Myelin genes and proteins in the WT (A,C,E) and IGF1 null (B,D,F) brains. The top 4 panels show dark-field micrographs depicting MBP (A&B) and PLP (C&D) mRNA expression patterns in the anterior forebrain. The lower 2 micrographs show MBP in the cerebellum. Most myelin tracts in the IGF1 null brain are modestly reduced in size, proportion to the ~25-30% reduction in brain size. The anterior commissure (ac in A-D) is reduced by about 60%, however, reflecting a major contribution by myelinated tracts from the OB. Western blots show that myelin-specific protein concentrations are not reduced in the IGF1 null brain, when equal amounts of total protein are loaded on the gels. This is because the decrease in myelin is proportionate to the decrease neuroaxonal mass.

are consistent with the current view[86] that myelination is induced by neuronal fiber growth and/or activity. IGF1 over-expression stimulates excessive growth in size and number of neuronal processes and possibly also the survival of additional neurons which, in turn, stimulates additional oligodendrocyte biosynthetic activity and myelination. The findings that myelination in IGF1 null and

IGF1-over-expressing mice is essentially matched to neuraxonal mass is best explained by the simple hypothesis that IGF1 stimulates neuronal process growth which in turn stimulates myelin formation.

The fact that IGF1 does not seem to have an essential role in developmental myelination does not mean that it is not important in repair processes after nervous system injury. IGF1 expression is induced in reactive astrocytes responding to demyelinating insults and IGF1 receptor expression is enhanced in injured oligodendrocytes.[32] Supporting the significance of these expression patterns, administration of exogenous IGF1 improves remyelination after injury.[34] IGF1's prominent effects on oligodendrocytes in vitro may actually reflect the fact that cell culture is essentially an injury model system.

Discussion

The brain requires enormous supplies of fuel and substrate to support neuroglial growth and process formation during early postnatal development. Murine and human brains consume over half the energy available to the organism as a whole during this critical period, characterized more by synapse formation than synaptic activity. Some neurons grow into gargantuan cells with surface areas exceeding all other cells in the body. How this remarkable anabolic feat is achieved when all brain cells are exposed to the same extracellular nutrient supply is unclear. This review has assembled evidence from in vivo studies of murine brain development suggesting that IGF1's role in normal brain development is to promote these extraordinary growth processes. IGF1 and its cognate receptor are most abundantly expressed in maturing projection neurons during a time of rapid process growth and synaptogenesis. IGF1 and IGF1 receptor local expression patterns parallel regional glucose utilization in the developing brain, which is profoundly diminished in the IGF1 null brain. Reduced glucose utilization in IGF1 null neurons is correlated with reduced/absent phosphorylation of Akt/PKB and GSK3β and reduced glycogen accumulation, specifically in projection neurons that normally express IGF1 (e.g., Purkinje and mitral cells) while phospho-forms of these enzymes and glycogen are abundant in WT, IGF1-expressing neurons.[50] These data indicate that IGF1 acts in an autocrine and/or paracrine (i.e., IGF1 secreted from Purkinje cells acting locally on the cell of origin and on neighboring Purkinje cells) manner to promote glucose utilization, using PI3K/Akt(PKB)/GSK3β pathways familiar from insulin signaling in peripheral tissues (Fig. 13).

The major effect of IGF1 action through these pathways is the promotion of cellular hypertrophy. This is clearly seen in IGF1-expressing neurons, and also in the epiphysial growth plate, where IGF1-induced chondrocyte hypertrophy promotes longitudinal bone growth.[87] IGF1 promotes hypertrophy of muscle cells using these same molecular signals, including GSK3β.[88] These observations in the mouse are supported by data from *Drososphila*, in which inactivation of paralogs of the insulin/IGF receptor, IRS, PI3K and Akt all result in globally reduced cell size resulting in proportionate dwarfism, while over expression of any of these molecules results in increased cell size and giantism.[89] As a further comment on IGF1 action in general, all of these studies in mice and in *Drosophila* suggest that IGF1 effects are neutral with respect to cellular differentiation.

It should be noted that these anabolic pathways are not involved in the ultra-rapid glucose fluxes associated with mature brain synaptic activity. IGF1-induced glucose utilization is not correlated with synaptic activity; it peaks during early postnatal development before synapses and synaptic activity mature. It is also observed in brain injury sites where synaptic activity is abolished due to neuronal injury or death, and in isolated nerve membrane preparations in which neural activity is absent.[50] Moreover, the glucose utilization accompanying neural transmission involves primarily glycolysis and glycogenolysis by glial cells involved in glutamate recycling,[90] while IGF1 promotes neuronal glycogen synthesis, not glycogenolysis. However, preferential allocation of glucose to IGF1-expressing neurons during postnatal development facilitates the extraordinary dendritic arborization characterizing these complex information-processing systems. These sensory processing centers continue to exhibit high-level glucose utilization in the mature brain, after IGF1 expression has receded, reflecting the extraordinary dendritic complexity and synaptic density achieved by these structures. Normal process growth and synaptogenesis are completed in the postnatal period, but synaptic remodeling occurs in structures such as the oxytocinergic system during nursing, and in other mature brain systems adapting to alterations in activity. It seems likely and indeed there is substantial evidence that IGF1 is involved in activity-induced synaptic remodeling in the mature brain.[91]

IGF1 and Brain Development

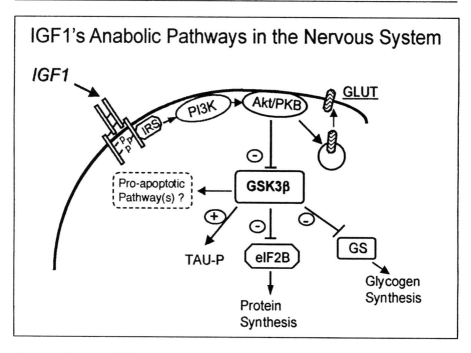

Figure 13. Schematic of IGF1 signaling pathways implicated in regulation of brain anabolic pathways and growth. IGF1 binding to the IGF1 receptor triggers receptor autophosphorylation and association with IRS docking proteins. Phosphtidylinositol 3 Kinase (PI3K) is then activated and generates phospholipids which activate Akt/PKB. Akt/PKB promotes translocation of GLUTs (1, 3 &4) from intracellular endosomal pools to the plasma membrane, thus augmenting glucose transport into the cell. Akt/PKB also serine phosphorylates GSK3 β, causing inhibition. Since GSK3β normally inhibits glycogen synthase and eIF2b, inactivation of GSK3β promotes both glycogen and protein synthesis. The microtubule associated protein tau is also a target for GSK3β and is hyperphosphorylated in the IGF1 null brain, providing further evidence that IGF1 normally inhibits brain GSK3β activity. Finally, GSK3β has been implicated as a pro-apoptotic factor in neurons, though there is no consensus on which cell death pathways are involved.

PKB/Akt and GSK3β appear to be central players in IGF signaling to the brain (Fig. 13). PKB/Akt phosphorylation in IGF1-expressing neurons is associated with increased GLUT4 expression and membrane localization. IGF1's apparent link with this 'insulin-sensitive' transporter may represent a specific anabolic pathway, distinct from the major brain glucose transport pathways involving GLUTs 1 and 3. PKB/Akt phosphorylation also leads to the inhibition of GSK3β in IGF1-expressing neurons (Figs. 7 and 13). This inhibition is expected to facilitate glycogen and protein synthesis, as GSK3β normally inhibits both glycogen synthase and eIF2B (Fig. 13). GSK3β also phosphorylates the microtubule associated protein tau, which is hyper-phosphorylated in the *IGF1* null brain, lending support to the view that GSK3β is a major player in IGF1 action in brain. Tau hyperphosphorylation is associated with the formation of neurofibrillary tangles and neuronal degeneration, as in Alzheimer's Disease.

Whereas most IGF1-expressing neurons survive IGF1 deletion essentially as dwarf neurons, selective populations in the dentate gyrus and olfactory bulb undergo extensive apoptotic cell death, during the process of neurogenesis. The reason for the selective loss of these neuronal populations is not clear, but an important and unique feature of both systems is continuing generation of new neurons during adult life.[92,93] A simple explanation for the dependence of these populations on IGF1 for survival is that these neurons are 'born' during a time when IGF1 is expressed, while most other neuronal system are born and differentiate before birth and before IGF1 expression is significant in brain, and thus must rely on other factors to survive critical developmental stages. It is not clear exactly how IGF1 promotes the survival of these nascent dentate and olfactory neurons. Both

Akt-dependent and-independent pathways are implicated in mediating IGF1's protective effect on non-neural cell lines.[94] IGF1 supports the survival of cultured cerebellar granule neurons via PI3K/Akt stimulated pathways. These cells are also protected from death by lithium treatment.[95,96] Since lithium inhibits GSK3β, these observations support the view that IGF1-induced inhibition of GSK3β may mediate IGF1's promotion of cerebellar granule cell survival. On the other hand, IGF1's neuroprotective role may also involve NF-kB.[97] GSK3β proapoptotic effects are still not well defined,[98] but tau hyper-phosphorylation is a clear result of GSKβ hyperactivity and may contribute to neurotoxicity. However, recent data suggest a coupling between increased glucose utilization, inhibition of GSK3β, and protection from cell death in endothelial cells—where tau presumably is not a player.[99] Further studies are required to define GSK3β 's potential role in IGF1-induced neuroprotection.

The identification of GSK3β as a major target of brain IGF1 signaling provides a unifying pathway for IGF1's well-established anabolic and anti-apoptotic functions, with IGF1-induced inhibition of GSK3β triggering multifaceted anabolic and neuroprotective effects. IGF1 deficiency during brain development results in smaller neurons with hypoplastic processes and decreased survival of dentate gyrus and olfactory neurons. In the human, global IGF1 deficiency due to homozygous IGF1 deletion results in a small brain and mental retardation, suggesting that these studies in mice are indeed relevant to human neurobiology.

Aknowledgements

We are grateful to Rickey Rinehardt, M.D., PhD., for his contributions to these studies, to Drs. Kazutomo Imahori and Koichi Ishiguro for providing the anti-phospho-tau antibodies, and to Ricardo Dreyfus for expert photomicrography.

References

1. Bondy CA. Transient IGF-I gene expression during the maturation of functionally related central projection neurons. J Neurosci 1991; 11(11):3442-3455.
2. Feldman EL, Sullivan KA, Kim B et al. Insulin-like growth factors regulate neuronal differentiation and survival. Neurobiol Dis 1997; 4(3-4):201-214.
3. Anlar B, Sullivan KA, Feldman EL. Insulin-like growth factor-I and central nervous system development. Horm Metab Res 1999; 31(2-3):120-125.
4. Bondy CA, Werner H, Roberts CT Jr et al. Cellular pattern of insulin-like growth factor-I (IGF-I) and type I IGF receptor gene expression in early organogenesis: comparison with IGF-II gene expression. Mol Endocrinol 1990; 4(9):1386-1398.
5. Bondy C, Werner H, Roberts CT Jr et al. Cellular pattern of type-I insulin-like growth factor receptor gene expression during maturation of the rat brain: comparison with insulin-like growth factors I and II. Neuroscience 1992; 46(4):909-923.
6. Werner H, Woloschak M, Adamo M et al. Developmental regulation of the rat insulin-like growth factor I receptor gene. Proc Natl Acad Sci USA 1989; 86(19):7451-7455.
7. Lesniak MA, Hill JM, Kiess W et al. Receptors for insulin-like growth factors I and II: autoradiographic localization in rat brain and comparison to receptors for insulin. Endocrinology Oct 1988; 123(4):2089-2099.
8. Bohannon NJ, Corp ES, Wilcox BJ et al. Localization of binding sites for insulin-like growth factor-I (IGF-I) in the rat brain by quantitative autoradiography. Brain Res 1988; 444(2):205-213.
9. Werther GA, Hogg A, Oldfield BJ et al. Localization and characterization of IGF-I receptors in rat brain and pituitary gland using in vitro autoradiography and computerized densitometry: distinct distribution from insulin receptors. J Endocrinol 1989; 1:369-377.
10. Devaskar S, Zahm DS, Holtzclaw L et al. Developmental regulation of the distribution of rat brain insulin- insensitive (Glut 1) glucose transporter. Endocrinology 1991; 129(3):1530-1540.
11. Bondy C, Bach M, Lee W. Mapping of brain insulin and insulin-like growth factor receptor gene expression by in situ hybridization. Neuroprotocols 1992; 1:240-249.
12. Rotwein P, Burgess SK, Milbrandt JD et al. Differential expression of insulin-like growth factor genes in rat central nervous system. Proc Natn Acad Sci USA 1988; 85:265-269.
13. Bach MA, Shen-Orr Z, Lowe WL Jr et al. Insulin-like growth factor I mRNA levels are developmentally regulated in specific regions of the rat brain. Brain Res Mol Brain Res 1991; 10(1):43-48.
14. Bartlett WP, Li XS, Williams M et al. Localization of insulin-like growth factor-1 mRNA in murine central nervous system during postnatal development. Dev Biol 1991; 147(1):239-250.
15. Ayer-Le Lievre C, Stahlbom PA, Sara VR. Expression of IGF-I and -II mRNA in the brain and craniofacial region of the rat fetus. Development 1991; 111:105-115.
16. Andersson IK, Edwall D, Norstedt G et al. Differing expression of insulin-like growth factor I in the developing rat cerebellum. Acta Physiol Scand 1988; 132:167-173.

17. Lee WH, Michels KM, Bondy CA. Localization of insulin-like growth factor binding protein-2 messenger RNA during postnatal brain development: correlation with insulin-like growth factors I and II. Neuroscience 1993; 53(1):251-265.
18. Bartlett WP, Li XS, Williams M. Expression of IGF-1 mRNA in the murine subventricular zone during postnatal development. Brain Res Mol Brain Res 1992; 12(4):285-291.
19. Hynes MA, Brooks PJ, Van Wyk JJ et al. Insulin-like growth factor II messenger RNAs are synthesized in the choroid plexus of the rat brain. Mol Endocrinol 1988; 2:47-54.
20. Stylianopoulou F, Herbert J, Soares MB et al. Expression of the insulin-like growth factor II gene in the choroid plexus and the leptomeninges of the adult rat central nervous system. Proc Natl Acad Sci USA 1988; 85:141-145.
21. Hetts SW, Rosen KM, Dikkes P et al. Expression and imprinting of the insulin-like growth factor II gene in neonatal mouse cerebellum. J Neurosci Res 1997; 50(6):958-966.
22. Clemmons DR. Role of insulin-like growth factor binding proteins in controlling IGF actions. Mol Cell Endocrinol 1998; 140(1-2):19-24.
23. Ye P, D'Ercole J. Insulin-like growth factor I (IGF-I) regulates IGF binding protein-5 gene expression in the brain. Endocrinology 1998; 139(1):65-71.
24. Logan A, Gonzalez AM, Hill DJ et al. Coordinated pattern of expression and localization of insulin-like growth factor-II (IGF-II) and IGF-binding protein-2 in the adult rat brain. Endocrinology 1994; 135(5):2255-2264.
25. Sullivan KA, Feldman EL. Immunohistochemical localization of insulin-like growth factor-II (IGF-II) and IGF-binding protein-2 during development in the rat brain. Endocrinology 1994; 135(2):540-547.
26. Lee WH, Bondy C. Insulin-like growth factors and cerebral ischemia. Ann N Y Acad Sci 1993; 679:418-422.
27. Bondy C, Lee WH. Correlation between insulin-like growth factor (IGF)-binding protein 5 and IGF-I gene expression during brain development. J Neurosci 1993; 13(12):5092-5104.
28. Brar AK, Chernausek SD. Localization of insulin-like growth factor binding protein-4 expression in the developing and adult rat brain: analysis by in situ hybridization. J Neurosci Res 1993; 35(1):103-114.
29. Lee WH, Wang GM, Seaman LB et al. Coordinate IGF-I and IGFBP5 gene expression in perinatal rat brain after hypoxia-ischemia. J Cereb Blood Flow Metab 1996; 16(2):227-236.
30. Johnston BM, Mallard EC, Williams CE et al. Insulin-like growth factor-1 is a potent neuronal rescue agent after hypoxic-ischemic injury in fetal lambs. J Clin Invest 1996; 97(2):300-308.
31. Cheng B, Mattson MP. IGF-I and IGF-II protect cultured hippocampal and septal neurons against calcium-mediated hypoglycemic damage. J Neurosci 1992; 12(4):1558-1566.
32. Komoly S, Hudson LD, Webster HD et al. Insulin-like growth factor I gene expression is induced in astrocytes during experimental demyelination. Proc Natl Acad Sci USA 1992; 89(5):1894-1898.
33. Gehrmann J, Yao DL, Bonetti B et al. Expression of insulin-like growth factor-I and related peptides during motoneuron regeneration. Exp Neurol 1994; 128(2):202-210.
34. Yao DL, West NR, Bondy CA et al. Cryogenic spinal cord injury induces astrocytic gene expression of insulin-like growth factor I and insulin-like growth factor binding protein 2 during myelin regeneration. J Neurosci Res 1995; 40(5):647-659.
35. Lee w, Clemens J, Bondy C. Insulin-like growth factors in the response to cerebral ischemia. Molec Cell Neuroscience 1992; 3:36-43.
36. Li XS, Williams M, Bartlett WP. Induction of IGF-1 mRNA expression following traumatic injury to the postnatal brain. Brain Res Mol Brain Res 1998; 57(1):92-96.
37. Walter HJ, Berry M, Hill DJ et al. Spatial and temporal changes in the insulin-like growth factor (IGF) axis indicate autocrine/paracrine actions of IGF-I within wounds of the rat brain. Endocrinology 1997; 138(7):3024-3034.
38. Beilharz EJ, Russo VC, Butler G et al. Co-ordinated and cellular specific induction of the components of the IGF/IGFBP axis in the rat brain following hypoxic-ischemic injury. Brain Res Mol Brain Res 1998; 59(2):119-134.
39. Walter HJ, Berry M, Hill DJ et al. Distinct sites of insulin-like growth factor (IGF)-II expression and localization in lesioned rat brain: possible roles of IGF binding proteins (IGFBPs) in the mediation of IGF-II activity. Endocrinology 1999; 140(1):520-532.
40. Coker GT 3rd, Studelska D, Harmon S et al. Analysis of tyrosine hydroxylase and insulin transcripts in human neuroendocrine tissues. Brain Res Mol Brain Res 1990; 8(2):93-98.
41. Havrankova J, Brownstein M, Roth J. Insulin and insulin receptors in rodent brain. Diabetologia 1981; 20(Suppl):268-273.
42. Young WS. Periventricular hypothalamic cells in rat brain contain insulin mRNA. Neuropeptides 1986; 8:93-97.
43. Bruning JC, Gautam D, Burks DJ et al. Role of Brain Insulin Receptor in Control of Body Weight and Reproduction. Science 2000; 289(5487):2122-2125.
44. Van Houten M, Posner BI, Kopriwa BM et al. Insulin binding in the rat brain: in vivo localization to the circumventricular organs by quantitative autoradiography. Endocrinology 1979; (105):666.
45. Reinhardt RR, Bondy CA. Insulin-like growth factors cross the blood-brain barrier. Endocrinology 1994; 135(5):1753-1761.
46. Laron Z, Klinger B. Comparison of the growth-promoting effects of insulin-like growth factor I and growth hormone in the early years of life. Acta Paediatr 2000; 89(1):38-41.

47. Guan J, Miller OT, Waugh KM et al. Insulin-like growth factor-1 improves somatosensory function and reduces the extent of cortical infarction and ongoing neuronal loss after hypoxia-ischemia in rats. Neuroscience 2001; 105(2):299-306.
48. Carro E, Trejo JL, Busiguina S et al. Circulating insulin-like growth factor I mediates the protective effects of physical exercise against brain insults of different etiology and anatomy. J Neurosci 2001; 21(15):5678-5684.
49. Soos MA, Field CE, Siddle K. Purified hybrid insulin/insulin-like growth factor-I receptors bind insulin-like growth factor-I, but not insulin, with high affinity. Biochem J 1993; 290(Pt 2):419-426.
50. Cheng CM, Reinhardt RR, Lee WH, Joncas G, Patel SC, Bondy CA. Insulin-like growth factor 1 regulates developing brain glucose metabolism. Proc Natl Acad Sci USA 2000; 97(18):10236-10241.
51. Gutierrez-Ospina G, Saum L, Calikoglu AS et al. Increased neural activity in transgenic mice with brain IGF-I overexpression: a [3H]2DG study. Neuroreport 1997; 8(13):2907-2911.
52. Haspel HC, Stephenson KN, Davies-Hill T et al. Effects of barbiturates on facilitative glucose transporters are pharmacologically specific and isoform selective. J Membr Biol 1999; 169(1):45-53.
53. Wang J, Zhou J, Powell-Braxton L et al. Effects of Igf1 gene deletion on postnatal growth patterns. Endocrinology 1999; 140(7):3391-3394.
54. Beck KD, Powell-Braxton L, Widmer HR et al. Igf1 gene disruption results in reduced brain size, CNS hypomyelination, and loss of hippocampal granule and striatal parvalbumin-containing neurons. Neuron 1995; 14(4):717-730.
55. Cheng CM, Joncas G, Reinhardt RR et al. Biochemical and morphometric analyses show that myelination in the insulin-like growth factor 1 null brain is proportionate to its neuronal composition. J Neurosci 1998; 18(15):5673-5681.
56. Cheng C, Mervis R, Niu S-L et al. Endogenous IGF1 is Essential for Normal Dendritic Growth. J Neuroscience (in press).
57. Niblock MM, Brunso-Bechtold JK, Riddle DR. Insulin-like growth factor I stimulates dendritic growth in primary somatosensory cortex. J Neurosci 2000; 20(11):4165-4176.
58. Woods KA, Camacho-Hubner C, Barter D et al. Insulin-like growth factor I gene deletion causing intrauterine growth retardation and severe short stature. Acta Paediatr 1997; 423(Suppl):39-45.
59. Woods KA, Camacho-Hubner C, Savage MO et al. Intrauterine growth retardation and postnatal growth failure associated with deletion of the insulin-like growth factor I gene. N Engl J Med 1996; 335(18):1363-1367.
60. Summers SA, Birnbaum MJ. A role for the serine/threonine kinase, Akt, in insulin-stimulated glucose uptake. Biochem Soc Trans 1997; 25(3):981-988.
61. Cho H, Mu J, Kim JK et al. Insulin resistance and a diabetes mellitus-like syndrome in mice lacking the protein kinase Akt2 (PKB beta). Science 2001; 292(5522):1728-1731.
62. Cho H, Thorvaldsen JL, Chu Q et al. Akt1/pkbalpha is required for normal growth but dispensable for maintenance of glucose homeostasis in mice. J Biol Chem 2001; 276(42):38349-38352.
63. Frame S, Cohen P. GSK3 takes centre stage more than 20 years after its discovery. Biochem J 2001; 359(Pt 1):1-16.
64. Borke RC, Nau ME. Glycogen, its transient occurrence in neurons of the rat CNS during normal postnatal development. Brain Res 1984; 318(2):277-284.
65. Lovestone S, Reynolds CH. The phosphorylation of tau: a critical stage in neurodevelopment and neurodegenerative processes. Neuroscience 1997; 78(2):309-324.
66. Takahashi M, Tomizawa K, Kato R et al. Localization and developmental changes of tau protein kinase I/glycogen synthase kinase-3 beta in rat brain. J Neurochem 1994; 63(1):245-255.
67. Hong M, Lee VM. Insulin and insulin-like growth factor-1 regulate tau phosphorylation in cultured human neurons. J Biol Chem 1997; 272(31):19547-19553.
68. Lesort M, Johnson GV. Insulin-like growth factor-1 and insulin mediate transient site-selective increases in tau phosphorylation in primary cortical neurons. Neuroscience 2000; 99(2):305-316.
69. Ishiguro K, Sato K, Takamatsu M et al. Analysis of phosphorylation of tau with antibodies specific for phosphorylation sites. Neurosci Lett 1995; 202(1-2):81-84.
70. Tong N, Sanchez JF, Maggirwar SB et al. Activation of glycogen synthase kinase 3 beta (GSK-3beta) by platelet activating factor mediates migration and cell death in cerebellar granule neurons. Eur J Neurosci 2001; 13(10):1913-1922.
71. Li M, Wang X, Meintzer MK et al. Cyclic AMP promotes neuronal survival by phosphorylation of glycogen synthase kinase 3beta. Mol Cell Biol 2000; 20(24):9356-9363.
72. Crowder RJ, Freeman RS. Glycogen synthase kinase-3 beta activity is critical for neuronal death caused by inhibiting phosphatidylinositol 3-kinase or Akt but not for death caused by nerve growth factor withdrawal. J Biol Chem 2000; 275(44):34266-34271.
73. Hetman M, Cavanaugh JE, Kimelman D et al. Role of glycogen synthase kinase-3beta in neuronal apoptosis induced by trophic withdrawal. J Neurosci 2000; 20(7):2567-2574.
74. Lucas JJ, Hernandez F, Gomez-Ramos P et al. Decreased nuclear beta-catenin, tau hyperphosphorylation and neurodegeneration in GSK-3beta conditional transgenic mice. Embo J 2001; 20(1-2):27-39.
75. Cheng CM, Cohen M, Tseng V et al. Endogenous IGF1 enhances cell survival in the postnatal dentate gyrus. J Neurosci Res 2001; 64(4):341-347.
76. Aberg MA, Aberg ND, Hedbacker H et al. Peripheral infusion of IGF-I selectively induces neurogenesis in the adult rat hippocampus. J Neurosci 2000; 20(8):2896-2903.

77. Lichtenwalner RJ, Forbes ME, Bennett SA et al. Intracerebroventricular infusion of insulin-like growth factor-I ameliorates the age-related decline in hippocampal neurogenesis. Neuroscience 2001; 107(4):603-613.
78. O'Kusky JR, Ye P, D'Ercole AJ. Insulin-like growth factor-I promotes neurogenesis and synaptogenesis in the hippocampal dentate gyrus during postnatal development. J Neurosci 2000; 20(22):8435-8442.
79. Chrysis D, Calikoglu AS, Ye P et al. Insulin-like growth factor-I overexpression attenuates cerebellar apoptosis by altering the expression of Bcl family proteins in a developmentally specific manner. J Neurosci 2001; 21(5):1481-1489.
80. Lee WH, Wang GM, Lo T et al. Altered IGFBP5 gene expression in the cerebellar external germinal layer of weaver mutant mice. Brain Res Mol Brain Res 1995; 30(2):259-268.
81. McMorris FA, Mozell RL, Carson MJ et al. Regulation of oligodendrocyte development and central nervous system myelination by insulin-like growth factors. Ann N Y Acad Sci 1993; 692:321-334.
82. McMorris FA, Smith TM, DeSalvo S et al. Insulin-like growth factor I/somatomedin C: a potent inducer of oligodendrocyte development. Proc Natl Acad Sci USA 1986; 83(3):822-826.
83. Carson MJ, Behringer RR, Brinster RL et al. Insulin-like growth factor I increases brain growth and central nervous system myelination in transgenic mice. Neuron 1993; 10(4):729-740.
84. Gao WQ, Shinsky N, Ingle G et al. IGF-I deficient mice show reduced peripheral nerve conduction velocities and decreased axonal diameters and respond to exogenous IGF-I treatment. J Neurobiol 1999; 39(1):142-152.
85. Ye P, Carson J, D'Ercole AJ. In vivo actions of insulin-like growth factor-I (IGF-I) on brain myelination: studies of IGF-I and IGF binding protein-1 (IGFBP-1) transgenic mice. J Neurosci 1995; 15(11):7344-7356.
86. Barres BA, Raff MC. axonal control of oligodendrocyte development. J Cell Biol 1999; 147(6):1123-1128.
87. Wang J, Zhou J, Bondy CA. Igf1 promotes longitudinal bone growth by insulin-like actions augmenting chondrocyte hypertrophy. Faseb J 1999; 13(14):1985-1990.
88. Rommel C, Bodine SC, Clarke BA et al. Mediation of IGF-1-induced skeletal myotube hypertrophy by PI(3)K/Akt/mTOR and PI(3)K/Akt/GSK3 pathways. Nat Cell Biol 2001; 3(11):1009-1013.
89. Potter CJ, Xu T. Mechanisms of size control. Curr Opin Genet Dev 2001; 11(3):279-286.
90. Shulman RG. Functional imaging studies: linking mind and basic neuroscience. Am J Psychiatry 2001; 158(1):11-20.
91. Torres-Aleman I. Serum growth factors and neuroprotective surveillance: focus on IGF-1. Mol Neurobiol 2000; 21(3):153-160.
92. Kempermann G, van Praag H, Gage FH. Activity-dependent regulation of neuronal plasticity and self repair. Prog Brain Res 2000; 127:35-48.
93. Kempermann G, Gage FH. Neurogenesis in the adult hippocampus. Novartis Found Symp 2000; 231:220-235; discussion 235-241, 302-226.
94. Kulik G, Weber MJ. Akt-dependent and -independent survival signaling pathways utilized by insulin-like growth factor I. Mol Cell Biol 1998; 18(11):6711-6718.
95. Kumari S, Liu X, Nguyen T et al. Distinct phosphorylation patterns underlie Akt activation by different survival factors in neurons. Brain Res Mol Brain Res 2001; 96(1-2):157-162.
96. Mora A, Sabio G, Gonzalez-Polo RA et al. Lithium inhibits caspase 3 activation and dephosphorylation of PKB and GSK3 induced by K+ deprivation in cerebellar granule cells. J Neurochem 2001; 78(1):199-206.
97. Koulich E, Nguyen T, Johnson K et al. NF-kappaB is involved in the survival of cerebellar granule neurons: association of IkappaBbeta [correction of Ikappabeta] phosphorylation with cell survival. J Neurochem 2001; 76(4):1188-1198.
98. Grimes CA, Jope RS. The multifaceted roles of glycogen synthase kinase 3beta in cellular signaling. Prog Neurobiol 2001; 65(4):391-426.
99. Hall JL, Chatham JC, Eldar-Finkelman H et al. Upregulation of glucose metabolism during intimal lesion formation is coupled to the inhibition of vascular smooth muscle cell apoptosis. Role of GSK3beta. Diabetes 2001; 50(5):1171-1179.

CHAPTER 9

IGFs and the Nervous System

Gina M. Leinninger, Gary E. Meyer and Eva L. Feldman

Introduction

The insulin-like growth factors-I and -II (IGF-I and IGF-II) are neurotrophic factors with sequence homology to pro-insulin. Through the type I IGF receptor (IGF-IR), the IGFs mediate proliferation, survival and differentiation of neuronal and non-neuronal cells.[1,2] IGFs are of particular interest in the nervous system because of their unique actions as survival agents for post-mitotic neurons. For this reason, IGFs are currently being studied in both the central and peripheral nervous systems as potential therapeutic agents for neurodegenerative diseases. This chapter will survey the multivariate roles of the IGFs in the nervous system, highlighting developmental expression, neurotrophic actions, regulation of myelination and how our knowledge of IGF in these systems is contributing to treatment paradigms in neurological disease.

Developmental Expression of the IGFs and IGFRs in the Nervous System

The IGFs and IGF receptors (IGFRs) are widely expressed in the vertebrate nervous system. Expression of the IGF system in the brain is outlined elsewhere in this volume. This chapter will highlight the expression of the IGF system in the other major component of the central nervous system (CNS), the spinal cord, and the peripheral nervous system (PNS).

Developmental Expression of the IGFs and IGF Receptors in the Spinal Cord

Most IGF system expression studies in the nervous system have been conducted in the developing rat. Expression of IGF-I in the rat spinal cord peaks around embryonic day 14 or 15 (E14-E15).[3,4] During this embryonic period, expression of IGF-I mRNA is 8-10 times more abundant in cervical-thoracic levels of the spinal cord than it is in whole brain.[5] IGF-I mRNA levels decrease just prior to birth, and again just after birth, suggesting that most of the neurotrophic developmental effects of IGF-I are mediated during embryogenesis.[3,4] IGF-I levels in the spinal cord and CNS as a whole diminish progressively during postnatal development, and are present only at low levels in the adult.[6] In contrast to IGF-I, expression of IGF-II in the spinal cord is mainly limited to leptomeninges and blood vessels.[7] This vascular pattern of IGF-II expression appears to be consistent throughout the CNS, as IGF-II expression is also limited to vascular structures of the brain.[8-10] The role of the IGFBPs in the spinal cord has not been studied in detail. It is known that there are low levels of IGFBP-2 and -5 in the spinal cord. IGFBP-6 mRNA is expressed more widely in the cord, particularly in meningeal cells, interneurons of the deep part of the dorsal horn and in some spinal cord motor neurons.[7]

There is little specific literature on the developmental expression of the type I and II IGFRs in the spinal cord. The IGF-IR is expressed in spinal motor neurons in the adult rat (Fig. 1). Unlike the IGF-IR, the role of the IGF-IIR, also known as the mannose-6-phosphate receptor, is more ambiguous. This receptor has a higher affinity for IGF-II than IGF-I, but its role in the nervous system is yet unknown. It may have a role in general cellular protein sorting, via the Golgi apparatus.[11,12] To date, there is no specific data on IGF-IIR expression in the spinal cord, or the nervous system as a whole.

Insulin-Like Growth Factors, edited by Derek LeRoith, Walter Zumkeller and Robert Baxter. ©2003 Eurekah.com and Kluwer Academic / Plenum Publishers.

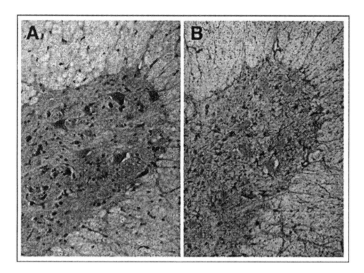

Figure 1. IGF-I and IGF-IR immunostaining in the spinal cord of an adult rat. A) IGF-IR immunostaining is localized in large motor neurons (arrows). B) and are surrounded by GFAP-positive astrocytes (arrowheads) in the ventral horn. The identification of neurons was confirmed by neurofilament immunostaining, data not shown. (Reprinted with permission from Anlar et al, Horm Metab Res 31:120-125, 1999.)

Developmental Expression of the IGFs and IGF Receptors in the Peripheral Nervous System

The IGF system also has potent developmental roles in the PNS, as suggested by its developmental expression pattern. In humans, IGF-IR is expressed within the perikaryon of DRG (Fig. 2D) and high levels of IGF-II are expressed in fetal spinal ganglia, as well as in muscle, during the formation of neuromuscular junctions.[13,14] IGF-IR protein is expressed in the growth cones of cultured embryonic rat DRG neurons (Fig. 2A-C). Adult rat motor neurons continue to express IGF-II protein, suggesting a protracted neurotrophic role for IGF-II in these neurons.[15] IGF-I mRNA is also expressed in the rat sciatic nerve from birth, through postnatal development and adulthood. Immunohistochemistry indicates that within the sciatic nerve, the IGF-I protein expression is localized to the cytoplasm of Schwann cells, the myelinating cells of the PNS.[16] Additionally, Schwann cells express IGF-IR, IGF-II and members of the IGF binding protein family, IGFBP-4 and -5. Both IGFBP-5 and IGFBP-1 are detected in embryonic rat peripheral nerves.[7,17,18] This expression data indicates that the IGF system is important in both neurons and glial support cells that compose the PNS.

Expression of the IGFs, IGF Receptors and IGF Binding Proteins During Neuronal Regeneration

CNS injury affects the expression of components of the IGF system. In particular, injury sustained in the CNS by hypoxia-ischemia (HI) immediately downregulates neuronal levels of IGF-I, IGFBP-3 and IGFBP-5, concurrent with an increase in the number of cells undergoing apoptosis.[19,20] The mRNA levels of IGFPB-2 and -4 are also decreased in the first 24 hours following HI.[21,22] While HI injury immediately downregulates IGFBP-3 in neurons, expression is increased within vascular endothelial cells of the affected hemisphere at 24 hours, a time when IGFBP-3 levels in neurons are steadily decreasing.[20] As IGFBP-3 is the main carrier protein for circulating IGFs in adults, this expression pattern following HI may represent a mechanism to increase IGF-I delivery from the blood to injured areas. Further, 3 days post HI-injury, IGF-I mRNA is expressed by reactive microglia and astrocytes surrounding the infarct, particularly those juxtaposed to surviving neurons. IGFBP-2 and -6 levels are also increased in the affected hemisphere.[23] In another CNS injury paradigm, a penetrating brain injury, expression of IGF-I, IGF-IR, and IGFBPs-1,-2,-3 and

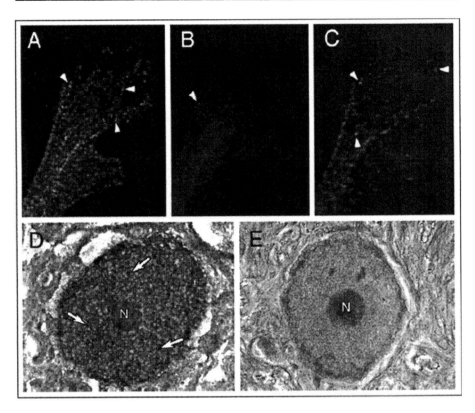

Figure 2. IGF-I receptor expression in the rat DRG neurite growth cone and in human DRG. A) The IGF-I receptor (white arrows) was abundant in the rat neurite growth cone. Rat DRG were incubated with chicken IGF-IR IgY, and viewed using confocal microscopy (X1500). The figure represents a z-series reconstruction of several image planes from the upper to the lower surface of the neurite growth cone. The IGF-I receptor was distributed throughout the neuron, but was most abundant in the DRG growth cone. B) Image from an upper plane from the z-series described in A. The image plane was adjusted to lie just below the upper surface of the neuron. Although the outline of the growth cone is seen, there are few IGF-I receptors present. C) Image from a lower plane (adjacent to the ECM) of the z-series described in A. More numerous IGF-I receptor staining (white arrows) is observed thoughout the growth cone. D) In-situ hybridization for IGF-I receptor in human DRG using anti-sense probe showing abundant punctate labeling of IGF-I receptor mRNA (white arrows) throughout the neuronal perikaryon (X500). N = DRG neuronal nucleus. There was less abundant labeling of satellite cells and fibroblasts. E) Sense control showing no labeling of the neuron or supporting cells (X500). N = DRG neuronal nucleus. Equal total concentrations of sense and antisense RNA probes were used on each tissue section. (Reprinted with permission from Russell et al, J Neurobiol 36:455-467, 1998.)

-6 are increased in astrocytes, neurons and monocytic cells during the acute phase (1-7 days post lesion).[24] IGF-II secretion from choriod plexus into the CSF is also upregulated during the acute phase, but these levels in the CSF and the wound neuropil decline by 7-10 days.[25] In contrast, IGF-II mRNA is decreased in the damaged tissue immediately after the wound is made, but levels recover to normal after 48 hours.[26] In another direct model of brain injury, lesioning of the unilateral entorhinal cortex, IGF-I expression is elevated around the lesion cavity in young and old rats. However, hippocampal IGF-I expression was more markedly increased in young rats after the lesion compared to old rats, suggesting that the IGF survival expression system declines during aging.[27] IGF-I and IGFBP-2 expression are also increased following cryogenic spinal cord injury, cytotoxic lesions and cerebral contusion.[28-30]

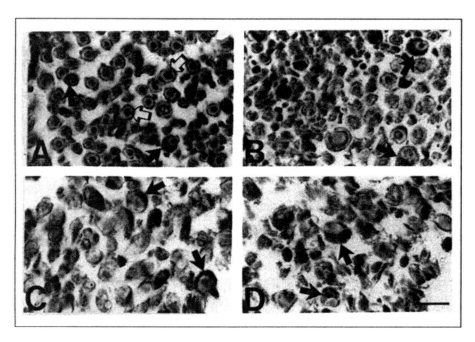

Figure 3. IGFBP-5 immunoreactivity in nerve transection model. IGFBP-5 immunohistochemistry studies were performed using a polyclonal antiserum against IGFBP-5. A) Intact nerve section. B) Proximal nerve stump of 3 days. C) Distal nerve stump of 3 days. D) Distal nerve stump of 7 days. The IGFBP-5 immunoreactivity was consistently detected in SC cytoplasm (black arrows, A, B, C, D) as well as axons in intact nerves (A, white arrows). Representative sections are from three animals. Bar = 20 μm. (Reprinted with permission from Cheng et al, J Neurochem 66:525-536, 1996.)

In contrast to the multiple CNS injury models, study of the IGF system in peripheral nerve injury has mainly focused on two similar injury models, crush or transection of the sciatic nerve. Sciatic nerve crush in the rat increases IGF-I and IGF-II mRNA distal to the crush site, while IGF-I expression at the crush site is not increased until 4 days post-injury.[31] IGF-II expression is not altered at the crush site, but rather in more distal, intramuscular branches of the nerves; this expression is lost after re-establishment of functional neuromuscular synapses.[31,32] Thus, IGF-II may be important in both regenerative and developmental formation of neuromuscular junctions.[14,32] In the transection model, IGFBP-6 mRNA levels increase in motor neurons and in the proximal nerve stump. IGFBP-4 and -5 are also upregulated in denervated sciatic nerve (Fig. 3).[7] Further, IGF-I, IGF-II and IGF-IR are observed following sciatic nerve transaction.[7,18] During the first 3-7 days following transection, the expressed IGF-I is localized mainly in Schwann cells of the intact nerve and the distal stump.[18,33] IGF-I expression has also been detected in transected facial nerve, 4-7 days post transection, where it is mainly localized in astrocyte processes.[34]

Regulation of the IGF system following injury to the CNS and PNS is presumably part of a survival and regeneration response. These in vivo responses not only implicate the IGF system in normal nervous system regeneration, but also suggest that additional clinical treatment with IGFs may have potent therapeutic effects. The expression pattern of the IGF system in nervous system development and regeneration, as well as the survival roles of the IGF system (addressed subsequently), have led to investigation of the effects of IGF-I on several models of neurodegenerative disease. (See section *IGFs in the Treatment of Neurological Diseases*).

Neurotrophic Roles of the IGFs in the Nervous System

IGF-I promotes the survival and differentiation of target neurons.[35,36] Numerous in vitro studies demonstrate the neurotrophic effects of IGF-I treatment in various CNS neurons, including those of the hippocampus,[37] cortex,[38] and retina.[39,40] IGF-I also exerts neurotrophic survival and differentiation effects in the PNS, as shown in peripheral sensory,[41] sympathetic,[41,42] and motor neurons.[43,44] The neurotrophic roles of IGFs in the PNS are particularly important because the IGFs are the only growth factors capable of supporting both sensory and motor nerve regeneration in adult animals (recently reviewed in ref. 45). In addition to survival and differentiation, IGFs are also implicated in neuronal neurite extension, motility and transformation. This section will address these neurotrophic roles of IGFs in the central and peripheral nervous systems.

IGFs Promote Survival by Inhibiting Apoptosis

The past two decades have yielded considerable evidence that IGFs promote neuronal survival by preventing apoptosis, or programmed cell death. Unlike necrotic death, which is characterized by an immune response, cellular swelling and lysis, apoptotic cell death is characterized by membrane blebbing, a specific pattern of DNA fragmentation, new transcription and cell shrinkage.[46] Apoptosis is mediated by a cast of cellular proteins that are normally present in healthy cells, but only promote apoptotic cellular changes and death in response to a toxin, stress, or a signal during injury or normal development. Among these apoptotic regulatory proteins are the Bcl family of proteins (bcl-2, bcl-xL, bad, bax, bid, bim), the cysteine protease family of caspases, and the c-Jun NH_2-terminal kinases (JNKs).[47-51] Thus, apoptotic death is colloquially termed "cellular suicide", because it represents a regulated and active process by which the cell dies in response to aversive or developmental stimuli.

IGFs, and in particular IGF-I, are able to rescue both neurons and glia from apoptotic death induced by various aversive stimuli. In vitro experiments demonstrate that IGF-I inhibits apoptosis in cultured cerebellar granule neurons exposed to low potassium, high potassium, serum withdrawal or okadaic acid.[52,53] IGF-I can also prevent apoptotic death in amyloid beta-treated neurons, suggesting it may have therapeutic benefits in treatment of Alzheimer's Disease.[54] In hippocampal and cortical neurons cultured from spontaneously hypertensive rats, treatment with N-methyl-D-aspartate (NMDA) or nitric oxide (NO) induces apoptosis, which is prevented by IGF-I treatment.[55,56] Neurons from the PNS are also rescued from apoptosis by IGF-I, including motor neurons after axotomy or spinal transection[57] and DRG after NGF withdrawal or exposure to high glucose (Fig. 4).[58,59] Further, neuroblastoma cells are protected from apoptotic death secondary to serum deficiency, doxorubicin treatment or hyperosmotic shock.[60-62] Interestingly, IGF-I protects neuroblastoma cells from hyperosmotic-induced apoptosis, but other growth factors, such as nerve growth factor (NGF), epidermal growth factor (EGF) and platelet-derived growth factor (PDGF) do not, suggesting that neuroprotection of these cells is specific to IGF-I.[63] The anti-apoptotic actions of IGF-I in different subtypes of CNS and PNS neurons suggests that IGF-I is widely neuroprotective, and this property of IGF-I may be exploited for therapeutic uses in the clinical setting.

IGF-I also prevents apoptosis in glial cells. In the CNS, IGF-I prevents apoptosis of mature oligodendrocytes.[64,65] IGF-I also protects the PNS myelinating counterpart of oligodendrocytes, Schwann cells.[66-68] The IGF-I mediated neuroprotection of myelinating cells may be important for remyelination following trauma or neurodegenerative disease (see section *IGF's Promote Myelination in the CNS and PNS*).

IGFs Inhibit Apoptosis Through Biochemical Signaling Mechanisms

The neuroprotective effects of IGF-I have prompted study of the intracellular mechanisms by which IGF-I inhibits apoptosis. The following section will provide a brief overview of the important key players in neuronal apoptotic signaling, and then will examine points at which IGF-I interrupts this signaling, and thereby inhibits apoptosis.

Key Mediators of Neuronal Apoptosis

Neuronal apoptosis has been widely studied in response to aversive stimuli, as well as in neuronal development, where excess neurons are pared down to create a functional nervous system.[48] Developmental and adult neuronal apoptosis both employ the same families of apoptotic proteins: the caspases, the bcl proteins and the JNKs. The cysteine protease family of caspases are critical for apoptotic progression, and their activation is considered a defining hallmark of the apoptotic process.[49,50] The 14 known members of the caspase family exist as inactive zymogens in the cell, but,

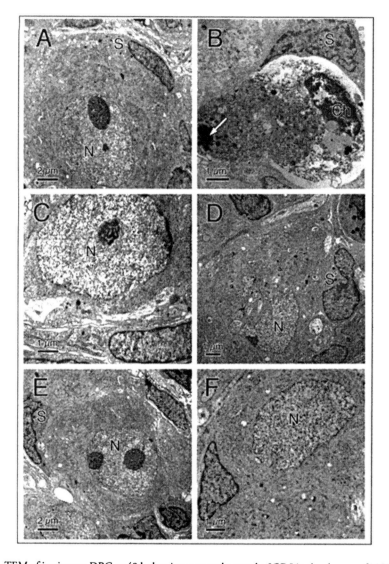

Figure 4. TEM of in vitro rat DRG at 48 h showing neuronal rescue by IGF-I in the absence of NGF. A) Control DRG with 10 ng/ml NGF showing normal chromatin staining in the nucleus (N), and an intact perikaryon. The neuroblast was encased in satellite cells (S). B) DRG after 48 h NGF withdrawal. There was severe chromatin clumping (Ch) observed in the nucleus, shrinkage of the perikaryon, although at higher magnification the plasma membrane remained intact, and single deletion of cells. At 48 h most of the neuroblasts were apoptotic. There was also a prominent collection of lipofuscin (white arrow) commonly observed in apoptotic DRG neurons. The satellite cells (S) were histologically unaffected by the NGF withdrawal. C) DRG with 1 nM IGF-I, but no NGF. The presence of IGF-I supported neuronal survival as indicated, although several neuroblasts continued to undergo apoptosis at this low concentration of IGF-I. D) DRG with 100 nM IGF-I. There was survival of most of the neuroblasts, with normal chromatin distribution in the nucleus (N). The normal neuronal perikaryon was surrounded by satellite cells (S). E) DRG with 1 nM IGF-I and 10 ng/ml NGF. F) DRG with 100 nM IGF-I and 10 ng/ml NGF. In both E and F, the neuronal nuclei (N) were normal. Apoptotic neurons were rare in the presence of both NGF and IGF-I, and survival was slightly greater than controls at higher concentrations of IGF-I.
(Reprinted with permission from Russell et al, J Neurobiol 36:455-467.)

when activated, cleave other caspases at specific aspartate residues. Thus, one activated caspase can unleash a catalytic caspase activation cascade.[46,50,69-71] Caspases are organized in a hierarchical manner, such that there are "initiator" caspases and "executor" caspases.[72] The apical "initiator caspases," caspases-8 and -9, respond selectively to different apoptotic stimuli.[73] Caspase-8 is activated by death-receptor activation, such as that induced by the Fas and TNFR1 receptors.[72,74] Alternatively, caspase-9 is activated in response to chemical stress stimuli, such as chemotherapeutic agents, and its activation requires cytochrome c release from the mitochondria.[75,76] Activated initiator caspases propagate an apoptotic signal by activating downstream effector caspases, such as caspases-3, -6 and -7. These 'executioner' caspases then enact further cleavage events and commit the cell to apoptotic death.[49,71,77]

The Bcl family of proteins are also intricately involved in neuronal apoptosis. This family consists of anti-apoptotic (bcl-2, bcl-xL, bcl-w, bim) and pro-apoptotic (bax, bad, bid) members.[78,79] The anti-apoptotic proteins bcl-2 and bcl-xL regulate mitochondrial integrity, and thereby inhibit apoptosis, by heterodimerizing with the pro-apoptotic bax protein.[80-83] Bcl-2 can be inactivated by phosphorylation, thus releasing bax.[82,84,85] Freed bax translocates to the mitochondria,[86] integrates into the mitochondrial membrane, increases mitochondrial membrane permeability and causes the release of cytochrome c into the cytoplasm.[87,88] The released cytochrome c binds to APAF-1 in the cytoplasm, which then facilitates the binding and activation of caspase-9.[89,90] Pro-apoptotic bid also acts to regulate cytochrome c translocation; bid is activated by caspase-8, translocates to the mitochondria, and then mediates cytochrome c release into the cytoplasm. This, again, allows for caspase-9 activation.[91] Like the sequestration of bax by bcl-2 or bcl-x, the pro-apoptotic protein bad is also normally sequestered by 14.3.3 proteins, and thus cannot mediate apoptotic effects. However, de-phosphorylated bad localizes to the mitochondrial membrane and antagonizes the anti-apoptotic functions of bcl-2.[92,93] Bcl-2, bcl-x, bax and bad are the most common bcl proteins known to function in neuronal apoptosis.

Additionally, the JNKs, also known as the stress-activated MAP kinases (SAPKs), are implicated in neuronal apoptosis.[94-96] The JNK cascade is initiated by apoptosis-signal regulated kinase (ASK1), which can activate the kinases MKK4 and MKK7 via phosphorylation; these go on to phosphorylate JNK. JNK can also be activated by MEKK1, cdc42 and rac.[97-99] Phosphorylated JNK activates several transcription factors, including c-Jun, which is critical to the execution of apoptosis.[100] Numerous apoptotic stimuli activate the JNK pathway, including cytokines, UV radiation, chemotherapeutic agents, osmolarity changes and oxidative stress.[96,101-104] Further, blocking the JNK pathway via dominant negative JNK expression or the CEP-1347 inhibitor prevents apoptosis in neuronal and non-neuronal cell-types.[103,105,106] Interestingly, some studies suggest that JNK phosphorylates Bcl-2, thereby inhibiting the ability of bcl-2 to preserve mitochondrial integrity.[107,108]

IGF-I Intervention Points in Apoptotic Signaling

IGF-I protects neurons from undergoing apoptosis by binding to the IGF-IR and activating the PI3K/Akt pathway (Reviewed in Schematic 1). First, IGF-I binding induces autophosphorylation of a cluster of tyrosines on the intracellular portion of the IGF-IR. This autophosphorylation activates the tyrosine kinase activity of the receptor, resulting in phosphorylation of additional intracellular receptor tyrosines.[109,110] Insulin receptor substrates -1 and -2 (IRS-1, -2) proteins bind to phosphotyrosine residues of the IGF-IR and act as docking proteins, allowing binding of additional downstream signaling molecules that contain Src homology 2 (SH2) domains.[111] This allows binding of the SH2-containing p85 regulatory subunit of PI3K.[112,113] Binding of IRS and p85 activates the catalytic subunit of PI3K, thereby increasing the production of phosphoinositides, particularly phosphatidylinositol 3,4,5-triphosphate (PIP3).[114] PI3K inhibitors abolish the anti-apoptotic effects of IGF-I, proving that PI3K is a critical mediator of IGF-I neuroprotection (Fig. 5).[66,113,115,116] Recent studies show that PI3K is critical for IGF-I mediated neuroprotection because it activates the serine/threonine kinase, Akt, in a two-part mechanism. Specifically, PI3K lipid products bind directly to Akt, inducing a conformational change that exposes an Akt threonine residue. PI3K lipid products also activate phosphoinositide-dependent-kinase (PDK-1), which phosphorylates the newly exposed Akt threonine.[117,118] Inhibitors of PI3K or expression of a kinase-inactive Akt mutant block the neuroprotective effects of IGF-I, proving that IGF-I activates Akt by way of PI3K.[117,119] Akt

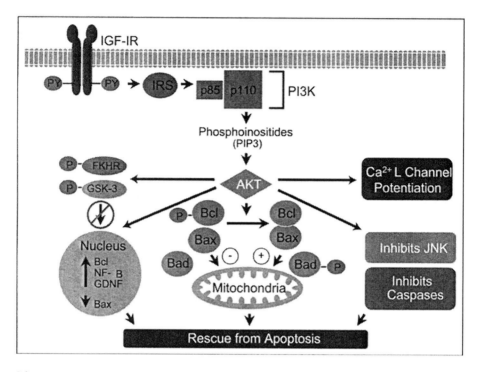

Schematic 1.

activity can also be inhibited by the dual phosphatase, PTEN, which prevents IGF-I mediated neuroprotection in neuroblastoma cells.[120]

IGF-I induced Akt activation inhibits apoptosis through several mechanisms. Akt directly regulates neuronal survival by potentiating L subtype calcium channels.[121] L-channel modulation likely inhibits apoptosis by regulating intracellular calcium levels in response to depolarizing KCl.[115,122] Akt also mediates survival effects by upregulating the anti-apoptotic bcl-2 and bcl-xL proteins.[123,124] Further, Akt downregulates pro-apoptotic bax levels in some neurons.[119] In other neuronal subtypes, Akt maintains bax protein levels, but prevents bax translocation to the mitochondria.[125] In this case, bax is likely bound to bcl proteins; the bcl-2/bax heterodimerization prevents bax translocation to the mitochondria, and thus prevents mitochondrial membrane changes that lead to apoptosis.[126] Additionally, Akt phosphorylates the pro-apoptotic bad protein, thereby antagonizing its apoptotic function.[52,127,128] Activation of caspases, the executioners of apoptotic death, are also inhibited by IGF-I/Akt signaling. In particular, IGF-I inhibits the cleavage, and therefore the activation, of caspase-3 (Fig. 6) and caspase -9 in neurons and Schwann cells.[59,66,129,130]

Akt also prevents apoptosis via regulation of transcription and transcription factors. For example, Akt phosphorylates forkhead transcription factors, thereby inhibiting their ability to translocate to the nucleus and activate transcription of apoptotic proteins, such as the Fas ligand.[131-133] Akt also phosphorylates and inactivates the apoptotic glycogen synthase kinase-3β (GSK-3β) and inhibits apoptotic induction by p53, though the mechanism of the latter is not well understood in neurons.[134,135] NF-κB, an anti-apoptotic transcription factor, is also induced by Akt.[136] Interestingly, glial cell line-derived neurotrophic factor (GDNF) is upregulated by IGF-I. Combined treatment of GDNF and IGF-I promotes cell survival in motor neurons,[137] suggesting that IGF-I may activate transcription of GDNF to mount a heightened neuroprotective response.

IGF-I inhibits activation of the JNKs, which are implicated in developmental and adult neuronal apoptosis.[138] IGF-I reduces JNK activation and prevents apoptosis in SH-SY5Y neuroblastoma cells, PC12 cells, and Schwann cells.[139-141] In neuroblastoma cells, IGF-I treatment also

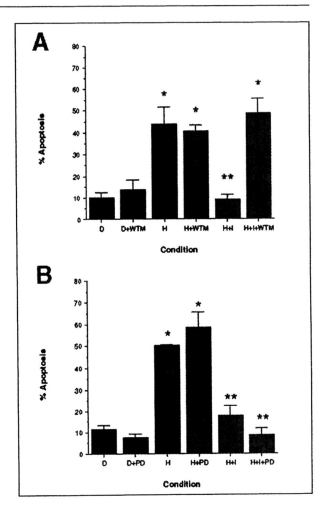

Figure 5. IGF-I rescues SHEP cells from hyperosmotic-induced apoptosis via PI-3K. A) Serum deprived SHEP cells are incubated with DMEM (D) + 100 nM wortmannin (WTM) or 300 mM mannitol (H) + 10 nM IGF-I (I) + 100 nM WTM. Cells exposed to conditions with WTM are pretreated with WTM for 1 h prior to addition of experimental conditions. WTM is replenished every 6 h, and IGF-I is replenished every 12 h. After 24 h, cells are collected and prepared for flow cytometry. % apoptotic cells represents the percentage of DNA in the sub-G_0 population as measured by propidium iodide staining for flow cytometry. B) Serum deprived SHEP cells were exposed to DMEM (D) + 10 μM PD98059 (PD) or 300 mM mannitol (H) + 10 nM IGF-I (I) + 10 μM PD. Cells exposed to PD are pretreated with the inhibitor for 1 h before addition of experimental conditions. After 24 h, cells are prepared for flow cytometry. For both (A) and (B), * = p < 0.01 compared to DMEM and ** = < 0.01 compared to 300 mM mannitol. (Reprinted with permission from van Golen et al, Cell Growth Differ 12:371-378, 2001.)

decreases JNK-induced activation of the c-Jun transcription factor (Fig. 7) but has no effect on P38 kinase (Fig. 8).[141] As c-Jun activation is required for apoptotic death,[100] this further implicates IGF-I as a neuroprotective agent. Inhibition of the PI3K pathway, but not the MAP Kinase (MAPK) pathway, reduces JNK activation in Schwann cells overexpressing JNK, suggesting that the IGF-I mediates its effects on JNK through the PI3K pathway.[141] IGF-I effects on JNK are likely exerted through PI3K-mediated activation of Akt, though the exact mechanism of this is now known.[140,142] Studies suggest that the JNK pathway is required for activation of caspases that lead to apoptotic execution,[143] and thus, IGF-I inhibition of JNK may prevent caspase activation.

Thus, IGF-I is capable of rescuing neurons and glia from apoptotic death at the cytoplasmic and nuclear levels. These actions are particularly intriguing in neurons, which are post-mitotic cells. Neuron depletion due to normal aging or neurodegenerative disease can have irreparable deleterious effects, as no new neurons can be generated after birth to replace those that have died. However, because IGF-I is neuroprotective, it may be a useful agent for promoting survival of degenerating neurons in disease or normal aging. Continued study of the effects of IGF-I and the IGF-IR will be important for understanding how IGF-I may be utilized therapeutically in these paradigms.

The IGF System in Cancers of the CNS: Loss of Apoptotic Signaling Systems

The IGF system has also been implicated in cancers of the CNS, where neurons have lost the ability to undergo apoptotic death.[144] For example, human medulloblastomas exhibit IGF-IR tyrosine

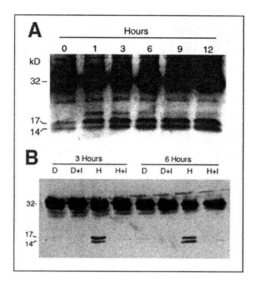

Figure 6. A) Caspase-3 is activated after hyperosmotic exposure in SHEP cells. Cells were serum deprived for 4 h, then treated with 300 mM mannitol. At 0, 1, 3, 6, 9, and 12 hours, whole cell lysates were collected and Western immunoblotting performed for caspase-3. Blot shown is 1 of 3 performed. B) IGF-I prevents hyperosmotic-induced caspase-3 activation in SHEP cells. Serum deprived cells were exposed to DMEM (D), DMEM + 10 nM IGF-I (D+I), 300 mM mannitol (H), or 300 mM mannitol + 10 nM IGF-I (H+I). At 3 and 6 h, whole cell lysates were collected and Western immunoblotting performed for caspase-3. Blot shown is 1 of 3 performed. (Reprinted with permission from van Golen et al, Cell Death Differ 7:654-665, 2000.)

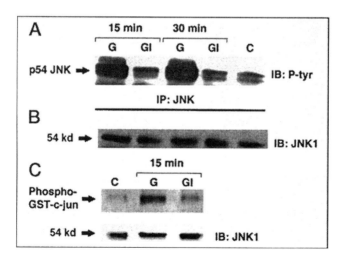

Figure 7. IGF-I inhibits JNK tyrosine phosphorylation. SH-SY5Y cells were incubated with DMEM containing various additions, and JNK tyrosine phosphorylation was assessed. A) Effect of IGF-I on glucose-induced JNK tyrosine phosphorylation. SH-SY5Y cells were treated with DMEM alone as a control (C) or treated with DMEM + 20 mM glucose (G) or DMEM + 20 mM glucose + 10 nM IGF-I (GI) for 15 and 30 min. B) The immunoblot was stripped and reprobed with anti-JNK serum. C) Effect of IGF-I on glucose-induced JNK kinase activity. Cells were treated for 15 min with DMEM (C), DMEM + 20 mM glucose (G) or DMEM + 20 mM glucose + 10 nM IGF-I (GI). JNK activity was determined by GST-c-Jun (1-169) phosphorylation. Blots from the JNK activity assay were stripped and immunoblotted with anti-JNK serum. Results are representative of three independent experiments. (Reprinted with permission from Cheng et al, JBC 273:14560-14565, 1998.)

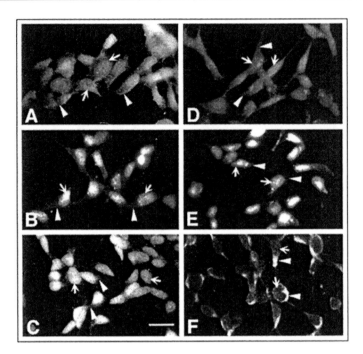

Figure 8. Effects of glucose and IGF-I on the nuclear translocation of p38 kinase and JNK. SH-SY5Y cells were treated for 30 min with DMEM alone (A and D), DMEM + 20 mM glucose (B and E), or DMEM + 20 mM glucose + 10 nM IGF-I (C and F). Cells were stained with p38 kinase (A-C) or JNK (D-F) antisera. Bar, 10 µM. Results are representative of three independent experiments. (Reprinted with permission from Cheng et al, JBC 273:14560-14565, 1998.)

phosphorylation, suggesting that continued receptor activation may induce transformation of cells of the medulla. In support of this, medulloblastoma cell lines overexpress IRS-I, and respond to IGF-I by proliferating in culture.[145] Though present in all astrocytoma tumor grades, IGF-I levels and expression patterns correlate with histopathologic grade, suggesting that IGF signaling is important in tumor cell proliferation.[146] Also, neuroblastoma cell lines overexpressing bcl-2 and n-myc upregulate IGF-IR levels. This is intriguing in light of the fact that human neuroblastoma tumors also exhibit increased bcl-2 and n-myc.[147] IGF-IR has also been implicated in increased tumorigenesis, loss of apoptotic signaling and resistance to the chemotherapeutic drug, doxirubicin, in neuroblastoma cell lines.[60,148] Additionally, in response to IGF-I or IGF-IR activation, neuroblastoma cells can resist hyperosmotic apoptosis induced by high mannitol, which is a vehicle for treatment of childhood tumors.[63,129,149] Further, IGFBP-2 expression is increased in high grade, versus low grade, human glial tumors, implicating IGFBP-2 in glioma malignancy.[150] These studies indicate that components of the IGF system are altered in cancerous cells resistant to apoptosis, and therefore that IGF signaling alterations may induce and stabilize the transformed phenotype.

In summary, these results demonstrate that IGF-I activation of the PI3K/Akt signaling pathway is a potent inhibitor of apoptotic signaling. As apoptosis is a presumed pathogenetic mechanism for several neurodegenerative diseases and in CNS malignancies, the further study of IGF-I and neuronal apoptosis is critical.[151,152] Continued exploration of IGF-I neuroprotection will enable further understanding of the apoptotic death mechanism and potential therapeutic intervention points in this process, which may be important for future clinical therapy.

IGFs Promote Axonal Growth and Regeneration

One of the defining morphological characteristics of differentiated neurons are their processes, namely dendrites and axons. These processes extend during development and must be re-established after neuronal injury.[153,154] Growth factors are important for the formation of dendrites and axons, as commonly associated with neuronal differentiation, as well as in their maintenance and regeneration. Functionality of these processes is critical for intact synapse formation and neuronal signaling, thus, growth factors are extremely important in this system. This section will address evidence that the IGF system is key for neurite extension and differentiation, as well as the biochemical signaling that mediates these events.

Cytoskeletal Mechanisms of Neurite Extension

In order to understand the components of IGF signaling that regulate neurite outgrowth, it is essential to understand the basic known mechanisms of neurite extension. Neurite growth begins with the extension of filopodia and lamellipodia. Filopodia, which are spike-like protrusions composed of parallel actin filaments, form in response to extracellular matrix proteins and soluble factors.[155-157] In contrast, actin containing lamellipodia are veil-like structures that mediate cellular protrusion.[158,159] Filopodia and lamellipodia are both present on structures at the neurite tips, called growth cones.[160] The composite filopodia and lamellipodia move the growth cone in response to extracellular cues, and thereby mediate axonal guidance and extension.[161,162] These extracellular cues induce movement by regulation of the actin cytoskeleton. The actin filaments of filopodia and lamellipodia undergo dynamic polymerization or de-polymerization, which mechanically propulses movement of these structures, and therefore allows for neurite movement.[163] This dynamic polymerization occurs at the leading edge of filopodial or lamellipodial structures, and depending on the extracellular cue, may induce attraction or repulsion of the extended growth cone.[164] Such dynamic cytoskeletal regulation in response to extracellular cues is critical for forming functional synapses in vivo.[165] Growth cone guidance and advance is also dependent on neurite adhesion and de-adhesion to the extracellular substrate.[166] Proteins such as focal adhesion kinase (FAK)[167] and the integrins[168] are important for regulating growth cone adhesion. Thus, neurite extension is mediated by a diverse cast of extracellular cues, adhesion proteins and cytoskeletal changes.

Examples of Neurite Extension Regulated by the IGF System

In vitro and in vivo experiments demonstrate IGF mediated neurite outgrowth in many neuronal subtypes of the central and peripheral nervous systems. In the CNS, IGFs promote neurite outgrowth in cultured cells from the rat cortex,[169] amacrine neurons of the retina,[39] and chick forebrain neurons.[170] Pyramidal neurons of the layer II somatosensory cortex respond to IGF-I treatment by producing more highly branched dendrites and longer apical and basal dendritic arbors.[171] IGF-I also increases the length of neurites generated by adult stem cells from the forebrains of mice, suggesting that the IGF system is important for optimal neurite extension of differentiating neurons.[172] Further, rats implanted with IGF-I containing polymeric implants in the ventral hypothalamus exhibit dramatic growth of oxytocinergic neuron fibers. This demonstrates that IGF-I can induce additional axonal sprouting even in mature hypothalamic neurons.[173] IGF-I is also important for the phasic remodeling of arcuate nucleus synapses between proestrus and estrus cycling. Antagonists of the IGF-IR block this IGF-I mediated synaptic remodeling, demonstrating that IGF-IR activation and signaling is crucial for axonal changes at the synapse in proestrus rats.[174] IGF-II is also important in neurite extension, as it stimulates neurite outgrowth of embryonic dopamine neurons isolated from the rat substantia nigra/vental tegmental area. However, the effects of IGF-II are cell type-dependent, as IGF-II did not stimulate neurite outgrowth of serotonergic neurons isolated from the dorsal raphe.[175] Additionally, the SH-SY5Y neuroblastoma cell line has been utilized as a model system of IGF-I and IGF-II-induced neurite outgrowth (Fig. 9).[113,176,177] IGF-I stimulated SH-SY5Y cells exhibit both PI-3K-dependent leading edge actin polymerization[178] and lamellipodial advance[179] (Fig. 10), thus demonstrating the hallmark mechanisms of cytoskeletal reorganization that underlie neurite extension.

IGFs also modulate neurite outgrowth in neurons of the PNS. IGF-I supports neurite outgrowth in guinea pig myenteric plexus neurons,[180] chick sympathetic and motor neurons[42,181,182] and rat sensory and sympathetic neurons.[183,184] IGF-II also supports outgrowth of cultured chick sympathetic neurons.[185] Most of the evidence for IGF promoted neurite extension comes from studies of transected peripheral nerves. For example, rats treated with IGF-I after sciatic nerve transection exhibit increased motor neuron survival and re-innervation of muscle, showing that IGF-I

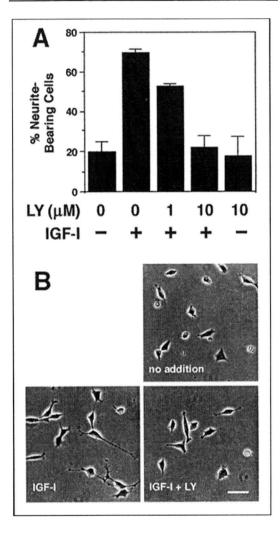

Figure 9. IGF-I induces neurite outgrowth via the PI3K pathway. A) SH-SY5Y neuroblastoma cells were serum starved and treated with 10 nM IGF-I for 24 hours in the presence of 0-10 μM of the PI3K inhibitor, LY29402 (LY). As a control, cells that did not receive LY were treated with 0.1% dimethylsulfoxide (DMSO), the same concentration of DMSO vehicle that LY is dissolved in. After 24 hours of treatment, neurite-bearing cells were counted from each treatment condition. LY treatment alone had no effect on neurite extension compared to control treatment. IGF-I stimulated neurite extension, but this effect was inhibited by inclusion of LY in a dose-dependent manner. B) Phase contrast microscopy images of SH-SY5Y cells treated with vehicle DMSO only (no addition), 10 nM (IGF-I) or 10 nM IGF-I + 10 μM LY. IGF-I induces neurite extension in SH-SY5Y cells, which is blocked by addition of LY, suggesting that IGF-I mediates its neurite enhancing effects via PI3K. (Reprinted with permission from Kim B et al, Endocrinol 139:4881-4889, 1998).

can mediate functional neurite regeneration in vivo.[43] IGF-II infusion in rats undergoing sciatic nerve crush increases the distance of motor axons in the regeneration phase.[186] Functional sciatic nerve regeneration is also promoted by IGF-I treatment in mice after sciatic nerve crush.[187] Furthermore, sciatic nerve damage upregulates IGF-I, IGF-II and IGF-IR,[18,188] but downregulates IGF-II after functional neuromuscular synapses have been re-established.[32] In another model of nerve transection, the end-to-side model, IGF-II yields higher axon counts and greater motor-end-plate counts in the re-innervated nerve.[91] These spatio-temporal regulation patterns of the IGFs and IGF-IR suggest a functional endogenous role of the IGF system in neurite regeneration after injury. Interestingly, neurite extension and regeneration may also be impaired in disease states. For example, diabetic nerves express lower amounts of IGF-I mRNA than normal nerves[189] along with delayed upregulation of IGF-I after sciatic nerve crush.[190] Further, IGFs are also important for developmental axon extension and resulting nerve conduction. This is demonstrated by transgenic mice lacking IGF-I (IGF$^{-/-}$), which have smaller axonal diameters within their peroneal nerves. These mice also exhibit decreased motor and sensory nerve conduction velocities compared to normal transgenic control mice, suggesting that IGF-I is imperative for both proper axonal formation and nerve conduction function.[191]

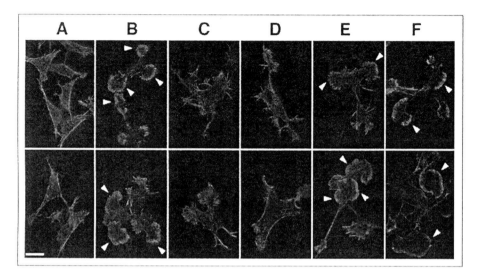

Figure 10. Inhibition of PI3K but not MAP kinase blocks IGF-I-mediated membrane ruffling. Serum-starved SH-SY5Y cells were treated without (A and B) or with 100 nM wortmannin (C), 10 μM LY294002 (D), or 10 μM PD98059 (E) for 1 h and then stimulated for 5 min without (A) or with (B-E) 10 nM IGF-I. p52 Shc mutant-transfected SH-SY5Y cells were treated with 10 nM IGF-I for 5 min (F). Cells were fixed and stained with rhodamine-phalloidin to visualize actin filaments. Arrowheads indicate the membrane ruffling. Duplicate panels are presented for each treatment. Results are representative of at least three different experiments. Bar, 20 μm. (Reprinted with permission from Kim et al, JBC 272:34543-34550, 1998.)

Intracellular Signaling Mechanisms in IGF Stimulated Neurite Outgrowth

Studies of SH-SY5Y neuroblastoma cells and other cultured neurons suggest that IGF-I stimulates neurite outgrowth through the IGF-IR receptor and activation of the PI3K and MAPK pathways. Schematic 2 details the current understanding of this signaling mechanism. The neurite extension properties of IGF-I are induced by IGF-I binding to, and activation of, the IGF-IR.[176,192,193] The activated IGF-IR proteins exhibit phosphorylated tyrosines on the intracellular portion of the receptor.[109,110] Insulin receptor substrate -1 and -2 (IRS-1, -2) proteins bind to phosphotyrosine residues of the IGF-IR and act as docking proteins, allowing binding of additional downstream signaling molecules that contain Src homology 2 (SH2) domains.[111] SH-2 containing proteins such as the p85 regulatory subunit of PI3K (which goes on to activate Akt) and Grb2 (which can go on to activate the MAPK pathway) bind to phosphorylated IRS proteins.[113,114] Interestingly, IRS-1 and IRS-2 may mediate the differential anti-apoptotic and neurite extension effects induced by IGF-I and the IGF-IR. For example, hippocampal rat cells expressing IGF-IR and IRS-I cannot differentiate and express neurites. However, cells expressing IGF-IR, along with a mutant IRS-I lacking a phosphotyrosine binding domain, can undergo differentiation and extend neurites.[194] Further, inhibitors of the PI3K/Akt pathway do not inhibit neurite outgrowth in these neurons, suggesting that the PI3K/Akt pathway is not imperative for this function of IGF-I.[193,194] However, studies in SH-SY5Y cells, which only express IRS-2, indicate that PI3K inhibition suppress IGF-I stimulated neurite outgrowth.[113] These results suggest that neurons expressing IRS-1 may signal predominantly through MAPK, while neurons expressing IRS-2 may utilize both PI3K and MAPK for neurite extension. Though there may be signaling differences between neuronal subtypes of IRS proteins, phosphorylation of IRS tyrosines has been linked to IGF-I stimulated actin polymerization and lamellipodial advance.[195] Thus, signaling through IRS proteins is one mechanism by which IGF-I stimulates neurite outgrowth.

IGF-I activation of the IGF-IR also regulates neurite extension through the binding of Shc and its activation of the MAPK pathway. IGF-I induces Shc binding to the activated IGF-IR, which causes sustained tyrosine phosphorylation of Shc.[196,197] IGF-I also sustains association of

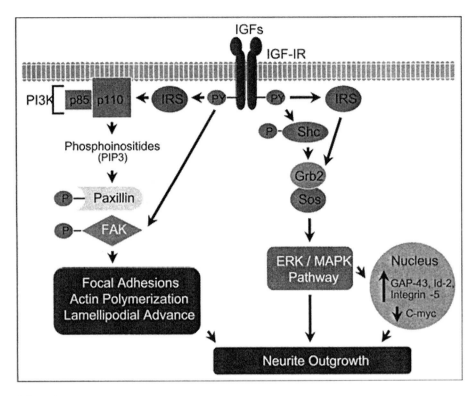

Schematic 2.

phosphorylated Shc with the adaptor protein, Grb2. The Shc/Grb2 complex is known to activate the Ras/MAPK pathway, via the Son of Sevenless (Sos) protein.[198] This association of Shc/Grb 2 is associated with an increase in ERK phosphorylation, a member of the MAPK pathway. Expression of a Shc mutant decreases the IGF-I induced phosphorylation of ERK and neurite outgrowth in SH-SY5Y cells.[197] Further, pharmacological inhibition of the MAPK pathway hinders IGF-I stimulated neurite outgrowth in SH-SY5Y cells and cultured DRG sensory neurons.[176,199] This suggests that the IGF-I stimulated Shc/Grb2 association is critical for activating the downstream MAPK signaling pathway that mediates neurite extension.

IGF-I also regulates the adhesion proteins FAK and paxillin that function in neurite extension. IGF-treated chick sympathetic neurons exhibit adhesion complexes in their growth cones, which include paxillin. These complexes also exhibit tyrosine-phosphorylated proteins, suggesting that the IGF-I/IGF-IR signaling cascade involves tyrosine phosphorylation of adhesion complex proteins.[200] Additionally, tyrosine kinase inhibitors prevent neurite outgrowth in SH-SY5Y cells by disrupting the actin cytoskeleton.[201] Further, IGF-I stimulated SH-SY5Y cells exhibit phosphotyrosine rich adhesion complexes at the leading edge of extending lamellipodia. These adhesion complexes contained phosphorylated paxillin and FAK, suggesting that IGF regulates growth cone extension by phosphorylation of these proteins.[178] In agreement with this theory, IGF-I also stimulates phosphorylation of FAK and paxillin in 3T3 cells.[202] Recent evidence indicates that IGF-induced FAK phosphorylation can be mediated through FAK interaction with the IGF-IR.[203] However, there may be some overlap of FAK phosphorylation by growth factors, as EGF can also promote neurite extension.[204]

In addition to regulation at the protein level, IGF-I regulates neurite extension via transcriptional activation. IGF-I induces transcription of GAP-43 mRNA, a marker of differentiated neurons and regulator of actin polymerization in filopodia.[205,206] In contrast, the c-myc transcription factor is downregulated in IGF-treated SH-SY5Y cells. As c-myc is associated with proliferating cells, downregulation of this transcription factor suggests that IGF-I promotes neurite extension by switching these cells from a proliferative to a differentiative phenotype.[205] Transcription of both GAP-43 and c-myc is blocked in IGF-I treated cells if the MAPK pathway is inhibited, indicating that this pathway is key for downstream nuclear events.[176] IGF-I also increases transcription of Id-2, a transcription factor that regulates proliferation or differentiation in many cell types, and thus may confer neurite extension phenotypes in neurons.[207] Transcription of Integrin α-5, which functions as the fibronectin receptor, is also increased by IGF-I treatment. Thus, IGF-I transcriptionally regulates a protein that mediates adhesion to extracellular substrate, suggesting a role in neurite outgrowth and guidance processes.[208] IGF-I also affects transcription by stabilizing tubulin mRNA, supporting IGF-I's potent regulation of structural proteins that function in neuronal shape and neurite extension.[209] Thus, IGF-I regulates both transcription and mRNA stabilization of several genes that are important for the extension of neurites and neuronal differentiation.

IGFs Promote Neuronal Motility: Relevance to Cancer

Movement of cells, or motility, is regulated by similar mechanisms to those employed in neurite extension. Cellular motility chiefly relies on cell membrane ruffling and the extension of lamellipodia, which are flat, organelle-free membranes composed of actin filaments.[210] As discussed earlier, the polymerization and de-polymerization of actin filaments at the leading edge of lamellipodia, allows for translocation of growth cones. Lamellipodia on cell bodies also mediate translocation via the same mechanism.[159] Translocation of whole cells is important in cancer, where motile cells are able to metastasize, or leave the primary tumor and propigate in a new environment.[211] IGFs have a role in regulating motility, as IGF treatment stimulates motility in breast,[212] lung,[213] prostate[214] and rhabdosarcoma cancer cells.[215] The role of IGFs and motility has also been studied in cells from neuroblastomas, which are brain tumors most commonly found in young children.[216] IGF-I stimulates motility of cultured neuroblastoma cells (Fig. 11), which is mediated through the IGF-IR and PI3K activation.[217,218] Inhibition of the PI3K pathway blocks IGF-I stimulated membrane ruffling in motile cells, though blocking the MAPK pathway does not block this effect.[219] Further, overexpression of PTEN, which de-phosphorylates the lipid products of PI3K, also attenuates IGF-I stimulated motility.[218] This evidence shows that IGF-I stimulation of the PI3K pathway activates motility in transformed cells. However, IGFs can also stimulate motility in primary glial cells, such as Schwann cells. IGF-I stimulated Schwann cell motility also activates PI3K and Akt; blocking Akt signaling by overexpression of dominant negative Akt inhibits this response.[68] Schwann cell motility also depends on the activity of Rac, a Rho GTPase that is instrumental in cytoskeleton dynamics.[220] Like inhibition of the PI3K pathway, blocking Rac signaling also blocks IGF-stimulated motility in Schwann cells. Blocking Rac additionally prohibits phosphorylation of FAK, suggesting that both the cytoskeleton and adhesion proteins are important in this motile process.[221] Thus, IGF-I can mediate motility in neuronal and glial cells, though the mechanisms of these processes are yet to be fully elucidated. Understanding the role of IGFs in transformed neuronal, and non-neuronal cells, may be important in the development of new chemotherapeutic agents for cancer treatment.

IGFs Promote Myelination in the CNS and PNS

Oligodendrocytes of the CNS and Schwann cells of the PNS ensheath the axons of neurons with myelin, a multilayered membrane structure.[222] This ensheathment increases conduction velocities, thereby allowing neuronal messages to be dispersed more rapidly. Thus, degradation of myelin sheaths, as seen in demyelinating diseases such as multiple sclerosis, have profound deleterious effects on nervous system function.[223] The discovery that IGF-I overexpressing mice have larger brains and express 30% more myelin, compared to transgenic controls, suggested that the IGF system might have therapeutic potential in demyelinating disorders.[224] Additionally, mice overexpressing IGF-I exhibit myelination of both large and small axons,[225] contrary to the concept that only large-diameter axons tend to be myelinated.[226] These roles of IGF are blocked in transgenic mice overexpressing IGFBP-1, which binds free IGF-I, such that these animals have fewer myelinated fibers and thinner myelin sheaths compared to transgenic controls.[227] Further, knockout mice lacking IGF-I have fewer axons and oligodendrocytes, resulting in reduced white matter of the brain and spinal cord.[228]

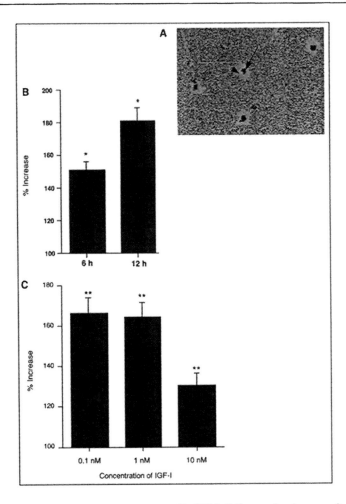

Figure 11. The motility of SH-SY5Y cells is increased by IGF-I. Cells were plated on coverslips coated with gold particles, treated with or without IGF-I for 6 or 12 h, then fixed and mounted. A) Example of tracks left in the gold-particle coated surface by migrating cells. Arrow indicates the gold internalized in the cell body. Arrowhead points to the area over which the cell moved during the incubation. B) Average increase in track area in response to 1 nM IGF-I treatment for 6 and 12 h. *P < 0.0001 in comparison to unstimulated controls for both conditions. C) Effect of different concentrations of IGF-I on SH-SY5Y motility. **P < 0.001 for each condition, in comparison to unstimulated controls. For all conditions, n = 150 cells. (Reprinted with permission from Meyer et al, Oncogene 20:7542-7550, 2001.)

Collectively, these results implicate IGF-I in myelination, though these in vivo studies did not clarify how IGF-I promoted these effects.

Animal models of neurodegenerative diseases of the CNS have proved useful for studying the effects of IGF-I and myelination. For example, carotid artery occlusion, a model of hypoxic ischemic injury, results in loss of oligodendrocytes in the intragyral white matter. IGF-I treatment after artery occlusion inhibits loss of oligodendrocytes, as well as reducing tissue swelling.[229] IGF-I, together with PDGF, enhances myelination of transected axons in a model of spinal cord transection, suggesting a role for IGF-I in spinal injury.[230] Additionally, animal models of demyelination that mimic multiple sclerosis, induced by cuprizone or experimental allergic encephalomyelitis (EAE), have provided insight to IGF-I regulation of myelination. In the cuprizone-induced demyelination model, IGF-IR mRNA is increased during early recovery stages,[231] though mRNA levels of other growth factors, such as PDGF and FGF, were unaltered.[232] Further, mice overexpressing IGF-I exhibit a nearly full recovery of myelination 5 weeks after cuprizone treatment, while control mice exhibit near complete demyelination at the same stage. This rapid recovery in IGF transgenic mice is due to reduced apoptotic death of mature oligodendrocytes, coupled with restoration of oligodendrocytes that had degenerated.[64] In the EAE model of demyelination, IGF-I decreases inflammation, demyelination and clinical deficit of treated animals.[233-235] However, the time of IGF-I treatment may be critical for these effects, as delivery during the acute phase of demyelination did not promote remyelination or enhancement of oligodendrocyte progenitors.[236] Finally, developing mice overexpressing IGF-I are protected from myelination defects due to nutritional deficit. Thus, IGF-I has potent effects on myelination in the developing and adult CNS.

IGF-I likely mediates myelination effects in the CNS through glia, particularly oligodendrocytes and astrocytes. Developing oligodendrocytes express IGF-I mRNA, suggesting a role for IGF-I regulation in formation of the CNS.[237] In the adult nervous system, IGF-I enhances survival of mature oligodendrocytes.[64] Also, culture models have also suggested that IGF-I can promote proliferation of oligodendrocytes, and thus increases total myelin levels.[238] These IGF-mediated effects in developing oligodendroctyes may be mediated by PI3K activation,[239] but little is known about IGF-I signaling in adult oligodendrocytes. Astrocytes also express IGF-I after injury to the brain. [231,240] Reduction of IGF-I in astroctyes, as well as microglia, correlates with improper re-myelination and delayed differentiation of oligodendrocyte precursors in cuprizone demyelination, suggesting that glia are a critical source of IGF-I for the myelination process of the CNS.[241]

IGF-I also plays a role in PNS myelination. When DRG and Schwann cells are cultured together, IGF-I promotes both Schwann cell attachment to (Fig. 12), and ensheathment of, DRG

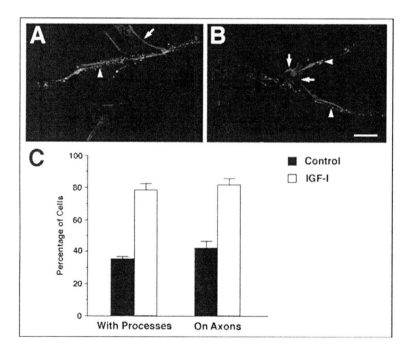

Figure 12. IGF-I Enhances Schwann cell Attachment on Axons. Schwann cells (SC) were cocultured with DRG neurons + 10 nM IGF-I for 6 (A) or 12 h (B). Cells were double immunostained for S-100 (rhodamine) and neurofilament (fluoroscein) to determine SC (arrows) and axons (arrowheads). B) After 12 h of IGF-I treatment, SC extend processes along axons. Pictures are representative of 3 separate experiments. C) IGF-I enhances numbers of SC with processes (processes longer than the width of the cell body) and SC attachment on axons by 2-fold compared to the serum free control. Bar = 20 μm. (Reprinted with permission from Cheng et al, JBC 275:27197-27204, 2000.)

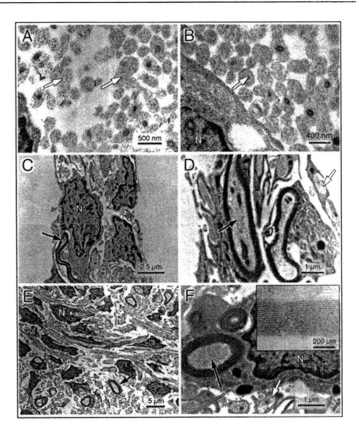

Figure 13. TEM photo-micrograph showing myelination of DRG axons in the presence of IGF-I. Dissociated DRG are cultured for 21d in SIFDM. A) SIFDM alone (control) showing healthy unmyelinated axons (white arrows). B) Higher magnification shows the nucleus (N) of an adjacent Schwann cell (SC). There are no SC processes extending and ensheathing the axons despite the close apposition of the SC and axon. C) SIFDM and 1 nM IGF-I show myelination of a few individual axons (black arrow) although many axons remain unmyelinated. D) Higher magnification of C shows myelinated axons (black arrow) and in addition extensive attachment and ensheathment of unmyelinated axons (white arrow) by SC processes. E) SIFDM and 10 nM IGF-I showing extensive axonal myelination (black arrow) in comparison with A and C. F) Higher magnification showing well myelinated large and small diameter axons with extensive SC (N-SC nucleus) ensheathment of remaining unmyelinated axons (white arrow). The insert shows normal periodicity of compact myelin and formation of major dense lines in the presence of IGF-I. Although IGF-I is seen to myelinate a small diameter axon in this figure, IGF-I does not preferentially myelinate or overmyelinate smaller axons (< 1 mm^2) compared to larger axons (> 1 mm^2). (Reprinted with permission from Russell et al, J Neuropathol Exp Neurol 59:575-584, 2000.)

axons (Fig. 13).[242,243] Interestingly, in the absence of IGF-I, Schwann cells survive in co-cultures, but do not myelinate DRG axons, indicating that IGF-I is an important factor for myelinogenesis in the PNS.[243] IGF-I also increases expression of P_o, which is a major component of myelin.[242,244,245] While there is much to be learned about how IGF-I regulates myelination in the PNS, these studies show that the role of IGF-I is important for the myelination process.

IGFs in the Treatment of Neurological Diseases

As IGFs are effective neurotrophic factors and promoters of CNS and PNS myelination, there has been much interest in the use of IGFs to treat muscle disorders[246] and diseases that result in loss of neurons, axons, or myelin. Thus far, the therapeutic potential of IGFs has been tested in animal models of stroke, trauma, neuropathies, and multiple sclerosis, and in human trials of patients with head trauma and amyotrophic lateral sclerosis

IGFs in the Treatment of Hypoxic-Ischemic Injury

Hypoxia/ischemia (HI) of nervous system tissue is a major cause of permanent disability and death. Neurons are lost in a selective and delayed fashion in perinatal hypoxia and focal ischemia (stroke). Both apoptotic[247,248] and necrotic mechanisms lead to neuronal death in HI. The ability of IGF-I to rescue neurons from apoptosis induced by a variety of insults has led to studies examining the effect of IGF-I on the outcome of HI. An animal model of HI involves the transient ligation of one of the carotid arteries. Several studies using this model have found that recombinant human IGF-I, administered centrally 1-2h after HI, reduces neuronal loss in the brain and decreases the incidence of cerebral infarct.[56,249-252] Reduction of neuronal apoptosis was partly responsible for this effect.[56] Centrally administered IGF-I was also found to decrease neuronal loss, incidence of seizures, and cerebral edema in a sheep model of HI.[253] Spinal cord ischemia benefits from IGF-I treatment as well. A single dose of IGF-I 30 min prior to the induction of cord HI in rabbits allowed for recovery of hind limb function 48h later, in comparison to untreated and insulin treated controls.[254] While single doses have effects on neuronal loss and functional outcome, longer durations of therapy may have a greater impact on infarct size. A single dose of IGF-I 2h after HI injury in rats resulted in long term neuroprotective and behavioral effects. Rats treated with IGF-I performed significantly better on a somatosensory test 3 d post-injury, and had less selective loss of neurons 20 d post-injury.[255] But, there was only a trend towards reduction in infarct size. IGF-I administered intraventricularly or subcutaneously daily for 3d post-HI decreased the size of cerebral infarction and the degree of neurological impairment.[256] Benefit from IGF-I treatment appears to be dependent upon comparatively higher doses (at least 20 µg/rat), as studies that have tried lower doses failed to demonstrate benefit.[256,257] Most studies have relied on intraventricular delivery of IGF-I to circumvent the blood-brain barrier. As this would be inconvenient in human patients, an intranasal delivery of IGF-I was tried in a rat model of stroke. Intranasal delivery successfully diminished infarct size and impairment, suggesting this may be a facile and efficient route for exposing the CNS to IGF-I following HI injury.[258] The potent effects of IGF-I on reducing nervous system damage caused by HI suggest that IGF-I may be a beneficial treatment for perinatal hypoxic injury or stroke.

IGFs as Promoters of Neuronal Regeneration Following Trauma

IGFs have also been studied in the treatment of traumatic nervous system injury, based on their increased endogenous expression following injury, and their ability to promote neurite growth. Administration of IGF-I to a crushed sciatic nerve in the rat promotes regeneration.[259] IGF-II causes similar regeneration in the frog sciatic nerve.[260] IGF-I improved motor neuron survival, functional re-innervation and morphometry of skeletal muscle following sciatic nerve transection in 2d old rats.[43] In rats subjected to a traumatic brain injury, IGF-I treatment helped restore cognitive and motor neuron functions.[261] To date, one human clinical trial has examined the benefits of IGF-I in the treatment of head trauma. Intravenous IGF-I was given continuously for a 14d period to patients with severe isolated traumatic head injuries. Six months later, eight of the eleven treated patients that achieved the highest serum concentrations of IGF-I received moderate to good outcome scores, versus only one of five patients that had low levels of IGF-I.[262] Thus, the initial evidence indicates that IGF-I may be clinically useful in the treatment of traumatic head and peripheral nerve injury, though more work is needed to confirm this.

IGFs in the Treatment of Peripheral Nerve Disease

Peripheral neuropathies are characterized by decreased tissue innervation, neuronal survival, and nerve regeneration. Peripheral neuropathies have a variety of etiologies, including advanced diabetes mellitus and toxicity associated with chemotherapeutic agents. IGFs have been considered in the treatment of peripheral neuropathy because of their effects on neuronal survival and axon regeneration. IGFs may be particularly important in the treatment of diabetic neuropathy, as some studies show that decreased IGF expression may contribute to diabetic neuropathy.[263-265] In rats with non-insulin-dependent diabetes mellitus, subcutaneous IGF-II restores pain and pressure thresholds towards normal.[263] IGF-I and IGF-II increase the ability of the sciatic nerve of the diabetic rat to regenerate after a crushing injury.[266] IGF-I treatment also reversed the neuroaxonal dystrophy seen in a rat model of diabetic autonomic neuropathy.[267] Studies have shown that IGFs can prevent peripheral neuropathies caused by vincristine, cisplatinum, and taxol.[268,269] Peripheral neuropathy is a dose-limiting effect of some anti-tumor drugs, so IGF treatment might allow higher doses of the drugs to be tolerated.

Motor neuron disease is a heterogeneous group of disorders that selectively affects upper and/or lower motor neurons. A common form of motor neuron disease is amyotrophic lateral sclerosis (ALS), or Lou Gehrig's disease. IGF-I was considered as a therapeutic for ALS because of its potential for protecting motor neurons. Expression of the IGF-IR is increased in the spinal cord of patients with ALS, suggesting that increased responsiveness to IGFs may be part of the body's adaptation to the disease.[270] Also, IGF-I preserves motor neurons in the *wobbler* mouse.[271] Two placebo-controlled trials of IGF-I in ALS patients have produced mixed results. The North American ALS/IGF-I Study Group found that patients receiving 9 months of daily IGF-I had slower disease progression, and reported a better quality of life, than placebo-treated controls.[272] However, The European ALS/IGF-I Study Group showed no benefit to IGF-I therapy in a similar paradigm.[273] IGF-I could potentially add to the benefits of the glutamate-release inhibitor Riluzole, the only currently effective drug for treating ALS. A study in rats with transected sciatic nerves showed no significant increased benefit when IGF-I was added to Riluzole treatment, in terms of motor neuron survival and axon diameter.[274] Despite the experimental data showing potential for IGF-I in the treatment of motor neuron disease, the usefulness of this treatment remains unclear.

IGFs in the Treatment of Demyelinating Diseases

IGFs are effective promoters of myelination in the PNS and CNS, and thus have been considered for treating demyelinating diseases like multiple sclerosis (MS). A number of studies have looked at the effect of IGF-I on the progression of the main animal model for MS, experimental autoimmune encephalomyelitis (EAE). The results of these studies have been contradictory. Some studies find that subcutaneous or intravenous IGF-I reduces the clinical signs of EAE, blood-brain barrier defects, and the number and size of lesions,[233,235,275] while at the same time increases remyelination and the proliferation of oligodendrocyte-like cells in the lesion.[234,275] IGF-I also stimulates the production of myelin components, including myelin basic protein, proteolipid protien, and 2',3'-cyclic nucleotide 3'-phosphodiesterase.[233,234] Other studies, however, suggest the effect of IGF-I on the clinical progression of the disease is transient,[236] and fail to detect a significant increase in remyelination or progenitor cell proliferation. In another study, IGF-I given prior to the onset of EAE had a positive effect, but given after disease onset had no effect. Administration of IGF-I along with IGFBP-3 led to severe relapse.[276] The contradictory results of these investigations leave the usefulness of IGF-I in the treatment of MS unanswered.

IGFs in the Treatment of Neurodegenerative Diseases

IGF-I may be neuroprotective in Huntington's disease by decreasing the toxicity of the mutant huntingtin protein. Striatal neurons cultured in vitro and transfected with huntingtin containing polyglutamine expansions die at a higher rate than cells transfected with normal huntingtin. IGF-I treatment preserves the neurons expressing mutant huntingtin, as does constitutively active Akt. Akt phosphorylates huntingtin on serine 421 in vitro and in vivo. This phosphorylation can be stimulated by IGF-I, through PI-3K signaling. Striatal neurons expressing an expanded huntingtin mutant lacking the phosphorylation site are no longer salvageable when treated with IGF-I. Conversely, mutating the phosphorylation site to mimic constitutive phosphorylation prevents neuronal death and decreases intracellular inclusions.[277] This single study is the first indication of a novel mechanism for IGF-I-mediated neuroprotection, where IGF-I may decrease the toxicity of a protein containing a polyglutamine expansion by phosphorylation of the protein.

Treating neuroblastoma cells with amyloid β-peptide, the main component of the amyloid plaques in Alzheimer's disease, causes cell death and a two- to three-fold increase in JNK activation. The cell death is blocked by a dominant-negative SAPK, suggesting that JNK activation is required for amyloid β-peptide induced death. IGF-I protects the neuroblastoma cells from amyloid β-peptide toxicity and decreases JNK activation via PI-3K signaling.[278] Thus there is potential for IGF-I to protect neurons from the effects of amyloid β-peptide in Alzheimer's disease.

Summary

Diseases of the nervous system may be among the most devastating to humans because they so radically affect quality of life, from memory to movement. As such, the search for neurological therapeutics is of utmost importance in medicine's quest to both extend and improve human life. The basic science of the last 20 years has illustrated the potent neuroprotective and regenerative properties of IGFs, and these effects are now being tested in the clinical setting. However, additional work is needed to understand how IGFs are utilized in neurological systems. Better understanding of the downstream intracellular and nuclear targets of IGF signals could allow for a more finely tuned therapy paradigm in future clinical applications. Further, understanding the intracellular controls that determine whether IGF will impart survival or neurite extension/regeneration are of vital interest for developing targeted therapies to divergent neurological disorders. In conclusion, continued research will expand the initial promise of IGFs and their roles both in the normal and diseased nervous system.

Acknowledgements

The authors would like to thank Judith Boldt for expert secretarial assistance. This work was supported by NIH NS43023 (GG), NIH CA09676 (GM), NIH NS38849, NIH NS36778, the Juvenile Diabetes Research Foundation Center for the Study of Complications in Diabetes and the Program for Understanding Neurological Diseases (ELF).

References

1. Daughaday WH, Rotwein P. Insulin-like growth factor I and II. Peptide, messenger ribonucleic acid and gene structures, serum, and tissue concentrations. Endocr Rev 1989; 10:68-91.
2. Sara VR, Hall K. Insulin-like growth factors and their binding proteins. Physiol Rev 1990; 70:591-614.
3. Bondy CA, Werner H, Roberts CT Jr et al. Cellular pattern of insulin-like growth factor-I (IGF-I) and type I IGF receptor gene expression in early organogenesis: comparison with IGF-II gene expression. Mol Endocrinol 1990; 4:1386-1398.
4. Devaskar SU, Sadiq HF, Holtzclaw L et al. The developmental pattern of rabbit brain insulin and insulin-like growth factor receptor expression. Brain Res 1993; 605:101-109.
5. Rotwein P, Burgess SK, Milbrandt JD et al. Differential expression of insulin-like growth factor genes in rat central nervous system. Proc Natl Acad Sci USA 1988; 85:265-269.
6. Lai M, Hibberd CJ, Gluckman PD et al. Reduced expression of insulin-like growth factor 1 messenger RNA in the hippocampus of aged rats. Neurosci Lett 2000; 288:66-70.
7. Hammarberg H, Risling M, Hokfelt T et al. Expression of insulin-like growth factors and corresponding binding proteins (IGFBP 1-6) in rat spinal cord and peripheral nerve after axonal injuries. J Comp Neurol 1998; 400:57-72.
8. Logan A, Gonzalez A-M, Hill DJ et al. Coordinated pattern of expression and localization of insulin-like growth factor-II (IGF-II) and IGF-binding protein-2 in the adult rat brain. Endocrinology 1994; 135:2255-2264.
9. Couce ME, Weatherington AJ, McGinty JF. Expression of insulin-like growth factor-II (IGF-II) and IGF-II/mannose-6-phosphate receptor in the rat hippocampus: An in situ hybridization and immunocytochemical study. Endocrinology 1992; 131:1636-1642.
10. Stylianopoulou F, Herbert J, Soares MB et al. Expression of the Insulin-Like Growth Factor II Gene in the Choroid Plexus and the Leptomeninges of the Adult Rat Nervous System. Proc Natl Acad Sci USA 1993; 85:141-145.
11. Kiess W, Yang Y, Kessler U et al. Insulin-like growth factor II (IGF-II) and the IGF-II/mannose-6-phosphate receptor: The myth continues. Horm Res 1994; 41 Suppl. 2:66-73.
12. Dahms NM, Lobel P, Kornfeld S. Mannose-6-phosphate receptors and lysosomal enzyme targeting. J Biol Chem 1989; 264:12115-12118.
13. Braulke T, Gotz W, Claussen M. Immunohistochemical localization of insulin-like growth factor binding protein-1, -3 and -4 in human fetal tissues and their analysis in media from fetal tissue explants. Growth Regul 1996; 6:55-65.
14. Ishii DN. Relationship of insulin-like growth factor II gene expression in muscle to synaptogenesis. Proc Natl Acad Sci USA 1989; 86:2898-2901.
15. Hansson H-A, Nilsson A, Isgaard J et al. Immunohistochemical localization of insulin-like growth factor I in the adult rat. Histochemistry 1988; 89:403-410.
16. Syroid DE, Zorick TS, Arbet-Engels C et al. A role for insulin-like growth factor-I in the regulation of Schwann cell survival. J Neurosci 1999; 19:2059-2068.
17. Cheng H-L, Sullivan KA, Feldman EL. Immunohistochemical localization of insulin-like growth factor binding protein-5 in the developing rat nervous system. Brain Res Dev Brain Res 1996; 92:211-218.
18. Cheng H-L, Randolph A, Yee D et al. Characterization of insulin-like growth factor-I (IGF-I), IGF-I receptor and binding proteins in transected nerves and cultured Schwann cells. J Neurochem 1996; 66:525-536.

19. Clawson TF, Vannucci SJ, Wang GM et al. Hypoxia-ischemia-induced apoptotic cell death correlates with IGF-I mRNA decrease in neonatal rat brain. Biol Signals Recept 1999; 8:281-293.
20. Lee WH, Wang GM, Yang XL et al. Perinatal hypoxia-ischemia decreased neuronal but increased cerebral vascular endothelial IGFBP3 expression. Endocrine 1999; 11:181-188.
21. Morano S, Sensi M, Di Gregorio S et al. Peripheral, but not central, nervous system abnormalities are reversed by pancreatic islet transplantation in diabetic Lewis rats. Eur J Neurosci 1996; 8:1117-1123.
22. Klempt M, Klempt ND, Gluckman PD. Hypoxia and hypoxia/ischemia affect the expression of insulin-like growth factor binding protein 2 in the developing rat brain. Brain Res Mol Brain Res 1993; 17:347-350.
23. Beilharz EJ, Russo VC, Butler G et al. Co-ordinated and cellular specific induction of the components of the IGF/IGFBP axis in the rat brain following hypoxic-ischemic injury. Brain Res Mol Brain Res 1998; 59:119-134.
24. Walter HJ, Berry M, Hill DJ et al. Spatial and temporal changes in the insulin-like growth factor (IGF) axis indicate autocrine/paracrine actions of IGF-I within wounds of the rat brain. Endocrinology 1997; 138:3024-3034.
25. Walter HJ, Berry M, Hill DJ et al. Distinct sites of insulin-like growth factor (IGF)-II expression and localization in lesioned rat brain: possible roles of IGF binding proteins (IGFBPs) in the mediation of IGF-II activity. Endocrinology 1999; 140:520-532.
26. Giannakopoulou M, Mansour M, Kazanis E et al. NMDA receptor mediated changes in IGF-II gene expression in the rat brain after injury and the possible role of nitric oxide. Neuropathol Appl Neurobiol 2000; 26:513-521.
27. Woods AG, Guthrie KM, Kurlawalla MA et al. Deafferentation-induced increases in hippocampal insulin-like growth factor-1 messenger RNA expression are severely attenuated in middle aged and aged rats. Neuroscience 1998; 83:663-668.
28. Yao D-L, West NR, Bondy CA et al. Cryogenic spinal cord injury induces astrocytic gene expression of insulin-like growth factor I and insulin-like growth factor binding protein 2 during myelin regeneration. J Neurosci Res 1995; 40:647-659.
29. Breese CR, D'Costa A, Rollins YD et al. Expression of insulin-like growth factor-1 (IGF-1) and IGF-binding protein 2 (IGF-BP2) in the hippocampus following cytotoxic lesion of the dentate gyrus. J Comp Neurol 1996; 369:388-404.
30. Nordqvist ACS, von Holst H, Holmin S et al. Increase of insulin-like growth factor (IGF)-1, IGF binding protein-2 and -4 mRNAs following cerebral contusion. Brain Res Mol Brain Res 1996; 38:285-293.
31. Pu SFw, Zhuang HX, Ishii DN. Differential spatio-temporal expression of the insulin-like growth factor genes in regenerating sciatic nerve. Brain Res Mol Brain Res 1995; 34:18-28.
32. Glazner GW, Ishii DN. Insulinlike growth factor gene expression in rat muscle during reinnervation. Muscle Nerve 1995; 18:1433-1442.
33. Zochodne DW, Cheng C. Neurotrophins and other growth factors in the regenerative milieu of proximal nerve stump tips. J Anat 2000; 196(Pt2):279-283.
34. Gehrmann J, Yao D-L, Bonetti B et al. Expression of insulin-like growth factor-I and related peptides during motoneuron regeneration. Exp Neurol 1994; 128:202-210.
35. Torres-Aleman I. Serum growth factors and neuroprotective surveillance: focus on IGF-1. Mol Neurobiol 2000; 21:153-160.
36. Connor B, Dragunow M. The role of neuronal growth factors in neurodegenerative disorders of the human brain. Brain Res Brain Res Rev 1998; 27:1-39.
37. Zhang Y, Tatsuno T, Carney JM et al. Basic FGF, NGF, and IGFs protect hippocampal and cortical neurons against iron-induced degeneration. J Cereb Blood Flow Metab 1993; 13:378-388.
38. Wilkins A, Chandran S, Compston A. A role for oligodendrocyte-derived IGF-1 in trophic support of cortical neurons. Glia 2001; 36:48-57.
39. Politi LE, Rotstein NP, Salvador G et al. Insulin-like growth factor-I is a potential trophic factor for amacrine cells. J Neurochem 2001; 76:1199-1211.
40. Kermer P, Klocker N, Labes M et al. Insulin-like growth factor-I protects axotomized rat retinal ganglion cells from secondary death via PI3-K-dependent akt phosphorylation and inhibition of caspase-3 In vivo. J Neurosci 2000; 20:722-728.
41. Ishii DN, Glazner GW, Pu S-F. Role of insulin-like growth factors in peripheral nerve regeneration. Pharmacol Ther 1994; 62:125-144.
42. Zackenfels K, Oppenheim RW, Rohrer H. Evidence for an important role of IGF-I and IGF-II for the early development of chick sympathetic neurons. Neuron 1995; 14:731-741.
43. Vergani L, Di Giulio AM, Losa M et al. Systemic administration of insulin-like growth factor decreases motor neuron cell death and promotes muscle reinnervation. J Neurosci Res 1998; 54:840-847.
44. Pu SF, Zhuang HX, Marsh DJ et al. Insulin-like growth factor-II increases and IGF is required for postnatal rat spinal motoneuron survival following sciatic nerve axotomy. J Neurosci Res 1999; 55:9-16.
45. Vincent AM, Feldman EL. Control of cell survival by IGF signaling pathways. Growth hormone and IGF Research 2002; 12(4):193.
46. Budihardjo I, Oliver H, Lutter M et al. Biochemical pathways of caspase activation during apoptosis. Annu Rev Cell Dev Biol 1999; 15:269-290.
47. Martin LJ. Neuronal cell death in nervous system development, disease, and injury (Review). Int J Mol Med 2001; 7:455-478.
48. Yuan J, Yankner BA. Apoptosis in the nervous system. Nature 2000; 407:802-809.
49. Cohen GM. Caspases: the executioners of apoptosis. Biochem J 1997; 326(Pt1):1-16.
50. Chang HY, Yang X. Proteases for cell suicide: functions and regulation of caspases. Microbiol Mol Biol Rev 2000; 64:821-846.

51. Ham J, Eilers A, Whitfield J et al. c-Jun and the transcriptional control of neuronal apoptosis. Biochem Pharmacol 2000; 60:1015-1021.
52. Gleichmann M, Weller M, Schulz JB. Insulin-like growth factor-1-mediated protection from neuronal apoptosis is linked to phosphorylation of the pro-apoptotic protein BAD but not to inhibition of cytochrome c translocation in rat cerebellar neurons. Neurosci Lett 2000; 282:69-72.
53. Villalba M, Bockaert J, Journot L. Concomitant induction of apoptosis and necrosis in cerebellar granule cells following serum and potassium withdrawl. Neuroreport 1997; 8:981-985.
54. Dore S, Bastianetto S, Kar S et al. Protective and rescuing abilities of IGF-I and some putative free radical scavengers against beta-amyloid-inducing toxicity in neurons. Ann N Y Acad Sci 1999; 890:356-364.
55. Tagami M, Yamagata K, Nara Y et al. Insulin-like growth factors prevent apoptosis in cortical neurons isolated from stroke-prone spontaneuosly hypertensive rats. Lab Invest 1997; 76:603-612.
56. Tagami M, Ikeda K, Fujino H et al. Insulin-like growth factor-1 attenuates apoptosis in hippocampal neurons caused by cerebral ischemia and reperfusion in stroke-prone spontaneously hypertensive rats. Lab Invest 1997; 76:613-617.
57. Lewis ME, Neff NT, Contreras PC et al. Insulin-like growth factor-I: Potential for treatment of motor neuronal disorders. Exp Neurol 1993; 124:73-88.
58. Russell JW, Windebank AJ, Schenone A et al. Insulin-like growth factor-I prevents apoptosis in neurons after nerve growth factor withdrawal. J Neurobiol 1998; 36:455-467.
59. Russell JW, Sullivan KA, Windebank AJ et al. Neurons undergo apoptosis in animal and cell culture models of diabetes. Neurobiol Dis 1999; 6:347-363.
60. Gil-Ad I, Shtaif B, Luria D et al. Insulin-like-growth-factor-I (IGF-I) antagonizes apoptosis induced by serum deficiency and doxorubicin in neuronal cell culture. Growth Horm IGF Res 1999; 9:458-464.
61. Matthews CC, Odeh H, Feldman EL. Insulin-like growth factor-I is an osmoprotectant in human neuroblastoma cells. Neuroscience 1997; 79:525-534.
62. Matthews CC, Feldman EL. Insulin-like growth factor I rescues SH-SY5Y human neuroblastoma cells from hyperosmotic induced programmed cell death. J Cell Physiol 1996; 166:323-331.
63. van Golen CM, Feldman EL. Insulin-like growth factor I is the key growth factor in serum that protects neuroblastoma cells from hyperosmotic-induced apoptosis. J Cell Physiol 2000; 182:24-32.
64. Mason JL, Ye P, Suzuki K et al. Insulin-like growth factor-1 inhibits mature oligodendrocyte apoptosis during primary demyelination. J Neurosci 2000; 20:5703-5708.
65. Ye P, D'Ercole AJ. Insulin-like growth factor I protects oligodendrocytes from tumor necrosis factor-alpha-induced injury. Endocrinology 1999; 140:3063-3072.
66. Delaney CL, Cheng H-L, Feldman EL. Insulin-like growth factor-I prevents caspase mediated apoptosis in Schwann cells. J Neurobiol 1999; 41:540-548.
67. Delaney CL, Russell JW, Cheng H-L et al. Insulin-like growth factor-I and over-expression of Bcl-xL prevent glucose-mediated apoptosis in Schwann cells. J Neuropathol Exp Neurol 2001; 60:147-160.
68. Cheng H-L, Steinway M, Delaney CL et al. IGF-I promotes Schwann cell motility and survival via activation of Akt. Mol Cell Endocrinol 2000; 170:211-215.
69. Wolf BB, Green DR. Suicidal tendencies: apoptotic cell death by caspase family proteinases. J Biol Chem 1999; 274:20049-20052.
70. Green DR. Apoptotic pathways: the roads to ruin. Cell 1998; 94:695-698.
71. Kumar S. Mechanisms mediating caspase activation in cell death. Cell Death Differ 1999; 6:1060-1066.
72. Bratton SB, MacFarlane M, Cain K et al. Protein complexes activate distinct caspase cascades in death receptor and stress-induced apoptosis. Exp Cell Res 2000; 256:27-33.
73. Sun XM, MacFarlane M, Zhuang J et al. Distinct caspase cascades are initiated in receptor-mediated and chemical-induced apoptosis. J Biol Chem 1999; 274:5053-5060.
74. Raoul C, Henderson CE, Pettmann B. Programmed cell death of embryonic motoneurons triggered through the Fas death receptor. J Cell Biol 1999; 147:1049-1062.
75. Li P, Nijhawan D, Budihardjo I et al. Cytochrome c and dATO-dependent formation of Apaf-1/caspase-9 complex initiates an apoptotic protease cascade. Cell 1997; 91:479-489.
76. Slee EA, Harte MT, Kluck RM et al. Ordering the cytochrome c-initiated caspase cascade: hierarchical activation of caspases-2,3,6,7,8, and 10 in a caspase-9 dependent manner. J Cell Biol 1999; 144:281-292.
77. Thornberry NA, Lazebnik Y. Caspases: enemies within. Science 1998; 281:1312-1316.
78. Cellerino A, Bahr M, Isenmann S. Apoptosis in the developing visual system. Cell Tissue Res 2000; 301:53-69.
79. Graham SH, Chen J, Clark RS. Bcl-2 family gene products in cerebral ischemia and traumatic brain injury. J Neurotrauma 2000; 17:831-841.
80. Hsu Y-T, Wolter KG, Youle RJ. Cytosol-to-membrane redistribution of Bax and Bcl-X$_L$ during apoptosis. Proc Natl Acad Sci USA 1997; 94:3668-3672.
81. Oltvai ZN, Milliman CL, Korsmeyer SJ. Bcl-2 heterodimerizes in vivo with a conserved homolog, Bax, that accelerates programmed cell death. Cell 1993; 74:609-619.
82. Yang J, Liu X, Bhalla K et al. Prevention of apoptosis by Bcl-2: Release of cytochrome c from mitochondria blocked. Science 1997; 275:1129-1132.
83. Minn AJ, Kettlun CS, Liang H et al. Bcl-xl regulated apoptosis by heterodimerization-dependent and independent mechanisms. EMBO J 1999; 18:632-643.
84. Shitashige M, Toi M, Yano T et al. Dissociation of bax from a bcl-2/bax heterodimer triggered by phosphorylation of serine 70 of bcl-2. J Biochem (Tokyo) 2001; 130:741-748.
85. Srivastava RK, Mi QS, Hardwick JM et al. Deletion of the loop region of Bcl-2 completely blocks paclitaxel-induced apoptosis. Proc Natl Acad Sci USA 1999; 96:3775-3780.

86. Cao G, Minami M, Pei W et al. Intracellular Bax translocation after transient cerebral ischemia: implications for a role of the mitochondrial apoptotic signaling pathway in ischemic neuronal death. J Cereb Blood Flow Metab 2001; 21:321-333.
87. Marzo I, Brenner C, Zamzami N et al. Bax and adenine nucleotide translocator cooperate in the mitochondrial control of apoptosis. Science 1998; 281:2027-2031.
88. Shimizu S, Narita M, Tsujimoto Y. Bcl-2 family proteins regulate the release of apoptogenic cytochrome c by the mitochondrial channel VDAC. Nature 1999; 399:483-487.
89. Putcha GV, Deshmukh M, Johnson EM Jr. BAX translocation is a critical event in neuronal apoptosis: regulation by neuroprotectants, BCL-2, and caspases. J Neurosci 1999; 19:7476-7485.
90. Robertson JD, Orrenius S. Molecular mechanisms of apoptosis induced by cytotoxic chemicals. Crit Rev Toxicol 2000; 30:609-627.
91. Caplan J, Tiangco DA, Terzis JK. Effects of IGF-II in a new end-to-side model. J Reconstr Microsurg 1999; 15:351-358.
92. Chao DT, Korsmeyer SJ. BCL-2 family: regulators of cell death. Annu Rev Immunol 1998; 16:395-419.
93. Tzivion G, Shen YH, Zhu J. 14-3-3 proteins; bringing new definitions to scaffolding. Oncogene 2001; 20:6331-6338.
94. Constantopoulos G, Rees S, Cragg BG et al. Suramin-induced storage disease. Mucopolysaccharidosis. Am J Pathol 1983; 113:266-268.
95. Kuan CY, Yang DD, Samanta RD et al. The Jnk1 and Jnk2 protein kinases are required for regional specific apoptosis during early brain development. Neuron 1999; 22:667-676.
96. Le-Niculescu H, Bonfoco E, Kasuya Y et al. Withdrawal of survival factors results in activation of the JNK pathway in neuronal cells leading to Fas ligand induction and cell death. Mol Cell Biol 1999; 19:751-763.
97. Kanamoto T, Mota M, Takeda K et al. Role of apoptosis signal-regulating kinase in regulation of the c-Jun N-terminal kinase pathway and apoptosis in sympathetic neurons. Mol Cell Biol 2000; 20:196-204.
98. Perona R, Montaner S, Saniger L et al. Activation of the nuclear factor-kappaB by Rho, CDC42, and Rac-1 proteins. Genes Dev 1997; 11:463-475.
99. Bazenet CE, Mota MA, Rubin LL. The small GTP-binding protein Cdc42 is required for nerve growth factor withdrawal-induced neuronal death. Proc Natl Acad Sci USA 1998; 95:3984-3989.
100. Estus S, Zaks WJ, Freeman RS et al. Altered gene expression in neurons during programmed cell death: identification of c-jun as necessary for neuronal apoptosis. J Cell Biol 1994; 6:1717-1727.
101. Iordanov MS, Magun BE. Different mechanisms of c-Jun NH(2)-terminal kinase-1 (JNK1) activation by ultraviolet-B radiation and by oxidative stressors. J Biol Chem 1999; 274:25801-25806.
102. Ho FM, Liu SH, Liau CS et al. High glucose-induced apoptosis in human endothelial cells is mediated by sequential activations of c-Jun NH(2)-terminal kinase and caspase-3. Circulation 2000; 101:2618-2624.
103. Maroney AC, Glicksman MA, Basma AN et al. Motoneuron apoptosis is blocked by CEP-1347 (KT 7515), a novel inhibitor of the JNK signaling pathway. J Neurosci 1998; 18:104-111.
104. Ip YT, Davis RJ. Signal transduction by the c-Jun N-terminal kinase (JNK)—from inflammation to development. Curr Opin Cell Biol 1998; 10:205-219.
105. Lee YJ, Galoforo SS, Sim JE et al. Dominant-negative Jun N-terminal protein kinase (JNK-1) inhibits metabolic oxidative stress during glucose deprivation in a human breast carcinoma cell line. Free Radic Biol Med 2000; 28:575-584.
106. Okubo Y, Blakesley VA, Stannard B et al. Insulin-like growth factor-I inhibits the stress-activated protein kinase/c-Jun N-terminal kinase. J Biol Chem 1998; 273:25961-25966.
107. Maundrell K, Antonsson B, Magnenat E et al. Bcl-2 undergoes phosphorylation by c-Jun N-terminal kinase stress-activated protein kinases in the presence of the constitutively active GTP-binding protein Rac1. J Biol Chem 1997; 272:25238-25242.
108. Deng X, Xiao L, Lang W et al. Novel role for JNK as a stress-activated Bcl2 kinase. J Biol Chem 2001; 276:23681-23688.
109. O'Connor R, Kauffmann-Zeh A, Liu YM et al. Identification of domains of the insulin-like growth factor I receptor that are required for protection from apoptosis. Mol Cell Biol 1997; 17:427-435.
110. De Meyts P, Wallach B, Christoffersen CT et al. The insulin-like growth factor-I receptor. Structure, ligand-binding mechanism and signal transduction. Horm Res 1994; 42:152-169.
111. Myers MG Jr, White MF. Insulin signal transduction and the IRS protein. Annu Rev Pharmacol Toxicol 1996; 36:615-658.
112. Giorgetti S, Ballotti R, Kowalski-Chauvel A et al. The insulin and insulin-like growth factor-I receptor substrate IRS-1 associates with and activates phosphatidylinositol 3-kinase in vitro. J Biol Chem 1993; 268:7358-7364.
113. Kim B, Leventhal PS, White MF et al. Differential regulation of insulin receptor substrate-2 and mitogen-activated protein kinase tyrosine phosphorylation by phosphatidylinositol 3-kinase inhibitors in SH-SY5Y human neuroblastoma cells. Endocrinology 1998; 139:4881-4889.
114. Carpenter CL, Cantley LC. Phosphoinositide 3-kinase and the regulation of cell growth. Biochim Biophys Acta 1996; 1288:M11-M16.
115. Miller TM, Tansey MG, Johnson EM Jr et al. Inhibition of phosphatidylinositol 3-kinase activity blocks depolarization- and insulin-like growth factor I-mediated survival of cerebellar granule cells. J Biol Chem 1997; 272:9847-9853.
116. Singleton JR, Dixit VM, Feldman EL. Type I insulin-like growth factor receptor activation regulates apoptotic proteins. J Biol Chem 1996; 271:31791-31794.
117. Dudek H, Datta SR, Franke TF et al. Regulation of neuronal survival by the serine-threonine protein kinase Akt. Science 1997; 275:661-665.

118. Vanhaesebroeck B, Leevers SJ, Panayotou G et al. Phosphoinositde 3-kinases: a conserved family of signal transducers. Trends Biochem Sci 1997; 22:267-272.
119. Matsuzaki H, Tamatani M, Mitsuda N et al. Activation of Akt kinase inhibits apoptosis and changes in Bcl-2 and Bax expression induced by nitric oxide in primary hippocampal neurons. J Neurochem 1999; 73:2037-2046.
120. van Golen CM, Schwab TS, Woods Ignatoski KM et al. PTEN/MMAC1 overexpression decreases insulin-like growth factor-I-mediated protection from apoptosis in neuroblastoma cells. Cell Growth Differ 2001; 12:371-378.
121. Blair LA, Bence-Hanulec KK, Mehta S et al. Akt-dependent potentiation of L channels by insulin-like growth factor- 1 is required for neuronal survival. J Neurosci 1999; 19:1940-1951.
122. Galli C, Meucci O, Scorziello A et al. Apoptosis in cerebellar granule cells is blocked by high KCl, forskolin, and IGF-I through distinct mechanisms of action: the involvement of intracellular calcium and RNA synthesis. J Neurosci 1995; 15:1172-1179.
123. Chrysis D, Calikoglu AS, Ye P et al. Insulin-like growth factor-I overexpression attenuates cerebellar apoptosis by altering the expression of Bcl family proteins in a developmentally specific manner. J Neurosci 2001; 21:1481-1489.
124. Pugazhenthi S, Nesterova A, Sable C et al. Akt/protein kinase B up-regulates Bcl-2 expression through cAMP-response element-binding protein. J Biol Chem 2000; 275:10761-10766.
125. Yamaguchi H, Wang HG. The protein kinase PKB/Akt regulates cell survival and apoptosis by inhibiting Bax conformational change. Oncogene 2001; 20:7779-7786.
126. Minshall C, Arkins S, Straza J et al. IL-4 and insulin-like growth factor-I inhibit the decline in Bcl-2 and promote the survival of IL-3-deprived myeloid progenitors. J Immunol 1997; 159:1225-1232.
127. Datta SR, Dudek H, Tao X et al. Akt phosphorylation of BAD couples survival signals to the cell-intrinsic death machinery. Cell 1997; 91:231-241.
128. Bai Hz, Pollman MJ, Inishi Y et al. Regulation of vascular smooth muscle cell apoptosis. Modulation of bad by a phosphatidylinositol 3-kinase-dependent pathway. Circ Res 1999; 85:229-237.
129. van Golen CM, Castle VP, Feldman EL. IGF-I receptor activation and Bcl-2 overexpression prevent early apoptotic events in human neuroblastoma. Cell Death Differ 2000; 7:654-665.
130. Kermer P, Ankerhold R, Klocker N et al. Caspase-9: involvement in secondary death of axotomized rat retinal ganglion cells in vivo. Brain Res Mol Brain Res 2000; 85:144-150.
131. Zheng WH, Kar S, Quirion R. Insulin-like growth factor-1-induced phosphorylation of the forkhead family transcription factor FKHRL1 is mediated by akt kinase in PC12 cells. J Biol Chem 2000; 275:39152-39158.
132. del Peso L, Gonzalez VM, Hernandez R et al. Regulation of the forkhead transcription factor FKHR, but not the PAX3-FKHR fusion protein, by the serine/threonine kinase Akt. Oncogene 1999; 18:7328-7333.
133. Suhara T, Kim HS, Kirshenbaum LA et al. Suppression of Akt signaling induces Fas ligand expression: involvement of caspase and Jun kinase activation in Akt-mediated Fas ligand regulation. Mol Cell Biol 2002; 22:680-691.
134. Crowder RJ, Freeman RS. Glycogen synthase kinase-3 beta activity is critical for neuronal death caused by inhibiting phosphatidylinositol 3-kinase or Akt but not for death caused by nerve growth factor withdrawal. J Biol Chem 2000; 275:34266-34271.
135. Sabbatini P, McCormick F. Phosphoinositide 3-OH kinase (PI3K) and PKB/Akt delay the onset of p53-mediated, transcriptionally dependent apoptosis. J Biol Chem 1999; 274:24263-24269.
136. Heck S, Lezoualc'h F, Engert S et al. Insulin-like growth factor-1-mediated neuroprotection against oxidative stress is associated with activation of nuclear factor kappaB. J Biol Chem 1999; 274:9828-9835.
137. Bilak MM, Kuncl RW. Delayed application of IGF-I and GDNF can rescue already injured postnatal motor neurons. Neuroreport 2001; 12:2531-2535.
138. Mielke K, Herdegen T. JNK and p38 stresskinases—degenerative effectors of signal-transduction-cascades in the nervous system. Prog Neurobiol 2000; 61:45-60.
139. Cheng H-L, Feldman EL. Bidirectional regulation of p38 kinase and c-Jun N-terminal protein kinase by insulin-like growth factor-I. J Biol Chem 1998; 273:14560-14565.
140. Levresse V, Butterfield L, Zentrich E et al. Akt negatively regulates the cJun N-terminal kinase pathway in PC12 cells. J Neurosci Res 2000; 62:799-808.
141. Cheng H-L, Steinway M, Xin X et al. Insulin-like growth factor-I and Bcl-X_L inhibit c-jun N-terminal kinase activation and rescue Schwann cells from apoptosis. J Neurochem 2001; 76:935-943.
142. Shimoke K, Yamagishi S, Yamada M et al. Inhibition of phosphatidylinositol 3-kinase activity elevates c-Jun N-terminal kinase activity in apoptosis of cultured cerebellar granule neurons. Brain Res Dev Brain Res 1999; 112:245-253.
143. Harada J, Sugimoto M. An inhibitor of p38 and JNK MAP kinases prevents activation of caspase and apoptosis of cultured cerebellar granule neurons. Jpn J Pharmacol 1999; 79:369-378.
144. Zumkeller W, Westphal M. The IGF/IGFBP system in CNS malignancy. Mol Pathol 2001; 54:227-229.
145. Wang JY, Del Valle L, Gordon J et al. Activation of the IGF-IR system contributes to malignant growth of human and mouse medulloblastomas. Oncogene 2001; 20:3857-3868.
146. Hirano H, Lopes MB, Laws ERJ et al. Insulin-like growth factor-1 content and pattern of expression correlates with histopathologic grade in diffusely infiltrating astrocytomas. Neuro-oncol 1999; 1:109-119.
147. Jasty R, van Golen C, Lin HJ et al. Bcl-2 and n-myc coexpression increases igf-ir and features of malignant growth in neuroblastoma cell lines. Neoplasia 2001; 3:304-313.
148. Singleton JR, Randolph AE, Feldman EL. Insulin-like growth factor I receptor prevents apotosis and enhances neuroblastoma tumorigenesis. Cancer Res 1996; 56:4522-4529.

149. Rapoport SI. Advances in osmotic opening of the blood-brain barrier to enhance CNS chemotherapy. Expert Opin Investig Drugs 2001; 10:1809-1818.
150. Elmlinger MW, Deininger MH, Schuett BS et al. In vivo expression of insulin-like growth factor-binding protein-2 in human gliomas increases with the tumor grade. Endocrinology 2001; 142:1652-1658.
151. Mattson MP. Apoptosis in neurodegenerative disorders. Nat Rev Mol Cell Biol 2000; 1:120-129.
152. Zumkeller W. IGFs and IGFBPs: surrogate markers for diagnosis and surveillance of tumor growth? Mol Pathol 2001; 54:285-288.
153. McFarlane S. Dendritic morphogenesis: building an arbor. Mol Neurobiol 2000; 22:1-9.
154. Kalil K, Szebenyi G, Dent EW. Common mechanisms underlying growth cone guidance and axon branching. J Neurobiol 2000; 44:145-158.
155. Goldberg DJ, Wu D-Y. Inhibition of formation of filopodia after axotomy by inhibitors of protein tyrosine kinases. J Neurobiol 1995; 27:553-560.
156. Ono K, Shokbuni T, Nagata I et al. Filopodia and growth cones in the vertically migrating granule cells of the postnatal mouse. Exp Brain Res 1997; 117:17-29.
157. Davenport DW, Dou P, Rehder V et al. A sensory role for neuronal growth cone filopodia. Nature 1993; 361:721-724.
158. Smith CL. Cytoskeletal movements and substrate interactions during initiation of neurite outgrowth by sympathetic neurons in vitro. J Neurosci 1994; 14:384-398.
159. Lauffenberger DA, Horwitz AF. Cell migration: a physically integrated molecular process. Cell 1996; 84:359-369.
160. Buettner HM. Nerve growth dynamics: Quantitative models for nerve development and regeneration. Ann N Y Acad Sci 1994; 745:210-221.
161. Zheng JQ, Wan JJ, Pooo MM. Essential role of filopodia in chemotropic turning of nerve growth cone induced by a glutamate gradient. J Neurosci 1996; 16:1140-1149.
162. Mueller BK. Growth cone guidance: first steps towards a deeper understanding. Annu Rev Neurosci 1999; 22:351-388.
163. Kuhn TB, Meberg PJ, Brown MD et al. Regulating actin dynamics in neuronal growth cones by ADF/ cofilin and rho family GTPases. J Neurobiol 2000; 44:126-144.
164. Korey CA, Van Vactor D. From the growth cone surface to the cytoskeleton: one journey, many paths. J Neurobiol 2000; 44:184-193.
165. Gavazzi I. Semaphorin-neuropilin-1 interactions in plasticity and regeneration of adult neurons. Cell Tissue Res 2001; 305:275-284.
166. Long KE, Lemmon V. Dynamic regulation of cell adhesion molecules during axon outgrowth. J Neurobiol 2000; 44:230-245.
167. Burgaya F, Menegon A, Menegoz M et al. Focal adhesion kinase in rat central nervous system. Eur J Neurosci 1995; 7:1810-1821.
168. Condic ML, Letourneau PC. Ligand-induced changes in integrin expression regulate neuronal adhesion and neurite outgrowth. Nature 1997; 389:852-856.
169. Aizenman Y, de Vellis J. Brain neurons develop in a serum and glial free environment: effects of transferrin, insulin, insulin-like growth factor-I and thyroid hormone on neuronal survival, growth and differentiation. Brain Res 1987; 406:32-42.
170. Robinson LJ, Leitner W, Draznin B et al. Evidence that p21ras mediates the neurotrophic effects of insulin and insulin-like growth factor I in chick forebrain neurons. Endocrinology 1994; 135:2568-2573.
171. Niblock MM, Brunso-Bechtold JK, Riddle DR. Insulin-like growth factor I stimulates dendritic growth in primary somatosensory cortex. J Neurosci 2000; 20:4165-4176.
172. Brooker GJ, Kalloniatis M, Russo VC et al. Endogenous IGF-1 regulates the neuronal differentiation of adult stem cells. J Neurosci Res 2000; 59:332-341.
173. Zhou X, Herman JP, Paden CM. Evidence that IGF-I acts as an autocrine/paracrine growth factor in the magnocellular neurosecretory system: neuronal synthesis and induction of axonal sprouting. Exp Neurol 1999; 159:419-432.
174. Fernandez-Galaz MC, Naftolin F, Garcia-Segura LM. Phasic synaptic remodeling of the rat arcuate nucleus during the estrous cycle depends on insulin-like growth factor-I receptor activation. J Neurosci Res 1999; 55:286-292.
175. Liu JP, Lauder JM. S-100b and insulin-like growth factor-II differentially regulate growth of developing serotonin and dopamine neurons in vitro. J Neurosci Res 1992; 33:248-256.
176. Kim B, Leventhal PS, Saltiel AR et al. Insulin-like growth factor-I-mediated neurite outgrowth in vitro requires MAP kinase activation. J Biol Chem 1997; 272:21268-21273.
177. Recio-Pinto E, Ishii DN. Effects of insulin, insulin-like growth factor-II and nerve growth factor on neurite outgrowth in cultured human neuroblastoma cells. Brain Res 1984; 302:323-334.
178. Leventhal PS, Shelden EA, Kim B et al. Tyrosine phosphorylation of paxillin and focal adhesion kinase during insulin-like growth factor-I-stimulated lamellipodial advance. J Biol Chem 1997; 272:5214-5218.
179. Leventhal PS, Feldman EL. Insulin-like growth factors as regulators of cell motility: signaling mechanisms. Trends Endocrinol Metab 1997; 8:1-6.
180. Mulholland MW, Romanchuk G, Simeone DM et al. Stimulation of myenteric plexus neurite outgrowth by insulin and IGF I and II. Life Sci 1992; 51:1789-1796.
181. Caroni P, Grandes P. Nerve sprouting in innervated adult skeletal muscle induced by exposure to elevated levels of insulin-like growth factors. J Cell Biol 1990; 110(4):1307-1317.
182. D'Costa AP, Prevette DM, Houenou LJ et al. Mechanisms of insulin-like growth factor regulation of programmed cell death of developing avian motoneurons. J Neurobiol 1998; 36:379-394.

183. Prager D, Melmed S. Insulin and insulin-like growth factor I receptors: Are there functional distinctions? Endocrinology 1993; 132:1419-1420.
184. Russell JW, Feldman EL. Insulin-like growth factor-I prevents apoptosis in sympathetic neurons exposed to high glucose. Horm Metab Res 1999; 31:90-96.
185. Recio-Pinto E, Rechler MM, Ishii DN. Effects of insulin, insulin-like growth factor-II, and nerve growth factor on neurite formation and survival in cultured sympathetic and sensory neurons. J Neurosci 1986; 6:1211-1219.
186. Near SL, Whalen LR, Miller JA et al. Insulin-like growth factor II stimulates motor nerve regeneration. Proc Natl Acad Sci USA 1992; 89:11716-11720.
187. Contreras PC, Steffler C, Yu EY et al. Systemic administration of rhIGF-I enhanced regeneration after sciatic nerve crush in mice. J Pharmacol Exp Ther 1995; 274:1443-1449.
188. Pu SF, Zhuang HX, Ishii DN. differential spatio-temporal expression of the insulin-like growth factor genes in regenerating sciatic nerve. Brain Res Mol Brain Res 1995; 34:18-28.
189. Wuarin L, Guertin DM, Ishii DN. Early reduction in insulin-like growth factor gene expression in diabetic nerve. Exp Neurol 1994; 130:106-114.
190. Xu G, Sima AA. Altered immediate early gene expression in injured diabetic nerve: implications in regeneration. J Neuropathol Exp Neurol 2001; 60:972-983.
191. Gao WQ, Shinsky N, Ingle G et al. IGF-I deficient mice show reduced peripheral nerve conduction velocities and decreased axonal diameters and respond to exogenous IGF- I treatment. J Neurobiol 1999; 39:142-152.
192. Feldman EL, Sullivan KA, Kim B et al. Insulin-like growth factors regulate neuronal differentiation and survival. Neurobiol Dis 1997; 4:201-214.
193. Morrione A, Romano G, Navarro M et al. Insulin-like growth factor I receptor signaling in differentiation of neuronal H19-7 cells. Cancer Res 2000; 60:2263-2272.
194. Morrione A, Navarro M, Romano G et al. The role of the insulin receptor substrate-1 in the differentiation of rat hippocampal neuronal cells. Oncogene 2001; 20:4842-4852.
195. Kotani K, Yonezawa K, Hara K et al. Involvement of phosphoinositide 3-kinase in insulin- or IGF-1-induced membrane ruffling. EMBO J 1994; 13:2313-2321.
196. Giorgetti S, Pelicci PG, Pelicci G et al. Involvement of Src-homology/collagen (SHC) proteins in signaling through the insulin receptor and the insulin-like-growth-factor-I-receptor. Eur J Biochem 1994; 223:195-202.
197. Kim B, Cheng H-L, Margolis B et al. Insulin receptor substrate 2 and Shc play different roles in insulin-like growth factor I signaling. J Biol Chem 1998; 273:34543-34550.
198. Ravichandran KS. Signaling via Shc family adapter proteins. Oncogene 2001; 20:6322-6330.
199. Kimpinski K, Mearow K. Neurite growth promotion by nerve growth factor and insulin-like growth factor-1 in cultured adult sensory neurons: role of phosphoinositide 3-kinase and mitogen activated protein kinase. J Neurosci Res 2001; 63:486-499.
200. Yagihashi S. Nerve structural defects in diabetic neuropathy: Do animals exhibit similar changes? Neurosci Res Commun 1997; 21:25-32.
201. Leventhal PS, Feldman EL. The tyrosine kinase inhibitor methyl 2,5-dihydroxycinnimate disrupts changes in the actin cytoskeleton required for neurite formation. Brain Res Mol Brain Res 1996; 43:338-340.
202. Casamassima A, Rozengurt E. Insulin-like growth factor I stimulates tyrosine phosphorylation of p130(Cas), focal adhesion kinase, and paxillin. Role of phosphatidylinositol 3'-kinase and formation of a p130(Cas).Crk complex. J Biol Chem 1998; 273:26149-26156.
203. Baron V, Calléja V, Ferrari P et al. p125^{Fak} focal adhesion kinase is a substrate for the insulin and insulin-like growth factor-I tyrosine kinase receptors. J Biol Chem 1998; 273:7162-7168.
204. Ivankovic-Dikic I, Gronroos E, Blaukat A et al. Pyk2 and FAK regulate neurite outgrowth induced by growth factors and integrins. Nat Cell Biol 2000; 2:574-581.
205. Sumantran VN, Feldman EL. Insulin-like growth factor I regulates c-myc and GAP-43 messenger ribonucleic acid expression in SH-SY5Y human neuroblastoma cells. Endocrinology 1993; 132:2017-2023.
206. Aigner L, Caroni P. Absence of persistent spreading, branching, and adhesion in GAP-43-depleted growth cones. J Cell Biol 1995; 128:647-660.
207. Navarro M, Valentinis B, Belletti B et al. Regulation of Id2 gene expression by the type 1 IGF receptor and the insulin receptor substrate-1. Endocrinology 2001; 142:5149-5157.
208. Dupont J, Khan J, Qu BH et al. Insulin and IGF-1 induce different patterns of gene expression in mouse fibroblast NIH-3T3 cells: identification by cDNA microarray analysis. Endocrinology 2001; 142:4969-4975.
209. Fernyhough P, Mill JF, Roberts JL et al. Stabilization of tubulin mRNAs by insulin and insulin-like growth factor I during neurite formation. Brain Res Mol Brain Res 1989; 6:109-120.
210. Mitchison TJ, Cramer LP. Actin-based cell motility and cell locomotion. Cell 1996; 84:371-379.
211. Woodhouse EC, Chuaqui RF, Liotta LA. General mechanisms of metastasis. Cancer 1997; 80:1529-1537.
212. Jackson JG, Zhang X, Yoneda T et al. Regulation of breast cancer cell motility by insulin receptor substrate-2 (IRS-2) in metastatic variants of human breast cancer cell lines. Oncogene 2001; 20:7318-7325.
213. Bredin CG, Liu Z, Hauzenberger D et al. Growth-factor-dependent migration of human lung-cancer cells. Int J Cancer 1999; 82:338-345.
214. Reiss K, Wang JY, Romano G et al. Mechanisms of regulation of cell adhesion and motility by insulin receptor substrate-1 in prostate cancer cells. Oncogene 2001; 20:490-500.
215. Kakizaki Y, Kraft N, Atkins RC. Differential control of mesangial cell proliferation by interferon-gamma. Clin Exp Immunol 1991; 85:157-163.

216. Brodeur GM, Maris JM, Yamashiro DJ et al. Biology and genetics of human neuroblastomas. J Pediatr Hematol Oncol 1997; 19:93-101.
217. Puglianiello A, Germani D, Rossi P et al. IGF-I stimulates chemotaxis of human neuroblasts. Involvement of type 1 IGF receptor, IGF binding proteins, phosphatidylinositol-3 kinase pathway and plasmin system. J Endocrinol 2000; 165:123-131.
218. Meyer GE, Shelden E, Kim B et al. Insulin-like growth factor I stimulates motility in human neuroblastoma cells. Oncogene 2001; 20:7542-7550.
219. Kim B, Feldman EL. Differential regulation of focal adhesion kinase and mitogen-activated protein kinase tyrosine phosphorylation during insulin-like growth factor-I-mediated cytoskeletal reorganization. J Neurochem 1998; 71:1333-1336.
220. Schmitz AA, Govek EE, Bottner B et al. Rho GTPases: signaling, migration, and invasion. Exp Cell Res 2000; 261:1-12.
221. Cheng H-L, Steinway M, Russell JW et al. GTPases and phosphatidylinositol-3 kinase are critical for insulin-like growth factor-I mediated Schwann cell motility. J Biol Chem 2000; 275:27197-27204.
222. Kursula P. The current status of structural studies on proteins of the myelin sheath (Review). Int J Mol Med 2001; 8:475-479.
223. Zumkeller W. The effect of insulin-like growth factors on brain myelination and their potential therapeutic application in myelination disorders. Europ J Paediatr Neurol 1997; 1:91-101.
224. Carson MJ, Behringer RR, Brinster RL et al. Insulin-like growth factor I increases brain growth and central nervous system myelination in transgenic mice. Neuron 1993; 10:729-740.
225. Ye P, Carson J, D'Ercole AJ. Insulin-like growth factor-I influences the initiation of myelination: Studies of the anterior commissure of transgenic mice. Neurosci Lett 1995; 201:235-238.
226. Windebank AJ, Wood P, Bunge RP et al. Myelination determines the caliber of dorsal root ganglion neurons in culture. J Neurosci 1985; 5:1563-1569.
227. Ye P, Carson J, D'Ercole AJ. In vivo actions of insulin-like growth factor-I (IGF-I) on brain myelination: Studies of IGF-I and IGF binding protein-1 (IGFBP-1) transgenic mice. J Neurosci 1995; 15:7344-7356.
228. Beck KD, Powell-Braxton L, Widmer H-R et al. Igf1 gene disruption results in reduced brain size, CNS hypomyelination, and loss of hippocampal granule and striatal parvalbumin-containing neurons. Neuron 1995; 14:717-730.
229. Guan J, Bennet L, George S et al. Insulin-like growth factor-1 reduces postischemic white matter injury in fetal sheep. J Cereb Blood Flow Metab 2001; 21:493-502.
230. Oudega M, Xu XM, Guénard V et al. A combination of insulin-like growth factor-I and platelet-derived growth factor enhances myelination but diminishes axonal regeneration into Schwann cell grafts in the adult rat spinal cord. Glia 1997; 19:247-258.
231. Komoly S, Hudson LD, Webster HD et al. Insulin-like growth factor I gene expression is induced in astrocytes during experimental demyelination. Proc Natl Acad Sci USA 1992; 89:1894-1898.
232. Mason JL, Jones JJ, Taniike M et al. Mature oligodendrocyte apoptosis precedes IGF-1 production and oligodendrocyte progenitor accumulation and differentiation during demyelination/remyelination. J Neurosci Res 2000; 61:251-262.
233. Yao DL, Liu X, Hudson LD et al. Insulin-like growth factor-I given subcutaneously reduces clinical deficits, decreases lesion severity and upregulates synthesis of myelin proteins in experimental autoimmune encephalomyelitis. Life Sci 1996; 58:1301-1306.
234. Yao D-L, Liu X, Hudson LD et al. Insulin-like growth factor I treatment reduces demyelination and up-regulates gene expression of myelin-related proteins in experimental autoimmune encephalomyelitis. Proc Natl Acad Sci USA 1995; 92:6190-6194.
235. Li W, Quigley L, Yao D-L et al. Chronic relapsing experimental autoimmune encephalomyelitis: effects of insulin-like growth factor-I treatment on clinical deficits, lesion severity, glial responses, and blood brain barrier defects. J Neuropathol Exp Neurol 1998; 57:426-438.
236. Cannella B, Pitt D, Capello E et al. Insulin-like growth factor-1 fails to enhance central nervous system myelin repair during autoimmune demyelination. Am J Pathol 2000; 157:933-943.
237. Shinar Y, McMorris FA. Developing oligodendroglia express mRNA for insulin-like growth factor-I, a regulator of oligodendrocyte development. J Neurosci Res 1995; 42:516-527.
238. Mozell RL, McMorris FA. Insulin-like growth factor I stimulates oligodendrocyte development and myelination in rat brain aggregate cultures. J Neurosci Res 1991; 30:382-390.
239. Vemuri GS, McMorris FA. Oligodendrocytes and their precursors require phosphatidylinositol 3-kinase signaling for survival. Development 1996; 122:2529-2537.
240. Garcia-Estrada J, Garcia-Segura LM, Torres-Aleman I. Expression of insulin-like growth factor I by astrocytes in response to injury. Brain Res 1992; 592:343-347.
241. Mason JL, Suzuki K, Chaplin DD et al. Interleukin-1beta promotes repair of the CNS. J Neurosci 2001; 21:7046-7052.
242. Cheng H-L, Russell JW, Feldman EL. IGF-I promotes peripheral nervous system myelination. Ann N Y Acad Sci 1999; 883:124-130.
243. Russell JW, Cheng H-L, Golovoy D. Insulin-like growth factor-I promotes myelination of peripheral sensory axons. J Neuropathol Exp Neurol 2000; 59:575-584.
244. Brockes JP, Raff MC, Nishiguchi DJ et al. Studies on cultured rat Schwann cells, III. Assays for peripheral myelin proteins. J Neurocytol 1980; 9:67-77.
245. Baron P, Shy M, Kamholz J et al. Expression of P0 protein mRNA along rat sciatic nerve during development. Brain Res Dev Brain Res 1994; 83:285-288.
246. Singleton JR, Feldman EL. Insulin-like growth factor-i in muscle metabolism and myotherapies. Neurobiol Dis 2001; 8:541-554.

247. Scott RJ, Hegyi L. Cell death in perinatal hypoxic ischemic brain injury. Neuropathol Appl Neurobiol 1997; 23:307-314.
248. Johnson EM, Greenlund LJ, Akins PT et al. Neuronal apoptosis: current understanding of molecular mechanisms and potential role in ischemic brain injury. J Neurotrauma 1995; 12:843-852.
249. Guan J, Williams CE, Skinner SJM et al. The effects of insulin-like growth factor (IGF)-1, IGF-2, and des-IGF-1 on neuronal loss after hypoxic-ischemic brain injury in adult rats: Evidence for a role for IGF binding proteins. Endocrinology 1996; 137:893-898.
250. Guan J, Williams C, Gunning M et al. The effects of IGF-1 treatment after hypoxic-ischemic brain injury in adult rats. J Cereb Blood Flow Metab 1993; 13:609-616.
251. Gluckman P, Klempt N, Guan J et al. A role for IGF-1 in the rescue of CNS neurons following hypoxic-ischemic injury. Biochem Biophys Res Commun 1992; 182:593-599.
252. Zhu CZ, Auer RN. Intraventricular administration of insulin and IGF-1 in transient forebrain ischemia. J Cereb Blood Flow Metab 1994; 14:237-242.
253. Johnston BM, Mallard EC, Williams CE et al. Insulin-like growth factor-1 is a potent neuronal rescue agent after hypoxic-ischemic injury in fetal lambs. J Clin Invest 1996; 97:300-308.
254. Nakao Y, Otani H, Yamamura T et al. Insulin-like growth factor 1 prevents neuronal cell death and paraplegia in the rabbit model of spinal cord ischemia. J Thorac Cardiovasc Surg 2001; 122:136-143.
255. Guan J, Miller OT, Waugh KM et al. Insulin-like growth factor-1 improves somatosensory function and reduces the extent of cortical infarction and ongoing neuronal loss after hypoxia-ischemia in rats. Neuroscience 1902; 105:299-306.
256. Schabitz WR, Hoffmann TT, Heiland S et al. Delayed neuroprotective effect of insulin-like growth factor-i after experimental transient focal cerebral ischemia monitored with mri. Stroke 2001; 32:1226-1233.
257. Bergstedt K, Wieloch T. Changes in insulin-like growth factor 1 receptor density after transient cerebral ischemia in the rat. Lack of protection against ischemic brain damage following injection of insulin-like growth factor 1. J Cereb Blood Flow Metab 1993; 13:895-898.
258. Liu XF, Fawcett JR, Thorne RG et al. Intranasal administration of insulin-like growth factor-I bypasses the blood-brain barrier and protects against focal cerebral ischemic damage. J Neurol Sci 2001; 187:91-97.
259. Kanje M, Skottner A, Sjoberg J et al. Insulin-like growth factor I (IGF-I) stimulates regeneration of the rat sciatic nerve. Brain Res 1989; 486:396-398.
260. Edbladh M, Fex-Svenningsen Å, Ekström PA et al. Insulin and IGF-II, but not IGF-I, stimulate the in vitro regeneration of adult frog sciatic sensory axons. Brain Res 1994; 641:76-82.
261. Saatman KE, Contreras PC, Smith DH et al. Insulin-like growth factor-1 (IGF-1) improves both neurological motor and cognitive outcome following experimental brain injury. Exp Neurol 1997; 147:418-427.
262. Hatton J, Rapp RP, Kudsk KA et al. Intravenous insulin-like growth factor-I (IGF-I) in moderate-to-severe head injury: A Phase II safety and efficacy trial. J Neurosurg 1997; 86:779-786.
263. Zhuang HX, Wuarin L, Fei ZJ et al. Insulin-like growth factor (IGF) gene expression is reduced in neural tissues and liver from rats with non-insulin-dependent diabetes mellitus, and IGF treatment ameliorates diabetic neuropathy. J Pharmacol Exp Ther 1997; 283:366-374.
264. Wuarin L, Namdev R, Burns JG et al. Brain insulin-like growth Factor-II mRNA content is reduced in insulin-dependent and non-insulin-dependent diabetes Mellitus. J Neurochem 1996; 67:742-751.
265. Migdalis IN, Kalogeropoulou K, Kalantzis L et al. Insulin-like growth factor-I and IGF-I receptors in diabetic patients with neuropathy. Diabet Med 1995; 12:823-827.
266. Ishii DN, Lupien SB. Insulin-like growth factors protect against diabetic neuropathy: Effects on sensory nerve regeneration in rats. J Neurosci Res 1995; 40:138-144.
267. Schmidt RE, Dorsey DA, Beaudet LN et al. Insulin-like growth factor I reverses experimental diabetic autonomic neuropathy. Am J Pathol 1999; 155:1651-1660.
268. Contreras PC, Vaught JL, Gruner JA et al. Insulin-like growth factor-I prevents development of a vincristine neuropathy in mice. Brain Res 1997; 774:20-26.
269. Songyang Z, Baltimore D, Cantley LC et al. Interleukin 3-dependent survival by the Akt protein kinase. Proc Natl Acad Sci USA 1997; 94:11345-11350.
270. Adem A, Ekblom J, Gillberg P-G et al. Insulin-like growth factor-1 receptors in human spinal cord: Changes in amyotrophic lateral sclerosis. J Neural Transm 1994; 97:73-84.
271. Vaught JL, Contreras PC, Glicksman MA et al. Potential utility of rhIGF-1 in neuromuscular and/or degenerative disease. CIBA Found Symp 1996; 196:18-27.
272. Lai EC, Felice KJ, Festoff BW et al. Effect of recombinant human insulin-like growth factor-I on progression of ALS. A placebo-controlled study. The North America ALS/IGF-I Study Group. Neurology 1997; 49:1621-1630.
273. Borasio GD, Robberecht W, Leigh PN et al. A placebo-controlled trial of insulin-like growth factor-I in amyotrophic lateral sclerosis. European ALS/IGF-I Study Group. Neurology 1998; 51:583-586.
274. Iwasaki Y, Ikeda K. Prevention by insulin-like growth factor-I and riluzole in motor neuron death after neonatal axotomy. J Neurol Sci 1999; 169:148-155.
275. Liu X, Yao DL, Webster J. Insulin-like growth factor I treatment reduces clinical deficits and lesion severity in acute demyelinating autoimmune encephalomyelitis. Mult Scler 1995; 1:2-9.
276. Lovett-Racke AE, Bittner P, Cross AH et al. Regulation of experimental autoimmune encephalomyelitis with insulin-like growth factor (IGF-1) and IGF-1/IGF-binding protein-3 complex (IGF-1/IGFBP3). J Clin Invest 1998; 101:1797-1804.
277. Humbert S, Bryson EA, Cordelieres FP et al. The IGF-1/Akt pathway is neuroprotective in Huntington's disease and involves Huntingtin phosphorylation by Akt. Dev Cell 2002; 2:831-837.
278. Wei W, Wang X, Kusiak JW. Signaling events in amyloid beta-peptide-induced neuronal death and insulin-like growth factor I protection. J Biol Chem 2002; 277:17649-17656.

CHAPTER 10

The Insulin-Like Growth Factors in Mammary Development and Breast Cancer

Teresa L. Wood, Malinda A. Stull, Dawn Kardash-Richardson, Michael A. Allar and Aimee V. Loladze

Abstract

Normal and abnormal growth of mammary or breast epithelium is coordinately controlled by circulating hormones and locally-produced growth factors. The insulin-like growth factors and their primary signaling receptor, the IGF type I receptor, have demonstrated roles in normal mammary gland development and are highly implicated in breast cancer. Genetic deletions of either IGF-I or the IGF-IR result in decreased epithelial growth during puberty in the mouse mammary gland. Conversely, overexpression of either IGF-I or IGF-II in mammary epithelium promotes growth and survival of mammary epithelial cells and leads to mammary tumor formation. Increased IGF levels also are correlated with breast cancers in humans. IGF signaling is critical for mammary/breast epithelial cell cycle progression mediated by both EGF-related ligands and by estrogen. Increasing evidence also supports important roles for the family of high-affinity IGF binding proteins that modulate IGF availability in the circulation and at the local tissue level. In addition, the IGFBPs may regulate epithelial-stromal interactions critical for growth during normal mammary development and in breast cancers through both IGF-dependent and IGF-independent mechanisms.

Introduction

The insulin-like growth factors, IGF-I and IGF-II, are essential mediators of fetal growth and have known roles in cell proliferation, survival and differentiation in a variety of tissues. The function of the IGFs and their primary signaling receptor, the IGF type I receptor (IGF-IR), has been of particular interest in the developing mammary gland. Postnatal growth of mammary epithelium is regulated by complex interactions between circulating hormones and locally-produced growth factors. In addition, the role of hormones and growth factors in mediating normal growth of mammary tissue also has significant implications for how these factors regulate neoplastic growth in breast cancers. Studies from several laboratories have contributed to our current knowledge of the IGFs in normal mammary development. It is now clear that the IGFs and IGF-IR are expressed and have essential functions in normal growth of mammary epithelium. However, there are still many unsolved questions about how the IGFs and IGF-IR are regulated and how the IGFs coordinate with hormones and other growth factors to mediate mammary epithelial growth. In addition to their demonstrated role in normal mammary gland development, previous and current work strongly supports an important role for the IGFs and IGF-IR in breast cancers. In addition to the IGF ligands and receptor, numerous studies support critical roles for the family of high-affinity IGF binding proteins (IGFBPs) in mammary gland development and breast cancer. There is increasing evidence that the IGFBPs can modulate mammary/breast epithelial growth through both IGF-dependent and IGF-independent mechanisms.

Insulin-Like Growth Factors, edited by Derek LeRoith, Walter Zumkeller and Robert Baxter. ©2003 Eurekah.com and Kluwer Academic / Plenum Publishers.

Postnatal Mammary Gland Development

Postnatal development of the rodent mammary gland occurs in two major phases, pubertal-induced elongation and branching of the ductal epithelium and pregnancy-induced alveolar formation and differentiation (for reviews, see refs. 1-4). Prior to puberty, the mammary epithelium consists of a rudimentary ductal structure. At the start of puberty under the influence of the ovarian hormone estrogen, highly proliferative terminal end buds (TEBs) form at the tips of the ductal structures, which then lengthen and branch to fill the mammary fat pad. While the majority of ductal growth and branching occurs during puberty, some cellular proliferation and branching continues with each estrous cycle in post-pubertal virgin glands.[5]

The second major period of growth in the postnatal mammary gland occurs during pregnancy. This growth phase involves massive proliferation and differentiation of the epithelium to form secretory alveoli. At the end of pregnancy, the gland reaches functional maturity and, at parturition, the alveoli actively secrete milk in response to suckling by the pups. After pup weaning, the epithelial structures regress through widespread apoptotic cell death, a process known as involution. With each subsequent pregnancy, the epithelial structures reinitiate the process of alveolar growth and differentiation in preparation for lactation. The continual growth and remodeling of mammary tissue in the adult concomitant with each estrous cycle and with each pregnancy is thought to provide the basis for the susceptibility of breast epithelial cells to carcinogenesis.

The ovarian and pituitary hormones, including estrogen, progesterone, prolactin and growth hormone (GH), are the major endocrine regulators of postnatal growth and differentiation of the mammary gland.[6-12] These hormones act through cognate receptors located in either, or both, the stromal and epithelial compartments of the developing gland. Analyses of mouse lines carrying deletions in receptors for estrogen, progesterone and prolactin have revealed critical functions for these hormones at specific postnatal stages of mammary development. These analyses have demonstrated an essential role for estrogen in TEB formation and ductal growth during puberty.[7] While progesterone and prolactin contribute to ductal branching during pubertal growth,[8,9,13] both hormones are essential for pregnancy-induced alveolar differentiation.[8,9,14,15] The function of GH in pubertal ductal growth through induction of IGF-I is discussed in detail in the subsequent sections.

In addition to circulating hormones, peptide growth factors function as local mediators of mammary gland growth and differentiation. Peptide growth factor families with demonstrated essential functions in the postnatal mammary gland include ligands and receptors related to the epidermal growth factor (EGF), fibroblast growth factor (FGF), transforming growth factor-beta (TGF-β and IGF families). Loss of function experiments in mice utilizing genetic deletions or dominant-negative transgenic approaches for specific receptors have provided evidence that the EGF and FGF families positively regulate epithelial growth during normal mammary development in vivo.[16-21] In contrast, a dominant negative approach to delete function of the TGF-β type II receptor demonstrated that members of the TGF-β family negatively regulate mammary growth and differentiation.[22,23] Recent genetic studies on the IGFs that will be reviewed below support the hypothesis that these ligands also are essential mediators of mammary epithelial growth.

The IGFs in GH-Mediated Mammary Development

GH is the major inducer of circulating IGF-I and is one of the lactogenic hormones that mediates mammary growth. When given either systemically or locally to gonadectomized, hypophysectomized and estrogen-treated male or female rats, GH induces ductal as well as alveolar growth.[11,12] Evidence that IGF-I can substitute for GH in this model provided the initial evidence that IGF-I directly mediates the effects of GH on mammary ductal growth.[24] Additional evidence that the actions of GH on pubertal mammary development are due to IGF-I induction was provided by studies demonstrating that GH induces stromal expression of IGF-I mRNA in a dose-dependent manner.[25] Moreover, IGF-I can synergize with estrogen to promote TEB formation and ductal elongation in hypophysectomized, ovariectomized rats.[26] Taken together, these experiments led to the hypothesis that IGF-I is produced in the stromal compartment in response to GH and that it synergizes with estrogen to promote TEB formation and ductal growth.

Function of the IGFs in Postnatal Mammary Development: Transgenic Mice

Transgenic and knockout mouse lines have provided critical information on the function of hormones and growth factors in mammary gland development in vivo. Mammary epithelial-specific promoters have been used to overexpress genes in the mammary epithelium during various stages of postnatal mammary gland development. The mouse mammary tumor virus long terminal repeat (MMTV-LTR) is activated in mammary epithelial cells during pubertal ductal morphogenesis as well as during pregnancy and lactation. The use of milk protein promoters such as those for the whey acidic protein (WAP) and beta-lactoglobulin (BLG) genes results in transgene expression in epithelial cells exclusively during pregnancy and lactation. The MMTV and milk-protein gene promoters have been used to overexpress a wide variety of proteins in mammary epithelial cells including the IGFs.[27-34]

Mice overexpressing IGF-I and IGF-II under mammary specific promoters have indicated that the IGF ligands can promote growth and survival of mammary epithelial cells during normal development. Overexpression of IGF-I in mammary epithelium during pubertal ductal growth results in precocious differentiation as seen by the formation of alveolar buds in 50-day old virgin mice.[33] Overexpression of either IGF-I or IGF-II during pregnancy and lactation results in disruption of the normal process of involution through blocking apoptotic cell death of the epithelium.[28,30,35,36] These studies indicate that overexpression of IGF-I during postnatal mammary gland development results in an increase in epithelial cell number, either due to precocious differentiation and formation of alveolar buds, or due to increased cell survival during involution.

An essential requirement for IGF-I in normal mammary gland development was demonstrated in recent studies on the IGF-I null mutant mice.[37] The mammary glands of these mice fail to undergo ductal morphogenesis, even upon treatment with estrogen.[37] TEB formation was induced when IGF-I null mice were treated for either 5 or 14 days with IGF-I alone or in combination with estrogen, and ductal penetration into the fat pad was detected after 14 days of treatment. Replacement with IGF-I resulted in a significant increase in TEB formation, number of ducts, and the area of the fat pad occupied by epithelial structures. In contrast, treatment of IGF-I null mice with GH failed to restore ductal morphogenesis confirming previous studies indicating that IGF-I is the primary mediator of GH actions in the developing mammary gland.

Analysis of postnatal mammary epithelial development in the absence of the IGF-IR has been difficult since the IGF-IR null mutation results in perinatal lethality.[38] However, a recent study reported the results of experiments using embryonic mammary gland transplantation to rescue the IGF-IR null epithelium.[39] Embryonic mammary buds from IGF-IR null mutant mice at embryonic day 16-18 were transplanted into wild-type mammary fat pads that had been cleared of endogenous epithelium. Analyses of the transplanted epithelium after 4-8 weeks of development demonstrated reduced growth of the IGF-IR null epithelium as compared to wild-type epithelium. Further analyses of the transplanted epithelium revealed decreased cell proliferation but no difference in apoptosis in the IGF-IR null grafts. Taken together, results of the transgenic overexpression and genetic deletion studies on the IGFs and IGF-IR in the developing mammary gland suggest that IGF signaling mediates diverse effects on mammary epithelial cells including proliferation, differentiation and survival.

Endogenous Expression of the IGFs and IGF-IR in Developing Mammary Tissue

Results from the studies discussed previously do not address whether the predominant effects of the IGFs on postnatal mammary development are due predominantly to circulating IGF-I or to local expression of IGF-I and IGF-II within the mammary tissue. The induction of IGF-I mRNA expression in mammary stroma in response to GH provided the initial evidence that IGF-I is produced locally in the rat mammary gland during postnatal stages of development.[25] Subsequent studies from our own and other laboratories now have provided clear evidence for endogenous expression of the IGFs and IGF-IR in the mammary gland of the mouse,[40,41] rat,[42] cow,[43] and sheep.[44] Utilizing in situ hybridization, we demonstrated that IGF-I, IGF-II and the IGF-IR are expressed in the epithelium as well as in the stroma in the mouse mammary gland during pubertal ductal growth and pregnancy-induced alveolar differentiation (Fig. 1).[41,45] During pubertal ductal growth, IGF-I

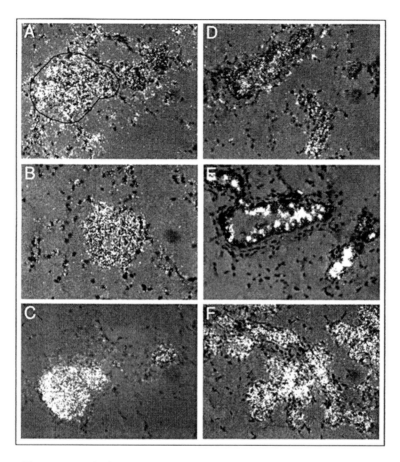

Figure 1. Photomicrographs showing ISH to IGF-I (A,D), IGF-II (B,E) and IGF-IR (C,F) mRNAs in mouse postnatal mammary gland. A,B,C are terminal end buds (TEBs) at 5.5 weeks during pubertal development. Note hybridization to both TEB epithelium and stromal cells for IGF-I (A) and IGF-II (B). D,E,F are sections from mammary glands taken at either day 13 of pregnancy (D,E) or at day 18 of pregnancy (F). Note non-uniform hybridization of IGF-II in the epithelium (E). Modified and reprinted with permission from: Richert MM, Wood TL. Endocrinology 1999; 140:454-461. © 1999 Endocrine Society and Wood TL, Richert MM, Stull MA, Allar MA. J Mam Gland Biol Neopl 2000; 5:31-42. © 2000 Plenum Publishing Corporation.

mRNA is expressed primarily in the TEBs and surrounding stromal cells while IGF-II mRNA is strongly expressed in the ductal structures as well as the TEBs and surrounding stroma.[41] During pregnancy-induced alveolar development, IGF-I mRNA expression is low in the epithelium until late pregnancy when its expression is induced throughout the epithelium. In contrast, IGF-II mRNA is expressed in mammary epithelium throughout pregnancy. The IGF-IR is expressed throughout the mammary epithelium during all stages of postnatal mammary development.[41] It is potentially important that the patterns of IGF-I and IGF-II expression are distinct throughout postnatal mammary development. While IGF-I mRNA expression is uniform in the ducts and alveoli in late pregnant glands, the expression of IGF-II is non-uniform throughout the epithelium by post-pubertal stages and during pregnancy. These data suggest that IGF-I and IGF-II are differentially regulated and that the two ligands may have distinct functions during mammary gland development.

The IGF-II expression pattern in the mammary epithelium is particularly interesting in the context of other studies on expression and function of hormone receptors in the developing mammary epithelium. A non-uniform pattern of expression in the mammary epithelium also has been described for the progesterone receptor (PR)[46-49] and the prolactin receptor[49] in post-pubertal virgin and pregnant glands. Several lines of evidence support the hypothesis that the non-uniform expression of hormone receptors is essential for normal proliferation and growth of the mammary epithelium. Investigations of steroid hormone receptor expression and sites of DNA synthesis demonstrated that proliferating cells are adjacent to cells expressing the estrogen receptor (ER) and PR in normal human breast tissue, but in breast cancers, the proliferating cells also are positive for the steroid hormone receptors.[50] Moreover, genetic deletion of the CCAAT/enhancer binding protein-beta (C/EBPβ) transcription factor in mice disrupts the normal expression of the PR such that PR expression remains uniform throughout the mammary epithelium.[48] The loss of C/EBPβ in the mammary gland also results in loss of epithelial proliferation and a severe deficit in ductal growth.[48,51] These data provide a basis for further investigations into the non-uniform expression of IGF-II in mammary epithelium as a possible mediator of epithelial proliferation that requires heterogeneous expression of several growth-promoting hormones and growth factors.

Interactions of IGF and EGF-Related Ligands in Normal Mammary Epithelial Cell Cycle Progression

Mammary epithelial cell culture and whole organ culture have been used extensively in an attempt to understand the factors involved in normal development of the mammary gland. The standard media for culturing either normal mammary epithelial cells, mammary explants, or intact mammary glands contains micromolar concentrations of insulin,[52-60] levels that are known to signal through the IGF-IR as well as the insulin receptor (IR).[61,62] In primary cultures of mammary or breast epithelial cells, micromolar levels of insulin are thought to be necessary both for cell viability as well as for the epithelial cells to respond to known growth promoting signals.[56,61] As a result, the role for IGF signaling has been overlooked in most in vitro experiments on mammary epithelial growth.

To examine the requirement for IGF-I in mammary organ culture, we conducted experiments using media containing nanomolar concentrations of insulin,[41,45] levels sufficient to stimulate the IR but not the IGF-IR. Mammary glands were obtained from 5-6 week old female C57BL6/J mice that had been primed with estrogen and progesterone for 9-14 days. The mammary glands were cultured in the presence of mammogenic hormones that included prolactin, aldosterone, hydrocortisone and nanomolar levels of insulin. Analysis of the intact glands following 5 days of culture indicated viable epithelial structures in control cultures and in cultures treated with IGF-I. In addition, analyses of induction of apoptotic genes also indicated no cell death in either group of glands.[41,63] These data suggest that even in the absence of signaling through the IGF-IR, the basal culture conditions provide sufficient cell survival signals likely through the combination of signaling through the prolactin receptor and the IR.

We have used the ex vivo organ culture system described above to study the effects of the IGFs and EGF-related ligands on the regulation of cell proliferation in the intact mammary gland (Fig. 2).[63] Glands were cultured in the presence of growth factors for 24 and 48 hours with the addition of ^{3}H-thymidine during the last 6 hours of culture. Analysis of glands cultured in various growth factor conditions showed that IGF-I significantly induced DNA synthesis in the cultured glands. In contrast, the EGF-related ligands, including EGF, amphiregulin and TGF-α failed to induce DNA synthesis in the cultured mammary glands. Interestingly, the combination of either EGF or TGF-α with IGF-I was synergistic in promoting DNA synthesis. Histological analyses of the cultured glands demonstrated that the sites of DNA synthesis induced by either IGF-I or IGF-I/EGF were primarily in epithelial cells at the tips of the ductal structures.[63]

To study the mechanisms by which IGF-I and the EGF-related ligands interact to promote DNA synthesis in mammary epithelium, we conducted experiments to investigate cell cycle regulation in mammary glands cultured in the presence of the growth factors.[63] Since it is well established that a critical point of cell cycle regulation is through induction of cell cycle regulatory proteins such as the cyclins, we examined expression levels of cyclin mRNAs in mammary glands after growth factor treatments (Fig. 3). Either IGF-I or EGF equally induced mRNA levels of the D type cyclins that are essential for early G_1 initiation and progression. In contrast, only IGF-I was able to induce cyclin E mRNA levels. Cyclin E is essential for cells to pass from the G_1 phase of the cell cycle past

The IGFs in Mammary Development and Breast Cancer 193

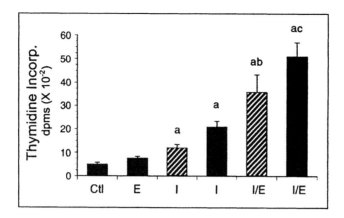

Figure 2. Analysis of ^{3}H-thymidine incorporation into pubertal mammary glands cultured in basal media (Ctl) or in basal media containing IGF-I (10 ng/ml or 100 ng/ml; I) and/or EGF (60 ng/ml; E), amphiregulin (60 ng/ml; A) or TGF-α (60 ng/ml; T). Values shown represent dpm in 25 μl of tissue homogenate (mean ± standard error). n = 4 glands from individual animals for each treatment group. Data was analyzed by ANOVA followed by Fisher's PLSD post-hoc test. Hatched bars (I and I/E): 10 ng/ml IGF-I; Solid bars (I and I/E): 100 ng/ml IGF-I. a: $P < 0.01$ versus Ctl, b: $P < 0.05$ versus 10 ng/ml IGF-I, c: $P < 0.01$ versus 100 ng/ml IGF-I. Reprinted with permission from Stull, MA, Richert, MM, Loladze, AV, Wood, TL. Endocrinology 2002; in press. © 2002 Endocrine Society.

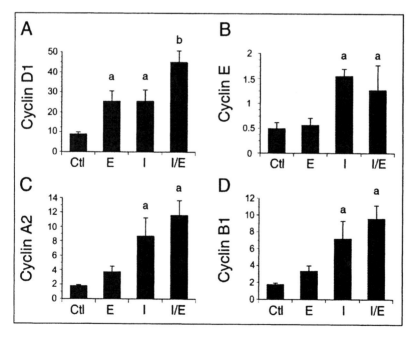

Figure 3. Expression of cyclin mRNA in cultured mammary glands after 48 hours of growth factor treatment. Graphs show changes in mRNA levels for cyclin D1 (A), cyclin E (B), cyclin A2 (C) and cyclin B1 (D). Values shown represent arbitrary optical density units (mean ± standard error) after adjustment to values obtained from hybridization to GAPDH mRNA within each sample. n = 4 glands from individual animals for each treatment group. Statistical analyses were performed using ANOVA followed by Fisher's PLSD post-hoc test. a: $P < 0.05$ vs Ctl; b: $P < 0.005$ vs IGF-I or EGF. Reprinted with permission from Stull, MA, Richert, MM, Loladze, AV, Wood, TL. Endocrinology 2002; in press. ©2002 Endocrine Society.

the G_1/S checkpoint and into the DNA synthetic phase. These results are consistent with the previous results showing that IGF-I but not EGF increased DNA synthesis in the cultured glands. IGF-I also induced mRNA expression of cyclin A2 and cyclin B1, which are required for the G_2-M transition. Analysis of cyclin B1 protein by western blot and immunohistochemical staining showed a similar increase in cyclin B1 protein in mammary epithelium following treatment with IGF-I compared to glands cultured in basal media or in the presence of EGF.[63] These results indicate that IGF-I mediates progression through the G_1-S and G_2-M checkpoints of the cell cycle. Additional analysis of the cyclin/cdk inhibitor p27 demonstrated no changes in p27 protein levels in response to either IGF-I or EGF in the cultured mammary glands (A. Loladze and T. Wood, unpublished data). This is in contrast to what has been shown previously in breast cancer cells where IGF-I reduces p27 levels.[64] In combination, these studies demonstrate that IGF-I induces cell cycle progression in normal mammary epithelial cells and provide support for the hypothesis that IGF-I is essential for EGF-mediated cell cycle progression in the intact mammary gland. Recent evidence demonstrating that the IGF-IR and ErbB-2 receptors form hetero-oligomers after induction by IGF-I and heregulin and that activation of the IGF-IR may be necessary for ErbB-2 phosphorylation in mammary tumor cells suggest that IGF signaling also is important for EGF-related growth in mammary tumors.[65]

The IGFBPs and Regulation of IGF Actions in the Developing Mammary Gland

The high-affinity IGF binding proteins (IGFBPs) are critical regulators of IGF bioavailability and biological actions (for reviews, see refs. 66-69). The IGFBPs are believed to have crucial roles during periods of growth and differentiation in a variety of tissues, however, the exact role of the IGFBPs in the developing mammary gland is unclear. Similar to the IGF ligands, several members of the IGFBPs are found in the circulation suggesting the possibility of endocrine actions on postnatal mammary gland development. At the circulating level, the IGFBPs extend the half-life of the IGFs and can transport the IGFs across endothelial barriers. However, results from our own and other laboratories also have shown local expression of specific IGFBPs in developing mammary tissue supporting a role for the IGFBPs as local autocrine/paracrine mediators of mammary epithelial growth.[40,43,45,70,71] Research from our laboratory has demonstrated that the six IGFBP family members each have a unique pattern of expression during both pubertal- and pregnancy-induced growth of the mouse mammary gland (Table 1).[40,45] Using in situ hybridization to analyze sites of mRNA synthesis in the developing gland, we determined that most of the IGFBPs are expressed in the TEBs, however, IGFBP-3 and IGFBP-5 are the most prominently expressed of the IGFBPs in the TEBs and throughout the ductal epithelium during pubertal growth. IGFBP-3 mRNA is expressed in the cap cells of the TEB while IGFBP-5 is expressed at high levels throughout the TEB. IGFBP-5 shows the strongest expression of all the IGFBPs throughout postnatal mammary gland development and is detected in stromal cells in addition to the epithelium. In contrast, IGFBP-2 and IGFBP-4 mRNAs are expressed predominantly in the stromal compartment. In recent studies, we analyzed tissue protein levels of the IGFBPs using western immunoblotting in the post-pubertal mammary gland and found that IGFBP-2, -3, -4, and -5 protein expression remains relatively stable throughout pubertal mammary growth (M. Allar and T. Wood, unpublished data). As was seen for mRNA expression, IGFBP-5 protein is the most abundant of the IGFBPs in developing mammary tissue (M. Allar and T. Wood, unpublished data). During pregnancy-induced alveolar development, expression of IGFBP-2 and IGFBP-4 mRNA is maintained in stromal cells. Interestingly, in both pubertal and pregnancy-induced growth stages, the most prominent expression of IGFBP-2 and IGFBP-4 is found in stroma adjacent to the epithelial structures. Similar to their prominent epithelial expression during pubertal growth, IGFBP-3 and IGFBP-5 mRNAs are highly expressed in the ducts and alveoli of the pregnant mammary gland. In addition to pregnancy stages, IGFBP protein expression also has been described in milk during lactation.[43,70,71]

Although the local expression of the IGFBPs is now well-documented in the developing mammary gland, the specific roles of these proteins in postnatal mammary development remain largely unclear. In the bovine mammary gland, IGFBP-3 is synthesized and released by mammary epithelial cells.[72] In vitro assays have demonstrated that IGFBP-3 inhibits the mitogenic activity of serum, mammary extracts or exogenous IGF-I on bovine mammary epithelial cells.[73] The most well-studied

Table 1. Summary of IGFBP mRNA expression patterns in virgin and pregnant mouse mammary glands

Age	Pubertal/Postpubertal Virgin Glands			Mid-Gestation Pregnant Glands		
Structure/ Compartment	TEBs	Ducts	Stroma	Ducts	Alveoli	Stroma
IGFBP-1	+	+	--	--	--	--
IGFBP-2	+[a]	--	+++[c]	--	--	+++[c]
IGFBP-3	++[b]	++	+	++	++	+
IGFBP-4	+	+	+++[c]	--	--	+++[c]
IGFBP-5	+++	+++	+++[d]	+++	+++	+
IGFBP-6	+	+	++[d]	--	--	++

Results of in situ hybridization analyses of IGFBP mRNA expression in sections of mouse mammary glands of pubertal (5-6 week), postpubertal virgin (15 week) and mid-pregnant (days 13-16) mice. Relative levels of expression (from + to +++) are based on hybridization intensities of sections with equivalent exposure times and with probes of equivalent specific activity. (--) = signal low to undetectable.
a: in cells at trailing edge of TEB
b: cap cells
c: stromal cells immediately surrounding ducts
d: distributed throughout stroma
Reprinted with permission from: Wood TL, Richert MM, Stull MA, Allar MA. J Mam Gland Biol Neopl 2000; 5:31-42. © 2000 Plenum Publishing Corporation.

period of postnatal mammary development in investigation of IGFBP function is involution. Transgenic mouse studies have revealed that IGFBP-3 overexpression results in inhibition of involution suggesting that IGFBP-3 promotes survival of mammary epithelial cells, possibly mediating the survival effects of IGF-I.[30] In contrast, IGFBP-5 has been proposed to have pro-apoptotic functions through blocking IGF-I survival effects during the process of involution.[71] In addition, IGFBP-5 likely contributes to remodeling of the extracellular matrix (ECM) during the process of involution by its ability to bind to plasminogen activator inhibitor-I resulting in activation of plasminogen cleavage.[74]

The studies discussed above have begun to elucidate the roles of the IGFBPs in the developing mammary gland, particularly during lactation and involution. However, little is known about these important regulators of growth during earlier stages of ductal growth and alveolar differentiation. One interesting finding from the mRNA expression patterns is that the IGFBPs exhibit expression patterns suggestive of a role in epithelial-stromal compartmentalization, and likely IGF compartmentalization, during mammary gland development. A variety of in vitro evidence has accumulated to support a model of local IGFBP action where the IGFBPs either inhibit or enhance IGF action depending on whether they are free in solution or associated with the cell surface or ECM.[66-68] The IGFBPs in solution have an affinity for the IGFs equal to or greater than the affinity of the IGF ligand for its receptor. In contrast, once the IGFBPs are associated with the cell surface or ECM, the affinity for the IGF is reduced thus allowing release of the IGF ligand to bind to its receptor. Consistent with this model of action, at least four of the IGFBPs have the ability to bind to the cell surface or ECM (for reviews, see refs. 66-68). Included in this group are IGFBP-1 and -3 which can associate with the cell surface and IGFBP-2,-3 and -5 which can associate with the ECM, at least in part through binding to heparin sulfate proteoglycans (HSPGs). The IGFBPs then could enhance the effects of the IGFs on mammary epithelial growth through association with the basement membrane that separates the stromal and epithelial compartments. When bound, the IGFBPs could enhance IGF accessibility to epithelial IGF receptors. Leading candidates for these actions are IGFBP-2, -3 and -5 that all have demonstrated ability to bind to HSPGs and are highly expressed in the mammary gland during ductal and alveolar growth. Regulation of IGF action by IGFBP tethering to the basement membrane is consistent with reports demonstrating association of other growth factors to the basement membrane around growing ducts.[75]

In contrast to other members of the high-affinity IGFBP family, IGFBP-4 is thought to be exclusively inhibitory for IGF actions. IGFBP-4 has no demonstrated ability to bind either cell surfaces or the ECM and consistently inhibits IGF actions on numerous cell types in vitro.[66,67] It is potentially relevant that IGFBP-4 is highly expressed in the stromal cells immediately surrounding the epithelial structures throughout postnatal mammary gland development.[40,45] These results suggest that IGFBP-4 is involved in inhibiting or restricting access of stromal IGFs to the epithelial compartment. Based on this model, the IGFs expressed in stroma are restricted to actions on stromal cells or adipocytes while IGFs expressed in the epithelium mediate proliferation and/or survival of epithelial cells. The proposed function of the IGFBPs in compartmentalization of the IGF ligands provides a model for how IGF actions are regulated locally in the mammary gland since both IGF-I and IGF-II are expressed in the epithelial as well as the stromal compartments. In addition, the distinct and spatially restricted patterns of expression for IGF-I and IGF-II, particularly in the epithelium, supports the hypothesis that local actions of the IGFs are restricted and highly regulated likely, in part, by the high-affinity IGFBPs.

The IGFs in Mammary Tumorigenesis and Breast Cancer

Breast cancer is among the most common forms of cancer and is the second leading cause of cancer death of women in the United States. A complete understanding of how and why breast tissue undergoes malignant transformation involves both epidemiological and genetic factors. Early investigations suggested that the underlying risks and susceptibility of developing breast cancer may be influenced by the breast developmental state.[76-81] Moreover, the pathways involved in normal breast proliferation and differentiation are related to those of carcinogenesis.[82]

Many of the genes involved in normal breast development also have a role in breast tumor formation including the BRCA genes,[83,84] genes important in cell cycle regulation such as cyclin D1,[32,85-87] cyclin E,[27,88] and p53,[89-92] and growth factor ligands and receptors of the EGF family.[16,21,29,31,34,93-95] In addition to their function in normal mammary epithelial growth, the IGF ligands and the IGF-IR also have demonstrated roles in abnormal growth of mammary epithelial cells. IGF-I and IGF-II are potent mitogens for breast cancer cell lines and for primary breast tumors in both humans and rodents.[1,96-99] In rodents, overexpression of either IGF-I or IGF-II in mammary glands results in tumors.[35,100] Moreover, mammary overexpression of IGF-I in combination with expression of a mutant form of the tumor suppressor p53 results in decreased mammary tumor latency demonstrating a cooperative interaction of these genes in mammary tumorigenesis.[100]

The IGFs also are highly implicated in breast cancer in humans (for reviews, see refs. 101,102). Serum IGF-I levels are 25% higher in women with breast cancer compared to normal age-matched controls.[103] In premenopausal women, high levels of biologically available IGF-I in plasma is associated with higher breast density and increased breast cancer risk and has been implicated in breast tumor growth.[104-106] While these studies suggest a correlation between breast cancer risk and circulating levels of IGF-I, IGF-I mRNA is not generally expressed in human breast cancer cell lines. Moreover, in breast cancers, IGF-I mRNA has been detected predominantly in the stromal cells adjacent to normal lobules.[107] Thus, IGF-I has been proposed as an paracrine stimulator of epithelial cells or as an autocrine stimulator of stromal cells in breast cancers.[104,108]

While both IGF-I and IGF-II are proposed to function as autocrine and/or paracrine growth promoters in breast cancers, at the local tissue level, IGF-II is more highly correlated with human breast cancers than IGF-I. Overexpression of IGF-II in breast epithelial cells in vitro results in changes associated with a malignant phenotype.[101,102,107,109] The induction of IGF-II has been correlated with invasive breast tumors,[110] and recent studies in breast cancer patients have shown abnormally high levels of IGF-II in 50% of breast cancers.[111] Moreover, IGF-II levels in tumor cytosols of women with breast cancer are inversely correlated with favorable prognostic indicators of breast cancer including the presence of the ER.[112] Initial studies on IGF-II in breast tissue showed IGF-II mRNA in pathologically normal and benign breast tissue.[113] IGF-II mRNA has been detected in normal breast fibroblasts as well as in the T47D breast cancer cell line.[113] However, like IGF-I, IGF-II is predominantly stromal in origin in breast cancers and is considered an important paracrine regulator of tumor epithelial growth.[114] It is also of interest that no significant relationship has been found between IGF-II RNA and ER expression in breast tissue; however, in one study, 73.5% of PR-positive breast cancer tissues were IGF-II positive.[114]

The IGF-IR in Breast Cancer

Actions of IGF-I and IGF-II in promoting breast tumor growth are thought to occur predominantly through the IGF-IR (for reviews, see refs. 115-117). H. Werner discusses the general function of the IGF-IR in human cancers elsewhere in this book, thus, in this section we review studies supporting a role for the IGF-IR specifically in breast cancers. The IGF-IR is generally overexpressed and demonstrates enhanced autophosphorylation and kinase activity in malignant breast tissue.[118-120] Clinical studies have suggested that IGF-IR overexpression is prognostic for breast cancer recurrence and reduced short-term survival.[121] In one study where 52% of breast cancers analyzed had high levels of IGF-IR, the patients with high IGF-IR levels relapsed more frequently within four years after diagnosis compared to patients without elevated expression of the IGF-IR. In another study, it was concluded that long-term survival is inversely correlated with IGF-IR levels.[122] This is consistent with results from other studies correlating IGF-IR expression with more differentiated tumors and a favorable clinical outcome.[118,123,124] Assessment of IGF-IR status in low risk cancer patients (ER and PR positive, low mitotic index, diploid); and high risk cancer patients (ER and PR negative, high mitotic index and aneuploid) also demonstrated a correlation between IGF-IR expression and favorable prognosis.[123] Unlike the IGF ligands that are highly expressed in stromal cells, high affinity IGF binding sites and IGF-IR mRNA have been localized primarily to breast epithelial cells in vivo.[118,124,125]

Signaling through the IGF-IR may contribute to the pathogenesis of breast cancer by stimulating proliferation as well as survival of breast tumor cells. Blockade of the IGF-IR inhibits growth of breast tumor cells both in vitro and in vivo.[97,98,122,126-129] Moreover, activation of the IGF-IR protects breast cancer cells from apoptosis induced by serum deprivation as well as by therapeutic agents and irradiation.[130] Additional investigations have suggested that IGF-IR signaling also contributes to the metastatic potential of breast cancer cells. A dominant-negative form of the IGF-IR suppressd adhesion, invasion and metastasis of metastatic breast cancer cells.[131] After injection into the mammary fat pad of mice, cells expressing the functionally impaired IGF-IR had no effect on growth of the primary tumor.[131] Similarly, metastatic breast cancer cells transfected with an antisense IGF-IR construct showed a decrease in metastasis when injected into nude mice.[132] In this latter study, however, the transfected breast cancer cells also showed a delay in tumor formation and in tumor size as well as decreased metastasis.[132]

Relationship between the IGF-IR and the Estrogen Receptor

Multiple lines of evidence now strongly support a role for the IGFs and IGF-IR in estrogen mediated regulation of breast cancer. High levels of ER and IGF-IR are correlated with susceptibility to breast cancer in women.[128,133] In the estrogen-responsive MCF-7 cell line, estrogen enhances the response to exogenously added IGF-I, while antiestrogens inhibit the response to IGF-I.[134] The mechanisms for estrogen and IGF-I synergistic actions in breast cancer cells include regulation of IGF-IR and IRS-1 expression and activation. Estrogen induces expression and phosphorylation of both the IGF-IR and IRS-1.[135-137] Conversely, antiestrogens, including tamoxifen, its derivatives, and the pure antiestrogens ICI 164,384 and ICI 182,780, inhibit IGF-IR-dependent proliferation through downregulation of the IGF-IR and IRS-1.[136,138-142] In vivo, antiestrogens also significantly decrease IGF-IR levels and activation in the normal rat mammary gland,[143] and estrogen induces IGF-IR mRNA levels in normal human breast xenografts.[144] The dependency of IGF-mediated growth on the ER was suggested by a recent study showing that both estrogen and IGF-mediated growth can be restored in ER-negative breast cancer cells by re-expression of the ER-alpha.[145] Estrogen-dependent growth of breast cancer cells is abolished in the presence of IGF-I when the IGF-IR is overexpressed.[141] Similarly, overexpression of IRS-1 depresses the ability of anti-estrogens to block IGF-dependent growth of hormone-dependent breast cancer cells.[142] In addition to the regulation of IGF-IR and IRS-1 by estrogen, the IGFs also can alter ER action through both transcriptional and post-transcriptional events.[146]

Synergism between the IGF-IR and estrogen is thought to involve predominantly the mitogenic function of the IGF-IR while the non-mitogenic functions of the IGF-IR are independent of ER status.[147] Recent studies on the interactions of IGF and estrogen signaling in breast cancer cells have revealed more information about activation of downstream targets by the two pathways. Subsequent to induction of IRS-1 expression, estradiol-enhanced IGF-IR signaling involves activation of the mitogen activated protein kinase, phosphoinositide 3' kinase, and Akt.[148] At the level of cell cycle regulation, estrogen enhances the effect of IGF-I on the expression of cyclin D1 and cyclin E and on the phosphorylation of the retinoblastoma protein.[64,148,149] An MCF-7-derived cell line containing a 50% reduction in IGF-IR expression showed no estrogen potentiation of the IGF-I effect on these cell cycle regulators.[64,148] Interestingly, IGF-I and estrogen both decreased the levels of the cell cycle inhibitor p27 but had differential effects on protein levels of the cell cycle inhibitor p21.[64,149] Estrogen abrogated the effect of IGF-I-induced p21 both by decreasing p21 levels[149] and by decreasing association of the inhibitors with the cyclinE-Cdk2 complex resulting in increased complex activation.[64,149]

The Insulin Receptor and IGF-II Receptor in Breast Cancer

While the IGF-IR is the primary signaling receptor for IGF-I and IGF-II, two other receptors with high affinity for IGF-II, but not IGF-I, are of potential interest in breast cancer. One of these, a splice variant of the IR known as IR-A, is found at proportionally higher levels in developing tissues and in cancers, including breast cancers, and may function as a second receptor by which IGF-II can promote growth of breast cancer cells.[150] The IR-A is of particular interest in the context of the data presented previously suggesting differential regulation and functions for IGF-I and IGF-II both in normal mammary development and in breast cancer.

IGF-II also binds with high affinity to the cation-independent mannose 6-phosphate/IGF-2 receptor (M6P/IGF2R).[151,152] A great deal of evidence suggests that the M6P/IGF2R does not function in biological signaling, rather, its function in binding IGF-II is to remove extracellular IGF-II for degradation in lysosomes.[153] In addition to its role in IGF-II degradation, the M6P/IGF2R has important functions in transport and endocytosis of M6P-containing proteins and in proteolytic activation of TGF-β. Thus, function of the M6P/IGF2R is directly relevant to cell physiology and growth regulation of many cell types and has been characterized as a candidate tumor suppressor gene.[153-155] In support of this hypothesis, loss of heterozygosity of the *m6p/igf2r* gene has been identified along with mutations in the remaining allele in both liver and breast cancers.[153,154,156,157]

The IGFBPs and Breast Cancer

The expression and regulation of all six high-affinity IGFBPs has been documented in breast cancer cells. In some cases, expression and secretion of the IGFBP family members is related to ER status of breast cancer cells.[158-161] IGFBP-2 and -5 were reported in ER-positive cell lines whereas IGFBP-1 and -6 were present in ER-negative cell lines. In contrast, IGFBP-3 and -4 expression was reported in both ER-positive and -negative cell lines. Estradiol markedly increases IGFBP-2, -4, and -5 concentrations but inhibits IGFBP-3 production in estrogen responsive cells.[161-163] In addition, progesterone agonists downregulates the IGF-IR mRNA and reduces IGFBP production in PR positive T47D human breast cancer cells.[163] In tamoxifen-resistant breast cancer cells, inhibition of cell growth by pure antiestrogen resulted in induction of IGFBP-5.[164] The expression of all six IGFBPs has been reported in breast cancer tissue in vivo.[161,165,166] The level of IGFBP-3 mRNA and protein is higher in ER negative tumors,[161,165-167] while IGFBP-4 and -5 expression is greater in ER positive cancer tissues.[166] Tamoxifen treatment in breast cancer patients elevated circulating levels of IGFBP-1[168-170] but did not effect total IGFBP-3 levels.[170]

The function of the IGFBPs in breast cancer is unclear. As discussed for their proposed function in normal tissues, individual IGFBPs have the ability to either augment or inhibit IGF action. Furthermore, IGF-independent actions of specific IGFBPs have been proposed particularly in relation to their function in breast cancer cells.[171-174] It is of interest that differential patterns of expression between the epithelial and stromal compartments for different IGFBPs in chemically-induced mammary tumors in the rat has led to the suggestion that the IGFBPs may be important for stromal-epithelial interactions in breast cancer cell proliferation.[175]

IGFBP-3 has been of particular interest in breast cancer. Increased serum levels of IGF-I and decreased IGFBP-3 serum levels are correlated with early-stage breast cancer[176]. More recently, a high IGF-I:IGFBP-3 ratio was correlated with increased breast density[105] and with an increased risk of breast cancer in premenopausal women.[177,178] Increased proteolysis of serum IGFBP-3 also has been associated with invasive breast cancers.[179] Moreover, in the tamoxifen studies cited previously, tamoxifen treatment had no effect on total levels of IGFBP-3 but did reduce proteolysis of circulating IGFBP-3.[170] At the cellular level, IGFBP-3 is thought to both inhibit IGF actions as well as promote apoptosis and block proliferation of breast cancer cells through IGF-independent mechanisms.[172-174,180-182] However, high IGFBP-3 levels in breast tumors has been associated with poor prognostic features (ER and PR negative, high S phase, aneuploidy and tumor size).[112,183-185] The significance of these findings are unclear since no correlation has been found between either circulating or breast tissue IGFBP-3 concentrations and disease recurrence.[184,185] Taken together, these data suggest that circulating IGFBP-3 is critical for regulating levels of free IGF-I, an increase in which is positively correlated with breast cancer. In contrast, the role of local IGFBP-3 expression in breast cancer is still unknown. It has been proposed that IGFBP-3 may antagonize tumor growth in the early stages of cancer but that during tumor progression, the tumor may become insensitive to the growth inhibitory effects of IGFBP-3.[173] In addition, recent evidence that IGFBP-3 may potentiate apoptosis of breast cancer cells induced by chemotherapeutic agents suggests a therapeutic potential for local IGFBP-3 expression.[186]

Conclusions

In summary, there is clear evidence that the IGF ligands, IGF receptors and IGFBPs are important regulators of epithelial growth both during normal mammary development and in breast cancer. It will be of great interest to continue investigations of how these molecules function specifically in normal and abnormal growth of this tissue and of how they coordinate with other growth factors and hormones to mediate proliferation, differentiation and survival of mammary/breast cells. The combination of genetic approaches in mice with in vitro approaches using both animal and human cells provide the basis for future studies to test function and regulation of the IGF system in normal, hyperplastic and neoplastic growth of breast tissue. These approaches will also be essential for developing and testing potential therapeutics designed to disrupt IGF- and IGFBP-mediated breast tumor growth.

Acknowledgements

The work in the authors' laboratory is supported by grants to TLW from the Department of Defense Breast cancer Research Program (CDA DAMD 17-99-1-9296) and the American Cancer Society (APG CNE-97721), to MAS from the National Cancer Institute (NRSA CA83174), and to DKR from the Susan G. Komen Breast cancer Foundation (PDF 0100718).

References

1. Russo I, Medado J, Russo J. Endocrine influences on the mammary gland. In: Jones T, Mohr U, Hunt R, eds. Integument and Mammary glands. New York: Springer-Verlag, 1989:252-266.
2. Russo IH, Russo J. Aging of the mammary gland. Pathobiology of the Aging Rat 1994; 2:447-458.
3. Imagawa W, Yang J, Guzman R et al. Control of mammary gland development. In: Knobil E, Neill JD, eds. The Physiology of Reproduction, Second Edition. New York: Raven Press, Ltd., 1994:1033-1063.
4. Richert M, Schwertfeger K, Ryder J et al. An atlas of mouse mammary gland development. J Mam Gland Biology Neoplasia 2000; 5:227-241.
5. Schedin P, Mitrenga T, Kaeck M. Estrous cycle regulation of mammary epithelial cell proliferation, differentiation, and death in the Sprague-Dawley rat: A model for investigating the role of estrous cycling in mammary carcinogenesis. J Mam Gland Biology Neoplasia 2000; 5:211-225.
6. Fendrick J, Raafat A, Haslam S. Mammary gland growth and development from the postnatal period to postmenopause: Ovarian steroid receptor ontogeny and regulation in the mouse. J Mam Gland Biology Neoplasia 1998; 3:7-22.
7. Bocchinfuso W, Korach K. Mammary gland development and tumorigenesis in estrogen receptor knockout mice. J Mam Gland Biology Neoplasia 1997; 2:323-334.
8. Humphreys R, Lydon J, O'Malley B et al. Use of PRKO mice to study the role of progesterone in mammary gland development. J Mam Gland Biology Neoplasia 1997; 2:343-354.

9. Ormandy C, Binart N, Kelly P. Mammary gland development in prolactin receptor knockout mice. J Mam Gland Biology Neoplasia 1997; 2:355-364.
10. Cunha G, Young P, Hom Y et al. Elucidation of a role for stromal steroid hormone receptors in mammary gland growth and development using tissue recombinants. J Mam Gland Biology Neoplasia 1997; 2:393-402.
11. Kleinberg D. Early mammary development: Growth hormone and IGF-1. J Mam Gland Biology Neoplasia 1997; 2:49-57.
12. Kleinberg D. Role of IGF-I in normal mammary development. Breast cancer Res Treat 1998; 47:201-208.
13. Atwood C, Hovey R, Glover J et al. Progesterone induces side-branching of the ductal epithelium in the mammary glands of peripubertal mice. J Endocrinol 2000; 167:39-52.
14. Ormandy CJ, Camus A, Barra J et al. Null mutation of the prolactin receptor gene produces multiple reproductive defects in the mouse. Genes & Development 1997; 11:167-178.
15. Brisken C, Park S, Vass T et al. A paracrine role for the epithlial progesterone receptor in mammary gland development. Proc Natl Acad Sci USA 1998; 95:5076-5081.
16. Luetteke N, Phillips H, Qiu T et al. The mouse waved-2 phenotype results from a point mutation in the EGF receptor tyrosine kinase. Genes and Development 1994; 8:263-278.
17. Fowler K, Walker F, Alexander W et al. A mutation in the epidermal growth factor receptor in waved-2 mice has a profound effect on receptor biochemistry that results in impaired lactation. Proc Natl Acad Sci USA 1995; 92:1465-1469.
18. Jackson D, Bresnick J, Rosewell I et al. Fibroblast growth factor signaling has a role in lobuloalveolar development of the mammary gland. J Cell Sci 1997; 110:1261-1268.
19. Xie W, Paterson A, Chin E et al. Targeted expression of a dominant negative epidermal growth factor receptor in the mammary gland of transgenic mice inhibits pubertal mammary duct development. Mol Endocrinol 1997; 11:1766-1781.
20. Wiesen J, Young P, Werb Z et al. Signaling through the stromal epidermal growth factor receptor is necessary for mammary ductal development. Development 1999; 126:335-344.
21. Luetteke N, Qiu T, Fenton S et al. Targeted inactivation of the EGF and amphiregulin genes reveals distinct roles for EGF receptor ligands in mouse mammary gland development. Development 1999; 126:2739-2750.
22. Gorska A, Joseph H, Derynck R et al. Dominant-negative interference of the transforming growth factor beta type II receptor in mammary gland epithelium results in alveolar hyperplasia and differentiation in virgin mice. Cell Growth Differ 1998; 9:229-238.
23. Joseph H, Gorska A, Sohn P et al. Overexpression of a kinase-deficient transforming growth factor-beta type II receptor in mouse mammary stroma results in increased epithelial branching. Mol Biol Cell 1999; 10:1221-1234.
24. Ruan W, Newman C, Kleinberg D. Intact and amino-terminally shortened forms of insulin-like growth factor I induce mammary gland differentiation and development. Proc Natl Acad Sci USA 1992; 89:10872-10876.
25. Walden P, Ruan W, Feldman M et al. Evidence that the mammary fat pad mediates the action of growth hormone in mammary gland development. Endocrinology 1998; 139:659-662.
26. Ruan W, Catanese V, Wieczorek R et al. Estradiol enhances the stimulatory effect of insulin-like growth factor (IGF-1) on mammary development and growth hormone-induced IGF-1 messenger ribonucleic acid. Endocrinology 1995; 136:1296-1302.
27. Bortner D, Rosenberg M. Induction of mammary gland hyperplasia and carcinomas in transgenic mice expressing human cyclin E. Mol Cell Biol 1997; 17:453-459.
28. Hadsell D, Greenberg N, Fligger J et al. Targeted expression of des(1-3) human insulin-like growth factor I in transgenic mice influences mammary gland development and IGF-binding protein expression. Endocrinology 1996; 137:321-330.
29. Matsui Y, Halter S, Holt J et al. Development of mammary hyperplasia and neoplasia in MMTV-TGFa transgenic mice. Cell 1990; 61:1147-1155.
30. Neuenschwander S, Schwartz A, Wood T et al. Involution of the lactating mammary gland is inhibited by the IGF system in a transgenic mouse model. J Clin Invest 1996; 97:2225-2232.
31. Sandgren E, Schroeder J, Qui T et al. Inhibition of mammary gland involution is associated with transforming growth factor a but not c-myc-induced tumorigenesis in transgenic mice. Cancer Res 1995; 55:3915-3927.
32. Wang T, Cardiff R, Zukerberg L et al. Mammary hyperplasia and carcinoma in MMTV-cyclin D1 transgenic mice. Nature 1994; 369:669-671.
33. Weber M, Boyle P, Corl B et al. Expression of ovine insulin-like growth factor-1 (IGF-1) stimulates alveolar bud development in mammary glands of transgenic mice. Endocrine 1998; 8:251-259.
34. Halter S, Dempsey P, Matsui Y et al. Distinctive patterns of hyperplasia in transgenic mice with mouse mammary tumor virus transforming growth factor-alpha. Am J Pathol 1992; 140:1131-1146.
35. Bates P, Fisher R, Ward A et al. Mammary cancer in transgenic mice expressing insulin-like growth factor II (IGF-II). British J Cancer 1995; 72:1189-1193.
36. Moorehead RA, Fata JE, Johnson MB et al. Inhibition of mammary epithelial apoptosis and sustained phosphorylation of Akt/PKB in MMTV-IGF-II transgenic mice. Cell Death Diff 2001; 8:16-29.
37. Ruan W, Kleinberg D. Insulin-like growth factor I is essential for terminal end bud formation and ductal morphogenesis during mammary development. Endocrinology 1999; 140:5075-5081.
38. Liu J-P, Baker J, Perkins AS et al. Mice carrying null mutations of the genes encoding insulin-like growth factor I (*Igf-1*) and type 1 IGF receptor (*Igf1r*). Cell 1993; 75:59-72.

39. Bonnette SG, Hadsell DL. Targeted disruption of the IGF-I receptor gene decreases cellular proliferation in mammary terminal end buds. Endocrinology 2001; 142:4937-4945.
40. Richert M, Wood T. Expression and regulation of insulin-like growth factors and their binding proteins in the normal breast. In: Manni A, eds. Endocrinology of Breast cancer. Totowa: Humana Press, 1999:39-52.
41. Richert M, Wood T. The insulin-like growth factors (IGF) and IGF type I receptor during postnatal growth of the murine mammary gland: Sites of messenger ribonucleic acid expression and potential functions. Endocrinology 1999; 140:454-461.
42. Kleinberg D, Ruan W, Catanese B et al. Non-lactogenic effects of growth hormone on growth and insulin-like growth factor-I messenger ribonucleic acid of rat mammary gland. Endocrinology 1990; 126:3274-3276.
43. Baumrucker CR, FErondu NE. Insulin-like growth factor (IGF) system in the bovine mammary gland and milk. J Mam Gland Biology Neoplasia 2000; 5:53-64.
44. Forsyth IA, Gabai G, Morgan G. Spatial and temporal expression of insulin-like growth factor-I, insulin-like growth factor-II and the insulin-like growth factor-I receptor in the sheep fetal mammary gland. J Dairy Res 1999; 66:35-44.
45. Wood T, Richert M, Stull M et al. The insulin-like growth factors (IGFs) and IGF binding proteins in postnatal development of murine mammary glands. J Mam Gland Biology Neoplasia 2000; 5:31-42.
46. Silberstein GB, Van Horn K, Shyamala G et al. Progesterone receptors in the mouse mammary duct: distribution and developmental regulation. Cell Growth Differ 1996; 7:945-952.
47. Shyamala G, Barcellos-Hoff MH, Toft D et al. In situ localization of progesterone receptors in normal mouse mammary glands: absence of receptors in the connective and adipose stroma and a heterogeneous distribution in the epithelium. J Steroid Biochem Mol Biol 1997; 63:251-259.
48. Seagroves TN, Lydon JP, Hovey RC et al. C/EBPbeta (CCAAT/enhancer binding protein) controls cell fate determination during mammary gland development. Mol Endocrinol 2000; 14:359-368.
49. Hovey RC, Trott JF, Ginsburg E et al. Transcriptional and spatiotemporal regulation of prolactin receptor mRNA and cooperativity with progesterone receptor function during ductal branch growth in the mammary gland. Dev Dyn 2001; 222:192-205.
50. Clarke RB, Howell A, Potten CS et al. Dissociation between steroid receptor expression and cell proliferation in the human breast. Cancer Res 1997; 57:4987-4991.
51. Seagroves TN, Krnacik S, Raught B et al. C/EBPβ, but not C/EBPα, is essential for ductal morphogenesis, lobuloalveolar proliferation, and functional differentiation in the mouse mammary gland. Genes & Development 1998; 12:1917-1928.
52. Ichinose R, Nandi S. Influence of hormones on lobulo-alveolar differentiation of mouse mammary glands in vitro. J Endocrinol 1966; 35:331-340.
53. Ganguly N, Ganguly R, Mehta N et al. Simultaneous occurrence of pregnancy-like lobuloalveolar morphognesis and casein-gene expression in a culture of the whole mammary gland. IN VITRO 1981; 17:55-60.
54. Imagawa W, Tomooka Y, Nandi S. Serum-free growth of normal and tumor mouse mammary epithelial cells in primary culture. Proc Natl Acad Sci USA 1982; 79:4074-4077.
55. Imagawa W, Tomooka Y, Hamamoto S et al. Stimulation of mammary epithelial cell growth in vitro: interaction of epidermal growth factor and mammogenic hormones. Endocrinology 1985; 116:1514-1524.
56. Gabelman BM, Emerman JT. Effects of estrogen, epidermal growth factor, and transforming growth factor-alpha on the growth of human breast epithelial cells in primary culture. Exp Cell Res 1992; 201:113-118.
57. Plaut K, Ikeda M, Vonderhaar B. Role of growth hormone and insulin-like growth factor-I in mammary development. Endocrinology 1993; 133:1843-1848.
58. Ip MM, Darcy KM. Three-dimensional mammary primary culture model systems. J Mam Gland Biology Neoplasia 1996; 1:91-110.
59. Barlow J, Casey T, Chiu J-F et al. Estrogen affects development of alveolar structures in whole-organ culture of mouse mammary glands. Biochem Biophys Res Commun 1997; 232:340-344.
60. Gingsburg E, Vonderhaar BK. Whole Organ Culture of the Mouse Mammary gland. In: Ip MM, Asch BB, eds. Methods in Mammary gland Biology and Breast cancer Research. New York: Kluwer Academic/Plenum Publishers, 2000:147-154.
61. Deeks S, Richards J, Nandi S. Maintenance of normal rat mammary epithelial cells by insulin and insulin-like growth factor. Exp Cell Res 1988; 174:448-460.
62. Imagawa W, Spencer E, Larson L et al. Somatomedin-C substitutes for insulin for the growth of mammary epithelial cells from normal virgin mice in serum-free collagen gel cell culture. Endocrinology 1986; 119:2695-2699.
63. Stull MA, Richert MM, Loladze AV et al. Requirement for insulin-like growth factor-I in epidermal growth factor-mediated cell cycle progression of mammary epithelial cells. Endocrinology 2002; in press.
64. Dupont J, Karas M, LeRoith D. The potentiation of estrogen on insulin-like growth factor I action in MCF-7 human breast cancer cells includes cell cycle components. J Biol Chem 2000; 275:35893-35901.
65. Balana ME, Labriola L, Salatino M et al. Activation of ErbB-2 via a hierarchical interaction between ErbB-2 and type I insulin-like growth factor receptor in mammary tumor cells. Oncogene 2001; 20:34-47.
66. Jones J, Clemmons D. Insulin-like growth factors and their binding proteins: Biological Actions. Endocrine Reviews 1995; 16:3-34.
67. Clemmons D, Busby W, Arai T et al. Role of insulin-like growth factor binding proteins in the control of IGF actions. Prog Growth Factor Res 1995; 6:357-366.

68. Clemmons D. Role of insulin-like growth factor binding proteins in controlling IGF actions. Molecular and Cellular Endocrinology 1998; 140:19-24.
69. Baxter RC. Insulin-like growth factor (IGF)-binding proteins: interactions with IGFs and intrinsic bioactivities. Am J Physiol Endocrinol Metab 2000; 278:E967-976.
70. Akers RM, McFadden TB, Purup S et al. Local IGF-I axis in peripubertal ruminant mammary development. J Mam Gland Biology Neoplasia 2000; 5:43-52.
71. Flint D, Tonner E, Allan G. Insulin-like growth factor binding proteins: IGF-dependent and -independent effects in the mammary gland. J Mam Gland Biology Neoplasia 2000; 5:65-73.
72. Gibson CA, Staley MD, Baumrucker CR. Identification of IGF binding proteins in bovine milk and the demonstration of IGFBP-3 synthesis and release by bovine mammary epithelial cells. Journal of Animal Science 1999; 77:1547-1557.
73. Weber M, Purup S, Vestergaard M et al. Contribution of insulin-like growth factor (IGF)-I and IGF-binding protein-3 to mitogenic activity in bovine mammary extracts and serum. J Endocrinol 1999; 161:365-373.
74. Tonner E, Allan G, Shkreta L et al. Insulin-like growth factor binding protein-5 (IGFBP-5) potentially regulates programmed cell death and plasminogen activation in the mammary gland. Advances in Experimental Medicine & Biology 2000; 480:45-53.
75. Silberstein G, Flanders K, Roberts A et al. Regulation of mammary morphogenesis: Evidence for extracellular matrix-mediated inhibition of ductal budding by transforming growth factor-β1. Dev Biol 1992; 152:354-362.
76. MacMahon B, Cole P, Lin TM et al. Age at first birth and breast cancer risk. Bulletin of World Health Organization 1970; 43:209-221.
77. MacMahon B, Tricholpoulos D, Brown J et al. Age at menarche, probability of ovulation and breast cancer risk. Int J Cancer 1982; 29:13-16.
78. Russo IH, Russo J. Developmental stage of the rat mammary gland as determinant of its susceptibility to 7,12-dimethylben(a)anthracene. J Nat Cancer Inst 1978; 61:1439-1449.
79. Russo J, Russo IH. Biological and molecular bases of mammary carcinogenesis. Lab Invest 1987; 57:112-137.
80. Lambe M, Hsieh C-C, Tricholpoulos D et al. Transient increase in the risk of breast cancer after giving birth. N Engl J Med 1994; 331:5-9.
81. Newcomb P, Storer B, Longnecker M et al. Lactation and a reduced risk of premenopausal breast cancer. N Engl J Med 1994; 330:81-87.
82. Chodosh LA, Gardner HP, Rajan JV et al. Protein kinase expression during murine mammary development. Developmental Biology 2000; 219:259-276.
83. Xu X, Wagner K-U, Larson D et al. Conditional mutation of Brca1 in mammary epithelial cells results in blunted ductal morphogenesis and tumor formation. Nature Genetics 1999; 22:37-43.
84. Deng C-X, Scott F. Role of the tumor suppressor gene Brca1 in genetic stability and mammary gland tumor formation. Oncogene 2000; 19:1059-1064.
85. Bartkova J, Lukas J, Muller H et al. Cyclin D1 protein expression and function in human breast cancer. Int J Cancer 1994; 57:353-361.
86. Said T, Luo L, Medina D. Mouse mammary hyperplasias and neoplasias exhibit different patterns of cyclins D1 and D2 binding to cdk4. Carcinogenesis 1995; 16:2507-2513.
87. Sicinski P, Donaher J, Parker S et al. Cyclin D1 provides a link between development and oncogenesis in the retina and breast. Cell 1995; 82:621-630.
88. Keyomarsi K, O'Leary N, Molnar G et al. Cyclin E, a potential prognostic marker for breast cancer. Cancer Res 1994; 54:380-385.
89. Li B, Murphy KL, Laucirica R et al. A transgenic mouse model for mammary carcinogenesis. Oncogene 1998; 16:997-1007.
90. Jerry D, Kuperwasser C, Downing S et al. Delayed involution of the mammary epithelium in BALB/c-p53 null mice. Oncogene 1998; 17:2305-2312.
91. Jerry D, Kittrell F, Kuperwasser C et al. A mammary-specific model demonstrates the role of the p53 tumor suppressor gene in tumor development. Oncogene 2000; 19:1052-1058.
92. Murphy KL, Rosen JM. Mutant p53 and genomic instability in a transgenic mouse model of breast cancer. Oncogene 2000; 19:1045-1051.
93. Kim H, Muller W. The role of the epidermal growth factor receptor family in mammary tumorigenesis and metastasis. Exp Cell Res 1999; 253:78-87.
94. Humphreys R, Hennighausen L. Transforming growth factor alpha and mouse models of human breast cancer. Oncogene 2000; 19:1085-1091.
95. Rose-Hellekant T, Sandgren E. Transforming growth factor alpha- and c-myc-induced mammary carcinogenesis in transgenic mice. Oncogene 2000; 19:1092-1096.
96. Karey K, Sirbasku D. Differential responsiveness of human breast cancer cell lines to growth factors and 17b-estradiol. Cancer Res 1988; 48:4083-4040.
97. Arteaga C, Kitten K, Coronado E. Blockade of the type I somatomedin receptor inhibits growth of human breast cancer cells in athymic mice. J Clin Invest 1989; 84:1418-1423.
98. Arteaga C, Osborne C. Growth inhibition of human breast cancer cells in vitro with an antibody against the type I somatomedin receptor. Cancer Res 1989; 49:6237-6241.
99. Bhalla V, Joshi K, Vohra H et al. Effect of growth factors on proliferation of normal, borderline, and malignant breast epithelial cells. Experimental & Molecular Pathology 2000; 68:124-132.

100. Hadsell DL, Murphy KL, Bonnette SG et al. Cooperative interaction between mutant p53 and des(1-3)IGF-I accelerates mammary tumorigenesis. Oncogene 2000; 19:889-898.
101. Yee D. Tyrosine kinase signaling in breast cancer: Insulin-like growth factors and their receptors in breast cancer. Breast cancer Research 2000; 2:170-175.
102. Sachdev D, Yee D. The IGF system and breast cancer. Endocr Relat Cancer 2001; 8:197-209.
103. Peyrat JP, Bonneterre J, Hecquet B et al. Plasma insulin-like growth factor -1 (IGF-1) concentrations in human breast cancer. European Journal of Cancer 1993; 29A:492-497.
104. Yee D, Paik S, Lebavic G et al. Analysis of IGF-I expression in malignancy-evidence for a paracrine role in human breast cancer. Mol Endocrinol 1989; 3:509-517.
105. Byrne C, Colditz GA, Willett WC et al. Plasma insulin-like growth factor (IGF) I, IGF-binding protein 3, and mammographic density. Cancer Res 2000; 60:3744-3748.
106. Stoll BA. Biological mechanisms in breast cancer invasiveness: relevance to preventive interventions. European Journal of Cancer Prevention 2000; 9:73-79.
107. Cullen KJ, Allison A, Martire I et al. Insulin-like growth factor expression in breast cancer epithelium and stroma. Breast cancer Research and Treatment 1992; 22:21-29.
108. Toropainen E, Lipponen P, Syrjanen K. Expression of insulin-like growth factor I (IGF-I) in female breast cancer as related to established prognostic factors and long-term prognosis. European Journal of Cancer 1995; 31:1443-1448.
109. Cullen K, Lippman M, Chow D et al. Insulin-like growth factor-II overexpression in MCF-7 cells induces phenotypic changes associated with malignant progression. Mol Endocrinol 1992; 6:91-100.
110. Giani C, Cullen K, Campani D et al. IGF-II mRNA and protein are expressed in the stroma of invasive breast cancers: an in situ hybridization and immunohistochemistry study. Breast cancer Research and Treatment 1996; 41:43-50.
111. Fichera E, Liang S, Xu Z et al. A quantitative reverse transcription and polymerase chain reaction assay for human IGF-II allows direct comparison of IGF-II mRNA levels in cancerous breast, bladder, and prostate tissues. Growth hormone & IGF Research 2000; 10:61-70.
112. Yu H, Levesque M, Khosravi M et al. Associations between insulin-like growth factors and their binding proteins and other prognostic indicators in breast cancer. Br J Cancer 1996; 74:1242-1247.
113. Yee D, Cullen KJ, Piak J et al. Insulin-like growth factor II mRNA expression in human breast cancer. Cancer Res 1988; 48:6691-6696.
114. Giani C, Pinchera A, Rasmussen A et al. Stromal IGF-II messenger RNA in breast cancer: Relationship with progesterone receptor expressed by malignant epithelial cells. Journal of Endocrinol Invest 1998; 21:160-165.
115. Ellis MJ, Jenkins S, Hanfelt J et al. Insulin-like growth factors in human breast cancer. Breast cancer Res Treat 1998; 52:175-184.
116. Surmacz E. Function of the IGF-I receptor in breast cancer. J Mam Gland Biology Neoplasia 2000; 5:95-105.
117. Zhang X, Yee D. Tyrosine kinase signalling in breast cancer: insulin-like growth factors and their receptors in breast cancer. Breast cancer Research 2000; 2:170-175.
118. Peyrat J, Bonneterre J. Type I IGF receptor in human breast diseases. Breast cancer Res Treat 1992; 22:59-67.
119. Athanassiadou P, Athanassiades P, Petrakakou E et al. Expression of insulin-like growth factor-I receptor and transferrin receptor by breast cancer cells in pleural effusion smears. Cytopathology 1996; 7:400-405.
120. Resnik JL, Reichart DB, Huey K et al. Elevated insulin-like growth factor I receptor autophosphorylation and kinase activity in human breast cancer. Cancer Res 1998; 58:1159-1164.
121. Turner BC, Haffty BG, Narayanan L et al. Insulin-like growth factor-I receptor overexpression mediates cellular radioresistance and local breast cancer recurrence after lumpectomy and radiation. Cancer Res 1997; 57.
122. Railo M, Smitten K, Pekonen F. The prognostic value of insulin-like growth factor-I in breast cancer patients. Results of a follow-up study on 126 patients. Eur J Can 1994; 30A:307-311.
123. Pezzino V, Papa V, Milazzo G et al. Insulin-like growth factor-I (IGF-I) receptors in breast cancer. Annals of the New York Academy of Sciences 1996; 784:189-201.
124. Schnarr B, Strunz K, Ohsam J et al. Down-regulation of insulin-like growth factor-I receptor and insulin receptor substrate-1 expression in advanced human breast cancer. Int J Cancer 2000; 89:506-513.
125. Jammes H, Peyrat JP, Ban E et al. Insulin-like growth factor 1 receptors in human breast tumor: localisation and quantification by histo-autoradiographic analysis. Br J Cancer 1992; 66:248-253.
126. Rohlik Q, Adams D, Kull F et al. An antibody to the receptor for insulin-like growth factor-I inhibits the growth of MCF-7 cells in tissue culture. Biochem Biophys Res Commun 1987; 149:276-281.
127. Arteaga C. Interference of the IGF system as a strategy to inhibit breast cancer growth. Breast cancer Res Treat 1992; 22:101-106.
128. Lee A, Yee D. Insulin-like growth factors and breast cancer. Biomed & Pharmacother 1995; 49:415-421.
129. Neuenschwander S, Roberts Jr. CT, LeRoith D. Growth inhibition of MCF-7 breast cancer cells by stable expression of an insulin-like growth factor I receptor antisense ribonucleic acid. Endocrinolgy 1995; 136:4298-4303.
130. Dunn SE, Hardman RA, Kari FW et al. Insulin-like growth factor 1 (IGF-1) alters drug sensitivity of HBL100 human breast cancer cells by inhibition of apoptosis induced by diverse anticancer drugs. Cancer Res 1997; 57:2687-2693.
131. Dunn SE, Ehrlich M, Sharp NJH et al. A dominant negative mutant of the insulin-like growth factor-I receptor inhibits the adhesion, invasion, and metastasis of breast cancer. Cancer Res 1998; 58.

132. Chernicky C, Yi L, Tan H et al. Treatment of human breast cancer cells with antisense RNA to the type I insulin-like growth factor receptor inhibits cell growth, suppresses tumorigenesis, alters the metastatic potential, and prolongs survival in vivo. Cancer Gene Therapy 2000; 7:384-395.
133. Stoll B. Breast cancer: further metabolic-endocrine risk markers? Br J Cancer 1997; 76:1652-1654.
134. Wakeling AE. Comparative studies on the effects of steroidal and nonsteroidal oestrogen antagonists on the proliferation of human breast cancer cells. Journal of Steroid Biochemistry 1989; 34:183-188.
135. Stewart A, Johnson M, May F et al. Role of insulin-like growth factors and the type I insulin-like growth factor receptor in the estrogen-stimulated proliferation of human breast cancer cells. J Biol Chem 1990; 265:21172-21178.
136. Surmacz E, Guvakova MA, Nolan MK et al. Type I insulin-like growth factor receptor function in breast cancer. Breast cancer Res Treat 1998; 47:255-267.
137. Lee AV, Jackson JG, Gooch JL et al. Enhancement of insulin-like growth factor signaling in human breast cancer: Estrogen regulation of insulin receptor substrate-1 expression in vitro and in vivo. Mol Endocrinol 1999; 10:787-796.
138. de Cupis A, Noonan D, Pirani P et al. Comparison between novel steroid-like and conventional nonsteroidal antioestrogens in inhibiting oestradiol- and IGF-I-induced proliferation of human breast cancer-derived cells. British Journal of Pharmacology 1995; 116:2391-2400.
139. Huynh H, Nickerson T, Pollak M et al. Regulation of insulin-like growth factor I receptor expression by the pure antiestrogen ICI 182780. Clinical Cancer Research 1996; 2:2037-2042.
140. de Cupis A, Favoni RE. Oestrogen/growth factor cross-talk in breast carcinoma: a specific target for novel antiestrogens. Trends Pharmcol Sci 1997; 18:245-251.
141. Guvakova M, Surmacz E. Tamoxifen interferes with the insulin-like growth factor I receptor (IGF-IR) signaling pathway in breast cancer cells. Cancer Res 1997; 57:2606-2610.
142. Salerno M, Sisci D, Mauor L et al. Insulin receptor substrate-1 (IRS-1) is a substrate for a pure antiestrogen ICI 182,780. Int J Cancer 1999; 81:299-304.
143. Chan TW, Pollak M, Huynh H. Inhibition of insulin-like growth factor signaling pathways in mammary gland by pure antiestrogen ICI 182,780. Clinical Cancer Research 2001; 7:2545-2554.
144. Clarke RB, Howell A, Anderson E. Type I insulin-like growth factor receptor gene expression in normal human breast tissue treated with oestrogen and progesterone. Br J Cancer 1997; 75:251-257.
145. Oesterreich S, Zhang P, Guler RL et al. Re-expression of estrogen receptor alpha in estrogen receptor alpha-negative MCF-7 cells restores both estrogen and insulin-like growth factor-mediated signaling and growth. Cancer Res 2001; 61:5771-5777.
146. Lee A, Weng C-N, Jackson J et al. Activation of estrogen receptor-mediated gene transcription by IGF-I in humna breast cancer cells. J Endocrinol 1997; 152:39-47.
147. Bartucci M, Morelli C, Mauro L et al. Differential insulin-like growth factor I receptor signaling and function in estrogen receptor (ER)-positive MCF-7 and ER-negative MDA-MB-231 breast cancer cells. Cancer Res 2001; 61:6747-6754.
148. Dupont J, LeRoith D. Insulin-like growth factor 1 and oestradiol promote cell proliferation of MCF-7 breast cancer cells: new insights into their synergistic effects. Molecular Pathology 2001; 54:149-154.
149. Lai A, Sarcevic B, Prall OWJ et al. Insulin/insulin-like growth factor and estrogen cooperate to stimulate cyclin E-Cdk2 activation and cell cycle progression in MCF-7 breast cancer cells through differential regulation of cyclin E and and p21$^{WAF1/Cip1}$. J Biol Chem 2001; 276:25823-25833.
150. Frasca F, Pandini G, Scalia P et al. Insulin receptor isoform A, a newly recognized high-affinity insulin-like growth factor II receptor in fetal and cancer cells. Mol Cell Biol 1999; 19:3278-3288.
151. Morgan DO, Edman JC, Standring DN et al. Insulin-like growth factor II receptor as a multifunctional binding protein. Nature 1987; 329:301-307.
152. MacDonald RG, Pfeffer SR, Coussens L et al. A single receptor binds both insulin-like growth factor II and mannose-6-phosphate. Science 1988; 1134-1137.
153. DaCosta SA, Schumaker LM, Ellis MJ. Mannose 6-phosphate /insulin-like growth factor 2 receptor, a bona fide tumor suppressor gene or just a promising candidate? J Mam Gland Biology Neoplasia 2000; 5:85-94.
154. Hankins GR, De Souza AT, Bentley RC et al. M6P/IGF2 receptor: a candidate breast tumor suppressor gene. Oncogene 1996; 12:2003-2009.
155. Oates AJ, Schumaker LM, Jenkins SB et al. The mannose 6-phosphate/insulin-like growth factor 2 receptor (M6P/IGF2R), a putative breast tumor suppressor gene. Breast cancer Res Treat 1998; 47:269-281.
156. De Souza AT, Hankins GR, Washington MK et al. Frequent loss of heterozygosity on 6q at the mannose 6-phosphate/insulin-like growth factor II receptor locus in human hepatocellular tumors. Oncogene 1995; 10:1725-1729.
157. Yamada Y, De Souza AT, Finkelstein S et al. Loss of the gene encoding mannose 6-phosphate/insulin-like growth factor II receptor is an early event in liver carcinogenesis. Proc Natl Acad Sci USA 1997; 94:10351-10355.
158. Clemmons DR, Camacho-Hubner C, Coronado E et al. Insulin-like growth factor binding protein secretion by breast carcinoma cell lines: Correlation with estrogen receptor status. Endocrinology 1990; 127:2679-2686.
159. Yee D, Favoni RE, Lippman ME et al. Identification of insulin-like growth factor binding proteins in breast cancer cells. Breast Cancer Research and Treatment 1991; 18:3-10.
160. Sheikh MS, Shao Z-M, Clemmons DR et al. Identification of the insulin-like growth factor-binding proteins 5 and 6 (IGFBP-5 and 6) in human breast cancer cells. Biochem Biophys Res Commun 1992; 183:1003-1010.

161. Figueroa J, Yee D. The insulin like growth factor binding proteins (IGFBPs) in human breast cancer. Breast cancer Research and Treatment 1992; 22:81-90.
162. Pratt SE, Pollak MN. Estrogen and antiestrogen modulation of MCF-7 human breast cancer cell proliferation is associated with specific alterations in accumulation of insulin-like growth factor-binding proteins in conditioned media. Cancer Res 1993; 53:519-5198.
163. Owens PC, Gill PG, DeYoung NJ et al. Estrogen and progesterone regulate secretion of insulin-like growth factor-binding proteins by human breast cancer cells. Biochem Biophys Res Commun 1993; 193:467-473.
164. Parisot JP, Leeding KS, Hu XF et al. Induction of insulin-like growth factor binding protein expression by ICI 182,780 in a tamoxifen-resistant human breast cancer cell line. Breast cancer Res Treat 1999; 55:231-242.
165. Pekonen F, Nyman T, Ilvesmaki V et al. Insulin-like growth factor-binding proteins in human breast cancer tissue. Cancer Res 1992; 52:5204-5207.
166. McGuire S, Hilsenbeck S, Figueroa J et al. Detection of insulin-like growth factor binding proteins (IGFBPs) by ligand blotting in breast cancer tissues. Cancer Letters 1994; 77:25-32.
167. Shao Z-M, Sheikh MS, Ordonez JV et al. IGFBP-3 gene expression and estrogen receptor status in human breast carcinoma. Cancer Res 1992; 52:5100-5103.
168. LØnning PE, Hall K, Aakvaag A et al. Influence of tamoxifen on the plasma levels of insulin-like growth factor-binding protein I in breast cancer patients. Cancer Res 1992; 52:4719-4723.
169. Pollak MN, Huynh HT, Lefebvre SP. Tamoxifen reduces serum insulin-like growth factor I (IGF-I). Breast cancer Res Treat 1992; 22:91-100.
170. Helle SI, Holly JM, Tally M et al. Influence of treatment with tamoxifen and change in tumor burden on the IGF-system in breast cancer patients. Int J Cancer 1996; 69:335-339.
171. Perks CM, Newcomb PV, Norman MR et al. Effect of insulin-like growth factor binding protein-1 on integrin signalling and the induction of apoptosis in human breast cancer cells. Journal of Molecular Endocrinology 1999; 22:141-150.
172. Perks CM, McCaig C, Holly JM. Differential insulin-like growth factor (IGF)-independent interactions of IGF binding protein-3 and IGF binding protein-5 on apoptosis in human breast cancer cells. Involvement of the mitochondria. Journal of Cellular Biochemistry 2000; 80:248-258.
173. Perks CM, Holly JMP. Insulin-like growth factor binding proteins (IGFBPs) in breast cancer. J Mam Gland Biology Neoplasia 2000; 5:75-84.
174. Baxter RC. Signalling pathways involved in antiproliferative effects of IGFBP-3: a review. Molecular Pathology 2001; 54:145-148.
175. Manni A, Badger B, Wei L et al. Hormonal regulation of insulin-like growth factor II and insulin-like growth factor binding protein expression by breast cancer cells in vivo: Evidence for stromal epithelial interactions. Cancer Res 1994; 54:2934-2942.
176. Bruning PF, Van Doorn JM, Bonfrer PA et al. Insulin-like growth factor-binding protein 3 is decreased in early-stage operable pre-menopausal breast cancer. Int J Cancer 1995; 62:266-270.
177. Bohlke K, Cramer DW, Trichopoulos D et al. Insulin-like growth factor-I in relation to premenopausal ductal carcinoma in situ of the breast. Epidemiology 1998; 9:
178. Li BD, Khosravi MJ, Berkel HJ et al. Free insulin-like growth factor-I and breast cancer risk. Int J Cancer 2001; 91:736-739.
179. Helle SI, Geisler S, Aas T et al. Plasma insulin-like growth factor binding protein-3 proteolysis is increased in primary breast cancer. Br J Cancer 2001; 85:74-77.
180. Rozen F, Zhang J, Pollak M. Antiproliferative action of tumor necrosis factor-alpha on MCF-7 breastcancer cells is associated with increased insulin-like growth factor binding protein-3 accumulation. International Journal of Oncology 1998; 13:865-869.
181. Butt AJ, Firth SM, King MA et al. Insulin-like growth factor-binding protein-3 modulates expression of Bax and Bcl-2 and potentiates p53-independent radiation-induced apoptosis in human breast cancer cells. J Biol Chem 2000; 275:39174-39181.
182. Fanayan S, Firth SM, Butt AJ et al. Growth inhibition by insulin-like growth factor-binding protein-3 in T47D breast cancer cells requires transforming growth factor-beta (TGF-beta) and the type II TGF-beta receptor. J Biol Chem 2000; 275:39146-39151.
183. Rocha RL, Hilsenbeck SG, Jackson JG et al. Correlation of insulin-like growth factor-binding protein-3 messenger RNA with protein expression in primary breast cancer tissues: Detection of higher levels in tumors with poor prognostic features. J Nat Cancer Inst 1996; 88:601-606.
184. Rocha RL, Hilsenbeck SG, Jackson JG et al. Insulin-like growth factor-binding protein-3 and insulin receptor substrate-1 in breast cancer: Correlation iwth clinical parameters and disease-free survival. Clinical Cancer Research 1997; 3:103-109.
185. Yu H, Levesque MA, Khosravi MJ et al. Insulin-Like Growth Factor-Binding protein-3 and Breast cancer Survival. Int J Cancer 1998; 79:624-628.
186. Fowler CA, Perks CM, Newcomb PV et al. Insulin-like growth factor binding protein-3 (IGFBP-3) potentiates paclitaxel-induced apoptosis in human breast cancer cells. Int J Cancer 2000; 88:448-453.

CHAPTER 11

The Insulin-Like Growth Factor System and Bone

Thomas L. Clemens and Clifford J. Rosen

Introduction

In the last several years, investigators have begun to unravel the role of insulin like growth factor-I (IGF-I), and its family of IGF binding proteins, in the building and maintenance of the adult skeleton. Although it was nearly a half a century ago that the presence of a 'somatomedin' peptide modulating growth hormone activity was first postulated, it has taken nearly five decades to determine how this factor regulates longitudinal growth and bone consolidation. The first steps in this discovery began with in vitro studies of primary murine calvarial cells, and transformed cell lines. Several classic experiments established the importance of IGF-I in bone cell differentiation. A decade later, studies of global transgenic and knockout mice were initiated. These studies confirmed that over or under- expression of IGF-I impacted bone acquisition and maintenance. More recently, the generation of mice using conditional mutagenesis, and the construction of congenics have provided investigators with unparalleled opportunities to study the mechanisms whereby IGF-I affects peak bone mass acquisition. Results from these studies have served an additional purpose; i.e., to propel IGF-I into the forefront as a potential peptide target for future therapeutics aimed specifically at building bone mineral and preventing osteoporosis. This chapter will examine three major aspects of the relationship between IGFs and bone, focusing particularly on IGF-I. These include:

1. the physiology of IGF-I in bone;
2. the role of IGF-I in modulating bone formation and bone resorption; and
3. in vivo model systems that have been used to understand IGF-I actions in the skeleton.

Rapid progress in these three areas of investigation promise to shed even more light on this complex regulatory circuit in bone.

Origins of Skeletal IGFs and Their Activity within Bone

The skeleton is a highly organized and physiologically active organ, continuously remodeling itself to preserve skeletal integrity, and provide a reliable and constant source of calcium for the circulation. Numerous growth factors and cytokines, each of which contributes to coupling bone dissolution (i.e., resorption) to new bone formation, orchestrate bone remodeling[1] (See Fig. 1). Pre-osteoblasts (pre-OBs), derived from mesenchymal stromal cells, represent key target cells for initiation of the remodeling cycle.[2] Systemic and local factors enhance pre-OB differentiation, and this, in turn, leads to the synthesis and release of m-CSF and RANK ligand.[3] These two peptides are necessary and sufficient for the recruitment of bone resorbing cells, i.e., the osteoclasts. Once bone resorption occurs, calcium, collagen fragments and growth factors such as the IGFs, and TGFs, are released from the bony matrix. The latter factors enhance the recruitment of osteoblasts to the bone surface, thereby setting the stage for collagen synthesis and matrix deposition/mineralization.[4] The entire remodeling cycle in humans takes approximately 90 days, with the majority of time consumed by the elaborate process of bone formation and subsequent mineralization.

As has been noted previously in this book, the circulation of most mammals contains large concentrations of IGF-I and IGF-II bound to high and low molecular weight IGFBPs.[5] Similarly, the skeletal matrix also is highly enriched with growth factors and non-collagenous proteins including IGFs, all six IGFBPs and several IGFBP proteases. In addition, the Type I IGF receptor is

Insulin-Like Growth Factors, edited by Derek LeRoith, Walter Zumkeller and Robert Baxter. ©2003 Eurekah.com and Kluwer Academic / Plenum Publishers.

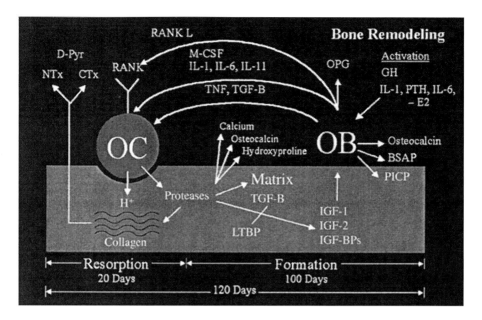

Figure 1. The bone remodeling unit is composed of osteoclasts and osteoblasts in an orchestrated cycle that is controlled by both systemic and local factors. IGF-I is one of many factors that are involved in the recruitment and differentiation of osteoblasts. IGF-I is produced by stromal cells and osteoblasts. It is also stored in the bone matrix and released during bone resorption. IGF-I may play a supplementary role in recruiting osteoclasts.

present on both osteoblasts and osteoclasts. It is reasonably certain that skeletal IGFs originate from two sources: 1) de novo synthesis by bone forming cells (i.e., pre-osteoblasts, and fully differentiated osteoblasts); and 2) the circulation. In fact some skeletal IGFs probably make their way into the matrix by way of canaliculi and sinusoids within the bone microcirculation.[1,4] IGFs, bound to IGFBPs, can also be found within the marrow milieu in close contact with the endosteal surface of bone. But, by most accounts, the vast majority of IGF-I in bone is derived from local synthesis. Yet during active bone resorption, as the matrix is dissolved, significant amounts of IGF-I and II are released from storage (i.e.,binding to IGFBP-5 and hydroxyapatite). Subsequently,both IGFs play an active role in the recruitment of precursor osteoblasts, and possibly early osteoclasts to the bone surface where remodeling is occurring.[1,4,6]

The IGFs act in diverse patterns via endocrine / autocrine / paracrine pathways to regulate differentiative functions of both osteoblasts and osteoclasts. Since the IGFs are stored within the skeletal matrix after osteoblast synthesis, and released during bone resorption, it has been suggested that IGF-I and -II are the critical coupling proteins that keep bone resorption closely linked to formation. On the other hand, several cytokines and differentiation factors also work in a manner analogous to the IGFs, some with greater potency on osteoblasts than others.[1,4] It seems more likely that the orchestration of bone remodeling requires the activity of both osteoblast-derived and systemic proteins, working through time and dose-dependent circuits to maintain a balanced bone turnover rate.

Although the autocrine/paracrine pathways for the IGFs are well delineated in the osteoblast, an area of active investigation is the endocrine role for IGF-I in the skeleton. Up to the present time, it has been virtually impossible to quantify the relative proportion of IGFs that are derived solely from the circulation vs those that originate directly from the skeleton. However, based on genetic engineering studies in mice, there may be a threshold effect for IGF-I, such that a certain concentration from the circulation is needed to augment skeletal sources for overall maintenance of bone remodeling.[7] Interestingly, the relative proportion of IGF-I:IGF-II is maintained in both the serum and skeleton of various species, suggesting that tissue specific activity may be related to the overall circulatory balance.[8]

IGF-I Gene Expression and Regulation in Bone

Much of the transcriptional regulation of the IGFs in hard tissue is obscure and illustrates the complexity underpinning the physiology of IGF actions in bone. For example, growth hormone (GH) is a potent regulator of skeletal and cartilage IGF-I synthesis, yet its mode of action is unknown. There are no identifiable GH responsive protein binding sites near promoter 1 or promoter 2 sites in the IGF-I gene for either osteoblasts or chondrocytes.[9-11] Several transcription factors have been identified which can bind to and enhance the activity of P1 in soft tissues, including a CAAT enhancing binding protein (C/EBP), HNF-1 and HNF-3).[12] Response elements within the IGF-I gene have also been identified in bone cells. For example, a cyclic AMP response element (CRE) and a glucocorticoid responsive region have been noted in close proximity to the P1 promoter of the IGF-I gene.[13] Glucocorticoids down regulate IGF-I expression in osteoblasts, through a steroid response element located approximately 100 bps upstream of P1.[14] Prostaglandins, in particular PGE2, have been shown to regulate osteoblast production of IGF-I by binding C/EBP, which, in turn, acts via CRE at a location approximately 200 base pairs 5' upstream of P1.[15] Similarly, although the IGF-I promoters lack estrogen response elements, 17 B estradiol suppresses IGF-I gene activation, by acting through receptor binding to C/EBP.[16] It seems certain there are other response elements within the IGF-I gene and tissue-specific transcription factors that regulate IGF-I expression in bone, although so far none have been mapped to either the P1 or P2 region of the IGF-I gene.

Systemic and Local Regulation of IGF-I in Bone Cells

Osteoblast-like (OB) cells from rodents and human in culture express both IGF-I and IGF-II mRNA transcripts. In vivo, studies have revealed that the major hormones regulating bone turnover, also affect IGF-I expression. These include parathyroid hormone (PTH), estrogen, glucocorticoids, and 1,25-dihydroxyvitamin D. There is substantial evidence from in vitro and in vivo studies that the anabolic effects of PTH on rat bone are mediated largely through increased local IGF-I expression.[15] PTH exerts its effect on IGF-I synthesis through increased cyclic AMP (cAMP) production.[17] PTH and other potent stimulators of cAMP in OB cells, such as prostaglandin E2, increase IGF-I synthesis via increases in gene transcription.[18] Estradiol also enhances IGF-I synthesis at the transcriptional level in rat bone cells transfected with estrogen receptors.[19] As indicated above there are no consensus estrogen responsive elements identified within the cloned promoter regions of the IGF-I gene. Hence, it is likely that estrogen acts through the c-AMP dependent C/EBP pathway either as an inhibitor in some cell lines and species, or as a stimulator of IGF-I transcription in rat and human osteoblast.

There is also unique genetic programming of skeletal IGF expression. Rosen et al demonstrated that for two healthy inbred strains of mice (C3H and C57BL6), of the same body length and size, serum and skeletal IGF-I content differed by as much as 30% and these interstrain differences in IGF-I expression were also observed in calvarial osteoblasts maintained in vitro.[20,21] There were, however, differences in promoter usage such that hepatic P2 promoter expression was nearly five fold greater in C3H than B6 mice, while P1 promoter transcripts were not different by strain (Adamo, personal communication). On the other hand, P1 IGF-I expression in the femurs of C3H were significantly greater than B6 without differences in P2 transcripts. Hence, there must be heritable regulators of IGF-I that are strain and tissue specific.

Besides genetic and systemic regulation of IGF-I, it is clear that there are numerous local factors, which contribute to the IGF regulatory circuit. Skeletal growth factors such as FGF-2 and cytokines such as the interleukins, regulate IGF expression in osteoblasts. BMP-2 increases IGF-I and II mRNA expression in rat osteoblasts, and may be a critical factor in early osteoblast recruitment within the remodeling unit.[15] BMP-7 also has a very potent effect on both IGF-I and II production in bone cells, and anti-sense IGF-I and IGF-II oligonucleotides block BMP-7 induced alkaline phosphatase expression. TGF-β also increases IGF-I and IGFBP-3 expression in human marrow stromal cells.[22] IL-6 up-regulates IGF-I expression mRNA in osteoblasts, while its effect on hepatic expression is the opposite.[15] Prostaglandins regulate IGF-I and -II expression and are produced locally by bone cells, thereby providing a major paracrine regulatory circuit in the skeleton. Mechanical loading is also a stimulus for enhanced IGF-I expression in bone cells, possibly through the induction of PGI2 and PGE2. Strain induced production of PGI2 has been shown to immunolocalize to osteocytes, where IGF-II is released. PGE2, also generated by strain, tends to localize to osteoblasts, and it can induce the generation of either IGF-I or IGF-II.[23]

IGF Binding Proteins (IGFBPs) and Bone

All cells involved in bone remodeling produce and/or respond to IGFs (pre-OBs, osteoblasts, osteocytes and osteoclasts). In addition, IGFs influence OB function at all stages of development (proliferation, differentiation, matrix production, and mineralization). But IGF peptides and receptors are relatively ubiquitous. Any consideration of IGF action must take into account binding proteins that modify IGF bioactivity. Six distinct yet structurally homologous IGFBPs have been characterized and designated IGFBP-1 through IGFBP-6. Wang recently documented that OB localized in trabecular bone of the postnatal growth plate express IGFBP-2, -4, -5, and -6 mRNAs during the course of skeletogenesis in rat and mouse.[24] All IGFBPs are expressed by bone cells in vitro, but, like IGF-I expression, IGFBP synthesis varies depending on cell type, cell density and cell culture conditions.[23,24]

In their native or recombinant state in solution, IGFBPs bind IGFs with high affinity, thereby preventing interaction with their receptor and effectively inhibiting IGF action. Posttranslational modifications produce dramatic changes in structure/ function of the IGFBPs, through phosphorylation, glycosylation or proteolytic cleavage of the native binding protein. These changes ultimately determine the biologic fate of the IGFs.

IGFBP-1 can inhibit or enhance IGF action dependent upon its phosphorylation state.[25] In addition, IGFBP-1 stimulates cell migration through interaction with integrins. Recent published data suggest that IGFBP-1 expression in hOB cells is directly stimulated by high dose glucocorticoid treatment and associated with suppressed type I collagen.[26] Since high levels of IGFBP-1 are noted in poorly controlled diabetics and in malnourished individuals, it is conceivable that suppression of bone formation noted in these conditions can be linked to locally high levels of IGFBP-1 expression in the skeleton.

IGFBP-2 is a 30 kD protein found in large quantities in the neonatal circulation. It is also secreted by rat and human OB cells. Addition of recombinant human IGFBP-2 inhibits the actions of IGF-I on fetal rat calvarial OB replication and matrix synthesis. However, when administered in combination with IGF-II, this complex actually stimulates new bone formation.[27]

IGFBP-3 is another Janus-faced IGFBP with both inhibitory and stimulatory potential. It is also the largest and most abundant circulating binding protein. In its intact form, exogenous IGFBP-3 is a potent inhibitor of bone cell growth.[4] However, Ernst and Rodan found that accumulation of endogenous IGFBP-3 correlated with enhanced IGF-I activity in OB cells.[28] The ability of cell-associated IGFBP-3 to modulate IGF action and the IGF-independent effects of IGFBP-3 described in other cell systems, have not yet been explored in pre-OBs or OBs in vitro.

IGFBP-4 was originally isolated from human bone cell culture media and found to have inhibitory properties.[29] In vitro, IGF-I IGFBP-4 blocks IGF-I stimulation of OB activity in all rodent cell lines. Circulating IGFBP-4 is increased with age and has been associated with fractures and secondary hyperparathyroidism But, not unlike other IGFBPs, when IGFBP-4 is administered in vivo, it can stimulate bone formation.[30]

Intact soluble or exogenous IGFBP-5 blocks IGF mediated bone growth in a variety of cell models. However, IGFBP-5 is not normally intact or in solution in the bone cell environment, but is preferentially located in the extracellular matrix due to its strong affinity for hydroxyapatite. In this form, IGFBP-5 serves to anchor IGF-I and IGF-II to the crystalline matrix of human bone. In fibroblasts, IGFBP-5 in the extracellular matrix is associated with enhancement of IGF action.[31] Preliminary studies using recombinant IGFBP-5 in intact animals and in vitro, demonstrate that independent of IGF-I, IGFBP-5 can enhance bone formation.[32] There is currently very little information about the role of IGFBP-6 in bone. However, message and protein expression of this binding protein can be detected in several rodent and human bone cell lines.

IGFBP Proteases

IGFBP bioavailability is determined not only by gene expression but also through limited proteolysis of secreted IGFBPs. Indeed, local IGF action may be largely controlled by this mechanism. IGFBP proteases that alter the high-affinity binding between IGFs and individual IGFBPs and are activated by particular physiological states, have been identified in several human bone cell systems.

IGFBP-4 Proteolysis in Bone Cells

It had been noted by several investigators that IGF-I treatment of normal hOB cells results in a loss of IGFBP-4 in serum free media, as determined by ligand blot analysis.[33-35] Further investigation of this phenomena revealed that the IGF-induced decrease in IGFBP-4 was not due to a decrease in IGFBP-4 mRNA expression or secretion. Rather, the effect could be reproduced in a cell free assay suggesting that hOBs secreted a protease that could cleave IGFBP-4 thereby enhancing the biologic activity of the bound IGFs. A novel IGFBP-4 specific protease was subsequently identified in conditioned media by hOB cells in 1994.[36] This protease was a calcium requiring metalloprotease that cleaves IGFBP-4, attenuating inhibition of IGF action by IGFBP-4 (an inhibitory IGFBP).[36] The IGFBP-4 protease was dependent on IGFs for its functional activity, with IGF-II being more effective than IGF-I. Overexpression of IGF-II conferred constitutive IGFBP-4 protease activity in a subset of hOB cells.[37] Subsequently it was found that TGF-β also regulated IGFBP-4 protease in hOB cells.[36] However, unlike IGF-II, TGF-β did not directly affect proteolysis in cell free assay, but rather treatment with TGF-β in hOB cells enhanced IGF-dependent IGFBP-4 protease activity in conditioned media. TGF-β may also stimulate hOB cell expression and/or secretion of the enzyme. In 1999 Conover and colleagues identified the protease synthesized by human fibroblasts and osteoblasts as pregnancy-associated plasma protein-A (PAPP-A).[38] PAPP-A is generated in various osteoblastic cell lines but its greatest expression is in osteoprogenitor and pre-OB cells. Interestingly, IGFBP-4 is the only IGFBP substrate for this protease, which is active in a broad pH range of 5.5- 9.0. Estrogen has been shown to decrease IGF dependent protease IGFBP-4 proteolysis in estrogen responsive cells, although it is unclear whether that works to decrease protease expression or increase inhibition[39] IGFBP-4 proteolysis can also be controlled by inhibitors produced by bone cells. Treatment of hOB cells with phorbol ester tumor promoters or transfection with SV40 T-antigen induces a cycloheximide-sensitive inhibitor of the IGFBP-4 proteolytic reaction, suggesting an association with early transformation processes.[36] As representative of the fully transformed OB phenotype, U-2, MG-63, and TE-85 human osteosarcoma cells secrete neither IGFBP-4 protease nor protease inhibitors. Thus, transformation appears to alter the IGFBP-4 protease system in bone cells.

IGFBP-5 Proteolysis in Bone Cells

U-2 osteosarcoma cell-conditioned medium readily degrades exogenous and endogenous IGFBP-5 due to a cation-dependent serine protease specific for IGFBP-5.[35] In contrast to their stimulatory role in IGFBP-4 proteolysis, IGFs attenuate IGFBP-5 proteolysis in U-2 cells.[33] IGF-regulated IGFBP-5 proteolysis has also been identified in hOB cell-conditioned media , and IGFBP-5 protease activity varied during murine OB development.[35,36,40]

IGFBP-5 may have numerous functions in bone. When intact and soluble, IGFBP-5 inhibits IGF-I action in bone cells in vitro. In osteoblasts, in which the IGFBP-5 protease has been identified, a truncated form of IGFBP-5 possesses intrinsic mitogenic activity, possibly acting through a putative IGFBP-5 receptor.[35] In addition, secreted IGFBP-5 that is not immediately proteolyzed appears to be preferentially localized in the extracellular matrix as the intact form, and in this state, is associated with enhanced IGF action. IGFBP-5 also serves the unique function of fixing the IGFs in the bone matrix by virtue of its high affinity to hydroxyapatite.

Other IGFBP Proteases in Bone Cells

MG-63 human osteosarcoma cells secrete an acid-activated IGFBP-3 protease identified as the aspartic protease, cathepsin D, based on its acidic pH optimum, inhibition by pepstatin, distinctive proteolytic fragment pattern, and immunoreactivity with cathepsin D antisera. Acid-activated cathepsin D is not IGFBP-specific and will proteolyze IGFBP-1 through -5. IGFs may influence this system as well, since IGF-II modulates type II IGF/M-6-P receptor-mediated binding and uptake of cathepsin D.[41,42] Plasmin is another highly active IGFBP protease in MG-63 osteosarcoma cells. Recently, Lalou et al demonstrated that IGF-I treatment of MG-63 cells decreased protease activity toward IGFBP-3 via inhibition of plasminogen conversion to plasmin,[42] which is also capable of degrading IGFBP-5.

Other IGFBP proteases identified in bone cell models include matrix metalloproteases, that are also under IGF control.[43] It is of note that the IGFs are major regulators of IGFBP proteolysis, acting at different levels and by various molecular mechanisms. By modulating IGFBP specific proteases, skeletal IGFs may autoregulate their biological activity. These highly regulated positive and negative feedback systems could ensure temporal and spatial specificity of bone response to critical growth factors.

Effects of IGF-I on Osteoblasts in Vitro

IGFs increase DNA synthesis and replication of cells of the OB lineage and play a major role in stimulating differentiated function of the mature OB. In vitro, human and rodent OB and osteosarcoma cells respond to ligand-activated type I IGF receptor stimulation with increases in DNA and protein synthesis.[44] Both IGF-I and IGF-II increase type I collagen expression and decrease collagen degradation in fetal rat OBs.[45] Whether bone cells register a mitogenic or a differentiated response to IGF stimulus may reflect receptor population and receptor cross-reactivity and depend on cell type and OB lineage.

IGF-I may act in a bimodal fashion. During in vitro development of fetal rat calvaria, IGF-I is an autocrine mitogen for pre-OBs, and, as these cells differentiate, IGF-I secretion decreases. A second rise in IGF-I secretion occurs later in OB development during matrix formation and mineralization. Part of this secondary enhancement in IGF-I synthesis may be tied to local and systemic factors that are necessary to complete the process of bone formation. If the pattern of late IGF-I expression exists in vivo, this increase in IGF-I secretion by mature OBs could lead to sequestration of IGF-I in the bone matrix for release and activity during subsequent remodeling cycles. Whether IGF-I has an autocrine effect in the mature osteoblast is unclear (see below). However, there are some data to suggest that in vitro, IGF-I is necessary for nodule formation. There is also conjecture that IGF-I is critical for mineralization of the newly synthesized bone matrix in vivo; but the evidence is indirect and inferential. Finally, it should be noted that IGF-I acting through the Type I IGF receptor, has a profound anti-apoptotic effect on osteoblasts undergoing differentiation.[7,45] In addition, IGF-I may also prevent osteocyte programmed cell death. In the overall scheme of osteoblast function and number, IGF-I play a critical role, not only in the differentiative pathway, but also by blocking programmed cell death.

Effects of IGF-I on Osteoclasts

IGFs may play a role in regulation of bone resorption. Middleton found that osteoclasts actively engaged in bone resorption expressed IGF-I, IGF-II, and type I IGF receptor mRNA.[6,46] In vitro, IGF-I has been shown to promote formation of osteoclasts from mononuclear precursors and to stimulate activity of preexisting osteoclasts.[47] However, a recent study suggests that these effects represent an indirect action of IGF on osteoclast activity via its effects on OB cells . In that same vein, it has recently been demonstrated that stromal cells produce osteoprotogerin (OPG) and its ligand, OPGL. OPGL is responsible for activating osteoclasts and coupling resorption to formation, while OPG is a member of the TNF receptor superfamily and serves as an extramembrane 'decoy receptor'. Very recent studies by Rubin et al demonstrate that physiologic doses of IGF-I significantly down regulates OPG expression and stimulates OPGL production.[48] Hence, increased osteoclast-mediated resorption by IGF-I may be a function of both direct activation of osteoclasts/ osteoclast precursors, and suppression of OPG synthesis, thereby making more OPGL available to its cognant receptor (RANK) on the osteoclast. This may also explain why the administration of rhGH or rhIGF-I to humans has been associated with a marked increase in bone resorption. How IGFs participate directly or indirectly in bone resorption in vivo remains an important issue to resolve.

Studies in Vivo and Genetically Altered Mice

In vivo studies using recombinant IGF-I in animal models have demonstrated that this growth factor can enhance longitudinal growth, periosteal circumference, and bone mineral density. Locally synthesized IGF-I is also anabolic to bone. Using in situ hybridization (ISH), Shinar et al found a close correlation between IGF-I expression and osteogenesis during rat development.[49] Also, estrogen treatment of ovariectomized rats resulted in decreased calvarial IGF-I mRNA that preceded reduction in bone formation. Similarly, Watson et al noted that IGF-I expression in osteoblasts by in situ hybridization(ISH) was markedly enhanced by PTH administration.[50] Lean et al undertook a novel study of genes expressed in rat osteocytes after a single, acute episode of dynamic loading to reproduce physiological strains in bone and found IGF-I mRNA expression in osteocytes preceded increases in IGF-I expression and matrix formation in overlying surface osteoblasts.[51]

Studies in mice with genetic modifications in specific components of the IGF system have provided additional insights into the actions of IGFs and Type I IGF receptor in vivo,[52] (Table 1). Mice lacking the *IGF-I* gene appear to develop normally but are smaller, have low bone density and frequently die in the postnatal period. The postnatal survivors do exhibit an interesting phenomenon. Although BMD and femur length are reduced in these animals, trabecular bone mass and connectivity are actually enhanced.[53] Whether this represents a compensation for the absence of IGF-I, or is instead due to the lack of IGF-I in cancellous bone, thereby reducing the recruitment of osteoclasts, still needs to be determined. Mice nullizygous for the IGF-I receptor demonstrate extreme organ hypoplasia, delayed skeletal calcification, severe growth retardation and invariably die postnatally. Cross-breeding of the IGF-I (-/-) and Type I IGF receptor (-/-) mice yields a phenotype indistinguishable to that observed in the Type I IGF receptor null mice. This suggests that IGF-I interacts exclusively with the Type I IGF receptor. By contrast, mice lacking the *IGF-II* gene show no delay in ossification and have normal sized skeletons.[54]

Ubiquitous overexpression of IGF-I in mice achieved using a metallothionein promoter resulted in an increased body weight and disproportionate overgrowth of some organs but normal skeletal size and morphology.[55-58] Likewise, overexpression of IGF-II does not cause major changes in skeletal growth and bone turnover in mice.[59]

The relative importance of circulating vs locally produced IGF-I in the process of bone acquisition has frequently been debated. Cross-sectional and cohort studies in various populations have implied that there is a strong correlation between serum levels of IGF-I and femoral or lumbar BMD.[60] Some insight can be gained into this critical question by examining recent work in mice by one investigative group. Yakar et al examined the effects of knocking out hepatic IGF-I expression and/or the acid-labile subunit protein(ALS)that bind circulating IGF-I IGFBP-3.[61,62] When hepatic IGF-I expression alone is deleted, serum IGF-I declines by 75%, but femur length and weight are only slightly reduced, while overall growth velocity is maintained.[62] On the other hand, the double knockout mice (hepatic IGF-I and ALS) had even lower concentrations of IGF-I (i.e., 90% reduction), significant growth retardation and low bone mineral density.[61] Moreover, the growth plates of these mice were significantly disordered. Selective knockout of the acid labile subunit also resulted in very low serum IGF-I concentrations, as well as shortened femurs, and reduced bone density. Interestingly, the level of free IGF-I was remarkably increased in these knockouts, although the circulating half life of IGF-I was significantly shortened. These data suggest there may be a major role for serum IGF-I in determining bone size and mass.

Although the original "somatomedin hypothesis" is no longer considered tenable, there is still uncertainty as to the exact relationship between the roles of GH and IGF-I on skeletal growth and acquisition.[7,62] Most recent data support the view that IGF-I is the mediator of GH actions on skeletal growth. Thus overexpression of IGF-I, but not IGF-I in the absence of GH, normalizes skeletal growth.[63-65] The availability of methods for bone-specific deletion of the GH receptor should provide a means to determine whether GH exerts effects on bone independent of IGF-I.

Other components of the IGF system have also been investigated in genetically altered mice. For example, mice lacking IRS-1, a key downstream effector of the IGF-IR, had severe osteopenia with low bone turnover.[66] Experiments using IRS-1 (-/-) osteoblasts demonstrated that IRS-1 deficiency impairs osteoblast proliferation, differentiation, and supports osteoclastogenesis, resulting in low-turnover osteopenia. Injection of a protease resistant IGFBP-4 directly into the parietal bones of mice inhibited the anabolic actions of IGF-I.[67] In addition, mice overexpressing IGFBP-4 in osteoblasts have marked growth retardation and disproportionally small bones.[68] These observations support the view that IGFBP-4 serves to sequester IGF-I and thereby inhibit its actions. Paradoxically, however, IGFBP-4 null mice have reduced weight at birth (10-15%).[69] One potential explanation is that

Table 1. Genetically altered IGF mouse models and their skeletal phenotypes

Mouse	IGF Alteration	Skeletal Phenotype	Reference
IGF-I Tg	Global Overexpression IGF-I	normal size,?BMD;Inc growth In GHD;inc tail length	63, 64
IGF-II Tg	Global Overexpression IGF-II	normal size;?BMD	59, 65
GH Tg	Overexpression of GH	Increased size, Inc BMD;Inc serum IGF-I	52
IGF-I-/-	Global knockout IGF-I	Short bones, low BMD Very low Serum IGF-I	53
IGF-IR-/-	Global knockout of Type I	lethal;growth impairment,poor Calcification	52
IRS-/-	Global knockout of IRS Signaling for IGF-I	severely impaired bone formation small bones,ostepenia	66
IGF-I Tg Targeted	Overexpression IGF-I in bone with OC promoter	increases in BMD,bone formation bone resorption;no size change	70
IGF R -/- Targeted	Conditional IGF receptor knockout in bone	decreases in BMD,bone formation no size change	71
Hepatic IGF-I -/-	Conditional KO of liver derived IGF-I	low serum IGF-I;slightly decrease bone size;BMD slightly reduced	62
Hepatic IGF-I -/-+ ALS -/-	Conditional KO of liver derived IGF-I and ALS	marked growth retardation,very low serum IGF-I, reduced BMD	61

* There is an IGF-I mutant called the Midi-IGF-I mutant. Attempts to ablate IGF-I gene function by homologous recombination resulted in a disrupted IGF-I gene which retained some function.

absence of IGFBP-4 diminishes tissue IGF storage capacity. This would predict that physiological levels of IGFBP-4 are required for normal growth, and that IGFs would be released through the action of IGFBP-4 proteases.

While studies of these mouse models have provided additional insights into our understanding of the actions of the IGF system, there are several limitations when interpreting skeletal findings in these animals. First, the high lethality and severe organ defects in the IGF-I and IGF-IR KO mice make it difficult to distinguish direct and indirect actions of IGF-I in bone and prohibit study of adult mice. Also, as alluded to above, systemic overexpression of IGF-I or II cannot unequivocally differentiate endocrine vs autocrine/paracrine actions in bone, nor does it necessarily give valid information on the changes in the availability of IGFs in bone microenvironment. However, tissue-specific modifications of IGF system components in mice should theoretically enable further understanding of the precise role of IGF-I in bone. Indeed, targeted overexpression of IGF-I to osteoblasts of transgenic mice increased cancellous bone formation rate and volume without any change in osteoblast number, suggesting locally delivered IGF-I exerts its anabolic actions primarily by increasing the activity of resident osteoblasts.[70] However, even this model has confounding variables. For example, the phenotype might result from the overexpression of a transgene in mature osteoblasts, and not necessarily reflect the global action of IGF-I at its physiological levels. In this regard, the feasibility of conditional mutagenesis of the *Igf1r* gene in osteoblasts using Cre- mediated recombination has recently been demonstrated and should theoretically provide a more powerful approach to further define the actions of IGF-I in bone.[71] Table 1 summarizes the skeletal phenotypes of genetically engineered mice that have targeted the IGF regulatory system.

Evidence for an interaction between IGF-I and other genes controlling skeletal acquisition has recently been obtained through the development of congenic mice. This strategy involves the donation of a quantitative trait locus (QTL) from one inbred strain to another using repetitive backcrossing[72] (Fig. 2). After ten generations, congenic mice carry the chromosomal region of interest on the homozygous background of the recipient strain. From there, investigators can define the full genetic

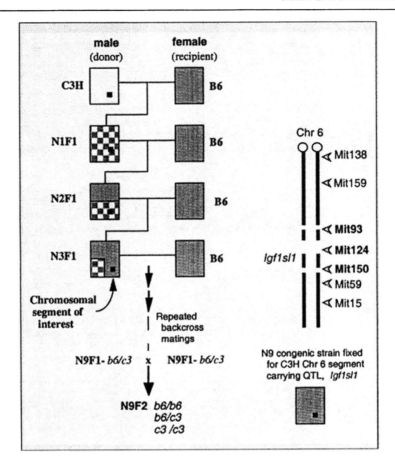

Figure 2. The development of a congenic mouse using a locus that contains an IGF-I regulatory gene. The B6.C3H-6T congenic strain was made by backcrossing F1 mice carrying the Chr 6 segment donated from C3H mice to B6 mice. The donated Chr segment (represented by a small black square) contains the serum IGF-I regulatory QTL, designated *Ifg1sl1*. DNAs from F1 carriers were identified by PCR as heterozygous for markers *D6Mit93 – D6Mit124 – D6Mit150*. As the backcrossing proceeds, the proportion of genomic DNA from C3H declines (represented by the checkerboard pattern). At the 6th generation of backcrossing, heterozygous carriers were intercrossed to produce N6F2 generation progeny that segregated as *c3/c3*, *c3/b6*, or *b6/b6* for the donated chromosomal region. Backcrossing was continued for 3 additional cycles, then N9F1 carriers were intercrossed and *c3/c3* homozygotes selected to begin the congenic strain designated B6.C3H-6T (6T). Major markers used for Chr 6 genotyping and the approximate location of *Ifg1sl1* are shown in the schematic chromosome diagram. The dashed line represents the donated Chr 6 segment made homozygous after 9 cycles of backcrossing (N9) in the 6T mice.

effect of a particular QTL, and determine whether there are gene x gene interactions. Two inbred strains, C3H and B6 are particularly useful models for creating congenic mice to study the interaction between BMD and serum IGF-I.[20,21,73a,b] These two strains have differences in peak bone acquisition that correlate very closely with rises in serum IGF-I.[20,21,73a,b] Bouxsein et al produced a congenic mouse by transferring a section of chromosome 6 from C3H mice (an inbred strain with high BMD and high serum IGF-I) to C57B6 (a strain with low BMD and low IGF-I) over ten generations.[20,2174] This 20 centimorgan segment contained a major QTL from C3H that regulated serum IGF-I in a negative direction (i.e., lowering IGF-I circulating concentrations in the high IGF-I strain), and accounted for nearly 15% of the variability in that phenotype. After 10 generations

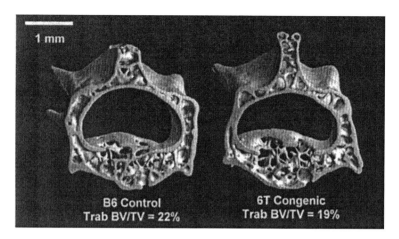

Figure 3. MicroCT image of vertebral body of female 6T congenic and B6 control mouse at 16 weeks of age showing the reduction in cancellous bone volume, accompanied by a reduction in trabecular number and increase in trabecular separation. These changes were accompanied by a significant decline in serum IGF-I (i.e., approximately 25%) in the congenic mouse compared to a B6 littermate, although skeletal expression of IGF-I did not differ by strain. Also noted were differences in cancellous bone architecture, with the B6 showing a more plate-like structure compared to the 6T congenic, which is relatively more rod-like. Overall, changes in serum IGF-I appeared to have a dramatic effect on trabecular microarchitecture.

of mice, both male and females carrying the Chr 6 QTL(c3/c3) but homozygous throughout the rest of the genome for B6 (b6/b6) exhibited 25% lower circulating IGF-I levels, and a nearly 50% reduction in trabecular bone density[74] (Fig. 3). These congenics established proof of principle that a QTL found through whole genome scanning of F2 mice, when transferred to a different background, strongly influenced IGF-I circulating concentrations. More intriguing, however, was the finding that IGF-I expression in the femur of the congenic did not differ from the background strain. This suggests that genetic determinants of IGF-I could have a major impact on bone mass, in the absence of changes in skeletal IGF-I expression. In summary, it is still too early to determine what role circulating IGF-I may play in bone acquisition, although it appears that adequate amounts of both circulating and skeletal IGF-I are critical for support of all the homeostatic mechanisms related to peak bone acquisition.

Clinical Aspects of IGF-I and Bone: Implications from Recent Findings

IGF-I is a potent anabolic agent for the skeleton, although its clinical use has been limited to some degree by other factors. First, IGF-I can reduce blood glucose and can cause significant edema when administered even at low doses. Second, the peptide can only be given subcutaneously and is expensive. Third, recent case control studies suggest that chronically high normal levels of IGF-I may place a person at risk for certain malignancies, including breast, prostate and colon.[75] Finally, except for specific clinical circumstances such as GH resistance, or growth hormone deficiency in children, the effects of IGF-I on bone density and fractures in humans have been relatively disappointing. In part this is a function of the lack of long term, large-scale randomized trials, as well as the fact that administration of recombinant human IGF-I enhances both bone resorption and bone formation, thereby compromising effects on bone mineral density. Indeed, several trials using rhIGF-I or rhGH in postmenopausal osteoporotic women have failed to demonstrate any enhancement in bone density. Interestingly, in those studies bone resorption and bone formation are equally activated suggesting that the entire remodeling sequence (not just bone formation) is stimulated by administration of IGF-I. These in vivo findings are also consistent with the in vitro suppressive effects of IGF-I on OPG and its induction of OPGL in marrow stromal cells.[48] On a positive note, one small trial using IGF-I and IGFBP-3 in combination did demonstrate that this complex could slow bone loss after hip fractures (Boonen, personal communication). Finally, trials with IGF-II coupled to IGFBP-2, could be initiated relatively soon.[27]

It should be noted however, that recombinant PTH has recently been approved for the treatment of osteoporosis. Since PTH induces IGF-I expression in bone cells, there is a strong rationale for modulating IGF-I expression locally rather than systemically. Still, there are potentially inherent dangers in using agents that can turn on skeletal IGF-I, since both the IGF-I ligand, and the Type I IGF receptor, are active in preventing osteoblast induced apoptosis.[76] Indeed, it is conceivable that the appearance of a genomically unstable marrow stromal cell progenitors, could be driven to expand and increase if exposed to an agent such as PTH which enhances IGF-I expression. This concern is heightened which during long term exposure to PTH theoretically could result in osteogenic sarcoma.[77,78] Clearly more work is needed to further understand the role of proliferative growth factors such as IGF-I in relation to the generation of bone tumors. Consequently, at the present time, there is sufficient concern on the part of the FDA to limit the use of PTH in humans for the treatment of osteoporosis to 24 months.[78]

In sum, recent molecular and genomic technologies have opened new avenues for investigation into the role of IGF-I in the acquisition and maintenance of peak bone mass. There seems to be little doubt that the IGFs have an important role in the proliferation and differentiation of osteoblasts; IGF-I might also be a permissive factor for osteoclast recruitment and may be important in the mineralization of newly formed matrix.[79] Future therapies aimed at manipulative individual components of the IGF regulatory system are currently under investigation and offer tremendous potential.

References

1. Hayden JM Mohan S, Baylink DJ. The IGF system and the coupling for formation to resorption. Bone 1995; 17:93S-98S.
2. Manolagas SC. Birth and death of bone cells: basic regulatory mechanisms and implications for the pathogenesis and treatment of osteoporosis 2000; 21:115-137.
3. Udgawa, N, Takahasi N, Jimi E et al. Osteoblasts/stromal cells stimulate osteoclast differentiation factor/RANKL but not mCSF. Bone 1999; 25:517-523.
4. Donahue LR, Rosen CJ. IGFs and Bone: The osteoporosis connection revisited. Proceeding of the Society for Experimental Biology and Medicine 1998;219:1-7.
5. Zapf J, Froesch E. IGFs/somatomedins: structure secretion biological actions and physiological role. Horm Res 1986; 24:121-130.
6. Mochizuki H, Hakeda Y, Wakatsuki N et al. IGF-I supports formation and activation of osteoclasts. Endocrinology 1992; 131:1075-1080.
7. LeRoith D, Bondy C, Yakar S et al.The Somatomedin Hypothesis: Endocrine Reviews 2001; 22:53-74.
8. Bautista CM, Mohan S, Baylink DJ. Insulin-like growth factors I and II are present in the skeletal tissues of ten vertebrates. Metabolism 1990; 39:96-100.
9. Rotwein P. Two insulin-like growth factor I messenger RNAs are expressed in human liver. Proc Natl Acad Sci USA 1986; 83(1):77-81.
10. Adamo M, Lowe WL Jr, LeRoith D et al. Insulin-like growth factor I messenger ribonucleic acids with alternative 5'-untranslated regions are differentially expressed during development of the rat. Endocrinology 1989; 124(6):2737-44.
11. Adamo ML, Ben-Hur H, Roberts CT Jr et al. Regulation of start site usage in the leader exons of the rat insulin- like growth factor-I gene by development, fasting, and diabetes. Mol Endocrinol 1991; 5(11):1677-86.
12. Nolten LA, Steenbergh PH, Sussenbach JS. Hepatocyte nuclear factor 1 alpha activates promoter 1 of the human insulin-like growth factor I gene via two distinct binding sites. Mol Endocrinol1995; 9:1488-99.
13. McCarthy TL, Thomas MJ, Centrella M et al. Regulation of insulin-like growth factor I transcription by cyclic adenosine -monophosphate (cAMP) in fetal rat bone cells through an element with exon 1: Protein kinase A-dependent control without a consensus AMP response element. Endocrinology 1995; 136:3901-3908.
14. Delany AM, Pash JM, Canalis E. Cellular and clinical perspectives on skeletal insulin-like growth factor I.J. Cell. Biochem 1994; 55:328-333
15. McCarthy TL, Ji C, Centrella M. Links among growth factors, hormones, and nuclear factors with essential roles in bone formation. Crit Rev Oral Biol Med 2000; 11(4):409-422.
16. McCarthy TL, Ji C, Shu H et al. 17beta-estradiol potently suppresses cAMP-induced insulin-like growth factor-I gene activation in primary rat osteoblast cultures." J Biol Chem 1997; 272(29):18132-9.
17. McCarthy TL, Centrella M, Canalis E. Parathyroid hormone enhances the transcript and polypeptide levels of insulin like growth factor I in osteoblast-enriched cultures from fetal rat bone. Endocrinology 1989; 124:1247-1253.
18. McCarthy TL, Centrella M, Canalis E. Cyclic AMP induces insulin-like growth factor I synthesis in osteoblast-enriched cultures. J Biol. Chem 1990; 265;15353-15356.
19. Ernst M, Rodan GA. Estradiol regulation of insulin-like growth factor-I expression in osteoblastic cells: Evidence for transcriptional control. Mol Endocrinol 1991; 5:1081-1089.

20. Rosen CJ, Dimai HP, Vereault D et al. Circulating and skeletal insulin-like growth factor-I (IGF-I) concentrations in two inbred strains of mice with different bone mineral densities. Bone 1997; 21(3):217-23.
21. Rosen CJ, Chruchill GA, Donahue LR et al. Mapping QTLs for serum IGF-I levels in mice. Bone 2000; 27:521-528.
22. Kveiborg M, Flyvbjerg A, Eriksen EF et al. Transforming growth factor-beta1 stimulates the production of insulin-like growth factor-I and insulin-like growth factor-binding protein-3 in human bone marrow stromal osteoblast progenitors. J Endocrinol 2001; 169(3):549-561.
23. Mohan S, Baylink DJ. IGF System Components and Their Role in Bone metabolism. In: Rosenfeld RG, Roberts CT, eds. The IGF System. Humana Press, 1999:457-496.
24. Wang E, Wang J, Chin E et al. Cellular patterns of insulin-like growth factor system gene expression in murine chondrogenesis and osteogenesis. Endocrinology 1995; 136:2741-2751.
25. Jones JI, D'Ercole AJ, Camacho-Hubner C et al. Phosphorylation of insulin-like growth factor (IGF)-binding protein 1 in cell culture and in vivo:Effects on affinity for IGF-I. Proc Natl Acad Sci USA 1991; 88:7481-7485.
26. Okazaki R, Riggs BL, Conover CA. Glucocorticoid regulation of insulin-like growth factor-binding protein expression in normal human osteoblast-like cells. Endocrinology 1994; 134:126-132.
27. Conover CA, Turner RT, Johnston EW et al. Subcutaneous administration of IGF-II/IGFBP-2 stimulates bone formation and prevents loss of bone mineral density in a rat model of disuse osteoporosis. J Bone Min Res 2001; 16:S148.
28. Ernst M, Rodan GA. Increased activity of insulin-like growth factor (IGF) in osteoblastic cells in the presence of growth hormone (GH): Positive correlation with the presence of the GH-induced IGF-binding protein BP-3. Endocrinology 1990; 127:807-814.
29. Mohan S, Bautista CM, Wergedal J et al. Isolation of an inhibitory insulin-like growth factor (IGF) binding protein from bone cell-conditioned medium: A potential local regulatory of IGF action. Proc Natl Acad Sci USA 1989; 86:8338-8342.
30. Miyakoshi N, Richman C, Qin X et al. Effects of recombinant insulin-like growth factor-binding protein-4 on bone formation parameters in mice. Endocrinology 1999 ; 140(12):5719-5728.
31. Andress DL, Birnbaum RS. A novel human insulin-like growth factor binding protein secreted by osteoblast-like cells. Biochem Biophys Res Commun 1991; 176:213-218.
32. Richman CD, Baylink DJ, Lang K et al. Recombinant human insulin-like growth factor-binding protein-5 stimulates bone formation parameters in vitro and in vivo. Endocrinology 1999; 140(10):4699-705.
33. Durham Sk, Kiefer MC, Riggs BL et al. Regulation of IGFBP-4 by a specific IGFBP4 protein proteases in normal human osteoblast-like cells:implications in bone cell physiology. J Bone Miner Res 1994; 111-117.
34. Durham SK, Riggs BL, Conover CA. The IGFBP-4 protease system in normal human osteoblast-like cells: regulation by TGF-b. J Clin Endocrinol metab 1994; 79:1752-1758.
35. Kanzaki S, Hilliker S, Baylink DJ et al. Evidence that human bone cells in cultures produce IGFBP-4 and −5 proteases. Endocrinology 1994; 134:383-392.
36. Durham SK, Riggs BL, Harris SA et al. Alterations in IGF dependent IGFBP-4 proteolysis in transformed osteoblastic cells. Endocrinology 1994; 136:1374-1380.
37. Durham SK, DeLeon DD, Okazaki R et al. Regulation of IGF binding protein −4 availability in normal human osteoblast like cells: role of endogenous IGFs. J Clin Endocrinol Metab 1995; 104-110.
38. Lawrence JB, Oxvig C, Overgaard MT et al. The IGF dependent IGF BP-4 protease secreted by human fibroblasts is pregnancy associated plasma protein A. Proc Natl Acad Sci USA 1999; 96:3149-3153.
39. Kassem M, Okazaki R, DeLeon D et al. Potential mechanism of estrogen mediated decrease in bone formation; estrogen increases production of inhibitory IGFBP-4 in human osteoblastic cell lines with high levels of estrogen receptors. Proc Assoc Am Physicians 1996; 1(108):151-164.
40. Thraillkill KM, Quarles LD, Nagase H et al. Characterization of IGFBP-5 degrading proteases produced throughout murine osteoblast differentiation. Endocrinology 1996; 3527-3533.
41. Nissley P, Kiess W, Sklar MM. The IGF-II/mannose 6-phosphate receptor. In: LeRoith D, ed. IGFs: Molecular and Cellular Aspects. Boca Raton: CRC Press, 1991:111-150.
42. LaLou C, Silve C, Rosato R et al. Interactions between IGF-I and the system of plasminogen activators and their inhibitors in the control of IGFBP-3 production and proteolysis in human osteosarcoma cells. Endocrinology 1994; 135:2318-2326.
43. Delany AM, Rydziel S, Cnalis E. Transcriptional repression of matrix metalloproteinase 1 by IGF in rat osteoblasts. J Bone Miner Res 1995; 10:S164.
44. Canalis E. Insulin like growth factors and the local regulation of bone formation. Bone 1993; 14:273-276.
45. Jia D, Heersche JN. Insulin-like growth factor-1 and -2 stimulate osteoprogenitor proliferation and differentiation and adipocyte formation in cell populations derived from adult rat bone. Bone 2000; 27(6):785-794
46. Middleton J, Arnott N, Walsh S et al. Osteoblasts and osteoclasts in adult human osteophyte tissue express the mRNAs for insulin-like growth factors I and II and the type 1 IGF receptor. Bone 1995; 16:287-293.
47. Slootweg CM, Hoogerbrugge CM, de Poorter TL et al. The presence of classical insulin-like growth factor (IGF) type-I and -II receptors on mouse osteoblasts: Autocrine/paracrine growth effect of IGFs. J Endocrinol 1990; 125:271-277.
48. Rubin JR, Rosen CJ. OPG/OPL Ratios are Increased in Bone Stromal Cells from C3H (High Bone Density) Compared to C57BL6 (Low Bone Density) Mice. Journal of Bone and Mineral Research 2001; 15(S1):S276.

49. Shinar DM, Endo N, Halperin D et al. Differential expression of insulin-like growth factor-I (IGF-I) and IGF-II messenger ribonucleic acid in growing rat bone. Endocrinology 1993; 132:1158-1167.
50. Watson P, Lazowski D, Han V et al. Parathyroid hormone restores bone mass and enhances osteoblast insulin-like growth factor I gene expression in ovariectomized rats. Bone 1995; 16:357-365.
51. Lean JM, Jagger CJ, Chambers TJ et al. Increased insulin-like growth factor I mRNA expression in rat osteocytes in response to mechanical stimulation. Am J Physiol 1995; 268:E318-E327.
52. Efstratiadis A. Genetics of mouse growth. Int J Dev Biol 1998; 42:955-976.
53. Bikle D, Majumdar S, Laib A et al. The Skeletal Structure of IGF-I Deficient Mice. J Bone Min Res 2001; 16:2320-2330.
54. DeChiara TM, Efstratiadis A, Robertson EJ. A growth-deficiency phenotype in heterozygous mice carrying an insulin- like growth factor II gene disrupted by targeting. Nature 1990; 345:78-80.
55. Quaife CJ, Mathews LS, Pinkert CA et al. Histopathology associated with elevated levels of growth hormone and insulin-like growth factor I in transgenic mice. Endocrinology 1989; 124:40-48.
56. Mathews LS, Hammer RE, Brinster RL et al. Expression of insulin-like growth factor I in transgenic mice with elevated levels of growth hormone is correlated with growth. Endocrinology 1988; 123:433-437.
57. Corsini A, Fantappiè S, Granata A et al. Binding-defective low-density lipoprotein in family with hypercholesterolaemia. Lancet 1989; 1:623.
58. Gotoda T, Senda M, Gamou T et al. Nucleotide sequence of human cDNA coding for a lipoprotein lipase (LPL) cloned from placental cDNA library. Nucleic Acids Res 1989; 17:2351.
59. Wolf E, Rapp K, Blum WF et al. Skeletal growth of transgenic mice with elevated levels of circulating insulin-like growth factor-II. Growth Regulation 1995; 5:177-183.
60. Langlois JA, Rosen CJ, Visser M et al. The association between IGF-I and bone mineral density in women and men: the Framingham heart study. J Clin Endocrinol Metab 1998; 83:4257-4262.
61. Yakar S, Ooi G, Boisclair Y et al. Inactivation of both liver IGF-I and the acid labile subunit genes caus a marked reduction in serum IGF-I and growth retardation. Endocrine Society 2001; 75.
62. Yakar S, Liu JL, Stannard B et al. Normal growth and development in the absence of hepatic insulin-like growth factor I. Proc Natl Acad Sci USA 1999; 96:7324-7329.
63. Matthews LS, Hammer RE, Behringer RR et al. Growth enhancement of transgenic mice expressing human IGF-I. Endocrinology 1988; 123:2827-2833.
64. Behringer RR, Lewin TM, Quaife CJ et al. Expression of insulin-like growth factor I stimulates normal somatic growth in growth hormone-deficient transgenic mice. Endocrinology 1990; 127:1033-1040.
65. Buul-Offers SC, de Haan K, Reijnen-Gresnigt MG et al. Overexpression of human insulin-like growth factor-II in transgenic mice causes increased growth of the thymus. J Endocrinol 1995; 144:491-502.
66. Ogata N, Chikazu D, Kubota N et al. Insulin receptor substrate-1 in osteoblast is indispensable for maintaining bone turnover. J Clin Invest 2000; 105(7):935-943.
67. Miyakoshi N, Qin X, Kasukawa Y et al. Systemic administration of insulin-like growth factor (IGF)-binding protein-4 (IGFBP-4) increases bone formation parameters in mice by increasing IGF bioavailability via an IGFBP-4 protease-dependent mechanism. Endocrinology 2001; 142:2641-2648.
68. Zhang M, Faugere MC, Malluche H. Paracrine overexpression of IGFBP-4 in osteoblasts of transgenic mice results in global growth retardation. J Bone Min Res 2003; in press.
69. Pintar JE, Cerro JA, Wood TL. Genetic approaches to the function of insulin-like growth factor- binding proteins during rodent development. Horm Res 1996; 45:172-177.
70. Zhao G, Monier-Faugere MC, Langub MC et al. Targeted overexpression of insulin-like growth factor I to osteoblasts of transgenic mice: increased trabecular bone volume without increased osteoblast proliferation. Endocrinology 2000; 141:2674-2682.
71. Zhang M, Xuan S, Bouxsein ML et al. Conditional mutagenesis of the IGF1 receptor gene in osteoblasts reduces cancellous bone volume and increases turnover. J Biol Chem 2002; 277:44005-44012.
72. Rosen CJ, Donahue LR, Beamer WG. Defining the genetics of osteoporosis: using the mouse to understand man. Osteoporosis Int 2001; 12:803-810.
73a. Beamer WG, Shultz KL, Donahue LR et al. Quantitativie trait loci for femoral and lumbar vertebral bone mineral density in C57BL/6J and C3H/HeJ Inbred Strains of Mice. J Bone Min Res 2001; 16:1195-1206.
73b. Richman C, Kutilek S, Miyakoshi N et al. Postnatal and pubertal skeletal changes contribute predominantly to the differences in peak bone density between C3H and C57BL/6J mice. J Bone Min Res 2001; 16:386-397.
74. Bouxsein ML, Rosen CJ, Turner CH et al. Generation of a new congenic mouse strain to test the relationship among serum IGF-I, bone mineral density and skeletal morphology in vivo. J Bone Min Res 2002; 17: in press.
75. Khandwala HM, McCutcheon IE, Flyvbjerg A et al. The Effects of IGFs on Neoplastic Growth 2000; 21:215-244.
76. LeRoith DM, Blakesley VA. The insulin-like growth factor-I receptor and apoptosis. Implications for the aging progress." Endocrine 1997; 7(1):103-5.
77. Whitfield JF. The parathyroid hormones (PTHs): Anabolic tools for mending fractures and treating osteoporosis. Medscape Women's Health Clinical Updates 2001. Available at: http://www.medscape.com/WomensHealth /ClinicalUpdate/2001/
78. Schneider BS. Food and Drug Administration Center for Drug Evaluation and Re-search; Endocrinologic and Metabolic Drugs Advisory Committee Meeting on July 27, 2001; p150. Available: http://www.fda.gov/ohrms/dockets/ac/01/transcripts/3761t2_01.pdf.
79. Zhang Q,Wastney ME, Rosen CJ et al. IGF-I infusion increases Bone Calcium Depostion in the Growing Rat Model. J Bone Min Res 2001; 16:S356

CHAPTER 12

IGFBP-5, a Multifunctional Protein, Is an Important Bone Formation Regulator

Subburaman Mohan, Yousef Amaar and David J Baylink

Abstract

Bone formation is essential to all aspects of bone physiology, including growth, remodeling, and repair. In terms of potential messenger molecules that regulate bone formation, recent studies have provided strong evidence for a physiological role for IGFBP-5 in the regulation of bone formation:
1. IGFBP-5 is the most abundant IGFBP stored in bone, which provides a mechanism for the fixation of IGFs in bone matrix for future actions;
2. IGFBP-5 production in osteoblasts is regulated by osteoregulatory agents that regulate bone formation, with serum levels of IGFBP-5 showing considerable changes in clinical disease states, correlating with bone formation changes;
3. IGFBP-5 has been shown to stimulate bone formation parameters in vitro and in vivo; and
4. IGFBP-5 can function as a growth factor besides its role as a traditional binding protein.

The growth factor effects of IGFBP-5 could involve IGFBP-5 binding to its putative receptor on the osteoblast cell surface to stimulate the intracellular signaling pathway and/or transcriptional activation of genes by IGFBP-5 that is transported into the nucleus of osteoblasts. In conclusion, IGFBP-5 has several functional features which are consistent with an important role for this binding protein in the regulation of bone formation.

Introduction

Bone formation is essential to all aspects of bone physiology, including growth, remodeling, and repair. Age-related impairment in bone formation, in part, contributes to the pathogenesis of senile osteoporosis.[1,2] Therefore, basic studies on the potential messenger molecules that regulate bone formation processes and their molecular mechanism of action are important to our future understanding of the causes of bone disease and the development of therapeutic drugs to stimulate bone formation in vivo. In terms of potential bone formation regulators, it has become increasingly clear that the IGF system plays a critical role in regulating many aspects of bone physiology, including growth, remodeling, and repair.[3-6] As one might expect from an important regulatory system involved in diverse actions, the IGF system is composed of many components, including receptors, binding proteins, binding protein proteases, and activators and inhibitors of binding protein proteases.[7-12] The focus of this review is on one of the components of the IGF system, namely IGFBP-5. We selected IGFBP-5 based on findings from our laboratory, as well as other laboratories, which demonstrate that IGFBP-5 has several unique features among the six high affinity IGFBPs and is an important player in the regulation of bone formation.[13-15] In this chapter, we will briefly review the work that led to the discovery of IGFBP-5 in bone and then describe what is known on the regulation and actions of IGFBP-5, with particular emphasis on bone formation.

Discovery of IGFBP-5

Animals treated systemically with agents that stimulate bone resorption show evidence of increased bone formation in addition to the expected increase in bone resorption.[16] This paradoxical

Insulin-Like Growth Factors, edited by Derek LeRoith, Walter Zumkeller and Robert Baxter. ©2003 Eurekah.com and Kluwer Academic / Plenum Publishers.

increase in bone formation, which represents a counter regulatory mechanism to maintain bone volume, was thought to be mediated by local mechanisms involving increased growth factor production.[17] Our subsequent studies on the purification of skeletal growth factor (SGF) activity in human bone extracts led to the discovery that IGFs were the most abundant growth factors stored in bone and produced by human osteoblasts.[17,18] During the course of our studies involving the extraction and purification of human SGF, now known as IGF-II, we found that human IGF-II was present as a large molecular weight form in non-dissociative extracts of human bone. However, the large molecular weight IGF-II could be dissociated from its carrier protein and purified as a low molecular weight form under strong dissociative conditions.[18,19] It was also found that the high molecular weight IGF-II complex bound to hydroxyapatite under native conditions and that low molecular weight purified IGF-II did not bind to hydroxyapatite columns. Based on these results, it was postulated that IGF-II was fixed in large amounts in bone by means of an IGF binding protein (IGFBP), which has a strong affinity for hydroxyapatite. Our subsequent studies on the purification of this binding protein led to the discovery of IGFBP-5 from human bone extract.[20] Simultaneously, Andress and Birnbaum[21] and Shimasaki et al[22] reported purification of IGFBP-5 from medium conditioned by U2 human osteosarcoma cells and rat serum respectively. Subsequently, Camacho-Hubner et al[23] and Roghani et al[24] also purified IGFBP-5 from media conditioned by T98G human glioblastoma cells and human cerebrospinal fluid respectively. The complete primary structure of the IGFBP-5 protein has been deduced from clones isolated from human placenta and osteosarcoma complementary DNA (cDNA) libraries.[22,25] The cDNA for human IGFBP-5 encodes a mature protein of 252 residues with no N-glycosylation sites or arginine-glycine-aspartic acid (RGD) sequences. IGFBP-5 exhibits several distinct structural features, including glycosylation sites, phosphorylation sites, extracellular matrix binding sites, acid labile subunit (ALS) binding site, hydroxyapaptite binding sites, heparin binding domain, and a nuclear localization signal.[3,8] Although IGFBP-5 is expressed by a variety of cell types, the findings that IGFBP-5, the most abundant IGFBP in human bone, exhibits significant biological effects on bone cells in vitro, and that the production of IGFBP-5 is regulated by key regulators of osteoblast cell proliferation, suggest that IGFBP-5 is an important component of the IGF system in bone.

Regulation of IGFBP-5 Levels by Proteases

During the course of purification of IGFBP-5 from human bone extract, we found evidence that IGFBP-5 in bone extract is subject to degradation by protease/s, resulting in fragments which bind IGF with much less affinity compared to intact IGFBP-5.[20] In subsequent studies, we and others found evidence for the presence of protease/s capable of degrading IGFBP-5 in a variety of biological fluids, including serum.[26-29] The findings that human osteoblasts in culture produce proteases capable of degrading IGFBP-5, and that the production and/or activity of IGFBP-5 protease can be regulated by a number of osteoregulatory agents, provide strong evidence that the rate of IGFBP-5 degradation by protease is one of the key control mechanisms regulating the effective concentration of IGFBP-5 in local areas of bone.[26,30] In terms of the identity of proteases that cleave IGFBP-5, it has been shown that matrix metalloprotease (MMP)-2 produced by mouse osteoblasts cleaves IGFBP-5.[31] In addition, Campbell and Andress[32] have shown that plasmin degrades IGFBP-5 and that the heparin binding domain in IGFBP-5 (201-218) inhibits plasmin proteolysis of intact IGFBP-5 via a mechanism involving inhibition of plasmin binding to substrate IGFBP-5. Both MMP-2 and plasmin have also been shown to degrade other IGFBPs besides IGFBP-5 and a host of other proteins, and are, therefore, considered relatively non-specific (Table 1). The physiological significance of IGFBP-5 regulation by MMP-2 and plasmin remains to be investigated.

In addition to the production of non-specific proteases capable of cleaving IGFBP-5, there is clear evidence for the production of IGFBP-5-specific protease by human osteoblasts. This conclusion is based on the findings that the fragmentation patterns, cleavage site, and inhibitor profiles for the major IGFBP-5 protease produced by human osteoblasts are different from those of non-specific IGFBP proteases known to cleave IGFBPs. In this regard, it has been recently shown that complement c1s is the major IGFBP-5-specific protease produced by human smooth muscle cells.[33] Furthermore, Overgaard et al[34] have shown that pregnancy-associated plasma protein (PAPP)-A2 cleaves IGFBP-5, but not other IGFBPs. Based on the findings that the cleavage site for PAPP-A2 and complement c1s for IGFBP-5 is different from the cleavage site for IGFBP-5 protease produced by

Table 1. Proteases produced by osteoblasts known to cleave IGFBP-5

Protease	Cleaves IGFBP-5	Cleaves Other IGFBPs	Cleaves Other Proteins	Reference
Matrix metalloprotease 2	Yes	Yes	Yes	31
Plasmin	Yes	Yes	Yes	32
Pregnancy associated plasma protein-A	Yes	Yes*	No	27
IGFBP-5 specific protease	Yes	No	No	26

* cleaves IGFBP-4 with higher potency

human osteoblasts, it can be concluded that the IGFBP-5 specific protease produced by human osteoblasts is not PAPP-A2 or complement c1s. Future studies on the identification of IGFBP-5 specific protease produced by human osteoblasts are essential to elucidate the role of the IGFBP-5 specific protease in the regulation of IGFBP-5 and, thereby, bone formation.

In order for IGFBP-5 protease to be physiologically significant in terms of regulating IGFBP-5 actions and, thereby, bone formation, one of the key requirements is that osteoregulatory agents regulate the action of this protease. In this regard, we have previously found that IGFBP-5 proteolysis in a human osteoblast cell-conditioned medium is subject to regulation by IGFs, BMPs, and dexamethasone.[3,26,30] Franchimont et al[35] have shown that interleukin-6 treatment increased IGFBP-5 proteolysis in the conditioned medium of fetal rat calvarial osteoblasts. Furthermore, Hakeda et al[36] have found evidence that treatment of neonatal mouse calavaria in culture with PTH and PGE2 increased accumulation of 21 kDa fragment of IGFBP-5 in the conditioned medium, which was due to increased proteolysis of intact IGFBP-5 and not due to increased synthesis. Based on these data, it can be concluded that IGFBP-5 proteolysis is subject to regulation by both local and systemic regulators of bone metabolism and is a key control mechanism regulating the effective concentration of intact IGFBP-5 in a variety of biological fluids.

Regulation of IGFBP-5 Production in Osteoblasts in Vitro

The rate of IGFBP-5 synthesis represents another key control mechanism regulating the effective concentration of IGFBP-5 in local areas of bone. Studies on IGFBP-5 production using osteoblasts derived from humans, rats, and mice demonstrate that a number of local and systemic factors influence the production of IGFBP-5 (Table 2). In addition, these studies also demonstrate that the regulation of IGFBP-5 production is complex, involving both transcriptional and post-transcriptional mechanisms.

Local Regulators

Treatment of osteoblasts derived from human and rat bone with IGF-I or IGF-II increased IGFBP-5 protein levels in the conditioned medium.[37-40] The increased IGFBP-5 levels appear to be mediated via both increased synthesis and decreased proteolysis. Similarly, bone morphogenetic protein (BMP)-7 treatment increased both IGFBP-5 protein and mRNA levels in a dose-dependent manner in TE85 and SaOS-2 human osteosarcoma cells.[30] In contrast to the stimulatory effect of BMP-7 on IGFBP-5 production in adult human osteoblasts, treatment of fetal rat calvarial osteoblasts with BMP-7 or BMP-2 decreased IGFBP-5 production.[41-43] The molecular mechanisms that contribute to the differential regulation of IGFBP-5 expression by BMPs in human versus rat osteoblasts remain to be established.

Treatment of fetal rat calvarial osteoblasts with basic FGF, TGFβ1, and PDGF BB caused a dose and time-dependent decrease in IGFBP-5 mRNA levels.[44] The effects of these growth factors on IGFBP-5 production was independent of cell division as hydroxyurea, an inhibitor of cell division, had no effect on the inhibitory effect of FGF, TGFβ1, and PDGF BB on IGFBP-5 expression. In contrast, treatment of fetal rat calvarial osteoblasts with interleukin-6, in the presence of a soluble

Table 2. Effects of systemic and local agents on IGFBP-5 production in osteoblasts in vitro

Effector	Effect on IGFBP-5 Production	Cell Type	Reference
Cortisol	Decrease	Fetal rat calvarial osteoblasts	55
Dexamethasone	Decrease	Normal human osteoblasts	53, 54
Growth hormone	Increase	Fetal rat calvarial osteoblasts	39
PTH	Increase	Fetal rat calvarial osteoblasts, UMR106 rat osteosarcoma	36, 48-51, 53
Progesterone	Increase	Normal human osteoblasts, U2 human osteosarcoma cells	58
Retinoic acid	Decrease	Normal human osteoblasts, SaOS-2 human osteosarcoma	56
Retinoic acid	Increase	Fetal rat calvarial osteoblasts	57
$1,25(OH)_2D_3$	Increase	Fetal rat calvarial osteoblasts	48
IGFs	Increase	Normal human osteoblasts, U2 human osteosarcoma, UMR 106 rat osteosarcoma, normal fetal rat calvarial osteoblasts	26, 38, 39, 47
BMP-7	Increase	TE85 and SaOS-2 human osteosarcoma cells	30
BMP-2, BMP-7	Decrease	Fetal rat calvarial osteoblasts	41-43
FGF	Decrease	Fetal rat calvarial osteoblasts	44
Interleukin-6	Increase	Fetal rat calvarial osteoblasts	35
PDGF BB	Decrease	Fetal rat calvarial osteoblasts	44
PGE2	Increase	Fetal rat calvarial osteoblasts	45-47
TGFβ1	Decrease	Fetal rat calvarial osteoblasts	44

interleukin-6 receptor, increased IGFBP-5 mRNA expression by transcriptional mechanism and this effect of interleukin-6 was blocked by the protein synthesis inhibitor cycloheximide.[35] Because prostaglandin E2 (PGE2) treatment also increased IGFBP-5 expression by transcriptional mechanism,[45-47] and because PGE2 has been shown to be an intracellular mediator of interleukins, the effect of interleukin-6 on IGFBP-5 expression was studied in the presence of indomethacin, an inhibitor of PGE2. It was found that indomethacin had no effect on an interleukin-induced increase in IGFBP-5 expression, thus suggesting that the stimulatory effect of interleukin-6 on IGFBP-5 production was mediated by a PGE2 independent mechanism. Although the above findings suggest that IGFBP-5 production in osteoblasts is subject to regulation by bone-derived growth factors, we are unable to construct a model of IGFBP-5 involvement in growth factor regulation of bone formation because our knowledge of role of growth factors in the regulation of bone formation is inadequate at the present time.

Systemic Regulators

Treatment of fetal rat calvarial osteoblasts with 1,25-dihydroxyvitamin D3 and parathyroid hormone (PTH) increased expression of IGFBP-5.[38,48-51] Similarly, PTH treatment increased IGFBP-5 expression in UMR106 rat osteosarcoma and SaOS-2 human osteosarcoma cells.[38,52] Growth hormone, another anabolic hormone for bone, caused an increase in IGFBP-5 production in rat osteoblasts.[39] In contrast to the stimulatory effects of growth hormone and PTH, treatment with dexamethasone, a bone formation inhibitor, decreased IGFBP-5 mRNA and protein levels in a dose-dependent manner in human osteoblasts derived from various skeletal sites.[53,54] Similarly, cortisol treatment decreased IGFBP-5 mRNA and protein levels in rat osteoblasts.[55] Retinoic acid, another inhibitor of bone formation, also caused a dose- and time-dependent decrease in IGFBP-5 mRNA levels in normal human osteoblasts by dual opposing mechanisms involving a stimulatory

Table 3. Potential control points for modulation of IGFBP-5 concentration in extracellular fluid by systemic and local factors

1. Transcriptional regulation
2. Post transcriptional (nuclear mRNA stability, cytoplasmic mRNA stability)
3. Translational modification (phosphorylation, glycosylation)
4. Synthesis/secretion
5. Degradation by specific and non-specific proteases
6. Binding to extracellular matrix proteins, cell surface

cytoplasmic stability mechanism and an inhibitory transcriptional or early post-transcriptional mechanism.[56,57] Since IGFBP-5 could itself function as a growth factor for osteoblasts (see section *Models of IGFBP-5 Action in Bone*), the issue of whether the effect of systemic polypeptide and steroid hormones on bone formation is mediated, in part, via modulation of IGFBP-5 production is an interesting hypothesis that deserves attention in future studies.

In contrast to dexamethasone and retinoic acid, progesterone treatment stimulated osteoblast cell proliferation and increased IGFBP-5 mRNA levels in human osteosarcoma cells, as well as in normal human osteoblasts.[37] Studies on the mechanism by which progesterone increases IGFBP-5 gene transcription suggest that progesterone may stimulate IGFBP-5 gene transcription via a novel mechanism involving progesterone receptor-A and CACCC binding factors.[58]

Based on the above findings, it can be concluded that the IGFBP-5 production in osteoblasts is regulated by both local and systemic regulators of bone metabolism and that the regulation of IGFBP-5 levels in the conditioned medium of osteoblasts appears to be complex, involving multiple control points (e.g., transcription, mRNA stability, proteolysis) (Table 3). If IGFBP-5 were an important mediator of local and systemic effectors of bone metabolism, we would expect that agents which increase bone formation would also increase production of IGFBP-5 while agents which decrease bone formation would decrease IGFBP-5 production. However, this does not appear to always be the case as evident from studies on the inhibitory effects of BMP-7 and TGFβ on IGFBP-5 production in fetal rat osteoblasts. This is further confounded by the observations that the same agents produce opposite effects on IGFBP-5 production in human versus rat osteoblast model systems. Future studies are needed to address the molecular mechanisms that contribute to these differences and the functional consequence of opposing effects of some agents on IGFBP-5 production in human versus rat osteoblasts.

Regulation of IGFBP-5 Levels in Vivo

Recent findings that serum IGFBP-5 levels are altered under various physiological and pathological situations, and that they change in the right direction which could explain the corresponding skeletal changes, are consistent with the general hypothesis that IGFBP-5 may play an important role in the regulation of bone formation (Table 4). The changes in IGFBP-5 levels during growth hormone treatment, and during certain physiological and pathological situations, are provided below as examples to illustrate the potential importance of the IGFBP-5 system in the regulation of bone formation.

Growth Hormone Effects on Serum IGFBP-5 Levels

In previous studies, it has been well established that GH administration increases serum levels of IGFBP-3 and that serum levels of IGFBP-3 are highly correlated with serum GH levels.[8] Thus, it is fairly well known that the serum IGFBP-3 level is under the influence of GH control. Based on the findings that serum levels of IGFBP-5 are significantly lower in both children and adults with GH deficiency,[59-61] and that serum levels of IGFBP-5 show a significant positive correlation with serum IGF-I and IGFBP-3 levels,[62,63] we predicted GH to be a major regulator of serum IGFBP-5 levels. To evaluate this possibility, GH deficient adults were treated with GH (mean dose of 0.23±0.01 IU/kg/week) for a period of three years. Serum IGFBP-5 levels increased by about two-fold after three months of therapy and remained elevated during the entire treatment. Similarly, GH treatment of

Table 4. Changes in serum IGFBP-5 levels during conditions known to influence bone formation

Condition	IGFBP-5 Level in Serum	Bone Formation	Reference
Puberty	Increase	Increase	98
Aging	Decrease	Decrease*	13, 99
Pregnancy	Decrease	Increase**	100, 101
Hip fracture	Decrease	Decrease	74
Growth hormone treatment	Increase	Increase	59-62
Prednisolone treatment	Decrease	Decrease	76

* Although bone formation increases during aging to compensate for the increased bone resorption, the magnitude of age-related increase in bone formation is less than that of bone resorption.
** Both bone formation and bone resorption parameters in serum were increased during pregnancy resulting in a high bone-turnover state.

GH-deficient children also increased serum levels of IGFBP-5.[61] Furthermore, treatment of healthy obese subjects with the GH secretagogue MK-677 increased serum IGFBP-5 levels by 44% after two weeks of treatment.[64] These studies are consistent with our position that GH is an important regulator of serum IGFBP-5 levels.

As with GH deficiency, GH insensitivity is characterized by abnormal growth caused by a lack of GH response due to abnormalities in GH receptor structure, GH receptor dimerization, and signal transduction.[59,65] In children with GH insensitivity from an inbred population in southern Ecuador, serum IGFBP-5 levels were decreased by more than 80% compared to corresponding age-matched normal children.[59] In addition, serum levels of IGF-I, IGF-II, and IGFBP-3 were also lower in children with GH insensitivity, despite elevated serum GH levels. Twice-daily subcutaneous administration of recombinant human IGF-I (120 ug/kg/day) had no significant effect on serum levels of IGFBP-5. If GH effect on IGFBP-5 were mediated via increased IGF-I production, we would anticipate IGF-I treatment to increase serum IGFBP-5 levels in children with GH receptor deficiency. The findings that IGF-I treatment failed to increase serum IGFBP-5 levels are consistent with the possibility that GH exerts a direct effect on IGFBP-5 production in target tissues.

Consistent with the idea that growth hormone, and not IGF-I, is the major regulator of IGFBP-5, we found that the serum level of IGFBP-5 was elevated compared to age-matched controls in a patient with primary IGF-I deficiency due to a partial deletion of the IGF-I gene.[66] In order to evaluate if the higher IGFBP-5 level was due to elevated growth hormone in this patient, the effect of IGF-I therapy on serum IGFBP-5 was determined based on the rationale that IGF-I therapy would decrease the growth hormone level and, correspondingly, serum IGFBP-5 level. Accordingly, we found that IGF-I therapy decreased the serum IGFBP-5 level and a significant negative correlation was found between serum IGF-I levels and IGFBP-5 levels during 6 months of IGF-I therapy.[66] These data suggested that growth hormone effects on serum IGFBP-5 is independent of IGF-I.

Regarding the source of increased serum IGFBP-5 levels in response to growth hormone, it is known that there are multiple growth hormone responsive tissues that produce IGFBP-5 (e.g., muscle, cartilage, and bone). In this regard, it has been shown that GH increases IGFBP-5 expression in rat osteoblasts,[39] thus suggesting that skeletal tissue could be one of the contributors to the increased circulating level of IGFBP-5 in response to growth hormone treatment. Future studies are needed to examine the mechanism by which growth hormone increases the production of IGFBP-5 in the target tissues, such as bone.

In regard to the mechanism(s) by which GH increases serum level of IGFBP-5, it has been shown by Baxter and his colleagues that IGFBP-5+IGF can form ternary complexes with the GH-dependent protein, ALS.[67] Because the half-life of IGFBP-5 in the ternary complex (IGF+IGFBP-5+ALS) is prolonged considerably compared to the binary complex (IGFBP-5+IGF), GH treatment could induce formation of ternary complex, which would stabilize IGFBP-5 in the circulation, and thereby increase circulating level of IGFBP-5. Thus, GH could mediate its effect on serum IGFBP-5 level by a post-translational effect involving the formation of ternary complex, in addition to the direct transcriptional effect of GH on IGFBP-5 expression. Future studies are needed to establish the relative contribution of transcriptional and post-translational mechanisms to GH-induced increase in serum IGFBP-5 levels.

Based on the above findings, and other findings on serum regulation of IGFBP-5, the following conclusions could be made:
1. GH is a major regulator of serum IGFBP-5 levels;
2. Both direct transcriptional effect of GH on IGFBP-5 expression and a post-translational effect involving the formation of ternary complex could contribute to GH-induced increase in serum IGFBP-5 levels; and
3. Decreased serum and skeletal IGFBP-5 levels during aging may be due to an age-related decline in serum GH levels (see below).

In terms of the significance of GH regulation of IGFBP-5, studies in our laboratories, and other laboratories as well, have provided evidence that IGFBP-5 may itself be a growth factor that can mediate its effects on bone, in part, via an IGF-independent mechanism.[68] Accordingly, the finding that GH stimulates the production of IGFBP-5 in bone raises the possibility that IGFBP-5 may mediate the IGF-independent effects of GH on target tissues, such as bone (Fig. 1). In this regard, it has been proposed that GH stimulates longitudinal bone growth directly by stimulating prechondrocytes in the growth plate in an IGF-independent manner.[69] Furthermore, it has been found that the growth rate achieved in patients with GH insensitivity in response to IGF-I therapy is less than the growth rate observed in children with GH deficiency receiving GH.[70] The issue of whether some of the IGF-independent effects of GH on longitudinal growth are mediated via IGFBP-5 acting as a growth factor independent of IGF-I is an interesting possibility that requires further studies.

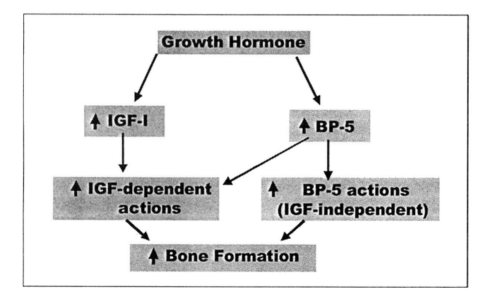

Figure 1. Model of Growth hormone Actions in Bone. Growth hormone may mediate its effects on bone by increasing circulating levels of IGF-I in an endocrine manner or local levels of IGF-I in an autocrine/paracrine manner. In addition, growth hormone has been shown to modulate production of IGFBP-5 in bone cells. Based on the findings that IGFBP-5 could itself function as a growth factor independent of IGFs, it is possible that a growth hormone-induced increase in IGFBP-5 may mediate growth hormone effects in an IGF-independent manner.

Skeletal and Serum Levels of IGFBP-5 Decrease with Age

It is well known that that bone loss occurs with age. In terms of the pathogenesis of age-related bone loss, it has been shown by calcium 47 kinetic studies that the rate of increase in bone resorption after the menopause is not compensated by a corresponding increase in bone formation, resulting in the net loss of bone.[71] To understand the mechanism that underlies an age-related deficiency in bone formation, we proposed the hypothesis that a deficiency in stimulatory IGFBP-5 function occurs, in addition to the deficiency in IGFs, as a consequence of age, which could lead to a decrease in osteoblast cell proliferation and a subsequent reduction in bone formation. As an indirect means of testing this hypothesis, we first measured serum levels of IGFBP-5 in 102 healthy men and women between ages 21-87 years. Consistent with our hypothesis, serum IGFBP-5 levels showed a negative correlation with age (r=-0.48, P<0.001). The mean serum level of IGFBP-5 was 35% lower in the 61-87 year age group compared to that of the 21-40 year age group.[72,73] In addition, serum levels of IGFBP-5 showed significant positive correlations with both IGF-I and IGF-II. In contrast to serum levels of IGFBP-5, serum levels of IGFBP-4 showed significant positive correlations with age (r=0.54, P<0.001). These data suggest that the mechanisms regulating serum levels of IGFBP-4 and IGFBP-5 are different.

Since multiple tissues contribute to circulating serum levels of IGFBP-5, we evaluated if similar age-related changes in IGFBP-5 levels occur in bone. We found that the skeletal content of IGFBP-5 decreased by 28% (P=0.02) between the ages 20-29 years and 54-64 years. In addition, we found that the skeletal contents of IGFBP-5 showed significant positive correlations with both bone IGF-I and IGF-II.[72,73] Based on these data, we speculate that the level of IGFBP-5 in bone is largely a reflection of its integrated secretion by osteoblasts and that the production of stimulatory IGFBP-5 by osteoblasts may decline with age.

Although the mechanisms that cause the age-related changes in IGFBP-5 remain unknown, the findings that serum levels of IGFBP-3 and IGFBP-5 correlate with serum levels of IGF-I and acid labide subunit (ALS), as well as with each other, suggest that serum levels of IGFBP-3 and IGFBP-5 may be GH dependent and that age-related changes in serum levels of these two IGFBPs may be due, in part, to the age-related decline in GH secretion.[8] Similarly, it is known that IGFBP-5 expression is under the influence of GH. Thus, it is conceivable that the decrease in skeletal content of IGFBP-5 could be, in part, contributed to by the decline in IGFBP-5 expression by osteoblasts in response to the GH deficiency that occurs with age.

Serum IGFBP-5 Levels and Clinical Bone Diseases

The findings that systemic administration of IGFBP-5 increased bone formation parameters, and that serum levels of IGFBP-5 decreases in clinical disease states, suggest that IGFBP-5 could play a pathogenic role in the development of osteoporosis and act as physiologic growth factor to maintain bone under normal conditions. Accordingly, we found that serum levels of IGFBP-5 were decreased 35% in patients with hip fractures compared to levels found in the control and that these levels correlated with bone density, thus suggesting that deficiency in IGFBP-5 could contribute to the pathogenesis of hip fractures.[74] Two additional clinical studies support our assumption that alterations in IGFBP-5 production could contribute to bone formation changes seen during clinical disease states. First, Jehle et al[75] have shown that serum levels of IGFBP-5 were decreased in chronic renal failure patients with impaired bone formation and that serum levels of IGFBP-5 showed significant, positive correlation with the bone formation rate as evaluated by histomorphometry. Second, glucocorticoid treatment caused an acute decline in serum IGFBP-5 levels, which correlated with a decrease in bone formation parameters.[76]

The findings that serum levels of IGFBP-5 show significant changes during clinical disease states and that serum regulation of IGFBP-5 levels in vivo and osteoblast cell regulation of IGFBP-5 production in vitro are under the influence of agents that affect bone metabolism are consistent with the speculation that IGFBP-5 could function to regulate bone formation. The extents to which local and systemic sources of IGFBP-5 contribute to regulation of bone formation remain to be established in future studies.

Actions of IGFBP-5

In Vitro Studies

IGFBP-5 is unique among IGFBPs in that it has been shown to stimulate both basal and IGF-induced cell proliferation in various bone cell types in vitro.[3,20,21,68,77-80] During the course of purification of IGFBP-5 from human bone extract, we found that purified IGFBP-5 stimulated basal and IGF-induced cell proliferation in osteoblasts derived from chick and mouse.[3,20,68] Similar potentiating effects of IGFBP-5 purified from U2 human osteosarcoma cell conditioned medium have been reported by Andress and Birnbaum.[21] Schmid et al[81] have also shown that IGFBP-5, but not IGFBP-3, stimulated thymidine incorporation into DNA, both in the absence and presence of IGF-I in rat osteoblasts. These results have been subsequently confirmed using purified recombinant human IGFBP-5 in various osteoblast cell types, including untransformed normal human osteoblasts.[77-80,82] Consistent with the stimulatory effects of IGFBP-5 on osteoblasts, Jones et al[83] have demonstrated that when IGFBP-5 was present in cell culture substrata, it potentiated the growth stimulatory effects of IGF-I on fibroblasts. Together, these studies suggest that IGFBP-5 modulates IGF action in a positive manner, not only in osteoblasts, but also in fibroblasts.

In addition to modulating both basal and IGF-induced cell proliferation, we have recently found that IGFBP-5 treatment increased osteoblast differentiation markers, namely alkaline phosphatase activity and osteocalcin production in MG63 human osteosarcoma cells.[84] Transgenic over-expression of IGFBP-5 by transfecting OS/50-K8 mouse osteosarcoma cells with an expression vector containing the osteocalcin promoter and the complete mouse IGFBP-5cDNA resulted in an inhibition of proliferation, but an increase in differentiation.[85] The growth inhibitory effect of IGFBP-5 over-expression could not be completely abolished by the exogenous addition of long [R3] IGF-I with a reduced affinity for IGFBP-5, thus suggesting that the growth inhibitory effect of IGFBP-5 overexpression is partly IGF-I-independent. In addition, Slootweg et al[86] have reported that IGFBP-5 treatment caused a dose-dependent increase in GH binding caused by up-regulation of GH mRNA levels in UMR106 rat osteosarcoma cells, and that this increase in GH binding by IGFBP-5 was associated with an increase in GH-induced cell proliferation. These data suggest that IGFBP-5 exerts a multitude of effects on osteoblasts in vitro.

In Vivo Studies

Consistent with the in vitro data, in vivo studies in several laboratories have shown that IGFBP-5 stimulates bone formation parameters in various experimental animal model systems. Richman et al[84] described, for the first time, that systemic administration of IGFBP-5, alone or in combination with IGF-I, stimulated bone formation parameters in mice in a dose and time-dependent manner. In this study, it was found that IGFBP-5 was as potent as IGF-I in stimulating bone formation parameters. In subsequent studies, we found that local single administration of IGFBP-5 increases bone formation parameters by more than 100% in mice.[68] This data has now been confirmed by two independent studies. First, Bauss et al[87] have demonstrated that local administration produced a clear synergistic effect of IGF-I/IGFBP-5, in comparison with IGF-I alone, on bone density and bone area in the calvariae of mice. Furthermore, it was also found that systemic administration of IGF-I/IGFBP-5 also increased cortical thickness, bone area, and mineral density in long bones after systemic administration. Second, Andress[15] recently reported that daily subcutaneous injections of intact or fragment (1-169) forms of IGFBP-5 for eight weeks increased bone mineral density in ovariectomized mice. Based on histomorphometric studies, it was shown in this study that IGFBP-5 effectively increases bone formation and bone accretion in ovariectomized mice by stimulating osteoblast activity. Thus, these data provide evidence that IGFBP-5 stimulates bone formation in vivo.

In contrast to the above reports from our group and other groups which have shown that intermittent administration of IGFBP-5 stimulated bone formation, Devlin et al[88] have shown recently that transgenic over-expression of IGFBP-5 in large amounts, using eight copies of the transgene, decreased bone formation in vivo. There are at least two potential explanations for this discrepancy: 1) IGFBP-5, like PTH, may stimulate bone formation when administered intermittently but not continuously; and 2) IGFBP-5, like other growth factors may be biphasic and may inhibit osteoblast activity at extremely high doses.[89,90] Future studies are needed to address if intermittent versus continuous administration of IGFBP-5 exert different effects on bone formation in mice.

Table 5. Potential mechanisms by which IGFBP-5 could stimulate IGF action in osteoblasts

1. Increase half-life of IGF in the local bone mileu
2. Increase concentration of IGF in the vicinity of signaling type I IGF receptor
3. Amplify IGF-I down stream signaling pathway by IGFBP-5 signaling pathway
4. Release IGF from inhibitory IGFBPs
5. Release IGF from the 150 kDa complex in the vascular compartment

Models of IGFBP-5 Action in Bone

The findings that IGFBP-5 exerts pleotrophic effects on osteoblasts raise the question as to how a single protein mediates multitude of these effects. In this regard, it is now known that locally produced IGFBP-5 could act to potentiate IGF actions, or it could act in an IGF-independent manner.[8] Furthermore, IGFBP-5 in serum could act as a hormone in a manner analogous to that of IGF-I. The potential mechanisms of actions of IGFBP-5 are discussed below.

IGFBP-5 treatment has been shown to potentiate IGF-I action in a variety of cell types, including osteoblasts, chondroblasts, and smooth muscle cells.[3,20,21,68,77-81,83-85] In terms of the mechanism of IGFBP-5 action, IGFBP-5 could increase IGF action by prolonging the half-life of IGF in the local mileu (Table 5). The association of IGFBP-5 with proteins on the cell surface or in the extracellular matrix could result in an increase in the local concentration of IGFs in the vicinity of IGF receptors, thereby allowing IGF to bind to IGF receptors.[10,79] Additionally, IGBP-5 could also release IGFs from inhibitory IGFBPs and thereby increase IGF bioavailability (Table 5).

Another way IGFBPs could potentiate IGF actions is by facilitating storage of IGFs in the extracellular matrices of certain tissues for future action. In this regard, Jones et al[83] have provided evidence for the fixation of IGFs via IGFBP-5 binding to extracellular matrix proteins (Fig. 2). We found evidence that IGFBP-5 may help fix IGFs in bone since the complex of IGFBP-5 and IGFs, but not IGFs alone, bind to hydroxylapatite.[73] In terms of the significance of fixation of IGFs in extracellular matrices, such as bone, it is speculated that the stored IGFs may be released during the bone resorption phase of remodeling to stimulate nearby osteoblasts during the bone formation phase of remodeling (Fig. 3). Similarly, IGFs stored in extracellular matrices of soft tissues may have a role in wound healing.

In addition to modulating IGF actions, IGFBP-5 could also function as a growth factor independent of its ligand. During the course of our studies on biological actions of IGFBP-5 in osteoblasts, we found that IGFBP-5 treatment increased differentiation in MG63 human osteosarcoma cells which produce detectable levels of neither IGF-I nor IGF-II.[84] In subsequent studies, we found that systemic administration of IGFBP-5 increased bone formation parameters but had no effect on circulating IGF-I levels in serum. Furthermore, Andress et al[78] showed that a carboxyterminal-truncated form of IGFBP-5, which binds IGFs with reduced affinity, increased osteoblast cell proliferation in the absence of IGF-I. Although these studies provided indirect evidence for an IGF-independent effect of IGFBP-5, it was only recently that direct evidence for an IGF-independent effect of IGFBP-5 was obtained, using osteoblasts from IGF-I knockout mice. In mutant osteoblasts that produce neither of the two IGFs, we found that IGFBP-5 promoted cell growth, alkaline phosphatase activity, and osteocalcin expression as robustly as wild type cells.[68] Furthermore, IGFBP-5, when administered locally to the outer periosteum of the parietal bone of IGF-I knockout mice, increased markers of bone formation to levels comparable to those seen in wild type mice.[68] Thus, these data provide direct evidence that IGFBP-5 could function as a growth factor that mediates its actions, in part, via an IGF-independent mechanism.

In terms of the mechanism by which IGFBP-5 mediate its effects via an IGF-independent mechanism, a number of possibilities exist. In this regard, we and others have shown that osteoblasts contain a putative receptor for IGFBP-5.[79,91] It is possible that binding of IGFBP to its putative receptor may stimulate a signaling pathway independent of an IGF receptor to mediate the effects of IGFBPs in certain target cell types.[91] IGFBP-5 putative receptor, a 420 kDa protein, was purified from membrane preparations of mouse osteoblasts using IGFBP-5[201-218] affinity column based on

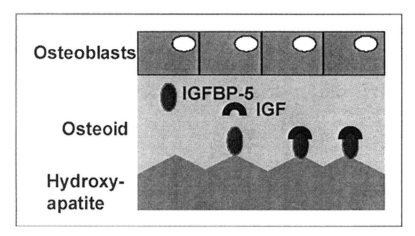

Figure 2. Model for fixation of IGFs by IGFBP-5 in bone. Osteoblasts produce IGF-I, IGF-II as well as IGFBP-5. IGF-I or IGF-II itself does not bind to either unmineralized osteoid or to hydroxylapaitite. On the other hand, IGFBP-5 either alone or complexed to IGF binds to hydroxyapaite. Accordingly, we propose that osteoblast cell produced IGF-I or IGF-II is fixed in bone via IGFBP-5 binding to hydroxyapatite. In terms of why there is much more IGF-II than IGF-I, there are two potential explanations. First, osteoblasts produce much more IGF-II than IGF-I. Second, IGFBP-5 binds to IGF-II with higher affinity than IGF-I.

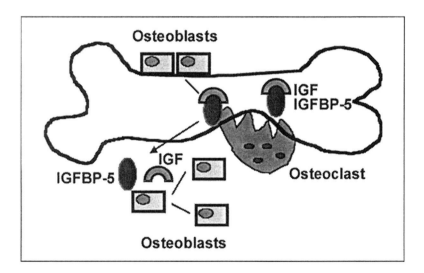

Figure 3. Model illustrating the delayed paracrine actions of IGFs fixed in bone by IGFBP-5. IGFs produced by osteoblasts are deposited in bone via IGFBP-5. Osteoclastic resorption of an area of bone releases IGFs and the IGFBP-5 both of which can act on osteoblast-line of cells to insure site-specific bone replacement.

the data that IGFBP-5[201-218] bound directly to the osteoblast surface.[90] Co-incubation of the affinity purified 420 kDa protein with ^{32}P-ATP resulted in auto-phosphorylation at serine residues which was enhanced by both intact IGFBP-5 and IGFBP-5 fragments,[90] suggesting that serine/threonine kinase activation may be important in mediating some of the IGF-independent effects of IGFBP-5 (Fig. 4).

Another mechanism by which IGFBP-5 could exert IGF-independent effects is by the transcriptional activation of genes by IGFBPs transported into the nucleus via their nuclear localization

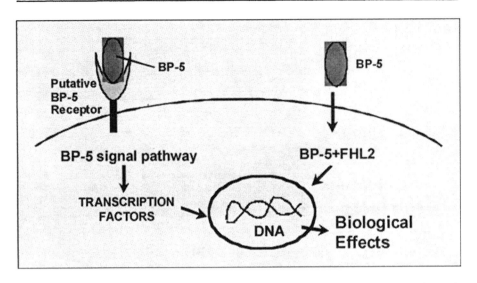

Figure 4. Model for IGF-independent action of IGFBP-5 in osteoblasts. Based on our current understanding, we have proposed two models for IGF-independent actions of IGFBP-5. In model 1, IGFBP-5 binds to putative IGFBP-5 receptor which has serine-threonine kinase activity. IGFBP-5 up on binding to its signaling receptor causes activation of transcription factors which binds to DNA to elicit biological effects. In model 2, IGFBP-5 from cytoplasm or extracellular enters the nucleus via its nuclear localization sequence. IGFBP-5 binds to FHL2, a potential transcription modulator, in the cytoplasm and shuttle FHL2 into the nucleus to stimulate transcription of target genes.

signal. In this regard, IGFBP-5 contains a nuclear localization sequence and there is evidence for nuclear uptake of IGFBP-5 by human osteoblasts and other cell types.[92-94] Based on these findings, we have proposed the concept that IGFBP-5 interacts with transcription factors that lead to increased osteoblast proliferation. As a means of testing this hypothesis, we recently undertook studies to identify proteins that bind to IGFBP-5 using IGFBP-5 as bait in a yeast two-hybrid screen of U2 human osteosarcoma cell cDNA library. Using this approach, we have identified FHL2 (four and a half LIM domain protein), a zinc finger protein, as a binding partner for IGFBP-5 in osteoblasts.[94] Based on the findings that FHL2 is expressed strongly in a variety of human osteoblast cell types and found to be localized in the nucleus as a complex with IGFBP-5, and based on the findings that FHL2 can function as transcription factor or coactivator in other cell types,[95] we propose that FHL2 is an intracellular mediator of IGFBP-5 action in osteoblasts (Fig. 4). In support of the above hypothesis, studies by Boden et al[96] have shown that LIM mineralization protein 1 (LMP-1), which is a LIM domain protein, mediates BMP-6 effects on osteoblast differentiation. Future studies are needed to evaluate the role of FHL2 in mediating the effects of IGFBP-5 on osteoblast proliferation and differentiation.

In addition to their local actions, IGFBP-5 in serum may play an important role in regulating the endocrine actions of IGFs. It is fairly well established that the presence of IGF in the 150-kDa complex (acid labile subunit+IGFBP-3+IGF) is critical for the prolonged half-life, as well as for the maintenance of, a large readily available IGF reserve in serum.[8] The biological activity of circulating IGFs to the tissues is determined by mechanisms involving a shifting of IGF from the 150 kDa complex to the 50 kDa complex and subsequent release of IGF from the 50 kDa complex.[14,97] Since IGFBP-5 is one of the IGFBPs that contribute to the circulating 50-kDa complex, the physiologic regulation of IGF in the IGFBP-5/IGF pool may influence the levels of serum IGF that is bioavailable to target tissues. Furthermore, the circulating serum level of IGFBP-5 could also act as an endocrine hormone since systemic administration of IGFBP-5 increased bone formation via a mechanism that did not involve increase in circulating IGF-I levels.

Conclusions

Several unique features suggest that IGFBP-5 is a multifunctional protein in bone:
1. it stimulates both proliferation and activity of osteoblast line cells in vitro;
2. it acts as a traditional binding protein whereby IGFBP-5 modulates actions of IGFs;
3. it acts as a growth factor to stimulate bone formation;
4. it acts both as a local factor, as well as a systemic factor, to regulate bone forming effects of osteoblasts; and
5. it binds to hydroxyapatite and extracellular matrix proteins with high affinity, which would be a means to fix IGF-I and IGF-II for future use during bone repair.

Studies on the molecular pathways that control the growth factor actions of IGFBP-5 have revealed evidence for the potential involvement of the IGFBP-5 receptor and IGFBP-5 binding protein in the nucleus of osteoblasts. Future studies are needed to clarify if one or both of these proteins are involved in mediating the IGF-independent effect of IGFBP-5 in osteoblasts. The above features of IGFBP-5, together with the findings that IGFBP-5 production is regulated by key osteoregulatory agents in vitro and in vivo in a manner consistent with their effects on bone formation, provide support to the concept that IGFBP-5 functions as a physiologic regulator of bone formation and is a key player in the local regulation of bone formation. Future studies are needed to evaluate if alterations in the production and/or actions of IGFBP-5 could, in part, contribute to why some people have impaired bone formation during aging. If so, alleles in IGFBP-5 pathway that are polymorphic could be used to predict people at risk for impaired bone formation during aging.

Acknowledgments

This work was supported by funds from NIH (AR31062), Veteran's Administration and Department of Medicine, Loma Linda University.

References

1. Baylink DJ, Strong DD, Mohan S. The diagnosis and treatment of osteoporosis: future prospects. Mol Med Today 1999; 5:133-40.
2. Rodan GA, Raisz LG, Bilezikian JP. Pathophysiology of osteoporosis. In: Bilezikian LGRJP, Rodan GA, eds. Principles of Bone Biology. San Diego: Academic Press, 1996:979-90.
3. Mohan S, Baylink DJ. IGF system components and their role in bone metabolism. In: Roberts CT, ed. IGFs in Health and Disease. New Jersey: Humana Press, 1999:457-96.
4. Mohan S, Baylink DJ. Role of growth hormone/insulin-like growth factor axis. In: Rosen CJ, Bilezikian JP, eds. The Aging Skeleton. San Diego: Academic Press, 1999:209-19.
5. Canalis E. Insulin-like growth factors and osteoporosis. Bone 1997; 21:215-6.
6. Conover CA. In vitro studies of insulin-like growth factor I and bone. Growth Horm IGF Res 2000; 10(Suppl B):S107-10.
7. Mohan S, Baylink DJ. Insulin-like growth factor (IGF)-binding proteins in serum—do they have additional roles besides modulating the endocrine IGF actions? [editorial]. J Clin Endocrinol Metab 1996; 81:3817-20.
8. Rajaram S, Baylink DJ, Mohan S. Insulin-like growth factor-binding proteins in serum and other biological fluids: regulation and functions. Endocr Rev 1997; 18:801-31.
9. Baxter RC. Insulin-like growth factor (IGF)-binding proteins: interactions with IGFs and intrinsic bioactivities. Am J Physiol Endocrinol Metab 2000; 278:E967-76.
10. Clemmons DR. Role of insulin-like growth factor binding proteins in controlling IGF actions. Mol Cell Endocrinol 1998; 140:19-24.
11. Wetterau LA, Moore MG, Lee KW et al. Novel aspects of the insulin-like growth factor binding proteins. Mol Genet Metab 1999; 68:161-81.
12. Hwa V, Oh Y, Rosenfeld RG. The insulin-like growth factor-binding protein (IGFBP) superfamily. Endocr Rev 1999; 20:761-87.
13. Mohan S, Baylink DJ. Serum insulin-like growth factor binding protein (IGFBP)-4 and IGFBP-5 levels in aging and age-associated diseases. Endocrine 1997; 7:87-91.
14. Miyakoshi N, Richman C, Qin X et al. Effects of recombinant insulin-like growth factor-binding protein-4 on bone formation parameters in mice. Endocrinology 1999; 140:5719-28.
15. Andress DL. IGF-binding protein-5 stimulates osteoblast activity and bone accretion in ovariectomized mice. Am J Physiol Endocrinol Metab 2001; 281:E283-8.
16. Mohan S, Baylink DJ. Bone growth factors. Clin Orthop 1991; 30-48.
17. Hayden JM, Mohan S, Baylink DJ. The insulin-like growth factor system and the coupling of formation to resorption. Bone 1995; 17:93S-8S.
18. Mohan S, Jennings JC, Linkhart TA et al. Primary structure of human skeletal growth factor: homology with human insulin-like growth factor-II. Biochim Biophys Acta 1988; 966:44-55.

19. Farley JR, Baylink DJ. Purification of a skeletal growth factor from human bone. Biochemistry 1982; 21:3502-07.
20. Bautista CM, Baylink DJ, Mohan S. Isolation of a novel insulin-like growth factor (IGF) binding protein from human bone: a potential candidate for fixing IGF-II in human bone. Biochem Biophys Res Commun 1991; 176:756-63.
21. Andress DL, Birnbaum RS. A novel human insulin-like growth factor binding protein secreted by osteoblast-like cells. Biochem Biophys Res Commun 1991; 176:213-8.
22. Shimasaki S, Ling N. Identification and molecular characterization of insulin-like growth factor binding proteins (IGFBP-1, -2, -3, -4, -5 and -6). Prog Growth Factor Res 1991; 3:243-66.
23. Camacho-Hubner C, Busby WH Jr, McCusker RH et al. Identification of the forms of insulin-like growth factor-binding proteins produced by human fibroblasts and the mechanisms that regulate their secretion. J Biol Chem 1992; 267:11949-56.
24. Roghani M, Segovia B, Whitechurch O et al. Purification from human cerebrospinal fluid of insulin-like growth factor binding proteins (IGFBPs). Isolation of IGFBP-2, an altered form of IGFBP-3 and a new IGFBP species. Growth Regul 1991; 1:125-30.
25. Kiefer MC, Ioh RS, Bauer DM et al. Molecular cloning of a new human insulin-like growth factor binding protein. Biochem Biophys Res Commun 1991; 176:219-25.
26. Kanzaki S, Hilliker S, Baylink DJ et al. Evidence that human bone cells in culture produce insulin-like growth factor-binding protein-4 and -5 proteases. Endocrinology 1994; 134:383-92.
27. Byun D, Mohan S, Yoo M et al. Pregnancy-associated plasma protein-A accounts for the insulin-like growth factor (IGF)-binding protein-4 (IGFBP-4) proteolytic activity in human pregnancy serum and enhances the mitogenic activity of IGF by degrading IGFBP-4 in vitro. J Clin Endocrinol Metab 2001; 86:847-54.
28. Fielder PJ, Pham H, Adashi EY et al. Insulin-like growth factors (IGFs) block FSH-induced proteolysis of IGF- binding protein-5 (BP-5) in cultured rat granulosa cells. Endocrinology 1993; 133:415-8.
29. Nam TJ, Busby WH Jr, Clemmons DR. Characterization and determination of the relative abundance of two types of insulin-like growth factor binding protein-5 proteases that are secreted by human fibroblasts. Endocrinology 1996; 137:5530-6.
30. Knutsen R, Honda Y, Strong DD et al. Regulation of insulin-like growth factor system components by osteogenic protein-1 in human bone cells. Endocrinology 1995; 136:857-65.
31. Thrailkill KM, Quarles LD, Nagase H et al. Characterization of insulin-like growth factor-binding protein 5- degrading proteases produced throughout murine osteoblast differentiation. Endocrinology 1995; 136:3527-33.
32. Campbell PG, Andress DL. Plasmin degradation of insulin-like growth factor-binding protein-5 (IGFBP-5): regulation by IGFBP-5-(201-218). Am J Physiol 1997; 273:E996-1004.
33. Busby WH Jr, Nam TJ, Moralez A et al. The complement component C1s is the protease that accounts for cleavage of insulin-like growth factor-binding protein-5 in fibroblast medium. J Biol Chem 2000; 275:37638-44.
34. Overgaard MT, Boldt HB, Laursen LS et al. Pregnancy-associated plasma protein-A2 (PAPP-A2), a novel insulin-like growth factor-binding protein-5 proteinase. J Biol Chem 2001; 276:21849-53.
35. Franchimont N, Durant D, Canalis E. Interleukin-6 and its soluble receptor regulate the expression of insulin-like growth factor binding protein-5 in osteoblast cultures. Endocrinology 1997; 138:3380-6.
36. Hakeda Y, Kawaguchi H, Hurley M et al. Intact insulin-like growth factor binding protein-5 (IGFBP-5) associates with bone matrix and the soluble fragments of IGFBP-5 accumulated in culture medium of neonatal mouse calvariae by parathyroid hormone and prostaglandin E2-treatment. J Cell Physiol 1996; 166:370-9.
37. Mohan S. Insulin-like growth factor binding proteins in bone cell regulation. Growth Regul 1993; 3:67-70.
38. Conover CA, Bale LK, Clarkson JT et al. Regulation of insulin-like growth factor binding protein-5 messenger ribonucleic acid expression and protein availability in rat osteoblast- like cells. Endocrinology 1993; 132:2525-30.
39. McCarthy TL, Casinghino S, Centrella M et al. Complex pattern of insulin-like growth factor binding protein expression in primary rat osteoblast enriched cultures: regulation by prostaglandin E2, growth hormone, and the insulin-like growth factors. J Cell Physiol 1994; 160:163-75.
40. Gabbitas B, Canalis E. Insulin-like growth factors sustain insulin-like growth factor-binding protein-5 expression in osteoblasts. Am J Physiol 1998; 275:E222-8.
41. Gabbitas B, Canalis E.5 Bone morphogenetic protein-2 inhibits the synthesis of insulin-like growth factor-binding protein-5 in bone cell cultures. Endocrinology 1995; 136:2397-403.
42. Yeh LC, Adamo ML, Duan C et al. Osteogenic protein-1 regulates insulin-like growth factor-I (IGF-I), IGF-II, and IGF-binding protein-5 (IGFBP-5) gene expression in fetal rat calvaria cells by different mechanisms. J Cell Physiol 1998; 175:78-88.
43. Yeh LC, Lee JC. Identification of an osteogenic protein-1 (bone morphogenetic protein- 7)-responsive element in the promoter of the rat insulin-like growth factor-binding protein-5 gene. Endocrinology 2000; 141:3278-86.
44. Canalis E, Gabbitas B. Skeletal growth factors regulate the synthesis of insulin-like growth factor binding protein-5 in bone cell cultures. J Biol Chem 1995; 270:10771-6.
45. McCarthy TL, Casinghino S, Mittanck DW et al. Promoter-dependent and -independent activation of insulin-like growth factor binding protein-5 gene expression by prostaglandin E2 in primary rat osteoblasts. J Biol Chem 1996; 271:6666-71.

46. Pash JM, Canalis E. Transcriptional regulation of insulin-like growth factor-binding protein-5 by prostaglandin E2 in osteoblast cells. Endocrinology 1996; 137:2375-82.
47. Ji C, Chen Y, Centrella M, McCarthy TL. Activation of the insulin-like growth factor-binding protein-5 promoter in osteoblasts by cooperative E box, CCAAT enhancer-binding protein, and nuclear factor-1 deoxyribonucleic acid-binding sequences. Endocrinology 1999; 140:4564-72.
48. Schmid C, Schlapfer I, Gosteli-Peter MA et al. 1 alpha,25-dihydroxyvitamin D3 increases IGF binding protein-5 expression in cultured osteoblasts. FEBS Lett 1996; 392:21-4.
49. Nasu M, Sugimoto T, Chihara K. Stimulatory effects of parathyroid hormone and 1,25-dihydroxyvitamin D3 on insulin-like growth factor-binding protein-5 mRNA expression in osteoblastic UMR-106 cells: the difference between transient and continuous treatments. FEBS Lett 1997; 409:63-6.
50. Nasu M, Sugimoto T, Kaji H et al. Carboxyl-terminal parathyroid hormone fragments stimulate type-1 procollagen and insulin-like growth factor-binding protein-5 mRNA expression in osteoblastic UMR-106 cells. Endocr J 1998; 45:229-34.
51. Kveiborg M, Flyvbjerg A, Eriksen EF et al. 1,25-Dihydroxyvitamin D3 stimulates the production of insulin-like growth factor-binding proteins-2, -3 and -4 in human bone marrow stromal cells. Eur J Endocrinol 2001; 144:549-57.
52. Nasu M, Sugimoto T, Kaji H et al. Estrogen modulates osteoblast proliferation and function regulated by parathyroid hormone in osteoblastic SaOS-2 cells: role of insulin-like growth factor (IGF)-I and IGF-binding protein-5. J Endocrinol 2000; 167:305-13.
53. Okazaki R, Riggs BL, Conover CA. Glucocorticoid regulation of insulin-like growth factor-binding protein expression in normal human osteoblast-like cells. Endocrinology 1994; 134:126-32.
54. Chevalley T, Strong DD, Mohan S et al. Evidence for a role for insulin-like growth factor binding proteins in glucocorticoid inhibition of normal human osteoblast-like cell proliferation. Eur J Endocrinol 1996; 134:591-601.
55. Gabbitas B, Pash JM, Delany AM et al. Cortisol inhibits the synthesis of insulin-like growth factor-binding protein-5 in bone cell cultures by transcriptional mechanisms. J Biol Chem 1996; 271:9033-8.
56. Zhou Y, Mohan S, Linkhart TA et al. Retinoic acid regulates insulin-like growth factor-binding protein expression in human osteoblast cells. Endocrinology 1996; 137:975-83.
57. Dong Y, Canalis E. Insulin-like growth factor (IGF) I and retinoic acid induce the synthesis of IGF-binding protein 5 in rat osteoblastic cells. Endocrinology 1995; 136:2000-6.
58. Boonyaratanakornkit V, Strong DD, Mohan S et al. Progesterone stimulation of human insulin-like growth factor-binding protein-5 gene transcription in human osteoblasts is mediated by a CACCC sequence in the proximal promoter. J Biol Chem 1999; 274:26431-8.
59. Burren CP, Wanek D, Mohan S et al. Serum levels of insulin-like growth factor binding proteins in Ecuadorean children with growth hormone insensitivity. Acta Paediatr Suppl 1999; 88:185-91; discussion 92.
60. Ono T, Kanzaki S, Seino Y et al. Growth hormone (GH) treatment of GH-deficient children increases serum levels of insulin-like growth factors (IGFs), IGF-binding protein-3 and -5, and bone alkaline phosphatase isoenzyme. J Clin Endocrinol Metab 1996; 81:2111-6.
61. Thoren M, Hilding A, Brismar T et al. 8 Serum levels of insulin-like growth factor binding proteins (IGFBP)-4 and -5 correlate with bone mineral density in growth hormone (GH)- deficient adults and increase with GH replacement therapy. J Bone Miner Res 1998; 13:891-9.
62. Powell DR, Durham SK, Brewer ED et al. Effects of chronic renal failure and growth hormone on serum levels of insulin-like growth factor-binding protein-4 (IGFBP-4) and IGFBP-5 in children: a report of the Southwest Pediatric Nephrology Study Group. J Clin Endocrinol Metab 1999; 84:596-601.
63. Ulinski T, Mohan S, Kiepe D et al. Serum insulin-like growth factor binding protein (IGFBP)-4 and IGFBP-5 in children with chronic renal failure: relationship to growth and glomerular filtration rate. The European Study Group for Nutritional Treatment of Chronic renal failure in Childhood. German Study Group for Growth hormone Treatment in Chronic renal failure. Pediatr Nephrol 2000; 14:589-97.
64. Svensson J, Ohlsson C, Jansson JO et al. Treatment with the oral growth hormone secretagogue MK-677 increases markers of bone formation and bone resorption in obese young males. J Bone Miner Res 1998; 13:1158-66.
65. Rosenbloom AL, Rosenfeld RG, Guevara-Aguirre J. Growth hormone insensitivity. Pediatr Clin North Am 1997; 44:423-42.
66. Camacho-Hubner C, Savage MO, Baylink DJ et al. Evidence that Growth hormone Regulation of IGFBP-5 is IGF-1 Independent: Studies from a patient with partial deletion of the IGF-1 gene. Growth hormone and IGF Research 1999; 9:370.
67. Twigg SM, Kiefer MC, Zapf J et al. Insulin-like growth factor-binding protein 5 complexes with the acid-labile subunit. Role of the carboxyl-terminal domain. J Biol Chem 1998; 273:28791-8.
68. Miyakoshi N, Richman C, Kasukawa Y et al. Evidence that IGF-binding protein-5 functions as a growth factor. J Clin Invest 2001; 107:73-81.
69. Ohlsson C, Bengtsson BA, Isaksson OG et al. Growth hormone and bone. Endocr Rev 1998; 19:55-79.
70. Guevara-Aguirre J, Rosenbloom AL, Vasconez O et al. Two-year treatment of growth hormone (GH) receptor deficiency with recombinant insulin-like growth factor I in 22 children: comparison of two dosage levels and to GH-treated GH deficiency. J Clin Endocrinol Metab 1997; 82:629-33.
71. Mohan S, Linkhart T, Farley J et al. Bone-derived factors active on bone cells. Calcif Tissue Int 1984; 36:S139-45.
72. Mohan S, Baylink DJ. Serum insulin-like growth factor binding protein (IGFBP)-4 and IGFBP-5 levels in aging and age-associated diseases. Endocrine 1997; 7:87-91.

73. Nicolas V, Mohan S, Honda Y et al. An age-related decrease in the concentration of insulin-like growth factor binding protein-5 in human cortical bone. Calcif Tissue Int 1995; 57:206-12.
74. Boonen S, Mohan S, Dequeker J et al. Down-regulation of the serum stimulatory components of the insulin-like growth factor (IGF) system (IGF-I, IGF-II, IGF binding protein [BP]-3, and IGFBP-5) in age-related (type II) femoral neck osteoporosis. J Bone Miner Res 1999; 14:2150-8.
75. Jehle PM, Ostertag A, Schulten K et al. Insulin-like growth factor system components in hyperparathyroidism and renal osteodystrophy. Kidney Int 2000; 57:423-36.
76. Libanati CR, Messina D, Lancioni G et al. Differential Effects of Deflaxacort Versus Prednisone On Bone Mineral Density, Bone formation, and Serum IGFS. Journal of Bone and Mineral Research Volume 1997; 12:S336.
77. Andress DL, Birnbaum RS. Human osteoblast-derived insulin-like growth factor (IGF) binding protein-5 stimulates osteoblast mitogenesis and potentiates IGF action. J Biol Chem 1992; 267:22467-72.
78. Andress DL, Loop SM, Zapf J et al. Carboxy-truncated insulin-like growth factor binding protein-5 stimulates mitogenesis in osteoblast-like cells. Biochem Biophys Res Commun 1993; 195:25-30.
79. Mohan S, Nakao Y, Honda Y et al. Studies on the mechanisms by which insulin-like growth factor (IGF) binding protein-4 (IGFBP-4) and IGFBP-5 modulate IGF actions in bone cells. J Biol Chem 1995; 270:20424-31.
80. Kiepe D, Andress DL, Mohan S et al. Intact IGF-binding protein-4 and -5 and their respective fragments isolated from chronic renal failure serum differentially modulate IGF-I actions in cultured growth plate chondrocytes. J Am Soc Nephrol 2001; 12:2400-10.
81. Schmid C, Schlapfer I, Gosteli-Peter MA et al. Effects and fate of human IGF-binding protein-5 in rat osteoblast cultures. Am J Physiol 1996; 271:E1029-35.
82. Andress DL, Birnbaum RS. A novel human insulin-like growth factor binding protein secreted by osteoblast-like cells [published erratum appears in Biochem Biophys Res Commun 1991 Jun 14;177(2):895]. Biochem Biophys Res Commun 1991; 176:213-8.
83. Jones JI, Gockerman A, Busby WH Jr et al. Extracellular matrix contains insulin-like growth factor binding protein-5: potentiation of the effects of IGF-I. J Cell Biol 1993; 121:679-87.
84. Richman C, Baylink DJ, Lang K et al. Recombinant human insulin-like growth factor-binding protein-5 stimulates bone formation parameters in vitro and in vivo. Endocrinology 1999; 140:4699-705.
85. Schneider MR, Zhou R, Hoeflich A et al. Insulin-like growth factor-binding protein-5 inhibits growth and induces differentiation of mouse osteosarcoma cells. Biochem Biophys Res Commun 2001; 288:435-42.
86. Slootweg MC, Ohlsson C, van Elk EJ et al. Growth hormone receptor activity is stimulated by insulin-like growth factor binding protein 5 in rat osteosarcoma cells. Growth Regul 1996; 6:238-46.
87. Bauss F, Lang K, Dony C et al. The complex of recombinant human insulin-like growth factor-I (rhIGF-I) and its binding protein-5 (IGFBP-5) induces local bone formation in murine calvariae and in rat cortical bone after local or systemic administration. Growth Horm IGF Res 2001; 11:1-9.
88. Devlin RD, Du Z, Jorgetti V et al. Insulin-like Growth Factor Binding protein (IGFBP)-5 Over-expression Decreases Bone formation in vivo. Journal of Bone and Mineral Research Volume 2001; 16:S200.
89. Jongen JW, Willemstein-van Hove EC, van der Meer JM et al. Down-regulation of the receptor for parathyroid hormone (PTH) and PTH- related peptide by PTH in primary fetal rat osteoblasts. J Bone Miner Res 1996; 11:1218-25.
90. Verrecchia F, Tacheau C, Schorpp-Kistner M et al. Induction of the AP-1 members c-Jun and JunB by TGF-beta/Smad suppresses early Smad-driven gene activation. Oncogene 2001; 20:2205-11.
91. Andress DL. Insulin-like growth factor-binding protein-5 (IGFBP-5) stimulates phosphorylation of the IGFBP-5 receptor. Am J Physiol 1998; 274:E744-50.
92. Radulescu RT. Nuclear localization signal in insulin-like growth factor-binding protein type 3. Trends Biochem Sci 1994; 19:278.
93. Schedlich LJ, Le Page SL, Firth SM et al. Nuclear import of insulin-like growth factor-binding protein-3 and -5 is mediated by the importin beta subunit. J Biol Chem 2000; 275:23462-70.
94. Amaar Y, Thompson G, Linkhart TA et al. FHL-2, A LIM Only Transcription factor, Is a Binding Partner for IGF binding protein (BP)-5 in Normal Human Osteoblasts (OB). Journal of Bone and Mineral Research 2001; 16:S175.
95. Muller JM, Isele U, Metzger E et al. FHL2, a novel tissue-specific coactivator of the androgen receptor. Embo J 2000; 19:359-69.
96. Boden SD, Liu Y, Hair GA et al. LMP-1, a LIM-domain protein, mediates BMP-6 effects on bone formation. Endocrinology 1998; 139:5125-34.
97. Miyakoshi N, Qin X, Kasukawa Y et al. Systemic administration of insulin-like growth factor (IGF)-binding protein-4 (IGFBP-4) increases bone formation parameters in mice by increasing IGF bioavailability via an IGFBP-4 protease-dependent mechanism. Endocrinology 2001; 142:2641-8.
98. Libanati C, Baylink DJ, Lois-Wenzel E et al. Studies on the potential mediators of skeletal changes occurring during puberty in girls. J Clin Endocrinol Metab 1999; 84:2807-14.
99. Mohan S, Libanati C, Dony C et al. Development, validation, and application of a radioimmunoassay for insulin-like growth factor binding protein-5 in human serum and other biological fluids. J Clin Endocrinol Metab 1995; 80:2638-45.
100. Baxter RC, Meka S, Firth SM. Molecular Distribution of IGF binding protein-5 in Human Serum. J Clin Endocrinol Metab 2002; 87:271-6.
101. Naylor KE, Iqbal P, Fledelius C et al. The effect of pregnancy on bone density and bone turnover. J Bone Miner Res 2000; 15:129-37.

CHAPTER 13

IGF Action and Skeletal Muscle

David T. Kuninger and Peter S. Rotwein

Abstract

The insulin-like growth factors, IGF-I and IGF-II, comprise a pair of structurally related, secreted proteins which control diverse cellular functions by regulating multiple signal transduction pathways. Gene targeting experiments have revealed essential roles for IGF action in normal muscle growth and development in vivo. In cultured muscle cells, IGFs can promote proliferation, sustain cell viability, and stimulate differentiation, depending on the cellular context. This chapter highlights recent information on IGF action in muscle and focuses on both biochemical signal transduction pathways and molecular mechanisms.

Introduction

The insulin-like growth factors (IGFs), IGF-I and IGF-II, are multifunctional growth factors that affect a range of biological processes in vivo and in vitro. By activating the cell surface type-I receptor (IGF-IR), both IGFs elicit effects on cell growth, development and metabolism. In skeletal muscle, IGF action is critical for normal growth and development. This chapter discusses the biological roles of IGF-I and IGF-II in skeletal muscle and examines recent information on potential signal transduction pathways and molecular mechanisms of IGF action in myoblasts and differentiated muscle.

Basics of Skeletal Muscle Development

Skeletal muscle in vertebrates, with the exception of most head muscles, is derived from progenitor cells present in somites. Somite formation occurs relatively early in embryonic development and requires a series of inductive cues derived from surrounding tissues, including the neural tube and notochord.[1,2] As differentiation proceeds, somites subsequently become divided into a dorsal dermomyotome, containing skin and muscle precursors, and a ventral sclerotome, containing precursors of cartilage and bone. Myogenic commitment requires induction of two muscle-specific transcription factors, Myf-5 and MyoD, as evidenced by gene ablation experiments in mice.[3] For a comprehensive review of the earliest steps in skeletal muscle commitment see references 1, 4.

In the embryo, two families of transcription factors play key roles in the developmental steps that culminate in muscle differentiation. The MyoD family consists of four proteins, MyoD, myogenin, MRF4, and Myf5. The MEF2 family also is composed of four members, MEF2 A, B, C, and D.[5] MyoD proteins are critical for muscle development in the embryo. They are structurally related to one another, and share amino acid similarities in two contiguous regions, the basic helix-loop-helix (bHLH) domains, that are necessary respectively, for DNA binding and interactions with ubiquitously expressed bHLH E proteins. All four MyoD family members are capable of inducing conversion of fibroblasts and other cell types to skeletal muscle when overexpressed in tissue culture cells.[6-12] MyoD family proteins are able to stimulate the transcription of muscle-specific genes after binding to DNA elements termed E boxes as heterodimers with bHLH E-proteins.[5] Targeted gene knockout studies in mice have demonstrated that MyoD and Myf5 together are essential for muscle determination, while myogenin controls differentiation of committed myoblasts.[5] In contrast, MRF4 appears to play an auxiliary role in differentiation, as its targeted inactivation resulted in grossly normal skeletal muscle, potentially because of up-regulation of myogenin expression.[13]

Insulin-Like Growth Factors, edited by Derek LeRoith, Walter Zumkeller and Robert Baxter. ©2003 Eurekah.com and Kluwer Academic / Plenum Publishers.

Unlike the MyoD family, MEF2 proteins are not expressed exclusively in skeletal muscle but are also found in heart muscle, brain, and other tissues.[14] They bind as dimers to A/T rich DNA sequences also found in the promoters of many muscle-specific genes. MEF2 binding can potentiate the transcriptional activity of MyoD proteins interacting with the same gene promoters. Loss of individual MEF2 proteins by targeted gene knock-out in muscle did not perturb skeletal muscle development, although disruption of the MEF2C gene caused embryonic death secondary to severe defects in cardiac morphogenesis.[15]

Many other proteins participate in control of gene expression in muscle. Transcriptional co-activators and co-repressors play modulatory roles in regulating the activity of members of the MyoD and MEF2 families as well as other transcription factors. The transcriptional co-activators p300/CBP and P/CAF collaborate with MyoD in enhancing muscle differentiation in cell culture models.[16,17] To date these observations have not been extended to muscle development or repair in vivo. Similarly, several transcriptional co-repressors have been found to inhibit actions of MEF2 and MyoD in cultured muscle cells.[18-21] Their effects also have not been evaluated yet in muscle in vivo.

Regulation and Expression of the IGF System in Skeletal Muscle

Components of the IGF System

The IGF system consists of two ligands, IGF-I and IGF-II, two receptors, the IGF-IR and the IGF-IL receptor (IGF-IIR), and six high-affinity binding proteins, IGFBPs 1-6.[22,23] The biological effects of both IGFs are mediated following binding to the IGF-IR, a transmembrane ligand-activated tyrosine protein kinase related to the insulin receptor.[22-24] In contrast, the IGF-IIR functions in the sequestration and degradation of IGF-II. This latter receptor also plays a major role in the targeting of lysosomal enzymes.[22] The six high-affinity IGFBPs bind IGF-I and IGF-II with affinities greater than those of the receptors, and function primarily to modulate IGF action by either promoting or restricting access to receptors or providing a storage pool of growth factors in the blood or extracellular fluids.[25] Recent studies have suggested that IGFBPs may additionally exert IGF-independent actions.[26,27] More extensive discussion of components of the IGF system may be found in other chapters in this volume.

IGF Action in Muscle—Lessons from in Vivo Model Systems

All components of the IGF system are expressed in skeletal muscle [28] and a variety of manipulations in experimental animals have confirmed important roles for IGF action in the normal growth and development of skeletal muscle in vivo. Mice engineered to lack the IGF-IR or deficient in both IGF-I and IGF-II showed generalized growth impairment during fetal life (45% and 30% normal birthweight, respectively), and exhibited marked muscle hypoplasia.[29,30] These animals died of respiratory failure in the perinatal period because the profound muscle weakness prevented normal inflation of the lungs.[29,30] The muscle hypoplasia observed in IGF-IR deficient mice was secondary to a decreased number of mature myofibers, but it has not been established whether this occurred secondary to impaired differentiation or to diminished myocyte viability. While the results discussed above demonstrate roles for IGF action in muscle growth and maturation, an interpretation of potential mechanisms is complicated by the disruption of IGF production and signaling in other tissues which may contribute to effects observed in muscle.

In order to specifically address the role of IGF action in skeletal muscle, transgenic mice have been engineered to express different components of the IGF system in muscle through the use of muscle-specific gene promoters. Using this strategy, over-expression of IGF-I in skeletal muscle has been shown to promote myofiber hypertrophy and to minimize the loss of muscle mass that occurs normally in mice during aging.[31-34] Conversely, muscle-specific impairment of IGF signaling by a dominant-negative form of the IGF-IR resulted in skeletal muscle hypoplasia early in postnatal life, which somewhat surprisingly, was followed by muscle hyperplasia as these mice grew to adulthood.[35] The molecular mechanisms of these in vivo actions have not been defined.

Local control of IGF gene and protein expression in muscle has been demonstrated in a variety of in vivo situations. In rats, IGF-I mRNA and protein levels increased in muscle and in satellite cells (muscle precursors) following ischemic or toxic injury.[36-38] Additionally, satellite cells isolated from mice over-expressing a muscle-targeted IGF-I transgene demonstrated enhanced in vitro replicative lifespan.[39] This increased proliferative potential was shown to be mediated by the PI3-kinase-Akt pathway (discussed in later sections). IGF-I and IGF-II were induced in rats during skeletal muscle regeneration following muscle fiber degeneration caused by bupivacaine.[40] Increases in IGF-I and IGF-II mRNA also accompanied work-induced muscle hypertrophy in rats, even in animals lacking growth hormone and thyroid hormone secondary to hypophysectomy.[41] Local over-expression of IGF-I in muscle also enhanced muscle hypertrophy in mice, indicating that IGF production within muscle facilitates this biological response.[34] Additionally, in fetal rats, increased IGF-II expression was seen during re-innervation of denervated skeletal muscle.[42] Taken together, these studies demonstrate local regulation of IGF expression during muscle growth and regeneration. These observations also imply roles for IGF action in muscle in response to a variety of cellular insults, although results to date have not identified specific mechanisms.

IGF Action in Muscle—Signaling Pathways and Molecular Mechanisms

Experiments in cultured muscle cells have demonstrated IGF actions on myoblast proliferation, viability, or differentiation depending on the cellular context. These potentially mutually exclusive outcomes appear to result from activation of different signal transduction pathways. In the following sections evidence will be presented supporting roles for different signaling pathways in muscle cell proliferation, survival, and differentiation.

IGF Action in Myoblast Proliferation

Like other peptide growth factors, IGFs are able to elicit mitogenic responses in a wide variety of cell types, including skeletal muscle.[43] As shown in Figure 1, IGF-I is as effective as PDGF in promoting proliferation of subconfluent myoblasts, although neither growth factor was able to stimulate proliferation beyond confluent cell density.

The use of specific pharmacological agents, which block different signaling pathways, has been instrumental in identifying pathways responsible for IGF-stimulated muscle cell proliferation. In both rat and mouse myoblasts, treatment of cells with inhibitors of MEK 1 and 2 blocked IGF-I mediated replication,[44,45] as depicted in Figure 2. These compounds prevent activation of the MAP kinases, Erks 1 and 2.[46] Events associated with cell proliferation, such induction of genes encoding c-fos and cyclin D, were also inhibited. Consistent with these observations, activation of Erks by over-expressed Raf-1 promoted proliferation of rat myoblasts in the absence of IGF-I.[47] In contrast, inhibitors of PI3-kinase did not significantly impair IGF-mediated myoblast replication.[44,47] Taken together, these results identify a prominent role for the MAP kinase pathway in mediating IGF-stimulated muscle cell proliferation.

IGF Action in Muscle Cell Viability

In addition to their ability to stimulate muscle cell proliferation, IGFs are able to sustain cell survival under conditions that would otherwise lead to apoptotic cell death. A direct role for IGF action in maintaining muscle cell viability has been shown using mouse myoblasts in which the expression of endogenous IGF-II was prevented.[48] These cells underwent progressive apoptotic cell death in low serum differentiation medium, but remained viable after addition of IGF-II or other IGF analogues which activated the IGF-IR. IGF-II also was able to maintain the viability of myoblasts derived from a mouse model of muscular dystrophy.[49] As illustrated in Figure 3, incubation of confluent IGF-II deficient myoblasts with IGF-I or PDGF sustained cell survival, while progressive myoblast death occurred in the absence of growth factor treatment. Subsequent studies identified the signaling intermediates PI3-kinase and Akt (also known as PKB) as being critical for IGF-mediated muscle cell survival while activation of the MEK-Erk pathway was required for PDGF-stimulated survival.[50,51]

Active Akt prevents programmed cell death in different cell types by a variety of mechanisms.[52,53] Akt can phosphorylate and inactivate several pro-apoptotic proteins including Bad,[54] caspase-9,[55] and the forkhead transcription factor FKHR-L1.[56] Akt also inhibits programmed cell death induced by several pro-apoptotic members of the Bcl-2 family by blocking the release of cytochrome c from mitochondria.[57]

In skeletal myoblasts, current evidence suggests that Akt uses another mechanism to promote cell survival. It induces the expression of the cyclin-dependent kinase inhibitor p21 (also known as

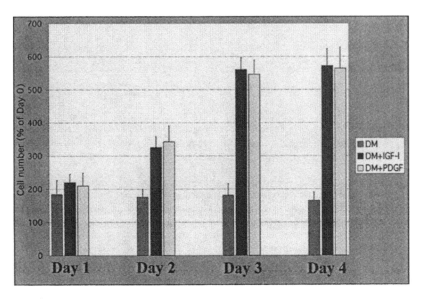

Figure 1. Growth factor-mediated myoblast proliferation. Subconfluent IGF-II deficient myoblasts were incubated in low serum differentiation medium (DM) with or without Des 1-3 IGF-I (2 nM) or PDGF-BB (0.4 nM) and cell number was measured for 4 days. Solid bars represent mean cell number expressed as a % of Day 0 +/- standard deviation. Note that neither factor stimulated growth beyond confluent density, which was reached on Day 3.

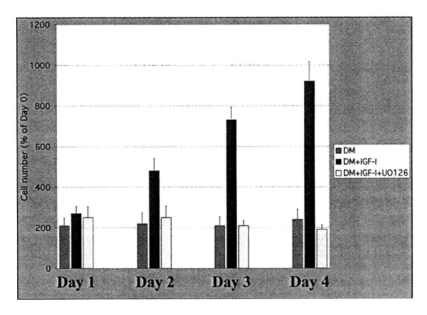

Figure 2. Pharmacological inhibition of MAP kinases blocks IGF-mediated myoblast proliferation. Subconfluent IGF-II deficient myoblasts were incubated in low serum differentiation medium (DM) alone or containing Des 1-3 IGF-I (2 nM) with or without UO126 (10 µM), a direct inhibitor of MEKs 1 and 2; OU126 thus prevents the activation of Erk 1 and 2. UO126 completely blocked IGF-mediated proliferation. Solid bars represent mean cell number expressed as % of Day 0 +/- standard deviation.

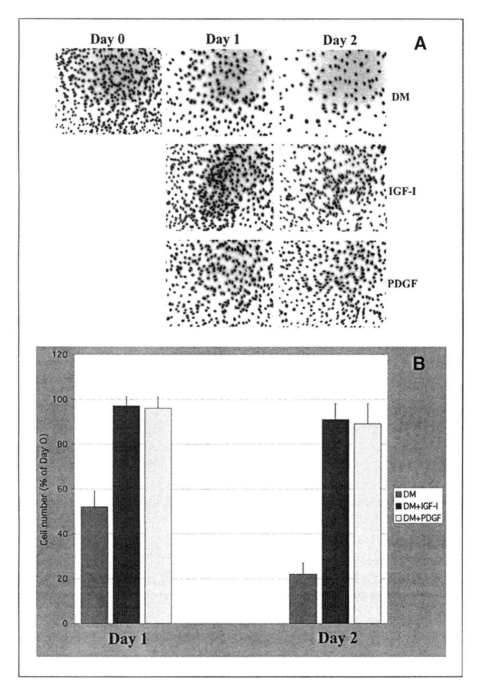

Figure 3. Growth factor-mediated myoblast cell survival. Confluent IGF-II deficient myoblasts were incubated in low serum differentiation medium (DM) supplemented with Des 1-3 IGF-I (2 nM) or PDGF-BB (0.4 nM) up to 2 days. A. Representative images of cells stained with a fluorescent nuclear dye (Hoechst) taken at 200x magnification. B. Quantification of myoblast survival, solid bars represent mean cell number expressed as a % of Day 0 +/- standard deviation. Note that both factors maintained near complete cell viability.

Wafl or Cipl). In addition, forced expression of p21 was found to maintain viability in the absence of IGF action while inhibition of p21 prevented IGF or Akt-mediated myoblast survival.[50,51] In developing mice, targeted deletion of the p21 gene and the related cyclin-dependent kinase inhibitor, p57, impaired muscle differentiation and promoted myoblast apoptosis,[58] thus confirming results in cultured muscle cells, where p57 is not expressed.[58] MyoD also can induce expression of p21 in differentiating myoblasts [59,60,61] but by a pathway distinct from that used by IGF-I or Akt,[50] indicating that redundant mechanisms act at the level of the p21 gene to mediate muscle cell survival.

IGF Signaling Pathways and Muscle Differentiation

Muscle differentiation in culture can be divided into several phases. The early events include permanent withdrawal from the cell cycle, and expression of the myogenic transcription factor, myogenin. Subsequent steps include induction of mRNAs and proteins encoding muscle structural proteins and muscle-specific enzymes. The final phase involves assembly of the multiprotein contractile apparatus and fusion of differentiating myoblasts to form multinucleated myotubes. IGF signaling appears to be involved in each of these stages. Illustrated in Figure 4 is a time course study using a skeletal muscle cell line dependent on exogenous growth factor stimulation for viability and differentiation. In this cell line PDGF maintains survival while IGF-I promotes survival and stimulates differentiation,[48,62] as evidenced in the figure by expression of the contractile protein, myosin heavy chain, and formation of myotubes.

Observations from several laboratories support the idea that endogenous expression of IGF-II and activation of IGF signaling pathways is required for early events in skeletal muscle differentiation in cell culture. IGF-II is produced by many myoblast cell lines as they differentiate,[63] secondary to stimulation of IGF-II gene transcription [64] and there is a correlation between the rate and extent of IGF-II expression and the rate of differentiation.[63] In contrast, inhibition of IGF-II impaired expression of myogenin and prevented differentiation.[50] In addition, inhibiting endogenous myogenin expression blocked IGF-mediated differentiation, demonstrating that induction of myogenin is a key step in the early actions of IGF signaling in muscle differentiation.[65] Current evidence indicates that IGF-mediated stimulation of myogenin gene expression involves the PI3-kinase-Akt pathway [66,67] although the precise molecular mechanisms have not been elucidated. Consistent with a critical role for IGF signaling in early events of differentiation, functional inhibition of the IGF-IR, using a dominant-negative approach, resulted in delayed differentiation and reduced myogenin expression in murine myoblasts.[68] Conversely, over-expression of either IGF-I, IGF-II, or the IGF-IR in cultured myoblasts caused accelerated and enhanced differentiation,[32,69-71] which in at least one experimental model correlated with early up-regulation of myogenin.[70]

Intermediate events in muscle differentiation include induction of genes encoding intermediate filaments (e.g., α skeletal actin), components of the contractile apparatus (e.g., troponins, desmin, myosin heavy and light chains) and muscle-specific enzymes (e.g., muscle creatine kinase). Experiments in which components of the IGF system have been overexpressed or functionally impaired support roles for IGF signaling in the intermediate aspects of myoblast differentiation but have not defined the specific biochemical pathways or mechanisms. For example, in C2C12 myoblasts, stimulation with IGF-I or stable expression of a IGF-I transgene resulted in increased expression of desmin and a skeletal actin mRNA compared with control myoblasts.[32] Stable expression of the IGF-IR in C2 murine myoblasts led to enhanced troponin T expression in the absence of exogenous IGF-I.[71] It remains possible that at least some of these actions may be attributed to the induction of myogenin by IGF signaling pathways and subsequent stimulation of muscle-specific genes by this transcription factor.

IGF signaling also appears to play a role in the final steps of muscle differentiation, myoblast fusion and formation of multinucleated myofibers. Several reports have demonstrated that ectopic overexpression of IGF-I or addition of IGF-I to the cell medium enhanced fusion and promoted myotube hypertrophy in cell culture.[72-75] While the hypertrophy-promoting effects of IGF stimulation have been consistently demonstrated, there is disagreement about the mechanisms or signaling pathways involved. For example, activation of calcium/calmodulin-dependent phosphatase, calcineurin, has been demonstrated in IGF-induced myotube hypertrophy.[72,73] A direct role for calcineurin was shown using pharmacological inhibition of its activation with the compound cyclosporin A, which blunted IGF-mediated myoblast hypertrophy.[72,73] In addition, a constituitively active form of calcineurin induced myoblast hypertrophy in the absence of added growth factors.[73]

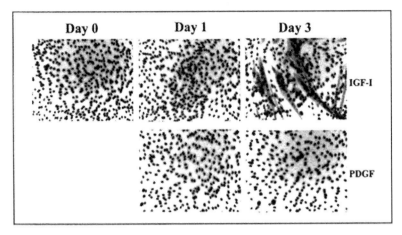

Figure 4. IGF action promotes myoblast differentiation. Confluent IGF-II deficient myoblasts were incubated in low serum differentiation medium (DM) containing either Des 1-3 IGF-I (2 nM) or PDGF-BB (0.4 nM) up to 3 days. Each panel shows a representative immunocytochemical images of cells double stained with a fluorescent nuclear dye (Hoechst) and with an antibody for the muscle-specific protein, myosin heavy chain (MHC). The formation of MHC positive, multinucleated myofibers is only seen in IGF-treated cells (Day 3).

However, other reports suggest that IGF-mediated muscle hypertrophy resulted from activation of the PI3-kinase—Akt pathway.[75,76] In these latter studies, IGF-induced myotube hypertrophy was impaired by the PI3-kinase inhibitor LY294002, and ectopic expression of an active form of Akt promoted hypertrophy in the absence of IGF action.[75] Interestingly, in this latter model neither over expression of calcineurin nor its inhibition altered myotube hypertrophy,[75] although cyclosporin A appeared to diminish overall myotube formation. Taken together, these results demonstrate key roles for IGF action in the last steps of muscle differentiation but do not clarify the molecular mechanisms required for cell fusion or myotube formation.

Perspectives

This chapter has outlined how IGF signaling pathways regulate different aspects of muscle biology. It is clear, from both in vivo and in vitro experiments, that IGF action is critical for normal muscle growth and development. Recent studies also have demonstrated a potential effect of IGF-I in maintaining muscle mass during aging.[33] For IGF-I or derivatives to become effective therapeutic agents for muscle disease, further investigation must focus on the specific biochemical pathways and molecular mechanisms through which IGF action maintains myoblast viability, and promotes and sustains muscle differentiation. A key challenge for the future thus will be to define and integrate these pathways, and then use this information to develop safe and effective treatments.

References

1. Buckingham M. Skeletal muscle formation in vertebrates. Curr Opin Genet Dev 2001; 11(4):440-8.
2. Perry RL, Rudnick MA. Molecular mechanisms regulating myogenic determination and differentiation. Front Biosci 2000; 5:D750-67.
3. Rudnicki MA, Schnegelsberg PN, Stead RH et al. MyoD or Myf-5 is required for the formation of skeletal muscle. Cell 1993; 75(7):1351-9.
4. Saga Y, Takeda H. The making of the somite: molecular events in vertebrate segmentation. Nat Rev Genet 2001; 2(11):835-45.
5. Sabourin LA, Rudnicki MA. The molecular regulation of myogenesis. Clin Genet 2000; 57(1):16-25.
6. Davis RL, Weintraub H, Lassar AB. Expression of a single transfected cDNA converts fibroblasts to myoblasts. Cell 1987; 51(6):987-1000.
7. Braun T, Buschhausen-Denker G, Bober E et al. A novel human muscle factor related to but distinct from MyoD1 induces myogenic conversion in 10T1/2 fibroblasts. EMBO J 1989; 8(3):701-9.
8. Wright WE, Sassoon DA, Lin VK. Myogenin, a factor regulating myogenesis, has a domain homologous to MyoD. Cell 1989; 56(4):607-17.

9. Edmondson DG, Olson EN. A gene with homology to the myc similarity region of MyoD1 is expressed during myogenesis and is sufficient to activate the muscle differentiation program. Genes Dev 1989; 3(5):628-40.
10. Rhodes SJ, Konieczny SF. Identification of MRF4: a new member of the muscle regulatory factor gene family. Genes Dev 1989; 3(12B):2050-61.
11. Braun T, Bober E, Winter B et al. Myf-6, a new member of the human gene family of myogenic determination factors: evidence for a gene cluster on chromosome 12. EMBO J 1990; 9(3):821-31.
12. Miner JH, Wold B. Herculin, a fourth member of the MyoD family of myogenic regulatory genes. Proc Natl Acad Sci USA 1990; 87(3):1089-93.
13. Zhang W, Behringer RR, Olson EN. Inactivation of the myogenic bHLH gene MRF4 results in up-regulation of myogenin and rib anomalies. Genes Dev 1995; 9(11):1388-99.
14. Black BL, Olson EN. Transcriptional control of muscle development by myocyte enhancer factor-2 (MEF2) proteins. Annu Rev Cell Dev Biol 1998; 14:167-96.
15. Lin Q, Schwarz J, Bucana C et al. Control of mouse cardiac morphogenesis and myogenesis by transcription factor MEF2C. Science 1997; 276(5317):1404-7.
16. Sartorelli V, Puri PL, Hamamori Y et al. Acetylation of MyoD directed by PCAF is necessary for the execution of the muscle program. Mol Cell 1999; 4(5):725-34.
17. Polesskaya A, Naguibneva I, Fritsch L et al. CBP/p300 and muscle differentiation: no HAT, no muscle. EMBO J 2001; 20(23):6816-25.
18. McKinsey TA, Zhang CL, Olson EN. Activation of the myocyte enhancer factor-2 transcription factor by calcium/calmodulin-dependent protein kinase-stimulated binding of 14-3-3 to histone deacetylase 5. Proc Natl Acad Sci USA 2000; 97(26):14400-5.
19. Dressel U, Bailey PJ, Wang SC et al. A dynamic role for HDAC7 in MEF2-mediated muscle differentiation. J Biol Chem 2001; 276(20):17007-13.
20. Singleton JR, Feldman EL. Insulin-like growth factor-I in muscle metabolism and myotherapies. Neurobiol Dis 2001; 8(4).541-54.
21. Puri PL, Iezzi S, Stiegler P et al. Class I histone deacetylases sequentially interact with MyoD and pRb during skeletal myogenesis. Mol Cell 2001; 8(4):885-97.
22. Jones JI, Clemmons DR. Insulin-like growth factors and their binding proteins: biological actions. Endocr Rev 1995; 16(1):3-34.
23. Stewart CE, Rotwein P. Growth, differentiation, and survival: multiple physiological functions for insulin-like growth factors. Physiol Rev 1996; 76(4):1005-26.
24. Baserga R, Resnicoff M, D'Ambrosio C, Valentinis B. The role of the IGF-I receptor in apoptosis. Vitam Horm 1997; 53:65-98.
25. Clemmons DR. Role of insulin-like growth factor binding proteins in controlling IGF actions. Mol Cell Endocrinol 1998; 140(1-2):19-24.
26. Schneider MR, Wolf E, Hoeflich A et al. IGF-binding protein-5: flexible player in the IGF system and effector on its own. J Endocrinol 2002; 172(3):423-40.
27. Hong J, Zhang G, Dong F et al. Insulin-like growth factor binding protein-3 (IGFBP-3) mutants that do not bind IGF-I or IGF-II stimulate apoptosis in human prostate cancer cells. J Biol Chem 2002.
28. Florini JR, Ewton DZ, Coolican SA. Growth hormone and the insulin-like growth factor system in myogenesis. Endocr Rev 1996; 17(5):481-517.
29. Powell-Braxton L, Hollingshead P, Warburton C et al. IGF-I is required for normal embryonic growth in mice. Genes Dev 1993; 7(12B):2609-17.
30. Liu JP, Baker J, Perkins AS et al. Mice carrying null mutations of the genes encoding insulin-like growth factor I (Igf-1) and type 1 IGF receptor (Igf1r). Cell 1993; 75(1):59-72.
31. Barton-Davis ER, Shoturma DI, Musaro A et al. Viral mediated expression of insulin-like growth factor I blocks the aging-related loss of skeletal muscle function. Proc Natl Acad Sci USA 1998; 95(26):15603-7.
32. Coleman ME, DeMayo F, Yin KC et al. Myogenic vector expression of insulin-like growth factor I stimulates muscle cell differentiation and myofiber hypertrophy in transgenic mice. J Biol Chem 1995; 270(20):12109-16.
33. Musaro A, McCullagh K, Paul A et al. Localized Igf-1 transgene expression sustains hypertrophy and regeneration in senescent skeletal muscle. Nat Genet 2001; 27(2):195-200.
34. Paul AC, Rosenthal N. Different modes of hypertrophy in skeletal muscle fibers. J Cell Biol 2002; 156(4):751-60.
35. Fernandez AM, Dupont J, Farrar RP et al. Muscle-specific inactivation of the IGF-I receptor induces compensatory hyperplasia in skeletal muscle. J Clin Invest 2002; 109(3):347-55.
36. Edwall D, Schalling M, Jennische E et al. Induction of insulin-like growth factor I messenger ribonucleic acid during regeneration of rat skeletal muscle. Endocrinology 1989; 124(2):820-5.
37. Caroni P, Schneider C. Signaling by insulin-like growth factors in paralyzed skeletal muscle: rapid induction of IGF1 expression in muscle fibers and prevention of interstitial cell proliferation by IGF-BP5 and IGF-BP4. J Neurosci 1994; 14(5 Pt 2):3378-88.
38. Jennische E, Hansson HA. Regenerating skeletal muscle cells express insulin-like growth factor I. Acta Physiol Scand 1987; 130(2):327-32.
39. Chakravarthy MV, Abraha TW, Schwartz RJ et al. Insulin-like growth factor-I extends in vitro replicative life span of skeletal muscle satellite cells by enhancing G1/S cell cycle progression via the activation of phosphatidylinositol 3'-kinase/Akt signaling pathway. J Biol Chem 2000; 275(46):35942-52.
40. Marsh DR, Criswell DS, Hamilton MT et al. Association of insulin-like growth factor mRNA expressions with muscle regeneration in young, adult, and old rats. Am J Physiol 1997; 273(1Pt2):R353-8.
41. DeVol DL, Rotwein P, Sadow JL et al. Activation of insulin-like growth factor gene expression during work-induced skeletal muscle growth. Am J Physiol 1990; 259(1Pt1):E89-95.
42. Ishii DN. Relationship of insulin-like growth factor II gene expression in muscle to synaptogenesis. Proc Natl Acad Sci USA 1989; 86(8):2898-902.

43. Florini JR, Ewton DZ, McWade FJ. IGFs, muscle growth, and myogenesis. Diabetes Rev 1995; 3:73-92.
44. Coolican SA, Samuel DS, Ewton DZ et al. The mitogenic and myogenic actions of insulin-like growth factors utilize distinct signaling pathways. J Biol Chem 1997; 272(10):6653-62.
45. Milasincic DJ, Calera MR, Farmer SR et al. Stimulation of C2C12 myoblast growth by basic fibroblast growth factor and insulin-like growth factor 1 can occur via mitogen-activated protein kinase-dependent and -independent pathways. Mol Cell Biol 1996; 16(11):5964-73.
46. Davies SP, Reddy H, Caivano M et al. Specificity and mechanism of action of some commonly used protein kinase inhibitors. Biochem J 2000; 351(Pt1):95-105.
47. Samuel DS, Ewton DZ, Coolican SA et al. Raf-1 activation stimulates proliferation and inhibits IGF-stimulated differentiation in L6A1 myoblasts. Horm Metab Res 1999; 31(2-3):55-64.
48. Stewart CE, Rotwein P. Insulin-like growth factor-II is an autocrine survival factor for differentiating myoblasts. J Biol Chem 1996; 271(19):11330-8.
49. Smith J, Goldsmith C, Ward A et al. IGF-II ameliorates the dystrophic phenotype and coordinately down-regulates programmed cell death. Cell Death Differ 2000; 7(11):1109-18.
50. Lawlor MA, Rotwein P. Coordinate control of muscle cell survival by distinct insulin-like growth factor activated signaling pathways. J Cell Biol 2000; 151(6):1131-40.
51. Lawlor MA, Rotwein P. Insulin-like growth factor-mediated muscle cell survival: central roles for Akt and cyclin-dependent kinase inhibitor p21. Mol Cell Biol 2000; 20(23):8983-95.
52. Datta SR, Brunet A, Greenberg ME. Cellular survival: a play in three Akts. Genes Dev 1999; 13(22):2905-27.
53. Khwaja A. Akt is more than just a Bad kinase. Nature 1999; 401(6748):33-4.
54. Datta SR, Dudek H, Tao X et al. Akt phosphorylation of BAD couples survival signals to the cell-intrinsic death machinery. Cell 1997; 91(2):231-41.
55. Cardone MH, Roy N, Stennicke HR et al. Regulation of cell death protease caspase-9 by phosphorylation. Science 1998; 282(5392):1318-21.
56. Brunet A, Bonni A, Zigmond MJ et al. Akt promotes cell survival by phosphorylating and inhibiting a Forkhead transcription factor. Cell 1999; 96(6):857-68.
57. Kennedy SG, Kandel ES, Cross TK et al. Akt/Protein kinase B inhibits cell death by preventing the release of cytochrome c from mitochondria. Mol Cell Biol 1999; 19(8):5800-10.
58. Zhang P, Wong C, Liu D et al. p21(CIP1) and p57(KIP2) control muscle differentiation at the myogenin step. Genes Dev 1999; 13(2):213-24.
59. Guo K, Walsh K. Inhibition of myogenesis by multiple cyclin-Cdk complexes. Coordinate regulation of myogenesis and cell cycle activity at the level of E2F. J Biol Chem 1997; 272(2):791-7.
60. Guo K, Wang J, Andres V et al. MyoD-induced expression of p21 inhibits cyclin-dependent kinase activity upon myocyte terminal differentiation. Mol Cell Biol 1995; 15(7):3823-9.
61. Halevy O, Novitch BG, Spicer DB et al. Correlation of terminal cell cycle arrest of skeletal muscle with induction of p21 by MyoD. Science 1995; 267(5200):1018-21.
62. Lawlor MA, Feng X, Everding DR et al. Dual control of muscle cell survival by distinct growth factor-regulated signaling pathways. Mol Cell Biol 2000; 20(9):3256-65.
63. Florini JR, Magri KA, Ewton DZ et al. "Spontaneous" differentiation of skeletal myoblasts is dependent upon autocrine secretion of insulin-like growth factor-II. J Biol Chem 1991; 266(24):15917-23.
64. Kou K, Rotwein P. Transcriptional activation of the insulin-like growth factor-II gene during myoblast differentiation. Mol Endocrinol 1993; 7(2):291-302.
65. Florini JR, Ewton DZ. Highly specific inhibition of IGF-I-stimulated differentiation by an antisense oligodeoxyribonucleotide to myogenin mRNA. No effects on other actions of IGF-I. J Biol Chem 1990; 265(23):13435-7.
66. Xu Q, Wu Z. The insulin-like growth factor-phosphatidylinositol 3-kinase-Akt signaling pathway regulates myogenin expression in normal myogenic cells but not in rhabdomyosarcoma-derived RD cells. J Biol Chem 2000; 275(47):36750-7.
67. Tureckova J, Wilson EM, Cappalonga JL et al. Insulin-like growth factor-mediated muscle differentiation: collaboration between phosphatidylinositol 3-kinase-Akt-signaling pathways and myogenin. J Biol Chem 2001; 276(42):39264-70.
68. Cheng ZQ, Adi S, Wu NY et al. Functional inactivation of the IGF-I receptor delays differentiation of skeletal muscle cells. J Endocrinol 2000; 167(1):175-82.
69. Quinn LS, Steinmetz B, Maas A et al. Type-1 insulin-like growth factor receptor overexpression produces dual effects on myoblast proliferation and differentiation. J Cell Physiol 1994; 159(3):387-98.
70. Stewart CE, James PL, Fant ME et al. Overexpression of insulin-like growth factor-II induces accelerated myoblast differentiation. J Cell Physiol 1996; 169(1):23-32.
71. Navarro M, Barenton B, Garandel V et al. Insulin-like growth factor I (IGF-I) receptor overexpression abolishes the IGF requirement for differentiation and induces a ligand-dependent transformed phenotype in C2 inducible myoblasts. Endocrinology 1997; 138(12):5210-9.
72. Semsarian C, Wu MJ, Ju YK et al. Skeletal muscle hypertrophy is mediated by a Ca^{2+}-dependent calcineurin signalling pathway. Nature 1999; 400(6744):576-81.
73. Musaro A, McCullagh KJ, Naya FJ et al. IGF-1 induces skeletal myocyte hypertrophy through calcineurin in association with GATA-2 and NF-ATc1. Nature 1999; 400(6744):581-5.
74. Musaro A, Rosenthal N. Maturation of the myogenic program is induced by postmitotic expression of insulin-like growth factor I. Mol Cell Biol 1999; 19(4):3115-24.
75. Rommel C, Bodine SC, Clarke BA et al. Mediation of IGF-1-induced skeletal myotube hypertrophy by PI(3)K/Akt/mTOR and PI(3)K/Akt/GSK3 pathways. Nat Cell Biol 2001; 3(11):1009-13.
76. Rommel C, Clarke BA, Zimmermann S et al. Differentiation stage-specific inhibition of the Raf-MEK-ERK pathway by Akt. Science 1999; 286(5445):1738-41.

CHAPTER 14

Insulin-Like Growth Factor-I and the Kidney

Franz Schaefer and Ralph Rabkin

Abstract

This review describes the multiple ways in which the kidney and the IGF-I system interact in both health and disease. IGF-I is an important physiological regulator of glomerular hemodynamics, renal growth and certain tubular functions. In several disease states the IGF-I system is an important determinant of compensatory renal growth and renal hypertrophy. Growth hormone and IGF-I also have a pathophysiological role in progressive glomerular and tubulointerstitial fibrosis. Renal IGF-1 levels are regulated by locally produced IGF-I, but more importantly by modulation of intrarenal IGF-1 receptor and binding protein expression affecting the trapping of circulating IGF-1 in the kidney. The intrarenal expression of IGF-1, its receptor and individual binding proteins is highly organized with a spatially distinct distribution, permitting differential effects of IGF-1 on different segments of the nephron. Besides the multiple renotropic actions of IGF-1, the kidney itself is an important regulator of the endocrine growth hormone-IGF-1 system. The kidney is the major degradation site for circulating growth hormone, IGF-1 and several of its binding proteins. The altered endocrine and metabolic milieu associated with chronic renal failure results in a state of multilevel growth hormone (GH) and IGF-I resistance that can partially be overcome by administration of recombinant human GH and IGF-I in pharmacological doses. This has led to the therapeutic use of GH to promote linear growth in children, while other potential clinical applications of GH and IGF-I treatment remain under investigation.

Introduction

The insulin-like growth factor system consisting of IGF-I, the IGF-I receptors and the six major binding proteins, is expressed in the kidney in an anatomically heterogeneous manner[1-3] and as elsewhere in the body, growth hormone is a major regulator of IGF-I expression. Locally produced and circulating IGF-I have profound effects on renal structure and function including modulation of tubular phosphate and sodium transport, renal blood flow and glomerular hemodynamics and stimulating growth. Animal studies have suggested that IGF-I may play a role in renal regeneration after acute tubular necrosis[4] and that it may promote renal hypertrophy in a number of conditions including diabetes and compensatory renal growth.[5-7] A more sinister role for growth hormone/IGF-I has been suggested in the pathogenesis of kidney diaease.[8] In contrast to our extensive understanding of the actions of IGF-I, far less is known about the function of the insulin-like growth factor binding proteins (IGFBPs) that are produced in the kidney. It appears that these proteins serve to trap IGF-I in the kidney, modulate its bioactivity and may also exhibit some limited IGF-I-independent actions.[3] Growth hormone receptors are also expressed in various renal structures, and most but not all of the actions transduced through this receptor are ultimately mediated by increasing the expression of IGF-I. While the growth hormone/IGF-I system in the kidney has major effects on renal structure and function in health and disease, loss of kidney function in turn has a major effect on the IGF-1 system throughout the body most notably in growing bone, skeletal muscle, the liver and various circulating components of this system.

In this chapter we will review the physiology of the growth hormone/ IGF-I system in the kidney, the changes and potential role of this system in kidney diseases and the effects of kidney failure on the growth hormone/IGF-I system in general. Finally we will briefly discuss the potential and

Insulin-Like Growth Factors, edited by Derek LeRoith, Walter Zumkeller and Robert Baxter. ©2003 Eurekah.com and Kluwer Academic / Plenum Publishers.

established therapeutic uses of recombinant human growth hormone and IGF-I for the treatment of kidney failure and some of its consequences.

Expression and Regulation of GH-IGF System Components in the Kidney

Studies in rats and humans have revealed a highly organized expression of the different members of the growth hormone (GH)-insulin-like growth factor-I (IGF-I) system in the kidney, with a distinct spatial distribution of individual components among the different anatomical and functional segments of the nephron (Fig. 1). Both IGF-I and IGF-II are found in the renal cortex and medulla.

Since the IGF-I gene is expressed in the kidney, and as IGF-I concentrations are greater in renal venous than in arterial blood, it is apparent that the kidney synthesizes this growth factor.[9] The amount of IGF-I peptide extractable from the kidney is comparable to liver IGF-I, whereas IGF-I mRNA is at least 10 fold greater in liver compared to kidney.[10,11] This suggests that a major fraction of the IGF-I found in the kidney is derived from the circulation and is trapped in the kidney by binding to cell surface receptors, cell associated IGFBPs and IGFBPs in the interstitial compartment.

In the glomerulus there is expression of the mRNAs for IGF-I, the IGF-I receptor, IGF-II receptor and IGFBPs-2, -4 and -5.[1-3,12,13] The expression of IGF-I receptors, which is more dense in the glomerulus than in any other part of the nephron, provides the basis for the biological effects of IGF-I on glomerular structure and function. IGF-I mRNA is not normally expressed in the proximal tubule, though it may transiently be expressed after acute tubular cell injury.[14] While not

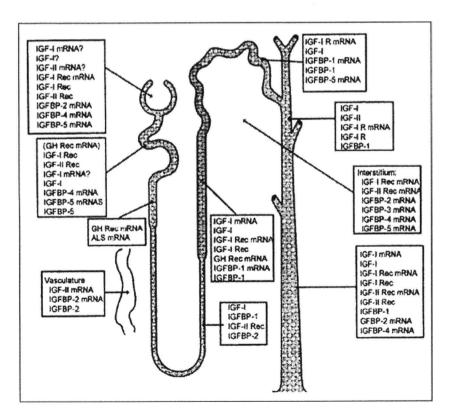

Figure 1. Expression of members of the GH/IGF-1 system in the rat nephron. Rec: receptor. Names with ? denote in vitro finding, not confirmed in vivo. The figure differentiates between proteins and mRNAs. The latter suggests in situ synthesis. Reproduced with permission from Feld & Hirschberg.[7]

synthesized in the proximal tubule, the IGF-I peptide is found both along the apical (luminal) and the basolateral membrane, colocalized with abundantly expressed IGF-I and IGF-II receptors.[2,3,14-16] This suggests that IGF-I derived from paracrine synthesis, from the peritubular circulation and from glomerular ultrafiltration exert physiological actions on proximal tubule cells. The GH receptor is also expressed in the proximal tubule.[1,17]

In the thick ascending limb of the loop of Henle of the rat the mRNAs for the GH receptor, IGF-I, IGF-1 receptor and IGFBP-1 are all expressed.[1,3,16] This colocalization suggests that circulating GH likely reacts with the GH receptors in this nephron segment and stimulates the local synthesis of IGF-I, which then may act through the IGF-I receptor in an autocrine or paracrine mode. Distal tubular cells do not synthesize IGF-1 or IGF-II, but express IGF-I receptor and IGFBP mRNA (BP-1 and -5 in rats, BP-2 in human kidney). Rat collecting ducts express mRNA for IGF-I and IGF-II receptors as well as for several IGF binding proteins (IGFBP-1 in rat cortical, BP-2 and -4 in rat medullary, IGFPB-2 in human cortical and medullary collecting duct).[1,3,14,16,18-20] IGF-I mRNA expression and/or cytoplasmic IGF-I peptide has been localized in the rat, but not human, medullary collecting duct.[14,16,18-23] Both IGF-I and IGF-II receptor, as well as IGFBP-2, -3, -4 and -5 mRNA are expressed in rat renal interstitium, probably by fibroblasts.[17,24,25] IGF-II mRNA and peptide expression has been demonstrated in human renal interstitial tissue.[26]

In summary, IGF-I receptors are expressed throughout the kidney, while IGF-I and IGF-II are expressed in a segmental manner, with the loop of Henle being the predominant site of expression. All six major IGFBPs are expressed in the kidneys, but in a strikingly heterogenous manner; they likely serve to trap IGF-I and modulate its bioactivity. IGFBPs may also possess intrinsic bioactivity though this does not appear to be a prominent property in the kidney. It is important to note that the IGF-I peptide found within the kidney is derived both from local synthesis and from the trapping of circulating IGF-I by the IGF-I receptors and local IGFBPs.

As anticipated in the rat, renal IGF-I expression is clearly regulated by GH,[11] which induces the IGF-I gene in collecting ducts and the medullary thick ascending limb of the loop of Henle.[22,23] Moreover, GH regulates the expression of some IGFBPs in the rat kidney; renal IGFBP-1 and -4 expression are inversely related to circulating GH concentrations.[25,27,28] Presumably GH also exerts some IGF-I independent direct actions via specific GH receptors in the proximal tubule, though many if not most effects of GH on the kidney appear to be mediated by local and systemic stimulation of IGF-I synthesis.

Nutritional status is another important regulatory element for the renal IGF system. In the rat, increased dietary protein intake positively regulates IGF-I mRNA and negatively regulates IGFBP-1 mRNA levels in the kidney as it does in the liver.[29,30] Hence, the kidney is affected by modifications of protein intake both via systemic and locally produced IGF-I. In addition to regulating IGF-I and IGFBP expression, reduced protein intake also upregulates IGF-I receptor and IGFBP-5 expression in proximal tubules. This in turn may serve to compensate for low IGF-I levels present in the protein restricted state. Potassium depletion and acidosis also affect the GH/IGF-I system.[31-33] As discussed later, both conditions stimulate renal growth while depressing body growth and this occurs in part by alterations in the GH/IGF-1 system. It has also been suggested that EGF, which in the kidney colocalizes with IGF-I in the thick ascending limb of Henle, may regulate IGF-I production.[34,35]

Renal Handling of GH, IGFs and IGFBPs

The glomerular capillary wall and basement membrane constitutes a size-selective ultrafiltration barrier, permitting free ultrafiltration of peptides less than 10 kD in size. Larger peptides and proteins are fractionally ultrafiltered according to their molecular size. Ultrafiltered proteins are reabsorbed in the proximal tubules via endocytosis and undergo endocytotic uptake and lysosomal degradation in tubular cells. Only minute fractions of filtered proteins are excreted in the urine.[36] With a molecular mass of 20 to 22 kD, growth hormone is ultrafiltered at the glomerulus with a sieving coefficient of 0.6 to 0.8[37-39] and is then reabsorbed by endocytosis in the proximal tubule.[40,41] This is an efficient process and the kidney serves as the major pathway of GH clearance from the circulation. In addition to being filtered at the glomerulus a small amount of circulating GH is taken up from the peritubular vessels and then presumably binds to receptors on the contraluminal aspect of the tubular cells.[40] However, since glomerular clearance is the major pathway of circulating GH removal, the metabolic clearance rate of GH is a linear function of the glomerular filtration rate.[38] In

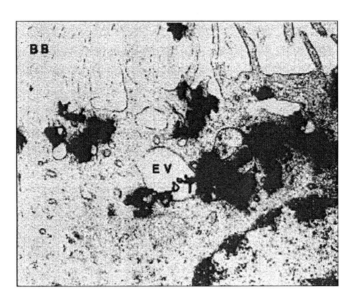

Figure 2. Endocytosis of IGF-1 from the lumen of the proximal renal tubule. Electron microscopic autoradiograph of luminal side of isolated rabbit proximal tubule (S2 segment) perfused with ^{125}I labeled IGF-1. Autoradiographic activity is located over endocytic vacuoles (EV) and lysosomes close to luminal brush border (BB). Reproduced with permission from Flyvbjerg et al.[43]

normal subjects the kidney is a major site of GH clearance from the circulation and accounts for about 50% of the metabolism of the circulating hormone.[38] This explains the increase in plasma half life and the elevated serum concentrations of GH seen in patients with chronic renal failure.

Although IGF-I and IGF-II are smaller molecules than GH (7.6 kD), glomerular ultrafiltration of IGF-I is extremely low under normal circumstances. This is explained by the fact that more than 98% of plasma IGF-I circulates bound to IGFBPs in complexes of 150 and 45 kD which because of their size are filtered through the glomerulus at much lower rates than free IGF-I peptide. Most of the bound IGF-I circulates complexed to IGFBP-3, that together with an acid-labile subunit forms a 150 kD complex.[42] A lesser amount of IGF-I circulates bound to IGFBP-1, -2, -4 and -6, and normally less than 2% is freely bioavailable. Reflecting their greater abundance in the serum, the filtered IGF binding proteins are predominantly IGFBP-1, IGFBP-2 and IGFBP-3 and appear in the urine in small quantities. Renal extraction and degradation of IGF-I occurs via glomerular filtration with receptor-mediated uptake of filtered IGF-I from the tubular lumen and also by uptake from peritubular capillaries through the basolateral tubular cell membranes (Fig. 2).[43] A small amount of the internalized IGF-I may also be transported into the cell nucleus.[44] Distribution studies in animals revealed that after an IV injection, labelled IGF-I reaches a higher concentration in the kidney compared to other organs.[45]

Interestingly, unlike the metabolic clearance of GH that is largely renal and thus impaired in CRF, the metabolic clearance rate of IGF-I is unaffected in kidney failure.[46,47] However because the IGFBPs levels are increased in the intravascular compartment in subjects with chronic renal failure, the volume of distribution of IGF-I is decreased and following the administration of IGF-I higher serum levels of IGF-I are achieved than in normals. Since the MCR of IGF-I is unaltered in CRF, it appears that the kidney may not be an important site of IGF-I disposal. However it is conceivable that other changes occurring in the uremic state such as an increase in skeletal muscle receptor number,[48] could cause an increase in extrarenal clearance compensating for the loss of renal clearance.

IGFBPs Have Direct and Indirect Actions on Kidney Function

As described earlier all six major IGFBPs are expressed in the kidney in an anatomically heterogenous distribution.[3] Since they have a similar or even higher affinity for IGF-1 than the IGF-1 receptor, the circulating and local IGFBPs form complexes with IGF-1. This influences the access of IGF-1 to the cells and thus modulates its bioactivity.[42] Most commonly the IGFBPs limit the immediate bioavailability of IGF-1and thus usually have a dampening effect on IGF-1 action. However, under selected conditions some IGFBP may actually enhance the activity of IGF-1 on some cell types or even initiate IGF-1-independent actions.[49-51] Despite their important role in modulating IGF-1 action, our understanding of their actions in the kidney is limited, and given the marked anatomical heterogeneity of IGFBP expression in the kidney these actions are likely to be diverse.[3]

Yap et al studied the effect of IGFBPs secreted by cultured rabbit proximal tubule cells on the mitogenic activity of IGF-1 and found that the overall effect is inhibitory.[52] Similarly, recombinant IGFBP-3, which has been shown to inhibit kidney cell IGF-I binding and internalization,[53] suppressed IGF-1-induced mitogenesis in cultured proximal tubular cells and in a proximal tubular cell line.[52] IGFBP-3 also inhibited DNA synthesis in the absence of added IGF-I, suggesting that IGFBP-3 may have an IGF-I independent action on kidney tubular cells. Interestingly, IGF-I and IGFBP-3 are both targeted to the nucleus of a cultured proximal tubular cell line, but only under conditions of cellular proliferation.[44] In resting cells IGF-I and IGFBP-3 are delivered to an endosomal compartment. These different sites of localization raise interesting questions as to the biological effects of IGF-I and IGFBP-3 on kidney tubular cells. Cultured mesangial cells are also affected by IGFBP-3 which accelerates apoptosis when growth factors are withdrawn from the culture medium.[54] Interestingly, recombinant IGFBP-5 stimulates cultured mesangial cell migration through an IGF-I independent action by binding to the putative serine/kinase IGFBP-5 receptor.[55]

The role of IGFBP on in vivo renal function has been more difficult to assess. Administration of IGF-1 complexed with IGFBP-3 causes an increase in the localization of IGF-I to the glomerulus, but the functional effect of this is unknown.[56] Infusions of IGFBP-1 alone or together with IGF-I stimulates kidney but not body growth in Snell dwarf mice, with a greater reponse when given together.[57,58] In contrast, overexpression of IGFBP-1 in transgenic mice leads to altered post-natal nephrogenesis with reduced nephron number and the development of glomerulosclerosis.[59,60]

As discussed later, profound and divergent changes in the expression of the renal IGFBPs occur in a variety of kidney diseases and physiologic processes. For example, kidney IGFBP-1 expression is increased in workload and in hypokalemia induced renal hypertrophy. This is associated with an increase in kidney IGF-1 peptide levels in both conditions, which has been attributed to trapping of IGF-I by the IGFBP since IGF-I gene expression is not increased.[32,61] Transient increases in IGF-I have also been described in other states of renal hypertrophy such as diabetes and compensatory renal growth, that are also attributable to an increase in kidney IGFBPs trapping IGF-I within the organ.[5-7,62] In all these conditions it has been proposed that the local increase in IGF-I contributes to the genesis of the renal hypertrophy. However until the intricate interactions between IGF-1, the various IGFBPs and the different cell types of the kidney are understood, the significance of changes in renal IGFBP expression remains largely conjectural.

GH and IGF-I Increase Renal Blood Flow and Glomerular Filtration Rate

In animals and in humans, glomerular hemodynamics are influenced by the GH and IGF-I status. Both GH[63-66] and IGF-I[67-71] cause an increase in renal plasma flow and glomerular filtration rate and a decrease in renal vascular resistance. In humans renal plasma flow and GFR increase by 20-30 % following their administration (Fig. 3). Whereas these effects are seen with GH only after several days of treatment at which time plasma IGF-I concentrations are increased, IGF-I induces the same changes within minutes to hours of administration. Taken together with the observation that rhIGF-I administration normalizes the reduced GFR in GH deficient rats and patients with GH receptor defects (Laron-type dwarfs), this all indicates that IGF-I is the mediator of the GH-induced increase in glomerular hemodynamics.[72-74] In single-nephron micropuncture studies carried out in the rat, IGF-I was demonstrated to decrease afferent and efferent arteriolar vascular resistance and, to increase the glomerular ultrafiltration coefficient, presumably via a relaxing effect on the mesangial cells determining the surface area available for ultrafiltration.[75] It appears that IGF-I mediated arteriolar vasodilatation is induced by metabolites of cyclooxygenase activity and through the generation of nitrous oxide which in turn acts through the generation of cyclic GMP.[67,68,76]

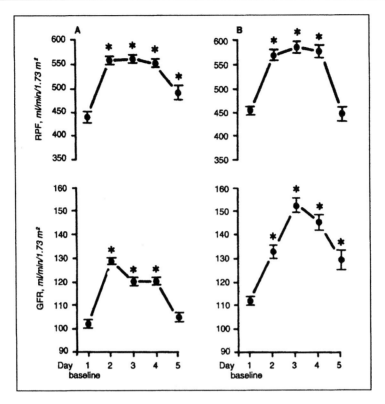

Figure 3. IGF-1 administration acutely raises renal plasma flow (RPF) and glomerular filtration rate (GFR) in healthy human subjects without (A) and with (B) volume expansion by salt loading. Recombinant human IGF-1 was administered subcutaneously on days 2 to 4. Reprinted with permission from Hirschberg et al.[69]

GH and IGF-1 Modulate Renal Tubular Function

The expression of IGF-I receptors along the different segments of the renal tubule system provides a mechanism for IGF-I to mediate its biologic effects on tubular function. Published literature examining the tubular effects of IGF-1 is largely limited to changes in the phosphate and sodium/water reabsorption, whereas much less is known regarding actions on the handling of other minerals and/or organic compounds.[73,74,77-79] Evidence from observational (patients with Laron-type dwarfism) as well as interventional studies in animals and humans clearly indicates that IGF-I, and GH via IGF-I, decrease renal phosphate losses by stimulating tubular reabsorption of phosphate.[80-84] This IGF-I action is immediate and is more efficiently mediated though apical (luminal) than basolateral membrane IGF-I receptors.[80] Absorption of ultrafiltered phosphate from the tubular lumen occurs via a specific sodium-phosphate co-transporter in the brush border membrane of proximal tubule cells. This sodium-phosphate co-transporter is activated independently by phosphate depletion, parathyroid hormone and IGF-I.[85-87]

IGF-1 has significant sodium retaining properties and this accounts for the edema formation observed in some patients receiving rhGH or rhIGF-I treatment. Sodium and water retention usually occurs during the first few days of treatment, is of modest degree and usually vanishes during continued therapy.[74,88-91] Fractional urinary excretion of filtered sodium transiently falls in patients treated short-term with rhIGF-I.[77,92] Neither GH nor IGF-I appear to affect sodium and water absorption in the proximal tubule,[80,93] but IGF-I may enhance distal tubular sodium absorption via activation of amiloride-sensitive apical sodium channels. Suppression of plasma atrial natriuretic peptide secretion and stimulation of renin release may be other mechanisms by which IGF-I therapy induces fluid retention.[94]

GH and IGF-I Increase Renal Mass

It is well established that growth hormone and IGF-I have renal growth promoting actions and while the actions of IGF-I are direct, it is likely that the renotropic effects of GH are largely mediated by IGF-I released into the circulation or produced in the kidney. On the other hand it cannot be excluded that GH may have direct growth promoting actions in those nephron segments where GH receptors are expressed. In humans the effect of GH is seen in patients with acromegaly for they have enlarged kidneys that reduce in size when the overproduction of GH, and hence IGF-I, is abolished.[95-97] In rats treated with IGF-I and in transgenic mice overexpressing either GH or IGF-I kidney and glomerular size is increased.[2,98-102] The stimulation of renal mass by GH and IGF-1 is due to hypertrophy in all segments of the nephron (most prominently in the proximal tubule), but also due to hyperplasia in the glomeruli and the interstitium.[100-102]

Since IGF-1 promotes renal growth, there has been intense interest in determining whether the intrarenal IGF-1 axis participates in compensatory renal growth (CRG) after loss of renal mass. There is good evidence showing that in the adult rat whole kidney, IGF-1 levels increase early after a uninephrectomy, returning to basal levels after 4 days.[18,18,103-105] Most but not all of the studies fail to show a parallel increase in IGF-1 gene expression.[105-107] An increase in IGF-I receptor and IGFBP-5 levels in glomerular and basolateral tubular membranes has also been described one month after a partial nephrectomy.[106,108,109] Administration of an IGF-1 receptor antagonist inhibits early CRG and the compensatory hyperfiltration after loss of renal tissue[110,111] (Fig. 4), indicating that IGF-I has an important role in promoting CRG.

The response to loss of renal mass differs in immature rats and in adult female rats from that described above in adult male rats.[112-114] First the renal response in adult male rats occurs predominantly by cellular hypertrophy, whereas in immature rats and adult females there is a large hyperplastic component. Second, early CRG in the adult male rat is GH dependent, while in the immature rat and in the adult female rat, CRG is GH independent. Third, immature and adult female rats exhibit an early increase in IGF-1 receptor and IGF-1 gene expression; IGF-1 receptor expression is unaltered in the adult male. Taken together this all suggests that IGF-1 plays an important role in early CRG. Furthermore it appears that the presence or absence of GH dependence determines whether CRG occurs predominantly through hyperplasia as in adult female and in immature rats, or through hypertrophy as in the adult male rat. Species differences also exist, for in adult female mice CRG is GH dependent.[115] Accordingly translation of findings obtained in rats to the clinical situation should be carried out with considerable circumspect.

Several reports suggest that GH and IGF-I may also be important mediators of early diabetic renal hypertrophy.[5,6] In rats with diabetes, kidney IGF-I content increases during the first three days of the disease and this initial increase precedes the earliest measured rise in kidney function and mass[116] Most studies have failed to demonstrate an increase in IGF-I mRNA levels to account for the increase in IGF-I peptide. Increased IGF-I production by mesangial cells cultured from non-obese diabetic mice and an increase in IGF-I receptors in mesangial cells from db/db mice have been reported, but whether this all occurs in vivo is unclear.[117,118] As observed with CRG, administration of an IGF-I receptor antagonist inhibits renal hypertrophy in diabetic rats.[110] GH has also been suggested to play a role in diabetic kidney disease largely based on the finding that somatostatin analogs and GH receptor antagonists inhibit diabetic renal growth, proteinuria and even morphologic changes.[8,119-122] Furthermore diabetic GH deficient rats have a smaller increase in kidney size and IGF-I content than diabetic non-GH deficient rats.[123] Most studies fail to show an increase in GH receptor expression,[5] though a recent study demonstrated increased GH receptor mediated signal transduction in the kidneys of STZ diabetic rats.[119] Significant and complex changes in kidney IGFBP expression, including an increase in cortical and a decrease in medullary IGFBP-1 mRNA levels,[124] are a feature of the diabetic kidney and are difficult to interpret. However it is likely that some of these changes may lead to increased trapping of IGF-I in the kidney.[116]

Potassium depletion is a complex condition in which there is renal hypertrophy while paradoxically, body growth is inhibited. In this model increased renal IGF-I peptide but not mRNA levels have been demonstrated. Again there are complex regional changes in kidney IGFBP expression, most notably an increase in IGFBP-1 mRNA and protein levels.[31,32] Metabolic acidosis is another condition in which there is renal hypertrophy, muscle wasting and growth retardation. Similarly

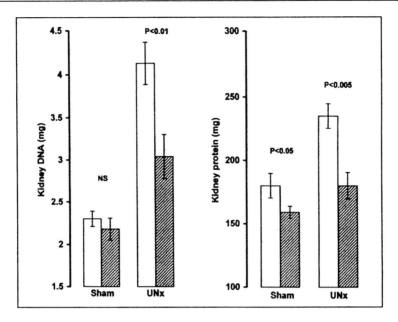

Figure 4. Treatment with IGF-1 receptor antagonist inhibits increase in total kidney DNA and protein content in rats 11 days after unilateral nephrectomy (UNx). Reprinted with permission from Haylor et al.[110]

there is an early and transient increase in kidney IGF-I peptide without an increase in mRNA levels.[33] The mRNA levels of several IGFBPs are elevated in the kidney especially IGFBP-1 and -4.

Common to all the causes of renal hypertrophy described above is an increase in IGF-I peptide concentration in the absence of an increase in gene expression. Several explanations have been proposed to account for this. First, elevated levels of low molecular weight IGFBPs as in potassium deficiency[31] could facilitate the movement of IGF-1 into the kidneys because of their relatively small size. Second, there could be increased trapping of circulating or locally produced IGF-1 within the kidney due to increased kidney membrane-associated IGFBP or IGF-1 receptor number. Third, reduced cellular breakdown of internalized IGF-1 may be involved as described in potassium depletion.[32] Finally, the disparity between mRNA and peptide levels could be caused by enhanced post-transcriptional mechanisms with increased generation of IGF-1 protein.

GH, IGF-I and Progression of Renal Disease

In rats, compensatory renal growth following reduction of functional renal mass is associated with progressive glomerulosclerosis and tubular interstitial fibrosis resulting in chronic renal failure. Administration of GH or transgenicity for GH accelerates glomerulosclerosis in rats and mice, respectively.[99,125,126] GH deficiency partially protects from glomerulosclerosis after partial nephrectomy and also in streptozotozin-induced diabetes.[127,128] It is currently unclear whether the sclerosis-inducing effect of GH is entirely mediated by IGF-I or not. Mice transgenic for IGF-I develop glomerular enlargement comparable to GH-transgenic mice, but do not develop accelerated glomerulosclerosis.[98,99] On the other hand, the increased glomerular synthesis of matrix proteins such as collagens type I and IV, laminin and heparin sulphate proteoglycan observed with GH over expression is mimicked by IGF-1 perfusion of isolated kidneys.[129] Because of the susceptibility of rodents to GH-induced development of progressive glomerulosclerosis, there has been concern that chronic GH or IGF-I treatment would accelerate the progression of chronic renal failure in humans. Fortunately, even extended GH treatment over several years had no detrimental effect on disease progression in children with chronic renal failure.[130] Also in adults with CRF treatment with IGF-I for several months has not been associated with a decline in renal function, indeed as discussed later, IGF-I treatment may actually increase the GFR.[131,132]

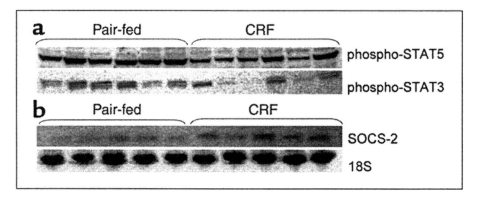

Figure 5. Upper panel: Deficient nuclear accumulation of tyrosine-phosphorylated STAT5 and STAT3 protein in rats with chronic renal failure (CRF) compared to pair-fed control animals. Lower panel: Increased mRNA abundance of Suppressor of Cytokine Signaling (SOCS-)2 in rhGH-treated rats with CRF. SOCS-2 inhibits STAT phosphorylation by binding to tyrosine kinase JAK2. Reproduced with permission from Schaefer et al.[135]

The GH-IGF System in Chronic Renal Failure

In chronic renal failure, circulating levels of GH are usually increased despite normal or even decreased pituitary secretion rates. This is due to the diminished renal metabolic clearance of the hormone.[38] GH-induced hepatic IGF-I synthesis is markedly reduced in rats with chronic renal failure.[133] This GH insensitivity may in part be due to deficient GH receptor expression, although this is controversial. Reduced hepatic GH receptor mRNA and receptor binding levels has been reported in some[133-135] but not all studies carried out with animals.[136,137] One explanation for the controversy is the effect of reduced nutritional intake. Indeed controlling for the anorexia of chronic uremia by pair feeding control animals abolished the difference in GH receptor binding to liver plasma membranes seen when ad-lib controls and uremic animals are compared.[135,136] On the other hand, reduced GH receptor protein expression was observed in the growth cartilage of rats with chronic renal failure even though nutritional intake was controlled.[138] In humans serum levels of GH binding protein, putatively reflecting hepatic GH receptor status, were decreased in some, but normal in other studies.[39,139,140]

Another mechanism accounting for the resistance to GH in uremia is provided by our recent demonstration of a marked post-receptor GH signaling defect in livers of chronically uremic rats resistant to growth hormone.[135] Despite GH receptor protein levels that were similar to those of pair fed normal controls, phosphorylation of the GH receptor associated tyrosine kinase janus kinase-2 (JAK-2) was diminished by 75% (Fig. 5). This resulted in a similar suppression of GH-dependent phosphorylation of major down stream signaling molecules, namely signal transducer and activator of transcription-1, -3 and -5. This defect was possibly caused by up-regulation of intracellular JAK-2 inhibitors, namely the suppressors of cytokine signaling-2 and -3.

As most metabolic effects of GH are mediated by IGF-I, GH insensitivity in uremia may also be due to IGF-I resistance. Indeed, numerous studies in the rat as well as in humans have clearly documented marked IGF-I resistance in chronic renal failure[48,141-143]. Whereas serum IGF-I levels are usually normal in rats and/or humans with chronic renal failure, the concentrations of circulating IGFBP-1 to -4 and -6 are increased in a manner inversely related to glomerular filtration rate[144,145] (Fig. 6). In addition, there is accumulation of a low-molecular weight fragment of IGFBP-3, which binds IGF-I albeit with markedly reduced affinity.[146] Experimental evidence suggests that the increase of IGFBP-1 and IGFBP-2 is not only due to reduced renal metabolic clearance but also to increased hepatic synthesis[147] Of note is the finding that IGFBP-1, -2, -4 and -5 plasma concentrations are inversely correlated with growth rates in children with chronic renal failure. This suggests an important role of IGFBP excess in the pathogenesis of clinical GH/IGF-1 insensitivity in uremia.[144,148] Apart from the increased plasma IGF-I binding capacity, a post-receptor IGF-I signaling

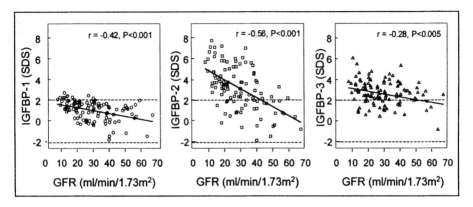

Figure 6. GFR-dependent increase in IGFBP-1, -2 and –3 serum concentrations in children with chronic renal failure. Age- and sex-related variation in IGFBP concentrations was accounted for by calculation of standard deviation scores. Normal range (-2 to 2) is indicated by dotted lines. Reproduced with permission from Tönshoff et al.[144]

defect may also contribute to IGF-1 resistance in chronic renal failure. In CRF rats the effect of IGF-I and various IGF-I analogs on protein turnover (synthesis and degradation) was suppressed (Fig. 7). The observation that the inhibitory effect of IGF-I and its analogs were affected to a similar degree indicates that the resistance arises because of a defect at a cellular level and not because of changes in the IGFBP levels. Deficient autophosphorylation of the IGF-I receptor in skeletal muscle of uremic rats has been reported,[48] but not confirmed in a later study.[149]

Therapeutic Manipulation of the GH/IGF-I System in Chronic Renal Failure

There are three potential target areas for the use of rhGH or rhIGF-1 in uremia; treatment of growth failure in children, reversal of catabolism and malnutrition in adults, and stimulation of residual renal function in patients with advanced, pre-endstage chronic renal failure. Fortunately despite the marked insensitivity to endogenous GH and IGF-I and to the administration of the recombinant molecules, several animal and clinical trials have documented significant therapeutic responses to recombinant GH and IGF-I in the setting of chronic renal failure.[150,151]

Mehls et al first demonstrated that pharmacological doses of GH can stimulate longitudinal growth in uremic rats.[152] Subsequently, numerous clinical trials have confirmed the efficacy of rhGH in children with chronic renal failure. RhGH therapy results in a persistent increase in body growth and tissue anabolism. Catch-up growth is observed during the first three treatment years, followed by a percentile-parallel growth pattern resulting in a significant improvement of final adult height[130,153] (Fig. 8). The effect of rhGH appears to be mediated by a marked stimulation of IGF-I production without major changes in the IGF binding protein profile, resulting in an increase of free, bioavailable IGF-I.[154] Notably, the relative growth response to pharmacological rhGH treatment remains dependent on the degree of chronic renal failure, the lowest efficacy being observed in patients with end-stage renal disease on long-term dialysis.[130,153,155]

Given its remarkable efficacy, rhGH treatment causes surprisingly few side effects in children with CRF. There has been concern that prolonged GH treatment might provoke diabetes mellitus, especially as patients with advanced renal failure develop insulin resistance and as carbohydrate intolerance is common. However GH administration stimulates insulin secretion and as this response continues as long as GH is administered,[156] the oral glucose tolerance remains unaltered in children with CRF treated with rhGH for up to five years.[156] Nevertheless the hyperinsulinemia and altered lipid profile induced by GH may potentially carry the risk of accelerated cardiovascular disease.[156,157] Concern has also been raised that the glomerular hyperfiltration induced by GH may result in glomerular sclerosis and accelerated deterioration of renal function during long-term treatment. However, several studies have shown that the GFR does not decrease more rapidly in CRF

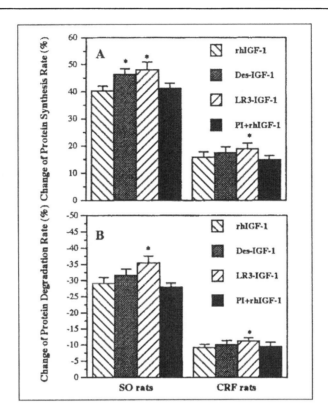

Figure 7. IGF-1 resistance in rats with chronic renal failure (CRF). Effects of rhIGF-1, IGF-1 analogues with low binding protein affinity (Des-IGF-1, LR3-IGF-1) and rhIGF-1 combined with five proteinase inhibitors (PI) were compared in CRF and sham-operated (SO) animals. Marked, binding protein and proteinase independent inhibition of the anabolic and anti-catabolic effects of IGF-1 in the CRF animals. Reproduced with permission from Ding et al.[48]

patients treated with rhGH than in untreated controls.[153,158] A possible explanation may be related to the renal resistance to GH that develops in kidney failure; for unlike the response in normal subjects GH fails to cause an increase in glomerular hemodynamics in CRF patients.[63]

RhGH has also been demonstrated in several short-term studies to improve nitrogen balance and nutritional status in adult patients on maintenance dialysis.[159-162] Similar anabolic effects were observed with rhIGF-I administration.[151] IGF-1 may also be effective in improving renal function in humans with advanced kidney failure.[132,163,164] In the earlier short-term studies in patients with advanced renal failure, rhIGF-1 was given in high doses and markedly stimulated glomerular filtration rate. Unfortunately this effect was lost during continued treatment, possibly due to the induction of compensatory changes in the IGFBP profile. Unfortunately the use of high dose IGF-I therapy was also associated with serious side effects. However, subsequent trials using lower dosage[131] and/or intermittent dosing (4 days on, 3 days off per week)[164] succeeded in stimulating an increase in GFR for the 4 to 6 weeks of treatment without major side effects.

The GH-IGF System and rhIGF Treatment in Acute Renal Failure

Cells of the proximal convoluted tubule and the thick ascending limb of the loop of Henle are most vulnerable to hypoxia and undergo necrosis and apoptosis in post-ischemic acute renal failure.[165-167] During the first few days after an ischemic insult, renal IGF-I mRNA and peptide levels as well as GH receptor and IGFBP-2, -3, -4 and -5 mRNA levels all decrease.[168] However, immunohistochemical studies have demonstrated a transient accumulation of IGF-I in proximal tubule cells

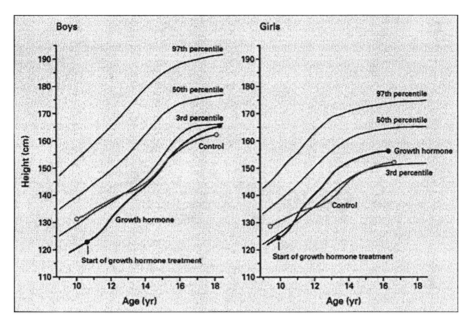

Figure 8. Successful reversal of uremic growth failure by growth hormone treatment in children with chronic renal failure. Mean growth curves during growth hormone treatment for 32 boys and 6 girls with chronic renal failure, compared with control children with chronic renal failure not treated with growth hormone, according to sex. The circles indicate the time of the first observation (the start of growth hormone treatment in the treated children) and the end of the pubertal growth spurt. Reproduced with permission from Haffner et al.[153]

in rats recovering from acute ischemic renal failure.[3] Since IGF-I has multiple biological actions that could accelerate recovery from acute renal failure including stimulation of renal blood flow and tubular cell proliferation and inhibition of apoptosis, several investigators have examined the efficacy of IGF-I for the treatment of acute renal failure.[4] Indeed a number of studies in rat models of postischemic and toxic acute renal failure demonstrated an impressive stimulation of proximal tubular cell proliferation and accelerated renal functional recovery by treatment with rhIGF-I or des(1-3)IGF-1.[169-173] These promising experimental results stimulated clinical trials of IGF-I. In one single-center study, IGF-I produced a modest, but clinically insignificant increase in GFR in patients with post-operative renal insufficiency.[174] In a second multi-center placebo-controlled study of 72 patients with severe acute renal failure, IGF-I failed to modify the course of the illness.[175] The reason for the lack of efficacy of IGF-I therapy in humans with acute renal failure against a background of success in animals is unclear and merits further study.

Summary and Conclusion

In summary, the components of the IGF-I system are expressed in the kidney in an anatomically heterogenous and physiologically regulated manner. IGF-I is involved in the regulation of glomerular hemodynamics, renal growth and the regulation of certain tubular functions. IGF-I is a major determinant of compensatory renal growth and renal hypertrophy in a number of disease states and growth hormone/IGF-I may be involved in the pathomechanism of progressive glomerular and interstitial fibrosis. The regulation of renal IGF-1 levels is complex and includes changes in the expression of locally produced IGF-I, increased trapping of circulating IGF-1 secondary to altered IGF-1 receptor and binding protein expression, increased delivery of IGF-I to the kidney because of an increase in circulating levels of low molecular weight IGFBPs. Finally, altered local IGF-I degrading activity may influence the level of IGF-I in the kidney. Chronic renal failure is associated with a state of multilevel GH and IGF-I resistance that can largely be overridden by the administration of rhGH and rhIGF-I in pharmacological doses. Whereas both these hormones stimulate growth and anabolism in chronic renal failure, and while IGF-I administration can increase glomerular filtration in advanced renal failure, the administration of IGF-1 to patients with acute renal failure has not been effective in promoting renal recovery. Taking together the major actions that IGF-I has on the kidney and its therapeutic potential, there remains a compelling need to improve our understanding of the renal growth hormone/ IGF-1 system, especially that of the insulin-like growth factor binding proteins.

References

1. Chin E, Zhou J, Bondy CA. Renal growth hormone receptor gene expression: relationship to renal insulin-like growth factor system. Endocrinology 1992; 131:3061-3066.
2. Chin E, Zhou J, Bondy C. Anatomical relationship in the patterns of insulin-like growth factor (IGF)-I, IGF binding protein-1, and IGF-I receptor gene expression in the rat kidney. Endocrinology 1992; 130:3237-3245.
3. Rabkin R, Brody M, Lu L et al. Expression of genes encoding the rat renal insulin-like growth factor-I system. J Am Soc Nephrol 1995; 6:1511-1518.
4. Rabkin R. Insulin-like growth factor-I treatment of acute renal failure. J Lab Clin Med 1995; 125:684-685.
5. Flyvbjerg A. Putative pathophysiological role of growth factors and cytokines in experimental diabetic kidney disease. Diabetologia 2000; 43:1205-1223.
6. Rabkin R, Fervenza FC. Renal hypertrophy and kidney disease in diabetes. Diabetes Metab Rev 1996; 12:217-241.
7. Feld S, Hirschberg R. Growth hormone, the insulin-like growth factor system, and the kidney. Endocr Rev 1996; 17:423-480.
8. Flyvbjerg A. Potential use of growth hormone receptor antagonist in the treatment of diabetic kidney disease. Growth Horm IGF Res 2001; 11(Suppl A):S115-S119.
9. Schimpff RM, Donnadieu M, Duval M. Serum somatomedin activity measured as sulphation factor in perpheral, hepatic and renal veins of mongrel dogs: basal levels. Acta Endocrinol 1980; 93:67-72.
10. D'Ercole AJ, Stiles AD, Underwood LE. Tissue concentrations of somatomedin C: further evidence for multiple sites of synthesis and paracrine or autocrine mechanisms of action. Proc Natl Acad Sci USA 1984; 81:935-939.
11. Mathews LS, Norstedt G, Palmiter RD. Regulation of insulin-like growth factor I gene expression by growth hormone. Proc Natl Acad Sci 1986; 83:9343-9347.
12. Landau D, Chin E, Bondy C et al. Expression of insulin-like growth factor binding proteins in the rat kidney: effects of long-term diabetes. Endocrinology 1995; 136:1835-1842.
13. Nakamura T, Fukui M, Ebihara I et al. mRNA expression of growth factors in glomeruli from diabetic rats. Diabetes 1993; 42:450-6.
14. Matejka GL, Eriksson PS, Carlsson B et al. Distribution of IGF-I mRNA and IGF-I binding sites in the rat kidney. Histochemistry 1992; 97:173-180.
15. Matejka GL, Jennische E. IGF-I binding and IGF-I mRNA expression in the post-ischemic regenerating rat kidney. Kidney Int 1992; 42:1113-1123.
16. Kobayashi S, Clemmons DR, Venkatachalam MA. Colocalization of insulin-like growth factor-binding protein with insulin-like growth factor I. Am J Physiol 1991; 261:F22-F28.
17. Chin E, Zhou J, Dai J et al. Cellular localization and regulation of gene expression for components of the insulin-like growth factor ternary binding protein complex. Endocrinology 1994; 134:2498-2504.
18. Evan AP, Henry DP, Connors BA et al. Analysis of insulin-like growth factors (IGF)-I, and -II, type II IGF receptor and IGF-binding protein-2 mRNA and peptide levels in normal and nephrectomized rat kidney. Kidney Int 1995; 48:1517-1529.
19. Andersson GL, Skottner A, Jennische E. Immunocytochemical and biochemical localization of insulin-like growth factor I in the kidney in rats before and after uninephrectomy. Acta Endocrinol 1988; 119:555-560.
20. Andersson GL, Ericson LE, Jennische E. Ultrastructural localization of IGF-I in the rat kidney; an immunocytochemical study. Histochemistry 1990; 94:263-267.
21. Bortz JD, Rotwein P, DeVol D et al. Focal expression of insulin-like growth factor I in rat kiney collecting duct. J Cell Biol 1988; 107:811-819.
22. Miller SB, Rotwein P, Bortz JD et al. Renal expression of IGF I in hypersomatrophic states. Am J Physiol 1990; 259:F251-F257.
23. Rogers SA, Miller SB, Hammerman MR. Growth hormome stimulates IGF I gene expression in isolated rat renal collecting duct. Am J Physiol 1990; 259:F474-F479.
24. D'Ercole A, Hill DJ, Strain AJ et al. Tissue and plasma somatomedin-C/insulin-like growth factor I concentrations in the human fetus during the first half of gestation. Pediatr Res 1986; 20:253-255.
25. Hise MK, Salmanullah M, Tannenbaum GS et al. mRNA expression of the IGF system in the kidney of the hypersomatotropic rat. Nephron 2001; 88:360-367.
26. Chin E, Michels K, Bondy CA. Partition of insulin-like growth factor (IGF)-binding sites between the IGF-I and IGF-II receptors and IGF-binding proteins in the human kidney. J Clin Endocrinol Metab 1994; 78:156-164.
27. Kobayashi S, Nogami H, Ikeda T. Growth hormone and nutrition interact to regulate expressions of kidney IGF-I and IGFBP mRNAs. Kidney Int 1995; 48:65-71.

28. Nogami H, Watanabe T, Kobayashi S. IGF-I and IGF-binding protein gene expressions in spontaneous dwarf rat. Am J Physiol 1994; 267:E396-E401.
29. Chin E, Bondy CA. Dietary protein-induced renal growth: correlation between renal IGF-I synthesis and hyperplasia. Am J Physiol 1994; 266:C1037-C1045.
30. Lemozy S, Pucilowska JB, Underwood LE. Reduction of insulin-like growth factor-I (IGF-I) in protein-restricted rats is associated with differential regulation of IGF-binding protein messenger ribonucleic acids in liver and kidney, and peptides in liver and serum. Endocrinology 1994; 135:617-623.
31. Hsu FW, Tsao T, Rabkin R. The IGF-I axis in kidney and skeletal muscle of potassium deficient rats. Kidney Int 1997; 52:363-370.
32. Tsao T, Fawcett J, Fervenza FC et al. Expression of insulin-like growth factor-I and transforming growth factor-beta in hypokalemic nephropathy in the rat. Kidney Int 2001; 59:96-105.
33. Fawcett J, Hsu FW, Tsao T et al. Effect of metabolic acidosis on the insulin-like growth factor-I system and cathepsins B and L gene expression in the kidney. J Lab Clin Med 2000; 136:468-475.
34. Rogers SA, Miller SB, Hammerman MR. Insulin-like growth factor I gene expression in isolated rat renal collecting duct is stimulated by epidermal growth factor. J Clin Invest 1991; 87:347-351.
35. Fervenza FC, Tsao T, Rabkin R. Response of the intrarenal insulin-like growth factor-I axis to acute ischemic injury and treatment with growth hormone and epidermal growth factor. Kidney Int 1996; 49:344-354.
36. Rabkin R, Haussman M. Renal metabolism of hormones. In: Becker KL, ed. Principles and Practice of Endocrinology and Metabolism Lippincott, 2000:1895-1901.
37. Johnson V, Maack T. Renal extraction, filtration, absorption, and catabolism of growth hormone. Am J Physiol 1977; 233:F185-196.
38. Haffner D, Schaefer F, Girard J et al. Metabolic clearance of recombinant human growth hormone in health and chronic renal failure. J Clin Invest 1994; 93:1163-1171.
39. Schaefer F, Baumann G, Haffner D et al. Multifactorial control of the elimination kinetics of unbound (free) growth hormone (GH) in the human: regulation by age, adiposity, renal function, an steady state concentrations of GH in plasma. J Clin Endocrinol Metab 1996; 81:22-31.
40. Rabkin R, Gottheiner TI, Fang VS. Removal and excretion of immunoreactive rat growth hormone by the isolated kidney. Am J Physiol 1981; 240:F282-287.
41. Rabkin R, Pimstone BL, Eales L. Autoradiographic demonstration of glomerular filtration and proximal tubular absorption of growth hormone 125-I in the mouse. Horm Metab Res 1973; 5:172-175.
42. Baxter RC. Insulin-like growth factor (IGF)-binding proteins: interactions with IGFs and intrinsic bioactivities. Am J Physiol Endocrinol Metab 2000; 278:E967-E976.
43. Flyvbjerg A, Nielsen S, Sheikh MI et al. Luminal and basolateral uptake and receptor binding of IGF-I in rabbit renal proximal tubules. Am J Physiol 1993; 265:F624-F633.
44. Li W, Fawcett J, Widmer HR et al. Nuclear transport of insulin-like growth factor-I and insulin-like growth factor binding protein-3 in opossum kidney cells. Endocrinology 1997; 138:1763-1766.
45. Bastian SE, Walton PE, Wallace JC et al. Plasma clearance and tissue distribution of labelled insulin-like growth factor-I (IGF-I) and an analogue LR3IGF-I in pregnant rats. J Endocrinol 1993; 138:327-336.
46. Rabkin R, Fervenza FC, Maidment H et al. Pharmacokinetics of insulin-like growth factor-1 in advanced chronic renal failure. Kidney Int 1996; 49:1134-1140.
47. Fouque D, Peng SC, Kopple JD. Pharmacokinetics of recombinant human insulin-like growth factor-1 in dialysis patients. Kidney Int 1995; 47:869-875.
48. Ding H, Gao XL, Hirschberg R et al. Impaired actions of insulin-like growth factor 1 on protein synthesis and degradation in skeletal muscle of rats with chronic renal failure. Evidence for a postreceptor defect. J Clin Invest 1996; 97:1064-1075.
49. Schneider MR, Wolf E, Hoeflich A et al. IGF-binding protein-5: flexible player in the IGF system and effector on its own. J Endocrinol 2002; 172:423-440.
50. Baxter RC. Signalling pathways involved in antiproliferative effects of IGFBP-3: a review. Mol Pathol 2001; 54:145-148.
51. Kelley KM, Oh Y, Gargosky SE et al. Insulin-like growth factor-binding proteins (IGFBPs) and their regulatory dynamics. Int J Biochem Cell Biol 1996; 28:619-637.
52. Yap J, Tsao T, Fawcett J et al. Effect of insulin-like growth factor binding proteins on the response of proximal tubular cells to insulin-like growth factor-I. Kidney Int 1997; 52:1216-1223.
53. Fawcett J, Rabkin R. The processing of insulin-like growth factor-I (IGF-I) by a cultured kidney celline is altered by IGF-binding protein-3. Endocrinology 1995; 136:1340-1347.
54. Verzola D, Villaggio B, Berruto V et al. Apoptosis induced by serum withdrawal in human mesangial cells. Role of IGFBP-3. Exp Nephrol 2001; 9:366-371.
55. Berfield AK, Andress DL, Abrass CK. IGFBP-5(201-218) stimulates Cdc42GAP aggregation and filopodia formation in migrating mesangial cells. Kidney Int 2000; 57:1991-2003.
56. Sandra A, Boes M, Dake BL et al. Infused IGF-I/IGFBP-3 complex causes glomerular localization of IGF-I in the rat kidney. Am J Physiol 1998; 275:E32-E37.
57. Van Buul-Offers SC, Van Kleffens M, Koster JG et al. Human insulin-like growth factor (IGF) binding protein-1 inhibits IGF-I-stimulated body growth but stimulates growth of the kidney in snell dwarf mice. Endocrinology 2000; 141:1493-1499.
58. Van Kleffens M, Lindenbergh-Kortleve DJ, Koster JG et al. The role of the IGF axis in IGFBP-1 and IGF-I induced renal enlargement in Snell dwarf mice. J Endocrinol 2001; 170:333-346.
59. Doublier S, Seurin D, Fouquerary B et al. Glomerulosclerosis in mice transgenic for human insulin-like growth factor-binding protein-1. Kidney Int 2000; 57:2299-2307.

60. Doublier S, Amri K, Seurin D et al. Overexpression of human insulin-like growth factor binding protein-1 in the mouse leads to nephron deficit. Pediatr Res 2001; 49:660-666.
61. Kobayashi S, Clemmons DR, Nogami H et al. Tubular hypertrophy due to work load induced by furosemide is associated with increases of IGF-1 and IGFBP-1. Kidney Int 1995; 47:818-828.
62. Hirschberg R, Adler S. Insulin-like growth factor system and the kidney: physiology, pathophysiology, and therapeutic implications. Am J Kidney Dis 1998; 31:901-919.
63. Haffner D, Zacharewicz S, Mehls O et al. The acute effect of growth hormone on GFR is obliterated in chronic renal failure. Clin Nephrol 1989; 32:266-269.
64. Parving H, Noer I, Mogensen C et al. Kidney function in normal man during short-term growth hormone infusion. Acta Endocrinol 1978; 89:796-800.
65. Christiansen JS, Gammelgaard J, Orskov H et al. Kidney function and size in normal subjects before and during growth hormone administration for one week. Eur J Clin Invest 1981; 11:487-490.
66. Hirschberg RR, Kopple JD. Increase in renal plasma flow and glomerular filtration rate during growth hormone treatment may be mediated by insulin-like growth factor I. Am J Nephrol 1988; 8:249-254.
67. Hirschberg R, Kopple JD. Evidence that insulin-like growth factor I increases renal plasma flow and glomerular filtration rate in fasted rats. J Clin Invest 1989; 83:326-330.
68. Baumann U, Eisenhauer T, Hartmann H. Increase of glomerular filtration rate and renal plasma flow by insulin-like growth factor-I during euglycaemic clamping in anaesthetized rats. Eur J Clin Invest 1992; 22:204-209.
69. Hirschberg R, Brunori G, Kopple DJ et al. Effects of insulin-like growth factor I on renal function in normal men. Kidney Int 1993; 43:387-397.
70. Guler HP, Eckardt KU, Zapf J et al. Insulin-like growth factor I increase glomerular filtration rate and renal plasma flow in man. Acta Endocrinol 1989; 121:101-106.
71. Guler HP, Schmid C, Zapf J et al. Effects of recombinant insulin-like growth factor I on insulin secretion and renal function in normal human subjects. Proc Natl Acad Sci 1989; 86:2868-2872.
72. Gargosky SE, Nanto-Salonen K, Tapanainen P et al. Pregnancy in growth hormone-deficient rats: assessment of insulin-like growth factors (IGFs), IGF-binding proteins (IGFBPs) and IGFBP protease acitivity. J Endocrinol 1993; 136:479-489.
73. Hirschberg R. Effects of growth hormone and IGF-I on glomerular ultrafiltration in growth hormone-deficient rats. Regul Pept 1993; 48:241-250.
74. Laron Z, Klinger B. IGF-I treatment of adult patients with Laron syndrome: preliminary results. Clin Endocrinol 1994; 41:631-638.
75. Hirschberg R, Kopple JD, Blantz RC et al. Effects of recombinant human insulin-like growth factor I on glomerular dynamics in the rat. J Clin Invest 1991; 87:1200-1206.
76. Hirschberg R, Kopple JD. The growth hormone insulin-like growth factor I axis and renal glomerular function. J Am Soc Nephrol 1992; 2:1417-1422.
77. Giordano M, DeFronzo R. Acute effects of human recombinant insulin-like growth factor I on renal function in humans. Nephron 1995; 71:10-15.
78. Corvilain J, Abramow M, Bergans A. Some effects of human growth hormone on renal hemodynamics and on tubular phosphate transport in man. J Clin Invest 1962; 41:1230-1235.
79. Mulroney SE, Lumpkin MD, Haramati A. Antagonist to GH-releasing factor inhibits growth and renal Pi reabsorption in immature rats. Am J Physiol 1989; 257:F29-F34.
80. Quigley R, Baum M. Effects of growth hormone and insulin-like growth factor I on rabbit proximal convoluted tubule transport. J Clin Invest 1991; 88:368-374.
81. Wada L, Don BR, Schambelan M. Hormonal mediators of amino acid-induced glomerular hyperfiltration in humans. Am J Physiol 1991; 260:F787-F792.
82. O'Shea M, Layish D. Growth hormone and the kidney: a case presentation and reviews of the literature. J Am Soc Nephrol 1992; 3:157-161.
83. Ogle GD, Rosenberg AR, Kainer G. Renal effects of growth hormone. II. Electrolyte homeostasis and body composition. Pediatr Nephrol 1992; 6:483-489.
84. Nishiyama S, Ikuta M, Nakamura T et al. Renal handling of phosphate can predict height velocity during growth hormone therapy for short children. J Clin Endocrinol Metab 1992; 74:906-909.
85. Bonjour J, Caverzasio J. IGF-I a key controlling element in phosphate homeostasis during growth. In: Spencer E, ed. Modern concepts of insulin-like growth factors. New York: Elsevier, 1991:193-198.
86. Caverzasio J, Bonjour J-P. Growth factors and renal regulation of phosphate transport. Pediatr Nephrol 1993; 7:802-806.
87. Ernest S, Coureau C, Escoubet B. Deprivation of phosphate increases IGF-II mRNA in MDCK cells but IGFs are not involved in phosphate transport adaptation to phosphate deprivation. J Endocrinol 1995; 145:325-331.
88. Bengtsson BA, Eden S, Lonn L et al. Treatment of adults with growth hormone (GH) deficiency with recombinant human GH. J Clin Endocrinol Metab 1993; 76:309-317.
89. Jabri N, Schalch DS, Schwartz SL et al. Adverse effects of recombinant human insulin-like growth factor I in obese insulin-resistant type II diabetic patients. Diabetes 1994; 43:369-374.
90. Clemmons DR. Use of growth hormone and insulin-like growth factor I in catabolism that is induced by negative energy balance. Horm Res 1993; 40:62-67.
91. Beck J, McGarry E, Dyrenfurth I et al. The metabolic effects of human and monkey growth hormone in man. Ann Intern Med 1958; 49:1090-1094.
92. Kopple J, Ding H, Hirschberg R. Effects of recombinant human insulin-like growth factor I on renal handling of phosphorus, calcium and sodium in normal humans. Am J Kidney Dis 1995; 26:818-824.

93. Gesek FA, Schoolwerth AC. Insulin increases Na(+)-H+ exchange activity in proximal tubules from normotensive and hypertensive rats. Am J Physiol 1991; 260:F695-F703.
94. Moller J, Jorgensen JO, Marqversen J et al. Insulin-like growth factor I administration induces fluid and sodium retention in healthy adults: possible involvement of renin and atrial natriuretic factor. Clin Endocrinol 2000; 52:181-186.
95. Gershberg H, Heinemann H, Stumpf H. Renal function studies and autopsy report in a patient with gigantism and acromegaly. J Clin Endocrinol Metab 1957; 17:377-385.
96. Dullaart RP, Meijer S, Marbach P et al. Effect of a somatostatin analogue, octreotide, on renal haemodynamics and albuminuria in acromegalic patients. Eur J Clin Invest 1992; 22:494-502.
97. Ikkos D, Ljunggren H, Luft R. Glomerular filtration rate and renal plasma flow in acromegaly. Acta Endocrinol 1956; 21:226-236.
98. Doi T, Striker LJ, Quaife C et al. Progressive glomerulosclerosis develops in transgenetic mice chonically expressing growth hormone and growth hormone releasing factor but not in those expressing insulin-like growth factor-1. Am J Pathol 1988; 131:398-403.
99. Doi T, Striker LJ, Gibson CC et al. Glomerular lesions in mice transgenetic for growth hormone and insulin-like growth factor-I. Relationship between increased glomerular size and mesangial sclerosis. Am J Pathol 1990; 137:541-552.
100. Ritz E, Tonshoff B, Worgall S et al. Influence of growth hormone and insulin-like growth factor-I on kidney function and kidney growth. Pediatr Nephrol 1991; 5:509-512.
101. Mehls O, Irzynjec T, Ritz E et al. Effects of rhGH and rhIGF-1 on renal growth and morphology. Kidney Int 1993; 44:1251-1258.
102. Pesce CM, Striker LJ, Peten E et al. Glomerulosclerosis at both early and late stages is associated with increased cell turnover in mice transgenetic for growth hormone. Lab Invest 1991; 65:601-605.
103. Stiles AD, Sosenko RS, D'Ercole AJ et al. Relation to kidney tissue somatomedin C/insulin-like growth factor I to postnephrectomy renal growth in the rat. Endocrinology 1985; 117:2397-2401.
104. Flyvbjerg A, Orskov H, Nyborg K et al. Kidney IGF-1 accumulation occurs in four different conditions with rapid initial kidney growth in rats. In: Spencer EM, ed. Modern Concepts of Insulin-like Growth Factors. Elsevier: Science Publishing Co, 1991:207-217.
105. Lajara R, Rotwein P, Bortz JD et al. Dual regulation of insulin-like growth factor I expression during renal hypertrophy. Am J Physiol 1989; 257:F252-F261.
106. Fagin JA, Melmed S. Relative increase in insulin-like growth factor I messenger ribonucleic acid levels in compensatory renal hypertrophy. Endocrinology 1987; 120:718-724.
107. Hise MK, Li L, Mantzouris N et al. Differential mRNA expression of insulin-like growth factor system during renal injury and hypertrophy. Am J Physiol 1995; 269:F817-F824.
108. Muchaneta-Kubara EC, Cope GH, el Nahas AM. Biphasic expression of IGF-I following subtotal renal ablation. Exp Nephrol 1995; 3:165-172.
109. Hise MK, Lahn JS, Shao ZM et al. Insulin-like growth factor-I receptor and binding proteins in rat kidney after nephron loss. J Am Soc Nephrol 1993; 4:62-68.
110. Haylor J, Hickling H, El Eter E et al. JB3, an IGF-I receptor antagonist, inhibits early renal growth in diabetic and uninephrectomized rats. J Am Soc Nephrol 2000; 11:2027-2035.
111. Haylor JL, McKillop IH, Oldroyd SD et al. IGF-I inhibitors reduce compensatory hyperfiltration in the isolated rat kidney following unilateral nephrectomy. Nephrol Dial Transplant 2000; 15:87-92.
112. Mulroney SE, Pesce C. Early hyperplastic renal growth after uninephrectomy in adult female rats. Endocrinology 2000; 141:932-937.
113. Mulroney SE, Woda C, Johnson M et al. Gender differences in renal growth and function after uninephrectomy in adult rats. Kidney Int 1999; 56:944-953.
114. Haramati A, Lumpkin MD, Mulroney SE. Early increase in pulsatile growth hormone release after unilateral nephrectomy in adult rats. Am J Physiol 1994; 266:F628-F632.
115. Flyvbjerg A, Bennett WF, Rasch R et al. Compensatory renal growth in uninephrectomized adult mice is growth hormone dependent. Kidney Int 1999; 56:2048-2054.
116. Flyvbjerg A, Thorlacius-Ussing O, Naeraa R et al. Kidney tissue somatomedin C and initial renal growth in diabetic and uninephrectomized rats. Diabetologica 1988; 31:310-314.
117. Oemer BS, Foellmer HG, Hodgdon-Anandant L et al. Regulation of insulin-like growth factor I receptors in diabetic mesangial cells. J Biol Chem 1991; 266:2369-2373.
118. Tack I, Elliot SJ, Potier M et al. Autocrine activation of the IGF-I signaling pathway in mesangial cells isolated from diabetic NOD mice. Diabetes 2002; 51:182-188.
119. Thirone AC, Scarlett JA, Gasparetti AL et al. Modulation of growth hormone signal transduction in kidneys of streptozotocin-induced diabetic animals: effect of a growth hormone receptor antagonist. Diabetes 2002; 51:2270-2281.
120. Landau D, Sergev Y, Afargan M et al. A novel somatostatin analogue prevents early renal complications in the nonobese diabetic mouse. Kidney Int 2001; 60:505-512.
121. Bellush LL, Doublier S, Holland AN et al. Protection against diabetes-induced nephropathy in growth hormone receptor/binding protein gene-disrupted mice. Endocrinology 2000; 141:163-168.
122. Sergev Y, Landau D, Rasch R et al. Growth hormone receptor antagonism prevents early renal changes in nonobese diabetic mice. J Am Soc Nephrol 1999; 10:2374-2381.
123. el Nahas AM, Le Carpentier JE, Bassett AH. Compensatory renal growth: role of growth hormone and insulin-like growth factor-I. Nephrol Dial Transplant 1990; 5:123-129.
124. Fervenza FC, Tsao T, Hoffman AR et al. Regional changes in the intrarenal insulin-like growth factor-I axis in diabetes. Kidney Int 1997; 51:811-818.

125. Allen DB, Fogo A, el-Hayek R et al. Effects of prolonged growth hormone adminstration in rats with chronic renal insufficiency. Pediatr Res 1992; 31:406-410.
126. Trachtman H, Futterweit S, Schwob N et al. Recombinant human growth hormone exacerbates chronic puromycin aminonucleoside nephropathy in rats. Kidney Int 1993; 44:1281-1288.
127. Yoshida H, Mitarai T, Kitamura M et al. The effect of selective growth hormone defect in the progression of glomerulosclerosis. Am J Kidney Dis 1994; 23:302-312.
128. Chen NY, Chen WY, Bellush L et al. Effects of streptozotocin treatment in growth hormone (GH) and GH antagonist transgenic mice. Endocrinology 1995; 136:660-667.
129. Haylor J, Johnson T, El Nahas M. Renal matrix protein mRNA elevated by insulin-like growth factor-I [Abstract]. J Am Soc Nephrol 6:896, 1995
130. Haffner D, Wühl E, Schaefer F et al. Factors predictive of the short- and long-term efficacy of growth hormone treatment in prepubertal children with chronic renal failure. German Study Group for Growth hormone Treatment in Children with Chronic renal failure. J Am Soc Nephrol 1998; 9:1899-1907.
131. Ike JO, Fervenza FC, Hoffman AR et al. Early experience with extended use of insulin-like growth factor-1 in advanced chronic renal failure. Kidney Int 1997; 51:840-849.
132. Miller SB, Moulton M, O'Shea M et al. Effects of IGF-I on renal function in end-stage chronic renal failure. Kidney Int 1994; 46:201-207.
133. Chan W, Valerie KC, Chan JCM. Expression of insulin-like growth factor-1 in uremic rats: Growth hormone resistance and nutritional intake. Kidney Int 1993; 43:790-795.
134. Tönshoff B, Eden S, Weiser E et al. Reduced hepatic growth hormone (GH) receptor gene expression and increase in plasma GH binding protein in experimental uremia. Kidney Int 1994; 45:1085-1092.
135. Schaefer F, Chen Y, Tsao T et al. Impaired JAK-STAT signal transduction contributes to growth hormone resistance in chronic uremia. J Clin Invest 2001; 108:467-475.
136. Villares SM, Goujon L., Maniar S et al. Reduced food intake is the main cause of low growth hormone receptor expression in uremic rats. Mol Cell Endocrinol 1994; 106:51-56.
137. Martínez V, Balbín M, Ordónez FA et al. Hepatic exression of growth hormone receptor/binding protein and insulin-like growth factor I genes in uremic rats. Influence of nutritional deficit. Growth Horm IGF Res 1999; 9:61-68.
138. Edmondson SR, Baker NL, Oh J et al. Growth hormone receptor abundance in tibial growth plates of uremic rats: GH/IGF-I treatment. Kidney Int 2000; 58:62-70.
139. Tönshoff B, Cronin MJ, Reichert M. Reduced concentration of serum growth hormone (GH)-binding protein in children with chronic renal failure: correlation with GH insensitivity. J Clin Endocrinol Metab 1997; 82:1007-1013.
140. Powell D, Liu F, Baker B et al. Modulation of growth factors by growth hormone in children with chronic renal failure. Kidney Int 1997; 51:1970-1979.
141. Phillips LS, Kopple JD. Circulating somatomedin activity and sulfate levels in adults with normal and impaired kidney function. Metabolism 1981; 30:1091-1095.
142. Fouque D. Insulin-like growth factor 1 resistance in chronic renal failure. Miner Electrolyte Metab 1995; 22:133-137.
143. Fouque D, Peng SC, Kopple JD. Impaired metabolic response to recombinant insulin-like growth factor-1 in dialysis patients. Kidney Int 1995; 47:876-883.
144. Tönshoff B, Blum WF, Wingen AM et al. Serum insulin-like growth factors (IGFs) and IGF binding proteins 1,2 and 3 in children with chronic renal failure: relationship to height and glomerular filtration rate. J Clin Endocrinol Metab 1995; 80:2684-2691.
145. Powell DR, Liu F, Baker BK et al. Effect of chronic renal failure and growth hormone therapy on the insulin-like growth factors and their binding proteins. Pediatr Nephrol 2000; 14:579-583.
146. Fouque D, Le Bouc Y, Laville M et al. Insulin-like growth factor-1 and its binding proteins during a low protein diet in uremic rats. J Am Soc Nephrol 1995; 6:1427-1433.
147. Tönshoff B, Powell DR, Zhao D et al. Decreased hepatic insulin-like growth factor (IGF)-I and increased IGF binding protein-1 and -2 gene expression in experimental uremia. Endocrinology 1997; 138:938-946.
148. Ulinski T, Mohan S, Kiepe D et al. Serum insulin-like growth factor binding protein (IGFBP)-4 and IGFBP-5 in children with chronic renal failure: Relationship to growth and glomerular filtration rate. Pediatr Nephrol 2000; 14:589-597.
149. Tsao T, Fervenza FC, Friedlaender M et al. Effect of prolonged uremia on insulin-like growth factor-I receptor autophosphorylation and tyrosine kinase activity in kidney and muscle. Exp Nephrol 2002; 10:285-292.
150. Rabkin R: Therapeutic use of growth factors in renal disease. In: Kopple JD, Massry SG, eds. Nutritional Management of renal disease. Lippincott, William and Wilkins, 2003:in press.
151. Chen Y, Fervenza FC, Rabkin R. Growth factors in the treatment of wasting in kidney failure. J Ren Nutr 2001; 11:62-66.
152. Mehls O, Ritz E, Hunziker EB et al. Improvement of growth and food utilization by human recombinant growth hormone in uremia. Kidney Int 1988; 33:45-52.
153. Haffner D, Schaefer F, Nissel R et al. Effect of growth hormone treatment on adult height of children with chronic renal failure. N Engl J Med 2000; 343:923-930.
154. Tönshoff B, Mehls O, Heinrich U et al. Growth-stimulating effects of recombinant human growth hormone in children with end-stage renal disease. J Pediatr 1990; 116:561-566.
155. Wühl E, Haffner D, Nissl R et al. Short children on dialysis treatment respond less to growth hormone than patients with chronic renal failure prior to dialysis. German Study Group for GH Treatment in children with CRF. Pediatr Nephrol 1996; 10:294-298.

156. Haffner D, Nissel R, Wuhl E et al. Metabolic effects of long-term growth hormone treatment in prepubertal children with chronic renal failure and after kidney transplantation. The German Study Group for Growth hormone Treatment in Chronic renal failure. Pediatr Res 1998; 43:209-215.
157. Anonymous. Critical evaluation of the safety of recombinant human growth hormone administration: Statement from the Growth hormone Research Society. J Clin Endocrinol Metab 2001; 86:1868-1870.
158. Tönshoff B, Tönshoff C, Mehls O et al. Growth hormone treatment over one year in children with preterminal chronic renal failure: no adverse effect on glomerular filtration rate. Eur J Pediatr 1992; 151:601-607.
159. Ziegler TR, Lazarus JM, Young LS et al. Effects of recombinant human growth hormone in adults receiving maintenance hemodialysis. J Am Soc Nephrol 1991; 2:1130-1135.
160. Ikizler TA, Wingard RL, Breyer JA et al. Short-term effects of recombinant human growth hormone in CAPD patients. Kidney Int 1994; 46:1178-1183.
161. Ikizler TA, Wingard RL, Hakim RM. Interventions to treat malnutrition in dialysis patients: the role of the dose of dialysis, intradialytic parenteral nutrition, and growth hormone. Am J Kidney Dis 1995; 26:256-265.
162. Schulman G, Wingard R, Hutchinson R et al. The effects of recombinant human growth hormone and intradialytic parenteral nutrition in malnourished hemodialyis patients. Am J Kidney Dis 1993; 21:527-534.
163. O'Shea MH, Miller SB, Hammerman MR. Effects of IGF-I on renal function in patients with chronic renal failure. Am J Physiol 1993; 264:F917-F922.
164. Vijayan A, Franklin SC, Behrend T et al. Insulin-like growth factor I improves renal function in patients with end-stage chronic renal failure. Am J Physiol 1999; 276:R929-R934.
165. Bonventre J. Mechanisms of ischemic acute renal failure. Kidney Int 1993; 43:1160-1178.
166. Yagil Y, Myers BD, Jamison RL. Course and pathogenesis of postischemic acute renal failure in the rat. Am J Physiol 1988; 255:F257-F264.
167. Schumer M, Colombel M, Sawczuk I et al. Morphologic, biochemical, and molecular evidence of apoptosis during the reperfusion phase after brief periods of renal ischemia. Am J Pathol 1992; 140:831-838.
168. Tsao T, Wang J, Fervenza FC et al. Renal growth hormone-insulin-like growth factor-I system in acute renal failure. Kidney Int 1995; 47:1658-1668.
169. Lin JJ, Cybulsky AV, Goodyer PR et al. Insulin-like growth factor-1 enhances epidermal growth factor receptor activation and renal tubular cell regeneration in postischemic acute renal failure. J Lab Clin Med 1995; 125:724-733.
170. Ding H, Kopple DJ, Cohen A et al. Recombinant human insulin-like growth factor-I accelerates recovery and reduces catabolism in rats with ischemic acute renal failure. J Clin Invest 1993; 91:2281-2287.
171. Miller SB, Martin DR, Kissane J et al. Insulin-like growth factor I accelerates recovery from ischemic acute tubular necrosis in the rat. Proc Natl Acad Sci 1992; 89:11876-11880.
172. Miller SB, Martin DR, Kissane J et al. Rat models for clinical use of insulin-like growth factor I in acute renal failure. Am J Physiol 1994; 266:F949-F956.
173. Clark R, Mortensen D, Rabkin R. Recovery from acute ischaemic renal failure is accelerated by des-(1-3)-insulin-like growth factor-1. Clin Sci 1994; 86:709-714.
174. Franklin SC, Moulton M, Sicard GA et al. Insulin-like growth factor I preserves renal function postoperatively. Am J Physiol 1997; 272:F257-F259.
175. Hirschberg R, Kopple J, Lipsett P et al. Multicenter clinical trial of recombinant human insulin-like growth factor I in patients with acute renal failure. Kidney Int 1999; 55:2424-2432.

CHAPTER 15

IGF-Independent Effects of the IGFBP Superfamily

Gillian E. Walker, Ho-Seong Kim, Yong-Feng Yang and Youngman Oh

Abstract

The proposed "insulin-like growth factor binding protein (IGFBP) superfamily" is said to consist of the high affinity binding proteins, IGFBP-1 to -6 and low affinity binding proteins (IGFBP-related proteins). A key mechanism for regulating IGF bioactivities is the high-affinity IGFBPs, which in either the circulation or the immediate extracellular environment modulate the activities of IGFs. All six high affinity IGFBPs inhibit IGF actions, while IGFBP-1, -3 and -5 also have IGF-potentiating effects. Interestingly, it is these same three proteins that exhibit dual IGF effects that have also been established to have a third biological function, distinct biological effects independent of the IGF axis. The inclusion of this important third classification for IGFBP biological actions, the "IGF-independent" actions, contributes to our comprehension of how the IGF axis can be so diverse in its biological outcomes.

Introduction

Insulin-like growth factor binding to and subsequent activation of the type I IGF-receptor (IGF-IR) for the most part results in diverse biological effects, including cellular proliferation and differentiation, an increase in metabolic activity, and cell survival via anti-apoptotic pathways in a wide range of cell types.[1] A key mechanism for regulating these diverse IGF bioactivities, are the high-affinity IGFBPs (IGFBP-1 to -6), which in either the circulation or the immediate extracellular environment modulate the activities of IGFs.[1-4] In recent years a further nine IGFBP-related proteins (IGFBP-rPs) have also been identified.[3] Although the IGFBP-rPs, by their biological activities, were identified initially independent of the IGF axis, based on their function, their nature as secreted proteins, their structural conservation of the IGFBP amino-terminal region and the ability of some of them to bind IGFs weakly, have led to the concept of an "IGFBP superfamily," consisting of the high affinity binding proteins and the IGFBP-rPs.[3,5]

All six high affinity IGFBPs have been shown to inhibit IGF actions in a range of cellular and metabolic environments, from the inhibition of cell type specific proliferation to the prevention of IGF "insulin-like" effects. In addition to inhibition, IGFBP-1, -3 and -5 have also been shown to have IGF-potentiating effects.[1-3] Interestingly, it is these same three proteins, IGFBP-1, -3 and -5, that exhibit dual IGF effects that have also been established to have a third function, very clear and distinct biological effects independent of the IGF axis. The inclusion of this third classification for IGFBP biological actions, the "IGF-independent" actions, contributes to our comprehension of how the IGF axis can be so diverse in its biological outcomes. In this chapter we will focus on the IGF-independent functions for the IGFBPs, from the first indications of the existence of independent actions, to the present day opinions selecting key papers that illustrate each point. Interestingly, we have observed during the preparation of this review that the present explosion in research for the high affinity IGF binders has been based on the characterization of IGF-independent actions, while it has been the IGF-dependent actions for the low affinity IGF binders that has been key to verifying the superfamily concept (Fig. 1).

Insulin-Like Growth Factors, edited by Derek LeRoith, Walter Zumkeller and Robert Baxter. ©2003 Eurekah.com and Kluwer Academic / Plenum Publishers.

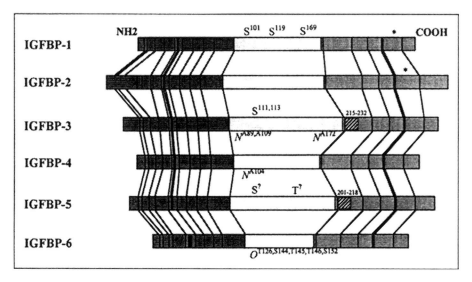

Figure 2. Characteristics of the human high affinity IGF binders of the IGFBP superfamily. Primary structures for IGFBP-1 to -6 are shown. Dark shaded box represents amino-terminal domain, light shaded box the carboxyl-terminal domain with the central white box representing the intermediate region. The locations of cysteines are shown by vertical lines.[3] The cross-hatched boxes indicate the heparin binding domain[29,54] while N- and O-linked glycosylation sites are shown on the lower side of each IGFBP structure.[28,29,40] The location of the phosphorylated serines (S) and threonines (T) are shown on the upper side of each IGFBP structure.[23,25,26] The * designate the location of the RGD sequence.[15,126]

modifications, such as phosphorylation,[23-26] glycosylation,[27-29] and proteolysis.[30-33] Of the high affinity IGFBPs, IGFBP-1, -3 and -5 are secreted as phosphoproteins[25,26] (Fig. 2). To date, a biological role for phosphorylated IGFBP-5 and its effect on IGF affinity has yet to be described, while direct mutagenesis of IGFBP-3 phosphorylated serines was found to have no effect on IGF binding characteristics.[25] The dephosphorylation of IGFBP-3 has however been suggested to cause a 2-fold increase in IGFBP-3 affinity for ALS.[34] In contrast to IGFBP-3 and -5, phosphorylation of human IGFBP-1, as opposed to rat IGFBP-1, has been shown to result in a 6-fold higher affinity for IGF-I, enabling the inhibition of its actions.[23,35] Recently, alpha(2)-macroglobulin has been identified as a binding protein for IGFBP-1, which modifies IGF-I/IGFBP-1 actions and is dependent on the phosphorylation status of IGFBP-1, overall resulting in enhanced IGF effects.[36] IGFBP-1[37,38] and phosphorylated IGFBP-1[39] are positively and negatively regulated by hormonal status, therefore suggesting that IGFBP-1 phosphorylation status is a key metabolic response.[37-39] IGFBP-3 is also posttranslationally modified through the glycosylation of consensus N-glycosyation sites at Asn[89] and Asn[109] and a variable site at Asn[27,29,172] (Fig. 2). IGFBP-4 contains a consensus N-glycosyation site at Asn,[28,104] while IGFBP-6 is O-glycosyated[40] (Fig. 2). To a large extent no significant differences have been observed in the affinity of IGFs for glycosylated and nonglycosylated IGFBP-3 and –4.[27,29,40] It has, however, been suggested that glycosylation of IGFBP-6 reduces its interaction with glycosaminoglycans and protects it from proteolytic cleavage, enabling it to maintain a high affinity interaction with IGF-II.[41] Proteolytic cleavage has been observed for IGFBP-1 to -5 of the high affinity IGFBPs and provides a key mechanism for regulating IGF bioavailabilty for IGF-receptors. Only a small number of the IGFBP specific proteases have been identified in serum and other biological fluids, including prostate-specific antigen[31] (PSA), plasmin[30,42] and thrombin,[43] cathepsin D[44,45] and the matrix metalloproteases[46] and disintegrin metalloproteases[47]. The IGFBP proteases all appear to cleave within the non-conserved central domains of the IGFBPs, with the resulting proteolysed fragments showing a significant decrease in their affinity for IGFs. In contrast, IGFBP- 4 and -5 mutants resistant to proteolytic degradation were found to cause an increase in IGFBP growth-inhibitory effects in the cell lines examined.[48-50]

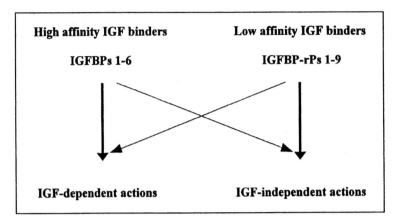

Figure 1. Schematic diagram of mechanisms for IGF-independent and -dependent biological actions of IGFBP superfamily.

IGF-Dependent Effects for the IGFBP Superfamily

Whether inhibiting or augmenting IGF bioactivities, these IGFBP functions are defined as "IGF-dependent":

1. their ability to control IGF "insulin-like" effects;
2. roles as IGF transporters, regulating cell and tissue localization;
3. roles as stabilizing factors, increasing IGF half-life and prolonging metabolic clearance;
4. the ability to modulate IGF/IGF receptor interactions, key for the IGF biological effects.[1,3,4]

These "IGF-dependent" functions have been shown to be key to the diverse biological outcomes of the IGF axis.

High-Affinity IGF Binders

In human serum the combined mean IGF concentration is 800µg/L, 1000-fold greater than the concentration of insulin, which theoretically could provide an insulin activity 50-fold higher than insulin alone.[6,7] However, blood glucose homeostasis remains controlled, and this is attributed to the presence of the high affinity IGFBPs. Both IGF-I and -II bind with a higher affinity to IGFBP-1 to -6, than to the IGF receptors, with binding affinity constants of k_d 10^{-10} M compared to k_d 10^{-8} - 10^{-9} M for the IGF receptors.[8-11] Affinity is key to the IGFBPs ability to regulate the IGFs. Serum contains the largest source of IGFs, where IGFBP-3 is the major circulating IGFBP. IGFBP-3 binds to >75% of serum IGFs in a ternary 150 kilodalton (kDa) complex with the acid-labile subunit[12,13,7] (ALS). In serum, IGFBP-5 also forms a ternary complex of approximately 130 kDa with IGFs and ALS,[14] while the remaining serum IGFs are bound to IGFBP-1, -2, -4 or -6, comprising lower molecular weight complexes of approximately 50 kDa.[15] The lower molecular weight complexes can cross the capillary barrier, enabling IGFs to access their target tissues; on the other hand, while bound in ternary complexes, IGFs remain stored within the vasculature, being unable to cross the capillary barrier.[7] As well as regulating access, ternary complexes also provide a means of stability, as the half-life of ternary complexed IGF-I, in comparison to free IGF-I, is increased from 10-12 minutes to 15 hours.[16] The functional significance of serum IGFs stored as a ternary complex is unknown, although, it has been hypothesized that proteases activated in response to stress, cleave the binding proteins at specific sites, decreasing their affinity and releasing IGFs, which can then cross the vasculature and access the IGF-receptors of their target tissues.[17,18] A range of pathological conditions associated with stress, such as starvation,[19] growth hormone (GH) deficiency,[20] hypoglyceamia[21] and surgery,[22] have been shown to have altered ternary complex to low molecular weight complexes and free IGF ratios.

As introduced, the regulation of IGFs and IGF/IGF receptor interactions by IGFBPs also involves modifications that alter IGF/IGFBP affinity, including specific IGFBP posttranslational

IGF regulation is also influenced by the interaction of IGFBPs with specific proteins on the cell surface or the extracellular matrix (ECM; Fig. 2). When IGFBPs are associated with the cell surface or ECM, a significant decrease in IGF affinity for the IGFBPs has been observed.[51-53] This has been substantiated by the identification of key basic residues within the carboxyl-terminal domain or heparin binding domain of IGFBP-3 and -5 responsible for cell surface and ECM binding, as well as IGF affinity (Fig. 2).[54-56] The physiological consequence of specifically localizing IGFBPs to the cell surface or ECM can be explained in the case of IGFBP-5, where it was shown that IGFBP-5 bound to the ECM could potentiate the effect of IGF-I on cell growth.[51,57] A more systemic example is in the case of cell surface proteoglycans, such as glycosaminoglycans, which have been shown to interfere with ternary complex formation involving IGFBP-3 or IGFBP-5, thus providing an additional mechanism for IGF release from the vasculature.[58,59]

Low-Affinity IGF Binders

"IGF-dependent" functions for the low-affinity IGFBPs of the IGFBP superfamily, the IGFBP-rPs, have yet to be elucidated. The fact that IGFBP-rPs have a 100-1000-fold lower affinity for IGFs than do the classic high affinity IGFBPs, indicates that the primary function of the IGFBP-rPs is not the regulation IGF action.[60-62] However, there is new evidence to show that, in fact, IGFBP-rPs may have IGF-dependent effects. Kofoed et al[63] have shown Mac25/IGFBP-rP1 and CTGF/IGFBP-rP2 differentially regulate IGF-I autophosphorylation of the IGF-IR in fibroblast cells engineered to over-express the IGF-IR. Studies such as this are now diversifying the scope of IGFBP/IGF-dependent functions and reinforcing the IGFBP-superfamily concept.

IGF-Independent Effects for the IGFBP Superfamily

From the inference and then identification of IGF carrier proteins in serum,[64] to the subsequent isolation and cloning of the high affinity IGFBPs[15] and the extension of the IGFBP superfamily concept with the identification of a family of low-affinity IGFBPs,[60-62,3] the major focus for IGFBP actions since their conception has been their stimulatory and inhibitory effects on IGFs, their "IGF-dependent" actions. One of the more dynamic areas in IGFBP research presently, is the now recognized third function of the IGFBPs, their "IGF-independent" functions; biological actions independent of IGF/IGF-receptor interactions. From the initial observation in 1989 by Blat et al[65] that IGFBP-3 could be a multifunctional protein exhibiting biological actions beyond its ability to regulate IGF/IGF-receptor interactions, it is now apparent that this concept is not limited to IGFBP-3. Although the high affinity IGFBPs have a high degree of structural and sequence based similarities, each IGFBP and IGFBP-rP also possesses unique characteristics that could be responsible for multifunctional effects, including IGF-independent actions. For example serine phosphorylation occurs for IGFBP-1,[24] and -3,[25,26] while IGFBP-1 and -2 contain the integrin receptor recognition sequence Arg-Gly-Asp (RGD sequence).[1] The binding proteins IGFBP-3, -5, and -6 contain heparin binding motifs[1]; IGFBP-3 and -5 and Mac25/IGFBP-rP1 have nuclear localization sequences (NLS), where in the case of IGFBP-3 and -5 they are known to be translocated to the nucleus.[60,66-68] These unique characteristics, in some cases, have been shown to be mechanisms that provide the means for IGF-independent actions. In this section of the review we will present the current knowledge of this expanding area of research, the IGF-independent actions of the IGFBP superfamily.

High Affinity IGF Binders

IGFBP-3

IGFBP-3 is a well-documented inhibitor of cell growth and/or promoter of apoptosis and while one theory is that this is achieved through the attenuation of IGF/IGF-receptor interactions,[69] it is becoming clearer that these bioactivities also occur via IGF-independent means.[70-74] The very first clear suggestion of IGF-independent effects for the IGFBPs was made for IGFBP-3. The protein inhibitory diffusible factor 45 (IDF45; IGFBP-3) was originally isolated as a novel inhibitory factor, being able to inhibit the stimulation of chick embryo fibroblasts (CEF) in the absence as well as presence of IGF-I, yet when bound to IGF-I, serum induced growth stimulation of CEF was attainable suggesting two activities for IDF45.[65] This study was followed quickly by another supporting this hypothesis, where fibroblast growth factor (FGF)-stimulated DNA synthesis in CEF and mouse embryo fibroblasts (MEF) was inhibited by mouse IGFBP-3, an action independent of IGFs, yet attenuated by IGFs.[75] Similarly, the over-expression of human IGFBP-3 in Balb/c mouse fibroblasts resulted in the inhibition of cellular proliferation in the presence/absence of IGFs or insulin.[76] The more definitive studies for the then new concept of IGFBP-3 IGF-independent actions came from evidence in human cell systems demonstrated by Oh et al[70,77] where IGFBP-3 inhibited the growth of Hs578T breast cancer cells in the presence of IGF analogs that had reduced affinity for IGFBPs but normal binding affinities for the IGF-receptors. This concept was supported in studies using a mouse embryo fibroblast cell line that contained a targeted disruption of the IGF-IR gene (R- cells). When transfected with human IGFBP-3 cDNA, R- cells showed a 10-fold slower growth rate and underwent apoptosis in comparison to wild type cells.[71,78] These studies and a recent investigation in a rat chondrocyte model system,[79] suggested that IGFBP-3 induced cell growth inhibition and apoptosis did not involve IGF/IGF-IR interactions, but cell signaling pathways independent of the IGF-IR.[71,78,79]

IGFBP-3 Is a Central Mediator

To understand the importance of independent actions for IGFBP-3 it is necessary to understand its scope of influence. IGFBP-3 directly regulates the biological outcomes of a number of growth factors and nuclear hormones. Agents that inhibit breast or prostate cancer cell proliferation in vitro via IGFBP-3-dependent mechanisms includes transforming growth factor-β[80,81] (TGF-β), retinoic acid[81] (RA), vitamin D,[82,83] tumor necrosis factor-α[84] (TNF-α) and antiestrogen compounds such as tamoxifen and ICI 182,780.[85,86] Treatment with these, as well the histone deacetylase inhibitors sodium butyrate (NaB) and trichostatin A[87] (TSA) as well as the expression of the tumor suppressor gene p53,[88] result in a significant increase in the synthesis and secretion of IGFBP-3. Conversely, mitogens of breast cancer cells such as epidermal growth factor (EGF) and estrogen inhibit the synthesis of IGFBP-3, therefore enabling the attenuation of its bioactivities.[69,84,86,89] The molecular mechanism for the increase or decrease in IGFBP-3 synthesis remains largely unknown, however in the case of NaB and TSA, IGFBP-3 up-regulation has been shown to occur through direct IGFBP-3 promoter interactions,[90] while p53 recognition sequences have been identified in the IGFBP-3 promoter[91] and IGFBP-3 introns.[88] As a consequence of its clear centralized role it has therefore been suggested that IGFBP-3, in most cases, is the or one of the downstream effector molecules for number of antiproliferative and proapoptotic agents.[92] This is not felt to be a possibility for p53, where pretreatment with IGFBP-3 enhances ionizing radiation-induced apoptosis in colon and esophageal cancer cell lines expressing wild type p53 in contrast to those expressing a mutant form.[93,94] In addition to these observations, pretreatment with IGFBP-3 increases the expression of p53 in the presence of ionizing radiation, therefore enhancing p53 bioactivities.[93] It seems that IGFBP-3 in this instance plays more a synergistic role with p53 than one as a downstream effector.

Downstream Targets for IGFBP-3

While the importance of IGFBP-3 in cell growth inhibition and apoptosis is no longer a question, the mechanism by which IGFBP-3 achieves these biological effects now is. While it appears that the sequestering of IGFs by IGFBP-3 is fundamental to this question, as observed for TNF-α and vitamin D in different cell systems,[82-84,95] a number of studies performed in the absence of endogenous and exogenous IGFs, have shown IGFBP-3 growth inhibitory and apoptotic effects suggesting mechanisms independent of IGF/IGF-receptor interactions. This was the case for TGF-β and RA in breast cancer cells in vitro lacking a functional IGF axis, where TGF-β and RA cell growth inhibition was attenuated using antisense oligonucleotides to block endogenous IGFBP-3 expression.[80,81] Recent experimental evidence now indicates specific IGFBP-3 downstream targets common to cell cycle control and apoptosis. The over-expression of IGFBP-3 has been found to increase the ratio of pro-apoptotic to anti-apoptotic proteins, independent of IGFs and the tumor suppressor p53. The p53 negative breast cancer cell line T47D when transfected with IGFBP-3 expressed higher levels of the pro-apoptotic proteins Bax and Bad and lower levels of the anti-apoptotic proteins Bcl-2 and Bcl-x(L) and subsequently underwent apoptosis.[96] This effect was further enhanced by treatment with ionizing radiation (Butt et al, 2000). To substantiate this finding it is of note that TGF-β,[97] RA[98] and vitamin D,[99] despite increasing the synthesis of IGFBP-3 also inhibit the expression of the anti-apoptotic protein Bcl-2 perhaps via IGFBP-3 actions.

In addition to pro- and antiapoptotic specific proteins, IGFBP-3 is now being shown to mediate a number of other downstream effectors, including a number of cell cycle specific proteins. Using an IGFBP-3 ecdysone inducible MCF-7 stable cell line, the expression of cell cycle related proteins specific to G1/S phase such as cyclin-D1, cyclin dependent kinase-4 (cdk4) and phosphorylated retinoblastoma protein (Rb) are regulated by elevated IGFBP-3, with a significant decrease in their expression.[100] The significance of such data is the observed clear regulation of cell cycle control by IGFBP-3. Similarly in LNCaP prostate cancer cells, IGFBP-3 has been shown to specifically regulate the induction of the CDK inhibitory protein p21/WAF/CIP1 by 1α, 25-dihydroxyvitamin D_3.[101] Links to caspases, intracellular proteases key for the induction of apoptosis, have also been implied, although have yet to be proven. Inactive or procaspase-3 expression levels were decreased while IGFBP-3 levels were increased in Hs578T breast cancer cells undergoing apoptosis in response to the chemotherapeutic agent paclitaxel.[102] Interestingly, an active Ras-mitogen-activated protein kinase pathway (MAPK) has been shown to block IGFBP-3 cell growth inhibition, which was restored by inhibiting the MAPK. These early investigations have implicated a large array of signaling proteins but for now, though, the identification of IGFBP-3 downstream effectors, key to its independent actions, is clearly only in its infancy.

Mechanisms for IGFBP-3 IGF-Independent Actions

Conceptually, IGFBP-3 can exert its actions on target cells in two ways:
1. activating cell surface receptors that initiate intracellular second messenger pathway/s or
2. by direct importation and translocation to the nucleus where it can induce its effect directly on gene expression.

Mechanisms for both concepts have been proposed for IGFBP-3, from the identification of cell surface proteins such as a novel putative high-affinity IGFBP-3 interacting protein,[70,76-78,80,81] to the previously identified TGF-β type V receptor[72] as well as influencing IGF-IR activation independent of IGFs.[103] Additionally, IGFBP-3 can be translocated directly to the nucleus by the importin β subunit[74] while it also interacts directly with the nuclear retinoid receptor-X-α subunit[73] (RXR-α). These recent studies show that IGFBP-3 does not achieve its biological effects by a single mechanism but appears to do so by a number of mechanisms, some of which still perhaps remain to be elucidated.

Novel Putative IGFBP-3 Cell Surface Interacting Protein

As described, IGFBP-3 is a key mediator for a number of growth factor and nuclear hormones, with treatment resulting in a significant increase in not only the synthesis but importantly the secretion of IGFBP-3. Secreted IGFBP-3 appears to have biological significance being that TGF-β cell growth inhibition of the IGF-nonresponsive Hs578T cells was attenuated by the presence of IGF-II.[80] Likewise the cell growth inhibitory effects of TGF-β and Vitamin D_3 on prostate cells could be negated by the immunodepletion of secreted IGFBP-3 from the culture medium.[71,101] The requirement of IGFBP-3 to be secreted prior to inducing its effects suggests that the presence of cell surface receptors/proteins is essential for IGFBP-3 independent actions. A number of groups have shown IGFBP-3 cell surface association in an array of cell lines.[104-106] Oh et al[70,77] have demonstrated that IGFBP-3 is capable of specific binding to the cell surface and acting as a growth inhibitor for both estrogen receptor responsive and nonresponsive human breast cancer cells. Using the yeast two-hybrid system to identify potential mediators of IGFBP-3-induced anti-proliferative effects, Ingermann et al[107] identified a novel putative IGFBP-3-interacting protein. The protein consists of 240 amino acids, with a predicted molecular weight of 26 kDa and a putative transmembrane domain near the C-terminus. The 0.9 kb mRNA of the IGFBP-3-interacting protein is widely expressed in human tissues. Over-expression of the IGFBP-3-interacting protein in the presence of IGFBP-3 resulted in significant inhibition of cell proliferation by visual assessment and MTS proliferation assay. This novel protein binds specifically to IGFBP-3, not other IGFBPs. This binding is competed in the presence of IGF peptides, indicating that the binding of IGFBP-3 to the IGFBP-3-interacting protein and IGF-I are mutually exclusive. The IGFBP-3-interacting protein was detected on the cell surface by immunostaining under conditions in which the integrity of the plasma membrane remains intact. In addition, cellular over-expression of the IGFBP-3-interacting protein resulted in a significant increase in IGFBP-3 binding to the cell surface as measured by monolayer binding assays. Finally, in breast cancer cells not expressing IGFBP-3, or in cells in which incubation with IGF-I or IGF-I analogs blocked IGFBP-3 interaction with the IGFBP-3-interacting protein, the effect of the IGFBP-3-interacting protein-mediated cell proliferation is abrogated. It has been proposed that this novel IGFBP-3 interacting protein may represent a functional IGFBP-3 receptor that mediates the biological function of IGFBP-3, such as cell growth inhibition, apoptosis and some yet to be revealed actions of IGFBP-3.[107]

Serine Threonine Kinase Type V TGF-β Receptor

As will be discussed in detail in a later chapter, IGFBP-3 has been proposed to be a functional ligand for the serine threonine kinase type V TGF-β receptor, causing cell growth inhibition.[72,108] The cell signaling pathway/s activated by IGFBP-3/type V TGF-β receptor interactions are thought to be distinct from those activated by TGF-β_1/type V TGF-β receptor, as phosphorylation of key signaling proteins for TGF-β_1/type V TGF-β receptor, Smad2 and Smad3 are not phosphorylated by IGFBP-3.[108] To date the cell signaling pathways for IGFBP-3/type V TGF-β receptor interactions remain to be elucidated, however the TGF-β receptor family consists of a number of receptor isoforms (type I to VI), with type I and type II TGF-b receptors involved in an interactive signal transduction pathway that involves the phosphorylation of Smad2 and Smad3.[109,110] It has recently been identified that the type I and type II TGF-b receptors, play a role in IGFBP-3 signaling.[111] Fanayan et al[111] showed that transient transfection of the type II TGF-b receptor into T47D cells which only express the type I TGF-b receptor and are unresponsive to TGF-β and IGFBP-3,[111,112] was required by both TGF-β and IGFBP-3 to induce growth inhibition, while IGFBP-3 alone had no effect. When they investigated the signaling intermediates of the interactive type I and -II TGF-β receptor second messenger pathway, they identified that IGFBP-3 enhanced TGF-β phoshorylation of Smad2 and Smad3 but also phosphorylated them independent of TGF-β, therefore suggesting overall that extracellular IGFBP-3 requires a functional TGF-β axis to cause cell growth inhibition.[111]

IGFBP-3/IGF-Receptor Interactions

Conceptually, IGFBP/IGF-receptor interactions cannot be excluded as a mechanism for IGF independent actions. Until recently, IGFBP-3 effects on the IGF-I receptor axis have remained elusive, however Ricort and Binoux[103] have now shown that IGFBP-3 can in fact influence the IGF-IR intracellular signaling independent of IGFs. Using the IGF-I analogs des(1-3) IGF-I and $Q^3A^4Y^{15}L^{16}$-IGF-I that have a significantly decreased affinity for IGFBPs, 100-fold and 1000-fold decrease in affinity respectively,[11] MCF-7 breast cancer cells preincubated with IGFBP-3 showed a similar inhibition of IGF-IR phosphorylation in the presence of IGF-I or IGF-I analogs. It was concluded that IGFBP-3 inhibition of the IGF-IR phosphorylation is independent of IGF sequestration. A direct interaction between IGFBP-3 and the IGF-IR could not be shown therefore suggesting that other intermediate proteins regulate the effects of IGFBP-3 on the IGF-IR, potentially putative cell surface associated proteins.

Importin-β-Nuclear Transport Protein

IGFBP-3 has been shown to have a positively charged NLS within the carboxy terminal domain that is essential for the translocation of IGFBP-3 to the nucleus.[74] These observations increase the scope of IGFBP-3's actions to potential direct nuclear effects, such as regulating gene transcription. The translocation of IGFBP-3 to the nucleus, is regulated by the importin-β-nuclear transport protein.[74] When the NLS of IGFBP-3 was mutated, binding to importin-β was diminished and the nuclear import of IGFBP-3 was ablated.[74] Nuclear translocation of IGFBP-3 has been reported for opossum kidney cells[66] and T47D breast cancer cells[68] while a number of cell lines have identified nuclear IGFBP-3,[67,113] however, to date a role for nuclear IGFBP-3 has yet to be established as IGFBP-3 remains to be shown to directly interact with DNA.

Nuclear Retinoid Receptor-X-α Subunit (RXR-α)

As will be discussed in detail in a later chapter, direct effects on transcription may come from the finding that IGFBP-3 can interact with an isoform of nuclear hormone receptor RXR, RXR-α, within the nucleus.[73] Retinoid-X receptors are nuclear receptor cofactors that play important roles in the regulation of gene transcription.[114] The RXR-α isoform heterodimerizes with other nuclear receptor subunits such as retinoic acid receptor (RAR),[115] the perixisome proliferator activating receptor,[116] thyroid receptor[117] and the vitamin-D receptor[118] where in turn these dimers recognize key DNA response elements following their regulation by an array of co-activators and inhibitors.[119] In the study by Liu et al,[73] IGFBP-3 enhanced RXR transactivation by an RXR-specific ligand through a DNA/protein complex involving RXR-α and an RXR response element, but dose dependently inhibited the transactivation of RAR by RA. They also observed that the IGFBP-3/RXR-α interaction had a functional consequence, since their interaction resulted in apoptosis of F9 embryonic cancer cells but not in a sister cell line containing RXR-α knockout gene. They also observed in prostate cancer cells that co-treatment with RXR-α agonist and IGFBP-3 enhanced the apoptotic effect caused by their independent treatment. From this study it seems clear that the RXR-α/IGFBP-3 interaction is important for RXR-α transcriptional activity and consequently IGFBP-3 IGF-independent influence on apoptosis.

IGF/IGF-Receptor Independent Actions for IGFBP-3 Proteolytic Fragments

It is now well recognized that IGFBP-3 is specifically proteolysed under a range of conditions that can increase with respect to different physiological and pathological circumstances.[120] As described, the primary role for IGFBP-3 proteolysis appears to be the reduction in its affinity for IGFs, allowing the sustained release of free IGF-I for the IGF receptor.[27,120] However, IGFBP-3 fragments derived from in vitro plasmin digestion or generated by Baculovirus and/or E.coli expression systems, have been shown to have secondary roles in being able to inhibit the mitogenic effects of IGF-I in CEF[121] and insulin, including insulin-induced autophosphorylation of the insulin receptor, and subsequent IRS-I phosphorylation in insulin receptor-overexpressing NIH-3T3 cells,[122,123] suggesting IGF-independent biological activities for the IGFBP-3 proteolytic fragments. Such an effect was also observed using R- cells, the mouse fibroblast cell line with a disrupted IGF-IR gene, where the plasmin generated IGFBP-3 amino-terminal proteolytic fragment of 16 kDa (IGFBP-3 1-85) inhibited FGF- and insulin-stimulated cell growth, while intact IGFBP-3 exhibited a little if any effect.[124] Yamanaka et al[125] (identified that only IGFBP-3 fragments that contain the intermediate region of the IGFBP-3 (88-148) could compete for IGFBP-3 cell surface binding, indicating that the intermediate region of the IGFBP-3 molecule is responsible for cell surface binding. Overall, studies of IGFBP-3 proteolytic fragments are key to the understanding of IGFBP-3 functional domains.

IGFBP-1 and -5

The high affinity IGFBPs have a high degree of structural and sequence based similarities, yet each also possesses unique characteristics that could be responsible for IGF-independent actions. This has been demonstrated for IGFBP-1 which like IGFBP-2, possesses an RGD-domain, an integrin binding motif within the carboxy terminal region which is responsible for specific binding of IGFBP-1 to the cell surface (Fig. 2). The first implication of IGF-independent functions for IGFBP-1 through the RGD-domain were described by Jones et al,[126] who demonstrated that IGFBP-1 could bind specifically to the α5β1-integrin receptor on the cell surface with the functional importance of this specific binding being the stimulation of cell migration of Chinese hamster ovary (CHO) cells in an IGF-independent manner. Via in vitro mutagenesis, the RGD sequence of IGFBP-1 was altered to Trp-Gly-Asp (WGD) and tested on the migration of CHO cells, which express a high level of α5β1-integrin receptors. Cells expressing or exogenously treated with the WT protein showed 3-fold and 2-fold greater migration respectively in 48 hr period, when compared with cells expressing the WGD mutant form of IGFBP-1. Likewise a synthetic peptide for RGD could compete with IGFBP-1 cell association while IGF-I had no effect thus concluding that IGFBP-1 cell surface binding to the α5β1-integrin receptor had a functional significance independent of IGFs.[126]

The affiliation of IGFBP-1 to the α5β1-integrin receptor appears to have a functional significance in the in vivo environment, where migration is critical to development. Human trophoblast implantation is an invasive process that is critically dependent on cellular migration. The binding of IGFBP-1 to the α5β1-integrin receptor is now known to be important for human trophoblast invasion because IGFBP-1 mediated trophoblast migration in vitro was blocked by antibodies specific to the α5- and β1-subunits of the α5β1-integrin receptor.[127]

It is well recognized that β1-integrins activate cytoplasmic signals.[128] Cell signaling events have been implied for the IGFBP-1/α5β1-integrin receptor interactions, where breast cancer cells grown in vitro were treated with IGFBP-1, a synthetic RGD-containing peptide (Gly-Arg-Gly-Asp-Thr-Pro) and a negative control peptide RGE (Arg-Gly-Glu-Ser). Both IGFBP-1 and the synthetic RGD-containing peptide dose dependently caused the dephosphorylation of focal adhesion kinase (FAK), which consequently influenced cell attachment and cell death.[129] These results are also applicable to the in vivo environment where IGFBP-1 IGF-independent stimulation of trophoblast cell migration regulated by binding of its RGD domain to the α5β1-integrin receptor, leads to the phosphorylation and activation of FAK and the subsequent stimulation of MAPK pathway.[130]

Of the IGFBPs, IGFBP-5 is probably the most closely related binding protein to IGFBP-3, exhibiting characteristics that are similar to IGFBP-3. Like IGFBP-3, IGFBP-5 is proteolysed into fragments of approximately 23-, 20- and 17-kDa[131] but the role of IGFBP-5 proteolysis is still unclear. It has been demonstrated that the proteolysed 23-kDa IGFBP-5 fragment, can stimulate mitogenesis in osteoblasts and in vivo bone formation, independently of IGFs,[132-134] suggesting that IGFBP-5 proteolysis could be an important mechanism for regulating cell growth and proliferation in an IGF-independent manner.

IGFBP-5, like IGFBP-3 has an 18-residue region (201-218) in the basic carboxy-terminal domain of the protein that is critical for cell attachment, matrix and glycosaminoglycan interactions and nuclear localization.[54,68] In addition to intact IGFBP-5, this IGFBP-5(201-218) heparin binding region is also able to stimulate cellular migration of mesangial cells (MC), a stimulation which can be inhibited by heparin suggesting this region is able to control direct IGFBP-5 effects on MC cells.[135]

As for IGFBP-3, it has been suggested that IGFBP-5 signals independently of the IGF-IR via a putative cell surface receptor located on the plasma membrane.[136] This putative 420kDa IGFBP-5 receptor was autophosphorylated on the serine residues when co-incubated with IGFBP-5 suggesting the possibility that serine/threonine kinase activation may be important in mediating some of the IGF-independent effects of IGFBP-5.[136] Interestingly, IGFBP-5 has also been identified as a ligand for the type V TGF-β receptor, although downstream activation and the biological significance for this interaction, has yet to be reported.[108] Interestingly, a signaling pathway has been proposed for the putative IGFBP-5 receptor. Apart from dose dependently stimulating cell migration of MC cells, Berfield et al[137] showed that intact IGFBP-5 and the IGFBP-5(201-218) heparin binding region, independently of IGFs induced dramatic morphological effects on MC cells, with the formation of multidirectional filopodial extensions. Of note was that this biological effect involved the intracellular activation of the Rho-related GTPase Cdc42, a specific mediator of cytoskeletal organization, suggesting the presence of a clear signaling pathway. Berfield et al[137] proposed that Cdc42 activation is as a result of IGFBP-5/IGFBP5-receptor interactions, as blocking serine/threonine phosphorylation inhibited Cdc42 activation.

The IGFBP-5(201-218) heparin binding region in IGFBP-5, like IGFBP-3, functions as a NLS and is essential for the translocation of IGFBP-5 to the nucleus in T47D breast cancer cells.[68] Interestingly, preincubation with IGFBP-3 blocked the nuclear translocation of both IGFBP-3 and -5 implying that the nuclear traslocation mechanism was via a common pathway.[68] In fact, as already described the translocation of IGFBP-5 to the nucleus was regulated by the same mechanism as IGFBP-3, the importin-β-nuclear transport protein.[74] The direct translocation of IGFBP-5 to the nucleus broadens the scope of IGFBP-5's actions to include potential direct nuclear effects such as transcriptional regulation, although the biological effects of nuclear IGFBP-5 as well as IGFBP-3 remain to be established.

IGFBP-2, -4 and -6

Of the remaining members of the high affinity IGFBPs, none to date have shown clear and convincing IGF-independent actions, despite unique characteristics that give each the potential for their own biological actions. IGFBP-2 is not glycosylated or phosphorylated but it does have an RGD-sequence in the carboxy-terminal domain like IGFBP-1, which serves as an integrin-binding motif and enables specific binding of IGFBP-1 to the cell surface.[126] Cell surface binding of IGFBP-2 to proteoglycans in rat olfactory bulb mitrial layer has been observed but no correlations to IGFBP-2 independent bioactivities have been shown.[53] It is also feasible that IGFBP-2 may bind to integrins and exert IGF-independent actions, but for now this remains to be demonstrated. Two truncated IGFBP-2 proteins of 14 and 16 kDa generated by proteolysis have been characterized in human milk.[138] Likewise, two truncated IGFBP-2 proteins of 14 and 22 kDa have been identified in the culture medium of SK-N-SH neuroblastoma cells with intact IGFBP-2 and the 14 kDa fragments showing cell surface binding.[139] Despite evidence implicating IGF regulation for these proteolysed IGFBP-2 fragments,[137,139] it remains plausible that their reduced affinity for IGFs and cell surface binding may be evidence for a pathway of independent actions.

Of the high affinity-IGFBPs, IGFBP-4 is the smallest and can be found in biological fluids in either a glycosylated or non-glycosylated form.[28] Proteolytic cleavage of IGFBP-4 by pregnancy associated plasma protein-A (PAPP-A) has been identified in a number of biological fluids, however this appears to be an IGF-dependent process.[140] There is no clear evidence for IGFBP-4 cell surface binding or its interaction with the ECM, however IGFBP-4 has been identified as a weak ligand for the type V TGF-β receptor, joining the ranks of IGFBP-3 and IGFBP-5 and therefore suggesting that IGFBP-4 may in fact induce bioactivities via IGF-independent signaling pathway.[108]

The observation that IGFBP-3, -5, and -6 bind to heparin- or heparin sulfate-containing proteoglycans in the ECM suggests IGF-dependent as well as IGF-independent actions for IGFBP-6.[141] The binding of IGFBP-6 to heparin occurs with a weak affinity and perhaps is the reason why IGFBP-6 cell associations have to date not been observed.[141] Like IGFBP-3, a range of growth inhibitory agents has been shown to increase the synthesis of IGFBP-6 including RA[142,143] and vitamin D.[144] Likewise, the stable transfection of IGFBP-6 into neuroblastoma cells inhibited the mitogenic effects of serum, IGF-I, IGF-II, and the IGF-I analog des (1-3) IGF-I, which has no affinity for IGFBP-6[145] while in non small cell lung carcinoma cell lines the adenoviral infection of IGFBP-6 induced apoptosis.[146] Additionally the sub cutaneous injection of IGFBP-6 transfected neuroblastoma cells into nude mice, reduced the tumorigenic potency of these cells.[146] These bioactivities for IGFBP-6, particularly when sustained in the presence of des (1-3) IGF-I which has no affinity for IGFBP-6, shows a great degree of similarity to the IGF-independent bioactivities of IGFBP-3 which supports the possibility that IGFBP-6 itself may induce its effects independently of IGFs.

Low Affinity IGF Binders

Since the identification of the low affinity IGF binders (IGFBP-rPs), the focus for a number of groups has been to establish "IGF-dependent" actions for the IGFBP-rPs, to extend the IGFBP superfamily concept. As previously described, the fact that the IGFBP-rPs are secreted and have a 100-1000-fold lower affinity for IGFs than the high affinity IGFBPs indicates that the primary function of the IGFBP-rPs is not the regulation IGF actions.[60-62,3] In fact, the IGFBP-rPs were discovered and isolated in systems independent of IGFs and it has only been through the high conservation of the amino-terminal sequence important for IGFBP binding to IGFs, that it is believed that the IGFBP-rPs are related, albeit distantly, to the high affinity IGFBPs and therefore comprise a sub-group of the IGFBP superfamily[3] (Figure3). As a result, the IGFBP-rP actions described presently in the literature are all independent of IGFs and we will briefly review their functions here.

Three groups independently isolated and characterized what IGF researchers have classified as Mac25/IGFBP-rP1.[147] This cysteine rich protein was first isolated and characterized as Mac25 and its structural relationship to the high affinity IGFBPs was described,[146-148] while tumor adhesion factor (TAF)[148] and prostacyclin stimulating factor (PSF)[149] were characterized later. The described functions for Mac25/IGFBP-rP1 seem to be diverse and the physiological significance of this protein remains to be defined. Mac25/IGFBP-rP1 has been detected in a range of biological fluids[150] and it is expressed in a wide range of normal tissues,[60,151] while it appears to be abrogated in various tumors such as breast cancer,[152] prostate cancer,[153] glioblastoma,[154] squamous cell carcinoma,[154] and in several cancer cell lines.[60,147] Such an expression pattern suggests that Mac25/IGFBP-rP1 may function as a growth-suppressing factor. In support of this hypothesis, a study by Sprenger et al[155] showed that the tumorigenic potential of malignant prostate cancer cells stably transfected with Mac25/IGFBP-rP1 was markedly diminished when compared with control cells. The legitimacy of such a biological function is supported by the finding that a number of growth suppressing factors significantly up-regulate Mac25/IGFBP-rP1, including RA,[151,153] TGF-β[153] and vitamin D3.[156]

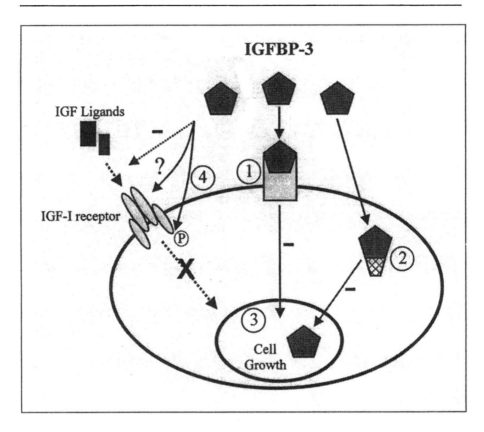

Figure 3. Mechanisms for IGFBP-3 IGF-independent actions. IGFBP-3 is a well-documented inhibitor of cell growth and/or promoter of apoptosis and while one theory is that this is achieved through the attenuation of IGF/IGF-receptor interactions it is becoming clearer that these bioactivities also occur via IGF-independent means. Several mechanisms have been proposed for IGFBP-3 IGF-independent actions including its direct interaction with the MEMBRANE (1) via a novel cell-surface associated protein[107] and also the TGF-β type V receptor.[72] IGFBP-3 interacts in the CYTOPLASM (2) with the importin β subunit where it is translocated to the nucleus,[74] however IGFBP-3 also has been shown potentially influence specific proteins of signaling pathways.[96,100-102,111] It has also been observed that IGFBP-3 has IGF-independent actions in the NUCLEUS (3) where it has been shown to interact with retinoid X receptor-alpha.[73] Finally IGFBP-3 has been shown to influence IGF-RECEPTOR (4) phosphorylation independent of IGF-I.[103]

However, Mac25/IGFBP-rP1 also appears to have another biological function. Mac25/IGFBP-rP1 was identified as being expressed at high levels in senescent normal human mammary epithelial cells (HMEC), suggesting that it could also function as a senescence factor.[151] This hypothesis was supported by independent studies performed in the prostate epithelium.[157] The down-regulation of Mac25/IGFBP-rP1 in human breast cancer, correlated to an inactivation of Rb and the over-expression of cyclin-E with a consequential increase in proliferation, therefore supporting its role as a suppression factor.[158]

The studies described are just a small representation of an increasing number that suggest a clear role for Mac25/IGFBP-rP1 in tumor suppression. In contrast, Mac25/IGFBP-rP1 has been shown to stimulate the growth of mouse fibroblasts, albeit in the presence of IGFs,[159] while it has also been described as an autocrine/paracrine factor that inhibits skeletal myoblast differentiation but permits proliferation in response to IGF.[160] In addition, a study examining co-factors to Mac25/IGFBP-rP1 identified a novel interacting protein, 25.1, which in collaboration with Mac25/IGFBP-rP1 caused neuroendocrine-like differentiation in M12 prostate cancer cells as well as a range of non small cell lung carcinoma (NSCLC) cell lines.[161,162] For now, it is clear from all of these studies that the biological role for Mac25/IGFBP-rP1 is complex, highly diverse and still requires understanding.

The next series of IGFBP-rPs include the immediate-early gene connective tissue growth factor/IGFBP-rP2[163] (CTGF/IGFBP-rP2), the human nephroblastoma overexpression gene NovH/IGFBP-rP3[164] and another immediate-early gene Cyr61/IGFBP-rP4,[165] all members of the CCN family of proteins[166] (CTGF, Cyr61 and NovH. Like Mac25/IGFBP-rP1, these proteins are also cysteine rich and possess a structural domain that is highly homologous to the amino-terminal domain of the IGFBPs.[166] In contrast to Mac25/IGFBP-rP1, they have biological functions that appear to clearly potentiate developmental events such as proliferation, cell migration, cell adhesion and cell survival.[167-170] It has been identified that CTGF/IGFBP-rP2 promotes cellular proliferation, terminal differentiation, and angiogenesis during follicular and luteal development, while Mac25/IGFBP-rP1 functions conversely, promoting the terminal differentiation of granulosa cells in preovulatory follicles.[171] The Cyr61/IGFBP-rP4 protein like CTGF/IGFBP-rP2 also induces angiogenesis in vivo,[172,173] promotes tumor growth[172] and potentiates the mitogenic effects of other growth factors, such as FGF on fibroblasts and endothelial cells[167] while CTGF/IGFBP-rP2 mediates the mitogenic activities of TGF-β.[174,175] The NovH/IGFBP-rP3 protein was first identified in myeloblastosis-associated virus type-I induced avian nephroblastomas, where its expression levels were higher than normal controls.[164] It was hypothesized that NovH/IGFBP-rP3 may be a protooncogene, a concept supported by studies in Wilm's tumors.[176] Interestingly, like Mac25/IGFBP-rP1, NovH/IGFBP-rP3 has divergent functions where the full-length protein can inhibit CEF but the amino terminal truncated version causes cellular transformation.[164] Interestingly, the members of the CCN family also possess similar key characteristics that make the high affinity IGFBPs independent proteins. Both CTGF/IGFBP-rP2 and Cyr61/IGFBP-rP4 bind heparin and can be found associated to the ECM,[177] while both are ligands for the $\alpha v \beta 3$ and $\alpha IIb \beta 3$-integrin receptors.[169,173,178] NovH/IGFBP-rP3 is N-glycosylated[164] while CTGF/IGFBP-rP2 has two potential glycosylation sites[163] However, a function for these post-translational modifications have yet to be described.

The CCN family of proteins has also recently expanded to include WISP-1/IGFBP-rP8,[179] WISP-2/IGFBP-rP7[180] and WISP-3/IGFBP-rP9.[180] Like the other members of the CCN family, these proteins are cysteine rich and possess the IGFBP amino-terminal domain and have thus been included as part of the IGFBP superfamily.[3] WISP-2/IGFBP-rP7 has also isolated independently and is recognized as rCop-1 and CTGF-like[181,182] (CTGF-L). Both WISP-2/IGFBP-rP7 and WISP-1/IGFBP-rP8, appear inhibit tumor growth while WISP-3/IGFBP-rP9 may potentiate tumorigenesis.[180,182] The final two proteins that have been proposed to be part of the IGFBP superfamily are L56/IGFBP-rP5[3,183] and endothelial cell-specific molecule-1/IGFBP-rP6[3,184] (ESM-1/IGFBP-rP6). L56/IGFBP-rP5 was originally isolated by subtraction cloning between normal human fibroblasts and transformed fibroblasts[183] and then independently by differential display in osteoarthritic cartilage where it was referred to as HtrA based on its similarity to the *Escherichia Coli* HtrA genes that are essential for bacterial survival at high temperatures.[185] Due to its sequence similarity to this bacterial serine protease it was postulated to also function as a serine protease for IGFBPs[183] and in fact exhibited proteolytic activity in osteoarthritic cartilage.[185] ESM-1/IGFBP-rP6 was cloned originally from human umbilical vein endothelial cells (HUVEC), but for now very little is known about this protein. Despite its conservation of the IGFBP amino-terminal domain, this protein is a class of its own and for now its biological function is largely unknown. It has been reported to be differentially regulated by cytokines: TNF-α and interleukin-1β up-regulate and interferon-gamma down-regulates in HUVEC cells while it associates with LFA-1 integrins on the cell surface of human blood lymphocytes, monocytes, and Jurkat cells.[185] Overall there is no question as to the "IGF-independent" functions of these nine proposed IGFBP-rP proteins based on their original isolation and the diversity of bioactivities. What remains is to establish their "IGF-dependent" functions.

Summary

The major focus for the biological actions of IGFBPs since their conception well over a decade ago has been their stimulatory and inhibitory effects on IGFs, their "IGF-dependent" actions. As has been described, one of the more dynamic areas in IGFBP research presently, is the now recognized third function of the IGFBPs, their "IGF-independent" functions; biological actions independent of IGF/IGF-receptor interactions. Since the initial observation in 1989 by Blat et al[65] that IGFBP-3 could be a multifunctional protein exhibiting biological actions beyond its ability to regulate IGF/IGF-receptor interactions, this concept has expanded rapidly amongst the IGFBP-superfamily. In the future the development of novel tools such as site-specific IGFBP mutants that fail to bind IGFs, as described for IGFBP-3 by Buckway et al[187] and IGFBP-3 and -5 by Imai et al[188] will enable more definitive characterizations of IGFBP IGF-independent functions.

Acknowledgment

Supported by the AACR-AstraZeneca-CRFA Grant and American Cancer Society Grant RPG-99-103-01-TBE.

References

1. Jones JI, Clemmons DR. Insulin-like growth factors and their binding proteins: biological actions. Endocr Rev 1995; 16:3-34.
2. Rajaram S, Baylink DJ, Mohan S. Insulin-like growth factor-binding proteins in serum and other biological fluids: regulation and functions. Endocr Rev 1997; 18:801-831.
3. Hwa V, Oh Y, Rosenfeld RG. The insulin-like growth factor-binding protein (IGFBP) superfamily. Endocr Rev 1999; 20:761-787.
4. Baxter RC. Insulin-like growth factor (IGF)-binding proteins: interactions with IGFs and intrinsic bioactivities. Am J Physiol Endocrinol Metab 2000; 278:E967-976.
5. Baxter RC, Binoux M, Clemmons DR et al. Recommendations for nomenclature of the insulin-like growth factor binding protein (IGFBP) superfamily. Growth Horm IGF Res 1998; 8:273-274.
6. Baxter RC. Insulin-like growth factor (IGF) binding proteins: the role of serum IGFBPs in regulating IGF availability. Acta Paediatr Scand Suppl 1991; 372:107-114.
7. Baxter RC. Insulin-like growth factor binding proteins in the human circulation: a review. Horm Res 1994; 42:140-144.
8. Elgin RG, Busby WH Jr, Clemmons DR. An insulin-like growth factor (IGF) binding protein enhances the biologic response to IGF-I. Proc Natl Acd Sci USA 1987; 84:3254-3258.
9. Ross M, Francis GL, Szabo L et al. Insulin-like growth factor (IGF)-binding proteins inhibit the biological activities of IGF-1 and IGF-2 but not des-(1-3)-IGF-1. Biochem J 1989; 258:267-272.
10. Clemmons DR, Cascieri MA, Camacho-Hubner C et al. Discrete alterations of the insulin-like growth factor I molecule which alter its affinity for insulin-like growth factor-binding proteins result in changes in bioactivity. J Biol Chem 1990; 265:12210-12216.
11. Oh Y, Muller HL, Lee DY et al. Characterization of the affinities of insulin-like growth factor (IGF)-binding proteins 1-4 for IGF-I, IGF-II, IGF-I/insulin hybrid, and IGF-I analogs. Endocrinology 1993; 132:1337-1344.
12. Baxter RC. Characterization of the acid-labile subunit of the growth hormone-dependent insulin-like growth factor binding protein complex. J Clin Endocrinol Metab 1988; 67:265-272.
13. Baxter RC. Circulating levels and molecular distribution of the acid-labile (alpha) subunit of the high molecular weight insulin-like growth factor-binding protein complex. J Clin Endocrinol Metab 1990; 70:1347-1353.
14. Twigg SM, Baxter RC. Insulin-like growth factor (IGF)-binding protein 5 forms an alternative ternary complex with IGFs and the acid-labile subunit. J Biol Chem 1998; 273:6074-6079.
15. Shimasaki S, Ling N. Identification and molecular characterization of insulin-like growth factor binding proteins (IGFBP-1, -2, -3, -4, -5 and -6). Prog Growth Factor Res 1991; 3:243-266.
16. Guler HP, Zapf J, Schmid C et al. Insulin-like growth factors I and II in healthy man. Estimations of half-lives and production rates. Acta Endocrinol (Copenh) 1989; 121:753-758.
17. Hossenlopp P, Segovia B, Lassarre C et al. Evidence of enzymatic degradation of insulin-like growth factor-binding proteins in the 150K complex during pregnancy. J Clin Endocrinol Metab 1990; 71:797-805.
18. Giudice LC, Farrell EM, Pham H et al. Insulin-like growth factor binding proteins in maternal serum throughout gestation and in the puerperium: effects of a pregnancy-associated serum protease activity. J Clin Endocrinol Metab 1990; 71:806-816.
19. Fukuda I, Hotta M, Hizuka N et al. Decreased serum levels of acid-labile subunit in patients with anorexia nervosa. J Clin Endocrinol Metab 1999; 84:2034-2036.
20. Aguiar-Oliveira MH, Gill MS, de A Barretto ES et al. Effect of severe growth hormone (GH) deficiency due to a mutation in the GH-releasing hormone receptor on insulin-like growth factors (IGFs), IGF-binding proteins, and ternary complex formation throughout life. J Clin Endocrinol Metab 1999; 84:4118-26.

21. Baxter RC, Daughaday WH. Impaired formation of the ternary insulin-like growth factor-binding protein complex in patients with hypoglycemia due to nonislet cell tumors. J Clin Endocrinol Metab 1991; 73:696-702.
22. Davenport ML, Isley WL, Pucilowska JB et al. Insulin-like growth factor-binding protein-3 proteolysis is induced after elective surgery. J Clin Endocrinol Metab 1992; 75:590-595.
23. Jones JI, D'Ercole AJ, Camacho-Hubner C et al. Phosphorylation of insulin-like growth factor (IGF)-binding protein 1 in cell culture and in vivo: effects on affinity for IGF-I. Proc Natl Acad Sci USA 1991; 88:7481-7485.
24. Frost RA, Tseng L. Insulin-like growth factor-binding protein-1 is phosphorylated by cultured human endometrial stromal cells and multiple protein kinases in vitro. J Biol Chem 1991; 266:18082-1808.
25. Hoeck WG, Mukku VR. Identification of the major sites of phosphorylation in IGF binding protein-3. J Cell Biochem 1994; 56:262-273.
26. Coverley JA, Baxter RC. Regulation of insulin-like growth factor (IGF) binding protein-3 phosphorylation by IGF-I. Endocrinology 1995; 136:5778-5781.
27. Conover CA. Glycosylation of insulin-like growth factor binding protein-3 (IGFBP-3) is not required for potentiation of IGF-I action: evidence for processing of cell-bound IGFBP-3. Endocrinology 1991; 129:3259-3268.
28. Ceda GP, Fielder PJ, Henzel WJ et al. Differential effects of insulin-like growth factor (IGF)-I and IGF-II on the expression of IGF binding proteins (IGFBPs) in a rat neuroblastoma cell line: isolation and characterization of two forms of IGFBP-4. Endocrinology 1991; 128:2815-2824.
29. Firth SM, Baxter RC. Characterisation of recombinant glycosylation variants of insulin-like growth factor binding protein-3. J Endocrinol 1999; 160:379-387.
30. Campbell PG, Novak JF, Yanosick TB et al. Involvement of the plasmin system in dissociation of the insulin-like growth factor-binding protein complex. Endocrinology 1992; 130:1401-1412.
31. Cohen P, Graves HC, Peehl DM et al. Prostate-specific antigen (PSA) is an insulin-like growth factor binding protein-3 protease found in seminal plasma. J Clin Endocrinol Metab 1992; 75:1046-1053.
32. Conover CA, Kiefer MC, Zapf J. Posttranslational regulation of insulin-like growth factor binding protein-4 in normal and transformed human fibroblasts. Insulin-like growth factor dependence and biological studies. J Clin Invest 1993; 91:1129-1137.
33. Booth BA, Boes M, Bar RS. IGFBP-3 proteolysis by plasmin, thrombin, serum: heparin binding, IGF binding, and structure of fragments. Am J Physiol 1996; 271(3 Pt 1):E465-470.
34. Coverley JA, Baxter RC. Phosphorylation of insulin-like growth factor binding proteins. Mol Cell Endocrinol 1997; 128:1-5.
35. Peterkofsky B, Gosiewska A, Wilson S et al. Phosphorylation of rat insulin-like growth factor binding protein-1 does not affect its biological properties. Arch Biochem Biophys 1998; 357:101-110.
36. Westwood M, Aplin JD, Collinge IA et al. Alpha(2)-Macroglobulin: A new component in the IGF/IGFBP-1 axis. J Biol Chem 2001; 276:41668-41674.
37. Lee PD, Conover CA, Powell DR. Regulation and function of insulin-like growth factor-binding protein-1. Proc Soc Exp Biol Med 1993; 204:4-29.
38. Thissen JP, Ketelslegers JM, Underwood LE. Nutritional regulation of the insulin-like growth factors. Endocr Rev 1994; 15:80-101.
39. Westwood M, Gibson JM, Williams AC et al. Hormonal regulation of circulating insulin-like growth factor-binding protein-1 phosphorylation status. J Clin Endocrinol Metab 1995; 80:3520-3527.
40. Bach LA, Thotakura NR, Rechler MM. Human insulin-like growth factor binding protein-6 is O-glycosylated. Biochem Biophys Res Commun 1992; 186:301-307.
41. Marinaro JA, Neumann GM, Russo VC et al. O-glycosylation of insulin-like growth factor (IGF) binding protein-6 maintains high IGF-II binding affinity by decreasing binding to glycosaminoglycans and susceptibility to proteolysis. Eur J Biochem 2000; 267:5378-5386.
42. Booth BA, Boes M, Dake BL et al. Isolation and characterization of plasmin-generated bioactive fragments of IGFBP-3. Am J Physiol 1999; 276:E450-E454.
43. Zheng B, Clarke JB, Busby WH et al. Insulin-like growth factor-binding protein-5 is cleaved by physiological concentrations of thrombin. Endocrinology 1998; 139:1708-1714.
44. Conover CA, De Leon DD. Acid-activated insulin-like growth factor-binding protein-3 proteolysis in normal and transformed cells. Role of cathepsin D. J Biol Chem 1994; 269:7076-7080.
45. Claussen M, Kubler B, Wendland M et al. Proteolysis of insulin-like growth factors (IGF) and IGF binding proteins by cathepsin D. Endocrinology 1997; 138:3797-3803.
46. Fowlkes JL, Enghild JJ, Suzuki K et al. Matrix metalloproteinases degrade insulin-like growth factor-binding protein-3 in dermal fibroblast cultures. J Biol Chem 1994; 269:25742-25746.
47. Shi Z, Xu W, Loechel F et al. ADAM 12, a disintegrin metalloprotease, interacts with insulin-like growth factor-binding protein-3. J Biol Chem 2000; 275:18574-18580.
48. Conover CA, Durham SK, Zapf J et al. Cleavage analysis of insulin-like growth factor (IGF)-dependent IGF-binding protein-4 proteolysis and expression of protease-resistant IGF-binding protein-4 mutants. J Biol Chem 1995; 270:4395-4400.
49. Imai Y, Busby WH Jr, Smith CE et al. Protease-resistant form of insulin-like growth factor-binding protein 5 is an inhibitor of insulin-like growth factor-I actions on porcine smooth muscle cells in culture. J Clin Invest 1997; 100:2596-2605.
50. Rees C, Clemmons DR, Horvitz GD et al. A protease-resistant form of insulin-like growth factor (IGF) binding protein 4 inhibits IGF-1 actions. Endocrinology 1998; 139:4182-4188.

51. Jones JI, Gockerman A, Busby WH Jr et al. Extracellular matrix contains insulin-like growth factor binding protein-5: potentiation of the effects of IGF-I. J Cell Biol 1993; 121:679-687.
52. McCusker RH, Clemmons DR. Use of lanthanum to accurately quantify insulin-like growth factor binding to proteins on cell surfaces. J Cell Biochem 1997; 66:256-267.
53. Russo VC, Bach LA, Fosang AJ et al. Insulin-like growth factor binding protein-2 binds to cell surface proteoglycans in the rat brain olfactory bulb. Endocrinology 1997; 138:4858-4867.
54. Booth BA, Boes M, Andress DL et al. IGFBP-3 and IGFBP-5 association with endothelial cells: role of C-terminal heparin binding domain. Growth Regul 1995; 5:1-17.
55. Firth SM, Ganeshprasad U, Baxter RC. Structural determinants of ligand and cell surface binding of insulin-like growth factor-binding protein-3. J Biol Chem 1998; 273:2631-2638.
56. Bramani S, Song H, Beattie J et al. Amino acids within the extracellular matrix (ECM) binding region (201-218) of rat insulin-like growth factor binding protein (IGFBP)-5 are important determinants in binding IGF-I. J Mol Endocrinol 1999; 23:117-123.
57. Parker A, Clarke JB, Busby WH Jr et alR. Identification of the extracellular matrix binding sites for insulin-like growth factor-binding protein 5. J Biol Chem 1996; 271:13523-13529.
58. Baxter RC. Glycosaminoglycans inhibit formation of the 140 kDa insulin-like growth factor-binding protein complex. Biochem J 1990; 271:773-777.
59. Twigg SM, Kiefer MC, Zapf J et al. Insulin-like growth factor-binding protein 5 complexes with the acid-labile subunit. Role of the carboxyl-terminal domain. J Biol Chem 1998; 273:28791-28798.
60. Oh Y, Nagalla SR, Yamanaka Y et al. Synthesis and characterization of insulin-like growth factor-binding protein (IGFBP)-7. Recombinant human mac25 protein specifically binds IGF-I and -II. J Biol Chem 1996; 271:30322-30325.
61. Kim HS, Nagalla SR, Oh Y et al. Identification of a family of low-affinity insulin-like growth factor binding proteins (IGFBPs): characterization of connective tissue growth factor as a member of the IGFBP superfamily. Proc Natl Acad Sci USA 1997; 94:12981-12986.
62. Burren CP, Wilson EM, Hwa V et al. Binding properties and distribution of insulin-like growth factor binding protein-related protein 3 (IGFBP-rP3/NovH), an additional member of the IGFBP Superfamily. J Clin Endocrinol Metab 1999; 84:1096-1103.
63. Kofoed EM, Hwa V, Twigg SM et al. CTGF/IGFBP-rP2 and Mac25/IGFBP-rP1 differentially regulate IGF-I-induced signal transduction. Endocrine Society's 83rd Annual Meeting, Abstract 2001; P2-253.
64. Zapf J, Waldvogel M, Froesch ER. Binding of nonsuppressible insulinlike activity to human serum. Evidence for a carrier protein. Arch Biochem Biophys 1975; 168:638-645.
65. Blat C, Delbe J, Villaudy J et al. Inhibitory diffusible factor 45 bifunctional activity. As a cell growth inhibitor and as an insulin-like growth factor I-binding protein. J Biol Chem 1989; 264:12449-12454.
66. Li W, Fawcett J, Widmer HR et al. Nuclear transport of insulin-like growth factor-I and insulin-like growth factor binding protein-3 in opossum kidney cells. Endocrinology 1997; 138:1763-1766.
67. Jaques G, Noll K, Wegmann B et al. Nuclear localization of insulin-like growth factor binding protein 3 in a lung cancer cell line. Endocrinology 1997; 138:1767-1770.
68. Schedlich LJ, Young TF, Firth SM et al. Insulin-like growth factor-binding protein (IGFBP)-3 and IGFBP-5 share a common nuclear transport pathway in T47D human breast carcinoma cells. J Biol Chem 1998; 273:18347-18352.
69. Martin JL, Coverley JA, Pattison ST et al. Insulin-like growth factor-binding protein-3 production by MCF-7 breast cancer cells: stimulation by retinoic acid and cyclic adenosine monophosphate and differential effects of estradiol. Endocrinology 1995; 136:1219-1226.
70. Oh Y, Muller HL, Lamson G et al. Insulin-like growth factor (IGF)-independent action of IGF-binding protein-3 in Hs578T human breast cancer cells. Cell surface binding and growth inhibition. J Biol Chem 1993; 268:14964-14971.
71. Rajah R, Valentinis B, Cohen P. Insulin-like growth factor (IGF)-binding protein-3 induces apoptosis and mediates the effects of transforming growth factor-beta1 on programmed cell death through a p53- and IGF-independent mechanism. J Biol Chem 1997; 272:12181-12188.
72. Leal SM, Liu Q, Huang SS et al. The type V transforming growth factor beta receptor is the putative insulin-like growth factor-binding protein 3 receptor. J Biol Chem 1997; 272:20572-20576.
73. Liu B, Lee HY, Weinzimer SA et al. Direct functional interactions between insulin-like growth factor-binding protein-3 and retinoid X receptor-alpha regulate transcriptional signaling and apoptosis. J Biol Chem 2000; 275:33607-33613.
74. Schedlich LJ, Le Page SL, Firth SM et al. Nuclear import of insulin-like growth factor-binding protein-3 and -5 is mediated by the importin beta subunit. J Biol Chem 2000; 275:23462-23470.
75. Liu L, Delbe J, Blat C et al. Insulin-like growth factor binding protein-3 (IGFBP-3), an inhibitor of serum growth factors other than IGF-I and -II. J Cell Physiol 1992; 153:15-21.
76. Cohen P, Lamson G, Okajima T et al. Transfection of the human insulin-like growth factor binding protein-3 gene into Balb/c fibroblasts inhibits cellular growth. Mol Endocrinol 1993; 7:380-386.
77. Oh Y, Muller HL, Pham H et al. Demonstration of receptors for insulin-like growth factor binding protein-3 on Hs578T human breast cancer cells. J Biol Chem 1993; 268:26045-26048.
78. Valentinis B, Bhala A, DeAngelis T et al. The human insulin-like growth factor (IGF) binding protein-3 inhibits the growth of fibroblasts with a targeted disruption of the IGF-I receptor gene. Mol Endocrinol 1995; 9:361-367.
79. Spagnoli A, Hwa V, Horton WA et al. Antiproliferative effects of insulin-like growth factor-binding protein-3 in mesenchymal chondrogenic cell line RCJ3.1C5.18. relationship to differentiation stage. J Biol Chem 2001; 276(8):5533-40.

80. Oh Y, Muller HL, Ng L et al. Transforming growth factor-beta-induced cell growth inhibition in human breast cancer cells is mediated through insulin-like growth factor-binding protein-3 action. J Biol Chem 1995; 270:13589-13592.
81. Gucev ZS, Oh Y, Kelley KM et al. Insulin-like growth factor binding protein 3 mediates retinoic acid- and transforming growth factor beta2-induced growth inhibition in human breast cancer cells. Cancer Res 1996; 56:1545-1550.
82. Colston KW, Perks CM, Xie SP et al. Growth inhibition of both MCF-7 and Hs578T human breast cancer cell lines by vitamin D analogues is associated with increased expression of insulin-like growth factor binding protein-3. J Mol Endocrinol 1998; 20:157-162.
83. Sprenger CC, Peterson A, Lance R et al. Regulation of proliferation of prostate epithelial cells by 1,25-dihydroxyvitamin D3 is accompanied by an increase in insulin-like growth factor binding protein-3. J Endocrinol 2001; 170:609-618.
84. Rozen F, Zhang J, Pollak M. Antiproliferative action of tumor necrosis factor-alpha on MCF-7 breastcancer cells is associated with increased insulin-like growth factor binding protein-3 accumulation. Int J Oncol 1998; 13:865-869.
85. Pratt SE, Pollak MN. Estrogen and antiestrogen modulation of MCF7 human breast cancer cell proliferation is associated with specific alterations in accumulation of insulin-like growth factor-binding proteins in conditioned media. Cancer Res 1993; 53:5193-5198.
86. Huynh H, Yang X, Pollak M. Estradiol and antiestrogens regulate a growth inhibitory insulin-like growth factor binding protein 3 autocrine loop in human breast cancer cells. J Biol Chem 1996; 271:1016-10121.
87. Tsubaki J, Choi WK, Ingermann AR et al. Effects of sodium butyrate on expression of members of the IGF-binding protein superfamily in human mammary epithelial cells. J Endocrinol 2001; 169:97-110.
88. Buckbinder L, Talbott R, Velasco-Miguel S et al. Induction of the growth inhibitor IGF-binding protein 3 by p53. Nature 1995; 377:646-649.
89. Skaar TC, Baumrucker CR. Regulation of insulin-like growth factor binding protein secretion by a murine mammary epithelial cell line. Exp Cell Res 1993; 209:183-188.
90. Walker GE, Wilson EM, Powell D et al. Butyrate, a histone deacetylase inhibitor, activates the human IGF binding protein-3 promoter in breast cancer cells: molecular mechanism involves an Sp1/Sp3 multiprotein complex. Endocrinology 2001; 142:3817-3827.
91. Bourdon JC, Deguin-Chambon V, Lelong JC et al. Further characterisation of the p53 responsive element—identification of new candidate genes for trans-activation by p53. Oncogene 1997; 14:85-94.
92. Oh Y. IGF-independent regulation of breast cancer growth by IGF binding proteins. Breast cancer Res Treat 1998; 47:283-293.
93. Hollowood AD, Lai T, Perks CM et al. IGFBP-3 prolongs the p53 response and enhances apoptosis following UV irradiation. Int J Cancer 2000; 88:336-341.
94. Williams AC, Collard TJ, Perks CM et al. Increased p53-dependent apoptosis by the insulin-like growth factor binding protein IGFBP-3 in human colonic adenoma-derived cells. Cancer Res 2000; 60:22-27.
95. Xie SP, James SY, Colston KW. Vitamin D derivatives inhibit the mitogenic effects of IGF-I on MCF-7 human breast cancer cells. J Endocrinol 1997; 154:495-504.
96. Butt AJ, Firth SM, King MA et al. Insulin-like growth factor-binding protein-3 modulates expression of Bax and Bcl-2 and potentiates p53-independent radiation-induced apoptosis in human breast cancer cells. J Biol Chem 2000; 275:39174-39181.
97. Tsukada T, Eguchi K, Migita K et al. Transforming growth factor beta 1 induces apoptotic cell death in cultured human umbilical vein endothelial cells with down-regulated expression of bcl-2. Biochem Biophys Res Commun 1995; 210:1076-1082.
98. Raffo P, Emionite L, Colucci L et al. Retinoid receptors: pathways of proliferation inhibition and apoptosis induction in breast cancer cell lines. Anticancer Res 2000; 20:1535-1543.
99. Diaz GD, Paraskeva C, Thomas MG et al. Apoptosis is induced by the active metabolite of vitamin D3 and its analogue EB1089 in colorectal adenoma and carcinoma cells: possible implications for prevention and therapy. Cancer Res 2000; 60:2304-2312.
100. Kim HS, Ingermann AR, Tsubaki J et al. Inducible expression of insulin-like growth factor binding protein-3 causes cell cycle arrest and apoptosis in MCF-7 human breast cancer cells. Growth hormone IGF Res 2000; 10:A28.
101. Boyle BJ, Zhao XY, Cohen P et al. Insulin-like growth factor binding protein-3 mediates 1 alpha,25-dihydroxyvitamin d(3) growth inhibition in the LNCaP prostate cancer cell line through p21/WAF1. J Urol 2001; 165:1319-1324.
102. Fowler CA, Perks CM, Newcomb PV et al. Insulin-like growth factor binding protein-3 (IGFBP-3) potentiates paclitaxel-induced apoptosis in human breast cancer cells. Int J Cancer 2000; 88:448-453.
103. Ricort JM, Binoux M. Insulin-like growth factor (IGF) binding protein-3 inhibits type 1 IGF receptor activation independently of its IGF binding affinity. Endocrinology 2001; 142:108-113.
104. Delbé J, Blat C, Desauty G et al. Presence of IDF45 (mIGFBP-3) binding sites on chick embryo fibroblasts. Biochem Biophys Res Commun 1991; 179:495-501.
105. Oh Y, Muller HL, Pham H et al. Non-receptor mediated, post-transcriptional regulation of insulin-like growth factor binding protein (IGFBP)-3 in Hs578T human breast cancer cells. Endocrinology 1992; 131:3123-3125.
106. Martin JL, Ballesteros M, Baxter RC. Insulin-like growth factor-I (IGF-I) and transforming growth factor-beta 1 release IGF-binding protein-3 from human fibroblasts by different mechanisms. Endocrinology 1992; 131:1703-1710.

107. Ingermann AR, Kim HS, Oh Y. Characterization of a functional receptor for insulin-like growth factor binding protein 3. Growth hormone IGF Res 2000; 10:A27.
108. Leal SM, Huang SS, Huang JS. Interactions of high affinity insulin-like growth factor-binding proteins with the type V transforming growth factor-beta receptor in mink lung epithelial cells. J Biol Chem 1999; 274:6711-6717.
109. Vivien D, Attisano L, Wrana JL et al. Signaling activity of homologous and heterologous transforming growth factor-beta receptor kinase complexes. J Biol Chem 1995; 270:7134-7141.
110. Nakao A, Imamura T, Souchelnytskyi S et al. TGF-beta receptor-mediated signalling through Smad2, Smad3 and Smad4. EMBO J 1997; 16:5353-5362.
111. Fanayan S, Firth SM, Butt AJ et al. Growth inhibition by insulin-like growth factor-binding protein-3 in T47D breast cancer cells requires transforming growth factor-beta (TGF-beta) and the type II TGF-beta receptor. J Biol Chem 2000; 275:39146-39151.
112. Kalkhoven E, Roelen BA, de Winter JP et al. Resistance to transforming growth factor beta and activin due to reduced receptor expression in human breast tumor cell lines. Cell Growth Differ 1995; 6:1151-1161.
113. Wraight CJ, Liepe IJ, White PJ et al. Intranuclear localization of insulin-like growth factor binding protein-3 (IGFBP-3) during cell division in human keratinocytes. J Invest Dermatol 1998; 111:239-242.
114. Solomin L, Johansson CB, Zetterstrom RH et al. Retinoid-X receptor signalling in the developing spinal cord. Nature 1998; 395:398-402.
115. Boylan JF, Lohnes D, Taneja R et al. Loss of retinoic acid receptor gamma function in F9 cells by gene disruption results in aberrant Hoxa-1 expression and differentiation upon retinoic acid treatment. Proc Natl Acad Sci USA 1993; 90:9601-9605.
116. Mukherjee R, Davies PJ, Crombie DL et al. Sensitization of diabetic and obese mice to insulin by retinoid X receptor agonists. Nature 1997; 386:407-410.
117. Collingwood TN, Butler A, Tone Y et al. Thyroid hormone-mediated enhancement of heterodimer formation between thyroid hormone receptor beta and retinoid X receptor. J Biol Chem 1997; 272:13060-13065.
118. Kliewer SA, Umesono K, Mangelsdorf DJ et al. Retinoid X receptor interacts with nuclear receptors in retinoic acid, thyroid hormone and vitamin D3 signalling. Nature 1992; 355:446-449.
119. Westin S, Kurokawa R, Nolte RT et al. Interactions controlling the assembly of nuclear-receptor heterodimers and co-activators. Nature 1998; 395:199-202.
120. Collett-Solberg PF, Cohen P. The role of the insulin-like growth factor binding proteins and the IGFBP proteases in modulating IGF action. Endocrinol Metab Clin North Am 1996; 25:591-614.
121. Lalou C, Lassarre C, Binoux M. A proteolytic fragment of insulin-like growth factor (IGF) binding protein-3 that fails to bind IGFs inhibits the mitogenic effects of IGF-I and insulin. Endocrinology 1996; 137:3206-3212.
122. Yamanaka Y, Wilson EM, Rosenfeld RG et al. Inhibition of insulin receptor activation by insulin-like growth factor binding proteins. J Biol Chem 1997; 272:30729-30734.
123. Devi GR, Yang DH, Rosenfeld RG et al. Differential effects of insulin-like growth factor (IGF)-binding protein-3 and its proteolytic fragments on ligand binding, cell surface association, and IGF-I receptor signaling. Endocrinology 2000; 141:4171-4179.
124. Zadeh SM, Binoux M. The 16-kDa proteolytic fragment of insulin-like growth factor (IGF) binding protein-3 inhibits the mitogenic action of fibroblast growth factor on mouse fibroblasts with a targeted disruption of the type 1 IGF receptor gene. Endocrinology 1997; 138:3069-3072.
125. Yamanaka Y, Fowlkes JL, Wilson EM et al. Characterization of insulin-like growth factor binding protein-3 (IGFBP-3) binding to human breast cancer cells: kinetics of IGFBP-3 binding and identification of receptor binding domain on the IGFBP-3 molecule. Endocrinology 1999; 140:1319-1328.
126. Jones JI, Gockerman A, Busby WH Jr et al. Insulin-like growth factor binding protein 1 stimulates cell migration and binds to the alpha 5 beta 1 integrin by means of its Arg-Gly-Asp sequence. Proc Natl Acad Sci USA 1993; 90:10553-10557.
127. Irving JA, Lala PK. Functional role of cell surface integrins on human trophoblast cell migration: regulation by TGF-beta, IGF-II, and IGFBP-1. Exp Cell Res 1995; 217:419-427.
128. Hynes RO. Integrins: versatility, modulation, and signaling in cell adhesion. Cell 1992; 69:11-25.
129. Perks CM, Newcomb PV, Norman MR et al. Effect of insulin-like growth factor binding protein-1 on integrin signalling and the induction of apoptosis in human breast cancer cells. J Mol Endocrinol 1999; 22:141-150.
130. Gleeson LM, Chakraborty C, McKinnon T et al. Insulin-like growth factor-binding protein 1 stimulates human trophoblast migration by signaling through alpha 5 beta 1 integrin via mitogen-activated protein Kinase pathway. J Clin Endocrinol Metab 2001; 86:2484-2493.
131. Nam TJ, Busby WH Jr, Clemmons DR. Human fibroblasts secrete a serine protease that cleaves insulin-like growth factor-binding protein-5. Endocrinology 1994; 135:1385-1391.
132. Andress DL, Birnbaum RS. Human osteoblast-derived insulin-like growth factor (IGF) binding protein-5 stimulates osteoblast mitogenesis and potentiates IGF action. J Biol Chem 1992; 267:22467-22472.
133. Andress DL. Heparin modulates the binding of insulin-like growth factor (IGF) binding protein-5 to a membrane protein in osteoblastic cells. J Biol Chem 1995; 270:28289-28296.
134. Andress DL. IGF-binding protein-5 stimulates osteoblast activity and bone accretion in ovariectomized mice. Am J Physiol Endocrinol Metab 2001; 281:E283-288.
135. Abrass CK, Berfield AK, Andress DL. Heparin binding domain of insulin-like growth factor binding protein-5 stimulates mesangial cell migration. Am J Physiol 1997; 273(6Pt2):F899-906.

136. Andress DL. Insulin-like growth factor-binding protein-5 (IGFBP-5) stimulates phosphorylation of the IGFBP-5 receptor. Am J Physiol 1998; 274(4 Pt 1):E744-750.
137. Berfield AK, Andress DL, Abrass CK. IGFBP-5(201-218) stimulates Cdc42GAP aggregation and filopodia formationin migrating mesangial cells. Kidney Int 2000; 57:1991-2003.
138. Ho PJ, Baxter RC. Characterization of truncated insulin-like growth factor-binding protein-2 in human milk. Endocrinology 1997; 138:3811-3818.
139. Russo VC, Rekaris G, Baker NL et al. Basic fibroblast growth factor induces proteolysis of secreted and cell membrane-associated insulin-like growth factor binding protein-2 in human neuroblastoma cells. Endocrinology 1999; 140:3082-3090.
140. Lawrence JB, Oxvig C, Overgaard MT et al. The insulin-like growth factor (IGF)-dependent IGF binding protein-4 protease secreted by human fibroblasts is pregnancy-associated plasma protein-A. Proc Natl Acad Sci USA 1999; 96:3149-3153.
141. Fowlkes JL, Thrailkill KM, George-Nascimento C et al. Heparin-binding, highly basic regions within the thyroglobulin type-1 repeat of insulin-like growth factor (IGF)-binding proteins (IGFBPs) -3, -5, and -6 inhibit IGFBP-4 degradation. Endocrinology 1997; 138:2280-2285.
142. Zhou Y, Mohan S, Linkhart TA et al. Retinoic acid regulates insulin-like growth factor-binding protein expression in human osteoblast cells. Endocrinology 1996; 137:975-983.
143. Gabbitas B, Canalis E. Retinoic acid stimulates the transcription of insulin-like growth factor binding protein-6 in skeletal cells. J Cell Physiol 1996; 169:15-22.
144. Gabbitas B, Canalis E. Cortisol enhances the transcription of insulin-like growth factor binding protein-6 in cultured osteoblasts. Endocrinology 1996; 137:1687-1692.
145. Grellier P, De Galle B, Babajko S. Expression of insulin-like growth factor-binding protein 6 complementary DNA alters neuroblastoma cell growth. Cancer Res 1998; 58:1670-1676.
146. Sueoka N, Lee HY, Wiehle S et al. Insulin-like growth factor binding protein-6 activates programmed cell death in non-small cell lung cancer cells. Oncogene 2000; 19:4432-4436.
147. Murphy M, Pykett MJ, Harnish P et al. Identification and characterization of genes differentially expressed in meningiomas. Cell Growth Differ 1993; 4:715-722.
148. Akaogi K, Okabe Y, Funahashi K et al. Cell adhesion activity of a 30-kDa major secreted protein from human bladder carcinoma cells. Biochem Biophys Res Commun 1994; 198:1046-1053.
149. Yamauchi T, Umeda F, Masakado M et al. Purification and molecular cloning of prostacyclin-stimulating factor from serum-free conditioned medium of human diploid fibroblast cells. Biochem J 1994; 303:591-598.
150. Wilson EM, Oh Y, Rosenfeld RG. Generation and characterization of an IGFBP-7 antibody: identification of 31kD IGFBP-7 in human biological fluids and Hs578T human breast cancer conditioned media. J Clin Endocrinol Metab 1997; 82:1301-1303.
151. Swisshelm K, Ryan K, Tsuchiya K et al. Enhanced expression of an insulin growth factor-like binding protein (mac25) in senescent human mammary epithelial cells and induced expression with retinoic acid. Proc Natl Acad Sci USA 1995; 92:4472-4476.
152. Burger AM, Zhang X, Li H et al. Down-regulation of T1A12/mac25, a novel insulin-like growth factor binding protein related gene, is associated with disease progression in breast carcinomas. Oncogene 1998; 16:2459-2467.
153. Hwa V, Tomasini-Sprenger C, Bermejo AL et al. Characterization of insulin-like growth factor-binding protein-related protein-1 in prostate cells. J Clin Endocrinol Metab 1998; 83:4355-4362.
154. Akaogi K, Okabe Y, Sato J et al. Specific accumulation of tumor-derived adhesion factor in tumor blood vessels and in capillary tube-like structures of cultured vascular endothelial cells. Proc Natl Acad Sci USA 1996; 93:8384-8389.
155. Sprenger CC, Damon SE, Hwa V et al. Insulin-like growth factor binding protein-related protein 1 (IGFBP-rP1) is a potential tumor suppressor protein for prostate cancer. Cancer Res 1999; 59:2370-2375.
156. Kanemitsu N, Kato MV, Bai F et al. Correlation between induction of the mac25 gene and anti-proliferative effects of 1alpha,25(OH)2-D3 on breast cancer and leukemic cells. Int J Mol Med 2001; 7:515-520.
157. Lopez-Bermejo A, Buckway CK, Devi GR et al. Characterization of insulin-like growth factor-binding protein-related proteins (IGFBP-rPs) 1, 2, and 3 in human prostate epithelial cells: potential roles for IGFBP-rP1 and 2 in senescence of the prostatic epithelium. Endocrinology 2000; 141:4072-4080.
158. Landberg G, Ostlund H, Nielsen NH et al. Downregulation of the potential suppressor gene IGFBP-rP1 in human breast cancer is associated with inactivation of the retinoblastoma protein, cyclin E overexpression and increased proliferation in estrogen receptor negative tumors. Oncogene 2001; 20:3497-3505.
159. Akaogi K, Sato J, Okabe Y et al. Synergistic growth stimulation of mouse fibroblasts by tumor-derived adhesion factor with insulin-like growth factors and insulin. Cell Growth Differ 1996; 7:1671-1677.
160. Haugk KL, Wilson HM, Swisshelm K et al. Insulin-like growth factor (IGF)-binding protein-related protein-1: an autocrine/paracrine factor that inhibits skeletal myoblast differentiation but permits proliferation in response to IGF. Endocrinology 2000; 141:100-110.
161. Wilson EM, Oh Y, Hwa V et al. Interaction of igf-binding protein-related protein 1 with a novel protein, neuroendocrine differentiation factor, results in neuroendocrine differentiation of prostate cancer cells. J Clin Endocrinol Metab 2001; 86:4504-4511.
162. Walker GE, Antoniono RJ, Ross HJ et al. Mac25/Insulin-like growth factor-related protein-1 (Mac25/IGFBP-rP1) and a Novel Associating Protein, 25.1, Induce Neuroendocrine-like Differentiation in Non-Small Cell Lung Carcinoma (NSCLC) cell lines. Endocrine Society's 83rd Annual Meeting, Abstract 2001; P1-221.

163. Bradham DM, Igarashi A, Potter RL et al. Connective tissue growth factor: a cysteine-rich mitogen secreted by human vascular endothelial cells is related to the SRC-induced immediate early gene product CEF-10. J Cell Biol 1991; 114:1285-1294.
164. Joliot V, Martinerie C, Dambrine G et al. Proviral rearrangements and overexpression of a new cellular gene (nov) in myeloblastosis-associated virus type 1-induced nephroblastomas. Mol Cell Biol 1992; 12:10-21.
165. O'Brien TP, Yang GP, Sanders L et al. Expression of cyr61, a growth factor-inducible immediate-early gene. Mol Cell Biol 1990; 10:3569-3577.
166. Bork P. The modular architecture of a new family of growth regulators related to connective tissue growth factor. FEBS Lett 1993; 327:125-130.
167. Kireeva ML, MO FE, Yang GP et al. Cyr61, a product of a growth factor-inducible immediate-early gene, promotes cell proliferation, migration, and adhesion. Mol Cell Biol 1996; 16:1326-1334.
168. Grotendorst GR. Connective tissue growth factor: a mediator of TGF-beta action on fibroblasts. Cytokine Growth Factor Rev 1997; 8:171-179.
169. Lau LF, Lam SC. The CCN family of angiogenic regulators: the integrin connection. Exp Cell Res 1999; 248:44-57.
170. Brigstock DR. The connective tissue growth factor/cysteine-rich 61/nephroblastoma overexpressed (CCN) family. Endocr Rev 1999; 20:189-206.
171. Wandji SA, Gadsby JE, Barber JA et al. Messenger ribonucleic acids for MAC25 and connective tissue growth factor (CTGF) are inversely regulated during folliculogenesis and early luteogenesis. Endocrinology 2000; 141:2648-2657.
172. Babic AM, Kireeva ML, Kolesnikova TV et al. CYR61, a product of a growth factor-inducible immediate early gene, promotes angiogenesis and tumor growth. Proc Natl Acad Sci USA 1998; 95:6355-6360.
173. Babic AM, Chen CC, Lau LF. Fisp12/mouse connective tissue growth factor mediates endothelial cell adhesion and migration through integrin alphavbeta3, promotes endothelial cell survival, and induces angiogenesis in vivo. Mol Cell Biol 1999; 19:2958-2966.
174. Kothapalli D, Frazier KS, Welply A et al. Transforming growth factor beta induces anchorage-independent growth of NRK fibroblasts via a connective tissue growth factor-dependent signaling pathway. Cell Growth Differ 1997; 8:61-68.
175. Kothapalli D, Hayashi N, Grotendorst GR. Inhibition of TGF-beta-stimulated CTGF gene expression and anchorage-independent growth by cAMP identifies a CTGF-dependent restriction point in the cell cycle. FASEB J 1998; 12:1151-1161.
176. Martinerie C, Huff V, Joubert I et al. Structural analysis of the human nov proto-oncogene and expression in Wilms tumor. Oncogene 1994; 9:2729-2732.
177. Yang GP, Lau LF. Cyr61, product of a growth factor-inducible immediate early gene, is associated with the extracellular matrix and the cell surface. Cell Growth Differ 1991; 2:351-357.
178. Jedsadayanmata A, Chen CC, Kireeva ML et al. Activation-dependent adhesion of human platelets to Cyr61 and Fisp12/mouse connective tissue growth factor is mediated through integrin alpha(IIb)beta(3). J Biol Chem 1999; 274:24321-24327.
179. Hashimoto Y, Shindo-Okada N, Tani M et al. Expression of the Elm1 gene, a novel gene of the CCN (connective tissue growth factor, Cyr61/Cef10, and neuroblastoma overexpressed gene) family, suppresses In vivo tumor growth and metastasis of K-1735 murine melanoma cells. J Exp Med 1998; 187:289-296.
180. Pennica D, Swanson TA, Welsh JW et al. WISP genes are members of the connective tissue growth factor family that are up-regulated in wnt-1-transformed cells and aberrantly expressed in human colon tumors. Proc Natl Acad Sci USA 1998; 95:14717-14722.
181. Zhang R, Averboukh L, Zhu W et al. Identification of rCop-1, a new member of the CCN protein family, as a negative regulator for cell transformation. Mol Cell Biol 1998; 18:6131-6141.
182. Kumar S, Hand AT, Connor JR et al. Identification and cloning of a connective tissue growth factor-like cDNA from human osteoblasts encoding a novel regulator of osteoblast functions. J Biol Chem 1999; 274:17123-17131.
183. Zumbrunn J, Trueb B. Primary structure of a putative serine protease specific for IGF-binding proteins. FEBS Lett 1996; 398:187-192.
184. Lassalle P, Molet S, Janin A et al. ESM-1 is a novel human endothelial cell-specific molecule expressed in lung and regulated by cytokines. J Biol Chem 1996; 271:20458-29464.
185. Hu SI, Carozza M, Klein M et al. Human HtrA, an evolutionarily conserved serine protease identified as a differentially expressed gene product in osteoarthritic cartilage. J Biol Chem 1998; 273:34406-34412.
186. Bechard D, Scherpereel A, Hammad H et al. Human endothelial-cell specific molecule-1 binds directly to the integrin CD11a/CD18 (LFA-1) and blocks binding to intercellular adhesion molecule-1. J Immunol 2001; 167:3099-3106.
187. Buckway CK, Wilson EM, Ahlsén M et al. Mutation of three critical amino acids of the N-terminal domain of insulin-like growth factor (IGF) binding protein-3 essential for high-affinity IGF binding. J Clin Endocrinol Metab 2001; 86:4943-4950.
188. Imai Y, Moralez A, Andag U et al. Substitutions for hydrophobic amino acids in the N-terminal domains of IGFBP-3 and -5 markedly reduce IGF-I binding and alter their biologic actions. J Biol Chem 2000; 275:18188-18194.

CHAPTER 16

Functional Relationships between Transforming Growth Factor-β and the Insulin-Like Growth Factor Binding Proteins

Susan Fanayan and Robert C. Baxter

Abstract

Insulin-like growth factor binding proteins (IGFBPs) modulate cell functions through IGF-dependent and independent mechanisms. The IGFBPs are subject to complex regulation, both inhibitory and stimulatory, by transforming growth factor-b (TGF-β) in various cell types, which may result in either decreased or increased cell proliferation. TGF-β commonly signals by binding to the type II TGF-b receptor, which interacts with the type I receptor, activating its serine/threonine kinase activity and leading to the phosphorylation and nuclear translocation of proteins of the Smad family. An alternative pathway, through the type V TGF-β receptor, has been proposed, but is less well characterized. IGFBP-3 is reported be a ligand for the type V receptor in mink lung cells and may signal through a Smad-independent mechanism. In human breast cancer cells, IGFBP-3 can signal through the type I/type II TGF-b receptor system, involving the phosphorylation of the type I receptor, Smad2 and Smad3. TGF-β thus interacts with the IGFBPs both at the level of regulating their protein levels, and through the commonality of their signaling systems.

Introduction

The insulin-like growth factor binding proteins (IGFBPs) are a family of six highly homologous proteins which are discussed in detail elsewhere in this volume. Structural aspects of these proteins have been reviewed recently.[1,2] The IGFBPs serve important functions in the circulation, providing a stable pool of IGFs which are believed to act in the regulation of growth and metabolism, while at the same time playing a more dynamic role in the acute regulation of IGF bioavailability and tissue targeting.[3-5] In the pericellular and intracellular environment, IGFBPs regulate cell function by modulating IGF interaction with the type 1 IGF receptor, and by a variety of mechanisms which appear independent of the type 1 IGF receptor. In different cell types the net effect of IGFBP action can be either pro- or anti-proliferative.[6,7]

The concentrations of IGFBPs in the circulation and the cellular environment are known to be modulated both transcriptionally, and posttranslationally by proteolytic degradation. Regulation may also occur at the level of mRNA stability, protein synthesis and secretion. A wide variety of factors involved in cell growth, differentiation, and death impact on the IGFBP system at different levels, including members of the transforming growth factor-b (TGF-β) family. This chapter will explore the regulation of IGFBPs by TGF-β, and pathways through which IGFBPs (principally IGFBP-3) and TGF-β might interact functionally.

The Transforming Growth Factor-β (TGF-β) Superfamily

The TGF-β superfamily consists of multifunctional cytokines including TGF-βs, activins, inhibins, and bone morphogenetic proteins. These are secreted signaling molecules that regulate a number of cellular responses, such as proliferation, differentiation, migration and apoptosis. TGF-β superfamily

Insulin-Like Growth Factors, edited by Derek LeRoith, Walter Zumkeller and Robert Baxter. ©2003 Eurekah.com and Kluwer Academic / Plenum Publishers.

members have critical roles during embryogenesis and in maintaining tissue homeostasis during adult life. Deregulated signaling by this group of proteins has been implicated in multiple developmental disorders and in various human diseases, including cancer, fibrosis and auto-immune diseases.[8]

The structural prototype for the TGF-β family is a disulfide-linked homodimer of 112- residue chains, which was first isolated from human platelets,[9] cloned from a human cDNA library,[10] and later named TGF-β1.[11] There are three human TGF-β proteins (TGF-β1, 2, and 3).[8] TGF-β displays a variety of biological activities, including the negative and positive regulations of cell growth, stimulation of extracellular matrix formation, stimulation of angiogenesis and induction of differentiation of several cell lineages.[12-14] Moreover, TGF-β has been implicated in several physiological and pathological processes such as wound repair, morphogenesis, and tissue fibrosis.[14]

Numerous cell types in culture express one or multiple forms of TGF-β, at least at the mRNA level.[15] The expression pattern of the different isoforms of TGF-β varies with each cell type, and is active throughout embryonic development and into adulthood.[16,17] Many human breast tumor cell lines secrete TGF-β,[18,19] and levels are elevated with increased malignancy.[20,21]

The nature of TGF-β action on its target cells depends not only on the cell type but also on its state of differentiation and on other growth factors present.[22,23] TGF-β proteins inhibit the growth of normal and transformed epithelial, endothelial, fibroblast, and hematopoietic cells,[14,24] but exert stimulatory effect on the growth of some fibroblast cell lines and osteoblasts under certain culture conditions.[14,24] Loss of autocrine TGF-β activity and/or responsiveness to exogenous TGF-β provides cells with a growth advantage leading to malignant progression.[25,26]

TGF-β is synthesized and secreted as a high molecular weight latent complex, which regulates its in vivo availability.[27,28] The complex is composed of three subunits, the mature 25 kDa TGF-β, the N-terminal remnant of the TGF-β precursor, and a component of 125-160 kDa denoted the latent TGF-β-binding protein (LTBP).[29] The N-terminal remnant of the TGF-β precursor is known as TGF-β-latency associated peptide (TGF-β-LAP).[30] When associated with LAP, TGF-β cannot interact with its receptors, and activation through the release of mature TGF-β from this complex is a necessary step before it can exert its actions. The in vitro conditions for activating latent TGF-β include pH extremes,[31] ionizing radiation,[32] deglycosylation[33] and protease action.[31] Certain cell types have been reported to secrete TGF-β in active form, such as breast cancer cells treated with antiestrogens,[34] and co-cultures of endothelial cells with smooth muscle cells.[35] However, the exact mechanism of activation in these systems remains to be elucidated. The IGF-II/mannose 6-phosphate receptor has been shown to have a role in latent TGF-β activation,[36] in a mechanism believed to involve urokinase plasminogen activator receptor binding and plasmin generation.[37]

TGF-β Receptors

Several putative or actual TGF-β receptors have been identified in cultured cells and tissues, designated type I ($M_r \sim 53,000$), type II ($M_r \sim 75,000$-110,000), type III ($M_r \sim 280,000$-310,000), type IV ($M_r \sim 60,000$), type V ($M_r \sim 400,000$) and type VI ($M_r \sim 180,000$).[38-40] The type I receptor is also classified as activin receptor-like kinase (ALK)-5. Among these receptors, types I, II, III, and V receptors are co-expressed in most cell types.[41] Other TGF-β receptors are found only in certain cell types and tissues.[39]

Type I (TβRI) and type II (TβRII) TGF-β receptors are transmembrane glycoprotein serine/threonine kinases which share a common overall molecular structure with some sequence similarity.[40,42,43] In the absence of TGF-β, the TβRII kinase is active and undergoes autophosphorylation on at least three serine residues that regulate receptor activity.[44] TβRII can bind extracellular TGF-β directly, allowing the receptor to recruit TβRI and phosphorylate it in a conserved glycine-serine-rich domain (GS domain).[45] This activates the type I receptor kinase, which subsequently recognizes and phosphorylates members of the intracellular Smad signal transduction pathway. Unresponsiveness to TGF-β in some cancer cell lines is associated with loss or low expression of TβRI and/or TβRII.[46,47] Re-expression of TβRII has been shown to restore responsiveness to the growth-inhibitory effects of TGF-β on cell proliferation and to reduce malignancy in several human cancer cell lines lacking this receptor.[26,47,48]

In many cell lines, the most abundant cell surface TGF-β-binding component is type III TGF-β receptor (TβRIII) or betaglycan,[43,49,50] an integral membrane proteoglycan of 280-310 kDa, which lacks a serine/threonine kinase domain. It consists of approximately 200 kDa of glycosaminoglycan (GAG) chain mass and ~10 kDa of N-linked glycans attached to a heterogeneous core polypeptide of ~100 kDa.[40,51] In contrast to receptors I and II, betaglycan shows similar affinity for TGF-β1, 2, and 3 in many cell lines tested.[52-54] Given its structural features, relative abundance, and secretory nature, betaglycan could function as a reservoir or clearance system for bioactive TGF-β.[55]

Smads

The discovery of Smad homologs in *Drosophila* [56] and *C. elegans*,[57] two organisms in which a TGF-β-like signaling pathway is present, led to a rapid unraveling of the TGF-β signaling pathway. Smads are molecules of relative molecular mass 42-60 kDa with two regions of homology at the amino and carboxy terminals, termed Mad-homology domains MH1 and MH2, respectively, which are connected with a proline-rich linker sequence.[58] The MH1 domain, together with part of the linker region, may be involved in direct DNA binding.[59] The MH2 domain may serve as an effector domain by activating transcription of target genes.[60,61]

Smads fall into three classes, based on sequence similarity and function. Class I or receptor-regulated Smads (R-Smads) couple to different receptors. Smad2 and Smad3 are phosphorylated and translocated to the nucleus after phosphorylation by activated TβRI[62] or type I activin receptor (Act-RI).[63] In the C-terminal regions, receptor-regulated Smads have a characteristic Ser-Ser-X-Ser motif, the two-most C-terminal serine residues of which are phosphorylated by type I receptors.[64,65] Class II Smad or co-Smad, represented by Smad4,[66] does not have the C-terminal Ser-Ser-X-Ser motif and therefore does not bind to, nor is it phosphorylated by, type I receptors.[62] Smad4 appears to be a general partner for the receptor-regulated Smads by bringing the cytoplasmic Smads into the nucleus, where they can activate transcriptional responses.[62,67] Class III Smads or inhibitory Smads, which include Smad6 and Smad7, are structurally different from other members of the Smad family.[68,69] They bind to type I receptors and interfere with the phosphorylation of the receptor-regulated Smads.[68,69]

The association between TGF-β and TβRII results in the recruitment of TβRI into a heteromeric complex in which TβRII phosphorylates and activates TβRI.[70] The R-Smads, Smad2 and Smad3, are then phosphorylated by active TβRI, followed by their association with Smad4 and the translocation of the heteromeric complex to the nucleus.[67,71] In the nucleus, they can potentially regulate the transcription of target genes either through binding to elements in the DNA, or indirectly by binding to other transcription factors. A protein called SARA (Smad anchor for receptor activation) regulates the interaction between receptors and Smads.[72] SARA binds unphosphorylated Smad2 and facilitates its phosphorylation by TβRI, resulting in its dissociation from SARA. Mutations in SARA that cause mislocalization of Smad2 inhibit TGF-β-dependent signaling, suggesting that the regulation of Smad localization is important for TGF-β signaling.[72]

Promoters of several genes induced by TGF-β signaling such as plasminogen activator inhibitor-1 (PAI-1) and type VII collagen, contain unique Smad binding elements (SBE) with which R-Smad/Smad4 complexes may physically interact.[73,74] In several TGF-β target genes, multiple copies of SBEs can be identified, often in close proximity to sites for other transcription factors.[75,76] However, these direct interactions are now regarded as being of relatively low affinity and specificity,[77] and transcriptional regulation by Smads is believed to require interaction with other, more specific DNA-binding partners. For example, Smad interaction with Sp1 has been suggested to activate the p21/WAF1/Cip1 gene promoter, which lacks SBEs, through Sp1 binding sites.[78] Members of the FAST (forkhead activin signal transducer) or FoxH1 family of transcription factors are now believed to be important in transcriptional regulation by Smads,[77] by forming DNA-binding complexes with R-Smad/Smad4 heteromers,[79] and a variety of other nuclear Smad-interacting parters have also been recently described.[77]

Alterations in TGF-β Signaling during Carcinogenesis

Many tumor cells and neoplastic cell lines are resistant to the growth-inhibitory effects of TGF-β.[80] Resistance to TGF-β-mediated growth inhibition may occur at many levels, including loss of latent TGF-β activation, loss of expression of TGF-β receptors, or inactivation of any member of the Smad family.[81,82] Resistance to TGF-β-mediated growth inhibition due to changes in the expression of functional TGF-β receptors has been shown in many tumors, including colorectal tumors.[83] liver tumors,[84] and prostate adenocarcinoma.[85] Re-expression of TβRII has been shown to restore TGF-β sensitivity and generate reversal of malignant properties in some of these systems.[47,86,87] There are also examples of TβRI mutations in human tumors,[46] and members of the Smad family, notably Smad4, may frequently be inactivated in a variety of tumor types.[66,88,89]

TGF-β Effects on IGFBP Production

A functional relationship between TGF-β and IGFBPs has been recognized for over a decade. Perhaps the earliest study implicating TGF-β in IGFBP regulation was the demonstration by Mondschein et al that TGF-β could down-regulate the level of IGFBP-3 in culture medium conditioned by porcine granulosa cells.[90] It was not clear from this study whether inhibition of production, or enhanced degradation, was responsible for this result, but it was speculated that TGF-β might influence IGF action through modifying IGFBPs.

The marked stimulatory effect of TGF-β on IGFBP-3 in fibroblasts was discovered as a result of investigating the identity of an acid-activated stimulator of IGFBP-3 production in fetal calf serum.[91] Inhibited by anti-TGF-β1 antibodies, its effect could be mimicked by exogenous TGF-β1. Subsequent to these studies, numerous reports have appeared which show a regulatory effect on the gene expression and/or production of IGFBP-3. In general, these effects are stimulatory to IGFBP-3 production, but in granulosa cells, endothelial cells and keratinocytes inhibition is seen (Table 1). The signaling pathway(s) involved in TGF-β regulation of IGFBP-3 have not been elucidated, but the observation that it can be inhibited by staurosporine, vanadate, or okadaic acid suggests the possible involvement of protein kinase C, and of both serine/threonine and tyrosine protein phosphatases.[92] As well as the many studies showing increased IGFBP-3 production in response to TGF-β, the appearance of an IGFBP-3 protease secreted by MCF-7 breast cancer cells is reported to be inhibited by TGF-β1, which might lead to increased active IGFBP-3.[93]

In addition to stimulatory and inhibitory effects on IGFBP-3, TGF-β has been found to either inhibit or increase the gene expression and protein production of IGFBP-2, -4, -5 and -6. Effects on IGFBP-2 all appear to be positive (Table 1), including one report showing the inhibition of IGFBP-2 proteolysis, which could increase unproteolyzed protein.[94] For IGFBP-4, TGF-β effects appear to be divided between inhibition and stimulation, with opposite effects reported in endothelial cells from two different sources (Table 1). IGFBP-5 and IGFBP-6 production are almost universally inhibited by TGF-β at the protein and/or mRNA level, with the exception of a report of increased IGFBP-5 in fetal chondrocytes.[95] Interestingly, TGF-β regulates IGFBP-3 and -5 in a reciprocal manner in human intestinal muscle, being stimulatory to IGFBP-3 and inhibitory to IGFBP-5.[96] Since IGFBP-3 was shown to suppress IGF-I-stimulated cell proliferation, and IGFBP-5 to augment it, the TGF-β-induced IGFBP changes would act together to inhibit proliferation.

With disparate effects of TGF-β on the production of various IGFBPs reported in different cell types, a unifying hypothesis relating TGF-β to cell growth control through IGFBP regulation does not emerge easily. This is complicated by the fact that several of the IGFBPs may themselves exert either growth-stimulatory or inhibitory effects, alone or in the presence of IGFs. One finding that does appear consistent for many cell types, however, is the role of IGFBP-3 induction in mediating at least some of the growth effects of TGF-β.

IGFBP-3 Induction in TGF-β Growth Regulation

IGFBP-3 has been identified as a cell senescence factor[97] and is a potent growth-inhibitory protein, acting to block cell cycle at the G1-S transition[98] and induce apoptosis,[99] either by impeding IGF access to the type I IGF receptor, or through IGF-receptor-independent mechanisms. Many effectors inhibitory to cell growth have been reported to induce IGFBP-3 gene expression in a range of cell types. These include — in addition to TGF-β — retinoic acid,[100] the antiestrogen ICI 182780,[101] TNF-α,[102] 1,25(OH)$_2$ vitamin D$_3$ and synthetic analogues,[103] dihydrotestosterone,[104] the histone deacetylase inhibitors butyrate[105] and trichostatin A,[106] a protein kinase Cα antisense oligonucleotide,[107] and the tumor suppressor p53.[108] In each case, growth inhibition resulting from the actions of these effectors has been attributed to their induction of IGFBP-3, but to date this has only been directly demonstrated in a limited number of cases.

TGF-β is the prototypic member of this diverse group of IGFBP-3 inducers. The observation that DNA synthesis in skin fibroblasts was inhibited concomitantly with the induction of IGFBP-3

Table 1. Effects of TGF-β on IGF binding proteins

IGFBP	Cell Type	TGF-β Effect	Ref.
IGFBP-1	human granulosa-luteal cells	no effect	128
IGFBP-2	lung alveolar epithelial cells	increases protein	129
	vascular smooth muscle cells	no effect	130, 131
	IEC-6 intestinal epithelial cells	increases protein and mRNA, → growth inhibition	132
	ovine fetal chondrocytes	increases mRNA	95
	broncheoalveolar cells in interstitial lung disease	association between elevated IGFBP-2 and TGF-β	133
	neuroblastoma cells	inhibits IGFBP-2 proteolysis	94
IGFBP-3	porcine granulosa cells	decreases protein	90
	skin fibroblasts	increases protein, different mechanism from IGF-I	91, 134, 135
	osteosarcoma cells	increases protein, synergistic with 1,25(OH)$_2$Vit D$_3$	136
	Hs578T breast cancer cells	increases protein; IGFBP-3 mediates growth inhibition	110
	endothelial cells (aorta, glomerular)	decreases protein and mRNA	137-139
	MDA-MD-231 breast cancer cells	increases protein	100
	mesangial cells	increases mRNA	140
	myogenic cells	increases mRNA	141
	PC-3 prostate cancer cells	increases protein; IGFBP-3 mediates growth inhibition	99, 142
	hepatic stellate cells	increases mRNA and protein	143
	human renal cortical fibroblasts	IGFBP-3 induced by TGF-β from proximal tubule cells	144
	human intestinal muscle	stimulates protein	96
	human keratinocytes	decreases protein and mRNA	145
	embyronic muscle cells	stimulates protein and mRNA, suppresses differentiation	146
	fetal chondrocytes	slightly increases protein and mRNA	95, 147
	prostatic stromal cells	increases protein in normal cells but much less in BPH cells	148
	human airway smooth muscle cells	increases protein, stimulates proliferation	113
	colon cancer cells	increased IGFBP-3 mediates proliferative response	112
	bone marrow stromal osteoblast progenitors	increases protein and mRNA	149
	MCF-7 human breast cancer cells	inhibits production of IGFBP-3 protease	93

continued on next page

by TGF-β[109] suggested the possibility that IGFBP-3 might mediate an inhibitory effect of TGF-β. This idea has been amply confirmed through the use of reagents to decrease IGFBP-3 levels or activity in cell culture. Oh et al[100,110] showed in Hs578T human breast cancer cells that IGFBP-3 antisense oligonucleotide blocked IGFBP-3 induction and cell growth inhibition in response to TGF-β, thus implicating IGFBP-3 induction in the TGF-β effect. Similarly, Rajah et al[99] showed

Table 1. Effects of TGF-β on IGF binding proteins (continued)

IGFBP	Cell Type	TGF-β Effect	Ref.
IGFBP-4	muscle cells	decreases protein	130
	human multiple myeloma cells	no effect	150
	bovine aorta endothelial cells	decreases mRNA	139
	human glomerular endothelial cells	increases protein	138
	bone marrow stromal osteoblast progenitors	no effect	149
	human osteoblasts	stimulates IGF-II-dependent IGFBP-4 protease	151
IGFBP-5	muscle cells	decreases protein	130
	rat fetal calvarial osteoblasts	decreases protein and mRNA	152
	human intestinal muscle	decreases protein	96
	hepatic stellate cells	decreases mRNA and protein	143
	ovine fetal chondrocytes	increases mRNA	95
IGFBP-6	human skin fibroblasts	decreases protein	153
	rat fetal calvarial osteoblasts	decreases protein	154

that TGF-β1-induced apoptosis in PC-3 prostate cancer cells was preceded by the induction of IGFBP-3, and could be blocked either by IGFBP-3 antisense oligonucleotide or IGFBP-3 antibody. Interestingly, while IGFBP-3 antisense oligonucleotides have also been used to show that antiestrogens can inhibit the growth of estrogen receptor-positive MCF-7 breast cancer cells by a similar mechanism involving IGFBP-3,[101] TGF-β does not appear to act as an intermediary in this process, since blockade of TGF-β signaling has no effect on the growth-inhibitory action of antiestrogen.[111]

While the examples above all relate to IGFBP-3 induction resulting in growth inhibition, TGF-β is known to be growth stimulatory in some cell types, and this effect may also depend on the induction of IGFBP-3. In colon carcinoma cell lines that proliferate in response to TGF-β1, IGFBP-3 antisense oligonucleotide was found to block the TGF-β effect, indicating that it was mediated by IGFBP-3. Exogenous IGFBP-3 also stimulated the growth of these cells.[112] Similarly in airway smooth muscle cells, both IGFBP-3 antisense oligonucleotide and IGFBP-3 antiserum were able to prevent an increase in cell number in response to TGF-β1 treatment.[113] These studies point to a common mechanism of IGFBP-3 induction that may contribute in some cell types to growth regulation by TGF-β.

Involvement of TGF-β Signaling Pathways in IGFBP-3 Action

Type V TGF-β Receptor

The induction of IGFBP-3 as a possible step in the growth-regulatory action of TGF-β does not imply any commonality between the signaling pathways directly activated by these proteins. After its induction in response to activation of a TGF-β pathway, IGFBP-3 might exert its cellular effect, whether stimulatory or inhibitory, through an entirely independent signaling process. However, there is considerable evidence that IGFBP-3 and TGF-β may indeed share a common intracellular pathway. This was first suggested by the observation that IGFBP-3 can compete with TGF-β1 for binding to the type V TGF-β receptor (TβRV).

The TβRV is a 400-kDa non-proteoglycan membrane glycoprotein that is expressed in many cell types.[41,114] First purified from bovine liver, TβRV shows a wide distribution in cultured cells, which may suggest a potential role for this receptor in mediating the growth inhibitory response to TGF-β.[114] It has been shown to undergo internalization following TGF-β binding,[114] but its exact role in TGF-β-induced growth inhibition has not been clearly defined. Liu et al demonstrated that, like type I and type II receptors, this receptor is also a serine/threonine-specific protein kinase, with casein kinase-like phosphorylation specificity, that can be stimulated by TGF-β and inhibited by heparin.[115] However, the significance of the kinase activity of the TβRV in the TGF-β-induced signal transduction, leading to various cellular responses, is poorly understood.

TβRV has been suggested to be the putative IGFBP-3 receptor,[116] mediating the IGF-independent action of IGFBP-3. This is based on the the observation of IGFBP-3 binding to the receptor in mink lung cells, its competition for TGF-β binding, and the fact that a TGF-β peptide antagonist blocks IGFBP-3-induced growth inhibition in these cells.[116,117] In addition to binding IGFBP-3, the receptor could be affinity labeled with IGFBP-4 and -5, though they are relatively weak ligands, and relatively poor inhibitors of DNA synthesis.[118] Interestingly, TGF-β but not IGFBP-3 was able to activate phosphorylation of the signaling intermediates Smad2 and Smad3 in mink lung cells.

Type II TGF-β Receptor

In breast cancer cells, the TβRI/TβRII system has been implicated in IGFBP-3 signaling. The possibility of a relationship between IGFBP-3 and TGF-β signaling was based on the observation that MCF-7 breast cancer cells, when transfected to express IGFBP-3, had enhanced sensitivity to growth inhibition by TGF-β1.[119] In a parallel system, growth inhibition by TGF-β1 in T47D cells required the addition of exogenous IGFBP-3, in addition to the introduction by cDNA transfection of TβRII, a protein normally expressed at undetectable levels in these cells. Conversely, TβRII expression, and the addition of exogenous TGF-β1, was necessary for IGFBP-3 to be growth-inhibitory.[119]

Examination of the signaling pathway used by IGFBP-3 in these cells showed that, when TβRII was expressed, IGFBP-3, like TGF-β, was able to stimulate the serine phosphorylation of TβRI, Smad2 and Smad3.[119,120] Figure 1 shows that IGFBP-3 enhances the effect of endogenous TGF-β on TβRI and Smad2 serine-phosphorylation, and can stimulate this pathway even when the endogenous TGF-β effect is abolished by immunoneutralization. Evidence supporting the formation of Smad2-Smad4 complexes in response to TGF-β1 was similarly observed in response to IGFBP-3.[120] In addition, transcription of a "classic" Smad-responsive gene, *PAI-1*, was activated by IGFBP-3, suggesting that the entire pathway from TβRI activation at the cell surface to nuclear transcription was responsive to IGFBP-3, similar to its responsiveness to TGF-β. Interestingly, a mutated form of IGFBP-3, lacking key basic carboxylterminal residues which are required for cell-surface binding[121] and nuclear translocation,[122] was as potent as wild-type IGFBP-3 in activating this pathway,[120] whereas IGFBP-5 was without effect. It would therefore appear unlikely that any nuclear action of IGFBP-3 is involved in activating the Smad pathway.

IGFBP-3 signaling through TβRI/TβRII is clearly distinguished from possible signaling through TβRV, which was reported (a) not to involve Smad phosphorylation and (b) to also be activated by IGFBP-5, albeit less potently than by IGFBP-3.[118] The requirement for an intact TGF-β signaling pathway for exogenous IGFBP-3 action can apparently be by-passed if the IGFBP-3 is expressed endogenously, e.g., induced by cDNA transfection. Under these circumstances, T47D cells show both cell-cycle blockade at the G1-S transition[98] and an increase in apoptosis, accompanied by an increase in the pro-apoptotic proteins Bax and Bad,[123] in response to IGFBP-3, even when TβRII is not restored by transfection. Whether these anti-proliferative responses to endogenously-produced IGFBP-3 involve the Smad signaling pathway at all is presently unknown.

Concluding Comments

Since the demonstration that IGFBP-3 is able to inhibit the proliferation of cells lacking the type I IGF receptor,[124] the search for IGFBP-3 signaling pathways has been intense. Although relatively little progress has been made, it is now evident that there is more than a single pathway, and that effects involving nuclear translocation, with an obligatory requirement for the nuclear localization signal residues 228-232,[122] and nuclear receptors,[125] are distinct from effects mediated through the TβRI/TβRII system, which do not require these residues. It seems equally possible that other IGFBPs will eventually be found to use multiple signaling pathways.

Although it is unclear whether TGF-β might be implicated in these pathways for other IGFBPs, the observation that TGF-β1 can upregulate $\alpha_5\beta_1$ integrin in hepatocarcinoma cells [126] suggests its involvement in IGFBP-1 action, since $\alpha_5\beta_1$ integrin mediates IGFBP-1 effects on cell migration.[4]

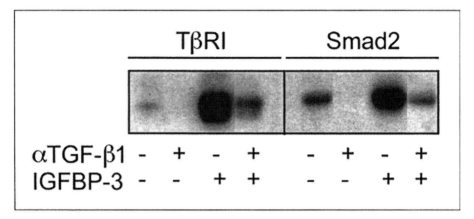

Figure 1. IGFBP-3 stimulates TβRI and Smad2 phosphorylation in human breast cancer cells. T47D cells transfected with TβRII cDNA to restore TGF-β1 signaling [119,120] were exposed to IGFBP-3 (500 ng/ml) in the presence or absence of TGF-β1-neutralizing antibody MAb240 (30 μg/ml). Bands representing serine-phosphorylated TβRI (left) and Smad2 (right) were measured as previously described.[120] The TGF-β1 antibody neutralized the basal effect of endogenous TGF-β1, and reduced the effect of exogenous IGFBP-3, suggesting that IGFBP-3 is active in the absence of TGF-β, and can synergize with endogenous TGF-β1.

TGF-β is known to affect multiple genetic programs within epithelial cells, involving up to 10% of all genes.[127] It therefore seems likely that many more interactions between TGF-β and IGFBP actions will eventually emerge.

Acknowledgement

Supported by the National Health & Medical Research Council, Australia and the Leo & Jenny Leukaemia and Cancer Foundation.

References

1. Baxter RC. Insulin-like growth factor (IGF) binding proteins: Interactions with IGFs and intrinsic bioactivities. Am J Physiol 2000; 278:E967-E976.
2. Hwa V, Oh Y, Rosenfeld RG. The insulin-like growth factor-binding protein (IGFBP) superfamily. Endocrine Rev 1999; 20:761-787.
3. Baxter RC. Circulating binding proteins for the insulinlike growth factors. Trends Endocrinol Metab 1993; 4:91-96.
4. Jones JI, Clemmons DR. Insulin-like growth factors and their binding proteins: biological actions. Endocrine Rev 1995; 16:3-34.
5. Rajaram S, Baylink DJ, Mohan S. Insulin-like growth factor binding proteins in serum and other biological fluids: Regulation and functions. Endocrine Rev 1997; 18:801-831.
6. Wetterau LA, Moore MG, Lee KW et al. Novel aspects of the insulin-like growth factor binding proteins. Mol Genet Metab 1999; 68:161-181.
7. Rechler MM, Clemmons DR. Regulatory actions of insulin-like growth factor-binding proteins. Trends Endocrinol Metab 1998; 9:176-183.
8. Massague J. TGF-β signal transduction. Ann Rev Biochem 1998; 67:753-791.
9. Assoian RK, Komoriya A, Meyers CA et al. Transforming growth factor-beta in human platelets. Identification of a major storage site, purification, and characterization. J Biol Chem 1983; 258:7155-7160.
10. Derynck R, Jarrett JA, Chen EY et al. Human transforming growth factor-β complementary DNA sequence and expression in normal and transformed cells. Nature 1985; 316:701-705.
11. Cheifetz S, Weatherbee JA, Tsang ML et al. The transforming growth factor-β system, a complex pattern of cross-reactive ligands and receptors. Cell 1987; 48:409-415.
12. Roberts AB, Flanders KC, Kondaiah P et al. Transforming growth factor-β: biochemistry and roles in embryogenesis, tissue repair and remodeling and carcinogenesis. Recent Prog Horm Res 1988; 44:157-197.
13. Moses HL, Yang EY, Pieterol JA. TGF-beta stimulation and inhibition of cell proliferation: new mechanistic insights. Cell 1990; 63:245-247.
14. Massague J. The transforming growth factor-β family. Ann Rev Cell Biol 1990; 6:597-641.
15. Derynck R. Transforming growth factor-β. Cell 1988; 54:593-595.
16. Heine U, Munoz EF, Flanders KC et al. Role of transforming growth factor-beta in the development of the mouse embryo. J Biol Chem 1987; 105:2861-2876.

17. Rappolee DA, Brenner CA, Schultz R et al. Developmental expression of PDGF, TGF-α, and TGF-β genes in preimplantation mouse embryos. Science 1988; 241:1823-1825.
18. Valverius EM, Walker-Jones D, Bates SE et al. Production of and responsiveness to transforming growth factor-β in normal and oncogene-transformed human mammary epithelial cells. Cancer Res 1989; 49:6269-6274.
19. Fynan TM, Reiss M. Resistance to inhibition of cell growth by transforming growth factor-β and its role in oncogenesis. Crit Rev Oncol 1993; 4:493-540.
20. Kasid A, Knabbe C, Lippman ME. Effect of v-ras oncogene transfection on estrogen-independent tumorigenicity of estrogen-dependent human breast cancer cells. Cancer Res 1987; 47:5733-5738.
21. Daly RJ, King RJB, Darbre PD. Interaction of growth factors during progression towards steroid independence in T47D human breast cancer cells. J Cell Biochem 1990; 43:199-211.
22. Roberts AB, Anzano MA, Wakefield IM et al. Type β transforming growth factor: a bifunctional regulator of cellular growth. Proc Natl Acad Sci USA 1985; 82:119-123.
23. Sporn MB, Roberts AB, Wakefield IM et al. Some recent advances in the chemistry and biology of transforming growth factor-beta. J Cell Biol 1987; 105:1039-1045.
24. Wright JA, Turley EA, Greenberg AH. Transforming growth factor beta and fibroblast growth factor as promoters of tumor progression to malignancy. Crit Rev Oncol 1993; 4:473-492.
25. Wu SP, Sun LZ, Willson KJV et al. Repression of autocrine transforming growth factor β1 and β2 in quiescent CBS colon carcinoma cells leads to progression of tumorigenic properties. Cell Growth Diff 1993; 4:115-123.
26. Kalkhoven E, Roelen BA, de Winter JP et al. Resistance to transforming growth factor β and activin due to reduced expression in human breast tumor cell lines. Cell Growth Diff 1995; 6:1151-1161.
27. Pircher R, Lawrence DA, Jullien P. Latent beta-transforming growth factor in nontransformed and Kirsten sarcoma virus-transformed normal rat kidney cells, clone 49F. Cancer Res 1984; 44:5538-5543.
28. Pircher R, Jullien P, Lawrence DA. Beta-transforming growth factor is stored in human blood platelets as a latent high molecular weight complex. Biochem Biophys Res Commun 1986; 136:30-37.
29. Miyazono K, Hellman U, Wernstedt C et al. Latent high molecular weight complex of transforming growth factor-b from human platelets; a high molecular weight complex containing precursor sequences. J Biol Chem 1988; 263:6407-6415.
30. Gentry LE, Nash BW. The pro domain of pre-pro-transforming growth factor beta 1 when independently expressed is a functional binding protein for the mature growth factor. Biochemistry 1990; 29:6851-6857.
31. Lyons RM, Keski-Oja J, Moses HL. Proteolytic activation of latent transforming growth factor-β from fibroblast conditioned medium. J Cell Biol 1988; 106:1659-1665.
32. Barcellos-Hoff MH, Derynck R, Tsang ML et al. Transforming growth factor-beta activation in irradiated murine mammary gland. J Clin Invest 1994; 93:892-899.
33. Miyazono K, Heldin CH. Role for carbohydrate structure in TGF-β1 latency. Nature 1989; 338:158-160.
34. Knabbe C, Lippman ME, Wakefield LM et al. Evidence that transforming growth factor-b is a hormonally regulated negative growth factor in human breast cancer cells. Cell 1987; 48:417-428.
35. Sato Y, Rifkin DB. Inhibition of endothelial cell movement by pericytes and smooth muscle cells: activation of a latent transforming growth factor-beta 1-like molecule by plasmin during co-culture. J Cell Biol 1989; 109:309-315.
36. Dennis PA, Rifkin DB. Cellular activation of latent transforming growth factor beta requires binding to the cation-independent mannose-6-phosphate/insulin-like growth factor type II receptor. Proc Natl Acad Sci USA 1991; 88:580-584.
37. Godar S, Horejsi V, Weidle UH et al. M6P/IGFII-receptor complexes urokinase receptor and plasminogen for activation of transforming growth factor-beta1. Eur J Immunol 1999; 29:1004-1013.
38. Tucker RF, Branum EL, Shipley GD et al. Specific binding to cultured cells of ^{125}I-labeled type beta transforming growth factor from human platelets. Proc Natl Acad Sci USA 1984; 81:6757-6761.
39. Cheifetz S, Ling N, Guiilemin R et al. A surface component on GH3 pituitary cells that recognizes transforming growth factor-beta, activin and inhibin. J Biol Chem 1988; 263:17225-17228.
40. Cheifetz S, Andres JL, Massague J. The transforming growth factor-β receptor type III is a membrane proteoglycan. Domain structure of the receptor. J Biol Chem 1988; 263:16984-16991.
41. O'Grady P, Kuo M-D, Baldassare JJ et al. Purification of a new high molecular weight receptor (type V receptor) of transforming growth factor β (TGF-β) from bovine liver. J Biol Chem 1991; 266:8583-8589.
42. Ebner R, Chen RH, Shum L et al. Cloning of a type I TGF-beta receptor and its effect on TGF-beta binding to the type II receptor. Science 1993; 260:1344-1348.
43. Cheifetz S, Like B, Massague J. Cellular distribution of type I and type II receptors for transforming growth factor-b. J Biol Chem 1986; 261:9972-9978.
44. Luo KX, Lodish HF. Positive and negative regulation of type II TGFβ receptor signal transduction by autophosphorylation on multiple serine residues. EMBO J 1997; 16:1970-1981.
45. Wrana JL, Attisano L, Wiesner R et al. Mechanism of activation of the TGF-β receptor. Nature 1994; 370:341-347.
46. Wang J, Han W, Zborowska E et al. Reduced expression of transforming growth factor β type I receptor contributes to the malignancy of human colon carcinoma cells. J Biol Chem 1996; 271:17366-17371.
47. Sun L, Wu G, Willson JKV et al. Expression of transforming growth factor β type II receptor leads to reduced malignancy in human breast cancer MCF-7 cells. J Biol Chem 1994; 269:26449-26455.
48. Chang J, Park K, Bang Y-J et al. Expression of transforming growth factor β type II receptor reduces tumorigenicity in human gastric cancer cells. Cancer Res 1997; 57:2856-2859.

49. Fanger BO, Wakefield LM, Sporn MB. Structure and properties of the cellular receptor for transforming growth factor type-beta. Biochemistry 1986; 25:3083-3091.
50. Massague J. Subunit structure of a high-affinity receptor for type β-transforming growth factor. J Biol Chem 1985; 260:7059-7066.
51. Segarini PR, Seyedin SM. The high molecular weight receptor to transforming growth factor-β contains glycosaminoglycan chains. J Biol Chem 1988; 263:8366-8370.
52. Cheifetz S, Hernandez H, Laiho M et al. Distinct transforming growth factor-β (TGF-β) receptor subsets as determinants of cellular responsiveness to three TGF-β isoforms. J Biol Chem 1990; 265:20533-20538.
53. Cheifetz S, Massague J. The TGF-β receptor proteoglycan. Cell surface expression and ligand binding in the absence of glycosaminoglycan chains. J Biol Chem 1989; 264:12025-12028.
54. Cheifetz S, Bassols A, Stanley K et al. Heteroimeric transforming growth factor-β. Biological properties and interaction with three types of cell surface receptors. J Biol Chem 1988; 263:10783-10789.
55. Massague J, Cheifetz S, Boyd FT et al. TGF-beta receptors and TGF-beta binding proteoglycans: recent progress in identifying their functional properties. Ann NY Acad Sci 1990; 593:59-72.
56. Sekelsky JJ, Newfeld SJ, Raftery LA et al. Genetic characterization and cloning of *Mothers against dpp*, a gene required for *decapentaplegic* function in *Drosophila melanogaster*. Genetics 1995; 139:1347-1358.
57. Savage C, Das P, Finelli AL et al. *Caenorhabditis elegans* genes Sma-2, Sma-3, and Sma-4 define a conserved family of transforming growth factor β pathway components. Proc Natl Acad Sci USA 1996; 93:790-794.
58. Heldin C-H, Miyazono K, ten Dijke P. TGF-β signalling from cell membrane to nucleus through SMAD proteins. Nature 1997; 390:465-471.
59. Kim J, Johnson K, Chen HJ et al. Drosophila Mad binds to DNA and directly mediates activation of vestigial by decapentaplegic. Nature 1997; 388:304-308.
60. Liu F, Hata A, Baker JC et al. A human Mad protein acting as a BMP-regulated transcriptional activator. Nature 1996; 381:620-623.
61. Baker JC, Harland RM. A novel mesoderm inducer, Madr2, functions in the activin signal transduction pathway. Genes Dev 1996; 10:1880-1889.
62. Nakao A, Imamura T, Souchelnytskyi S et al. TGF-β receptor-mediated signaling through Smad2, Smad3 and Smad4. EMBO J 1997; 16:5353-5362.
63. Chen Y, Lebrun JJ, Vale W. Regulation of transforming growth factor-β and activin-induced transcription by mammalian Mad proteins. Proc Natl Acad Sci USA 1996; 93:12992-12997.
64. Kretzschmar M, Liu F, Hata A et al. The TGF-β family mediator Smad1 is phosphorylated directly and activated functionally by the BMP receptor kinase. Genes Dev 1997; 11:984-995.
65. Souchelnytskyi S, Tamaki K, Engstrom U et al. Phosphorylation of Ser465 and Ser467 in the C-terminus of Smad2 mediates interaction with Smad4 and is required for TGF-β signalling. J Biol Chem 1997; 272:28107-28115.
66. Hahn SA, Schutte M, Shamsul Hoque ATM et al. DPC4, a candidate tumor suppressor gene at human chromosome 18q21.1. Science 1996; 271:350-353.
67. Lagna G, Hata A, Hemmati-Brivanlou A et al. Partnership between DPC4 and SMAD proteins in TGF-β signalling pathways. Nature 1996; 383:832-836.
68. Imamura T, Takase M, Nishihara A et al. Smad6 inhibits signalling by the TGF-beta superfamily. Nature 1997; 389:622-626.
69. Nakao A, Afrakhte M, Moren A et al. Identification of Smad7, a TGF-beta-inducible antagonist of TGF-beta signaling. Nature 1997; 389:631-635.
70. ten Dijke P, Miyazono K, Heldin CH. Signalling via heter-oligoneric complexes of type I and type II serine/threonine kinase receptors. Curr Opin Cell Biol 1996; 8:139-145.
71. Wu RY, Zhang Y, Feng XH et al. Heteromeric and homomeric interactions correlate with signaling activity and functional cooperativity of SMAD3 and SMAD4/DPC4. Mol Cell Biol 1997; 17:2521-2528.
72. Tsukazaki T, Chiang TA, Davison AF et al. SARA, a FYVE domain protein that recruits Smad2 to the TGF β receptor. Cell 1998; 95:779-791.
73. Dennler S, Itoh S, Vivien D et al. Direct binding of Smad3 and Smad4 to critical TGF-b-inducible elements in the promoter of human plasminogen activator inhibitor-type 1 gene. EMBO J 1998; 17:3091-3100.
74. Vindevoghel L, Lechleider RJ, Kon A et al. SMAD3/SMAD4-dependent transcriptional activation of the human type VII collagen (COL7A1) promoter by transforming growth factor β. Proc Natl Acad Sci USA 1998; 95:14769-14774.
75. Labbe E, Silvestri C, Hoodlss PA et al. Smad2 and Smad3 positively and negatively regulate TGFβ-dependent transcription through the forkhead DNA-binding protein FAST2. Mol Cell 1998; 2:109-120.
76. Zhang Y, Feng XH, Derynck R. Smad3 and Smad4 cooperate with c-Jun/c-Fos to mediate TGF-β-induced transcription. Nature 1998; 394:909-913.
77. Attisano L, Silvestri C, Izzi L et al. The transcriptional role of Smads and FAST (FoxH1) in TGFbeta and activin signalling. Mol Cell Endocrinol 2001; 180:3-11.
78. Li JM, Datto MB, Shen X et al. Sp1, but not Sp3, functions to mediate promoter activation by TGF-beta through canonical Sp1 binding sites. Nucleic Acids Res 1998; 26:2449-2456.
79. Liu B, Dou CL, Prabhu L et al. FAST-2 is a mammalian winged-helix protein which mediates transforming growth factor beta signals. Mol Cell Biol 1999; 19:424-430.
80. Markowitz SD, Roberts AB. Tumor suppressor activity of the TGF-β pathway in human cancers. Cytokine Growth Factor Rev 1996; 7:93-102.

81. Kelly DL, Rizzino A. Growth regulatory factors and carcinogenesis: the roles played by transforming growth factor β, its receptors and signaling pathways. Anticancer Res 1999; 19:4791-4808.
82. Kretzschmar M. Transforming growth factor-beta and breast cancer: Transforming growth factor-beta/SMAD signaling defects and cancer. Breast cancer Res 2000; 2:107-115.
83. Parsons R, Myeroff LL, Liu B et al. Microsatellite instability and mutations of the transforming growth factor beta type II receptor gene in coloretal cancer. Cancer Res 1995; 55:5548-5550.
84. Bedossa P, Peltier E, Terris B et al. Transforming growth factor-β1 (TGF-β1) and TGF-β1 receptors in normal, cirrhotic, and neoplastic human livers. Hepatology 1995; 21:760-766.
85. Guo Y, Jacobs SC, Kyprianou N. Down-regulation of protein and mRNA expression for the transforming growth factor β (TGF-β1) type I and type II receptors in human prostate cancer. Int J Cancer 1997; 71:573-579.
86. Inagaki M, Maustakas A, Lin HY et al. Growth inhibition by transforming growth factor beta (TGF-beta) type I is restored in TGF-beta-resistant hepatoma cells after expression of TGF-beta receptor type II cDNA. Proc Natl Acad Sci USA 1993; 90:5359-5363.
87. Park K, Kim SJ, Bang YJ et al. Genetic changes in the transforming growth factor beta (TGF-beta) type II receptor gene in human gastric cancer cells correlation with sensitivity to growth inhibition by TGF-beta. Proc Natl Acad Sci USA 1994; 91:8772-8776.
88. Nagatake M, Takagi K, Osada H et al. Somatic in vivo alterations of the DPC4 gene at 18q21 in human lung cancers. Cancer Res 1996; 56:2718-2720.
89. Schutte M, Hruban RH, Hedrick L et al. DPC4 gene in various tumor types. Cancer Res 1996; 56:2527-2530.
90. Mondschein JS, Smith SA, Hammond JM. Production of insulin-like growth factor binding proteins (IGFBPs) by porcine granulosa cells: identification of IGFBP-2 and -3 and regulation by hormones and growth factors. Endocrinology 1990; 127:2298-2306.
91. Martin JL, Baxter RC. Transforming growth factor-beta stimulates production of insulin-like growth factor-binding protein-3 by human skin fibroblasts. Endocrinology 1991; 128:1425-1433.
92. Srinivasan N, Baylink DJ, Sampath K et al. Effects of inhibitors of signal transduction pathways on transforming growth factor beta1 and osteogenic protein-1-induced insulinlike growth factor binding protein-3 expression in human bone cells. J Cell Physiol 1997; 173:28-35.
93. Salahifar H, Baxter RC, Martin JL. Differential regulation of insulin-like growth factor-binding protein-3 protease activity in MCF-7 breast cancer cells by estrogen and transforming growth factor-beta1. Endocrinology 2000; 141:3104-3110.
94. Menouny M, Binoux M, Babajko S. Role of insulin-like growth factor binding protein-2 and its limited proteolysis in neuroblastoma cell proliferation: modulation by transforming growth factor-beta and retinoic acid. Endocrinology 1997; 138:683-690.
95. De Los Rios P, Hill DJ. Expression and release of insulin-like growth factor binding proteins in isolated epiphyseal growth plate chondrocytes from the ovine fetus. J Cell Physiol 2000; 183:172-181.
96. Bushman TL, Kuemmerle JF. IGFBP-3 and IGFBP-5 production by human intestinal muscle: reciprocal regulation by endogenous TGF-beta1. Am J Physiol 1998; 275:G1282-1290.
97. Goldstein S, Moerman EJ, Baxter RC. Accumulation of insulin-like growth factor binding protein-3 in conditioned medium of human fibroblasts increases with chronologic age of donor and senescence in vitro. J Cell Physiol 1993; 156:294-302.
98. Firth SM, Fanayan S, Benn D et al. Development of resistance to insulin-like growth factor binding protein-3 in transfected T47D breast cancer cells. Biochem Biophys Res Commun 1998; 246:325-329.
99. Rajah R, Valentinis B, Cohen P. Insulin-like growth factor (IGF)-binding protein-3 induces apoptosis and mediates the effects of transforming growth factor-β1 on programmed cell death through a p53- and IGF-independent mechanism. J Biol Chem 1997; 272:12181-12188.
100. Gucev ZS, Oh Y, Kelley KM et al. Insulin-like growth factor binding protein 3 mediates retinoic acid- and transforming growth factor b2-induced growth inhibition in human breast cancer cells. Cancer Res 1996; 56:1545-1550.
101. Huynh H, Yang X, Pollak M. Estradiol and antiestrogens regulate a growth inhibitory insulin-like growth factor binding protein 3 autocrine loop in human breast cancer cells. J Biol Chem 1996; 271:1016-1021.
102. Rozen F, Zhang J, Pollak M. Antiproliferative action of tumor necrosis factor-a on MCF-7 breast cancer cells is associated with increased insulin-like growth factor binding protein-3 accumulation. Int J Oncology 1998; 13:865-869.
103. Colston KW, Perks CM, Xie SP et al. Growth inhibition of both MCF-7 and Hs578T human breast cancer cell lines by vitamin D analogues is associated with increased expression of insulin-like growth factor binding protein-3. J Mol Endocrinol 1998; 20:157-162.
104. Martin JL, Pattison SL. Insulin-like growth factor binding protein-3 is regulated by dihydrotestosterone and stimulates deoxyribonucleic acid synthesis and cell proliferation in LNCaP prostate carcinoma cells. Endocrinology 2000; 141:2401-2409.
105. Walker GE, Wilson EM, Powell D et al. Butyrate, a histone deacetylase inhibitor, activates the human IGF binding protein-3 promoter in breast cancer cells: molecular mechanism involves an Sp1/Sp3 multiprotein complex. Endocrinology 2001; 142:3817-3827.
106. Gray SG, Kytola S, Lui WO et al. Modulating IGFBP-3 expression by trichostatin A: potential therapeutic role in the treatment of hepatocellular carcinoma. Int J Mol Med 2000; 5:33-41.
107. Shen L, Dean NM, Glazer RI. Induction of p53-dependent, insulin-like growth factor-binding protein-3-mediated apoptosis in glioblastoma multiforme cells by a protein kinase Cα antisense oligonucleotide. Mol Pharmacol 1999; 55:396-402.

108. Buckbinder L, Talbott R, Velasco-Miguel S et al. Induction of the growth inhibitor IGF-binding protein 3 by p53. Nature 1995; 377:646-649.
109. Baxter RC, Martin JL. Regulation and actions of insulin-like growth factor binding protein-3. In: Raizada MK, LeRoith D, eds. Molecular Biology and Physiology of Insulin and Insulin-like Growth Factors. New York: Plenum Press, 1991:125-135.
110. Oh Y, Muller HL, Ng L et al. Transforming growth factor β-induced cell growth inhibition in human breast cancer cells is mediated through insulin-like growth factor-binding protein-3 action. J Biol Chem 1995; 270:13589-13592.
111. Koli KM, Ramsey TT, Ko V et al. Blockade of transforming growth factor-β signalling does not abrogate antiestrogen-induced growth inhibition of human breast carcinoma cells. J Biol Chem 1997; 272:8296-8302.
112. Kansra S, Ewton DZ, Wang J et al. IGFBP-3 mediates TGF beta 1 proliferative response in colon cancer cells. Int J Cancer 2000; 87:373-378.
113. Cohen P, Rajah R, Rosenbloom J et al. IGFBP-3 mediates TGF-β1-induced cell growth in human airway smooth muscle cells. Am J Physiol 2000; 278:L545-L551.
114. O'Grady P, Huang SS, Huang JS. Expression of a new high molecular weight receptor (type V receptor) of transforming growth factor β in normal and transformed cells. Biochem Biophys Res Commun 1991; 179:378-385.
115. Liu Q, Huang SS, Huang JS. Kinase activity of the type V transforming growth factor β receptor. J Biol Chem 1994; 269:9221-9226.
116. Leal SM, Liu Q, Huang SS et al. The type V transforming growth factor β receptor is the putative insulin-like growth factor-binding protein 3 receptor. J Biol Chem 1997; 272:20572-20576.
117. Wu HB, Kumar A, Tsai WC et al. Characterization of the inhibition of DNA synthesis in proliferating mink lung epithelial cells by insulin-like growth factor binding protein-3. J Cell Biochem 2000; 77:288-297.
118. Leal SM, Huang SS, Huang JS. Interactions of high affinity insulin-like growth factor-binding proteins with the type V transforming growth factor-β receptor in mink lung epithelial cells. J Biol Chem 1999; 274:6711-6717.
119. Fanayan S, Firth SM, Butt AJ et al. Growth inhibition by insulin-like growth factor binding protein-3 in T47D breast cancer cells requires transforming growth factor-b (TGF-β) and the type II TGF-b receptor. J Biol Chem 2000; 275:39146-39151.
120. Fanayan S, Firth SM, Baxter RC. Signaling through the Smad pathway by insulin-like growth factor binding protein-3 in breast cancer cells: Relationship to transforming growth factor-β1 signaling. J Biol Chem 2002; 277:7255-7261.
121. Firth SM, Ganeshprasad U, Baxter RC. Structural determinants of ligand and cell surface binding of insulin-like growth factor binding protein-3. J Biol Chem 1998; 273:2631-2638.
122. Schedlich LJ, Young TF, Firth SM et al. Insulin-like growth factor-binding protein (IGFBP-3)-3 and IGFBP-5 share a common nuclear transport pathway in T47D human breast carcinoma cells. J Biol Chem 1998; 273:18347-18352.
123. Butt AJ, Firth SM, King MA et al. Insulin-like growth factor-binding protein-3 modulates expression of Bax and Bcl-2 and potentiates p53-independent radiation-induced apoptosis in human breast cancer cells. J Biol Chem 2000; 275:39174-39181.
124. Valentinis B, Bhala A, De Angelis T et al. The human insulin-like growth factor (IGF) binding protein-3 inhibits the growth of fibroblasts with a targeted disruption of the IGF-1 receptor gene. Mol Endocrinol 1995; 9:361-367.
125. Liu B, Lee H-Y, Weinzimer SA et al. Direct functional interactions between insulin-like growth factor binding protein-3 and retinoid X receptor-α regulate transcriptional signaling and apoptosis. J Biol Chem 2000; 275:33607-33613.
126. Cai T, Lei QY, Wang LY et al. TGF-beta 1 modulated the expression of alpha 5 beta 1 integrin and integrin-mediated signaling in human hepatocarcinoma cells. Biochem Biophys Res Commun 2000; 274:519-525.
127. Zavadil J, Bitzer M, Liang D et al. Genetic programs of epithelial cell plasticity directed by transforming growth factor-beta. Proc Natl Acad Sci USA 2001; 98:6686-6691.
128. Yap OW, Chandrasekher YA, Giudice LC. Growth factor regulation of insulin-like growth factor binding protein secretion by cultured human granulosa-luteal cells. Fertil Steril 1998; 70:535-540.
129. Cazals V, Mouhieddine B, Maitre B et al. Insulin-like growth factors, their binding proteins, and transforming growth factor-beta 1 in oxidant-arrested lung alveolar epithelial cells. J Biol Chem 1994; 269:14111-14117.
130. McCusker RH, Clemmons DR. Effects of cytokines on insulin-like growth factor-binding protein secretion by muscle cells in vitro. Endocrinology 1994; 134:2095-2102.
131. Cohick WS, Gockerman A, Clemmons DR. Regulation of insulin-like growth factor (IGF) binding protein-2 synthesis and degradation by platelet-derived growth factor and the IGFs is enhanced by serum deprivation in vascular smooth muscle cells. J Cell Physiol 1995; 164:187-196.
132. Guo YS, Townsend CM Jr, Jin GF et al. Differential regulation by TGF-beta 1 and insulin of insulin-like growth factor binding protein-2 in IEC-6 cells. Am J Physiol 1995; 268:E1199-1204.
133. Chadelat K, Boule M, Corroyer S et al. Expression of insulin-like growth factors and their binding proteins by bronchoalveolar cells from children with and without interstitial lung disease. Eur Respir J 1998; 11:1329-1336.

134. Martin JL, Ballesteros M, Baxter RC. Insulin-like growth factor-I (IGF-I) and transforming growth factor-beta 1 release IGF-binding protein-3 from human fibroblasts by different mechanisms. Endocrinology 1992; 131:1703-1710.
135. Yateman ME, Claffey DC, Cwyfan Hughes SC et al. Cytokines modulate the sensitivity of human fibroblasts to stimulation with insulin-like growth factor-I (IGF-I) by altering endogenous IGF-binding protein production. J Endocrinol 1993; 137:151-159.
136. Nakao Y, Hilliker S, Baylink DJ et al. Studies on the regulation of insulin-like growth factor binding protein 3 secretion in human osteosarcoma cells in vitro. J Bone Miner Res 1994; 9:865-872.
137. Erondu NE, Dake BL, Moser DR et al. Regulation of endothelial IGFBP-3 synthesis and secretion by IGF-I and TGF-beta. Growth Regul 1996; 6:1-9.
138. Giannini S, Cresci B, Pala L et al. Human glomerular endothelial cells IGFBPs are regulated by IGF-I and TGF-beta1. Mol Cell Endocrinol 1999; 154:123-136.
139. Dahlfors G, Arnqvist HJ. Vascular endothelial growth factor and transforming growth factor-beta1 regulate the expression of insulin-like growth factor-binding protein-3, -4, and -5 in large vessel endothelial cells. Endocrinology 2000; 141:2062-2067.
140. Grellier P, Sabbah M, Fouqueray B et al. Characterization of insulin-like growth factor binding proteins and regulation of IGFBP3 in human mesangial cells. Kidney Int 1996; 49:1071-1078.
141. Hembree JR, Pampusch MS, Yang F et al. Cultured porcine myogenic cells produce insulin-like growth factor binding protein-3 (IGFBP-3) and transforming growth factor beta-1 stimulates IGFBP-3 production. J Anim Sci 1996; 74:1530-1540.
142. Hwa V, Oh Y, Rosenfeld RG. Insulin-like growth factor binding protein-3 and -5 are regulated by transforming growth factor-beta and retinoic acid in the human prostate adenocarcinoma cell line PC-3. Endocrine 1997; 6:235-242.
143. Gentilini A, Feliers D, Pinzani M et al. Characterization and regulation of insulin-like growth factor binding proteins in human hepatic stellate cells. J Cell Physiol 1998; 174:240-250.
144. Johnson DW, Saunders HJ, Baxter RC et al. Paracrine stimulation of human renal fibroblasts by proximal tubule cells. Kidney Int 1998; 54:747-757.
145. Edmondson SR, Murashita MM, Russo VC et al. Expression of insulin-like growth factor binding protein-3 (IGFBP-3) in human keratinocytes is regulated by EGF and TGFbeta1. J Cell Physiol 1999; 179:201-207.
146. Johnson BJ, White ME, Hathaway MR et al. Decreased steady-state insulin-like growth factor binding protein-3 (IGFBP-3) mRNA level is associated with differentiation of cultured porcine myogenic cells. J Cell Physiol 1999; 179:237-243.
147. Garcia-Ramirez M, Audi L, Andaluz P et al. Effects of TGF-beta1 on proliferation and IGFBP-3 production in a primary culture of human fetal epiphyseal chondrocytes (HFEC). J Clin Endocrinol Metab 1999; 84:2978-2981.
148. Cohen P, Nunn SE, Peehl DM. Transforming growth factor-beta induces growth inhibition and IGF-binding protein-3 production in prostatic stromal cells: abnormalities in cells cultured from benign prostatic hyperplasia tissues. J Endocrinol 2000; 164:215-223.
149. Kveiborg M, Flyvbjerg A, Eriksen EF et al. Transforming growth factor-beta1 stimulates the production of insulin-like growth factor-I and insulin-like growth factor-binding protein-3 in human bone marrow stromal osteoblast progenitors. J Endocrinol 2001; 169:549-561.
150. Feliers D, Woodruff K, Abboud S. Potential role of insulin-like growth factor binding protein-4 in the uncoupling of bone turnover in multiple myeloma. Br J Haematol 1999; 104:715-722.
151. Durham SK, Riggs BL, Conover CA. The insulin-like growth factor-binding protein-4 (IGFBP-4)-IGFBP-4 protease system in normal human osteoblast-like cells: regulation by transforming growth factor-beta. J Clin Endocrinol Metab 1994; 79:1752-1758.
152. Canalis E, Gabbitas B. Skeletal growth factors regulate the synthesis of insulin-like growth factor binding protein-5 in bone cell cultures. J Biol Chem 1995; 270:10771-10776.
153. Martin JL, Coverley JA, Baxter RC. Regulation of immunoreactive insulin-like growth factor binding protein-6 in normal and transformed human fibroblasts. J Biol Chem 1994; 269:11470-11477.
154. Gabbitas B, Canalis E. Growth factor regulation of insulin-like growth factor binding protein-6 expression in osteoblasts. J Cell Biochem 1997; 66:77-86.

CHAPTER 17

Insulin-Like Growth Factor Binding Proteins (IGFBPs) and Apoptosis

Claire M. Perks and J. M. P. Holly

Abstract

Apoptosis is a physiologically regulated mode of cell death, which plays a critical role during development and maturation. Apoptosis is tightly controlled both within the cell and by extracellular cues from soluble survival factors and from cell contacts with its extracellular environment. For many cell types the most prevalent and potent survival factor is IGF-I. The family of six high affinity IGFBPs modulate the availability and hence the anti-apoptotic actions of the IGFs. It has become increasingly clear that IGFBPs can also modulate cell survival in a manner, which is not dependent upon interaction with IGF-I. Although there is much still to be learned regarding the mechanisms of actions of the IGFBPs, it is already clear that they may provide very specific strategies for new therapeutic interventions in disorders where apoptosis is dysregulated. We have demonstrated that the IGFBPs alter integrin receptor function and that their effects on cell survival are preceded by altered intracellular signaling from integrin receptors. We believe that the intrinsic actions of IGFBPs on cell survival are via modulation of the integrin-mediated survival cues from the ECM. This enables the IGFBPs to impact on both the soluble and the matrix signals that control cell survival.

What Is Apoptosis?

Apoptosis or programmed cell death (PCD) is a highly regulated physiological mode of cell death, which is characterized by a number of morphological, biochemical and immunochemical alterations. The cells undergo intense membrane blebbing, the cell volume decreases, the chromatin condenses and the cell's DNA undergoes internucleosomal degradation due to the activation of endogenous endonucleases. This is followed by budding of both the nucleus and cytoplasm into multiple, small, membrane-bound apoptotic bodies, which are rapidly phagocytosed by neighboring cells or macrophages. In vivo the whole process can be completed within a couple of hours. In contrast to apoptosis, necrosis is a nonspecific form of cell death, which is distinguished by swelling of the mitochondria and immediate loss of plasma membrane integrity. These events culminate in the release of cellular contents leading to an inflammatory response which is not seen during apoptotic death.[1,2]

Apoptosis during Development and Maturation

It is now clear that apoptosis is an essential component of animal development, important for the establishment and maintenance of tissue architecture.[3,4] In addition to playing a critical role in development, programmed cell death continues to have widespread importance throughout life, where it occurs naturally to maintain a balance between cell proliferation and death ensuring tissue homeostasis. It is involved in a wide variety of physiological processes including follicular development in the ovary[5] and involution of the mammary gland.[6] Dysfunctions in apoptosis not only manifest themselves in developmental abnormalities but also in a wide variety of pathological conditions including cancer, AIDS and neurodegenerative disorders.[7]

Insulin-Like Growth Factors, edited by Derek LeRoith, Walter Zumkeller and Robert Baxter. ©2003 Eurekah.com and Kluwer Academic / Plenum Publishers.

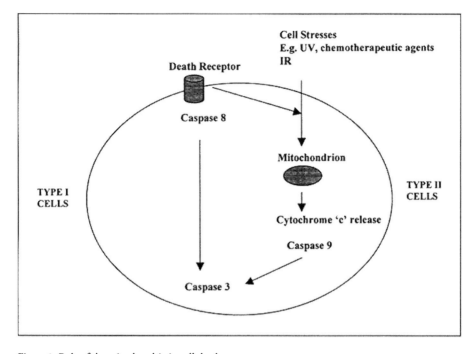

Figure 1. Role of the mitochondria in cell death.

Activation of Apoptotic Signaling Pathways

The components required for apoptotic signaling are constitutively expressed in most cells, but in inactive forms(1). Apoptosis can be triggered by a variety of physiological and stress stimuli. Cell death signals are mediated by two major signaling routes in mammalian cells. The 'intrinsic' pathway, which can be activated by many cell stresses is modulated by Bcl-2 and Bax family members involving the mitochondrion. The 'extrinsic' pathway is triggered by ligation of 'death receptors' belonging to the TNF receptor superfamily. These receptors associate with signaling proteins to form a death inducing signaling complex (DISC).[8] It has been demonstrated that different cell types use distinct death receptor signaling pathways. Type 1 cells use a mitochondria-independent pathway, accompanied by high levels of caspase 8 activation by the DISCs, whereas in type II cells DISC formation is greatly reduced and activation of caspase 9 and subsequently caspase 3 is mediated via the mitochondrion (See Fig. 1).[9]

The Sphingomyelin Pathway

The sphingomyelin pathway is a ubiquitous, evolutionary conserved signaling system. This pathway is associated with the generation of ceramide, by the action of either acidic or neutral sphingomyelinases on sphingomyelin.[10] It was originally assumed that ceramide was produced exclusively at the plasma membrane. However, there is evidence to suggest that it may also be generated at other sites within the cell including the endoplasmic reticulum,[11] mitochondria,[12,13] endosomes and lysosomes[14] and in the nuclear membrane.[15] Neutral sphingomyelinases are primarily located in the plasma membrane, whereas acid sphingomyelinases are found in endosomal and lysosomal compartments.[16,17] Ceramide acts as a second messenger for a number of different agents in activating a plethora of cellular functions including proliferation, inflammation, cell survival and apoptosis.[18] Ceramide levels are elevated in response to a number of diverse challenges including chemotherapeutic agents,[19] ionizing radiation[20] or treatment with pro-death ligands such as TNF-alpha.[21] Ceramide in turn activates multiple signaling targets (summarized in Fig. 2) including ceramide activated protein kinase (CAPK; 97kda),[22] ceramide activated protein phosphatase (CAPP),[23]

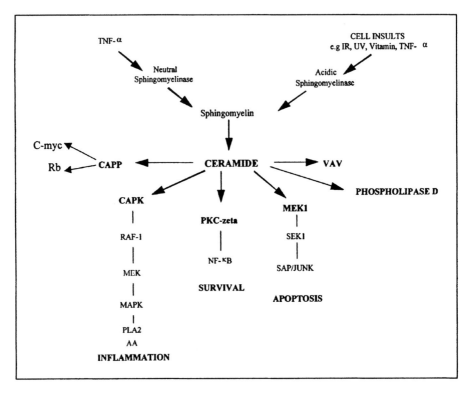

Figure 2. Ceramide-induced signaling cascades.

protein kinase C,[24,25] phospholipase D,[26] and the guanine nucleotide exchange factor, Vav.[27] In most cell systems, ceramide-induced apoptosis is thought to be mediated via activation of the SAPK/JNK cascade[28] which culminates in caspase activation.[29] It has however been demonstrated that the pro-apoptotic Bcl-2 family member, BAD mediates ceramide-induced apoptosis in COS-7 cells, and this is activated via the MAPK pathway, which is normally associated with proliferative and anti-apoptotic signaling.[30] This study highlighted that alternative pathways to the SAPK/JNK cascade can be utilized by ceramide to induce cell death and that the mitochondria is often central to these apoptotic actions of ceramide. It has also been shown that C2-ceramide can act by causing the formation of reactive oxygen intermediates which may affect mitochondrial components directly[31,32] and can cause cytochrome c release from the mitochondria.[33] The various upstream signaling cascades converge downstream to activate a common final effector mechanism for dismantling the dying cell.

Extracellular Control of Apoptosis

In mammalian cells, apoptosis is regulated by signals from other cells in a coordinated manner throughout development.[34] Apoptosis can be induced by inadequate or inappropriate cell-matrix interactions; integrin-mediated cell adhesion signals play a critical role in controlling apoptotic machinery.[35] Cell death can also be induced through death signals or lack of survival signals from surrounding cells. It has been proposed that most mammalian cells are programmed to undergo apoptosis when stimulated to proliferate unless they are exposed to adequate levels of survival factors.[34] Cell death is considered to be the default pathway. In order to prevent inappropriate cell proliferation, the growth factor providing the survival signal should be in limited supply and tightly controlled. If inappropriate cell proliferation occurred, the survival signal would then be outstripped and the default pathway to apoptosis activated. The IGFs are potent stimulators of cell survival and growth in many cell types[36] and fulfil all the necessary requirements to play such a role in preventing inappropriate tissue growth. The IGFs are the most prevalent and potent survival factors in the body and their supply is strictly controlled by the family of high affinity binding proteins (IGFBPs).

Intracellular Control

The extracellular controls on apoptosis act via a number of intracellular mediators. There are a number of interactions between IGFs and p53, which is a key regulator of cell death, that triggers apoptosis in response to DNA damage.[37] In addition to its direct actions on the cell death pathways, p53 can also accentuate cell death by downregulating IGF-II and the type 1 IGF receptor, hence reducing their survival effects.[38,39] Conversely it has been demonstrated that IGF-I can inhibit the transcriptional activity of p53.[40] The vital role of p53 is evident when considering that it is the most commonly found mutated gene in human cancer. The survival effects of IGFs can be mediated by a large number of intracellular signaling pathways, the most common being via P-I3 kinase activation of PKB which in turn phosphorylates BAD resulting in its sequestration by 14-3-3 proteins preventing its proapoptotic effects on the mitochondria.[41]

IGF-Dependent Modulation of Apoptosis by IGFBPs

The IGFBPs 1-6 have a greater affinity for the IGFs than do the type 1 IGF receptors and therefore have a large influence on IGFs binding to their receptor and hence on their consequent survival actions.[42] The IGFBPs have the potential to be either stimulatory[43] or inhibitory[44] to IGF action. As outlined above, Raff (1992)[34] postulated that an important trigger for apoptotic cell death is a reduction in the availability of survival signals. The IGFBPs have been shown to play a major role in regulating the survival effects of IGF-I.

In normal physiology, upregulation of IGFBP-5 has been associated with initiation of apoptosis during involution of the mammary gland,[45] involution of the prostate[46] and atresia of ovarian follicles.[5] Similarly, in pathological conditions such as prostate cancer, interventions such as the vitamin D analogue, EB1089,[47] casodex[48] and finasteride[49] can accentuate apoptosis due to sequestration of IGF-I as a result of induced increases in the production of IGFBPs –2, -3, -4 and –5. Similarly, IGFBPs –2 and –3 have been implicated in the cell death of lung epithelial cells following oxidant injury.[50] In MCF-7 breast cancer cells, the antiestrogen ICI 182,780 induced programmed cell death by upregulation of IGFBP-3[51] and in non-small cell lung cancer cells IGFBP-6 also acted similarly in a pro-apoptotic manner.[52]

IGF-Independent Modulation of Apoptosis by IGFBPs

It was originally believed that the IGFBPs were simply present to modulate the biological actions of the IGFs. However, evidence is accumulating indicating that the IGFBPs have direct 'intrinsic' actions of their own, affecting aspects of cell growth, cell adhesion, migration and apoptosis. It has been reported that apoptosis can be induced directly by addition of IGFBP-3 to PC-3 human prostate cancer cells[53] and also by overexpressing IGFBP-3 in human breast cancer cells.[54] However, accumulating evidence suggests that in most cell types, at physiological concentrations, IGFBP-3 alone is unable to effect cell death but can potently accentuate a variety of apoptotic triggers including C2-ceramide,[55] paclitaxel[56] and antimycin A[57] in Hs578T human breast cancer cells, gamma radiation in colorectal carcinoma cells[58] and UV-radiation in KYSE oesophageal carcinoma cells.[59] All of these studies consistently indicate that IGFBP-3 can affect cell survival; there is however discrepancy between whether IGFBP-3 can induce apoptosis or modulate apoptosis triggered by other cell insults. There are major differences in the physiological and pathological implications of these two different actions. There are a number of potential explanations for the discrepancy in reported actions. There may well be differences in biological potency between different preparations of IGFBP-3 used in various laboratories. We have generally reported findings using a widely available preparation of recombinant nonglycosylated IGFBP-3. We have however compared the actions of 8 different preparations of IGFBP-3, including 3 different preparations of recombinant glycosylated IGFBP-3 (produced in different laboratories) and one preparation of IGFBP-3 purified from human serum; all of these preparations produced comparable effects modulating apoptosis (unpublished observations). The different reported actions could also be attributed to the different cell lines that have been used; as with any other cell regulator, it should be expected that there will be a range of sensitivity to IGFBP-3 actions between different cells. A further contributing factor to the different observed cell responses could be the experimental design. There is always a degree of stress induced by culturing cells in the laboratory and this can vary considerably dependent upon the substrata, the culture media and whether this is serum-supplemented or serum-free. In addition some studies have examined effects of addition of various preparations of IGFBP-3 and others have relied upon transfecting cells to over-express IGFBP-3. It has been reported that transfecting cells to express IGFBP-3 resulted in induction of apoptosis and that this was associated with an increase in cell conditioned-media concentrations of IGFBP-3 from undetectable to 20-25 ng/ml.[54] It is difficult to reconcile the physiological significance of cell death induction associated with such low concentrations of IGFBP-3 when considering that concentrations of IGFBP-3 in the body are normally more than a 100 fold higher than this. It is evident that IGFBP-3 can interact with many proteins on the cell surface, in the cytoplasm and in the nucleus, the functional implications of all of these interactions remains unclear. It should be considered that transfected IGFBP-3 expressed within a cell, not under the normal transcriptional controls, may result in a different pattern of IGFBP-3 interactions within the cell compared to that induced by IGFBP-3 encountered as an extracellular signal. The machinery required for induction and execution of apoptosis are constitutively present within cells, but they are normally kept in check by a series of balances. If IGFBP-3 interacts with components that are involved in maintaining these balances, then employing some experimental paradigms, IGFBP-3 may result in apoptosis. Care is however required before extrapolating such observations to infer that IGFBP-3 acts as direct trigger for apoptosis in normal physiology.

In the case of IGFBP-3, it is not only regulated by p53[60] but also appears to act like p53—which does not induce death of normal healthy cells but is activated in situations of cell stress, arresting growth allowing repair or initiating apoptosis. Similarly IGFBP-3 does not compromise normal healthy cells but acts on stressed cells; in our studies the response to even a small stress, such as a change in media, can be enhanced in the presence of IGFBP-3. This could explain some of the apparent discrepancy in the literature regarding induction or enhancement of apoptosis. Although the IGFBP-3 gene is a target of the tumor suppressor protein p53[60] the effects of IGFBP-3 on cell death can involve interactions with p53 but can also clearly act independently. The direct induction of apoptosis by IGFBP-3 has been reported in p53 null PC3 cells;[53] in addition accentuating actions of IGFBP-3 on cell death occurred in both a p53 dependent[58,59] and independent[54,55] manner.

We have also found that IGFBPs -4 and -5 were unable to directly trigger cell death, but in contrast to IGFBP-3 they acted as potent survival factors against C2-induced apoptosis in Hs578T cells.[55] Additionally in this model, IGFBP-5 was also a potent survival factor against apoptosis induced by an RGD (-arg-gly-asp-) disintegrin, whereas IGFBP-3 was without effect.[55] These studies indicated that despite sharing a high degree of structural similarity, IGFBP-3 and -5 exerted opposite effects on apoptosis; whereas IGFBP-4 and -5 had identical actions despite being relatively more dissimilar in structure. These data suggested that IGFBP-3 was acting via a different pathway to IGFBPs -4 and -5 to modulate cell death. This was confirmed using a further alternative trigger of cell death, antimycin A, which initiates apoptosis by inhibiting mitochondrial respiratory chain transport. In Hs578T cells antimycin A-induced apoptosis was accentuated by IGFBP-3, whereas IGFBP-5 was without effect.[57] In T47D human breast cancer cells it has been reported that the enhancement of radiation-induced apoptosis by IGFBP-3 occurred via upregulation of the mitochondria-associated proteins Bax and Bad.[54] Based on these data, it appears that IGFBP-3 may only accentuate apoptosis that has been induced via pathways involving the mitochondria. In contrast IGFBP-5, could inhibit RGD and C2-induced death but had no effect on apoptosis induced by antimycin A, indicating that IGFBP-5 was exerting effects upstream or simply independently of the mitochondria. These results would fit into a model analogous to that described by Scaffidi (1998)[9] (see Fig. 3) for two distinct intracellular signaling pathways for apoptosis outlined in the previous section (1.3). Further attempts using the Hs578T cells to dissect the pathways by which IGFBPs -3 and -5 exerted their IGF-independent effects on cell death, determined that inhibiting MAPK blocked the ability of IGFBP-3 to accentuate apoptosis but had no effect on the ability of IGFBP-5 to confer survival.[61] In contrast, inhibition of PKC prevented IGFBP-5 conferring survival but had no effect on the accentuating actions of IGFBP-3.[62] These studies provide further evidence that IGFBP -3 and -5 are modulating apoptosis via distinct signaling pathways.

While it has become accepted to refer to the actions of IGFBPs that are not dependent upon them binding IGFs as "IGF-independent", this term is not strictly correct. It is evident that the

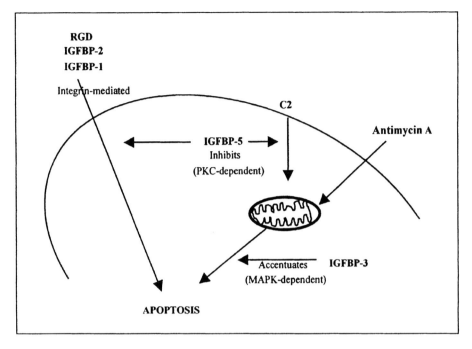

Figure 3. IGF-independent modulation of apoptosis.

binding of IGFs to IGFBP-3 can block its intrinsic action on cells[63] and this appears to be due to reducing its binding to the cell surface.[64] There is evidence that the binding of IGFs similarly alters the affinity of IGFBP-5 for proteoglycans and the cell surface[65] and it is likely that IGF will affect the intrinsic actions of this and other IGFBPs. As a consequence there will be complex balances in vivo with IGFBPs modulating the potent survival actions of the IGFs, acting directly on cell survival and with these actions in turn being modulated by the IGFs.

Mechanism of IGFBP Intrinsic Actions

There have been reports of putative IGFBP-3 and –5 cell surface receptors.[66,67] The differences in cellular responses to IGFBP-3 and –5 could be dependent upon variations in the concentrations of these receptors on different cell types and differences in intracellular pathways activated. It has been reported, for example that Hs578T cells have 30 fold more IGFBP-3 receptors than MCF-7 cells.[68] An association of IGFBP-3 with the TGF-beta V receptor[69] has also been reported. All of these associations are at present putative 'IGFBP-receptors' as links to intracellular signaling with any of these receptors have not been clearly established and confirmation of these receptors by independent laboratories is eagerly awaited.

A clearly established mechanism has been determined for actions of two of the IGFBPs. The only high affinity IGFBPs to possess an RGD integrin-binding motif are IGFBP-1 and -2. Direct IGF-independent actions of IGFBP-1 on cell migration have been demonstrated[70,71] and were mediated by interaction of this RGD sequence with the alpha 5 beta 1 integrin receptor.[72] It has also been shown that IGFBP-1 and -2 induced focal adhesion kinase (FAK) dephosphorylation, cell detachment and subsequent apoptosis of Hs578T and T47D human breast cancer cells and these actions were mimicked by a synthetic RGD-containing disintegrin.[73,74] Interactions of other IGFBPs with integrin receptors have also been proposed as mechanisms by which they could modulate cellular responses independently of IGF-I. We have demonstrated that, despite not possessing a classical RGD sequence, IGFBP-3 could modulate the phosphorylation and localization of one of the key integrin signaling molecules, FAK in Hs578T cells completely independent of IGF involvement.[75] Furthermore, IGFBP-3 and –5 could also modulate one of the main functions of integrin receptors, that of cell attachment; we have shown that IGFBP-3 inhibited, whereas IGFBP-5 had the opposite effect and promoted adhesion of Hs578T cells onto extracellular matrix.[61] These contrasting actions on cell attachment are identical to the opposing actions of these IGFBPs on the survival of these cells. In addition, IGFBP-3 and -5 have been shown to bind to known integrin ligands such as fibronectin,[76] ADAM-12[77] and thrombospondin and osteopontin.[78] Furthermore, the IGF-independent modulation of both apoptosis and cell attachment by IGFBP-3 were effectively blocked via modulation of integrin receptor status.[61] These data suggest that these IGF-independent actions of IGFBPs -3 and -5 may be mediated via as yet uncharacterized interactions with integrin receptors. Interestingly the CCN family of proteins, that are closely related to IGFBPs[79] and share some structural homology, have similarly been shown to affect cell function via modulation of integrin signaling despite not possessing classical integrin recognition sequences.[80]

In an attempt to elucidate which region of IGFBP-3 is responsible for exerting these IGF-independent actions, we synthesized a number of different 15-20 amino acid peptides spanning the structure of IGFBP-3. We determined that only one, a 15 amino acid peptide sequence spanning the two mid-region serines of IGFBP-3, was able to mimic the accentuating actions of IGFBP-3 on cell death.[81] This peptide corresponds to the region of the IGFBP-3 molecule, which is also responsible for binding to the putative IGFBP-3 receptor.[68] Furthermore, we have demonstrated that the ability of IGFBP-3 to accentuate apoptosis was negated following phosphorylation of these mid-region serines.[81] The six high affinity IGFBPs are all composed of three major domains; the N, C and mid-region.[42] The N and C terminal regions are highly homologous for all six IGFBPs and contain sites involved in IGF binding. These homologous regions thus provide a common mechanism for controlling IGF function. Specific post-translational modifications (including phosphorylation and proteolysis) to each of the IGFBPs occur at residues found within the variable mid-region of the IGFBPs. These alterations in this domain could lend specificity to the actions of the individual IGFBPs and provide a very specific way of modulating them. We have also demonstrated that the IGFBPs have differential actions on cell attachment and cell survival[55] and the evidence suggests that the variable mid-region will be responsible for these specific actions.

Two of the IGFBPs, IGFBP-3 and -5, possess nuclear localization sequences[82] and can traverse to the nucleus, suggesting that they may be able to modulate transcriptional activity; indeed it has recently been shown that IGFBP-3 can bind with the retinoid RXR receptor.[83] Translocation of IGFBP-3 and -5 to the nucleus and subsequent effects on gene transcription, which has also been proposed as a possible mechanism by which they could exert IGF-independent actions, could not however explain the very rapid cellular responses such as IGFBP-3 modulating the phosphorylation status of signaling molecules such as FAK[75] and SMAD-2 and -3.[84]

Therapeutic Interventions and Potential Modulation by IGFBPs

An understanding of the mechanisms by which the IGFBPs can modulate apoptosis in an IGF-independent manner may provide novel therapeutic strategies. It is clear that programmed cell death is a critical underlying component of many developmental abnormalities and human diseases. Cancer is associated with insufficient cell death. As outlined in section 1.7, a number of current therapeutic interventions to induce apoptosis of tumor cells are known to act at least in part via an increase in endogenous IGFBPs and subsequent sequestration of the survival signal IGF-I.

Chemotherapy and radiotherapy used for the treatment of cancer induce apoptosis at least partly via an increase in endogenous ceramide, which may also be accentuated by the IGF-independent actions of IGFBP-3. The combination of these current treatments with agents that could mimic or activate IGFBP-3 actions could be a way of maximizing the levels of apoptosis within tumors.

Conversely, other diseases including AIDS, osteoporosis and neurodegenerative diseases are associated with excessive cell death. In these disorders actions of IGFBPs, such as IGFBP-5, which could suppress apoptosis, may be of value to restore tissue homeostasis. It is now clear that the IGFBPs have many effects on cell death, via both IGF-dependent and independent actions. Although the mechanisms underlying these latter actions are only beginning to be understood, it is already clear that they may provide very specific strategies for fine-tuning therapeutic interventions.

References

1. Wyllie AH. Cell death: the significance of apoptosis. Int Rev Cytology 1980; 60:251-306.
2. Kerr JFR, Wyllie AH, Currie AR. Apoptosis:a basic biological phenomenon with wide-ranging implication in tissue kinetics. Br J Cancer 1972; 26:239-257.
3. Rakic S, Zecevic N. Programmed cell death in the developing human telencephalon. Eur J Neurosci 2000; 12(8):2721-34.
4. Schaller SA, Li S, Ngo-Muller V et al. Cell biology of limb patterning. Int Rev Cytol 2001; 203:483-517.
5. Kaipia A, Hsueh AJ. Regulation of ovarian follicle atresia. Annu Rev Physiol 1997; 59:349-63.
6. Kumar R, Vadlamudi RK, Adam L. Apoptosis in mammary gland and cancer. Endocr Relat Cancer 2000; 7(4):257-69.
7. Thompson CB. Apoptosis in the pathogenesis and treatment of disease. Science 1995; 267(5203):1456-62.
8. Kischkel FC, Hellbardt S, Behrmann I et al. Cytotoxicity-dependent APO-1 (Fas/CD95)-associated proteins form a death-inducing signaling complex (DISC) with the receptor. Embo J 1995; 14(22):5579-88.
9. Scaffidi C, Fulda S, Srinivasan A et al. Two CD95 (APO-1/Fas) signaling pathways. Embo J 1998; 17(6):1675-87.
10. Kolesnick RN. Sphingomyelin and derivatives as cellular signals. Prog Lipid Res 1991; 30(1):1-38.
11. Bose R, Verheij M, Haimovitz-Friedman A et al. Ceramide synthase mediates daunorubicin-induced apoptosis: an alternative mechanism for generating death signals. Cell 1995; 82(3):405-14.
12. El Bawab S, Roddy P, Qian T et al. Molecular cloning and characterization of a human mitochondrial ceramidase. J Biol Chem 2000; 275(28):21508-13.
13. Shimeno H, Soeda S, Sakamoto M et al. Partial purification and characterization of sphingosine N-acyltransferase (ceramide synthase) from bovine liver mitochondrion-rich fraction. Lipids 1998; 33(6):601-5.
14. Heinrich M, Wickel M, Schneider-Brachert W et al. Cathepsin D targeted by acid sphingomyelinase-derived ceramide. Embo J 1999; 18(19):5252-63.
15. Tsugane K, Tamiya-Koizumi K, Nagino M et al. A possible role of nuclear ceramide and sphingosine in hepatocyte apoptosis in rat liver. J Hepatol 1999; 31(1):8-17.
16. Hannun YA. The sphingomyelin cycle and the second messenger function of ceramide. J Biol Chem 1994; 269(5):3125-8.
17. Schutze S, Potthoff K, Machleidt T et al. TNF activates NF-kappa B by phosphatidylcholine-specific phospholipase C-induced " acidic" sphingomyelin breakdown. Cell 1992; 71(5):765-76.
18. Spiegel S, Foster D, Kolesnick R. Signal transduction through lipid second messengers. Curr Opin Cell Biol 1996; 8(2):159-67.
19. Kaufmann SH. Induction of endonucleolytic DNA cleavage in human acute myelogenous leukemia cells by etoposide, camptothecin, and other cytotoxic anticancer drugs: a cautionary note. Cancer Res 1989; 49(21):5870-8.
20. Haimovitz-Friedman A, Kan CC, Ehleiter D et al. Ionizing radiation acts on cellular membranes to generate ceramide and initiate apoptosis. J Exp Med 1994; 180(2):525-35.
21. Joseph CK, Byun HS, Bittman R et al. Substrate recognition by ceramide-activated protein kinase. Evidence that kinase activity is proline-directed. J Biol Chem 1993; 268(27):20002-6.
22. Mathias S, Dressler KA, Kolesnick RN. Characterization of a ceramide-activated protein kinase: stimulation by tumor necrosis factor alpha. Proc Natl Acad Sci USA 1991; 88(22):10009-13.
23. Dobrowsky RT, Kamibayashi C, Mumby MC et al. Ceramide activates heterotrimeric protein phosphatase 2A. J Biol Chem 1993; 268(21):15523-30.
24. Huwiler A, Fabbro D, Pfeilschifter J. Selective ceramide binding to protein kinase C-alpha and -delta isoenzymes in renal mesangial cells. Biochemistry 1998; 37(41):14556-62.
25. Lozano J, Berra E, Municio MM et al. Protein kinase C zeta isoform is critical for kappa B-dependent promoter activation by sphingomyelinase. J Biol Chem 1994; 269(30):19200-2.
26. Gomez-Munoz A, Martin A, O'Brien L et al. Cell-permeable ceramides inhibit the stimulation of DNA synthesis and phospholipase D activity by phosphatidate and lysophosphatidate in rat fibroblasts. J Biol Chem 1994; 269(12):8937-43.
27. Gulbins E, Coggeshall KM, Baier G et al. Direct stimulation of Vav guanine nucleotide exchange activity for Ras by phorbol esters and diglycerides. Mol Cell Biol 1994; 14(7):4749-58.
28. Kyriakis JM, Banerjee P, Nikolakaki E et al. The stress-activated protein kinase subfamily of c-Jun kinases. Nature 1994; 369(6476):156-60.
29. Lievremont JP, Sciorati C, Morandi E et al. The p75(NTR)-induced apoptotic program develops through a ceramide-caspase pathway negatively regulated by nitric oxide. J Biol Chem 1999; 274(22):15466-72.
30. Basu S, Bayoumy S, Zhang Y et al. BAD enables ceramide to signal apoptosis via Ras and Raf-1. J Biol Chem 1998; 273(46):30419-26.
31. Gudz TI, Tserng KY, Hoppel CL. Direct inhibition of mitochondrial respiratory chain complex III by cell-permeable ceramide. J Biol Chem 1997; 272(39):24154-8.
32. Garcia-Ruiz C, Colell A, Mari M et al. Direct effect of ceramide on the mitochondrial electron transport chain leads to generation of reactive oxygen species. Role of mitochondrial glutathione. J Biol Chem 1997; 272(17):11369-77.
33. Ghafourifar P, Klein SD, Schucht O et al. Ceramide induces cytochrome c release from isolated mitochondria. Importance of mitochondrial redox state. J Biol Chem 1999; 274(10):6080-4.
34. Raff MC. Social controls on cell survival and cell death. Nature 1992; 356(6368):397-400.
35. Frisch SM, Screaton RA. Anoikis mechanisms. Curr Opin Cell Biol 2001; 13(5):555-62.

36. Barres BA, Schmid R, Sendnter M et al. Multiple extracellular signals are required for long-term oligodendrocyte survival. Development 1993; 118(1):283-95.
37. Levine AJ. p53, the cellular gatekeeper for growth and division. Cell 1997; 88(3):323-31.
38. Zhang L, Zhan Q, Zhan S et al. p53 regulates human insulin-like growth factor II gene expression through active P4 promoter in rhabdomyosarcoma cells. DNA Cell Biol 1998; 17(2):125-31.
39. Ohlsson C, Kley N, Werner H et al. p53 regulates insulin-like growth factor-I (IGF-I) receptor expression and IGF-I-induced tyrosine phosphorylation in an osteosarcoma cell line: interaction between p53 and Sp1. Endocrinology 1998; 139(3):1101-7.
40. Yamaguchi A, Tamatani M, Matsuzaki H et al. Akt activation protects hippocampal neurons from apoptosis by inhibiting transcriptional activity of p53. J Biol Chem 2001; 276(7):5256-64.
41. Kulik G, Weber MJ. Akt-dependent and -independent survival signaling pathways utilized by insulin-like growth factor I. Mol Cell Biol 1998; 18(11):6711-8.
42. Jones JI, Clemmons DR. Insulin-like growth factors and their binding proteins: biological actions. Endocr Rev 1995; 16(1):3-34.
43. Blum WF, Jenne EW, Reppin F et al. Insulin-like growth factor I (IGF-I)-binding protein complex is a better mitogen than free IGF-I. Endocrinology 1989; 125(2):766-72.
44. Ritvos O, Ranta T, Jalkanen J et al. Insulin-like growth factor (IGF) binding protein from human decidua inhibits the binding and biological action of IGF-I in cultured choriocarcinoma cells. Endocrinology 1988; 122(5):2150-7.
45. Flint DJ, Tonner E, Allan GJ. Insulin-like growth factor binding proteins: IGF-dependent and -independent effects in the mammary gland. J Mammary gland Biol Neoplasia 2000; 5(1):65-73.
46. Thomas LN, Cohen P, Douglas RC et al. Insulin-like growth factor binding protein 5 is associated with involution of the ventral prostate in castrated and finasteride-treated rats. Prostate 1998; 35(4):273-8.
47. Nickerson T, Huynh H. Vitamin D analogue EB1089-induced prostate regression is associated with increased gene expression of insulin-like growth factor binding proteins. J Endocrinol 1999; 160(2):223-9.
48. Nickerson T, Pollak M. Bicalutamide (Casodex)-induced prostate regression involves increased expression of genes encoding insulin-like growth factor binding proteins. Urology 1999; 54(6):1120-5.
49. Thomas LN, Wright AS, Lazier CB et al. Prostatic involution in men taking finasteride is associated with elevated levels of insulin-like growth factor-binding proteins (IGFBPs)-2, -4, and -5. Prostate 2000; 42(3):203-10.
50. Besnard V, Corroyer S, Trugnan G et al. Distinct patterns of insulin-like growth factor binding protein (IGFBP)-2 and IGFBP-3 expression in oxidant exposed lung epithelial cells. Biochim Biophys Acta 2001; 1538(1):47-58.
51. Huynh H, Alpert L, Alaoui-Jamali MA et al. Co-administration of finasteride and the pure anti-oestrogen ICI 182,780 act synergistically in modulating the IGF system in rat prostate. J Endocrinol 2001; 171(1):109-18.
52. Sueoka N, Lee HY, Wiehle S et al. Insulin-like growth factor binding protein-6 activates programmed cell death in non-small cell lung cancer cells. Oncogene 2000; 19(38):4432-6.
53. Rajah R, Valentinis B, Cohen P. Insulin-like growth factor (IGF)-binding protein-3 induces apoptosis and mediates the effects of transforming growth factor-beta1 on programmed cell death through a p53- and IGF-independent mechanism. J Biol Chem 1997; 272(18):12181-8.
54. Butt AJ, Firth SM, King MA et al. Insulin-like growth factor-binding protein-3 modulates expression of Bax and Bcl-2 and potentiates p53-independent radiation-induced apoptosis in human breast cancer cells. J Biol Chem 2000; 275(50):39174-81.
55. Perks CM, Bowen S, Gill ZP et al. Differential IGF-independent effects of insulin-like growth factor binding proteins (1-6) on apoptosis of breast epithelial cells. J Cell Biochem 1999; 75(4):652-64.
56. Fowler CA, Perks CM, Newcomb PV et al. Insulin-like growth factor binding protein-3 (IGFBP-3) potentiates paclitaxel-induced apoptosis in human breast cancer cells. Int J Cancer 2000; 88(3):448-53.
57. Perks CM, McCaig C, Holly JM. Differential insulin-like growth factor (IGF)-independent interactions of IGF binding protein-3 and IGF binding protein-5 on apoptosis in human breast cancer cells. Involvement of the mitochondria. J Cell Biochem 2000; 80(2):248-58.
58. Williams AC, Collard TJ, Perks CM et al. Increased p53-dependent apoptosis by the insulin-like growth factor binding protein IGFBP-3 in human colonic adenoma-derived cells. Cancer Res 2000; 60(1):22-7.
59. Hollowood AD, Lai T, Perks CM et al. IGFBP-3 prolongs the p53 response and enhances apoptosis following UV irradiation. Int J Cancer 2000; 88(3):336-41.
60. Buckbinder L, Talbott R, Velasco-Miguel S et al. Induction of the growth inhibitor IGF-binding protein 3 by p53. Nature 1995; 377(6550):646-9.
61. Holly SP, Larson MK, Parise LV. Multiple roles of integrins in cell motility. Exp Cell Res 2000; 261(1):69-74.
62. McCaig C, Perks CM, Holly JMP. IGF-independent effects of IGFBP-5 on breast epithelial cell survival and attachment are blocked by a synthetic RGD-containing peptide. Proceedings of the Endocrine Society's 83rd Annual Meeting 2001; Denver:Abstract.
63. Maile LA, Gill ZP, Perks CM et al. The role of cell surface attachment and proteolysis in the insulin-like growth factor (IGF)-independent effects of IGF-binding protein-3 on apoptosis in breast epithelial cells. Endocrinology 1999; 140(9):4040-5.
64. Oh Y, Muller HL, Lamson G et al. Insulin-like growth factor (IGF)-independent action of IGF-binding protein-3 in Hs578T human breast cancer cells. Cell surface binding and growth inhibition. J Biol Chem 1993; 268(20):14964-71.

65. Parker A, Rees C, Clarke J et al. Binding of insulin-like growth factor (IGF)-binding protein-5 to smooth-muscle cell extracellular matrix is a major determinant of the cellular response to IGF-I. Mol Biol Cell 1998; 9(9):2383-92.
66. Andress DL. Insulin-like growth factor-binding protein-5 (IGFBP-5) stimulates phosphorylation of the IGFBP-5 receptor. Am J Physiol 1998; 274(4 Pt 1):E744-50.
67. Oh Y, Muller HL, Pham H et al. Demonstration of receptors for insulin-like growth factor binding protein-3 on Hs578T human breast cancer cells. J Biol Chem 1993; 268(35):26045-8.
68. Yamanaka Y, Fowlkes JL, Wilson EM et al. Characterization of insulin-like growth factor binding protein-3 (IGFBP- 3) binding to human breast cancer cells: kinetics of IGFBP-3 binding and identification of receptor binding domain on the IGFBP-3 molecule. Endocrinology 1999; 140(3):1319-28.
69. Leal SM, Liu Q, Huang SS et al. The type V transforming growth factor beta receptor is the putative insulin-like growth factor-binding protein 3 receptor. J Biol Chem 1997; 272(33):20572-6.
70. Irving JA, Lala PK. Functional role of cell surface integrins on human trophoblast cell migration: regulation by TGF-beta, IGF-II, and IGFBP-1. Exp Cell Res 1995; 217(2):419-27.
71. Jones JI, Doerr ME, Clemmons DR. Cell migration: interactions among integrins, IGFs and IGFBPs. Prog Growth Factor Res 1995; 6(2-4):319-27.
72. Jones JI, Gockerman A, Busby WH et al. Insulin-like growth factor binding protein 1 stimulates cell migration and binds to the alpha 5 beta 1 integrin by means of its Arg-Gly-Asp sequence. Proc Natl Acad Sci USA 1993; 90(22):10553-7.
73. Perks CM, Newcomb PV, Norman MR et al. Effect of insulin-like growth factor binding protein-1 on integrin signalling and the induction of apoptosis in human breast cancer cells. J Mol Endocrinol 1999; 22(2):141-50.
74. Schutt BS, Langkamp M, Ranke MB et al. Intracellular Signalling Of Insulin-like Growth Factor Binding protein-2. Growth hormone & IGF research 2000; 10(4):A30 P35.
75. Perks CM, McCaig C, Laurence NJ et al. IGFBP-3 modulates integrin signalling independently of IGF-I in Hs578T human breast cancr cells. Growth hormone & IGF research 2000; 10(4):A26 O6.4.
76. Gui Y, Murphy LJ. Insulin-like growth factor (IGF)-binding protein-3 (IGFBP-3) binds to fibronectin (FN): demonstration of IGF-I/IGFBP-3/fn ternary complexes in human plasma. J Clin Endocrinol Metab 2001; 86(5):2104-10.
77. Shi Z, Xu W, Loechel F et al. ADAM 12, a disintegrin metalloprotease, interacts with insulin-like growth factor-binding protein-3. J Biol Chem 2000; 275(24):18574-80.
78. Nam TJ, Busby WH Jr, Rees C et al. Thrombospondin and osteopontin bind to insulin-like growth factor (IGF)- binding protein-5 leading to an alteration in IGF-I-stimulated cell growth [In Process Citation]. Endocrinology 2000; 141(3):1100-6.
79. Hwa V, Oh Y, Rosenfeld RG. The insulin-like growth factor-binding protein (IGFBP) superfamily. Endocr Rev 1999; 20(6):761-87.
80. Perbal B. NOV (nephroblastoma overexpressed) and the CCN family of genes: structural and functional issues. Mol Pathol 2001; 54(2):57-79.
81. Hollowood AD, Stewart CEH, Lai T et al. The role of serines in the mid region of IGFBP-3 in its autonomous actions. Growth hormone and IGF research 2000; 10(4):A27 P30.
82. Schedlich LJ, Young TF, Firth SM et al. Insulin-like growth factor-binding protein (IGFBP)-3 and IGFBP-5 share a common nuclear transport pathway in T47D human breast carcinoma cells. J Biol Chem 1998; 273(29):18347-52.
83. Liu B, Lee HY, Weinzimer SA et al. Direct functional interactions between insulin-like growth factor-binding protein-3 and retinoid X receptor-alpha regulate transcriptional signaling and apoptosis. J Biol Chem 2000; 275(43):33607-13.
84. Fanayan S, Firth SM, Butt AJ et al. Growth inhibition by insulin-like growth factor-binding protein-3 in T47D breast cancer cells requires transforming growth factor-beta (TGF-beta) and the type II TGF-beta receptor. J Biol Chem 2000; 275(50):39146-51.

CHAPTER 18

The Role of Insulin-Like Growth Factor-1 and Extracellular Matrix Protein Interaction in Controlling Cellular Responses to This Growth Factor

David R. Clemmons

Abstract

Insulin-like growth factor-I (IGF-I) is a small polypeptide growth factor that is ubiquitously present in physiologic fluids. IGF-I is a potent stimulant of extracellular matrix protein (ECM) synthesis by all connective tissue cell types and it can control synthesis of basement membrane proteins by epithelial cells. Similarly changes in extracellular matrix protein composition result in changes in the amount of IGF-I or II that are localized in the ECM. One specific mechanism by which this occurs is through insulin-like growth factor binding proteins. These high affinity proteins can bind to specific ECM components and act as a site of focal localization of IGF-I and II. Factors that regulate the amount of IGF-I that is localized include proteolytic cleavage of these binding proteins as well as alterations in ECM components that control their rate of deposition and affinity for IGF-I and II. In addition to ECM, interactions and binding to IGF binding proteins, the IGF actions are regulated by ECM protein binding to integrin receptors. Changes in ligand occupancy of integrin receptors can influence cell responsiveness to IGF-I by directly altering IGF-I signaling. One important means by which the integrin $\alpha V \beta 3$ regulates IGF-I signaling is by controlling the rate of IGF-I receptor dephosphorylation. This occurs by integrin mediated regulation of the rate of transfer of the phosphatase SHP-2 to the IGF-I receptor. Such specific interactions are important modulators of the ability of the ECM proteins to regulate cellular responsiveness to IGF-I.

Introduction

Although cells derived from all three embryonic lineages possess IGF receptors, IGF-I was originally discovered based on its ability to stimulate growth and metabolism of cartilage cells.[1] In humans IGF-I is an essential growth factor for stimulating the growth of the mesenchymal cell types that account for increased statural growth such as, chondrocytes, osteoblasts and fibroblasts. These cell types are normally enveloped in the extracellular matrix (ECM). ECM anchorage is an important growth regulatory variable for each of these cell types and when they are not attached to the ECM they are less sensitive to mitogens and have markedly reduced rates of proliferation. In general following a proapoptotic stimulus mesenchymal cells in suspension are less responsive to the antipoptotic and growth stimulating effects of IGF-I. Furthermore, in states of abnormal rates of proliferation such as, wound repair after injury, organ hypertrophy or malignancy, the normal relationship between cell contacts and the ECM is perturbed and in many cases these alterations have been associated with alterations in responsiveness to peptide growth factors, such as IGF-I. However the interaction between changes in ECM components and the effects that these changes have on cellular responsiveness to IGF-I are complex. This chapter will review the findings that have been published to date and their implication for the research into this important area of cell biology.

Insulin-Like Growth Factors, edited by Derek LeRoith, Walter Zumkeller and Robert Baxter. ©2003 Eurekah.com and Kluwer Academic / Plenum Publishers.

IGFs and Extracellular Matrix Component Composition

IGF-I and II are important stimuli of extracellular matrix production. Several cell types have been shown to be responsive to these two growth factors and to secrete abundant amounts of ECM proteins following IGF exposure. Important substituents of matrix such as type I and type IV collagen, elastin, tenascin-C and chondroitin sulfate to name a few have been shown to be synthesized and secreted in response to IGF-I.[2-4] IGF-I has also been shown to stimulate the synthesis of ECM components in several tumor cell types and this has been postulated to provide a selective growth advantage for these cells. IGF-I administration to animals is associated with increased type I fibrillar collagen synthesis, increased collagen crosslinking and stimulation of lysyl hydroxylase and which enhances matrix formation.[5]

Often IGF-I works in conjunction with other growth factors. Smith et al showed that in articular chondrocytes IGF-I stimulated proteoglycan synthesis by 2-3 fold and that its effect was additive with bone morphogenic protein-2.[2] Similarly following inhibition of proteoglycan synthesis in bovine articular cartilage by retinoic acid treatment, IGF-I could restore proteoglycan synthesis to normal and that its effect was additive with TGFβ-1.[6] Hill et al demonstrated that IGF-I had additive effects with TGFβ-1 in stimulating sulphated glycosaminoglycan synthesis by bovine growth plate chrondrocytes but that their actions were not additive in terms of total protein or collagen synthesis.[7] PDGF and IGF-I had additive effects in stimulating monocyte-ECM protein interactions.[8]

IGF-I can also stimulate changes in ECM proteins that result in changes in cell architecture. Van Osch et al demonstrated that IGF-I in conjunction with TGFβ-2 increased the synthesis of several components that are necessary for proteoglycan assembly in bovine articular chondrocytes.[9] IGF-I induced a 24% increase and TGFβ induced a 36% increase while the combination led to an 82% increase in proteoglycan and an 87% increase in total ECM component synthesis. Similarly IGF-I increased collagen gel contraction by fibroblasts, suggesting that it stimulated the interaction between cell surface and ECM components.[10] In cardiac fibroblasts IGF-I stimulated ECM synthesis and stretching of these cells resulted in increased autocrine synthesis of IGF-I which led to further ECM deposition.[11] IGF-I has also been shown to be a potent stimulatory factor for limbud development and is important stimulant for condensation of ECM during this process.[12]

While IGF-I and II had been shown to globally increase proteoglycan synthesis as well as important components of bone and cartilage ECM, such as type I collagen, recent studies have focused on the ability of these two growth factors to increase the synthesis of specific ECM components. IGF-I and TGFβ-1 increased the synthesis of tenascin-C, an important ECM glycoprotein by ovarian stromal cells and this increase in tenascin-C led to enhanced ovarian cell adhesion and migration.[13] Gahary et al showed that IGF-I stimulated the mRNA expression of the pro alpha-1 chain of type 1 procollagen, the proalpha-3 chain of type III prollagen and tissue inhibitor of metalloprotease.[14]

Connective tissue cells are an important source of IGF-I.[15] In general within organs the stromal cells contain abundant IGF-I mRNA and in situations where adequate growth hormone receptors are present these connective tissue cells have been shown to be growth hormone responsive and to increase their IGF-I synthesis following GH administration.[16] Since this locally secreted IGF-I can act in an autocrine/paracrine manner it represents an important mechanism by which the synthesis of ECM components may be regulated within the cellular microenvironment. Factors that enhance connective tissue cell IGF-I synthesis have been shown to modulate ECM composition. Madry and coworkers demonstrated that overexpression of IGF-I in articular cartilage resulted in increased glycosaminoglycan and DNA synthesis.[17] This role for the paracrine stimulatory effects of IGF-I is supported by the observation that IGF-I stimulated bovine intervertebral disk cells to increase ECM protein synthesis and that blocking the effects of autocrine/paracrine produced IGF-I could inhibit this activity.[18] Utilizing a strain of mice that had spontaneously developed high insulin-like growth factor concentrations, Reiser et al showed these mice had increased collage deposition in normal tissue as well as granulation tissue and increased collagen crosslinks as well as activation of lysyl oxidases.[19] However, the effect of IGF-I on tendon ECM protein synthesis is regionalized and certain areas of the tendon respond much better presumably due to changes in local IGF-I concentrations. Similarly the mitogenic effect of IGF-I is not uniform throughout.[20] IGF-I stimulates extracellular matrix and proteoglycan synthesis in embryonic kidney cells and antisense IGF-I oligonucleotides can retard this effect suggesting that autocrine/paracrine released IGF-I is capable of mediating these effects.[21] In summary, IGF-I has been shown to be a potent stimulant of both global ECM component synthesis as well as synthesis of specific proteins that are required for normal extracellular matrix function.

Interaction between IGF-I and ECM Components and Subsequent Cellular Responses

In addition to direct stimulation of IGF-I synthesis several factors have been shown to increase the concentration of ECM components that act to enhance IGF-I localization and this helps to sustain mesenchymal cell viability. For example Slater et al demonstrated that estradiol increased the osteoblast production of IGF-I and II as well as ECM components that can bind IGF-I.[22] Similarly lactogenic hormones increase ECM protein expression as well as IGF-I synthesis and deposition by mammary epithelial cells in culture.[23] Estrogen increases thrombospondin-1 synthesis by osteoblast cells and this leads to more accumulation of IGF-I within the ECM.[24] Ghahary et al showed that in posthypertrophic scars in burn patients there was increased fibroblast expression of ECM components and that this was associated with increased expression and deposition of IGF-I within the ECM.[14] The composition of ECM is an important determinant of IGF-I receptor expression in rat intestinal epithelial cells and switching cells from matrigel to plastic results in a marked diminution of IGF-I expression and localization as well as IGF-I receptor expression.[25] This effect has also been noted with other specific ECM proteins. Plating ovarian epithelial cells on a matrix containing laminin and fibronectin enhances the ability of IGF-I to stimulate progesterone synthesis suggesting an interaction between factors that increase IGF-I synthesis and ECM synthesis concominantly.[26]

Several biologic responses of cells are influenced by cell-ECM interactions following IGF-I receptor activation. These include increased cell adhesion, increased cell-cell interactions, cell migration, protein and DNA synthesis. Brooker et al demonstrated that IGF-I enhanced neuronal cell differentiation and that this process required proteoglycan interaction with the IGF-I.[27] Likewise adherent MCF-7 breast carcinoma cells respond to IGF-I with activation of MAP kinase without PI-3 kinase activation however if cells are placed in suspension, they respond to IGF-I only if the PI-3 kinase signaling pathway is functional.[28] This suggests that anchorage of cells enhances their ability to respond to IGF-I by lowering the requirement for activation of PI-3 kinase. ECM attachment enhances the ability of IGF-I to stimulate proliferation and to inhibit apoptosis in ovarian follicular and granulosa cells and these effects can be abrogated by RGD containing peptides suggesting that they are integrin mediated.[29] ECM derived from human intestinal epithelial cells enhances the ability of IGF-I to stimulate their proliferation and paracrine-derived ECM factors are required for IGF-I to enhance the epithelial cell response.[30] This finding has been extended by Woodard et al who showed that ECM proteins derived from mammary fibroblasts would enhance the proliferative effect of EGF and IGF-I on mammary epithelial cells.[31] ECM factors further upregulated IGF-I receptor expression and down regulated the expression of inhibitory forms of IGF binding proteins. Valentinis et al showed that over expression of the IGF-I receptor resulted in abrogation of the requirement for ECM attachment to protect mouse embryo fibroblasts from apoptosis induced by growth factor withdrawal suggesting an interaction between ligand occupancy of the IGF-I receptor and ECM protein deposition.[32] IGF-II stimulates the adherence of colon cancer cells to type IV collagen, fibronectin and laminin.[33] Similarly IGF-I increases cell/cell adhesion and that this is mediated through increased activity of the cell surface protein adherence.[34] This finding has been confirmed by Bracke et al who showed that IGF-I also increased the adherence of mammary epithelial carcinoma cells.[35]

Stimulation of Cell Migration

An important ECM dependent cell function that is modulated by IGF-I is the stimulation of cell migration. Manes et al showed that IGF-I stimulated migration of MCF-7 breast carcinoma cells plated on vitronectin and collagen and importantly that it promoted the association of the docking protein IRS-1 with focal adhesion kinase (FAK) resulting in the FAK dephosphorylation.[36] Since FAK was dephosphorylated by the tyrosine phosphatase SHP-2, abrogation of this phosphatase activity resulted in the inability of IGF-I to stimulate cell migration suggesting that dephosphorylation was an important event in the stimulation of this process. IGF-I stimulates keratinocyte migration and this effect is additive with EGF.[37] This effect could not be demonstrated in the absence of coating the plates with either type IV collagen or fibronectin suggesting that ECM protein interaction is facilitated by exposure to either one of these growth factors. IGF-I induces dephosphorylation of E cadherin and this is necessary for it to stimulate colonic epithelial cells to migrate.[38] Similarly IGF-I stimulates chemotaxis of breast carcinoma cells and that this requires ligand engagement of the αVβ3 integrin.[39] Further interaction with ECM is suggested by the work of Mira et al who showed that IGF-I stimulated cell migration in breast epithelial cells but that this required activation of matrix metalloprotease-9 which appeared to be important for this reaction to proceed.[40] IGF-I is required for chemotactic stimulation of preosteoclasts in response to migrating factors secreted by bone endothelial cells.[41] Similarly IGF-I has been shown to stimulate colonic carcinoma cell migration by reorganizing the cellular localization of the α2β1 integrin.[38] Specifically it promotes relocalization to the leading edge and this relocalization appeared to be required for increased motility. The IGF-I receptor is required for breast carcinoma cell migration and overexpression of a dominant negative form of the receptor that does not signal results in inhibition of this process.[42] Both IGF-I and platelet derived growth factor have been shown to be potent stimuli of smooth muscle cell migration[43] and this process requires deposition of vitronectin in the extracellular matrix and ligand engagement of αVβ3 integrin (see below).

Role of IGF Binding Proteins in ECM Localization

ECM localization of IGF-I and II has been proposed to be an important determinant of cell and tissue growth and development. Although most ECM proteins do not directly bind to the IGFs (a major exception being vitronectin which binds IGF-II)[44] IGF binding proteins have been shown to bind to several ECM proteins. Therefore IGFBPs can function to alter the amount of IGF-I and II that are present in ECM. There are six members of the IGFBP family. Although the forms of IGFBPs that circulate as soluble binding proteins can partition IGFs in extracellular fluids away from receptor and thus act to inhibit IGF-I action, forms of IGFBPs that can bind to ECM can provide a reservoir for IGFs thus making a low level concentration of IGF-I constantly available to receptors.[45] Therefore changes in IGFBP deposition in ECM have the potential to significantly influence cell behavior.

In several situations the increase in IGF-I within the ECM is associated with an increase in IGF binding protein abundance. IGFBPs 2-5 have been shown in various circumstances to bind to ECM.[46-51] By acting as a binding entities for IGFs within the ECM, IGFBPs can indirectly serve as an abundant source of either IGF-I or IGF-II. Factors that regulate IGFBP association within the matrix can also regulate the amount of IGF-I that is present in this tissue compartment. Since a principal role of IGFBPs is to determine the distribution of IGF-I and II among cells and tissues and to provide controlled access to receptors, the factors that regulate the abundance of IGFBPs within the ECM and that control their affinities are important determinants of IGF-I cell surface receptor accessibility. Because IGFBPs usually exist in concentrations in interstitial fluids that are greater than the IGF-I and IGF-II concentrations, a mechanism needs to exist for IGF release to receptors in the pericellular environment. ECM proteins provide an important means by which IGFBPs can accomplish this function.

Jones et al demonstrated that IGFBP-5 was the principle form of IGF binding protein in ECM of mesenchymal cell types such as fibroblasts, chondrocytes and osteoblasts.[51] ECM-associated IGFBP-5 was shown to act as an reservoir for IGF-I and II.[52] Since fibroblasts synthesize more IGFBP-3 than IGFBP-5 it was notable that ECM prepared from these cultures contained predominantly IGFBP-5 and that IGFBP-3 was much less abundant.[45] This suggested that IGFBP-5 had a much higher affinity for fibroblast ECM than did IGFBP-3. A further point that was noted in that study was that the IGFBP-5 that was detected in the ECM was intact. In contrast, the IGFBP-5 in the culture medium was entirely degraded. This suggested that ECM-associated IGFBP-5 was protected from proteolysis and that it could form an important reservoir for localizing IGF-I and II within the ECM. This property has been further studied extensively in osteoblasts. Specifically IGF-II is localized in bone extracellular matrix and forms an important reservoir of IGF-II that is necessary for normal osteoblast function.[53] Mohan et al determined that IGFBP-5 is the principle binding protein component of bone extracellular matrix and that this protein is a reservoir for IGF-II.[54] Although these investigators determined that a small amount of IGFBP-3 and IGFBP-4 were also localized in bone ECM, IGFBP-5 was clearly predominant. They also showed that hydroxyappatite was the principal moiety in bone ECM that bound to IGFBP-5. Campbell and Andres showed that IGFBP-5 bound to hydroxyappatite within bone and that this enhanced IGF binding to bone extracellular matrix and that this interaction could be inhibited by heparin.[55] Wirtz et al showed that IGFBP-5 specifically associated with trabecular mesh work within the retina and that the association of IGFBP-5 with the trabecular meshwork resulted in increased ECM deposition of IGF-I.[56] Ingman et al showed that stimulation of IGFBP-5 synthesis by ovarian granulosa cells resulted in its increased deposition of ECM which resulted in concomitant increase in ECM associated IGF-I.[57] Similarly up regulation of IGFBP-5 synthesis by mammary epithelial cells in culture resulted in increased IGF-I deposition in ECM and inhibition of apoptosis.[58]

Other forms of IGFBPs have been detected in ECM. Human osteoblasts produce IGFBP-2 which focally localizes IGF-II in bone ECM and this has been postulated to enhance its ability to stimulate osteoblast growth and division.[59] IGFBP-2 also localizes to cell surface proteoglycans of the rat olfactory bulb and this localization is dependent upon IGFBP-2 binding to IGF-II.[60] This IGFBP-proteoglycan interaction had been demonstrated earlier by Arai et al who showed that soluble glycosaminoglycans such as heparin would bind to IGFBP-2 only in the presence of high concentrations of IGF-II suggesting that IGF-II binding induces a conformational change in IGFBP-2 that makes it more likely to bind to proteoglycan ECM components.[47] Russo et al showed that exposure of olfactory bulb neuronal cells to FGF resulted in proteolytic cleavage of IGFBP-2 within the ECM and this resulted in increased release of IGF-I to cell receptors thus proposing an ECM binding protein deposition associated mechanism for releasing IGF-I to neural tissue.[61] Similarly systemic administration of IGFBP-4 to bone results in localization of this binding protein to bone extracellular matrix and if it is cleaved locally by a protease this results in enhanced IGF-I bioavailability.[62] In contrast if these investigators infused IGFBP-5, proteolytic cleavage was not required for this binding protein to enhance the ability of IGF-I to stimulate bone cell activation and IGFBP-5 appeared to have some ability to directly stimulate cell proliferation even in the absence of IGF-I.[63] In contrast IGFBP-4 has been shown to be consistently inhibitory when deposited in myoblast ECM and it inhibits IGF-I and II's ability to stimulate myocyte differentiation, whereas ECM associated IGFBP-5 has both inhibitory and stimulatory actions.[64]

Knutson, et al showed that perfusion of rat heart with IGFBP-3 or IGFBP-4 resulted in ECM deposition of both proteins. IGFBP-4 was localized primarily around microvascular endothelial cells whereas IGFBP-3 was retained in the interstitium.[48] Martin and Buckwalter showed that IGFBP-3 that was present in chondrocyte ECM was colocalized with fibronectin but not with tenascin-C or type VI collagen.[5] They proposed that the IGF-I that was associated with IGFBP-3 functions as a reservoir increasing local matrix synthesis following tissue damage. IGFBP-6 inhibits the binding of IGF-II to type IV collagen within colonic cell extracellular matrix.[65] IGF-I treatment of mouse mammary epithelial cells decreases ECM associated IGFBP-2 and IGFBP-3 and this change is associated with an increase in the response of the cells to IGF-I.[66]

Several epithelial cell types have also been shown to deposit IGFBPs in their basement membrane. Those that have been definitively localized include IGFBP-2, IGFBP-3, IGFBP-5 and IGFBP-6. However very few functional studies either adding or deleting these forms of IGFBPs from ECM have been reported for epithelial tissue.

Other mesenchymal cell types that have been shown to contain predominately IGFBP-5 in their ECM include skeletal muscle cells, smooth muscle cells and chondrocytes. The ECM components that result in localization of IGFBPs to ECM have been partially defined. Proteoglycans are clearly a major component that is involved in localization of IGFBPs to both ECM and to cell surfaces.[53,66] Both IGFBP-3 and IGFBP-5 have been extensively studied in this regard. Heparan sulfate containing proteoglycans appear to be the most abundant component of ECM that binds to these forms of IGFBPs. The preference of highly charged proteoglycans for these two proteins involves specific charged sequences within the midregion of both molecules.[67] Specifically the amino acids within IGFBP-5 that are located between position 131 and 141 and between position 201 and 218 contain several charged amino acids. These two sequence motifs with a high charge density form the basis for proteoglycan binding which can be easily displaced with heparan sulfate. Jones et al utilized synthetic peptides containing the 131 to 141 or the 201 to 218 sequence to show that both peptides would compete with intact IGFBP-5 for binding to proteoglycans and to ECM.[51] Addition of the 201-218 peptide to fibroblast cultures also inhibited IGF-I stimulated DNA synthesis.[68]

Subsequent studies also showed that heparan sulfate containing proteoglycans such as tenascin-C would not bind to IGFBP-5 if these peptides were used in competitive binding assays.[67] Similarly the 201-218 IGFBP-5 peptide inhibited IGFBP-5 binding to osteoblasts.[55]

The region of IGFBP-3 that is homologous to the 201-218 region within IGFBP-5 has been shown to account for IGFBP-3 binding to proteoglycans.[69] Using an in vivo perfusion model Booth et al showed that IGFBP-3 bound to ECM components within the cardiac interstitium and that heparin could inhibit binding.[69] The only other IGFBP that has been shown to associate with proteoglycans is IGFBP-2. This occurs under very specialized circumstances wherein high concentrations in IGF-I or II are present.[47] Presumably these high concentrations result in a conformational change in IGFBP-2 which allows it to bind to highly charged proteoglycans both in extracellular matrix and on cellular surfaces. This important property of IGFBP-2 helps it to localize high concentrations of IGF-II in the olfactory bulb and on the cell surfaces of several neuronal cell types wherein high concentrations of proteoglycans are also present.[60]

To delineate structural domains of IGFBP-3 and –5 that account for ECM and proteoglycan association and to analyze the effect of loss of IGFBP-2 reassociation on IGF actions investigators have utilized in vitro mutagenesis.[70] These studies have altered specific amino acids within the 201-218 sequence of IGFBP-5 and thereby determined the exact residues that mediate proteoglycan and ECM binding. Mutation of 4 residues at positions 211, 214, 217 and 218 was shown to markedly attenuate IGFBP-5 binding to heparan sulfate containing proteoglycans.[67] Additionally mutants involving substitutions for residues at positions 201, 206, 208, or 202, 206, 207 also showed marked reduction in proteoglycan binding. Since the 206-211 sequence contained a BBBXXB motif a sequence that confers heparin binding it is likely that residues 206, 207, 208 and 211 are very important for mediating this effect. Mutant forms of IGFBP-5 that contained these substitutions were also analyzed for their capacity to bind to ECM.[71] The ability of IGFBP-5 to associate with ECM was markedly attenuated if specific residues within the 201-218 sequence motif were mutated from charged to neutral residues. The most important residue in this regard was R214 which accounted for at least 50% of the binding activity and when this single residue was mutated a marked reduction in the association of IGFBP-5 with either cultured fibroblast or smooth muscle cell ECM was noted.[71,72] A second group of residues located located at positions 202, 206 and 207 are also important for ECM binding. The requirement for these specific residues is best explained using helicalwheel alignment of the 201-218 sequence. Assuming that this area forms an alpha helix then residues 211 and 214 align with 207 and 218 to form an asymmetric charge cluster on one side of the wheel suggesting that substitution for any of these residues is likely to result in the greatest disruption of binding.[72]

To analyze the functional significance of loss of ECM binding two types of experiments have been conducted. In the first ECM is prepared from fibroblast cultures by cell removal and then IGFBP-5 is layered onto the ECM. Cells are replated on this matrix and the ability of IGF-I to stimulate cell growth over a 48 hour period is quantified. Using this experimental paradigm it was shown that addition of IGFBP-5 resulted in a 2.2 fold enhancement in the ability of IGF-I to stimulate cell division.[51] This was also demonstrated for cultured smooth muscle cells.[72] To more definitively confirm the importance of IGFBP-5 association with ECM components for optimum IGF-I actions, in vitro mutagenesis was utilized. Mutants containing substitutions for 211, 214, 217 and 218 or 202, 206 and 207 were transfected into smooth muscle cells.[72] The expressed proteins were shown to have minimal association with ECM. When the cells that expressed the mutants that bound poorly to ECM were compared to cells expressing wild type IGFBP-5, there was a 7 fold difference in the amount of IGFBP-5 incorporated into the ECM. Cells expressing the mutant forms with these disruptions showed a markedly attenuated DNA synthesis response to IGF-I. Specifically when the ability of IGF-I to stimulate ^{3}H-thymidine incorporation into DNA was analyzed the cells that expressed wild type IGFBP-5 had a 2.5 fold greater increase compared to cells expressing either of the two ECM binding defective mutants.

In summary, the region of amino acids between R201 and R 218 within IGFBP-5 determines ECM binding. Constitutive expression of mutants that have neutral substitutions for these basic amino acids is associated with a reduced cellular proliferation response to IGF-I. This confirms the importance of ECM-associated IGFBP-5 for enhancing IGF-I actions in these mesenchymal cell types. The mechanism accounting for this change has to do with IGFBP affinity. Specifically when

IGFBP-5 is associated with ECM its affinity for IGF-I is reduced 8 fold.[51] Since its affinity is 10 fold greater than the receptor, this 8 fold reduction allows IGF-I that is bound to ECM-associated IGFBP-5 to enter into a more favorable equilibrium with the IGF-I receptor leading to enhancement of IGF-I actions. In vivo studies have also shown the essential role of IGFBP-5 in modulating IGF-I action. Miyakoshi et al injected IGFBP-5 into bone ECM. They demonstrated that in addition to potentiating IGF-I action IGFBP-5 also had independent effects that were trophic for bone if it associated with ECM in vivo suggesting that both IGF-I enhancing effects and non-IGF dependent effects were important for its trophic effect on bone.[63] The molecular mechanisms by which these independent actions might be mediated by ECM associated IGFBP-5 in vivo have not been elucidated.

Role of Integrin Receptor Activation in Modulating IGF-I Biologic Actions

A second mechanism by which changes in ECM microenvironment has been shown to alter IGF-I actions is ligand occupancy of integrin receptors. Integrins are heterodimeric transmembrane proteins that contain one alpha and one beta subunit.[73] These proteins bind to ECM proteins such as fibronectin or vitronectin and then transmit signals from the ECM to their cytoplasmic tail which associates with multiple different cytoplasmic components including cytoskeletal proteins. Intermediary proteins that link the cytoplasmic domains of integrins with cytoskeletal proteins include talin, paxcillin and vinculin.[73] Importantly following ligand attachment integrins can activate specific tyrosine kinases which phosphorylate substrates that can interact with multiple intracellular signaling pathways. The most extensively studied of these kinases are focal adhesion kinase and integrin-linked kinase which have been shown to localize to integrin cytoplasmic tails that are activated following integrin mediated cell attachment.[74,75] Several integrins activate focal adhesion kinase during cell attachment. Once activated, FAK phosphorylates multiple substrates that are important for cell attachment and motility. Although growth factors have been shown to alter the ability of integrins to mediate attachment, IGF-I has limited ability to alter this process.[76]

In contrast to its effect on cell attachment, IGF-I is a potent stimulant of cell migration.[52] This has been shown for cells migrating through an ECM such as a boyden chamber assay and for cells migrating across the lateral surface.[77] For both types of experiments, integrins receptors are required not only to attach to their ECM ligands but also for these ligands to activate cell motility. Thus IGF-I has the specific property of stimulating chemokinesis through integrin receptors. IGF-I integrin interaction can be modulated by deposition of IGFBPs within the ECM. IGFBPs have also been shown to interact with integrins and modulate IGF-I signaling. Initially it was demonstrated that IGFBP-1 could stimulate the migration of CHO cells to binding of the $\alpha 5\beta 1$ integrin and that mutation of its RGD sequence destroyed this property suggesting that it was a direct effect of IGFBP-1 binding to this integrin.[78] Perks et al extended this observation to demonstrate that IGFBP-1 binding to $\alpha 5\beta 1$ could induce apoptosis and induce FAK expression in breast cancer cells.[79]

The molecular mechanisms by which IGF-I stimulates cell motility through integrin-ECM protein interaction are not entirely understood. It is clear in several systems however the PI-3 kinase pathway has to be activated for IGF-I to stimulate cell migration.[80] Similarly, membrane ruffling has been shown to be an IGF-I dependent effect for several cell types and IGF-I stimulated increases can be blocked with PI-3 kinase inhibitors.

In some cell types activation of the MAP kinase pathway is also required to stimulate migration. To determine the specific integrin(s) on cell surfaces that were required for IGF-I to stimulate motility, Jones et al identified all of the integrins that were present on fibroblasts and smooth muscle cell surfaces.[52,77-78] They then determined that IGF-I had no effect on the abundance of $\alpha 5\beta 1$ or $\alpha V\beta 3$, the two most prevalent integrins. However when their affinities were analyzed, $\alpha 5\beta 1$ affinity was unchanged but $\alpha V\beta 3$ affinity was increased between 6 and 8 fold in response to IGF-I.[52] The significance of this finding was confirmed by showing that these cells would migrate on an ECM containing only vitronectin, a known ligand for $\alpha V\beta 3$.[77] Because the affinity of this particular integrin was altered these investigators then determined whether $\alpha V\beta 3$ antagonists would inhibit IGF-I stimulated cell motility. To determine if ligand occupancy was required a competitive antagonist, echistatin was utilized. This is a small peptide that binds to $\alpha V\beta 3$ with an affinity that is similar to its ECM ligands such as vitronectin and thrombospondin but its binding does not result in integrin activation.[81,82] Integrin heterodimers have to change confirmation and enter high activation state before cells can be stimulated to attach or migrate. This activation step requires multidomain ECM proteins, such as vitronectin and fibronectin. In contrast, small peptides such as echistatin do not have multiple epitopes that are capable of activating the integrin receptor but they do contain a high affinity binding site that is necessary for vitronectin or fibronectin mediate to adherence to $\alpha V\beta 3$. Therefore they serve as excellent competitive antagonists.[83,84] When increasing concentrations of two disintegrins, echistatin or kistrin were added to SMC cultures, the SMC migration response to IGF-I was inhibited in a dose dependent manner.[52] This occurred at relatively low concentrations of disintegrins which were physiologically relevant, i.e., 10^{-8} M.

Molecular Mechanism Mediating Signaling between the $\alpha V\beta 3$ Integrin and the IGF-I Receptor

To further elucidate the molecular mechanism by which integrin ligand occupancy was modulating IGF-I action, it was first determined whether this property was generalizable to other IGF-I stimulated biologic effects. Specifically, protein synthesis, inhibition of apoptosis and stimulation of cell replication were analyzed. For all three of these processes it was determined that echistatin significantly inhibited each biologic action and this could be demonstrated both in fibroblasts and smooth muscle cells. Since $\alpha V\beta 3$ antagonists globally inhibited IGF-I responsiveness this suggested that blocking ligand occupancy of the $\alpha V\beta 3$ integrin was leading to inhibition of a relatively proximal step in IGF-I receptor linked signal transduction. Since biologic actions that required IGF-I stimulation of both the MAP and PI3 kinase pathways were inhibited by $\alpha V\beta 3$ antagonists this suggested blocking ligand occupancy of $\alpha V\beta 3$ interfered with an IGF-I receptor linked signaling step that was proximal to activation of the point of divergence of these two pathways. That narrowed the possibilities to stimulation of phosphorylation of IRS-1 or IRS-2 and/or stimulation of IGF-I receptor autophosphorylation. To test these possibilities, cells were incubated with $\alpha V\beta 3$ receptor antagonists, then stimulated with IGF-I and IRS-1 phosphorylation was analyzed. Echistatin completely inhibited the ability of IGF-I to stimulate IRS-1 phosphorylation.[85] This suggested that ligand occupancy of $\alpha V\beta 3$ was required for IRS-1 to be activated. Zheng et al determined if this block occurred at the most proximal step by examining receptor autophosphorylation. Following exposure to IGF-I, there is a peak of receptor autophosphorylation activity at 10-15 minutes. This could be inhibited by 80% with prior exposure to echistatin. This strongly suggested that ligand occupancy of $\alpha V\beta 3$ by one of its ECM binding substituants, such as fibronectin or vitronectin was resulting in the ability of the receptor to be normally activated. To determine if differential ligand occupancy of integrin receptors would alter this response, smooth muscle cells were plated on the matrix containing fibronectin and allowed to differentiate. These cells were then examined to determine if integrin ligand occupancy altered IGF-I stimulated IGFBP-5 synthesis.[86] Cells that were plated on fibronectin resulting in $\alpha V\beta 3$ occupancy had a marked increase in IGFBP-5 synthesis. In contrast if the cells were plated on a different ECM substituant, eg type IV collagen and laminin which bind to the $\alpha 2\beta 1$ integrin, they did not synthesize IGFBP-5 in response to IGF-I. This inhibitory effect of the $\alpha 2\beta 1$ activation could be blocked with an anti $\alpha 2\beta 1$ specific antibody and the IGFBP-5 synthesis response to IGF-I could be reactivated. This strongly suggests that differential occupancy of integrins by specific extracellular matrix substitituents controls the cellular responsiveness to IGF-I.

To further analyze the mechanism by which ligand occupancy of the $\alpha V\beta 3$ integrin was altering the ability of IGF-I to activate IGF-I receptor phosphorylation, they analyzed the time course of the receptor phosphorylation response with or without prior echistatin exposure. These studies showed that after 5 min the receptor was fully activated whether it had been preexposed to echistatin or not and that the rate of receptor dephosphorylation was markedly accelerated in the presence of echistatin.[87] This strongly suggested that following echistatin exposure, a phosphatase was being activated prematurely resulting in abortive IGF-I receptor linked signaling. Failure to maintain the phosphorylated signal response for the appropriate period of time, that is 20-30 min resulted in impaired activation of IRS-1 and downstream signaling. SHP-2 is a tyrosine phosphatase that had been shown to dephosphorylate the IGF-I receptor in vitro.[88] This phosphatase binds to Src homology (SH2) domains once they are phosphorylated and specifically dephosphorylates them. Once SHP-2 dephosphorylates its substrates it no longer binds to them and is free to bind to other substrates. However SHP-2 is normally present in the cytoplasm and it has to be focally concentrated on the membrane in close approximation to the IGF-I receptor to function as a phosphatase. Recruitment is achieved as with other growth factor receptors, such as the insulin receptor, by phosphorylation of a transmembrane termed protein SHPS-1.[89] SHPS-1 has been shown to be an inhibitor of growth factor receptor activation if it is overexpressed, suggesting that enhanced recruitment of SHP-2 might be responsible for dephosphorylation of tyrosine kinase growth factor receptors and growth inhibition in response to SHPS-1 overexpression. SHPS-1 contains a cytoplasmic tail with four tyrosine residues that are located within SH2 binding domains and are consensus binding sites for SHP-2.[90] To test the hypothesis that IGF-I receptor stimulated SHPS-1 phosphorylation on those tyrosines, Maile et al incubated cells expressing wild type SHPS-1 and cells containing a mutant SHPS-1 that had had these two sites of tyrosine phosphorylation deleted. Cells expressing a wild type SHPS-1 showed that the activated IGF-I receptor phosphorylated SHPS-1 on these two tyrosines and stimulated recruitment of SHP-2 to SHPS-1 from the cytosol. In contrast cells expressing SHPS-1 mutant showed a reduced phosphorylation response and no recruitment of SHP-2.[87] They then examined whether SHP-2 could be recruited to the IGF-I receptor. Following a 10 min stimulation with IGF-I there was a marked increase in the amount of SHP-2 bound to the IGF-I receptor in cells expressing wild type SHPS-1. In contrast, SHP-2 could not be recruited to the IGF-I receptor in cells expressing the SHPS-1 mutant that did not contain the SHP-2 binding sites. This strongly implied that SHPS-1 phosphorylation by the IGF-I receptor was required for SHP-2 to be recruited to the plasma membrane and transferred to the IGF-I receptor. To further definitively determine that this resulted in a change in the IGF-I receptor phosphorylation response Maile et al examined IGF-I receptor autophosphorylation in the presence of the SHPS-1 mutant and in cells transfected with a SHP-2 mutant that could bind to the receptor but had no intrinsic phosphatase activity. Both types of transfected cells showed prolonged IGF-I receptor phosphorylation following IGF-I binding and dephopshorylation did not occur at the 10 and 20 min time points. They then showed that exposure of cells expressing mutant SHP-2 to echistatin resulted in no effect on their ability to alter their autophosphorylation response to IGF-I and did not alter IGF-I signaling. These findings strongly suggest that ligand occupancy of $\alpha V\beta 3$ by ECM proteins is required for the normal rate of transfer of SHP-2 from SHPS-1 and to the IGF-I receptor which results in a normal rate of receptor dephosphorylation and down regulation of IGF-I linked signaling. If this system is perturbed and integrin ligand occupancy is not present, then there is premature recruitment of SHP-2 to the receptor and premature dephosphorylation resulting in abortive signaling.[91] This is of significant importance because the SHPS-1-SHP-2 relay system is the same mechanism that is used to downregulate growth hormone receptor signaling and suggests that conservation of this attenuating mechanism for both responses may be an important means of regulating balanced cell growth. The fact that this relies on ligand occupancy of $\alpha V\beta 3$ suggests that this may be an important way by which ECM regulates growth and development in a coordinate manner. Furthermore, since SHPS-1 can be activated by cell surface proteins that are present on adjacent cell surfaces such as integrin activating protein, a ligand for SHPS-1, this suggests that potential means for regulating cell density in coordination with changes in extracellular matrix composition.

IGF-I Stimulation of Integrin Activation

IGF-I binding to its receptor also alters the activity of integrins. Gohel et al demonstrated that exposure to IGF-I for 72 hours increased cell process formation by osteoblasts and the number of osteocytes per unit area and this was associated with activation of the beta 1 integrin.[92] Similarly Losher showed that IGF-I exposure to chondrocytes resulted in enhanced $\beta 1$ in expression as well as increased expression of $\alpha 3$ and $\alpha 5$ integrin.[93] This resulted in enhanced cell adhesion to type IV collagen. Likewise Chandrsekaran et al showed that IGF-I stimulated activation of the $\alpha 3\beta 1$ on breast epithelial cells and that this was required for IGF-I stimulated migration.[39]

IGF-I exposure to smooth muscle cells increases the affinity of the $\alpha V\beta 3$ integrin for ligand. Specifically vitronectin binding is increased approximately 2.4 fold after 14 hour exposure to IGF-I.[52] The molecular mechanism accounting for this increased activity of this integrin involves a second cell surface protein termed "integrin associated protein" (IAP). IAP has been shown in several cell types to bind to $\alpha V\beta 3$ and change its conformation thereby increasing its activity for ligands.[52] In smooth muscle cells a 14 hr exposure to IGF-I results in translocation of IAP from the raft compartment of plasma membrane to the nonraft domains.[94] Since $\alpha V\beta 3$ receptors are localized almost exclusively in the nonraft domain, this enables IAP to be present in the same compartment as αVβ3 resulting in their association. For reasons that are poorly understood, this process requires approximately 14 hrs and therefore is no change seen in IAP-αVβ3 binding prior to this time and no change in the affinity of αVβ3 for ligands. This process requires activation of the PI-3 kinase pathway and increases up to a maximum at 24 hrs. Exposure to a specific monoclonal antibody that inhibits IAP-αVβ3 association inhibits the ability of IGF-I to stimulate this process and inhibits the IGF-I induced enhancement of αVβ3 affinity.[94] Whether this or a similar mechanism applies to the changes noted previously in β1 activity following cellular exposure to IGF-I has not been determined.

A similar mechanism has been shown for IGF-I stimulated enhancement of α3β1 integrin activation in breast carcinoma epithelial cells and that this was associated with HSP-60 binding to the α3β1 integrin.[95]

In summary IGF-I and IGF-II regulate synthesis of ECM components and this has important consequences for regulating cellular response to these mitogens. Similarly ECM components can act to increase IGF-I and IGFBP localization within the ECM and variables such as proteases and glycosaminoglycans may function to modulate IGF-I
bioavailability by altering its binding to these components. Thus ECM substituents may use several types of pathways to alter cellular response to the IGFs.

Acknowledgements

The author wishes to thank Ms. Laura Lindsey for her help in preparing the manuscript. This work was supported by grants from the National Intitutes of Health AG-02331 and HL-56850.

References

1. Salmon Jr WD, Daughaday WH. A hormonally controlled serum factor which stimulates sulfate incorporation by cartilage in vitro. J Lab Clin Med 1957; 49:825-836.
2. Smith P, Shuler FD, Georgescu HI et al. Genetic enhancement of matrix synthesis by articular chondrocyltes:comparison of different growth factor genes in the presence and absence of interleukin-1. Arthritis Rheum 2000; 43:1156-1164.
3. Pricci F, Pulisese G, Romano G et al. Insulin-like growth factors I and II stimulate extracullar matrix production in human glomerular cells. Comparison with transforming growth factor-beta. Endocrinology 1996; 137:879-885.
4. Heidenreich S, Tepel M, Lang et al. Differential effects of insulin-like growth factor I and platelet-derived growth factor on growth response, matrix formation and cystolic free calcium of glomerular mesangial cells of spontaneously hypertensive and normotensive rates. Nephron 1994; 68:481-488.
5. Martin JA, Buckwalter JA. The role of chondrocyte-matrix interactions in maintaining and repairing articular cartilage. Biorheology 2000; 37:129-140.
6. Jendraschak E, Kaminski WE, Kiefl R et al. IGF-1, PDGF and CD18 are adherence-responsive genes: regulation during monocyte differentiation. Biochimica et Biophysica Acta. 1998; 1396:320-335.
7. Morales TI. Transforming growth factor-beta and insulin-like growth factor-1 restore proteoglycan metabolism of bovine articular cartilage after depletion by retinoic acid. Arch Biochem Biophys 1994; 315:190-198.
8. Hill DJ, Logan A, McGarry M et al. Control of protein and matrix-molecule synthesis in isolated ovine fetal growth-plate chondrocytes by the interactions of basic fibroblast growth factor, insulin-like growth factors-I and II, insulin and transforming growth factor-beta 1. J Endocrinol 1992; 133:363-373.
9. van Osch GJ, van den Berg WB, Hunziker EB et al. Differential effects of IGF-1 and TGF beta-1 on the assembly of proteoglycans in pericellular and territorial matrix by cultured bovine articular chondrocytes. Osteo Cart 1998; 6:187-195.
10. Kanekar S, Borg TK, Terracio L et al. Modulation of heart fibroblast migration and collagen gel contraction by IGF-I. Cell Adh Comm 2000; 7:5134-523.
11. MacKenna D, Summerour SR, Villarreal FJ. Role of mechanical factors in modulating cardiac fibroblast function and extracellular matrix synthesis. Cardio Res 2000; 46:257-263.
12. Geduspan JS, Solursh M. Effects of mesonephros and insulin-like growth factor I on chondrogenesis of limb explants. Dev Biol 1993; 156:500-508.
13. Wilson KE, Bartlett JM, Miller ET et al. Regulation and function of the extracellular matrix protein tenascin-C in ovarian cancer cell lines. Brit J Can 1999; 80:685-692.
14. Ghahary A, Shen YJ, Nedelec B et al. Enhanced expression of mRNA for insulin-like growth factor-1 in post-burn hypertrophic scar tissue and its fibrogenic role by dermal fibroblasts. Mol Cell Biochem 1995; 148:25-32.
15. Han VKM, Hill DJ, Strain AJ et al. Identification of somatomedin/insulin-like growth factor immunoreactive cells in the human fetus. Pediatr Res 1987; 22:245-249.
16. Lowe WL, Adam OM, Werner H et al. Regulation by fasting of insulin-like growth factor I and its receptor: Effects on gene expression and binding. J Clin Invest 1989; 84:619-626.

17. Madry H, Zurakowski D, Trippel SB. Overexpression of human insulin-like growth factor-I promotes new tissue formation in an ex vivo model of articular chondrocyte transplantation. Gene Therapy 2001; 1443-1449.
18. Osada R, Ohshima H, Ishihara H et al. Autocrine/paracrine mechanism of insulin-like growth factor-1 secretion, and the effect of insulin-like growth factor-1 on proteoglycan synthesis in bovine intervertebral discs. J Orth Res 1996; 14:690-699.
19. Reiser K, Summers P, Medrano JF et al. Effects of elevated circulating IGF-1 on the extracullar matrix in "high-growth" C57BL/6J mice. Amer J Physiol 1996; 271:R696-703.
20. Abrahamsson SO, Lohmander S. Differential effects of insulin-like growth factor-I on matrix and DNA synthesis in various regions and types of rabbit tendons. J Ortho Res 1996; 14:370-376.
21. Liu ZZ, Kumar A, Wallner EI et al. Trophic effect of insulin-like growth factor-I on metanephric development: relationship to proteoglycans. Eur J Cell Biol 1994; 65:378-391.
22. Slater M, Patava J, Kingham K et al. Modulation of growth factor incorporation into ECM of human osteoblast-like cells in vitro by 17 beta-estradiol. Amer J Physiol 1994; 267:E990-1001.
23. Romagnolo D, Akers RM, Wong EA. Lactogenic hormones and extracellular matrix regulate expression of IGF-1 linked to MMTV-LTR in mammary epithelial cells. Mol Cell Endocrinol 1993; 96:147-157.
24. Slater M, Patava J, Mason RS. Thrombospondin co-localises with TGF beta and IGF-I in the extracellular matrix of human osteoblast-like cells and is modulated by 17 beta estradiol. Experientia 1995; 51:235-244.
25. Benya RV, Duncan MD, Mishra L et al. Extracellular matrix composition influences insulin-like growth factor I receptor expression in rat IEC-18 cells. Gastroentrol 1993; 1705-1711.
26. Aten RF, Kolodecik TR, Behrman HR. A cell adhesion receptor antiserum abolishes, whereas laminin and fibronectin glycoprotein components of extracellular matrix promote, luteinization of cultured rat granulosa cells. Endocrinology 1995; 1753-1758.
27. Brooker GJ, Kalloniatis M, Russo VC et al. Endogenous IGF-1 regulates the neuronal differentiaion of adult stem cells. J Neurosci Res 2000; 59:332-341.
28. Suzuki K, Takahaski K. Anchorage-independent activation of mitrogen-activated protine kinase through phosphatidylinositol-1 kinase by insulin-like growth factor I. Biochem Biophys Res Comm 2000; 272:111-115.
29. Huet C, Pisselet C, Mandon-Pepin B et al. Extracellular matrix regulates granulosa cell survival, proliferation and steroidogenesis: relationships between cell shape and function. J Endocrinology 2001; 169:347-360.
30. Simmons JG, Pucilowska JB, Lund PK. Autocrine and paracrine actions of intestinal fibroblast-derived insulin-like growth factors. Amer J Physiol 1999; 276:G817-827.
31. Woodward TL, Xie J, Fendrick JL et al. Proliferation of mouse mammary epithelial cells in vitro: interactions among epidermal growth factor, insulin-like growth factor-I, ovarian hormones, and extracellular matrix proteins. Endocrinology 2000; 141:3578-3586.
32. Valentinis B, Morrione A, Peruzzi F et al. Anti-apoptotic signaling of the IGF-I receptor in fibroblasts following loss of matrix adhesion. Oncogene 1999; 18:1827-1836.
33. Leng SL, Leeding KS, Whitehead RH et al. Insulin-like growth factor (IGF)-binding protein-6 inhibits IGF-II-induced but not basal proliferation and adhesion of LIM 1215 colon cancer cells. Mol Cell Endocrinol 2001; 174:121-127.
34. Guvakova MA, Surmacz E. Overexpressed IGF-I receptors reduce estrogen growth requirements, enhance survival, and promote E-cadherin-mediated cell-cell adhesion in human breast cancer cells. Exper Cell Res 1997; 231:149-162.
35. Bracke ME, Vyncke BM, Bruyneel EA et al Insulin-like growth factor I activates the invasion suppressor function of E-cadherin in MCF-7 human mammary carcinomal cells in vitro. Brit J Can 1993; 68:282-289.
36. Manes S, Mira E, Gomez-Mouton C et al. Concerted activity of tyrosine phosphatase SHP-2 and focal adhesion kinase in regulation of cell motility. Mol Cell Biol 1999; 19:3125-3135.
37. Ando Y, Jensen PJ. Epidermal growth factor and insulin-like growth factor I enhance keratinocyte migration. J Invest Dermatol 1993; 100:633-639.
38. Andre F, Rigot V, Thimonier J et al. Integrins and E-cadherin cooperate with IGF-I to induce migration of epithelial colonic cells. Inter J Can 1999; 83:497-505.
39. Chandrasekaran S, Guo NH, Rodriques RG. Pro-adhesive and chemotactic activities of thrombospondin-1 for breast carcinoma cells are mediated by alph3beta1 integrin and regulated by insulin-like growth factor-1 and CD98. J Biol Chem 1999; 274:11408-11416.
40. Mira E, Manes S, Lacalle RA. Insulin-like growth factor I-triggered cell migration and invasion are mediated by matrix metalloproteinase-9. Endocrinology 1999; 140:1657-1664.
41. Formigli L, Fiorelli G, Benvenuti S et al. Insulin-like growth factor-I stimulates in vitro migration of preosteroclasts across bone endothelial cells. Cell Tissue Res 1997; 288:101-110.
42. Dunn SE, Ehrlich M, Sharp NJ et al. A dominant negative mutant of the insulin-like growth factor-I receptor inhibits the adhesion, invasion, and metastasis of breast cancer. Cancer Res 1998; 58:3353-3361.
43. Cospedal R, Abedi H, Zachary I. Platelet-derived growth factor-BB (PDGF-BB) regulation of migration and focal adhesion kinase phosphorylation in rabbit aortic vascular smooth muscle cells: roles of phosphatidylinositol 3-kinase and mitrogen-activated protein kinases. Cardio Vas Res 1999; 41:708-721.
44. Upton Z, Webb H, Hale K et al. Identification of vitronectin as a novel insulin-like growth factor-II binding protein. Endocrinology 1999; 140 (6):2928-31.
45. Camacho-Hubner C, Busby WH, McCusker RH et al. Identification of the forms of insulin-like growth factor binding proteins produced by human fibroblasts and the mechanisms that regulate their secretion. J Biol Chem 1992; 267:11949-11956.

46. Grulich-Henn J, Ritter J, Mesewinkel S et al. Transport of insulin-like growth factor-I across endothelial cell monolayers and its binding to the subendothelial matrix. Exp Clin Endocrinol Diabetes 2002; 110(2): 67-73.
47. Arai T, Busby W Jr, Clemmons DR. Binding of insulin-like growth factor (IGF) I or II to IGF-binding protein-2 enables it to bind to heparin and extacellular matrix. Endocrinology 1996; 137(11): 4571-5.
48. Knudtson KL, Boes M, Sandra A et al. Distribution of chimeric IGF binding protein (IGFBP)-3 and IGFBP-4 in the rat heart: importance of C-terminal basic region. Endocrinology 2001; 142(9):3749-55.
49. Gosiewska A, Yi CF, Brown LJ et al. Differential expression and regulation of extracellular matrix-associated genes in fetal and neonatal fibroblasts. Wound Repair Regeneration 2001; 9(3):213-22.
50. Remacle-Bonnet MM, Garrouste FL, Pommier GJ. Surface-bound plasmin induces selective proteolysis of insulin-like-growth-factor (IGF)-binding protein-4 (IGFBP-4) and promotes autocrine IGF-II bio-availability in human colon-carcinoma cells. Int J Cancer 1997; 72 (5):835-43.
51. Jones JI, Gockerman A, Busby WH et al. Extracellular matrix contains insulin-like growth factor binding protein-5: Potentiation of the effects of IGF-I. J Cell Biol 1993; 121:679-687.
52. Jones JI, Prevette T, Gockerman A et al. Binding of vitronectin to an aVB3 integrin is necessary for smooth muscle cells to migrate in response to IGF-I. Proc Natl Acad Sci USA 1996; 93:2462-2467.
53. Andress DL, Birnbaum RS. Human osteoblast-derived insulin-like growth factor (IGF) binding protein-5 stimulates osteoblast mitogenesis and potentiates IGF action. J Biol Chem 1992; 267:22467-22472.
54. Mohan S, Nakao Y, Honda Y et al. Studies on the mechanisms by which insulin-like growth factor (IGF) binding protein-4 (IGFBP-4) and IGFBP-5 modulate IGF actions in bone cells. J Biol Chem 1995; 270:20424-20431.
55. Campbell PG and Andress DL. Insulin-like growth factor (IGF)-binding protein-5-(201-218) region regulates hydroxyapatite and IGF-I binding. Amer J Physiol 1997: 273:E1005-1013.
56. Wirtz MK, Xu H, Rust K et al. Insulin-like growth factor binding protein-5 expression by human trabecular network. Invest Ophthalmol Vis Sci 1998; 39:45-53.
57. Ingman WV, Owens PC, Armstrong DT. Differential regulation by FSH and IGF-I of extracellular matrix IGFBP-5 in bovine granulosa cells: effect of association with the oocyte. Mol Cell Endocrinol 2000: 164:53-58.
58. Blatchford DR, Quarrie LH, Tonner E. Influence of microenvironment on mammary epithelial cell survival in primary culture. J Cell Physiol 1999; 181:304-311.
59. Khosla S, Hassoun AA, Baker BK et al. Insulin-like growth factor system abnormalities in hepatitis C-associated osteosclerosis. Potential insights into increasing bone mass in adults. J Clin Invest 1998; 101:2165-2173.
60. Russo VC, Bach LA, Fosang AJ et al. Insulin-like growth factor binding protein-2 binds to cell surface proteoglycans in the rat brain olfactory bulb. Endocrinology 1997; 138:4858-4867.
61. Russo VC, Rekaris G, Baker NL. Basic fibroblast growth factor induces proteolysis of secreted and cell membrane-associated insulin-like growth factor binding protein-2 in human neuroblastoma cells. Endocrinology 1999; 140:3082-3090.
62. Miyakoshi N, Richman C, Qin X et al. Effects of recombinant insulin-like growth factor-binding protein-4 on bone formation parameters in mice. Endocrinology 1999; 140:5719-5720.
63. Miyakoshi N, Richman C, Kasukawa Y et al. Evidence that IGF-binding protein-5 functions as a growth factor. J Clin Invest 2001; 107:73-81.
64. Ewton DZ, Collican SA, Mohan S et al. Modulation of insulin-like growth factor actions in L6A1 myoblasts by insulin-like growth factor binding protein (IGFBP)-4 and IGFBP-5: a dual role for IGFBP-5. J Cell Physiol 1998; 177:47-57.
65. Leng SL, Leeding KS, Whitehead RH et al. Insulin-like growth factor (IGF)-binding protein-6 inhibits IGF-II-induced but not basal proliferation and adhesion of LIM 1215 colon cancer cells. Mol Cell Endocrinol 2001; 174: 121-127.
66. Woodward TL, Xie J, Fendrick JL et al. Proliferation of mouse mammary epithelial cells in vitro: interactions among epidermal growth factor, insulin-like growth factor I, ovarian hormones, and extracellular matrix proteins. Endocrinology 2000; 141:3578-3586.
67. Arai T, Clarke JB, Parker A et al. Substitution of specific amino acids in insulin-like growth factor-binding protein-5 alters heparin binding and its change in affinity for IGF-I in response to heparin. J Biol Chem 1996; 271:6099-6106.
68. Rees C, Clemmons DR. Inhibition of IGFBP-5 binding to extracellular matrix and IGF-I stimulated DNA synthesis by a peptide fragment of IGFBP-5. J Cell Biochem 1998; 71:375-381.
69. Booth BA, Boes M, Andress DL et al. IGFBP-3 and IGFBP-5 association with endothelial cells: role of c-terminal heparin binding domain. Growth Regul 1995; 5:1-17.
70. Clemmons DR. Use of mutagenesis to probe IGF binding protein structure function in relationship. Endo Rev 2001; 22:800-871.
71. Parker A, Busby WH, Clemmons DR. Identification of the extracellular matrix binding site for insulin like growth factor binding protein-5. J Biol Chem 1996; 271:13523-13529.
72. Parker A, Rees C, Clarke JB et al. Binding of insulin-like growth factor binding protein-5 to smooth muscle cell extracellular matrix is a major determinant of the cellular response to IGF-I. Mol Biol Cell 1998; 9:2383-2392.
73. Rouslahti E. Structure and biology of proteoglycans. Ann Rev Cell Biol 1988; 4:229-255.
74. Ishii T, Satoh E, Nishimura M. Integrin-linked kinase controls neurite outgrowth in N1E-115 neuroblastoma cells. J Biol Chem 2001; 276(46): 42994-43003.

75. Carloni V, Defranco RM, Caligiuri A et al. Cell adhesion regulates platelet-derived growth factor-induced MAP kinase and PI-3 kinase activation in stellate cells. Hepatology 2002; 36(3): 582-91.
76. Miyamoto S, Teramoto H, Gutkind IS et al. Integrins can collaborate with growth factors for phosphorylation of receptor tyrosine kinases and MAP kinase activation: roles of integrin aggregation and occupancy of receptors. J Cell Biol 1996; 135:1633-1642.
77. Jones, JI, Doerr ME, Clemmons DR. Cell migration: interactions among integrins, IGFs, and IGFBPs. Progress in Growth Factor Research 1995; 6:319-327.
78. Jones JI, Gockerman A, Busby WH Jr et al. Insulin-like growth factor binding protein 1 stimulates cell migration and binds to the αVβ3 integrin by means of its Arg-Gly-Asp sequence. Proc Natl Acad Sci USA 1993; 90:10553-10557.
79. Perks CM, Newcomb PV, Norman MR et al. Effect of insulin-like growth factor binding protein-1 on integrin signaling and the induction of apoptosis in human breast cancer cells. J Mol Endocrinol 1999; 11:141-150.
80. Imai Y, Clemmons DR. Roles of phosphatidylinositol 3-kinase and mitogen-activated protein kinase pathways in stimulation of vascular smooth muscle cell migration and deoxyribonucleic acid synthesis by insulin-like growth factor-I. Endocrinology 1999; 140:4228-4235.
81. Yahalom D, Wittelsberger A, Mierke DF et al. Identification of the principal binding site for RGD-containing ligands in the alpha (V) beta(3) integrin: a photoaffinity cross-linking study. Biochemistry 2002; 41(26):8321-31.
82. Scheibler L, Mierke DF, Bitan G et al. Identification of a contact domain between echistatin and the integrin alpha(v) beta(3) by photoaffinity cross-linking. Biochemistry 2001; 40(50):15117-26.
83. Haas Ta, Plow EF. Integrin-ligand interactions: a year in review. Curr Opin Cell Biol 1994; 6(5)656-62.
84. Niewiarowski S, McLane MA, Kloczewiak M, Stewart GJ. Disintegrins and other naturally occurring antagonists of platelet fibrinogen receptors. Semin Hematol 1994; 31(4):289-300.
85. Zheng B, Clemmons DR. Blocking ligand occupancy of the αVβ3 integrin inhibits IGF-I signaling in vascular smooth muscle cells. Proc Natl Acad Sci USA 1998; 95:11217-11222.
86. Zheng B, Duan C, Clemmons DR. The effect of extracellular matrix protein on porcine smooth muscle cell insulin like growth factor binding protein-5 synthesis and responsiveness to IGF-I. J Biol Chem 1998; 273:8994-9000.
87. Maile LA, Clemmons DR. Regulation of insulin-like growth factor I receptor dephosphorylation by SHPS-1 and the tyrosine phosphatase SHP-2. J Biol Chem 2002; 277:8955-8960.
88. Seely BL, Reichart DR, Staubs PA et al. Localization of the insulin-like growth factor I receptor binding sites for the SH2 domain proteins p85, Syp, and GTPase activating protein. J Biol Chem 1995; 270:19151-19157.
89. Takada T, Matozaki T, Takeda H et al. Roles of the complex formation of SHPS-1 with SHP-2 in insulin-stimulated mitogen-activated protein kinase activation. J Biol Chem 1998; 173:9234-9242.
90. Oshima K, Ruhul Amin AA, Suzuki A et al. SHPS-1, a multifunctional transmembrane glycoprotein. FEBS Letters 2002; 519:1-7.
91. Maile LA and Clemmons DR. The αVβ3 integrin regulates IGF-I receptor phosphorylation by altering the rate of SHP-2 recruitment to the activated IGF-I receptor. Endocrinology 2002; 143:4259-4264.
92. Gohel AR, Hand AR, Gronowicz GA. Immunogold localization of beta 1-intergrin in bone: effect of glucocorticoids and insulin-like growth factor I on integrins and osteocyte formation. J Histochem Cytochem 1995: 43:1085-1096.
93. Loeser RF. Growth factor regulation of chondrocyte integrins. Differential effects of insulin-like growth factor 1 and transforming growth factor beta on alpha 1 beta 1 integrin expression and chondrocyte adhesion to type VI collagen. Arth Rheum 1997; 40:270-276.
94. Maile LA, Imai Y, Clark JB et al. Insulin-like growth factor I increases alpha V beta 3 affinity by increasing the amount of integrin-assocaited protein that is associated with non-raft domains of the cellular membrane. J Biol Chem 2002; 277:1800-1805.
95. Barazi HO, Zhou L, Templeton NS et al. Identification of heat shock protein 60 as a molecular mediator of alpha 3 beta 1 integrin activation. Cancer Res 2002; 62:1541-1548.

CHAPTER 19

Epidemiologic Approaches to Evaluating Insulin-Like Growth Factor and Cancer Risk

Eva S. Schernhammer and Susan E. Hankinson

Epidemiologic Methods in Studying Insulin-Like Growth Factor

Epidemiologic studies that simply observe the natural cause of events often are referred to as observational studies. In contrast, studies in which the investigator intervenes to change some participants' behavior, assigning the exposure status of each participant, are referred to as intervention studies. Both types of studies seek to investigate relationships between a certain exposure (such as serum levels of insulin-like growth factor-I [IGF-I]) and an outcome (such as cancer). Often both design strategies are used in an effort to answer a particular research question; for example, the observation of higher plasma estrogen levels in women with breast cancer helped motivate a subsequent intervention study that investigated the effects of a selective estrogen receptor modulator (SERM), tamoxifen,[1] as a breast-cancer chemopreventive. In this chapter we focus on the evidence regarding IGF-I and cancer risk that was generated through observational studies, as few relevant intervention studies have been conducted.

Epidemiologists make observations on small, representative portions of the total population they wish to characterize using two main sampling strategies: the cohort study and the case-control study. In a cohort study, the investigator enrolls groups of people who are representative of different levels or types of a naturally occurring exposure. For example, a cohort of women might include some with higher and others with lower serum levels of IGF-I, thus representing the natural range of IGF-I in a population. By following this population over time, the epidemiologist can measure the incidence (or occurrence) of disease, such as breast cancer, separately for women in each of the exposure groups, defined according to their IGF-I level. For rare diseases, however, a large number of subjects must be observed over many years. Because of the high cost of large cohort studies (such as the Nurses' Health Study), investigators often use a study design in which disease has occurred before the start of the observation—the case-control study. In a case-control study, researchers first identify two different groups: one with people who already have the disease of interest (such as breast cancer) and one with people who do not have the disease (controls). They then evaluate the characteristics of the case and control groups (e.g., proportion with high serum IGF-I levels) and see whether the exposure of interest is more prevalent in one group than in the other.

These two study designs have different strengths and weaknesses that are important to bear in mind when reviewing scientific evidence. The case-control design, for instance, is particularly efficient, in terms of both time and cost, for investigating a relatively rare disease, because the case group selected has already developed the disease. Case-control studies allow evaluation of a wide range of exposures, making such studies a useful first step in the identification of risk factors for a disease. They can be prone, however, to selection bias arising from the occurrence of both exposure and disease before the participants' entry into the study. Another concern that is particularly relevant to studies on biomarkers arises from errors in the measurements of the biomarker (e.g., IGF-I). If the disease being studied (e.g., breast cancer), or its treatment, influences IGF-I levels, we must be concerned that the IGF-I levels assessed after diagnosis of the disease will not accurately reflect those before the disease state. Such "information bias" or misclassification can substantially distort or bias study results. For all these reasons, particularly when evaluating biomarkers such as IGF-I, case-control

Insulin-Like Growth Factors, edited by Derek LeRoith, Walter Zumkeller and Robert Baxter. ©2003 Eurekah.com and Kluwer Academic / Plenum Publishers.

studies are more prone to bias and results need to be interpreted with care. Conversely, cohort studies, which enroll initially healthy individuals, are best suited for relatively common diseases that will affect sufficiently large numbers over a reasonably short time. Although cohort studies minimize selection and information bias, they are very time consuming and expensive. However, with large sample sizes and high follow-up rates, data from prospective cohort studies can be viewed as particularly reliable and informative.

A study design called a nested case-control study is especially relevant for the analysis of IGF-I data. Here, a case-control study is nested within a cohort; blood samples are taken from the cohort at baseline and frozen and stored. Participants are followed up until occurrence of disease. When a sufficient number of cases have accrued, the blood samples from these individuals will be analyzed along with samples from a comparison group (the controls) of individuals without cancer selected from within the cohort. This is an extremely efficient approach that minimizes assay costs while still maintaining the strengths of a prospective study design.

Because of both logistic and cost reasons, most epidemiologic studies use a single blood level of IGF-I to define exposure. Thus, how well this single measurement represents the exposure time period of interest is an important issue. Intra-class-correlations (ICC) calculated from repeated IGF-I measures from samples, collected from the same subjects over time can elucidate this issue. Although only limited ICC data for IGF-I and insulin-like growth factor binding protein-3 (IGFBP-3) are available, to date they range from 0.94 to 0.97 for samples measured over 8 weeks[2] and 0.81 for IGF-I and 0.60 for IGFBP-3 measured over a 1-year period,[3] suggesting that a single measure reflects average levels over at least a 1-year period. This level of reproducibility appears similar to that found for other biologic variables, such as blood pressure and serum cholesterol measurements (ICC, 0.6-0.8 over several years), parameters considered reasonably well measured and consistent predictors of disease in epidemiologic studies. However, one should bear in mind one important caveat: IGF-I levels decrease with increasing age; therefore, unless an individual maintains his or her population rank (which is unknown), a single IGF measure will become increasingly misclassified over many years.

Finally, although the vast majority of reports have been investigations of plasma IGF levels, these levels may serve as an indirect marker of tissue bioactivity. In one report, tissue IGF levels did not appear to be related to circulating IGF levels in humans,[4] although further assessments are needed. Moreover, recent studies of mice with a liver-specific deletion of the IGF-I gene (LID mice, which have IGF-I levels that are only 25% of those in wild-type animals) support a relationship of both circulating and tissue IGF-I levels to the enhancement of growth and metastasis of both mammary tumor[5] and colon cancer.[6]

Introduction to Insulin-Like Growth Factor and Cancer

IGF Family

Insulin-like growth factor type I (IGF-I, also known as somatomedin C) and insulin-like growth factor II (IGF-II) are peptide hormones that control cellular proliferation and differentiation. IGF-I is regulated primarily by human growth hormone (GH), which is a hormone produced by the pituitary gland. GH is the primary regulator of hepatic production of IGF-I, which in turn is the main source of circulating IGF-I.[7] It has also been established that IGFs, acting in a paracrine and possibly autocrine manner, are synthesized locally in the breast,[8] colon,[9,10] prostate,[11] and the lung.[12] Insulin-like growth factor binding protein-3 (IGFBP-3) is one of six currently identified IGFBPs that, by binding IGF peptides, prolong their half-lives and maintain the reservoir of IGF. IGFBP-3 not only has IGF-modulating activity, it appears to have independent effects on DNA synthesis and cell apoptosis. Together with IGF-I, IGF-II, cell-surface receptors, and other IGFBP-interacting molecules, the IGFBPs constitute a comprehensive regulation system of cell survival and death.

Acromegaly

Acromegaly is a rare pituitary disorder. Its clinical symptoms result from excessive secretion of GH and are mediated in part by IGF-I. Clinical features include acral enlargement, bone and soft tissue overgrowth, and visceromegaly. Early follow-up studies of patients with acromegaly demonstrated that they have an increased mortality from cardiovascular disease as compared with the normal population.[13,14] In addition, male patients frequently suffer from secondary hypogonadism and benign prostate hyperplasia[15] and women commonly have menstrual irregularities[16] and hypogonadism.[17] These reports emphasize the complexity of using data from persons with acromegaly, who have multiple hormonal alterations, to assess the relationship between IGF and cancer risk specifically. For example, exposure to estrogens is well confirmed to increase risk of breast cancer. In women with acromegaly, any increase in breast-cancer risk due to elevated IGF might be offset, at least in part, by lower estrogen exposure.[16-18] Studies have consistently reported an increased incidence of adenomatous colon polyps and colon cancer in persons with acromegaly,[19-23] especially in those with particularly high serum levels of IGF-I.[24] The increased proliferation of epithelial cells in the colon of patients with acromegaly[25] and the positive correlation of this proliferation with serum levels of IGF-I and GH suggest a direct effect of the IGFs on the colonic epithelium. While prostatic hyperplasia is a frequent consequence of acromegaly,[15] currently there is limited evidence for a higher prostate-cancer risk among acromegalics.[26] Ongoing efforts to follow up larger numbers of acromegalics, coupled with improved treatment, which results in survival of these patients to the older ages when prostate cancer occurs, should help resolve this issue. Similarly, evidence linking acromegaly with breast cancer and lung cancer is limited because only a few, small studies have examined the issue (standardized mortality ratio [SMR] for breast cancer, 1.60; 95% confidence interval [CI], 0.85-2.74; SMR for lung cancer, 0.69; 95% CI, 0.37-1.18).[22] The largest study to date included only 13 cases of breast cancer and 13 cases of lung cancer. The evidence, taken together, supports a relationship between acromegaly and higher rates of colon cancer; data regarding other cancers are too sparse to permit any conclusions to be drawn.

In the following, we will outline and summarize the scientific evidence from observational studies published through 2001 that relate serum IGF levels to certain demographic, lifestyle, and dietary predictors and to cancer at various sites.

Demographic, Lifestyle, and Dietary Predictors of Levels of Insulin-Like Growth Factor

Few studies have investigated demographic, lifestyle, or dietary predictors of levels of insulin-like growth factors. These predictors are of special interest, however, as they may help elucidate the biologic basis of known or suspected cancer risk factors and identify factors that may alter IGF levels.[27]

Age and Gender

It is now well established that serum levels of IGF-I decrease with increasing age.[28] However, there does not appear to be any substantial independent relation between circulating levels of IGFBP-3 and age.[28,29] In two studies,[2,30] men were found to have higher IGF-I levels and lower IGFBP-3 levels than women.

Race and Ethnicity

The most common cancers diagnosed in men (prostate cancer) and in women (breast cancer) show striking racial/ethnic variations in their incidence. According to data from the Surveillance, Epidemiology, and End Results (SEER) program registries, the incidence rate for prostate cancer among American black men (181 per 100,000) is more than seven times that among U.S. Koreans (24 per 100,000) and still distinctly higher than among U.S. whites (136 per 100,000). Similarly, the highest incidence rates for breast cancer occur among white, Hawaiian, and black women, with Koreans being the group with the lowest incidence rate.[31] African-American men appear to have lower IGFBP-3 levels than white men, which might contribute to their greater risk of prostate cancer.[32,33] In the only study to date in women, mean serum IGF-I levels were significantly higher among black women than among white women.[34] Such findings suggest that racial variations in serum IGF levels exist and need to be taken into consideration in planning potential preventive strategies.

Lifestyle and Diet

Because of the more complicated, because not uniform effects of different kinds of physical activity on both tissue and circulating levels of IGF, an association with IGF is difficult to assess.

However, in a recent report that evaluated physical activity by calculating metabolic equivalent task (MET)-hours per week, with MET-hour defined as the energy expended in sitting quietly for an hour, higher levels of physical activity were associated with a higher IGF-I/IGFBP-3 ratio.[35] Smoking history has only inconsistently been linked with serum levels of IGF-I.[2,30] Although height is strongly correlated with IGF levels in childhood,[28] levels have not been consistently correlated with height in adults.[35-37] Higher body mass index (BMI) was significantly related to increased expression of IGF-I in the breast tissue of postmenopausal women with breast cancer.[38] In a detailed evaluation among healthy women, BMI was weakly associated with a lower IGF-I/IGFBP-3 ratio.[35] Recently, an inverse association of circulating IGF-I levels and the IGF-I/IGFBP-3 ratio with parity[35] and an inverse relation of serum levels of IGFBP-3 with higher alcohol intake [p for trend, 0.02][2,35] were reported. Current oral contraceptive use was associated with lower IGF-I levels.[34] In addition, current estrogen replacement therapy is associated with substantially (20-30%) lower levels of IGF-I and a lower IGF-I/IGFBP-3 ratio,[35,39,40] and one study found that levels decreased with increasing duration of estrogen replacement therapy.[40]

Protein-calorie malnutrition has been well documented to lower IGF levels;[41] however, there is little information about how diet affects IGF levels in well-nourished populations. The only two studies of this issue to date, observed a modest positive correlation between energy intake and IGF levels.[30,42] Consumption of red meat and total fat intake were positively associated with IGF-I levels in one study,[30] but this was not confirmed in a second assessment.[42] Finally, there are data, although sparse, that relate IGF levels to several dietary factors thought to be related to prostate cancer. Tomato intake (a consistent protective dietary factor for prostate cancer) was inversely related to serum IGF-I levels.[43] Recent studies suggest that vitamin D has a protective role against prostate cancer,[44] potentially through an increase in IGFBP-3.[45] Both increased bone resorption and decreased IGF-I levels have been linked to polymorphisms in the vitamin D receptor gene;[46] this polymorphism has also been associated with risk of prostate cancer in several studies in Japan.[47,48] However, all these associations require confirmation.

Milk and Dairy Products

The strongest and, together with age, the most consistent association observed with IGF levels is the amount of milk and dairy products in the diet. Both the Nurses' Health Study (NHS), and the Physicians' Health Study (PHS) reported a positive association between intake of dairy foods and both IGF-I levels and the IGF-I/IGFBP-3 ratio.[42,49] Furthermore, a randomized trial that investigated the association between higher milk consumption and bone remodeling found a significant increase in average serum IGF-I levels among the subjects in the intervention group (i.e., those with higher milk consumption).[50]

Etiologic Studies of Insulin-Like Growth Factor and Cancer

Insulin-Like Growth Factor and Breast Cancer

The potent mitogenic and anti-apoptotic actions of IGF-I influence both normal and transformed breast epithelial cells.[51,52] In rodents, the overexpression of GH increased the frequency of breast tumors,[53] and treatment with GH or IGF-I led to mammary gland hyperplasia in monkeys.[51] The expression of IGFBP-3 in many tissues, on the other hand, suggests that it locally modulates the action of IGF peptides. Retinoic acid-induced expression of IGFBP-3, for example, inhibits the growth-promoting effects of IGF-I in breast-cancer cells.[54] IGFBP-3 may also have other, not yet fully understood, physiologic roles. In addition, several groups have reported a positive association between plasma IGF-I levels in premenopausal women and the percent breast density observed on a mammogram.[55,56] Mammographic breast density is strongly and consistently related to breast-cancer risk.[57]

The proliferative effect of IGF-I and evidence that increased turnover of epithelial cells is associated with a greater risk of neoplastic transformation have fueled several epidemiologic studies of IGF-I levels and breast-cancer risk (Table 1). To date, 11 retrospective case-control studies have evaluated the relationship between plasma IGF-I levels and breast-cancer risk.[58-68] Six noted a positive relationship,[58-60,63,68] whereas five of the smaller studies reported no relationship[61,64-66] or an inverse[62] relationship with risk. In one of the largest of these studies,[60] which had 99 breast-cancer

Table 1. Serum IGF-1 and IGFBP-3 levels and the risk of breast cancer

Study	Study Population	Adjustment Covariates	Breast Cancer Risk Related to IGF-1 Level in ng/ml (highest vs. lowest) RR (95% CI)		Breast Cancer Risk Related to IGFBP-3 Level in ng/ml (highest vs. lowest) RR (95% CI)	
			Premenopausal	Postmenopausal	Premenopausal	Postmenopausal
Prospective Nested Case-Control Studies						
Hankinson et al., 1998[69]	Nurses' Health Study, 397ca/620co, matched by year of birth, day blood drawn, fasting status, month of blood sampling, menopausal status, postmenopausal hormone use; 4 years follow-up (1990-1994)	Age at first birth, age at menarche, age at menopause, parity, family history of breast cancer, serum IGF-1 and IGFBP-3 levels (quintiles)	2.33 (1.06-5.16) 2.88 (1.21-6.85)*	0.85 (0.53-1.39) 0.89 (0.51-1.55)*	-- --	-- --
Toniolo et al., 2000[70]	New York University Women's Health Study, 287ca/706co, matched by age, menopausal status, date of blood sampling, day of menstrual cycle at blood sampling; up to 10 years follow-up after recruitment between 1985 and 1991	History of benign breast disease, parity, serum IGFBP-3 levels (quartiles)	2.30 (1.07-4.94) 1.90 (0.82-4.42)*	0.95 (0.49-1.86) --	2.17 (0.99-4.76) --	1.08 (0.54-2.16) --
Retrospective Case-Control Studies						
Peyrat et al., 1993[58]	44ca/92co women age >35 years	None	Median concentrations significantly different		--	
Bruning et al., 1995[68]	150ca/441co, matched by age, geographic region, SES	Age, BMI, family history of breast cancer, height, menopausal status, serum CRP, albumin, testosterone levels (quintiles)	7.34 (1.67-32.16) [IGF-1/IGFBP-3 ratio] all women combined			

continued on next page

Table 1. Serum IGF-1 and IGFBP-3 levels and the risk of breast cancer (continued)

Study	Study Population	Adjustment Covariates	Breast Cancer Risk Related to IGF-1 Level in ng/ml (highest vs. lowest) RR (95% CI) Premenopausal	Breast Cancer Risk Related to IGF-1 Level in ng/ml (highest vs. lowest) RR (95% CI) Postmenopausal	Breast Cancer Risk Related to IGFBP-3 Level in ng/ml (highest vs. lowest) RR (95% CI) Premenopausal	Breast Cancer Risk Related to IGFBP-3 Level in ng/ml (highest vs. lowest) RR (95% CI) Postmenopausal
Bohlke et al., 1998[59]	94ca (ductal carcinoma in situ)/76co, matched by age, residence	Age, age at first birth, age at menarche, BMI, ethnicity, family history of breast cancer, height, parity, serum estradiol, IGF-1 and IGFBP-3 levels (tertiles)		1.8 (0.7-4.6)*		0.7 (0.3-1.7)§
Del Giudice et al., 1998[60]	99ca/99co, matched by age	Age, weight (quintiles)	1.47 (0.66-3.27)	--	2.05 (0.93-4.53)	--
Ng et al., 1998[65]	63ca/27co, matched by time period	Age, age at menarche, BMI, family history of breast cancer, history of breast biopsy, menopausal status, oral contraceptive use, parity, postmenopausal hormone use (median)		NS (p=0.23)†		0.18 (0.05-0.55)§
Jernstrom et al., 1999[64]	Rancho Bernardo Study, 45ca/393co, 2 years of follow-up (1992-1994)	Alcohol, exogenous hormone use, height, parity, physical activity, smoking status, weight (mean)		NS (p=0.31)†		--

continued on next page

Table 1. Serum IGF-1 and IGFBP-3 levels and the risk of breast cancer (continued)

Study	Study Population	Adjustment Covariates	Breast Cancer Risk Related to IGF-1 Level in ng/ml (highest vs. lowest) RR (95% CI)		Breast Cancer Risk Related to IGFBP-3 Level in ng/ml (highest vs. lowest) RR (95% CI)	
			Premenopausal	Postmenopausal	Premenopausal	Postmenopausal
Holdaway et al., 1999[66]	31 breast cancer cases/31co, 12 cases with benign breast disease/12co, matched by age	(mean)		NS (p=0.85)[†]		NS (p=0.051)[†]
Mantzaros et al., 1999[61]	83 ca (carcinoma in situ)/69co, matched by age and residence	None (tertiles)	NS (p=0.70)[‡]	--	NS (p=0.75)[‡]	--
Agurs-Collins et al., 2000[67]	30ca/30co, African-American women only, postmenopausal matched for age	BMI, serum SHBG (per 10 ng/ml of IGF-1 concentration)	--	1.18 (1.17-1.20)	--	--
Petridou et al., 2000[62]	75ca/75co, matched by age, residence	BMI, education, time-interval between blood collection and analysis (1 SD)	0.4 (0.1-1.4)	1.1 (0.7-1.7)	--	--
Li et al., 2001[63]	40ca/40co, matched by age, race (>50% African-American)	Menopausal status, serum IGFBP-3 levels (median)	2.00 (0.43-9.28)[*] 6.31 (1.03-38.72)[‡]		0.65 (0.16-2.58)[§]	

[*] Further adjustment for IGFBP-3 ; [§] Further adjustment for IGFBP-3; [†] two-sample t-test with unequal variances; [‡] Free IGF-I, adjusted for IGFBP-3 and menopausal status
Abbreviations: ca, cases; co, controls; SD, standard deviation; NS, not significantly different; SES, socioeconomic status; BMI, body mass index; CRP, C-reactive protein; SHBG, sex hormone-binding globulin.

cases and an equal number of age-matched control subjects, a nonsignificantly elevated breast-cancer risk (relative risk [RR] for top versus bottom quintile 1.47; 95 % CI, 0.66-3.27) was observed among premenopausal women. A case-control study of similar size[59] of premenopausal women with carcinoma in situ reported an elevated RR of 1.8 (95% CI, 0.7-4.6). Only two of the case-control studies provided separately results for postmenopausal women.[62,67] No association between IGF levels and risk was noted in one study (RR, 1.1; 95% CI 0.7-1.7),[62] while a strong association was seen in the second (RR, 1.18 per 10 ng/ml of IGF-I concentration; 95% CI, 1.17-1.20).[67]

One of the first, and still the largest, case-control studies to assess the ratio of IGF-I to IGFBP-3 levels[68] reported a strong association between the ratio and breast cancer risk. Among 150 cases and 441 matched controls, the RR was 7.34 for the highest compared with the lowest quintile of IGF-I/IGFBP3 ratio (95% CI, 1.67-32.16); this association tended to be strongest in premenopausal women. The control subjects were matched by age, geographic region, and socioeconomic status, and the authors controlled for family history of breast cancer, height, BMI, body fat distribution, menopausal status, and serum C-reactive protein, albumin, and testosterone in their analysis. Bohlke et al[59] also evaluated the IGF-I/IGFBP-3 ratio and reported an RR of 1.6 (95% CI, 0.7-3.8, top versus bottom tertile).

Two prospective analyses have been published to date.[69,70] The larger of the two collected blood samples from 32,826 women in the NHS from 1989 to 1990.[69] A total of 397 incident cases of breast cancer were diagnosed during follow-up (1990-1994); one or two control subjects were selected for each case subject matched by year of birth, day of blood draw, fasting status, month of blood sampling, menopausal status, and postmenopausal hormone use. Among premenopausal women, a significant relation was seen between plasma IGF-I levels and breast cancer risk (top versus bottom tertile of levels: RR, 2.33; 95% CI, 1.06-5.16). Additional adjustment for IGFBP-3 levels further elevated the risk for breast cancer among premenopausal women (RR, 2.88; 95% CI, 1.21-6.85), and in the small subset of women 50 years or younger, the breast-cancer risk associated with IGF-I levels was increased up to sevenfold (RR, 7.28; 95% CI, 2.40-22.0). In contrast, for postmenopausal women, there was no apparent association (quintiles, RR, 0.89; 95% CI, 0.51-1.55) and adjustment for IGFBP-3 did not alter the estimate further. The second prospective study[70] was conducted among 14,275 women followed up for up to 10 years, and findings were similar. A modest increase in breast-cancer risk with increasing IGF-I levels was found among women who were premenopausal at blood collection. The association appeared strongest among those who were also premenopausal at diagnosis (top versus bottom quartile of levels: RR, 1.90; 95% CI, 0.82-4.42). No association was noted in postmenopausal women (comparable RR, 0.95; 95 % CI, 0.49-1.86).

Few studies have investigated the relation between IGFBP-3 specifically and breast-cancer risk. The only prospective study published[70] noted no significant association for either IGFBP-3 or the IGF-I/IGFBP-3 ratio among either premenopausal or postmenopausal women (RR for premenopausal women, 1.18 [95% CI, 0.66-2.08]; RR for postmenopausal women, 1.08 [95% CI, 0.54-2.16]). The results of case-control studies have been quite inconsistent, with elevated risk,[60] reduced risk,[59,63,65] or no statistically difference in risks[61] reported (Table 1).

To date, there is increasing evidence of a relation between IGF-I and breast-cancer risk. Findings from the retrospective case-control studies have not been consistent, likely due in part to the small size of many of the studies. However, the larger and more carefully conducted studies have tended to observe a positive association. The two prospective studies published to date had similar findings: a relatively strong, positive association among the subset of premenopausal women and no apparent association with risk among postmenopausal women. These findings do require confirmation in additional large prospective studies, however. The reason for a difference in RRs between premenopausal and postmenopausal women is unknown but may reflect an effect of the GH/IGF axis that occurs after breast development early in life. Alternatively, IGF-I concentrations may be particularly relevant to the risk of breast cancer in premenopausal women because estradiol enhances the action of IGF-I in the breast.[71] The recent reports of a correlation between IGF levels and mammographic density in premenopausal women also provide indirect support for a relation with breast cancer. In summary, the potential relation between levels of IGF in plasma and breast-cancer risk may point to new and promising strategies for prevention of breast cancer.

Insulin-Like Growth Factor and Prostate Cancer

Several observations suggest that the regulatory effects of IGF-I and IGFBP-3 on cell proliferation are relevant not only for the breast but also for prostate cells and tissue. IGF-I has been shown to stimulate prostate cell growth,[72] and the inhibition of IGF action appears to slow proliferation of prostate cells.[73] IGFBP-3 reduces the proliferative effect of IGF-I on prostate cell growth.[72] Furthermore, prostate-specific antigen (PSA), used for the early detection of prostate cancer, appears to function as an IGFBP protease, nullifying the inhibitory effect of IGFBP-3 on IGF-I in vitro.[72,74]

A recent meta-analysis found a doubling of the risk of prostate cancer in men with higher IGF-I levels when the highest quartile of serum hormones was compared to the lowest.[75] The evidence from observational studies has been accruing rapidly since the publication of this meta-analysis and strongly supports its results (Table 2).

Nine retrospective case-control studies have evaluated the relationship between IGF-I and prostate cancer.[76-84] Typically, in a case-control study of IGF-I and prostate cancer, IGF-I levels in men with prostate cancer are compared with levels in men without a diagnosis of prostate cancer (controls). One smaller such study compared IGF-I levels in 52 men with prostate cancer to those in 52 controls with benign prostatic hyperplasia and 52 healthy controls.[81] IGF-I levels were significantly higher among patients with prostate cancer (RR, 1.91; 95% CI, 1.00-3.73) but not among men with benign prostatic hyperplasia (RR, 0.99; 95% CI, 0.48-2.06) when compared with the healthy control group. Also Khosravi et al[83] reported a statistically significant difference in mean IGF-I levels between patients with prostate cancer and those with benign prostatic hyperplasia. Two other case-control studies did not observe a significant difference in IGF-I levels; however, these studies were compromised by their very small number of study participants.[79,80] Another small case-control study investigated whether IGF-I levels varied among groups of men with prostate cancer, healthy men with elevated PSA values, and healthy men with normal PSA values and found no difference in serum levels in any of these groups.[77] Of the three larger case-control studies,[76,82,84] only one found no significant difference in the mean serum levels of IGF-I in men with and without prostate cancer.[76] However, in this report, the authors did not adjust for IGFBP-3 levels and the number of controls (n=67) was less than 50% of the number of cases (n=171). Wolk et al[82] described elevated serum IGF-I levels among their patients with prostate cancer (RR, 1.43; 95%CI, 0.88-2.33), and in another report that also included IGFBP-3 levels in the statistical model, the RR associated with higher IGF-I serum levels was 2.63 (95% CI, 1.19-5.79).[84] Four cross-sectional studies have been published to date (Table 2). Two of them report higher IGF-I levels among men who developed prostate cancer,[85,86] and the other two reported no statistically significant difference.[87,88] However, the analysis of Finne et al[87] was restricted to men with elevated PSA (≥ 4 µg/l) and the controls in the study by Koliakos et al[88] were men with benign prostatic hyperplasia rather than healthy subjects, factors that potentially could account for the different results.

In a prospective nested case-control study, serum IGF-I levels measured in men before diagnosis of prostate cancer were compared with levels in men without a diagnosis of prostate cancer. In contrast to previous studies, these studies are not hospital-based but are nested in a cohort consisting of healthy, random individuals who are followed up over time until the occurrence of prostate cancer. To date, four nested case-control studies have reported on this relationship, and the larger three described an elevated risk of prostate cancer associated with higher serum IGF-I levels (Table 2). In findings from the Physicians' Health Study (PHS), in which 152 cases were matched with 152 controls,[89] the RR associated with the highest versus the lowest quartile of serum IGF-I was 4.32 (95% CI, 1.76-10.6). The Baltimore Longitudinal Study of Aging reported similar findings: the association of higher IGF-I levels with an RR of 3.11 (95% CI, 1.11-8.74) for prostate cancer.[90] Finally, Stattin et al observed a modest, nonsignificant RR of 1.32 (95% CI, 0.73-2.39) among men with IGF-I levels in the highest quartile compared with those in the lowest quartile.[91] All three prospective studies adjusted their estimates for IGFBP-3 levels (changes in RRs due to controlling for IGFBP-3 are shown in Table 2). The smallest study, with only 30 cases, reported no association between IGF-I levels and prostate cancer risk (RR, 0.6; 95% CI, 0.1-2.9).[92]

Many of the studies that investigated serum IGF-I levels and their relation to prostate-cancer risk also measured serum levels of IGFBP-3. Overall, IGFBP-3 does not appear to be positively associated with prostate-cancer risk in these studies. Moreover, controlling for IGF-I, IGFBP-3 was inversely related to prostate-cancer risk in two of the three prospective studies [RR, 0.41; 95% CI,

Table 2. Serum IGF-1 and IGFBP-3 levels and the risk of prostate cancer

Study	Study Population	Adjustment Covariates (Comparison Mode)^5	Prostate Cancer Risk Related to IGF-1 Level in ng/ml (highest vs. lowest) RR (95% CI)	Prostate Cancer Risk Related to IGFBP-3 Level in ng/ml (highest vs. lowest) RR (95% CI)
Prospective Nested Case-Control Studies				
Chan et al., 1998[89]	Physicians' Health Study, 152ca/152co, matched by age, duration of follow-up, and smoking; 10 years follow-up	Androgen receptor gene, BMI, CAG polymorphism, height, serum levels of lycopene, estrogen, testosterone, PSA, DHT, SHBG, prolactin, AAG, IGF-1 and IGFBP-3, weight (quartiles)	2.41 (1.23-4.74) 4.32 (1.76-10.6)*	1.07 (0.54-2.11) 0.41 (0.17-1.03)**
Harman et al., 2000[90]	Baltimore Longitudinal Study of Aging, 72ca/127co, matched by age; 9 years follow-up	Sample date (visit date) (tertiles)	1.65 (0.71-3.86) 3.11 (1.11-8.74)*	0.71 (0.30-1.70) 0.76 (0.30-1.94)**
Stattin et al., 2000[91]	Northern Sweden Health and Disease Cohort, 149ca/298co, matched by age, date the survey was completed, residence; up to 14 years of follow-up	BMI, serum levels of IGF-1, IGFBP-3 and insulin, smoking status (quartiles)	1.72 (0.93-3.19) 1.32 (0.73-2.39)*	1.83 (0.98-3.24) 1.19 (0.66-2.15)**
Lacey et al., 2001[92]	The Washington County Serum Bank, 30ca/60co, matched by age, race, and date of blood draw; 2 to 15 years of follow-up	Age at blood draw	0.6 (0.1-2.9)*	1.1 (0.3-3.8)

continued on next page

Table 2. Serum IGF-1 and IGFBP-3 levels and the risk of prostate cancer (continued)

Study	Study Population	Adjustment Covariates (Comparison Mode)[§]	Prostate Cancer Risk Related to IGF-1 Level in ng/ml (highest vs. lowest) RR (95% CI)	Prostate Cancer Risk Related to IGFBP-3 Level in ng/ml (highest vs. lowest) RR (95% CI)
Cross-Sectional Studies				
Cutting et al., 1999[85]	37 ca/57 co	(mean)	2.47 (0.57-10.67)	--
Djavan et al., 1999[86]	71ca/174co	(mean)	Significantly higher in prostate cancer patients ($p<0.001$)[†]	--
Finne et al., 2000[87]	179 prostate cancer/174 benign prostatic hyperplasia, 268 healthy controls (restricted to men with elevated serum PSA ≥4 μg/l)	Age, prostate volume, serum PSA, IGF-1, and IGFBP-3 (quartiles)	0.60 (0.34-1.04) 0.50 (0.26-0.97)[*]	1.24 (0.68-2.24)[**]
Koliakos et al., 2000[88]	34 prostate cancer cases/131 benign prostatic hyperplasia	(mean)	NS ($p=0.35$)[†]	--
Retrospective Case-Control Studies				
Kanety et al., 1993[77]	14ca/4co with elevated PSA/6 healthy co	(mean)	NS ($p=0.14$)[†]	Significantly lower in cases ($p<0.02$)[†]
Cohen et al., 1993[78]	32ca/16co, matched by age	(mean)	NS ($p=0.45$)[†]	NS ($p=0.84$)[†]
Tellez Martinez-Fornes et al., 1996[79]	Tissue from 8 prostate cancer patients/5 patients with benign prostatic hyperplasia/5 healthy controls	(mean)	NS ($p=0.24$)[†]	--

continued on next page

Table 2. Serum IGF-1 and IGFBP-3 levels and the risk of prostate cancer (continued)

Study	Study Population	Adjustment Covariates (Comparison Mode)§	Prostate Cancer Risk Related to IGF-1 Level in ng/ml (highest vs. lowest) RR (95% CI)	Prostate Cancer Risk Related to IGFBP-3 Level in ng/ml (highest vs. lowest) RR (95% CI)
Ho and Baxter, 1997[80]	16 prostate cancer/8 benign prostatic hyperplasia/ 7 controls	(mean)	NS (p=0.48)†	NS (p=0.27)†
Mantzoros et al., 1997[81]	52 prostate cancer /52 benign prostatic hyperplasia/52 controls, matched by age, residence	Age, BMI, education, height, serum levels of steroid hormones and SHBG (per 60 ng/ml⁻¹ increment of IGF-1)	1.91 (1.00-3.73)	--
Wolk et al., 1998[82]	210ca/224co, matched by age	BMI, height, total energy intake (quartiles)	1.43 (0.88-2.33)	1.21 (0.75-1.93)
Kurek et al., 2000[76]	171ca/67co, matched by age	(mean)	NS (p=0.96)†	--
Khosravi et al., 2001[83]	84 prostate cancer/75 benign prostatic hyperplasia, matched by age and total PSA	(mean)	Significantly higher in cases (p<0.001)†	NS (p=0.09)†
Chokkalingam et al., 2001[84]	128ca/306co, matched by age	BMI, height, serum levels of IGFBP-3, 3-alpha-diol G, SHBG, and estradiol, waist-to-hip-ratio, weight (quartiles)	1.58 (0.87-2.88) 2.63 (1.19-5.79)*	0.90 (0.49-1.66) 0.54 (0.26-1.15)**

* Further adjustment for IGFBP-3; ** Further adjustment for IGF-1; † two sample t-test with unequal variances
Abbreviations: ca, cases; co, controls; NS, not statistically different; PSA, prostate-specific antigen; BMI, body mass index; SHBG, sex hormone binding globulin; DHT, dihydrotestosterone; CAG, CAG polymorphism of the androgen receptor gene; AAG, 3alpha-androstendiol glucuronide
§ Category contrast for relative risks presented (e.g., tertiles, quintiles, or means, where means compared)

0.17-1.03;[89] RR, 0.76; 95%CI, 0.30-1.94,[90]]. In a current update of the results from the PHS (379 cases and 383 controls), the RRs for both IGF-I (RR, 2.86; 95% CI, 1.13-7.23) and IGFBP-3 (RR, 0.33; 95% CI, 0.13-0.86) were confined to high-grade/stage prostate cancer, whereas the results for low-grade/stage prostate cancer were generally null.[93]

Differences in tissue levels of IGF-I in normal, hyperplastic, and neoplastic prostate tissue have been evaluated in one small study. This study did not compare these measurements with serum IGF-I levels, and the reported tissue levels were not significantly different in the three groups.[79]

In summary, there is fairly consistent evidence from observational studies that higher IGF-I serum levels in men are associated with a significantly increased risk (two-fold or greater) for developing prostate cancer. While serum IGFBP-3 levels generally show an inverse relation to risk of prostate cancer, results are less consistent. Furthermore, early evidence suggests that IGF-I and IGFBP-3 may be strongly associated with clinically important risk of prostate cancer but not with early-stage or low-grade tumors.

Insulin-Like Growth Factor and Colon Cancer

IGF-I receptors are expressed by colorectal cancer cells,[94] and exogenous IGF-I promotes the differentiation of human colorectal cancer cells.[95,96] It has been shown that blockade of the IGF-I receptor inhibits growth of tumor cells in the colon.[97] Moreover, persons with acromegaly appear to have an increased risk of colon cancer (See Introduction).

Few epidemiologic studies have evaluated associations between IGF levels and colon-cancer risk. However, three of a total of four studies were cohort studies, and results are quite consistent (Table 3). The association with IGF-I tended to increase after IGFBP-3 levels were taken into account. In the Physicians' Health Study (PHS), a strong positive association was found between baseline serum IGF-I levels and subsequent risk of colon cancer among men,[98] and after controlling for serum IGFBP-3 levels, the RR for the top (vs. bottom) 25% of levels increased from 1.36 to 2.51 (95% CI, 1.15-5.46). In the Nurses' Health Study (NHS), the RR associated with higher IGF-I levels was 2.18 (95% CI, 0.94-5.08), after controlling for known colon-cancer risk factors and IGFBP-3.[29] In the third of the cohort studies, the New York University Women's Health Study (NYUWHS), however, the RR was attenuated after serum IGFBP-3 levels were taken into account (RR, 1.23; 95% CI, 0.47-3.22).[99] A similar positive association was observed in the only retrospective case-control study.[100]

For IGFBP-3, the findings are somewhat less consistent. Although the NHS and the PHS observed almost identical inverse associations between IGFBP-3 levels and colon-cancer risk (RR, 0.28; 95% CI, 0.10-0.83 for women; RR, 0.28; 95%CI, 0.12-0.66 for men, respectively),[29,98] the third cohort study (NYUWHS) reported a weak positive association (RR, 1.23; 95% CI, 0.51-2.95, top versus bottom quintile).[99] The only published case-control study supports the latter finding.[100] Although different types of assays were used in the studies, a recent report[101] indicated that the assay methods were unlikely to be the primary cause of this difference.

Studies that investigated adenomas, precursors of colon cancer, and their association with IGF levels found that particularly high-risk adenomas were related to IGF-I and IGFBP-3 levels in a manner similar to that seen with colon cancer in the NHS and the PHS. Mean serum IGF-I levels were significantly increased, and IGFBP-3 levels were decreased in individuals with high-risk adenomas.[102]

In summary, cohort studies to date provide fairly strong and consistent evidence for a positive relation between higher IGF-I levels and colon cancer risk. However, the existing data on IGFBP-3 levels and colon cancer are less consistent. The somewhat unexpected differences between the study results for IGFBP-3 in two very similar cohorts (NHS versus NYUWHS) require further exploration. To help delineate when during carcinogenesis IGF may be important, additional studies of both colon cancer and colon polyps are needed.

Insulin-Like Growth Factor and Lung Cancer

There is some evidence for the local synthesis and expression of the IGF-I system in the lung and its action on the bronchial epithelium.[12] The presence of IGF-I receptors in the bronchial epithelial cells of both normal lung and primary lung cancer cells has been documented.[103] In vitro, evidence of the strong mitogenic and anti-apoptotic activity of IGF-I has been extended to lung cancer cell lines.[104,105]

Table 3. Serum IGF-1 and IGFBP-3 levels and the risk of colon cancer

Study	Study Population	Adjustment Covariates (Comparison Mode)[§]	Colon Cancer Risk Related to IGF-1 Level in ng/ml (highest vs. lowest) RR (95% CI)	Colon Cancer Risk Related to IGFBP-3 Level in ng/ml (highest vs. lowest) RR (95% CI)
Prospective Nested Case-Control Studies				
Ma et al., 1999[98]	Physicians' Health Study, 193ca/318co, matched by age, smoking 13 years follow-up (1982-1995)	Age, alcohol, BMI, serum levels of IGFBP-3, smoking status (quintiles)	1.36 (0.72-2.55) 2.51 (1.15-5.46)[*]	0.47 (0.23-0.95) 0.28 (0.12-0.66)[**]
Giovannucci et al., 2000[29]	Nurses' Health Study, 79ca/158co, matched by age, month of blood draw, fasting status 6 years follow-up (1990-1996)	Age, alcohol, aspirin use, BMI, intake of fat, carbohydrates, protein, folate, methionine, and red meat, physical activity, postmenopausal hormone use, serum levels of IGFBP-3 and IGF-1, smoking status (tertiles)	1.21 2.18 (0.94-5.08)[*]	0.53 0.28 (0.10-0.83)[**]
Kaaks et al., 2000[99]	New York University Women's Health Study, 102ca/200co, matched by age, menopausal status (at enrollment), date of recruitment, number of blood donations, time of day when blood was drawn; up to 13 years of follow-up	Age, BMI, day of menstrual cycle (for premenopausal women), menopausal status, serum levels of IGFBP-3 and IGF-1, smoking status, time of blood donation, time of last food consumption (quintiles)	1.88 (0.72-4.91) 1.23 (0.47-3.22)[*]	2.46 (1.09-5.57) 1.23 (0.51-2.95)[**]
Retrospective Case-Control Studies				
Manousos et al., 1999[100]	41ca/50co, matched by age, gender, education	BMI, blood sampling date, height (tertiles)	2.3 (0.6-9.1)	0.5 (0.1-1.7)

[*] Further adjustment for IGFBP-3; [**] Further adjustment for IGF-1
[§] Category contrast for relative risks presented (e.g., tertiles, quintiles, or means, where means compared)
Abbreviations: ca, case; co, control; BMI, body mass index

Of the four epidemiologic studies published through 2001 (Table 4), only one is prospective.[106] Three reasonably large case-control studies[4,107,108] observed a statistically significant association between higher IGF-I levels and lung-cancer risk (RR, 2.06; 95% CI, 1.19-3.56;[107] and RR, 2.13; 95% CI, 1.2-3.78,[108] respectively, top versus bottom). The cohort study did not report a positive association but noted instead a nonsignificant inverse association between serum IGF-I levels and lung-cancer risk.[106]

To date, no substantial association has yet been noted between IGFBP-3 and lung-cancer risk. In the cohort study, there was no association of IGFBP-3 with lung-cancer risk, either with or without adjustment for potential confounding factors, including serum IGF-I.[106] The case-control studies noted a small reduction in risk associated with high levels of IGFBP-3, but none of the results were statistically significant.[106-108]

Minuto et al described higher tissue levels of IGF-I in neoplastic lung tissue compared with normal lung tissue.[4] However, the serum IGF-I values measured in the same 10 subjects were unrelated to either normal or neoplastic tissue content.

In summary, there is only preliminary and weak evidence supporting a positive association of lung-cancer risk with serum IGF-I levels and a modest inverse association with serum IGFBP-3 levels.

Insulin-Like Growth Factor and Other Cancers

Laboratory investigations of the past few years suggest that IGF has important influences on the biology and carcinogenesis of several other systems. The few epidemiologic studies that have evaluated these relationships are summarized below.

Childhood Leukemia

Because both childhood leukemia[109] and prostate cancer[37] were associated with increased height in previous observational studies and because IGF-I was linked to prostate cancer,[89] it was postulated that IGF-I and IGFBP-3 also may play a role in the pathogenesis of childhood leukemia. Studies of the effects of chemotherapy on the growth of children with leukemia have consistently demonstrated decreased IGF-I and IGFBP-3 serum levels at diagnosi.s[110-112] In the largest observational study published to date,[113] 122 patients with childhood leukemia and 122 gender- and age-matched controls were evaluated. In this study, an increment in IGFBP-3 of 1 ng/ml corresponded to a statistically significant reduction in risk for childhood leukemia (odds ratio [OR], 0.72; 95% CI, 0.55-0.93), whereas there appeared to be no relation between IGF-I levels and leukemia (OR, 0.95; 95% CI, 0.80-1.13). However, levels of free IGF-I were reported to be significantly increased in patients with acute lymphatic leukemia at diagnosis and throughout therapy.[112] If free IGF-I is the biologically active fraction, leukemic children thus may not have low levels of biologically active IGF-I, despite their lower overall levels of IGF.[114]

Thyroid Cancer

Both IGF-I and its receptor are expressed in thyroid-cancer tissues.[115,116] IGF-I may also be produced locally in thyroid-cancer tissue.[117] Such overactivation of the IGF system was observed particularly in tissue of adenomas and carcinomas of the thyroid gland as compared with normal thyroid tissue.[118,119] More research, however, is needed to confirm these potentially important findings.

Other Cancers

Preliminary data indicate a significant downregulation of the expression of IGFBP-3 mRNA in human hepatocellular carcinoma tissues[120,121] and of IGF-I mRNA in endometrial carcinomas.[122] Serum IGF-I levels were found to be significantly lower in patients with gastric cancer than in control subjects.[123] Other IGF factors have been described as potential markers for cervical cancer[124] and for the occurrence of metastases in hepatocellular carcinoma.[125] Furthermore, the levels of IGF-I in cyst fluid from epithelial ovarian cancer were significantly higher than those in cyst fluid in benign neoplasms.[126]

In conclusion, preliminary data relate changes in IGF levels with liver, endometrial, gastric, and ovarian cancers. However, all of these reports require further exploration and confirmation.

Table 4. Serum IGF-1 and IGFBP-3 levels and the risk of lung cancer

Study	Study Population	Adjustment Covariates (Comparison Mode)[§]	Lung Cancer Risk Related to IGF-1 Level in ng/ml (highest vs. lowest) RR (95% CI)	Lung Cancer Risk Related to IGFBP-3 Level in ng/ml (highest vs. lowest) RR (95% CI)
Prospective Nested Case-Control Studies				
Lukanova et al., 2001[106]	New York University Women's Health Study, 93ca/186co, matched by age, date of blood sampling, menopausal status, day of menstrual cycle, and questionnaire data of smoking status at the time of blood donation; up to 14 years of follow-up	BMI, serum cotinine, IGF-1, and IGFBP-3 levels, time since last meal (quartiles)	0.79 (0.29-2.19) 0.54 (0.14-2.07)[*]	0.90 (0.36-2.25) 0.90 (0.28-2.85)[**]
Retrospective Case-Control Studies				
Minuto et al., 1986[4]	10 patients: neoplastic and normal lung tissue	(mean)	Significantly higher content in lung tumors (p=0.01)[†]	--
Yu et al., 1999[107]	204ca/218co, matched by age, gender, smoking status	Serum levels of IGFBP-3, IGF-1 (quartiles)	2.06 (1.19-3.56)	0.73 (0.43-1.26)
Wu et al., 2000[108]	183ca/227co, matched by age, gender, ethnicity, smoking status	BMI, bleomycin sensitivity, BPDE sensitivity, family history of cancer, serum levels of IGFBP-3 and IGF-1, smoking status (high versus low, cutoff 75th percentile values for IGF-1 and 25th percentile values for IGFBP-3 —quartiles)	2.13 (1.20-3.78)[*]	0.59 (0.33-1.05)[**]

[*] Further adjustment for IGFBP-3; [**] Further adjustment for IGF-1; [†] Two-sample t-test with unequal variances
[§] Category contrast for relative risks presented (e.g., tertiles, quintiles, or means, where means compared)
Abbreviations: ca, cases; co, controls; BMI, body mass index; BPDE, benzo[a]pyrene diol epoxide.

Summary

Our knowledge about what demographic and lifestyle factors influence IGF levels and our understanding of the pathophysiology responsible for the relation of the IGF family to carcinogenesis has grown steadily over the past several years. There is fairly substantial evidence from observational studies linking IGF levels with the risk of both prostate cancer and premenopausal breast cancer in particular. Although fewer epidemiologic studies have evaluated IGF and colon cancer, results have been quite consistent, and both laboratory data and studies of persons with acromegaly are supportive of a relation. However, these associations still need to be confirmed in additional large prospective studies. Moreover, many important aspects, such as the magnitude of the relation and the relation with the binding proteins, still need to be settled. In addition, genetic polymorphisms, such as the recently described promotor polymorphism in IGFBP-3^{127} and others yet to be identified, are likely to contribute to variation in endogenous levels of, or responsiveness to, members of the IGF family. Also, the influence of dietary and other lifestyle factors on the IGF system are incompletely defined. Finally, future studies need to define the role of certain members of the IGF family in cancer treatment and their relationship to cancer prognosis.

In summary, observational studies have both generated new hypotheses and helped confirm laboratory findings relating the IGF system to cancer risk. With growing evidence, future intervention studies that investigate potential measures for prevention, diagnosis, and treatment of cancer may be warranted and could ultimately have an important impact on cancer incidence.

Acknowlegment

This research was supported in part by National Cancer Institute Grant CA49449.

References

1. Fisher B, Costantino JP, Wickerham DL et al. Tamoxifen for prevention of breast cancer: report of the National Surgical Adjuvant Breast and Bowel Project P-1 Study. J Natl Cancer Inst 1998; 90:1371-88.
2. Goodman-Gruen D, Barrett-Connor E. Epidemiology of insulin-like growth factor-I in elderly men and women. The Rancho Bernardo Study. Am J Epidemiol 1997; 145:970-76.
3. Muti PC, Quattrin T, Misciagna G et al. Fasting serum glucose and insulin and subsequent risk of breast cancer in premenopausal and postmenopausal women; the ORDET Study. American Association for Cancer Research, 2001.
4. Minuto F, Del Monte P, Barreca A et al. Evidence for an increased somatomedin-C/insulin-like growth factor I content in primary human lung tumors. Cancer Res 1986; 46:985-88.
5. Yakar S, Green J, LeRoith D. Serum IGF-1 levels as a risk marker in mammary tumors. Denver: Pro ENDO 2001, 2001:P1-242.
6. Wu Y, Yakar S, LeRoith D. Colon cancer growth and metastasis is dependent on circulating IGF-1 levels. Denver: Pro ENDO 2001, 2001:P1-197.
7. Jones JI, Clemmons DR. Insulin-like growth factor binding proteins and their role in controlling IGF actions. Endocrinol Rev 1995; 16:3-34.
8. Oh Y, Muller HL, Lamson G et al. Insulin-like growth factor (IGF)-independent action of IGF-binding protein-3 in Hs578T human breast cancer cells. Cell surface binding and growth inhibition. Biol Chem 1993; 268:14964-71.
9. Singh P, Rubin N. Insulinlike growth factors and binding proteins in colon cancer. Gastroenterology 1993; 105:1218-37.
10. Jones JI, Clemmons DR. Insulin-like growth factors and their binding proteins: biological actions. Endocr Rev 1995; 16:3-34.
11. Kaicer EK, Blat C, Harel L. IGF-I and IGF-binding proteins: stimulatory and inhibitory factors secreted by human prostatic adenocarcinoma cells. Growth Factors 1991; 4:231-37.
12. Stiles AD, D'Ercole AJ. The insulin-like growth factors and the lung. Am J Respir Cell Mol Biol 1990; 3:93-100.
13. Wright AD, Hill DM, Lowy C et al. Mortality in acromegaly. Q J Med 1970; 39:1-16.
14. Bengtsson BA, Eden S, Ernest I et al. Epidemiology and long-term survival in acromegaly. A study of 166 cases diagnosed between 1955 and 1984. Acta Med Scand 1988; 223:327-35.
15. Colao A, Marzullo P, Ferone D et al. Prostatic hyperplasia: an unknown feature of acromegaly. J Clin Endocrinol Metab 1998; 83:775-9.
16. Kaltsas GA, Mukherjee JJ, Jenkins PJ et al. Menstrual irregularity in women with acromegaly. J Clin Endocrinol Metab 1999; 84:2731-5.
17. Katznelson L, Kleinberg D, Vance ML et al. Hypogonadism in patients with acromegaly: data from the multi-centre acromegaly registry pilot study. Clin Endocrinol 2001; 54:183-88.
18. Escobar-Morreale HF, Serrano-Gotarredona J, Garcia-Robles R et al. Abnormalities in the serum insulin-like growth factor-1 axis in women with hyperandrogenism. Fertil Steril 1998; 70:1090-100.

19. Delhougne B, Deneux C, Abs R et al. The prevalence of colonic polyps in acromegaly: a colonoscopic and pathological study in 103 patients. J Clin Endocrinol Metab 1995; 80:3223-6.
20. Ezzat S, Strom C, Melmed S. Colon polyps in acromegaly. Ann Intern Med 1991; 114:754-5.
21. Jenkins PJ, Fairclough PD, Richards T et al. Acromegaly, colonic polyps and carcinoma. Clin Endocrinol (Oxf) 1997; 47:17-22.
22. Orme SM, McNally RJ, Cartwright RA et al. Mortality and cancer incidence in acromegaly: a retrospective cohort study. United Kingdom Acromegaly Study Group. J Clin Endocrinol Metab 1998; 83:2730-4.
23. Terzolo M, Tappero G, Borretta G et al. High prevalence of colonic polyps in patients with acromegaly. Influence of sex and age. Arch Intern Med 1994; 154:1272-6.
24. Jenkins PJ, Frajese V, Jones AM et al. Insulin-like growth factor I and the development of colorectal neoplasia in acromegaly. J Clin Endocrinol Metab 2000; 85:3218-21.
25. Fehmann HC, Goke B. [Intestinal cell proliferation, acromegaly and colon tumors]. Z Gastroenterol 1997; 35:305-6.
26. Jenkins PJ, Besser M. Acromegaly and cancer: a problem. J Clin Endocrinol Metab 2001; 86:2935-41.
27. Harrela M, Koistinen H, Kaprio J et al. Genetic and environmental components of interindividual variation in circulating levels of IGF-I, IGF-II, IGFBP-1, and IGFBP-3. J Clin Invest 1996; 98:2612-15.
28. Juul A, Main K, Blum WF et al. The ratio between serum levels of insulin-like growth factor (IGF)-I and the IGF binding proteins (IGFBP-1, 2 and 3) decreases with age in healthy adults and is increased in acromegalic patients. Clin Endocrinol 1994; 41:85-93.
29. Giovannucci E, Pollak MN, Platz EA et al. A prospective study of plasma insulin-like growth factor-1 and binding protein-3 and risk of colorectal neoplasia in women. Cancer Epidemiol Biomarkers Prev 2000; 9:345-9.
30. Kaklamani VG, Linos A, Kaklamani E et al. Dietary fat and carbohydrates are independently associated with circulating insulin-like growth factor 1 and insulin-like growth factor-binding protein 3 concentrations in healthy adults. J Clin Oncol 1999; 17:3292-98.
31. Miller BA, Kolonel LN, Bernstein L et al. Racial/Ethnic Patterns of Cancer in the United States 1988-1992. Bethesda: National Cancer Institute, 1996:NIH96-4101.
32. Platz EA, Pollak MN, Rimm EB et al. Racial variation in insulin-like growth factor-1 and binding protein-3 concentrations in middle-aged men. Cancer Epidemiol Biomarkers Prev 1999; 8:1107-10.
33. Tricoli JV, Winter DL, Hanlon AL et al. Racial differences in insulin-like growth factor binding protein-3 in men at increased risk of prostate cancer. Urology 1999; 54:178-82.
34. Jernstrom H, Chu W, Vesprini D et al. Genetic factors related to racial variation in plasma levels of insulin-like growth factor-1: implications for premenopausal breast cancer risk. Mol Gen Metab 2001; 72:144-54.
35. Holmes MD, Pollak MN, Hankinson SE. Lifestyle predictors of insulin-like growth factors. 2001; in press.
36. Vaessen N, Heutink P, Janssen JA et al. A polymorphism in the gene for IGF-I: functional properties and risk for type 2 diabetes and myocardial infarction. Diabetes 2001; 50:637-42.
37. Signorello LB, Kuper H, Lagiou P et al. Lifestyle factors and insulin-like growth factor 1 levels among elderly men. Eur J Cancer Prev 2000; 9:173-8.
38. Suga K, Imai K, Eguchi H et al. Molecular significance of excess body weight in postmenopausal breast cancer patients, in relation to expression of insulin-like growth factor I receptor and insulin-like growth factor II genes. Jpn J Cancer Res 2001; 92:127-34.
39. Campagnoli C, Biglia N, Altare F et al. Differential effects of oral conjugated estrogens and transdermal estradiol on insulin-like growth factor 1, growth hormone and sex hormone binding globulin serum levels. Gynecol Endocrinol 1993; 7:251-58.
40. Goodman-Gruen D, Barrett-Connor E. Effect of replacement estrogen on insulin-like growth factor-I in postmenopausal women: the Rancho Bernardo Study. J Clin Endocrinol Metab 1996; 81:4268-71.
41. Thissen J, Ketelsleger J, Underwood L. Nutritional regulation of the insulin-like growth factors. Endocr Rev 1994; 15:80-101.
42. Holmes MD, Pollak M, Willett WC et al. Dietary predictors of insulin-like growth factors. 2001; in press.
43. Mucci LA, Tamimi R, Lagiou P et al. Are dietary influences on the risk of prostate cancer mediated through the insulin-like growth factor system? Br J Urol Int 2001; 87:814-20.
44. Tuohimaa P, Lyakhovich A, Aksenov N et al. Vitamin D and prostate cancer. J Steroid Biochem Mol Biol 2001; 76:125-34.
45. Martin JL, Pattison SL. Insulin-like growth factor binding protein-3 is regulated by dihydrotestosterone and stimulates deoxyribonucleic aicd synthesis and cell proliferation in LNCaP prostate carcinoma cells. Endocrinology 2000; 141:2401-09.
46. Dresner Pollack R, Rachmilewitz E, Blumenfeld A et al. Bone mineral metabolism in adults with beta-thalassaemia major and intermedia. Br J Haematol 2000; 111:902-07.
47. Habuchi T, Liqing Z, Suzuki T et al. Increased risk of prostate cancer and benign prostatic hyperplasia associated with a CYP17 gene polymorphism with a gene dosage effect. Cancer Res 2000; 60:5710-13.
48. Habuchi T, Suzuki T, Sasaki R et al. Association of vitamin D receptor gene polymorphism with prostate cancer and benign prostatic hyperplasia in a Japanese population. Cancer Res 2000; 60:305-08.
49. Ma J, Giovannucci E, Pollak M et al. Milk intake, circulating levels of insulin-like growth factor-1, and risk of colorectal cancer in men. J Natl Cancer Inst 2001; 93:1330-36.
50. Heaney RP, McCarron DA, Dawson-Hughes B et al. Dietary changes favorably affect bone remodeling in older adults. J Am Diet Assoc 1999; 99:1228-33.

51. Ng ST, Zhou J, Adesanya OO et al. Growth hormone treatment induces mammary gland hyperplasia in aging primates. Nat Med 1997; 3:1141-44.
52. Yang XF, Beamer WG, Huynh H et al. Reduced growth of human breast cancer xenografts in hosts homozygous for the lit mutation. Cancer Res 1996; 56:1509-11.
53. Bates P, Fisher R, Ward A et al. Mammary cancer in transgenic mice expressing insulin-like growth factor II (IGF-II). Br J Cancer 1995; 72:1189-93.
54. Shang Y, Baumrucker CR, Green MH. Signal relay by retinoic acid receptors alpha and beta in the retinoic acid-induced expression of insulin-like growth factor-binding protein-3 in breast cancer cells. J Biol Chem 1999; 274:18005-10.
55. Byrne C, Colditz GA, Willett WC et al. Plasma insulin-like growth factor (IGF) I, IGF-binding protein 3, and mammographic density. Cancer Res 2000; 60:3744-8.
56. Guo YP, Martin LJ, Hanna W et al. Growth factors and stromal matrix proteins associated with mammographic densities. Cancer Epidemiol Biomarkers Prev 2001; 10:243-8.
57. Byrne C, Schairer C, Wolfe J et al. Mammographic features and breast cancer risk: effects with time, age, and menopause status. Journal of the National Cancer Institute 1995; 87:1622-29.
58. Peyrat JP, Bonneterre J, Hecquet B et al. Plasma insulin-like growth factor-1 (IGF-1) concentrations in human breast cancer. Europ J Cancer 1993; 29A:492-97.
59. Bohlke K, Cramer DW, Trichopoulos D et al. Insulin-like growth factor-I in relation to premenopausal ductal carcinoma in situ of the breast. Epidemiology 1998; 9:570-3.
60. Del Giudice ME, Fantus IG, Ezzat S et al. Insulin and related factors in premenopausal breast cancer risk. Breast cancer Res Treat 1998; 47:111-20.
61. Mantzoros CS, Bolhke K, Moschos S et al. Leptin in relation to carcinoma in situ of the breast: a study of pre- menopausal cases and controls. Int J Cancer 1999; 80:523-6.
62. Petridou E, Papadiamantis Y, Markopoulos C et al. Leptin and insulin growth factor I in relation to breast cancer (Greece). Cancer Causes Control 2000; 11:383-8.
63. Li BD, Khosravi MJ, Berkel HJ et al. Free insulin-like growth factor-I and breast cancer risk. Int J Cancer 2001; 91:736-9.
64. Jernstrom H, Barrett-Connor E. Obesity, weight change, fasting insulin, proinsulin, C-peptide, and insulin-like growth factor-1 levels in women with and without breast cancer: the Rancho Bernardo Study. J Womens Health Gender-Based Med 1999; 8:1265-72.
65. Ng EH, Ji CY, Tan PH, Lin V et al. Altered serum levels of insulin-like growth-factor binding proteins in breast cancer patients. Ann Surg Oncol 1998; 5:194-201.
66. Holdaway IM, Mason BH, Lethaby AE et al. Serum levels of insulin-like growth factor binding protein-3 in benign and malignant breast disease. Aust N Z J Surg 1999; 69:495-500.
67. Agurs-Collins T, Adams-Campbell LL, Kim KS et al. Insulin-like growth factor-1 and breast cancer risk in postmenopausal African-American women. Cancer Detect Prev 2000; 24:199-206.
68. Bruning PF, Van Doorn J, Bonfrer JM et al. Insulin-like growth-factor-binding protein 3 is decreased in earlystage operable pre-menopausal breast cancer. Int J Cancer 1995; 62:266-70.
69. Hankinson SE, Willett WC, Colditz GA et al. Circulating concentrations of insulin-like growth factor-I and risk of breast cancer. Lancet 1998; 351:1393-96.
70. Toniolo P, Bruning PF, Akhmedkhanov A et al. Serum insulin-like growth factor-I and breast cancer. Int J Cancer 2000; 88:828-32.
71. Ruan W, Catanese V, Wieczorek R et al. Estradiol enhances the stimulatory effect of insulin-like growth factor-I (IGF-I) on mammary development and growth hormone-induced IGF-I messenger ribonucleic acid. Endocrinology 1995; 136:1296-302.
72. Cohen P, Peehl DM, Graves HC et al. Biological effects of prostate specific antigen as an insulin-like growth factor binding protein-3 protease. J Endocrinol 1994; 142:407-15.
73. Pietrzkowski Z, Wernicke D, Porcu P et al. Inhibition of cellular proliferation by peptide analogues of insulin-like growth factor 1. Cancer Res 1992; 52:6447-51.
74. Cohen P, Graves HC, Peehl DM et al. Prostate-specific antigen (PSA) is an insulin-like growth factor binding protein-3 protease found in seminal plasma. J Clin Endocrinol Metab 1992; 75:1046-53.
75. Shaneyfelt T, Husein R, Bubley G et al. Hormonal predictors of prostate cancer: a meta-analysis. J Clin Oncol 2000; 18:847-53.
76. Kurek R, Tunn UW, Eckart O et al. The significance of serum levels of insulin-like growth factor-1 in patients with prostate cancer. Br J Urol Int 2000; 85:125-9.
77. Kanety H, Madjar Y, Dagan Y et al. Serum insulin-like growth factor-binding protein-2 (IGFBP-2) is increased and IGFBP-3 is decreased in patients with prostate cancer: correlation with serum prostate-specific antigen. J Clin Endocrinol Metab 1993; 77:229-33.
78. Cohen P, Peehl DM, Stamey TA et al. Elevated levels of insulin-like growth factor-binding protein-2 in the serum of prostate cancer patients. J Clin Endocrinol Metab 1993; 76:1031-5.
79. Tellez Martinez-Fornes M, Balsa J, Maganto Pavon E et al. [Insulin growth factor I (IGF-I) in normal, hyperplastic, and tumor prostatic tissue [Spanish]. Acta Urol Esp 1996; 20:409-13.
80. Ho PJ, Baxter RC. Insulin-like growth factor-binding protein-2 in patients with prostate carcinoma and benign prostatic hyperplasia. Clin Endocrinol (Oxf) 1997; 46:333-42.
81. Mantzoros CS, Tzonou A, Signorello LB et al. Insulin-like growth factor 1 in relation to prostate cancer and benign prostatic hyperplasia. Br J Cancer 1997; 76:1115-8.
82. Wolk A, Mantzoros CS, Andersson SO et al. Insulin-like growth factor 1 and prostate cancer risk: a population- based, case-control study. J Natl Cancer Inst 1998; 90:911-5.

83. Khosravi J, Diamandi A, Mistry J et al. Insulin-like growth factor I (IGF-I) and IGF-binding protein-3 in benign prostatic hyperplasia and prostate cancer. J Clin Endocrinol Metab 2001; 86:694-9.
84. Chokkalingam AP, Pollak M, Fillmore CM et al. Insulin-like growth factors and prostate cancer: a population-based case-control study in China. Cancer Epidemiol Biomarkers Prev 2001; 10:421-7.
85. Cutting CW, Hunt C, Nisbet JA et al. Serum insulin-like growth factor-1 is not a useful marker of prostate cancer. Br J Urol Int 1999; 83:996-9.
86. Djavan B, Bursa B, Seitz C et al. Insulin-like growth factor 1 (IGF-1), IGF-1 density, and IGF-1/PSA ratio for prostate cancer detection. Urology 1999; 54:603-6.
87. Finne P, Auvinen A, Koistinen H et al. Insulin-like growth factor I is not a useful marker of prostate cancer in men with elevated levels of prostate-specific antigen. J Clin Endocrinol Metab 2000; 85:2744-7.
88. Koliakos G, Chatzivasiliou D, Dimopoulos T et al. The significance of PSA/IGF-1 ratio in differentiating benign prostate hyperplasia from prostate cancer. Dis Markers 2000; 16:143-6.
89. Chan JM, Stampfer MJ, Giovannucci E et al. Plasma insulin-like growth factor-I and prostate cancer risk: a prospective study. Science 1998; 279:563-6.
90. Harman SM, Metter EJ, Blackman MR et al. Serum levels of insulin-like growth factor I (IGF-I), IGF-II, IGF- binding protein-3, and prostate-specific antigen as predictors of clinical prostate cancer. J Clin Endocrinol Metab 2000; 85:4258-65.
91. Stattin P, Bylund A, Rinaldi S et al. Plasma insulin-like growth factor-I, insulin-like growth factor-binding proteins, and prostate cancer risk: a prospective study. J Natl Cancer Inst 2000; 92:1910-7.
92. Lacey JVJ, Hsing AW, Fillmore CM et al. Null association between insulin-like growth factors, insulin-like growth factor-binding proteins, and prostate cancer in a prospective study. Cancer Epidemiol, Biomarkers Prev 2001; 10:1101-02.
93. Chan JM, Stampfer M, Giovannucci E et al. Insulin-like growth factor I (IGF-I), IGF-binding protein-3 and prostate cancer risk: epidemiological studies. J Natl Cancer Inst 2002; in press.
94. Pollak MN, Perdue JF, Margolese RG et al. Presence of somatomedin receptors on primary human breast and colon carcinomas. Cancer Lett 1987; 38:223-30.
95. Lahm H, Suardet L, Laurent PL et al. Growth regulation and co-stimulation of human colorectal cancer cell lines by insulin-like growth factor I, II and transforming growth factor alpha. Br J Cancer 1992; 65:341-46.
96. Remacle-Bonnet M, Garrouste F, el Atiq F et al. des-(1-3)-IGF-I, an insulin-like growth factor analog used to mimic a potential IGF-II autocrine loop, promotes the differentiation of human colon-carcinoma cells. Int J Cancer 1992; 52:910-17.
97. Lahm H, Amstad P, Wyniger J et al. Blockade of the insulin-like growth-factor-I receptor inhibits growth of human colorectal cancer cells: evidence of a functional IGF-II-mediated autocrine loop. Int J Cancer 1994; 58:452-59.
98. Ma J, Pollak MN, Giovannucci E et al. Prospective study of colorectal cancer risk in men and plasma levels of insulin-like growth factor (IGF)-I and IGF-binding protein-3. J Natl Cancer Inst 1999; 91:620-5.
99. Kaaks R, Toniolo P, Akhmedkhanov A et al. Serum C-peptide, insulin-like growth factor (IGF)-I, IGF-binding proteins, and colorectal cancer risk in women. J Natl Cancer Inst 2000; 92:1592-600.
100. Manousos O, Souglakos J, Bosetti C et al. IGF-I and IGF-II in relation to colorectal cancer. Int J Cancer 1999; 83:15-7.
101. Kaaks R, Rinaldi S, Lukanova A et al. Correspondence re: Giovannucci et al., A Prospective Study of Plasma Insulin-like Growth Factor-1 and Binding protein-3 and Risk of Colorectal Neoplasia in Women. Cancer Epidemiol Biomark Prev 2000; 9:345-349. Cancer Epidemiol Biomarkers Prev 2001; 10:1103-04.
102. Renehan AG, Painter JE, Atkin WS et al. High-risk colorectal adenomas and serum insulin-like growth factors. Br J Surg 2001; 88:107-13.
103. Werner H, LeRoith D. The role of the insulin-like growth factor system in human cancer. Adv Cancer Res 1996; 68:183-223.
104. Macaulay VM, Everard MJ, Teale JD et al. Autocrine function for insulin-like growth factor I in human small cell lung cancer cell lines and fresh tumor cells. Cancer Res 1990; 50:2511-17.
105. Nakanishi Y, Mulshine JL, Kasprzyk PG et al. Insulin-like growth factor-I can mediate autocrine proliferation of human small cell lung cancer cell lines in vitro. J Clin Invest 1988; 82:354-59.
106. Lukanova A, Toniolo P, Akhmedkhanov A et al. A prospective study of insulin-like growth factor-I, IGF-binding proteins-1, -2 and -3 and lung cancer risk in women. Int J Cancer 2001; 92:888-92.
107. Yu H, Spitz MR, Mistry J et al. Plasma levels of insulin-like growth factor-I and lung cancer risk: a case-control analysis. J Nat Cancer Inst 1999; 91:151-56.
108. Wu X, Yu H, Amos CI et al. Joint effect of insulin-like growth factors and mutagen sensitivity in lung cancer risk. Growth hormone IGF Res 2000; 10(Suppl A):S26-S27.
109. Broomhall J, May R, Lilleyman JS et al. Height and lymphoblastic leukaemia. Arch Dis Child 1983; 58:300-01.
110. Crofton PM, Ahmed SF, Wade JC et al. Effects of intensive chemotherapy on bone and collagen turnover and the growth hormone axis in children with acute lymphoblastic leukemia. J Clin Endocrinol Metab 1998; 83:3121-29.
111. Mohnike KL, Kluba U, Mittler U et al. Serum levels of insulin-like growth factor-I, -II and insulin-like growth factor binding proteins -2 and -3 in children with acute lymphoblastic leukaemia. Eur J Pediatr 1996; 155:81-86.
112. Arguelles B, Barrios V, Pozo J et al. Modifications of growth velocity and the insulin-like growth factor system in children with acute lymphoblastic leukemia: a longitudinal study. J Clin Endocrinol Metab 2000; 85:4087-92.

113. Petridou E, Dessypris N, Spanos E et al. Insulin-like growth factor-I and binding protein-3 in relation to childhood leukaemia. Int J Cancer 1999; 80:494-96.
114. Hasegawa T, Hasegawa Y, Takada M et al. The free form of insulin-like growth factor I increases in circulation during normal human pregnancy. J Clin Endocrinol Metab 1995; 80:3284-86.
115. van der Laan BF, Freeman JL, Asa SL. Expression of growth factors and growth factor receptors in normal and tumorous human thyroid tissues. Thyroid 1995; 5:67-73.
116. Vannelli GB, Barni T, Modigliani U et al. Insulin-like growth factor-I receptors in nonfunctioning thyroid nodules. J Clin Endocrinol Metab 1990; 71:1175-82.
117. Vella V, Sciacca L, Pandini G et al. The IGF system in thyroid cancer: new concepts. Mol Pathol 2001; 54:121-24.
118. Takahashi MH, Thomas GA, Williams ED. Evidence for mutual interdependence of epithelium and stromal lymphoid cells in a subset of papillary carcinomas. Br J Cancer 1995; 72:813-17.
119. Maiorano E, Ciampolillo A, Viale G et al. Insulin-like growth factor 1 expression in thyroid tumors. Appl Immunohistochem Mol Morphol 2000; 8:110-19.
120. Gong Y, Cui L, Minuk GY. The expression of insulin-like growth factor binding proteins in human hepatocellular carcinoma. Mol Cell Biochem 2000; 207:101-04.
121. Stuver SO, Kuper H, Tzonou A et al. Insulin-like growth factor 1 in hepatocellular carcinoma and metastatic liver cancer in men. Int J Cancer 2000; 87:118-21.
122. Maiorano E, Loverro G, Viale G et al. Insulin-like growth factor-I expression in normal and diseased endometrium. Int J Cancer 1999; 80:188-93.
123. Lee DY, Yang DH, Kang CW et al. Serum insulin-like growth factors (IGFs) and IGF binding protein (IGFBP)-3 in patients with gastric cancer: IGFBP-3 protease activity induced by surgery. J Korean Med Sci 2000; 12:32-39.
124. Mathur SP, Mathur RS, Young RC. Cervical epidermal growth factor-receptor (EGF-R) and serum insulin-like growth factor II (IGF-II) levels are potential markers for cervical cancer. Am J Reproductive Immunol 2000; 44:222-30.
125. Song BC, Chung YH, Kim JA et al. Association between insulin-like growth factor-2 and metastases after transcatheter arterial chemoembolization in patients with hepatocellular carcinoma: a prospective study. Cancer 2001; 91:2386-93.
126. Karasik A, Menczer J, Pariente C et al. Insulin-like growth factor-I (IGF-I) and IGF-binding protein-2 are increased in cyst fluids of epithelial ovarian cancer. J Clin Endocrinol Metab 1994; 78:271-76.
127. Deal C, Ma J, Wilkin F et al. Novel promotor polymorphism in imsulin-like growt factor-binding protein-3: correlation with serum levels and interaction with known regulators. J Clin Endocrinol Metab 2001; 86:1274-80.

CHAPTER 20

IGF-I, Insulin and Cancer Risk:
Linking the Biology and the Epidemiology

Michael Pollak

Introduction

Insulin and insulin-like growth factor signaling systems arose early in evolution and play key roles in regulating cellular proliferation and survival, energy utilization at both the cellular and whole organism levels, and body size and longevity. In higher organisms, more specific regulatory roles have evolved in the context of the physiology of specific tissues, including, for example, control of ovarian follicle development. These broad areas have been the subject of recent reviews (for example, refs. 1-5).

Recent laboratory and epidemiologic research has led to increased interest in the relationship of insulin and IGF signaling to neoplasia, and to the related issue of aging at the cellular and whole-organism levels.[6,7] Broadly speaking, this research involves on one hand efforts to identify and therapeutically exploit novel molecular targets for cancer treatment, and on the other hand the characterization of novel cancer risk factors with potential implication for risk reduction strategies. This review will summarize current concepts concerning the biological basis for the relationship between circulating IGF-I levels and cancer risk observed in recent epidemiologic research. The reader is referred to the chapter by Hankinson and Scherhammer in this volume[8] for a summary of the epidemiologic data, and to other recent reviews for information regarding IGF research relevant to cancer treatment[9,10] and to relevant aging research.[6,7]

It has long been appreciated that there is substantial inter-individual variation in circulating levels of IGF-I and IGFBP-3 in human populations.[11,12] There is evidence that the circulating concentration of IGF-I and IGFBP-3, its major binding protein, are determined by both genetic and non-genetic factors;[13] these determinants are being defined in ongoing research (for example, refs. 14-19). The measurement of circulating levels of IGF-I and/or IGFBP-3 to aid in the diagnosis of growth hormone deficiency and acromegaly is well established in clinical endocrinology. Recent results have challenged the traditional view that there is no biological or medical significance to the inter-individual variation in IGF-I and IGFBP-3 levels that fall within the broad normal range. Dozens of reports (for example, refs. 20-26), recently reviewed[8,27-30] have provided evidence that risk of a subsequent diagnosis of certain cancers varies directly with IGF-I levels, and there is also evidence linking risk of cardiac disease inversely to these levels.[31]

Variation of Cancer Risk according to Circulating Level of IGF-I

Are the Epidemiological Results Reproducible?

Initial prospective studies (for example, ref. 20) linking IGF-I levels within the broad normal range to cancer risk were greeted with both interest and skepticism. Subsequent prospective studies from different laboratories performed on different populations (for example, refs. 25,26) provided further support for the hypothesis. Nevertheless, not all studies have detected correlations between IGF-I levels and cancer risk, and much work remains to be done to confirm and better describe the relationships.

Insulin-Like Growth Factors, edited by Derek LeRoith, Walter Zumkeller
and Robert Baxter. ©2003 Eurekah.com and Kluwer Academic / Plenum Publishers.

Inconsistencies may result from technical and/or biological factors. It is clear that technical factors such as suboptimal storage of serum samples and/or assay imprecision could easily result in false negative conclusions. On the other hand, relations observed in certain populations may in fact be absent in other populations due to interactions that are currently incompletely described. The strength of the relationship may be modified by racial, dietary, body mass index, or other factors. For example, oral estrogen administration is known to reduce circulating levels.[32,33] It is unclear if the resulting concentrations have the same biological significance as an individual's original levels: if not, estrogen use might obscure confound any relationship between IGF-I level and risk.

Some confusion has arisen due to confusion between the concepts of 'tumor marker' and 'risk factor'. There is evidence from many investigators that circulating IGF-I level does not have the characteristics of a tumor marker (for example, ref. 34). Indeed, patients with advanced cancer who are in negative nitrogen balance tend to have IGF-I levels lower than healthy controls. However, this in no way provides evidence against the hypothesis that higher levels are associated with increased risk for a future cancer diagnosis among healthy subjects. An analogy may be drawn to cholesterol levels in the context of risk of heart disease: elevated levels of cholesterol are associated with an increased probability of a future cancer diagnosis, but one cannot use a cholesterol level to determine if a patient with chest pain is having a myocardial infarction.

Are the Epidemiological Results Plausible?

Epidemiologic observations are more convincing if they make biologic sense. Does the relationship of IGF-I level to cancer risk pass this test? The a priori prediction that IGF-I levels would relate to cancer risk was clearly speculative. However, in retrospect, the positive relationship between IGF-I level and cancer risk seems plausible based on data from both the laboratory and the clinic. These data are circumstantial in the sense that formal cause-and-effect relationships are usually not demonstrated, yet in each example the data are at least consistent with a positive relation between IGF-I level and cancer risk.

Laboratory Models

Many animal carcinogenesis models have yielded data in keeping with the epidemiologic results, in that cancer risk and/or behavior is influenced by experimental manipulation of IGF-I levels and/or signaling. Examples include:

1. the demonstration of increased prostate cancer risk in mice with increased IGF-I gene expression in the prostate,[35]
2. the demonstration of decreased number of carcinogen-induced mammary gland cancers in transgenic mice that express a growth hormone antagonist, resulting in decreased IGF-I levels,[36]
3. decreased neoplastic progression of colon cancer in mice genetically manipulated to have reduced hepatic IGF-I expression and circulating levels,[37]
4. reduced proliferation of human breast cancer xenografts[38] or human osteosarcoma xenografts[39] in mice homozygous for the lit mutation, which is associated with reduced circulating IGF-I levels,[38] and
5. the demonstration that the classic protective effect of decreased caloric intake in rodent chemical carcinogenesis models is correlated with the suppressive effect of decreased intake on IGF-I levels, and reversed by IGF-I supplementation.[40]

An interesting non-laboratory model concerns osteosarcoma risk in dogs. IGF-I levels are higher in larger breeds[41,42] and these breeds are much more susceptible to this neoplasm.[43,44] Osteosarcoma in humans also may be IGF-I responsive.[39,45]

Clinical Data

Circumstantial clinical data consistent with the epidemiologic observations comes from evidence that several previously identified cancer risk factors for which no clear mechanism was known are each related to higher IGF-I levels. This implies that these risk factors may at least in part be surrogates for IGF-related risk. High birth weight has been associated with increased cancer risk,[46-51] and is umbilical cord blood IGF-I level is positively correlated with birthweight.[52-54] Taller individuals have slightly increased risk,[55] and height in adolescence is correlated with IGF-I levels.[11] Parity is associated with reduced risk of breast cancer and lower IGF-I levels,[18] and mammographic density, an important breast cancer risk factor, is positively correlated with circulating IGF-I level.[56]

What about Insulin, IGF-II, and IGF Binding Proteins?

Despite recent progress, we believe the relationship between circulating IGF-I level and cancer risk remains incompletely described and requires further confirmation. We are at even an earlier stage with respect to investigation of possible relations between insulin, IGF-II, and IGF binding proteins and cancer risk.

Recent data provide support for the view that insulin resistance is associated with higher cancer risk (for example, ref. 57), and poorer prognosis (for example, ref. 58). The syndrome of insulin resistance is complex, with many metabolic alterations, besides elevated insulin levels. If insulin resistance is shown in future research to indeed be associated with elevated risk of neoplasia, it will be of interest to determine if this is pathophysiologically related to elevated insulin levels, or to other metabolic abnormalities.

Circulating IGF-II levels have not been associated with risk in most epidemiologic studies, which is perhaps surprising as IGF-II concentration in the circulation is higher than that of IGF-I, and IGF-II is a ligand for the IGF-I receptor. One potential explanation is that serum IGF-II levels are poorly correlated with IGF-II levels or bioactivity at the tissue level. Another proposal relates to the IGF-II receptor, which appears to be a decoy receptor that binds IGF-II but does not transducer a mitogenic signal. The presence of this receptor may serve to attenuate IGF-II action. Interestingly, there is evidence that the IGF-II receptor has properties of a tumor suppressor gene, in that overexpression tends to reduce proliferation.[59,60] A recent paper[61] provides early evidence that loss of imprinting of the IGF-II gene represents an important risk factor for colorectal cancer. While this important observation requires confirmation, it may imply that IGF-II expression at the tissue level is indeed related to risk, even if circulating levels are not.

Data concerning IGF binding proteins and risk are inconsistent. Early observations[20] suggested that IGFBP-3 levels are inversely related to risk. This was a remarkable observation, as IGF-I and IGFBP-3 were in general correlated with each other, but oppositely related to risk. The risk relationship arose due to influence of 'outlier' individuals who have unusually high or low IGF-I to IGFBP-3 ratios. Subsequent studies have yielded conflicting results ranging from positive to no relation to an inverse relation. One of the issues that is limiting progress in this area concerns assay methodology. IGFBP-3 is present in the circulation in different glycosylation and phosphorylation states. The circulation contains not only intact IGFBP-3, but also a variety of proteolytic fragments, some of which retain IGF binding properties. Current assays are incompletely characterized with respect to their specificity. Epidemiology of IGFBP-3 and risk will progress as the measurement tools are refined. Other IGF binding proteins have not been studied with respect to risk. There evidence, however, for a role of IGFBP-2 as a tumor marker in prostate cancer,[62] although of less clinical utility than PSA.

Hypotheses to Explain the Epidemiologic Observations

IGF Signaling and Early Steps of Carcinogenesis: Cellular Renewal Dynamics of At-Risk Tissues

The life span of the individual epithelial cells of organs such as the colon, the breast, and the prostate is far shorter than that of the organism, usually only a few days. The control systems that regulate the balance between proliferation and apoptosis in these tissues are incompletely understood, but there is evidence that IGF signaling is one of several control systems involved.

Most concepts of stepwise carcinogenesis view sequential acquisition of genetic damage due to somatic cell mutations as an underlying principle. Obvious variables involved include the rate of mutations and the efficacy of DNA repair. Also important is the probability of survival of cells with a few genetic 'hits'. If such cells undergo apoptosis, the carcinogenic process is deleted. If they survive, they remain possible targets for further mutational damage that could lead to full transformation. Given the potent cell-survival signal conferred by activation of the IGF-I receptor, it is reasonable to speculate that even small differences in availability of ligands for the receptor at the tissue level might influence this process. This hypothesis assumes that the influence may be small; yet with millions of cell divisions per second over decades of life, even minute differences in probability of survival of damaged cells could influence stepwise carcinogenesis and explain the epidemiologic observations. This model represents an extension of prior speculation that emphasized cellular proliferation per se, as distinct from survival, as a key factor in the cellular mechanisms contributing to carcinogenesis.[63]

Of course, the subtle differences in proliferation and apoptosis rates between individuals hypothesized to be attributable to differences in 'tuning' of the IGF-I control system are only one of many factors that influence carcinogenesis. One topic for future research concerns the huge difference in risk of neoplasia between different tissues of an individual, as distinct from differences between individuals. The best example of this perhaps is the enormous difference in small bowel and colon cancer risk. Presumably host hormonal influences are identical. Both organs have huge rates of cell renewal. It is possible that the difference relates to different rates of carcinogen exposure or DNA repair, but the possibility that subtle differences in cell renewal regulatory systems exist between the small and large bowel has not been excluded.

IGF Signaling and Neoplastic Progression

The above hypothesis links IGF-I signaling to the epidemiologic observations concerning IGF-I levels and cancer risk at the level of early carcinogenesis. A separate model to account for the epidemiologic observations concerns IGF influences on neoplastic progression rather than on carcinogenesis. The models are not mutually exclusive.

Neoplastic progression refers to the process by which cancers have a tendency to become more aggressive with time, exhibiting increased invasiveness, higher proliferation rates, and greater propensity to metastasize. The process is likely driven by a Darwinian type of natural selection for rapidly growing cells. If a single cell within a tumor undergoes a somatic mutational event that confers increased growth rate, the resulting clonal population will over time dominate the cancer and determine its clinical characteristics. This is natural selection based on somatic cell genetics within a single organism, which is analogous to but proceeds much more rapidly than the more widely discussed natural selection based on variation of germ line genetics within individuals of a species competing within an ecosystem.

Recent data provides some support for the view that by age 50, most humans harbour some early neoplastic lesions. An often cited example concerns the unexpectedly high rate of prostate neoplasia seen when careful autopsies are performed.[64,65] In view of the data from animal models cited above, as well as the known IGF responsivity of neoplastic cells seen in tissue culture assays, it is possible that subtle differences between individuals with respect to IGF-I levels may influence the rate of neoplastic progression, and hence the probability of a clinical diagnosis of cancer. In contrast to the 'early carcinogenesis' model, the concept here is that individuals with IGF-I levels at the lower end of the normal range do not actually have fewer transforming events. Rather, they have a tendency to undergo neoplastic progression at a slower rate, simply because their proliferation is to a certain extent IGF-dependent, and the concentration of IGF ligands in the tumor microenvironment is limiting. This constraint would lengthen the time between transformation and a clinically detectable cancer, a process that could account for the epidemiologic results.

While these models are plausible and non-mutually exlusive, they remain speculative at this time. Ongoing epidemiologic research as well as laboratory research will provide data that may support these models or suggest others.

How Can Small Differences in Concentration Have Important Consequences?

It is puzzling that the absolute differences in circulating IGF-I concentration that are associated with varying cancer risk are small, often in the range of 10%. Such small concentration differences generally have small or undetectable effects on tissue culture bioassays of IGF-I bioactivity. One current hypothesis concerning this issue is that circulating level represents a surrogate for tissue IGF bioactivity, which may show interpersonal variability that is parallel in direction but larger in magnitude than interpersonal variability in circulating levels. Although technically challenging, ongoing efforts to determine if there is an experimentally detectable relationship between IGF-I level in the circulation and IGF-I receptor activation in vivo should address the issue.

Alternatively, it is conceivable that small differences in fact are associated with differences in proliferation or apoptosis that are biologically important, particularly if one assumes that the differences, though small in magnitude, are present over a lifetime. A conditional knock-out model provides an example of intact autocrine and paracrine IGF-I expression outside the liver, but reduced circulating level.[37] This model demonstrates important differences in neoplastic behavior, indicating that circulating IGF-I may indeed influence carcinogenesis and neoplastic progression, especially for cancers that do not express IGF-I or IGF-II in an autocrine manner.

Challenges

Measurement Issues: Serum Analytes

Recent experience has revealed that there is substantial room for improvement in assay methodology for IGF-I and related analytes in the circulation. Many different assays for these analytes are available from commercial sources and academic laboratories. Most of these were originally designed to distinguish pathologic (acromegalic or growth hormone deficient) states from normal, and clearly are sufficient for this application. However, not all assays have the precision, sensitivity, or long-term stability and lack of drift required for epidemiologic research that seeks to compare individuals whose levels fall within the normal range. Agreement in absolute values of individual samples assayed for IGF-I or IGFBP-3 by different methods is often poor, with differences as high as 30% commonly seen. Variability of results even between different production lots from the same supplier is common. Additional complexities result from the fact that varying storage conditions can interact with the variability related to antibodies employed in immunoassays. For example, methods based on highly specific antibodies and those that utilize less specific antibodies that recognize intact molecules as well as certain cleavage products may yield similar results on freshly obtained or optimately stored specimens, but very different results if specimens have been suboptimately collected or stored. These analytic issues need to be addressed to facilitate future epidemiologic research.

Measurement Issues: Cell Renewal Dynamics

Direct experimental approaches to test the hypothesis that interpersonal variability in circulating IGF-I levels is correlated with interpersonal variability in cell proliferation rates or apoptosis rates in at-risk renewing epithelial cell populations are difficult. This relates to the fact that the hypothesized differences in cell renewal are small enough that they approach the measurement error associated with currently available techniques such as Ki67 or Apotag labeling. There are reports, however, of differences in colonic epithelial proliferation according to IGF-I levels among acromegalic subjects.[66]

Potential Medical Relevance

Medical relevance of the recent observations concerning the relation between IGF-I and related analytes and cancer risk remains speculative at this time, although many investigators are actively exploring specific hypotheses in this area.

Population Attributable Risk

The variation of risk related to IGF-I level is small in magnitude relative to other well-known risks such as tobacco use or mutations in the *BRCA1* gene. However, the proportion of the population with IGF-I levels in the higher part of the normal range is large, particularly in comparison to the *BRCA1* example, so that even a modest increase in risk associated with higher levels might lead to a considerable IGF-I related cancer burden in the population.

Risk Reduction and Intermediate Endpoints for Prevention Strategies

Another issue concerns the possibility that baseline IGF-I levels and/or related analytes might predict the probability of success of certain proposed pharmacologic approaches to risk reduction, including (but not confined to) retinoids, SERMs, and deltanoids, all of which are known to influence IGF physiology.[67-69] It is not suggested that the perturbations of IGF physiology account for all the antiproliferative actions of these compounds. However, it is possible that induction of effects on IGF-related analytes, conveniently measured in serum, may correlate with longer-term efficacy with respect to risk reduction. Current data are insufficient to evaluate the hypothesis that certain individuals (such as *BRCA1* mutation carriers) might benefit from pharmacological efforts to target IGF signaling, either by lowering ligand levels by growth hormone or GHRH antagonists, or by receptor blockade .

Secular Trends

Another speculative issue concerns the possibility that secular trends towards increase in cancer incidence associated with a 'western' lifestyle may somehow relate to secular trends in insulin and/or IGF-I levels. Secular trends in certain countries involve increases in height, BMI (and presumably insulin levels), and cancer risk. It is impossible to evaluate the hypothesis that age-specific IGF-I levels are in general higher now than a century ago, but this is certainly possible, particularly in relationship to changing nutritional factors. If so, it is conceivable that this represents one (of many) factors that account for increasing incidence of cancer.

Growth Hormone Therapy

Growth hormone replacement therapy for growth hormone deficiency represents an important medical advance that offers major benefits to patients. On the other hand, growth hormone supplementation to achieve IGF-I levels higher than age-specific population means is more controversial.[70] The evidence concerning the relationship between cancer risk and IGF-I levels would suggest that a cautious approach to growth hormone therapy is wise, and that efforts should be made to titrate the growth hormone dose to achieve levels near the age-specific mean levels in an appropriate reference population.

Conclusion

There is now substantial data to support the view that individuals with IGF-I levels at the higher end of the normal range have modestly elevated risk of neoplasia than those at the lower end the range. These data are biologically plausible in the context of laboratory models. We hypothesize that the variation in risk relates to variation in epithelial cell renewal dynamics (proliferation rates and apoptosis rates) in at risk tissues. Additional research is required to confirm the relationship between IGF-I and other analytes related to insulin and IGF-I physiology and risk, and to investigate hypothesized mechanisms underlying the relationship.

References

1. Giudice LC. Insulin-like growth factor family in Graafian follicle development and function. J Soc Gynecol Investig 2001; 8(1 Suppl):S26-S29.
2. Firth SM, Baxter RC. Cellular actions of the insulin-like growth factor binding proteins. Endocr Rev 2002; 23:824-54.
3. Nakae J, Kido Y, Accili D. Distinct and overlapping functions of insulin and IGF-I receptors. Endocr Rev 2001; 22:818-35.
4. Khandwala HM, McCutcheon IE, Flyvbjerg A et al. The effects of insulin-like growth factors on tumorigenesis and neoplastic growth. Endocr.Rev 2000; 21(3):215-44.
5. Jones JI, Clemmons DR. Insulin-like growth factors and their binding proteins: biological actions. Endocr Rev 1995; 16(1):3-34.
6. Tatar M, Bartke A, Antebi A. The endocrine regulation of aging by insulin-like signals. Science 2003; 299:1346-51.
7. Longo VD, Finch CE. Evolutionary medicine: from dwarf model systems to healthy centenarians? Science 2003; 299:1342-6.
8. Schernhammer ES, Hankinson SE. Epidemiologic approaches to evaluating insulin-like growth factor and cancer risk. In: LeRoith D, Zumkeller W, Baxter RC, eds. Insulin-like growth factors. Georgetown: Landes Bioscience/Eurekah.com, 2003.
9. De Meyts P, Whittaker J. Structural biology of insulin and IGF-1 receptors: implications for drug design. Nat Rev Drug Discov 2002; 1:769-83.
10. Wang Y, Sun Y. Insulin-like growth factor receptor-1 as an anti-cancer target: blocking transformation and inducing apoptosis. Curr Cancer Drug Targets 2002; 2:191-207.
11. Juul A, Dalgaard P, Blum WF et al. Serum levels of insulin-like growth factor (IGF)-binding protein-3 (IGFBP-3) in healthy infants, children, and adolescents: the relation to IGF-I, IGF-II, IGFBP-1, IGFBP-2, age, sex, body mass index, and pubertal maturation. J Clin Endocrinol Metab 1995; 80(8):2534-42.
12. Juul A, Bang P, Hertel N et al. Serum insulin-like growth factor I in 1030 healthy children, adolescents, and adults: relation to age, sex, stage of puberty, testicular size, and body mass index. J Clin Endocrinol Metab 1994; 78:744-52.

13. Harrela M, Koinstinen H, Kaprio J et al. Genetic and environmental components of interindividual variation in circulating levels of IGF-I, IGF-II, IGFBP-1, and IGFBP-3. J Clin Invest 1996; 98:2612-5.
14. Deal C, Ma J, Wilkin F, Paquette J et al. Novel promoter polymorphism in insulin-like growth factor-binding protein-3: correlation with serum levels and interaction with known regulators. J Clin Endocrinol Metab 2001; 86:1274-80.
15. Jernstrom H, Wilkin F, Deal C et al. Genetic and non-genetic factors associated with variation of plasma levels of insulin-like growth factor-I and insulin-like growth factor binding protein-3 in healthy premenopausal women. Cancer Epidemiol Biomarkers Prev 2001; 10:377-84.
16. Jernstrom H, Chu W, Vesprini D et al. Genetic factors related to racial variation in plasma levels of insulin-like growth factor-I: implications for pre-menopausal breast cancer risk. Mol Genet Metab 2001; 72:144-54.
17. Holmes MD, Pollak MN, Willett WC et al. Dietary correlates of plasma insulin-like growth factor-I and insulin-like growth factor binding protein-3 concentrations. Cancer Epidemiol Biomarkers Prev 2002; 11:852-61.
18. Holmes MD, Pollak MN, Hankinson SE. LIfestyle correlates of plasma insulin-like growth factor-I and insulin-like growth factor binding protein-3 concentrations. Cancer Epidemiol Biomarkers Prev 2002; 11:862-7.
19. Giovannucci E, Pollak M, Liu Y et al. Nutritional predictors of insulin-like growth factor-I and their relationships to cancer in men. Cancer Epidemiol Biomarkers Prev 2003; 12:84-9.
20. Chan JM, Stampfer MJ, Giovannucci E et al. Plasma insulin-like growth factor-I and prostate cancer risk: a prospective study. Science 1998; 279:563-6.
21. Hankinson SE, Willett WC, Colditz GA et al. Circulating concentrations of insulin-like growth factor-I and risk of breast cancer. Lancet 1998; 351:1393-6.
22. Ma J, Pollak M, Giovannucci E. et al. Prospective study of colorectal cancer risk in men and plasma levels of insulin-like growth factor (IGF)-I, and IGF-binding protein-3. JNCI 1999; 91:620-5.
23. Chan JM, Stampfer MJ, Ma J, Gann P, Gaziano JM, Pollak M et al. Insulin-like growth factor-I (IGF-I) and IGF binding protein-3 as predictors of advanced-stage prostate cancer. J Natl Cancer Inst 2002; 94(14):1099-106.
24. Pollak M. Insulin-like growth factors (IGFs) and prostate cancer. Epidemiological Reviews 2001; 23:59-66.
25. Harman SM, Metter EJ, Blackman MR et al. Serum levels of insulin-like growth factor I (IGF-I), IGF-II, IGF-binding protein-3, and prostate-specific antigen as predictors of clinical prostate cancer. J Clin.Endocrinol Metab 2000; 85:4258-65.
26. Stattin P, Bylund A, Inaldi S et al. Plasma insulin-like growth factor-I, insulin-like growth factor-binding proteins, and prostate cancer risk: a prospective study. J Natl Cancer Inst 2000; 92:1910-7.
27. Moschos SJ, Mantzoros CS. The role of the IGF system in cancer: from basic to clinical studies and clinical applications. Oncol 2002; 63:317-32.
28. Yu H, Rohan T. Role of the insulin-like growth factor family in cancer development and progression. J Natl Cancer Inst 2000; 92:1472-89.
29. Giovannucci E. Insulin, insulin-like growth factors and colon cancer: a review of the evidence. J Nutr 2001; 131:3109S-20S.
30. Furstenberger G, Senn HJ. Insulin-like growth factors and cancer. Lancet Oncol 2002; 3:298-302.
31. Juul A, Scheike T, Davidsen M et al. Low serum insulin-like growth factor-1 is associated with increased risk of ischemic heart disease: a population-based case-control study. Circulation 2002; 106:939-44.
32. O'Sullivan AJ, Ho KK. Route-dependent endocrine and metabolic effects of estrogen replacement therapy. J Pediatr Endocrinol Metab 2000; 13(Suppl 6):1457-66.
33. Cardim HJ, Lopes CM, Giannella-Neto D et al. The insulin-like growh factor-I system and hormone replacement therapy. Fertil Steril 2001; 75:282-7.
34. Ismail AH, Pollak M, Behlouli H et al. Insulin-like growth factor-I and insulin-like growth factor binding protein-3 for prostate cancer detection in patients undergoing prostate biopsy. J Urol 2002; 168:2426-30.
35. DiGiovanni J, Kiguchi K, Frijhoff A et al. Deregulated expression of insulin-like growth factor 1 in prostate epithelium leads to neoplasia in transgenic mice. Proc Natl Acad Sci USA 2000; 97(7):3455-60.
36. Pollak M, Blouin MJ, Zhang JC et al. Reduced mammary gland carcinogenesis in transgenic mice expressing a growth hormone anatgonist. Brit J Cancer 2001; 85(3):428-430.
37. Wu Y, Yakar S, Zhao L et al. Circulating insulin-like growth factor-1 levels regulate colon cancer growth and metastasis. Cancer Res 2002; 62:1030-5.
38. Yang XF, Beamer W, Huynh HT et al. Reduced growth of human breast cancer xenografts in hosts homozygous for the 'lit' mutation. Cancer Res 1996; 56:1509-11.
39. Deitel K, Dantzer D, Ferguson P et al. Reduced growth of human sarcoma xenografts in hosts homozygous for lit mutation. J Surgical Oncology 2002; 81:75-9.
40. Dunn SE, Kari FW, French J et al. Dietary restriction reduces IGF-I levels, which modulates apoptosis, cell proliferation, and tumor progression in p53 deficient mice. Cancer Res 1997; 57:4667-72.
41. Eigenmann JE, Amador A, Patterson DF. Insulin-like growth factor I levels in proportionate dogs, chondrodystrophic dogs and in giant dogs. Acta Endocrinol (Copenh) 1988; 118:105-8.
42. Eigenmann JE, Patterson DF, Zapf J et al. Insulin-like growth factor -I in the dog: a study in different dog breeds and in dogs with growth hormone elevation. Acta Endocrinol (Copenh) 1984; 105:294-301.
43. Withrow SJ, Powers BE, Straw RC et al. Comparative aspects of osteosarcoma. Dog versus man. Clin Orthop 1991; 270:159-68.

44. Tjalma RA. Canine bone sarcoma: estimation of relative risk as a function of body size. J Natl Cancer Inst 1966; 36:1137-50.
45. Pollak M, Polychronakos C, Richard M. Insulin like growth factor 1: a potent mitogen for human osteogenic sarcoma. JNCI 1990; 82:301-5.
46. McCormack VA, dos Santos Silva I, De Stavola BL et al. Fetal growth and subsequent risk of breast cancer: results from long term follow up of Swedish cohort. BMJ 2003; 326:248.
47. Sandhu MS, Luben R, Day NE et al. Self-reported birth weight and subsequent risk of colorectal cancer. Cancer Epidemiol Biomarkers Prev 2002; 11:935-8.
48. Vatten LJ, Maehle BO, Lund Nilsen TI et al. Birth weight as a predictor of breast cancer: a case-control study in Norway. Br J Cancer 2002; 86:89-91.
49. Andersson SW, Bengtsson C, Hallberg L et al. Cancer risk in Swedish women: the relation to size at birth. Br J Cancer 2001; 84:1193-8.
50. Stavola BL, Hardy R, Kuh D et al. Birthweight, childhood growth and risk of breast cancer in British cohort. Br J Cancer 2000; 83(7):964-8.
51. Tibblin G, Eriksson M, Cnattingius S et al. High birthweight as a predictor of prostate cancer risk. Epidemiology 1995; 6:423-4.
52. Lo HC, Tsao LY, Hsu WY et al. Relation of cord serum levels of growth hormone, insulin-like growth factors, insulin-like growth factor binding proteins, leptin, and interleukin-6 with birth weight, birth length, and head circumference in term and preterm neonates. Nutrition 2002; 18:604-8.
53. Vatten LJ, Nilsen ST, Odegard RA et al. Insulin-like growth factor-I and leptin in umbilical cord plasma and infant birth size at term. Pediatrics 2002; 109:1131-5.
54. Christou H, Connors JM, Ziotopoulou M et al. Cord blood leptin and insulin-like growth factor levels are independent predictors of fetal growth. J Clin Endocrinol Metab 2001; 86:935-8.
55. Gunnell D, Okasha M, Smith GD et al. Height, leg length, and cancer risk: a systematic review. Epidemiol Rev 2001; 23:313-42.
56. Byrne C, Colditz GA, Willett WC et al. Plasma insulin-like growth factor-I, insulin-like growth factor-binding protein-3 and mammographic density. Cancer Res 2000; 60(14):3744-8.
57. Hsing AW, Gao YT, Chua S Jr et al. Insulin-resistance and prostate cancer risk. J Natl Cancer Inst 2003; 95:67-71.
58. Goodwin PJ, Ennis M, Pritchard KI et al. Fasting insulin and outcome in early-stage breast cancer: results of a prospective cohort study. J Clin Oncol 2002; 20:42-51.
59. Kong FM, Anscher MS, Washington MK et al. M6P/IGF2R is mutated in squamous cell carcinoma of the lung. Oncogene 2000; 19:1572-8.
60. O'Gorman DB, Weiss J, Hettiaratchi A et al. Insulin-like growth factor-II/mannose 6-phosphate receptor overexpression reduces growth of choriocarcinoma cells in vitro and in vivo. Endocrinol 2002; 143:4287-94.
61. Cui H, Cruz-Correa M, Giardiello FM et al. Loss of IGF2 imprinting: a potential marker of colorectal cancer risk. Science 2003; 299:1753-5.
62. Shariat SF, Lamb DJ, Kattan MW et al. Association of preoperative plasma levels of insulin-like growth factor-I and insulin-like growth factor binding proteins-2 and -3 with prostate cancer invasion, progression, and metastasis. J Clin Oncol 2002; 20:833-41.
63. Cohen SM, Ellwein LB. Cell proliferation in carcinogenesis. Science 1990; 249(4972):1007-11.
64. Neal DE, Donovan JL. Prostate cancer: to screen or not to screen. Lancet Oncol 2000; 1:17-24.
65. Whitmore WF. Localized prostate cancer: management and detection issues. Lancet 1994; 343:1263-7.
66. Cats A, Dullaart RP, Kleinbeuker JH et al. Increased epithelial cell proliferation in the colon of patients with acromegaly. Cancer Res 1996; 56(3):523-6.
67. Huynh HT, Yang XF, Pollak M. Estradiol and antiestrogens regulate a growth inhibitory insulin-like growth factor binding protein 3 autocrine loop in human breast cancer cells. J Biol Chem 1996; 271:1016-21.
68. Rozen F, Yang X, Huynh HT et al. Antiproliferative action of vitamin-D-related compounds and insulin-like growth factor binding protein 5 accumulation. J Natl Cancer Inst 1997; 89:652-6.
69. Gucev ZS, Oh Y, Kelley KM et al. Insulin-like growth factor binding protein 3 mediates retinoic acid and transforming growth factor beta 2-induced growth inhibition in human breast cancer cells. Cancer Res 1996; 56:1545-50.
70. Giovannucci E, Pollak M. Risk of cancer after growth-hormone treatment. (editorial). Lancet 2002; 360:268-9.

CHAPTER 21

The Molecular Basis of IGF-I Receptor Gene Expression in Human Cancer

Haim Werner

Abstract

The insulin-like growth factor-I receptor (IGF-IR) has a central role in normal cellular proliferation as well as in transformation processes. Transcription factors have been identified that modulate the activity of the *IGF-IR* gene. Transcription factors with tumor suppressor activity, such as p53 and WT1, were shown to inhibit transcription of the *IGF-IR* gene. Loss-of-function mutation of these genes in certain malignancies results in transcriptional derepression of the *IGF-IR* gene, with ensuing increases in the levels of IGF-IR. Likewise, the mechanisms of action of many oncogenic agents depend on their ability to *trans*activate the IGF-IR promoter and/ or to phosphorylate the cytoplasmic domain of the receptor. The expression of the *IGF-IR* gene is, ultimately, the net result of complex interactions between positive and negative nuclear factors, as well as between stimulatory and inhibitory secreted factors. The proliferative status of the cell is a direct consequence of this level of expression.

Introduction

The central role of the IGF system in a variety of growth and differentiation processes has been well established. Similarly well accepted is the notion that the vast majority of the biological actions of both IGF-I and IGF-II are mediated via activation of the IGF-IR heterotetramer. The role of the IGF-IR in cell cycle progression and apoptosis, as well as the signal transduction pathways responsible for these activities, are described in large detail in other Chapters of this book. The present Chapter will focus on the molecular mechanisms that are responsible for the expression of the *IGF-IR* gene during normal development, on understanding the events and factors that govern its levels of expression (and therefore determine, to a large extent, the proliferative status of the cell) and, in particular, on analyzing the mechanisms that underlie the pathological expression of the IGF-IR in the transformed cell.

Overexpression of the *IGF-IR* Gene as a Common Theme in Malignancy

Most human cancers and transformed cell lines express increased levels of IGF-IR on their cell surface, as well as augmented levels of IGF-IR mRNA (Table 1). These tumors include ovarian, colon, thyroid, lung, pheochromocytoma, breast, glioblastoma, astrocytoma, hepatoma, gastric, renal, rhabdomyosarcoma, and others (for an extensive review see ref. 1). In addition, amplification of the IGF-IR locus at band 15q26 has been reported in a small number of breast and melanoma cases.[2,3] It is generally assumed that the tumor IGF-IR is capable of responding to circulating/ endocrine IGFs, and to IGFs produced locally by neighboring (stromal) cells or by the cancer cells themselves.

In view of the abundant expression of the *IGF-IR* gene in most cancers, it is relevant to ask what are the mechanisms employed by the cell in order to control the synthesis and function of this important receptor. To gain an understanding on these basic questions it may be helpful to briefly review the role and regulation of the *IGF-IR* gene during ontogenesis.

Insulin-Like Growth Factors, edited by Derek LeRoith, Walter Zumkeller
and Robert Baxter. ©2003 Eurekah.com and Kluwer Academic / Plenum Publishers.

Table 1. IGF-IR in human cancers

Ovary	Astrocytoma
Colon	Hepatoma
Thyroid	Stomach
Lung	Kidney
Pheochromocytoma	Rhabdomyosarcoma
Breast	Glioblastoma
Pancreas	Endometrium
Leukemia	Ewings

IGF-IR have been characterized using competitive binding assays, affinity cross-linking, Northern blots, RNase protection assays, RT-PCR, or a combination of them.

The *IGF-IR* gene is constitutively expressed at each and every developmental stage, although its levels may vary over a wide range. IGF binding and IGF-IR mRNA can be detected as early as the oocyte stage.[4] In preimplantation mouse embryos the receptor mediates preferentially the effects of IGF-II, since no insulin or IGF-I transcripts can be detected at this early stage.[5] Following implantation, the IGF-IR is expressed by virtually every cell. The widespread distribution of the IGF-IR underscores its fundamental role as a survival factor required by most cells during ontogeny.[6,7] Further evidence for a survival role for the IGF-IR is provided by the fact that the majority of cultured cells constitutively express the *IGF-IR* gene during their proliferative stages.

In sharp contrast to the universal expression of the *IGF-IR* gene, the IGF ligands are expressed during ontogeny following distinct spatial and temporal patterns, suggesting that IGF-I and IGF-II have different roles during embryonic development. It appears, however, that the IGF-IR is able to mediate the endocrine effects of circulating IGFs (present both in the bloodstream and in the cerebrospinal fluid) as well as the effects of locally produced IGFs.

Late developmental stages, characterized by a reduction in the number of rapidly-proliferating cells and by an increment in the proportion of postmitotic, terminally differentiated cells, are associated with a generalized reduction in the levels of IGF-IR mRNA and binding in most organs.[8] These findings are consistent with the results of experiments demonstrating that cultured cells induced to differentiate exhibit a reduction in the levels of IGF-IR.

Dedifferentiation states associated with malignancy, similar to early developmental stages, exhibit high levels of IGF-IR, consistent with the increased proliferative activity of the transformed cell.[9,10] Furthermore, augmented receptor levels correlate with a large reduction in apoptosis. The implication of these observations is that activation of the IGF-IR may rescue from apoptosis cell populations that are otherwise tagged for elimination. Induction of apoptosis, on the other hand, seems to be the common theme of a number of approaches that are aimed at targeting the IGF-IR as a potential anticancer therapy.

An interesting novel paradigm of IGF-IR-independent proliferation has been recently described in metastatic prostate and breast cancer cells.[11,12] The implications of these observations, suggesting that the IGF-IR is required during the early stages of transformation, but not at the metastatic stage, are discussed in this book by C.T. Roberts.

The Role of the IGF-IR in the Transformation Process

Because the IGF-IR is expressed at very high levels in most naturally-occurring cancers, as well as in experimentally-induced tumors, it is important to examine its role in different cellular systems.

Artificial overexpression of the IGF-IR in fibroblasts results in a ligand-dependent, highly transformed phenotype, which includes the formation of tumors in nude mice.[13] On the other hand, abrogation of the IGF-IR signaling pathway using specific anti-IGF-IR antibodies resulted in a drastic reduction in cellular proliferation of melanoma,[14] breast,[15] hematopoietic,[16] colorectal,[17] neuroblastoma[18] and Wilms' tumor cells.[19] Likewise, inhibition of IGF-I-mediated growth and clonogenicity in soft agar of human T98G and rat C6 glioblastoma cells was achieved by introducing antisense oligodeoxynucleotides against the IGF-IR mRNA, or by transfection with plasmids encoding antisense cDNA fragments.[20,21]

The central role of the IGF-IR in the transformation process is further illustrated by the results of experiments showing that fibroblast cell lines established from mouse embryos in which the IGF-IR was disrupted by homologous recombination (R⁻) cannot be transformed by any of a number of oncogenes (including the SV40 large T antigen, activated *ras*, bovine papillomavirus E5 protein, and others).[22-23] Reintroduction of a functional receptor renders R⁻ cells susceptible to the transforming activities of these oncogenes. However, certain exceptions to this general paradigm have been reported. For instance, stable transfection of the GTPase-deficient mutant human $G_{\alpha13}$ resulted in transformation of R⁻ cells, as tested using the soft agar assay. These results demonstrate that $G_{\alpha13}$ can induce cellular transformation through pathways apparently independent of the IGF-IR.[24] Cooperation between oncogenes, in addition, may activate additional survival pathways which may potentially result in the transformation of IGF-IR-null cells. Thus, while human papillomavirus-16 E6 and E7 proteins were unable to induce colony formation in R⁻ cells when transfected separately, combined transfection of both oncogenes resulted in a transformed phenotype.[25]

The transforming activity of the IGF-IR depends, to a large extent, on its strong antiapoptotic activity. The ability of the IGF-IR to protect cells from apoptosis (thus conferring them an increased survivability) has been demonstrated in many different cell types, including fibroblasts, neural-derived, hematopoietic, and others.[26-29] Furthermore, a highly significant correlation was established between the number of cell surface IGF-IRs and the in vivo and in vitro survival capacity of the cell.[30]

Mapping of Receptor Domains Involved in Transformation

Early studies indicated that an intact tyrosine kinase domain is required for the transduction of the proliferative actions of the IGF-IR.[31] Specifically, mutations in the ATP binding site and triple tyrosine residues at positions 1250, 1251, and 1316, either individually or in combination, totally abrogated tumor formation.[32] Essential roles in IGF-I-mediated mitogenesis were also associated with tyrosine residues 1131, 1135, and 1136.[33] Transfection of rat-1 fibroblasts with a truncated β-subunit mutant (952 STOP) resulted in cells which were unable to grow in soft agar and to induce tumors in athymic mice.[34] Furthermore, mutation of a series of four serine residues at the C-terminal domain which are involved in specific binding to the 14-3-3 protein, a potential substrate of IGF-IR action, similarly affected tumorigenesis.[35]

Mapping of functional domains in the cytoplasmic portion of the IGF-IR revealed that the domains required for its antiapoptotic function are distinct from those required for its proliferative or transforming activities.[36] Furthermore, the domains of the receptor required for inhibition of apoptosis are necessary but not sufficient for transformation.

While most research focused on elucidating the structure-function relationship of the cytoplasmic portion of the receptor, an important modulatory role has been ascribed to the 36-amino acid extracellular segment of the IGF-IR β-subunit. Using N-terminally truncated IGF-IR fused to avian sarcoma virus UR2 gag p19 it was shown that the 20 residues located immediately upstream of the transmembrane domain have an inhibitory effect on the transforming and tumorigenic potential of the fusion protein, whereas N-linked glycosylation within this region had a positive effect.[37,38]

Phosphorylation of the IGF-IR in Malignancy

As previously indicated, the presence of a functional IGF-IR is an essential prerequisite for oncogenic transformation. Furthermore, an intact tyrosine kinase domain is fundamental in order for the receptor to exert its potent mitogenic, antiapoptotic, and transforming activities. The mechanisms of action of a number of oncogenic agents depend, in fact, on their ability to efficiently phosphorylate the receptor. Thus, transformation by pp60^{src}, the protein encoded by the *src* oncogene of Rous sarcoma virus, results in the constitutive tyrosine phosphorylation of the IGF-IR β subunit.[39,40] It has been estimated that between 10-50% of the receptors are phosphorylated in the unstimulated *src*-transformed cell. Addition of IGF-I synergistically increased the extent of phosphorylation of the receptor. These results raise the possibility that pp60^{src} alters growth regulation by rendering the cells constitutively subject to a mitogenic signal. Moreover, they suggest that the IGF-IR kinase is more active as an autokinase in transformed than in nontransformed cells.

Likewise, the EWS-FLI-1 chimeric protein that results from the reciprocal translocation t(11;22)(q24;q12) and that is characteristic of the Ewings family of tumors, requires the presence of an intact IGF-IR for transformation. Fibroblasts which were stably transfected with the fusion protein exhibited a larger degree of IGF-I-stimulated IRS-1 phosphorylation, suggesting that expression of the EWS-FLI-1 oncogene may sensitize the IGF-IR signaling pathway to the action of IGF-I.[41]

Transcriptional Regulation of the *IGF-IR* Gene

Transcriptional regulation of the *IGF-IR* gene constitutes one of the most important mechanisms employed by the cell to control expression of this receptor during normal development and in response to physiological and pathological stimuli. Cloning and characterization of the IGF-IR promoter region revealed a number of features that are shared by a family of genes which are constitutively expressed by most cells and that are referred to as *housekeeping genes*. The IGF-IR regulatory region is very rich in G and C nucleotides and lacks TATA or CAAT motifs, two regulatory elements that are required for efficient transcription initiation of most eukaryotic genes. Accurate transcription of the *IGF-IR* gene is directed from an "initiator" sequence, a control element that is present in the promoters of genes that are highly regulated during differentiation and development, and which is able to assemble a functional transcription complex in the absence of a TATA box.[42-45]

Similar to other widely expressed genes, the promoter region of the *IGF-IR* gene contains a number of GC boxes (GGGCGG), which are putative binding sites for members of the Sp1 family of transcription factors.[46] Sp1 is an ubiquitous zinc-finger nuclear protein that stimulates transcription from a group of RNA polymerase II-dependent promoters. Using transient cotransfections, electrophoretic mobility shift assays (EMSA), and DNaseI footprinting experiments it was demonstrated that Sp1 is a potent *trans*activator of the IGF-IR promoter. The capacity of Sp1 to *trans*activate this gene is consistent with its ability to bind with high affinity to consensus sites present in the promoter region.[47]

Although ubiquitously expressed, levels of Sp1 fluctuate during development.[48] Similar to the *IGF-IR* gene, lowest Sp1 mRNA and protein levels are seen in terminally differentiated cells, suggesting that Sp1 constitutes a positive activator of IGF-IR transcription in most physiological states. In addition, the mechanism of action by which certain tumor suppressors, including p53, inhibit transcription of the *IGF-IR* gene involves specific interaction with Sp1.[49] This point will be discussed below in more detail.

Regulation of the *IGF-IR* Gene by Oncogenes

Cellular and viral oncogenes can induce transformation by "recruiting" and activating the IGF signaling pathway. In a previous section I described the mechanism of action of the *src* oncogene of the Rous sarcoma virus, and showed that it involves the ligand-independent phosphorylation of the IGF-IR β subunit. Additional oncogenes were shown to directly *trans*activate the IGF-IR promoter.

An example of this class of oncogenes is c-*myb* (the cellular equivalent of the viral transforming oncogene v-*myb*). Overexpression of c-*myb* in Balb/c-3T3 cells induced an increase in the levels of both the IGF-IR and IGF-I ligand transcripts.[50,51] This event resulted in abrogation of the requirement for IGF-I in the growing media, which in itself constitutes one of the distinctive hallmarks of a malignantly-transformed cell.

An additional oncoprotein shown to stimulate the transcription of the *IGF-IR* gene is the hepatitis B virus X (HBx) gene product. In hepatocellular carcinoma-derived cell lines containing HBx protein, the endogenous levels of IGF-IR mRNA were ~5-fold higher than in cells that do not express HBx transcripts. Similarly, transfection of HepG2 cells with an expression vector encoding the HBx cDNA induced an ~2-3-fold increase in the levels of IGF-IR promoter activity, mRNA, and IGF binding.[52] These findings clearly demonstrate that the mechanism of action of oncogene HBx in the pathophysiology of hepatocellular carcinoma involves the *trans*activation of the *IGF-IR* gene.

In summary, the requirement for a functional IGF-IR in order for a cell to undergo oncogenic transformation can be explained, at a molecular level, by the fact that many oncogenes "adopt" the IGF-IR signaling pathway as *their* mechanism of transformation. Certain oncogenes achieve this goal by directly *trans*activating the IGF-IR promoter and thus drastically increasing the concentration of receptors in the preneoplastic cell. Additional oncogenes induce a large increase in the level of phosphorylation of the IGF-IR β subunit. Regardless of the molecular mechanism employed, the transformed cell displays essentially identical phenotypes, including the ability to proliferate in the absence of exogenous IGFs.

Regulation of the *IGF-IR* Gene by Tumor Suppressor p53

Given the strong proliferative action of the IGF-IR, it may be asked how does the adult, terminally differentiated cell succeed in remaining out of the cell cycle. One potential mechanism that may be responsible of keeping IGF-IR levels below a certain threshold involves its transcriptional suppression by a family of negative growth regulators, collectively referred to as *tumor suppressors*. We may predict that beneath these receptor concentrations, the cell will remain at the G_0 stage and will not engage in any mitogenic activity. Mutation, deletion, or chromosomal rearrangement of tumor suppressor genes in transformed cells may lead to transcriptional derepression of the *IGF-IR* gene.

P53 is a tumor suppressor that, in its hyperphosphorylated state, blocks progression of cells through the cell cycle.[53,54] P53 is involved in the etiology of many human tumors, and mutations of the *p53* gene are the most frequent mutations in human cancers.[55,56] The p53 protein functions as a transcription factor that binds specifically to DNA sequences in various promoters and stimulates their transcriptional activity. It can also function as a transcriptional repressor of many growth-regulated genes. Transient expression of wild type (wt) p53 in osteosarcoma- and rhabdomyosarcoma-derived cell lines suppressed the activity of cotransfected IGF-IR promoter constructs by 75-90 %. On the other hand, cotransfection of tumor-derived, mutant versions of p53 (encoding point mutations at codons 143, 248, and 273) stimulated promoter activity by ~2.3- to 4-fold of control values.[57] In addition, wt p53 decreased the IGF-I-induced tyrosine phosphorylation of the IGF-IR and of IRS-1, whereas mutant p53 stimulated their phosphorylation.[49] Although the mechanism/s for transcriptional suppression by wt p53 are not fully understood, results of EMSA assays suggest that wt p53 can bind to the TATA-binding protein (TBP), thus preventing this protein from binding to the initiator region of the *IGF-IR* gene and assembling a functional initiation complex. An additional potential mechanism of action of p53 involves its interaction with Sp1. Wild-type and mutant forms of p53 were shown to physically interact with Sp1, which counteracted the inhibitory effect of wt p53 in a dose-dependent manner. Further support for a physiological role for the *IGF-IR* gene as a target for p53 action was provided by experiments performed in murine hemopoietic cells using a temperature-sensitive mutant of p53. Expression of p53 in its wt conformation reduced the number of IGF-IRs in cells in which the transfected receptor was under the control of the IGF-IR, but not the cytomegalovirus, promoter.[58] Taken together, these results suggest that, at least part of, the effects of wt p53 on apoptosis and cell cycle arrest are mediated via suppression of the IGF-IR promoter. Lack of inhibition by mutant p53 may accelerate tumor growth and inhibit apoptosis, thus providing an increased survival capacity to malignant cell populations (Fig. 1).

In addition to controlling the activity of the *IGF-IR* gene, p53 has been shown to modulate additional components of the IGF signaling system. Thus, the expression of IGF-II transcripts is reduced by wt p53.[59] On the other hand, the activity of the *IGFBP3* gene is stimulated by wt, but not mutant, p53.[60] Because IGFBP3 is an inhibitor of mitogenic signaling by IGFs, it may be inferred that p53 can regulate the IGF system both at the level of availability of IGF ligands, and at the level of activity of the IGF-IR promoter.[61]

A similar paradigm of tumor suppressor modulation of *IGF-IR* gene expression was reported for BRCA1, the breast and ovarian cancer susceptibility gene.[62] The *BRCA1* gene encodes a 220-kDa phosphorylated protein that functions as a transcription factor with tumor suppressor activity.[63] Mutations at the *brca1* locus are linked to a large proportion of familial breast and/or ovarian cancer. Transient expression of a BRCA1 expression vector in a number of cell lines resulted in the dose-dependent suppression of cotransfected IGF-IR promoter constructs.[64] Although the molecular targets of BRCA1 have not yet been identified, it is possible that part of the proapoptotic activity of BRCA1 is achieved via suppression of the strongly antiapoptotic *IGF-IR* gene. Mutant versions of BRCA1 lacking *trans*activational activity can potentially *de*repress the IGF-IR promoter, resulting in augmented levels of IGF-IR mRNA and IGF-I binding in breast cancer.

A link between radiosensitivity and the *IGF-IR* gene has been recently reported.[65] Ataxia telangiectasia (AT) cells, displaying a mutant *ATM* gene, express low levels of IGF-IR and show

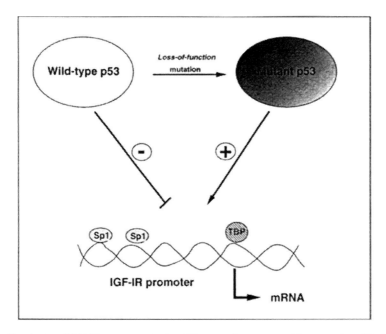

Figure 1. Regulation of *IGF-IR* gene expression by wild-type and mutant p53. The expression of the *IGF-IR* gene in the normal cell appears to be under inhibitory regulation by wt p53. As a result of this negative control, cellular proliferation is reduced and apoptosis is increased. *Loss-of-function* mutation of p53 in malignant cells can *de*repress the IGF-IR promoter, with ensuing increases in the levels of cell surface receptors. The mechanism of action of p53 involves interaction with transcription factor Sp1 and with the TATA-binding protein, TBP.

decreased IGF-IR promoter activity compared with wild type cells. Complementation of AT cells with the ATM cDNA results in increased IGF-IR promoter activity and protein levels. These results suggest that reduced expression of the IGF-IR may contribute to the radiosensitivity of AT cells.

Regulation of the *IGF-IR* Gene by Tumor Suppressor WT1

An additional tumor suppressor whose mechanism of action has been well characterized is the Wilms' tumor suppressor, WT1. Wilms' tumor is a pediatric kidney cancer that arises from metanephric blastema cells and whose etiology is associated with deletion or mutation of the WT1 gene. The WT1 gene encodes a 52-54 kDa protein that contains four zinc fingers of the C2-H2 class in its C-terminus.[66] WT1, via its zinc finger domain, binds to target DNAs containing versions of the consensus sequence GCGGGGGCG. Among other promoters, this specific motif is present in the regulatory regions of the *IGF-II* and *IGF-IR* genes.[67,68]

During normal kidney development, WT1 functions as a transcription factor with important roles in the differentiation of the metanephric blastema to renal epithelium.[69] IGF-II and the IGF-IR are also involved in kidney development, being their expressions negatively regulated by the WT1 gene product. In addition, the IGF-II-IGF-IR axis has an important role in Wilms' tumor progression, as illustrated by the fact that administration of anti-IGF-IR antibodies to nude mice bearing Wilms' tumor heterotransplants prevented tumor growth and resulted in partial tumor remission.[19] Furthermore, the levels of IGF-II and IGF-IR mRNAs are significantly elevated in the tumors in comparison to normal adjacent tissue.[70,71]

The molecular mechanisms responsible for transcriptional repression of the *IGF-IR* gene by WT1 were revealed by means of transient and stable transfections, EMSA and DNase footprinting assays. WT1 represses IGF-IR promoter activity in a dose-dependent manner, binding to sites in both the 5'-flanking and 5'-untranslated regions.[68] The DNA-binding capacity of WT1 is critical for maximal repression of the IGF-IR promoter, but some effects may be mediated through protein-protein interactions involving the N-terminal domain.[72] Stable expression of the WT1 cDNA in kidney tumor-derived G401 cells resulted in a decreased rate of cellular proliferation, decreased levels of IGF-IR mRNA and binding, and reduced activity of a transfected IGF-IR promoter. In addition, WT1-expressing cells exhibited a reduction in IGF-I-stimulated proliferation, thymidine incorporation, and anchorage-independent growth.[73]

In summary, Wilms' tumors and other nephropathies are cases of aberrant *de*differentiation that are characterized, in many instances, by underexpression, deletion, or mutation of the WT1 gene. Loss of WT1 activity may result in *de*repression of the IGF-IR and IGF-II promoters. Increased transcription and expression of both ligand and receptor genes may induce a mitogenic event with important consequences in tumor progression.

Regulation of the *IGF-IR* Gene by Disrupted Transcription Factors

As mentioned above, certain human malignancies are characterized by recurrent chromosomal translocations, frequently resulting in the fusion of genes. These fusion gene products (or chimeras) often comprise domains derived from unrelated transcription factors and nucleic acid-binding proteins. Furthermore, the chimeras usually display altered transcriptional activities that confer upon them a *gain-of-function* type of action.

Desmoplastic small round cell tumor (DSRCT) is an aggressive primitive pediatric tumor associated with the recurrent translocation t(11;22)(p13;q12). This rearrangement joins the N-terminal (activation) domain of the Ewings' sarcoma gene, EWS (an ubiquitously expressed RNA-binding protein), to the C-terminal (DNA-binding) domain of WT1, including 3 out of 4 zinc fingers.[74,75] The fusion protein, EWS-WT1, is capable of binding consensus WT1 sites in the IGF-IR promoter region with an affinity comparable to that of native WT1, and of *trans*activating the IGF-IR promoter in transient transfection assays.[76] Hence, fusion of EWS to WT1 abrogates the tumor suppressor function of WT1 and the RNA-binding capacity of EWS, and generates an oncogene capable of binding and *trans*activating the IGF-IR promoter (Fig. 2). Augmented levels of IGF-IR may constitute an important prerequisite in progression of DSRCT.

A similar paradigm of *trans*activation of the *IGF-IR* gene by a disrupted transcription factor was recently reported for the PAX3-FKHR oncoprotein.[77] This chimeric protein results from the recurrent translocation t(2;13)(q35;q14), the cytogenetic event characteristic of alveolar rhabdomyosarcoma (ARMS).[78] PAX3-FKHR includes the N-terminal domain of PAX3 (a developmentally-regulated transcription factor that contains a *paired-box* and an *homeodomain* DNA-binding motifs) fused in-frame to the C-terminal domain of FKHR (a member of the forkhead family of transcription factors). Transfection of sarcoma-derived cell lines with expression vectors encoding PAX3-FKHR resulted in *trans*activation of a cotransfected IGF-IR promoter construct, whereas PAX3 exhibited a reduced potency in comparison to the chimera. These results can be interpreted to suggest that the *IGF-IR* gene constitutes a molecular target for aberrant transcription factor PAX3-FKHR. Increased levels of IGF-IR are potentially critical in the etiology of ARMS and other pediatric sarcomas.

Regulation of the *IGF-IR* Gene by Growth Factors, Cytokines and Steroid Hormones

In addition to cellular factors such as oncogenes and tumor suppressors, the expression of the *IGF-IR* gene can be also modulated by various secreted factors, including peptide and steroid hormones, growth factors, and cytokines. Humoral regulation of the *IGF-IR* gene is important in many physiological processes. For example, cell cycle progression occurs only in the presence of two families of growth factors: competence factors (such as PDGF and FGF) and progression factors (such as IGF-I). It has been postulated that the main role of competence factors is to induce the production of sufficient amounts of progression factors and their receptors that will allow the cell to engage in mitogenesis.[79] In fact, both FGF and PDGF have been shown to stimulate transcription of the *IGF-IR* gene.[80,81] PDGF increased the activity of the IGF-IR promoter via an ~100-bp promoter fragment located immediately upstream of the initiator element that includes a consensus *c-myc* binding site.[82] On the other hand, IGF-I negatively regulates the expression of the *IGF-IR* gene.

The expression of the *IGF-IR* gene depends also on the ambient concentration of steroid hormones. Estrogens, for instance, were shown to stimulate the levels of IGF-IR mRNA and binding in

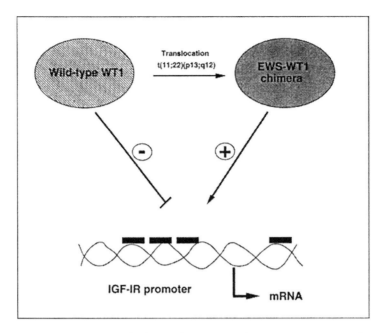

Figure 2. Regulation of the *IGF-IR* gene by disrupted transcription factor EWS-WT1. Desmoplastic small round cell tumor is characterized by the recurrent chromosomal translocation t(11;22)(p13;q12), that fuses the N-terminal domain of the Ewings' gene product, EWS, to the C-terminal, DNA-binding domain, of WT1. This event abrogates the tumor suppressor activity of WT1 and generates a chimeric oncoprotein, EWS-WT1, whose mechanism of action involves binding to, and transactivation of, the IGF-IR promoter. WT1 binding elements (denoted as black boxes) are located both upstream and downstream of the *IGF-IR* gene transcription start site.

MCF7 cells by ~7-fold, suggesting that a potential mechanism by which estrogens stimulate breast cancer proliferation involves sensitization to the mitogenic effects of IGFs by augmenting receptor concentration.[83] Progestins, on the other hand, induced a reduction in the levels of IGF-IR mRNA and binding in estrogen-responsive breast cancer cell lines. It appears that this effect is mediated by IGF-II, whose secretion is stimulated by progestins and which, in turn, down-regulates the *IGF-IR* gene.[84]

Finally, certain cytokines were also shown to control the activity of the *IGF-IR* gene. Tumor necrosis factor-α (TNF-α) and interferon-γ (IFN-γ) are multifunctional cytokines which are produced mainly by activated macrophages and lymphocytes, respectively, although TNF-α is also synthesized by a number of non-hematopoietic cells. TNF-α controls cellular proliferation by inducing apoptosis, alone or in combination with other cytokines. Furthermore, TNF-α and IFN-γ reportedly inhibited a number of IGF-mediated biological actions, as well as the expression of the *IGF-I* and *IGF-II* genes. It has been recently demonstrated that both cytokines suppressed the activity of the IGF-IR promoter, resulting in a drastic reduction in the levels of IGF-IR mRNA and protein.[85] TNF-α, in addition, decreased the stability of mRNA molecules. Regulation of *IGF-IR* gene expression at both transcriptional and posttranscriptional levels may constitute a potential mechanism by which TNF-α and IFN-γ (and probably other cytokines) affect cellular proliferation.

Conclusions

The IGF-IR plays a critical role in normal and pathological growth processes. Controlling the expression of this gene appears to be an important mechanism that allows the cell to "decide" whether to adopt proliferative or apoptotic pathways. The expression of this gene can be tightly regulated by secreted factors of endocrine or local (autocrine/paracrine) origin that can either stimulate or inhibit the synthesis of the IGF-IR. In addition, a number of nuclear proteins displaying either oncogenic or antioncogenic activities have been also shown to regulate the activity of the *IGF-IR* gene at the transcriptional level. Transcription factors with tumor suppressor activity, such as p53 and WT1, negatively regulate the expression of the *IGF-IR* gene. The etiology of neoplasias associated with *loss-of-function* mutation of tumor suppressors is linked to the inability of the mutant forms to suppress the activity of their molecular targets, including the *IGF-IR* gene. On the other hand, *gain-of-function* mutations of oncogenes are associated with increased *trans*activational activity of the IGF-IR promoter and/or augmented phosphorylation of the cytoplasmic domain of the receptor. Interactions between stimulatory and inhibitory factors may ultimately determine the level of expression of the *IGF-IR* gene.

Acknowledgments

The work in the authors' laboratory is supported by grants from The Israel Cancer Association, The Israel Science Foundation, The U.S.-Israel Binational Science Foundation, and The Fogarty International Center.

References

1. Werner H, LeRoith D. The role of the insulin-like growth factor system in human cancer. Adv Cancer Res 1996; 68:183-223.
2. Almeida A, Muleris M, Dutrillaux B et al. The insulin-like growth factor I receptor gene is the target for the 15q26 amplicon in breast cancer. Genes Chrom Cancer 1994; 11:63-65.
3. Zhang J, Trent JM, Meltzer PS. Rapid isolation and characterization of amplified DNA by chromosome microdissection: Identification of IGF1R amplification in malignant melanoma. Oncogene 1993; 8:2827-2831.
4. Scavo L, Shuldiner AR, Serrano J et al. Genes encoding receptors for insulin and insulin-like growth factor I are expressed in *Xenopus* oocytes and embryos. Proc Natl Acad Sci USA 1991; 88:6214-6218.
5. Schultz GA, Hahnel R, Arcellana-Panlilio M et al. Expression of IGF ligand and receptor genes during preimplantation mammalian development. Mol Reprod Dev 1993; 35:414-420.
6. Bondy CA, Werner H, Roberts Jr CT et al. Cellular pattern of insulin-like growth factor I (IGF-I) and type I IGF receptor gene expression in early organogenesis: comparison with *IGF-II* gene expression. Mol Endocrinol 1990; 4:1386-1398.
7. Bondy CA, Werner H, Roberts CT Jr et al. Cellular pattern of Type I insulin-like growth factor receptor gene expression during maturation of the rat brain: Comparison with insulin-like growth factors I and II. Neuroscience 1992; 46:909-923.
8. Werner H, Woloschak M, Adamo M et al. Developmental regulation of the rat insulin-like growth factor I receptor gene. Proc Natl Acad Sci USA 1989; 86:7451-7455.
9. Werner H , LeRoith D. The insulin-like growth factor-I receptor signaling pathways are important for tumorigenesis and inhibition of apoptosis. Crit Rev Oncogenesis 1997; 8:71-92.
10. Baserga R. The insulin-like growth factor I receptor: A key to tumor growth? Cancer Res 1995; 55:249-252.
11. Damon SE, Plymate SR, Carroll JM et al. Transcriptional regulation of insulin-like factor-I receptor gene expression in prostate cancer cells. Endocrinology 2001; 142:21-27.
12. Schnarr B, Strunz K, Ohsam J et al. Down-regulation of insulin-like growth factor-I receptor and insulin receptor substrate-1 expression in advanced human breast cancer. Int J Cancer 2000; 89:506-513.
13. Kaleko M, Rutter WJ, Miller AD. Overexpression of the human insulin-like growth factor I receptor promotes ligand-dependent neoplastic transformation. Mol Cell Biol 1990; 10:464-473.
14. Furlanetto RW, Harwell SE, Baggs RB. Effects of insulin-like growth factor receptor inhibition on human melanomas in culture and in athymic mice. Cancer Res 1993; 53:2522-2526.
15. Peyrat JP, Bonneterre J. Type 1 IGF receptor in human breast diseases. Breast cancer Res. Treat. 1992; 22:59-67.
16. McCubrey JA, Steelman LS, Mayo MW et al. Growth-promoting effects of insulin-like growth factor-I (IGF-I) on hematopoietic cells: Overexpression of introduced IGF-I receptor abrogates interleukin-3 dependency of murine factor-dependent cells by a ligand-dependent mechanism. Blood 1991; 78:921-929.
17. Lahm H, Amstad P, Wyniger J et al. Blockade of the insulin-like growth factor I receptor inhibits growth of human colorectal cancer cells: evidence of a functional IGF-II mediated autocrine loop. Int J Cancer 1994; 58:452-459.
18. El-Badry OM, Romanus JA, Helman LJ et al. Autonomous growth of a human neuroblastoma cell line is mediated by insulin-like growth factor II. J Clin Invest 1989; 84:829-839.
19. Gansler T, Furlanetto R, Gramling TS et al. Antibody to Type I insulin-like growth factor receptor inhibits growth of Wilms' tumor in culture and in athymic mice. Am J Pathol 1989; 135:961-966.
20. Resnicoff M, Sell C, Rubini M et al. Rat glioblastoma cells expressing an antisense RNA to the insulin-like growth factor I (IGF-I) receptor are non-tumorigenic and induce regression of wild type tumors. Cancer Res 1994; 54:2218-2222.
21. Ambrose D, Resnicoff M, Coppola D et al. Growth regulation of human glioblastoma T98G cells by insulin-like growth factor I (IGF-I) and the IGF-I receptor. J Cell Physiol 1994; 159:92-100.
22. Morrione A, DeAngelis T, Baserga R. Failure of the bovine papillomavirus to transform mouse embryo fibroblasts with a targeted disruption of the insulin-like growth factor I receptor gene. J Virol 1995; 69:5300-5303.

23. Sell C, Rubini M, Rubin R et al. Simian virus 40 large tumor antigen is unable to transform mouse embryonic fibroblasts lacking type 1 insulin-like growth factor receptor. Proc Natl Acad Sci USA 1993; 90:11217-11221.
24. Liu J-L, Blakesley VA, Gutkind JS et al. The constitutively active mutant $G_{\alpha13}$ transforms mouse fibroblast cells deficient in insulin-like growth factor-I receptor. J Biol Chem 1997; 272:29438-29442.
25. Steller MA, Zou Z, Schiller JT et al. Transformation by human papillomavirus 16 E6 and E7: Role of the insulin-like growth factor 1 receptor. Cancer Res 1996; 56:5087-5091.
26. Harrington EA, Bennett MR, Fanidi A et al. c-Myc-induced apoptosis in fibroblasts is inhibited by specific cytokines. EMBO J 1994; 13:3286-3295.
27. Rodriguez-Tarduchy G, Collins MKL, Garcia I et al. Insulin-like growth factor-I inhibits apoptosis in IL-3-dependent hemopoietic cells. J Immunol 1992; 149:535-540.
28. Sell C, Baserga R, Rubin R. Insulin-like growth factor I (IGF-I) and the IGF-I receptor prevent etoposide-induced apoptosis. Cancer Res 1995; 55:303-306.
29. Singleton JR, Randolph AE, Feldman EL. Insulin-like growth factor I receptor prevents apoptosis and enhances neuroblastoma tumorigenesis. Cancer Res 1996; 56:4522-4529.
30. Resnicoff M, Burgaud J-L, Rotman HL et al. Correlation between apoptosis, tumorigenesis, and levels of insulin-like growth factor I receptors. Cancer Res 1995; 55:3739-3741.
31. Kato H, Faria TN, Stannard B et al. Role of tyrosine kinase activity in signal transduction by the insulin-like growth factor-I (IGF-I) receptor. J Biol Chem 1993; 268:2655-2661.
32. Blakesley VA, Kalebic T, Helman LJ et al. Tumorigenic and mitogenic capacities are reduced in transfected fibroblats expressing mutant insulin-like growth factor (IGF)-I receptors. The role of tyrosine residues 1250, 1251, and 1316 in the carboxy-terminus of the IGF-I receptor. Endocrinology 1996; 137:410-417.
33. Kato H, Faria TN, Stannard B et al. Essential role of tyrosine residues 1131, 1135, and 1136 of the insulin-like growth factor-I (IGF-I) receptor in IGF-I action. Mol Endocrinol 1994; 8:40-50.
34. Prager D, Li H-L, Asa S et al. Dominant negative inhibition of tumorigenesis in vivo by human insulin-like growth factor I receptor mutant. Proc Natl Acad Sci USA 1994; 91:2181-2185.
35. Craparo A, Freund R, Gustafson TA. 14-3-3 interacts with the insulin-like growth factor-I receptor and insulin receptor substrate-I in a phosphoserine-dependent manner. J Biol Chem 1997; 272:11663-11669.
36. O'Connor R, Kauffmann-Zeh A, Liu Y et al. Identification of domains of the insulin-like growth factor I receptor that are required for protection from apoptosis. Mol Cell Biol 1997; 17:427-435.
37. Liu D, Rutter WJ, Wang L-H. Enhancement of transforming potential of human insulin-like growth factor I receptor by N-terminal truncation and fusion to avian sarcoma virus UR2 gag sequence. J Virology 1992; 66:374-385.
38. Liu D, Zong CS, Wang L-H. Distinctive effects of the carboxy-terminal sequence of the insulin-like growth factor I receptor on its signaling functions. J Virology 1993; 67:6835-6840.
39. Kozma LM, Weber MJ. Constitutive phosphorylation of the receptor for insulinlike growth factor I in cells transformed by the src oncogene. Mol Cell Biol 1990; 10:3626-3634.
40. Peterson JE, Jelinek T, Kaleko M et al. C phosphorylation and activation of the IGF-I receptor in src-transformed cells. J Biol Chem 1994; 269:27315-27321.
41. Toretsky JA, Kalebic T, Blakesley V et al. The insulin-like growth factor-I receptor is required for EWS-FLI-1 transformation of fibroblasts. J Biol Chem 1997; 272:30822-30827.
42. Mamula PW, Goldfine ID. Cloning and characterization of the human insulin-like growth factor-I receptor gene 5'-flanking region. DNA Cell Biol 1992; 11:43-50.
43. Cooke DW, Bankert LA, Roberts Jr CT et al. Analysis of the human type I insulin-like growth factor receptor promoter region. Biochem. Biophys Res Comm 1991; 177:1113-1120.
44. Werner H, Stannard B, Bach MA et al. Cloning and characterization of the proximal promoter region of the rat insulin-like growth factor I (IGF-I) receptor gene. Biochem Biophys Res Comm 1990; 169:1021-1027.
45. Werner H, Bach MA, Stannard B et al. Structural and functional analysis of the insulin-like growth factor I receptor gene promoter. Mol Endocrinol 1992; 6:1545-1558.
46. Courey AJ, Tjian R. Mechanisms of transcriptional control as revealed by studies of human transcription factor Sp1. In: McKnight S, Yamamoto K, eds. Transcriptional Regulation. Cold Spring Harbor: Cold Spring Harbor Laboratory, 1992:743-769.
47. Beitner-Johnson D, Werner H, Roberts CT, Jr. et al. Regulation of insulin-like growth factor I receptor gene expression by Sp1: Physical and functional interactions of Sp1 at GC boxes and at a CT element. Mol Endocrinol 1995; 9:1147-1156.
48. Saffer JD, Jackson SP, Annarella MB. Developmental expression of Sp1 in the mouse. Mol Cell Biol 1991; 11:2189-2199.
49. Ohlsson C, Kley N, Werner H et al. p53 regulates IGF-I receptor expression and IGF-I induced tyrosine phosphorylation in an osteosarcoma cell line: Interaction between p53 and Sp1. Endocrinology 1998; 139:1101-1107.
50. Reiss K, Ferber A, Travali S et al. The protooncogene c-myb increases the expression of insulin-like growth factor I and insulin-like growth factor I receptor messenger RNAs by a transcriptional mechanism. Cancer Res 1991; 51:5997-6000.
51. Travali S, Reiss K, Ferber A et al. Constitutively expressed c-myb abrogates the requirement for insulinlike growth factor I in 3T3 fibroblasts. Mol Cell Biol 1991; 11:731-736.
52. Kim SO, Park JG, Lee YI. Increased expression of the insulin-like growth factor I (IGF-I) receptor gene in hepatocellular carcinoma cell lines: Implications of IGF-I receptor gene activation by hepatitis B virus X gene product. Cancer Res 1996; 56:3831-3836.
53. Oren M. p53: The ultimate tumor suppressor gene? FASEB J 1992; 6:3169-3176.

54. Kern SE, Kinzler KW, Bruskin A et al. Identification of p53 as a sequence-specific DNA-binding protein. Science 1991; 252:1708-1711.
55. Hollstein M, Sidransky D, Vogelstein B et al. p53 mutations in human cancers. Science 1991; 253:49-53.
56. Harris CC, Hollstein M. Clinical implications of the p53 tumor suppressor gene. N Eng J Med 1993; 329:1318-1327.
57. Werner H, Karnieli E, Rauscher FJ, III et al. Wild type and mutant p53 differentially regulate transcription of the insulin-like growth factor I receptor gene. Proc Natl Acad Sci USA 1996; 93:8318-8323.
58. Prisco M, Hongo A, Rizzo MG et al. The insulin-like growth factor I receptor as a physiologically relevant target of p53 in apoptosis caused by interleukin-3 withdrawal. Mol Cell Biol 1997; 17:1084-1092.
59. Zhang L, Kashanchi F, Zhan Q et al. Regulation of insulin-like growth factor II P3 promoter by p53: A potential mechanism for tumorigenesis. Cancer Res 1996; 56:1367-1373.
60. Buckbinder L, Talbott R, Velasco-Miguel S et al. Induction of the growth inhibitor IGF-binding protein 3 by p53. Nature 1995; 377:1367-1373.
61. Werner H , LeRoith D. New concepts in regulation and function of the insulin-like growth factors: implications for understanding normal growth and neoplasia. Cell Mol Life Sci 2000; 57:932-942.
62. Miki Y, Swensen J, Shattuck-Eidens D et al. A strong candidate for the breast and ovarian cancer susceptibility gene *BRCA1*. Science 1994; 266:66-71.
63. Holt JT, Thompson ME, Szabo C et al. Growth retardation and tumor inhibition by BRCA1. Nature Gen 1996; 12:298-301.
64. Maor SB, Abramovitch S, Erdos MR et al. BRCA1 suppresses insulin-like growth factor-I receptor promoter activity: Potential interaction between BRCA1 and Sp1. Mol Gen Metab 2000; 69:130-136.
65. Peretz S, Jensen R, Baserga R et al. ATM-dependent expression of the insulin-like growth factor-I receptor in a pathway regulating radiation response. Proc Natl Acad Sci USA 2001; 98:1676-1681.
66. Rauscher FJ, III. The WT1 Wilms tumor gene product: A developmentally regulated transcription factor in the kidney that functions as a tumor suppressor. FASEB J 1993; 7.896-903.
67. Drummond IA, Madden SL, Rohwer-Nutter P et al. Repression of the insulin-like growth factor II gene by the Wilms' tumor suppressor WT1. Science 1992; 257:674-678.
68. Werner H, Rauscher FJ, III, Sukhatme VP et al. Transcriptional repression of the insulin-like growth factor I receptor (*IGF-I-R*) gene by the tumor suppressor WT1 involves binding to sequences both upstream and downstream of the *IGF-I-R* gene transcription start site. J Biol Chem 1994; 269:12577-12582.
69. Kreidberg J, Sariola H, Loring JM et al. WT1 is required for early kidney development. Cell 1993; 74:679-691.
70. Yun K, Fidler AE, Eccles MR et al. Insulin-like growth factor II and WT1 transcript localization in human fetal kidney and Wilms' tumor. Cancer Res 1993; 53:5166-5171.
71. Werner H, Re GG, Drummond IA et al. Increased expression of the insulin-like growth factor-I receptor gene, *IGFIR*, in Wilms' tumor is correlated with modulation of IGFIR promoter activity by the WT1 Wilms' tumor gene product. Proc Natl Acad Sci USA 1993; 90:5828-5832.
72. Tajinda K, Carroll J, Roberts CT, Jr. Regulation of insulin-like growth factor I receptor promoter activity by wild-type and mutant versions of the WT1 tumor suppressor. Endocrinology 1999; 140:4713-4724.
73. Werner H, Shen-Orr Z, Rauscher FJ III et al. Inhibition of cellular proliferation by the Wilms' tumor suppressor WT1 is associated with suppression of insulin-like growth factor I receptor gene expression. Mol Cell Biol 1995; 15:3516-3522.
74. Gerald WL, Rosai J, Ladanyi M. Characterization of the genomic breakpoint and chimeric transcripts in the EWS-WT1 gene fusion of desmoplastic small round cell tumor. Proc Natl Acad Sci USA 1995; 92:1028-1032.
75. Rauscher FJ, III, Benjamin LE, Fredericks WJ et al. Novel oncogenic mutations in the WT1 Wilms' tumor suppressor gene: A t(11;22) fuses the Ewings' sarcoma gene, *EWS1*, to WT1 in desmoplastic small round cell tumor. Cold Spring Harbor Symp Quant Biol 1994; 59:137-146.
76. Karnieli E, Werner H, Rauscher FJ, III et al. The IGF-I receptor gene promoter is a molecular target for the Ewings' sarcoma-Wilms' tumor 1 fusion protein. J Biol Chem 1996; 271:19304-19309.
77. Ayalon D, Glaser T, Werner H. Transcriptional regulation of IGF-I receptor gene expression by the PAX3-FKHR oncoprotein. Growth hormone IGF Res 2001; in press.
78. Shapiro DN, Sublett JE, Li B et al. Fusion of PAX3 to a member of the forkhead family of transcription factors in human alveolar rhabdomyosarcoma. Cancer Res 1993; 53:5108-5112.
79. Baserga R , Rubin R. Cell cycle and growth control. Crit Rev Euk Gene Exp 1993; 3:47-61.
80. Rosenthal SM, Brown EJ, Brunetti A et al. Fibroblast growth factor inhibits insulin-like growth factor II (*IGF-II*) gene expression and increases IGF-I receptor abundance in BC3H-1 muscle cells. Mol Endocrinol 1991; 5:678-684.
81. Hernandez-Sanchez C, Werner H, Roberts Jr CT et al. Differential regulation of IGF-I receptor gene expression by IGF-I and basic fibroblast growth factor. J Biol Chem 1997; 272:4663-4670.
82. Rubini M, Werner H, Gandini E et al. Platelet-derived growth factor increases the activity of the promoter of the IGF-I receptor gene. Exp Cell Res 1994; 211:374-379.
83. Stewart AJ, Johnson MD, May FEB et al. Role of insulin-like growth factors and the type I insulin-like growth factor receptor in the estrogen stimulated proliferation of human breast cancer cells. J Biol Chem 1990; 265:21172-21178.
84. Goldfine ID, Papa V, Vigneri R et al. Progestin regulation of insulin and insulin-like growth factor I receptors in cultured human breast cancer cells. Mol Endocrinol 1992; 6:1665-1672.
85. Shalita-Chesner M, Katz J, Shemer J et al. Regulation of insulin-like growth factor-I receptor gene expression by tumor necrosis factor-α and interferon-γ. Mol Cell Endocrinol 2001; 176:1-12.

CHAPTER 22

Antisense and Triple Helix Strategies in Basic and Clinical Research:
Challenge for Gene Therapy of Tumors Expressing IGF-I

L. C. Upegui-Gonzalez, J. C. Francois, L. A. Trojan, A. Ly, R. Przewlocki, C. Malvy and Jerry Trojan

Introduction

In prokaryotes and eukaryotes, genetic information is supported by double-stranded DNA, in which only one strand is usually transcribed in messenger RNA. Nevertheless, transcription could occur from both strands or from complementary strands, leading sometimes to the synthesis of complementary antisense and sense RNAs. For several years, it has been shown that natural antisense RNA which is transcribed from one strand could hybridize to the sense RNA. This natural physiological regulation represents the basis of the antisense approach to artificially inhibit gene expression of particular genes involved in human diseases. Using this antisense strategy, the translation of messenger RNA (sense RNA) can be blocked by binding of a complementary strand to mRNA. The achievement of this artificial regulation could be done using either plasmid constructs transcribing intracellular antisense RNA or short oligonucleotides delivered to cells via appropriate carriers. In the latter case, a specific ribonuclease (RNAse H) recognizes the oligonucleotide/mRNA hybrid and cleaves the RNA moiety, leading to a specific inhibition of translation. Several examples of both strategies, antisense RNA and antisense oligonucleotides, i.e., utilization of antisense RNA to inhibit the intracellular insulin-like growth factor I expression, are presented in relation to future applications in anti-tumor gene therapy. Gene therapy and gene immunotherapy provide new approaches for clinical trials. Among these are strategies using either antisense or triple helix technologies which lead to activation of the host immune system. In the latter technology the oligonucleotides that block gene expression are the triple-helix forming oligonucleotides (TFOs). They block transit of RNA polymerases by forming a triple-helical structure on DNA. These TFOs promise to be a new class of sequence-specific DNA-binding drugs which will target malignancies at the transcriptional level. TFOs may prove to be the basis for effective chemotherapy drugs for different cancers. Numerous cancers express IGF-I, a protein commonly associated with cellular normal and neoplastic growth and differentiation. IGF-I antisense RNA and IGF-I triple helix RNA-DNA strategies have been shown to induce the arrest and rejection of tumors in vivo such as rat glioma, mouse teratocarcinoma and rat and mouse hepatoma. These results have formed the basis for a gene therapy clinical trial for human primary glioblastoma (USA and Poland) and hepatocellular cancer (China).

General View of Antisense Strategy

The untranscribed DNA strand was considered as a stabilizer and a protector of genetic information. Recently, regulating the activity of this "antisense" strand has been suggested.[1] It has also been widely proven that a lot of genes present an open reading frame on the antisense strand; it is a frequent and nonrandom phenomenon, which depends on codons, and to a lesser degree, on the percentage of GC residues. Open reading frames on the antisense strand have been found in all

---

Insulin-Like Growth Factors, edited by Derek LeRoith, Walter Zumkeller and Robert Baxter. ©2003 Eurekah.com and Kluwer Academic / Plenum Publishers.

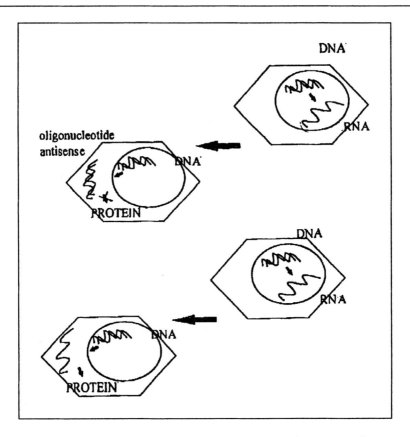

Figure 1. Basic mechanism of antisense oligonucleotides strategy to stop the protein synthesis.

genomes studied, as well in prokaryotes as in eukaryotes.[2-3] For example, in Wilms tumors some RNA is transcribed from the WT1 gene, which is encoded by the antisense strand of the first exon.[4] The role of this natural antisense RNA is not yet understood. More recently, it was found that the mouse thymidine kinase (Tk) gene expression is regulated by antisense transcription: a putative promoter in intron 3 of the murine Tk gene will transcribe this antisense RNA.[5]

However, concerning natural antisense RNAs in prokaryotes, it has been shown that they could play a regulatory role in replication, transcription or translation steps of some genes; it was demonstrated that the translation of the bacterial enzyme transposase was controlled by an antisense RNA.[6]

An antisense RNA, hybridized on its complementary sequence in mRNA, blocks the ribosome progression during the translation of the mRNA. This observation constitutes the " start point " of the antisense or non-sense approach[7] based on antisense RNA or antisense oligonucleotides (Fig. 1) to modulate artificially and specifically the expression of genes involved in important cellular processes. The mRNA complementary sequence is introduced in the cell either by a plasmid vector (dsDNA) coding for an antisense RNA or by a single stranded oligonucleotide form. The plasmid vector allows the intracellular transcription of antisense RNA which can strongly hybridize to the mRNA and stop the translation. Generally, an effective inhibition demands a high copy number of antisense RNA relative to mRNA. The antisense oligodeoxynucleotides, once in the cell, can stimulate the ribonuclease H after hybridization with target RNA. This enzyme, which is implicated in DNA replication, damages the RNA moiety of the hybrids formed in the cell (Fig. 2). On the other hand, the antisense oligonucleotide can remain nondegraded, hybridizing to another messenger and inducing the degradation of this mRNA. In this way, in the presence of RNAse H, the antisense oligonucleotide acts in a catalytic marrow, with the enzyme potentiating the antisense effect.[8]

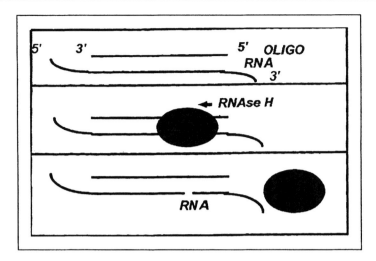

Figure 2. Implication of ribonuclease H in the dégradation of cellular RNA hybridized with antisense oligonucleotides.

The chemical stability of plasmid-derived antisense RNA seems much more efficient than that of antisense oligonucleotides delivered directly into cells. The antisense oligonucleotides are exposed to intra- and extracellular nuclease activity. In order to avoid nuclease-induced degradation and to maintain cellular penetration, the antisense primers should be chemically modified using phosphodiester backbone (phosphorothioate and phosphoramidate), sugars (morpholinos) or bases (C5 propyne).

The action of antisense oligomers can be reinforced by association with polycations like polyethyleneimine (PEI), polylysine or cationic lipids (DOTMA, DOTAP) facilitating endocytosis of oligomers.[9] These positively charged molecules are also used for transfection of cells with plasmids encoding antisense RNA. The difficulties of cellular penetration for antisense RNA or oligonucleotides are frequently similar. However, in spite of those problems, antisense RNA and antisense oligonucleotides are being used more and more in gene therapy assays. The first antisense oligonucleotide used in clinical pharmacology, anti-cytomegalovirus (Vitravene™), introduced the antisense strategy in human therapy, and especially in anti-tumor treatment.

Antisense Strategy in Protein Function Studies

The modulation of gene expression by antisense oligonucleotide or RNA approaches has been used in basic research and for clinical therapy purposes. These approaches are also useful for investigation of protein function in normal metabolism or in diseases such as cancer.

The antisense technology was used to study several protein actions: the alpha subunit of human chorionic gonadotrophin in choriocarcinoma cells;[10] the regulating protein E2F-1, in S phase of cell cycle, and its action on genes linked to proliferation;[11] nerve growth factor (NGF) in skin of transgenic mice, and its relationship with response to mechanical stimuli.[12] Lately, the antisense strategy is "classically" used to analyze gene expression and intron splicing. The same technology was employed to study the function of the heat shock protein hsp70, overexpressed in mouse fibrosarcoma cells; a direct correlation was found between hsp70 overexpression and tumorigenicity of cells. Cells which express high levels of hsp70 are resistant in vitro to cytotoxic cells and macrophages.[13]

The action of insulin-like growth factor I-binding protein 4 (IGFBP-4) has been also studied using antisense strategy. IGF BP-4 was shown to inhibit the mitogenic effect of exogenous IGF-I on HT29 tumor cells.[14] The same antisense strategy was applied to study p27kipl protein. The quiescent state of cells requires p27. The inhibition of p27 expression induces the progression of cell cycle and stimulates the cyclin D1 promoter activity. Hamster fibroblasts transformed in this way grow faster than nontreated cells, even in serum free medium.[15]

Antisense Strategy in Tumor Gene Therapy

Antisense Oncogenes

Antisense therapy targets genes implicated directly in the etiology of disease states, like proto-oncogenes or oncogenes. Antisense oligonucleotides were used to analyze the effect of negative regulation of c-Myc protein on chronic myeloid leukaemia (CML) cells. The decreasing expression of c-Myc protein, due to the antisense strategy, resulted in a total inhibition of cell proliferation (BV173 line) and a 50% inhibition in colony formation.[16] In this case, the c-myc antisense suppresses the malignancy of cells in vitro. c-myc expression is necessary for the stability of NIH3T3 cells transformed by ras oncogene.[17] c-myc antisense, transcribed in NIH3T3 cells transformed with ras, deprives the cells of the c-Myc protein. The cells lose their tumorigenic potential in nude mice and their capacity for growth in soft agar. These changes depend on the c-myc antisense expression . The antisense approach does not however modify the proliferation rate in vitro and the immortality of cells.

The c-myc antisense oligo-phosphorothioate inhibits tumor development in a Burkitt lymphoma transgenic mouse model. The antisense c-myc therapy could be used for prevention of lymphoma in asymptomatic patients carrying a c-myc translocation.[18] In melanoma tumor cells transfected in vitro with antisense c-myc oligonucleotide, the decreased c-Myc protein level inhibits the cell growth. These cells also present apoptotic characteristics. These antisense c-myc cells cause tumor growth inhibition and a decrease number of metastases in nude mice.[19]

Chromosomal translocation, that induces myeloid leukaemia, creates the junction of two genes: bcr/abl. This junction has been targeted using antisense oligonucleotides; the complementary antisense oligomers of the mRNA bcr/abl showed an antiproliferative effect on Philadelphia leukaemia cell line (Ph-1) in culture. Moreover, they inhibited the normal transformation induced by transfection with P210 bcr/abl oncoprotein P210 bcr/abl in mouse hematopoietic cells.[20] In normal hematopoietic cells, the block of c-abl proto-oncogene activity by antisense oligonucleotides revealed its importance in the S phase of the cell cycle and in the process of differentiation.[21]

In another in vivo experiment, leukaemic mice were injected with Philadelphia leukaemia cells transformed with antisense bcr/abl oligomers or c-myc oligomers, or with both simultaneously. The mice injected with the doubly transformed cells survived longer than those injected with the "single" oligomers, and the leukemia development was significantly retarded. The results suggest that the simultaneous targeting of several oncogenes may constitute an efficient therapy in cancer prevention.[22]

In spite of encouraging results, there still exists the possibility that certain leukaemia cells do not express the bcr/abl or escape the antisense therapy. For this reason, the c-myb proto-oncogene has also been targeted (Myb protein is expressed preferentially in hematopoietic cells). The studies in vitro, using antisense c-myc approach, showed that Myb is necessary for proliferation of normal and malignant cells, and its cellular level inversely correlates with differentiation. The anti-proliferative effect of c-myb antisense on leukaemia cells has been proven in vitro and in vivo in a SCID mouse model.[23] A clinical trial in patients suffering from chronic myeloid leukaemia using c-myb antisense approach has been started.

Antisense oligonucleotide strategy is also used in leukaemia and in lymphoma treatment to eliminate malignant blood cells. Thus, an oligonucleotide targeted bcr/abl RNA was used to purify a patient's bone marrow. This type of treatment in leukemia patients gave a total hematological remission.[9]

The importance of p53 gene alterations playing a role in etiology of different types of cancer is clearly described in literature. Ovary cancer cell lines show that only 50% of the lines have mutations in p53 gene ; thus, a mutation in p53 is not necessary for development of ovary cancer. The antisense oligonucleotides synthesized specifically for the p53 mutated sequence, inhibit the proliferation of cancer cells, whereas those targeted wild type gene do not inhibit lines which carry an intact p53 gene.[24] This observation could indicate that p53 antisense approach selectively inhibits the growth of mutated cells.

Among other genes implicated in ovary cancer, c-erbB2 proto-oncogene was also a target for antisense strategy. c-erbB2 phosphothiorate oligonucleotides induced a decrease in protein production and in the proliferation capacity of ovary cancer cells.[25]

Antisense of Genes Encoding Enzymes

The protein kinase C-alpha, PKC, is implicated in the regulation of cell proliferation and differentiation. In addition, it is associated with multiple drug resistance, related to the activity of the P-glycoprotein. The P-glycoprotein, a product of the "multidrug resistance" MDR gene, constitutes the evacuation pump of cell membranes. This action, amplified in transformed cells, is the principal cause of anticancer chemotherapy failures, i.e., the MCF-7 carcinoma cells are resistent to chemotherapy due to P-glycoprotein action. PKC antisense decreases PKC activity (up to 75%), and induces a two-fold increase in drug retention and a three-fold increase in doxorubicin cytotoxicity. Thus the PKC antisense strategy should become a good adjuvant for anticancer therapy.[26]

In a similar way, the antisense PKC gene was transfected into a U-87 glioblastoma cell line. After transfection, the total PKC activity decreased by 95%, the transfected cells showed a doubling in division time, and presented a decreased sensitivity to Ro 31-8220, the PKC inhibitor. These cells fail to produce tumors in nude mice. This strategy could be considered as one possible treatment for glioblastomas.[27]

The inhibition of protein kinase C delta expression (PKC delta) was studied using antisense oligomers in a model of mouse erythroleukaemia. PKC delta seems to be an essential enzyme for the maintenance of the undifferentiated phenotype and/or for the regulation of the initiation of the differentiation state.[28]

A phosphorothioate oligonucleotide antisense of the 3' untranslated portion of the PKC alpha gene showed its efficiency by producing the regression of tumors in an animal model of lung carcinoma. As a consequence, a clinical study using this oligo has been approved and clinical positive results have been observed in 3 of 17 patients treated for ovary cancer.[9]

In a study concerning the proliferation of bone marrow hematopoietic stem cells, Raf-1 protein kinase was recognized as an essential element for multiplication of growth factor-dependent cells; Raf-1 kinase is activated by the phosphorylation induced by hematopoietic growth factors. c-raf antisense oligonucleotides inhibited T lymphocytes and myeloid cells proliferation and could constitute a tool for prevention of proliferation of hematopoietic malignant cells.[29]

Antisense of Proteins Related to MHC Expression

The antisense technology was also employed to study mechanisms which allow tumor cells expressing HER-2/neu to resist the cytotoxic action of cytotoxic lymphocytes, natural killer cells (NK) and lymphokine-activated killer cells (LAK). Carcinoma cells expressing HER-2/neu present high levels of MHC-I antigen. Beta-2-microglobulin antisense oligonucleotides induce a decrease of class 1 molecules in selected cells (Raji or U937 lines) accompanied by an increase in sensitivity to NK and LAK effectors. Moreover, cells not expressing HER-2/neu, become more sensitive to the cytotoxicity generated by NK and LAK after the negative regulation of MHC-I is achieved by antisense approach. As NK and LAK are concerned, the resistance of HER-2/neu cells to cytotoxicity seems to be linked to MHC-I overexpression.[30]

The Ii protein blocks the antigen binding site in MHC-II molecules. In consequence, the suppression of Ii should increase the presentation of cancer epitopes by the CD4+ helper cells. Indeed, Ii antisense oligonucleotides eliminate the expression of Ii protein in SaI murine sarcoma cells. The antisense effect is observable in 35%-55% of cells. The inoculation of cells missing Ii into mice protects the animals from tumor development after subsequent challenges with SaI cells.[31]

Antisense of Genes Encoding Growth Factors

Rat 9L glioma, similar to human gliomas, secretes transforming growth factor beta (TGF beta) which is an immunosuppressor. In order to stimulate the immune response against glioma cells, 9L cells have been transfected with a plasmid containing TGF beta gene inserted in the antisense orientation. The 100% survival was observed in rats bearing glioma tumors and injected with "antisense TGF beta" cells. The tumors regressed and the cytolytic activity of cytotoxic cells increased.[32]

The role of the epidermal growth factor receptor (EGFR) in malignant glioma development has been studied using a vector containing complementary antisense oligonucleotides to EGFR cDNA. In C6 glioma cells transfected with this vector, a dramatic decrease of endogenous EGFR RNA was observed. The transfected cells injected into the brains of Wistar rats did not induce tumors. The plasmid encoding the antisense EGFR injected into established tumors induced tumor regression.[33]

The vascular-endothelial growth factor (VEGF) antisense approach has been also used to analyze the effect on the development of glioma tumors. VEGF is an endothelial mitogen factor which induces angiogenesis in solid tumors. Rat C6 glioma cells transfected with a VEGF antisense expression vector showed in an in vivo experiment weak development of tumors when compared to tumors developed after injection of non-transfected cells.[34]

On the other hand, the same C6 glioma cells transfected with antisense IGF-I receptor expression vector and injected into rats induced the regression of pre-established tumors in rats, probably by the stimulation of cytotoxic lymphocyte proliferation.[35] The same antisense IGF-I-R cells injected into athymic mice did not induce tumor regression.[36]

The successful transfer of antisense technology to clinical gene therapy was obtained using antisense IGF-I strategy. The IGF-I cDNA was inserted into a plasmid containing the Epstein-Barr virus origin of replication. The transfer of the plasmid into either rat glioma or rat hepatoma or mouse teratocarcinoma cells suppressed cell tumorigenicity. Moreover, the transfected cells injected subcutaneously in animals bearing the corresponding tumors induced T CD8 mediated immune response and suppressed the established tumors.[37-40] The IGF-I antisense-expressing hepatoma cells also demonstrated the phenomenon of apoptosis.[40,41] The IGF-I antisense approach was introduced in a clinical trial of glioblastoma treatment (Cleveland, USA) and in the treatment of hepatocarcinoma (Shanghai, China).[42]

The expression of PTHrP peptide synthesized by rat mGH3 pituitary metastatic tumor cells was suppressed in the tumor cells transfected with antisense oligonucleotides of PTHrP. The injection of transfected cells into animals prevented the formation of metastasis and decreased the size of the primary tumor. Thus, the application of antisense technology could be a plausible strategy for the treatment of metastases of somatotropic tumor.[43]

In conclusion, although several aspects of antisense oligonucleotides or antisense RNA strategies need to be improved, the results obtained are promising and the multiple applications appear appropriate with basic biological research, pharmacology or gene therapy as examples. Continued genomic investigation brings sequences of new molecules, often encoding proteins with unknown functions, and the expression of these proteins can be modulated by an antisense (RNA or oligonucleotide). Thus new drug receptors, viruses, or signalling pathways will be discovered and be validated.

Triple Helix Strategy in Tumor Gene Therapy

The aim of anti-tumor chemotherapy is to block the abnormal proliferation of malignant cells. Unfortunately, some tumoral cells can escape chemotherapy by developing multidrug resistance via the over-expression and sometimes amplification of a transmembrane glycoprotein (mdrl gene). This protein is likely responsible for the rapid efflux of many anti-tumor chemicals. The human mdrl gene was chosen as target for triplex-forming oligonucleotides (TFO). It was shown that this TFO downregulated specifically the mdrl expression in a human drug-resistant cell line.[44]

Triple helix strategy was applied to the ras oncogenes which are the most frequently activated oncogenes in human cancer. In vitro transcription of human Ha-ras was inhibited by triplex-forming oligonucleotides targeted to sequences recognized by the Sp I trancription factor.[45] Using transient transfection assays, it was demonstrated that a purine-rich TFO could also inhibit the transcription of murine c-Kras gene in NIH 3T3 cells.[46] At present, no data is available concerning the effect of these TFOs on the expression of endogeneous ras oncogenes in cell lines.

C-myc protooncogene was chosen as target for triplex-forming oligonucleotides (an overexpression of c-myc protein appears in a variety of malignant cells). Endogeneous c-myc gene expression was inhibited by TFO treatment in a human cervical carcinoma cell line (Hela)[47] and in a breast carcinoma cell line (MCF-7).[48] Triplex-forming oligonucleotides were also shown to bind in vitro to the human EGF receptor promoter,[49] and to inhibit in vitro transcription of HER2/neu gene.[50] The transcription of endogenous human HER-2/neu oncogene, which is overexpressed in breast cancer and other human malignancies, has been inhibited by TFO treatment of a breast carcinoma cell line (MCF-7).[51]

Growth factors are known to play a role in tumorigenesis, and thereby represent relevant targets for antigene therapies. The synthesis of human tumor necrosis factor (TNF), which acts as an autocrine growth factor in various tumor cell lines including neuroblastoma and glioblastoma, has been blocked

```
                    ANTIPARALLEL TRIPLE HELIX

       3'...AGAAGAGGGAGAGAGAGAGAAGG...5'   third strand
            ************************
       5'...AGAAGAGGGAGAGAGAGAGAAGG...3'   first strand
            IIIIIIIIIIIIIIIIIIIIIII
       3'...TCTTCTCCCTCTCTCTCTCTTCC...5'   second strand
```

Figure 3. Potential structure of antiparallel RNA – DNA – DNA triple helix complex : first strand and second strand constitute the genomic DNA (IGF-I); third strand constitutes the homopurine RNA.
***** Hoogsteen hydrogen bonds
IIIIIIII Watson-Crick bonds

by triplex-forming oligonucleotide treatment.[52] Growth of the human glioblastoma tumor cells decreased in a dose-dependant manner with the TFO treatment. These results demonstrate the potential therapeutic utility of triplex-mediated strategy in modulating the expression of oncogenes or growth factors.

The examples of the inhibitory activity of triplex-forming oligonucleotides on target genes involved in tumorigenesis are becoming more numerous.[53,54,55,56] Most of the TFOs are targeted to polypurine-polypyrimidine sequences located in control regions of the gene of interest and are delivered via transfection with various chemical carriers. An alternative way to introduce TFOs in cells is to use a plasmid vector that can drive the synthesis of an RNA triplex-forming oligonucleotide within the cells. This TFO generated in situ, is therefore protected from degradation by nucleases and could reach its DNA target without being trapped in lysosomal vesicles. Obviously, it could be transfected in cells via either standard cell transfection procedures or via ways similarly used in virus-based gene therapy. An application of this triplex-based approach has been used for the inhibition of the insulin-like growth factor-I protein which plays a major role in tumorigenesis of glioblastoma and hepatocarcinoma[57,58,59] (Fig. 3).

Gene Therapy of Tumors Expressing IGF-I

IGF-I and its receptor (IGF-IR) seem to play a major role in various tumors such as glioblastoma, prostatic carcinoma, hepatocarcinoma or cancers of bone and of mammary gland. Glioblastoma is the most frequent brain tumor in man and is usually fatal. Both human and rat glioma cells express high amounts of IGF-I.[60,61] When C6 rat glioma cells were transfected with a vector containing an antisense IGF-I cDNA transcriptional cassette, they lost their tumorigenicity.[37] Moreover, when antisense-transfected cells were injected into syngeneic animals with existing glioblastoma tumors, the established tumors regressed.[62] IGF-I seems to act by enabling cancer cells to evade immune recognition.[62,63] Down-regulation in the expression of IGF-I is coincident with a reappearance of B-7 and MHC class I antigens at the surface of transfected cells.[64] When injected subcutaneously, the transfected cancer cells initiate an immune reaction involving CD8+ lymphocytes which results in regression of tumors.[38,39,62]

IGF-I expression was also inhibited following stable transfection of C6 rat glioblastoma cells with a DNA plasmid encoding a triplex forming RNA (oligopurine sequence); this RNA is targeted to the IGF-I promoter via triple helix formation. The ability of this vector to inhibit in vitro IGF-I gene expression as well as in vivo tumorigenesis was successfully tested.[57] The C6 glioma cells transfected with this triple-helix vector displayed morphological changes, upregulation of MHC class I antigens and B7 antigen accompanied by apoptosis,[58] and increased expression of protease nexin I. Dramatic inhibition of tumor growth occurred in nude mice following injection of transfected C6 cells.[57] Similar results of IGF-I triple helix strategy were obtained using syngeneic model of PCC3 derived mouse tearatocarcinoma.[65] Another utilization of triplex-forming oligonucleotide delivered inside to cells via transcription of a plasmid vector was recently described with the receptor of the IGF-1 gene in glioblastoma cells.[66] This RNA TFO was targeted to the IGF-1 receptor gene (IGF-I-R).

Hepatocellular carcinoma (HCC) causes more than 1 million deaths per year in the world. Prognosis depends on the possibility of resection of tumor. Cellular gene therapy offers new possibilities for treatment. Inhibition of cellular IGF-I expression by transfection with the vectors that continuously synthesize either antisense RNA to IGF-I cDNA or RNA strands forming a triple helix with IGF-I cDNA, causes tumor arrest when the transfected mhAT1F1 hepatoma cells are injected into mice bearing hepatoma.[40,59] These cells also stopped producing IGF-I and recovered MHC-I expression accompanied by apoptosis but showed a reduction in IL-10 and TNF-alpha production. The injection of transfected cells into mice with terminal-phase significantly prolonged their survival. The results suggest that the injection of transfected hepatoma cells using antisense or triple helix approaches could constitute a vaccine against murine hepatoma. The triple-helix strategy for clinical gene therapy of hepatoma has currently been introduced at the University Hospital of Krakow, Poland.[59]

References

1. Ring BZ, Roberts JW. Function of a nontranscribed DNA strand site in transcription elongation. Cell 1994; 78(2):317-24.
2. Merino E, Balbas P, Puente JL et al. Antisense overlapping open reading frames in genes from bacteria to humans. Nucleic Acids Res 1994; 22(10):19-38
3. Yomo T, Urabe I. A frame-specific symmetry of complementary strands of DNA suggests the existence of genes on the antisense strand. J Mol Evol 1994; 38(2):113-20.
4. Campbell CE, Huang A, Gurney AL et al. Antisense transcripts and protein binding motifs within the Wilms tumor (WT1) locus. Oncogene 1994; 9(2):583-95.
5. Sutterluety H, Bartl S, Doetzlhofer et al. Growth-regulated antisense transcription of the mouse thymidine kinase. Nucleic Acids Res 1998; 26:4989-95.
6. Weintraub H, Izant JG, Harland RM. Antisense RNA as a molecular tool for genetic analysis. Trends Gen 1985; 1(1):23-5.
7. Rubinstein JL, Nicolas JF, Jacob F. L'ARN non sens (nsARN) : un outil pour inactiver spécifiquement l'expression d'un gène donné in vivo. C R Acad Sci Paris 1984; 299(8):271-4.
8. Hélène C. Specific regulation of gene expression by antisense, sense and antigene nucleic acids. Biochem Biophys Acta 1990; 1049:99-125.
9. Galderisi U, Cascino A, Giordano A. Antisense oligonucleotides as therapeutic agents. J Cell Phys 1999; 181:251-7.
10. Cao H, Lei ZM, Rao CV. Consequences antisense human chorionic gonadotrophin-alpha subunit cDNA expression in human choriocarcinoma JAR cells. J Mol Endocrinol 1995; 14(3):337-47.
11. Sala A, Nicolaides NC, Engelhard A et al. Correlation between E2F-1 requirement in the S phase and E2F-1 transactivation of cell cycle-related genes in human cells. Cancer Res 1994; 54(6):1402-6.
12. Davis BM, Lewin GR, Mendell LM et al. Altered expression of nerve growth factor in the skin of transgenic mice leads to changes in response to mechanical stimuli. Neurosciences 1993; 56(4):789-92.
13. Jaattela M. Over-expression of hsp70 confers tumorigenicity to mouse fibrosarcoma cells. Int J Cancer 1995; 60(5):689-93.
14. Singh P, Dai B, Dhruva B et al. Episomal sense and antisense insulin like growth factor (IGF)-binding protein-4 complementary DNA alters the mitogenic response of a human colon cancer cell line (HT-289) by mechanisms that are independent of and dependent upon IGF-I. Cancer Res 1994; 54(24):6563-70.
15. Rivard N, Allemain G, Bartek J et al. Abrogation of p27kipl by cDNA antisense suppresses quiescence in fibroblasts. J Biol Chem 1996; 271(31):18337-41.
16. Nieborowska-Skorska M, Ratajczak MZ, Calabretta B et al. The role of c-Myc protooncogene in chronic myelogenous leukemia. Folia Histochem Cytobiol 1994; 32(4):231-4.
17. Sklar MD, Thompson E, Welsh MJ et al. Depletion of c-myc with specific antisense sequences reverses the transformed phenotype in ras oncogene-transformed NIH 3T3 cells. Mol Cell Biol 1991; 11(7):3699-710.
18. Huang Y, Snyder R, Kligshteyn M et al. Prevention of tumor formation in a mouse model of Burkitt's lymphoma by 6 weeks of treatment with anti-c-myc DNA phosphorothioate. Mol Med 1995; 1(6):647-58.
19. Leonetti C, D'Agnano I, Lozupone F et al. Antitumor effect of c-myc antisense phosphorothioate oligodeoxynucleotides on human melanoma cells in vitro and in mice. J Natl Cancer Inst 1996; 88(7):419-29.
20. Okabe M, Kunieda Y, Miyagishima T et al. BCR/ABL oncoprotein-targeted antitumor activity of antisense oligodeoxynucleotides complementary to bcr/abl mRNA and herbimycin A, an antagonist of protein tyrosine kinase: inhibitory effects on in vitro growth of Ph1- positive leukemia cells and BCR/ABL oncoprotein-associated transformed cells. Leuk Lymphoma 1993; 10(4-5):307-16.
21. Rosti V, Bergamaschi G, Lucotti C et al. Oligodeoxynucleotides antisense to c-abl specifically inhibit entry into S phase of CD34+ hematopoietic cells and their differentiation to granulocyte—macrophage progenitors. Blood 1995; 86(9):3387-93.
22. Skorski T, Nieborowska-Skorska M, Campbell K et al. Leukemia treatment in severe combined immunodeficiency mice by antisense oligodeoxynucleotides targeting cooperating oncogenes. J Exp Med 1995; 182(6):1645-53.

23. Calabretta B, Skorski T, Ratajczak M et al. Antisense strategies in the treatment of leukemias. Sem Oncology 1996; 23(1):78-87.
24. Skilling JS, Squatrito RC, Connor JP et al. p53 gene mutation analysis and antisense mediated growth inhibition of human ovarian carcinoma cell lines. Gynecol Oncol 1996; 60(1):72-80.
25. Wiechen K, Dietel M. c-erbB-2 anti-sense phosphorothioate oligodeoxynucleotides inhibit growth and serum-induced cell spreading of P185c-erbB-2-overexpressing ovarian carcinoma cells. Int J Cancer 1995; 63(4):604-8.
26. Ahmad S, Glazer RI. Expression of the antisense cDNA for protein kinase C alpha attenuates resistance in doxorubicin-resistant MCF-7 breast carcinoma cells. Mol Pharmacol 1993; 43(6):858-62.
27. Ahmad S, Mineta T, Martuza RL et al. Antisense expression of protein kinase C alpha inhibits the growth and tumorigenicity of human glioblastoma cells. Neurosurgery 1994; 35(5):904-8.
28. Pessino A, Passalacqua M, Sparatore B et al. Antisense oligodeoxynucleotide inhibition of delta protein kinase C expression accelerates induced differentiation of murine erythroleukaemia cells. Biochem J 1995; 312(Pt 2):549-54.
29. Keller JR, Ruscetti FW, Heidecker G et al. The effect of c-raf antisense oligonucleotides on growth factor-induced proliferation of hematopoietic cells. Curr Top Microbiol Immunol 1996; 211:43-53.
30. Lichtenstein A, Fady C, Gera JF et al. Effects of beta-2 microglobulin anti-sense oligonucleotides on sensitivity of HER2/neu oncogene-expressing and non expressing target cells to lymphocyte-mediated lysis. Cell Immunol 1992; 141(1):219-32.
31. Qiu G, Goodchild J, Humphreys R et al. Cancer immunotherapy by antisense suppression of Ii protein in MHC-II positive tumor cells. Cancer Immunol Immunother 1999; 48(9):499-506.
32. Fakhrai H, Dorigo O, Shawler DL et al. Eradication of established intracranial rat gliomas by transforming growth factor beta antisense gene therapy. Proc Natl Acad Sci USA 1996; 93(7):2909-14.
33. Pu P, Liu X, Liu A et al. Inhibitory effect of antisense epidermal growth factor receptor on the proliferation of rat C6 glioma cells in vitro and in vivo. J Neurosurg 2000; 92(1):132-9.
34. Saleh M, Stacker SA, Wilks AF. Inhibition of growth of C6 glioma cells in vivo by expression of antisense vascular endothelial growth factor sequence. Cancer Res 1996; 56(2):393-401
35. Resnicoff M, Sell C, Rubini M et al. Rat glioblastoma cells expressing an antisens RNA to the insulin-like growth factor receptor are non tumorigenic and induce regression of wild-type tumors. Cancer Res 1994; 54:2218-22.
36. Resnicoff M, Li W, Basak S et al. Inhibition of rat C6 glioblastoma tumor growth by expression of insulin-like growth factor I receptor antisense mRNA. Cancer Immunol Immunother 1996; 42(1):64-8.
37. Trojan J, Blossey B, Johnson T et al. Loss of tumorigenicity of rat glioblastoma directed by episome-based antisense cDNA transcription of insulin-like growth factor I. Proc Natl Acad Sci USA 1992; 89:4874-8.
38. Trojan J, Johnson T, Rudin S et al. Gene therapy of murine teratocarcinoma: separate functions for insulin-like growth factors I and II in immunogenicity and differentiation. Proc Natl Acad Sci USA 1994; 91:6088-92.
39. Lafarge-Frayssinet C, Sarasin A, Duc HT et al. Gene therapy for hepatocarcinoma: antisense IGF-I transfer into a rat hepatoma cell line inhibits tumorigenesis into syngeneic animal. Cancer Gene Ther 1997; 4:276-85.
40. Upegui-Gonzalez LC, Duc H, Buisson Y et al. Use of antisense strategy in the treatment of the hepatocarcinoma. Adv Exp Med Biol 1998; 451:35-42.
41. Szabo I, Tarnawski AS. Apoptosis in the gastric mucosa: molecular mechanisms, basic and clinical implications. J Physiol Pharmacol 2000; 51(1):3-15.
42. Anthony D, Pan Y, Wu S et al. Ex vivo and in vivo IGF-I antisense RNA strategies for treatement of cancers in humans. Adv Exp Med Biol 1998; 45:27-34.
43. Akino K, Ohtsuru A, Yano H et al. Antisense inhibition of parathyroid hormone-related peptide gene expression reduces malignant pituitary tumor progression and metastases in the rat. Cancer Res 1996; 56(1):77-86.
44. Scaggiante B, Morassutti C, Tolazzi G et al. Effect of unmodified triple helix-forming oligodeoxyribonucleotide targeted to human multidrug-resistance gene mdr1 in MDR cancer cells. FEBS Lett 1994; 352:380-4.
45. Mayfield C, Ebbinghaus S, Gee I et al. Triplex formation by the human Ha-ras promoter inhibits Sp1 binding and in vitro transcription. J Biol Chem 1994; 269:18232-8.
46. Alunni-Fabbroni M, Pirolli D, Manzini G et al. (A,G)-oligonucleotides form extraordinary stable triple helices with a critical R.Y sequence of the murine c-Ki-ras promoter and inhibit transcription in transfected NIH 3T3 cells. Biochemistry 1996; 35:16361-9.
47. Postel EH, Flint SJ, Kessler DJ et al. Evidence that a triplex-forming oligodeoxyribonucleotide binds to the c-myc promoter in HeLa cells, thereby reducing c-myc mRNA levels. Proc Natl Acad Sci USA 1991; 88:8227-31.
48. Thomas T, Faaland C, Gallo M et al. Suppression of c-myc oncogene expression by a polyamine-complexed triplex forming oligonucleotide in MCF-7 breast cancer cells. Nucleic Acids Res 1995; 23:3594-9.
49. Durland R, Kessler D, Gunnell S et al. Binding of triple helix forming oligonucleotides to sites in gene promoters. Biochemistry 1991; 30:9246-55.
50. Ebbinghaus S, Gee J, Rodu B et al. Triplex formation inhibits HER-2/neu transcription in vitro. J Clin Invest 1993; 92:2433-9.
51. Porumb H, Gousset H, Letellier R et al. Temporary ex vivo inhibition of the expression of the human oncogene HER2 (NEU) by a triple helix-forming oligonucleotide. Cancer Res 1996; 56:515-22.

52. Aggarwal B, Schwarz L, Hogan M et al. Triple helix-forming oligodeoxyribonucleotides targeted to the human tumor necrosis factor (TNF) gene inhibit TNF production and block the TNF dependent growth of human glioblastoma tumor cells. Cancer Res 1996; 56:5156-64.
53. Chan PR, Glazer RM. Triplex DNA: Fundamentals, advances and potential applications for gene therapy. J Mol Med 1997; 75:267-82.
54. Giovannangeli C, Hélène C. Progress in developments of triplex-based strategies. Antisense Nucleic Acid Drug Dev 1997; 7:413-21.
55. Maher III LJ. Prospects for the therapeutic use of antigene oligonucleotides. Cancer Investigation 1996; 14:66-82.
56. Vasquez KM, Wilson IH. Triplex-directed modification of genes and gene activity. Trends in Biochem Sci 1998; 23:4-9.
57. Shevelev A, Burfeind P, Schulze E et al. Potential triple helix-mediated inhibition of IGF-I gene expression significantly reduces tumorigenicity of glioblastoma in an animal model. Cancer Gene Ther 1997; 4:105-12.
58. Ly A, Duc HT, Kalamarides M et al. Human glioma cells transformed by IGF-I triple-helix technology show immune and apoptotic characteristics determining cell selection for gene therapy of glioblastoma. J Clin Pathol: Mol Pathol 2001; 54(4):230-9.
59. Upegui-Gonzalez LC, Ly A, Sierzega M et al. IGF-I triple helix strategy in hepatoma treatment Hepato-Gastroentero 2001; 48:656-62.
60. Kiess W, Lee L, Graham DE et al. Rat C6 glial cells synthesize insulin-like growth factor I (IGF-I) and express IGF-I receptors and IGF-II/mannose 6-phosphate receptors. Endocrinology 1989; 124:1727-36.
61. Antoniades HN, Galanopoulis T, Nevile-Golden J et al. Expression of insulin like growth factor I and II and their receptor mRNAs in primary human astrocytomas and meningiomas : In vivo studies using in situ hybridization and immunocytochemistry. Int J Cancer 1992; 50:215-22.
62. Trojan J, Johnson TR, Rudin SD et al. Treatment and prevention of rat glioblastoma by immunogenic C6 cells expressing antisense insulin-like growth factor I RNA. Science 1993; 259:94-7.
63. Baserga R. The insulin-like growth factor I receptor: a key to tumor growth?. Cancer Res 1995; 55 :249-52.
64. Trojan J, Duc H, Upegui-Gonzalez L et al. Presence of MHC-I and B-7 molecules in rat and human glioma cells expressing antisense IGF-I mRNA. Neurosci Lett 1996 212:9-12.
65. Ly A, François JC, Duc HT et al. IGF-I triple helix technology changes tumorigenicity of embryonal carcinoma cells by immune and apoptotic effects. Life Sciences 2000; 68:307-9.
66. Rininsland F, Johnson T, Chernicky C et al. Suppression of insulin-like growth factor type I receptor by a triple-helix strategy inhibits IGF-I transcription and tumorigenic potential of rat C6 glioblastoma cells. Proc Natl Acad Sci USA 1997; 94:5854-9.

CHAPTER 23

The IGF System in Breast Cancer

Janet L. Martin

Abstract

Breast cancer remains one of the most common causes of mortality in women worldwide, despite intensive research aimed at identifying the factors involved in its establishment and progression. Many lines of evidence support a key role for the IGF system in the development and progression of breast cancer. IGFs are potent stimulators of breast cancer cell proliferation, and animal models of breast cancer have shown that strategic blockade of IGF action can reduce the formation, growth and metastatic spread of breast tumors in vivo. Dissection of the underlying mechanisms involved in the proliferative and antiapoptotic effects of IGFs continues to reveal a high level of complexity within the IGF axis itself, and interactions between the IGF axis and multiple other growth regulatory systems in the breast. Understanding of the IGFBPs as regulatory factors in IGF action is now being expanded to encompass an appreciation of their intrinsic bioactivity in modulating breast cancer cell proliferation and apoptosis. Recent epidemiological studies indicating that there may be prognostic and diagnostic value in evaluating the IGF axis in breast cancer highlight the significance of this growth regulatory system in breast cancer. In order to fully appreciate the role of the IGF axis in the tumorigenic process and the therapeutic value which may be afforded by modulation of components of this system, a clear understanding of the mechanisms and regulation of IGF action in breast tumors is essential.

Introduction

Breast cancer affects one in ten women, and is one of the most common causes of mortality among women worldwide. A number of factors are known to increase an individual's risk of developing breast cancer, including family history of the disease, the use of oral contraceptives, alcohol consumption and other dietary influences, but the relationship between these risk factors and the molecular basis of malignant breast disease remains largely unknown. Like other cancers, breast cancer is characterized by uncontrolled proliferation and spread of abnormal cells, a process facilitated by derangement of the regulatory mechanisms which function in normal cells to balance proliferation and death, and thus ensure tissue homeostasis.

Numerous observations in vitro and in vivo support an important role for the IGF system in breast tumorigenesis. Multiple components of the IGF axis are changed in the circulation of breast cancer patients, and the expression of IGFs, their receptors and binding proteins is altered in malignant compared with non-malignant breast tissue. Experiments have shown that antibody blockade of the type 1 IGF receptor (IGFR1) may result in reduced breast cancer cell proliferation and anchorage-independent growth in vitro, and inhibition of the growth of breast cancer cell-derived tumors as xenografts in vivo.[1,2] Collectively these observations provide a sound rationale for targeting the IGF system in the treatment of breast cancer.

The IGF System in Breast Cancer: Clinical Studies

The Clinical Significance of Endocrine IGFs in Breast Tumorigenesis

The role of circulating IGFs in the development and progression of breast cancer is controversial. Many clinical studies over the years have demonstrated significantly elevated IGF-I in the serum of

---

Insulin-Like Growth Factors, edited by Derek LeRoith, Walter Zumkeller and Robert Baxter. ©2003 Eurekah.com and Kluwer Academic / Plenum Publishers.

Table 1. Changes in the IGF system in the circulation during treatment of breast cancer

	Increased in Circulation	Decreased in Circulation
IGF-I	Aromatase inhibitor glutethimide[8,20] Aromatase inhibitor 4-hyroxyandrostenedione[19] Aromatase inhibitor letrozole[21] Medroxyprogesterone acetate[164] megestrol acetate[20] growth hormone[165]	Tamoxifen[7-10,12,164] droloxifene[11] Goserelin[8] diethylstilboestrol[166] bromocriptine/octreotide[167] fenretinide[13-15] melatonin[168]
IGF-II		diethylstilboestrol[166]
IGFBP-1	tamoxifen[12,164,169] droloxifene[11] diethylstilboestrol[166]	Medroxyprogesterone acetate[164]
IGFBP-2	glutethimide[20]	diethylstilboestrol[166]
IGFBP-3	fenretinide[15]	diethylstilboestrol[166] megestrol acetate[20]
IGFBP-4	diethylstilboestrol[166]	

breast cancer patients compared with age-matched healthy women,[3-6] with apparently no relationship with other prognostic indicators such as estrogen receptor or nodal status.[4] Treatment of the disease, in particular by administration of antiestrogens[7-12] or retinoids[13-15] is frequently accompanied by a decrease in the level of IGF-I in the circulation (Table 1), and this has been suggested to contribute to the efficacy of these agents.[10] Changes in the levels of IGFBPs in response to some of these agents has also been reported (Table 1), and collectively, these clinical observations have given rise to the hypothesis that a high concentration of free IGF-I in the circulation promotes tumor growth because of its increased bioavailability compared with IGFs in binary or ternary complexes. Decreased serum levels of IGFBP-1 and IGFBP-6,[16] and increased proteolyzed IGFBP-3 in the circulation of breast cancer patients compared to women with benign breast disease have also been reported,[17] and these changes might also increase the proportion of free IGFs in the circulation.

However there is no direct evidence that circulating IGFs contribute to the growth of breast tumors. Furthermore, it must be remembered that the levels of IGF-I and IGFBPs in the circulation are presumably a reflection of their expression by tissues, and as will be discussed in greater detail later, it appears likely that autocrine/paracrine IGF and IGFBP activity may be of greater importance in breast tumorigenesis than the proteins deriving from the circulation. There are, however, clearly important ramifications if indeed endocrine IGF-I is involved in breast tumor growth. Some current and proposed antitumor therapies, including suramin[18] and aromatase inhibitors[19-21] increase rather than decrease IGF-I in the circulation. Such changes may be accompanied by modulation of other members of the axis which would theoretically limit the impact of increased endocrine IGFs; for example the rise in plasma IGF-I associated with use of aromatase inhibitors is accompanied by an increase in IGFBP-2 and a decrease in IGFBP-3 proteolysis.[20] Nevertheless, until the role of endocrine IGF-I in breast tumorigenesis is delineated, careful monitoring of patients receiving treatments which modulate this axis is imperative.

Prognostic Significance of IGF-I and IGFBP-3 in Breast Cancer

Although the functional significance of elevated IGF-I in the circulation of breast cancer patients remains unclear, a number of studies indicate that there is increased risk of breast cancer associated with specific changes in the endocrine IGF system, but only for pre-menopausal women. Cross-sectional studies comprising premenopausal women with newly diagnosed breast cancer and age-matched controls have shown that a high concentration of IGF-I in the circulation is associated with increased risk of breast cancer,[6,22,23] and, conversely, high IGFBP-3 is associated with decreased risk.[6] More extensive analysis of these relationships has indicated that the level of IGFBP-3, the primary carrier of IGF-I and IGF-II in the circulation, is an important consideration when assessing IGF-I and relative risk. A number of case-control studies have now shown that the relative risk associated with a high concentration of IGF-I is further increased if IGFBP-3 is low,[5,6,23] suggesting that a high IGF:IGFBP-3 ratio is positively associated with risk of breast cancer. A cross-sectional analysis which examined associations between IGF-I, IGFBP-3 and mammographic density, a very strong predictor of breast cancer risk, gave results consistent with this;[24] IGF-I was independently positively associated with mammographic density, and the IGF:IGFBP-3 ratio showed the strongest positive association with breast density, and by inference breast cancer risk, in pre-menopausal women.

The possibility that measurement of IGF-I and IGFBP-3 in the circulation could be of predictive significance was addressed in a prospective study within the Nurses' Health Study cohort.[25] Consistent with the findings of the case-control studies, this study identified a positive correlation between circulating IGF-I and subsequent risk of developing breast cancer in pre-menopausal but not post-menopausal women, and also indicated that although IGFBP-3 itself showed a non-significant inverse association with breast cancer risk, a low circulating concentration of IGFBP-3 increased the relative risk associated with a high level of IGF-I. Thus despite a lack of a clearly defined functional role for circulating IGF-I and IGFBP-3 in breast cancer, it appears that there may be prognostic value in their assessment for pre-menopausal women. Future studies will no doubt help to further delineate the significance of endocrine IGFs and IGFBPs in breast cancer.

Mechanisms of IGF Action in Breast Cancer

The Autocrine/Paracrine Model of IGF Action in Breast Tumorigenesis

Early studies using breast cancer cell lines derived from tumors indicated that these cells express IGF-like peptides[26-30] and cell-surface IGFR1 capable of ligand binding,[27,31-35] raising the possibility that a functional autocrine IGF growth loop exists in breast tumors in vivo. Such autocrine IGF action was postulated to mediate the proliferative effects of other growth factors and hormones such as estrogen, and oncogenic transformation was also suggested to be mediated, at least in part, by changes in the expression of growth-stimulatory IGFs by breast epithelial cells.[26,28,36-38]

However clear evidence that IGF peptides are secreted by breast cancer cells and contribute to autocrine growth has been made difficult by a number of confounding factors. IGFs present in fetal bovine serum used in cell growth media will cross-react in many IGF radioimmunoassays, and IGFBPs—originating from fetal bovine serum and the cancer cells—may also interfere in the measurement of IGFs.[39] The results of many studies have now indicated that IGF-I is not produced by malignant breast epithelial cells, with no IGF-I mRNA or secreted protein detected in any breast cancer cell line in vitro.[35] Although IGF-II mRNA has been reported in some breast cancer cells,[35] it is not clear that expression of the IGF-II transcript is always followed by secretion of bioactive IGF-II at concentrations sufficient to stimulate cell proliferation, particularly in the presence of endogenous IGFBPs which would compete with receptors for ligand binding. A small amount of "big" IGF-II, a 15 kDa precursor form of the protein, may be secreted in addition to mature IGF-II[40,41] but again, its bioactivity and significance has not been demonstrated in breast cancer cells.

Immunohistochemical and in situ analyses of breast tissue have confirmed little expression of IGFs by malignant breast epithelial cells, revealing instead that IGF-I and IGF-II mRNA are localised to the stromal compartment of the breast,[34,42,43] thereby implicating stromal fibroblasts as the primary source of growth stimulatory IGFs in the breast. A number of lines of evidence also suggest that although both IGF-I and IGF-II are present in breast tumor tissue,[44] IGF-II may be the more important of the two in breast malignancy. IGF-II mRNA is highly expressed in stroma associated with malignant tissue, while IGF-I mRNA predominates in stromal cells adjacent to normal epithelial cells.[43,45-47] The possibility that soluble, cancer cell-derived factors are involved in promoting IGF-II expression by stromal cells in breast tumors is suggested by a study showing that IGF-II expression in fibroblasts from normal breast tissue is induced by co-culture with malignant breast epithelial cells.[46] An animal model of breast cancer also indicated that expression of IGF-II occurs in malignant cells but not normal epithelial cells adjacent to the tumor, and a high level of expression correlates with enhanced tumor growth.[48] Together these data provide support for a paracrine model of IGF action in breast cancer proliferation, with malignant epithelial cells inducing expression of stimulatory IGFs, in particular IGF-II, by adjacent stromal cells. These growth factors can enhance the proliferation of both stromal and epithelial cells, thereby promoting tumor growth.

The Expression of the Type I IGF Receptor (IGFR1) in Breast Cancer

The IGFR1 is highly expressed in malignant breast tissue[49-53] and virtually all breast cancer cell lines[27,31-35] implying that it confers a significant growth advantage on malignant breast epithelial cells. A recent report has shown that BRCA1, a tumor suppressor gene which is mutated in a large proportion of familial breast cancers, may suppress IGFR1 promoter activity.[54] This suggests that BRCA1 mutants have the potential to de-repress the IGFR1 promoter, which may be one mechanism by which overexpression of the receptor occurs in some breast cancers. Similarly, p53 regulation of insulin receptor and IGFR1 expression has also been proposed[55] and a dominant-negative p53 mutant is able to de-repress the insulin receptor promoter in cells containing normal p53. Thus p53 inactivation, which is common in breast and other tumors, may also result in overexpression of the insulin receptor, and perhaps also the IGFR1.

Elevated levels of IGFR1 in breast tumors are highly correlated with early tumor recurrence following lumpectomy and radiation therapy[56] suggesting that overexpression of the IGFR1 is an indicator of poor prognosis. Other prognostic indicators, such as expression of progesterone receptors, lymph node status, tumor size and patient age, show no consistent association with IGFR1.[50,57,58] Surprisingly however, IGFR1 expression is highly correlated with expression of the estrogen receptor (ER),[49-51,57] which is perhaps the most well-recognized indicator of favorable outcome. This implies that a high level of IGFR1 expression in breast tumors is in fact a favorable prognostic indicator, and may predict longer disease-free survival.[51,57] In support of this, immunohistochemical data indicate that expression of IGFR1 and its signalling intermediate IRS-1 is high in well and moderately differentiated carcinomas and much lower in poorly differentiated carcinomas, suggesting down-regulation of both proteins during disease progression.[59]

It is not clear why overexpression of the IGFR1, a transducer of mitogenic and antiapoptotic signals, should be associated with favorable prognosis. Lee et al have suggested that because the IGFR1 is normally down-regulated upon IGF-binding, a high level of IGFR1 expression may reflect absence of ligand and therefore a low level of receptor activation.[60] Furthermore, IGFR1 expression per se is not necessarily an indicator of its level of activity—there must also be ligand activation and downstream signaling for its overexpression to be of significance. Experimental overexpression of IGFR1 or its substrate IRS-1 in breast cancer cell models is not necessarily accompanied by an IGF-stimulated increase in proliferation, anchorage-independent growth, or invasiveness,[61-63] consistent with the concept that downstream signaling pathways must also be highly activated for a maximal response from overexpressed IGFR1.

Functional Effects of IGFR1 Activation in Breast Cancer

Activation of IGFR1 by ligand binding can initiate a number of responses in breast cancer cells, each of which may contribute to the tumorigenic process. Cell cycle progression, or the mitogenic response to IGFs, involves inter alia induction of cyclins D1 and E, increased expression of p21, decreased expression of p27, phosphorylation and nuclear exclusion of p53 and hyperphosphorylation of the retinoblastoma protein pRb.[64-68] These changes may be accompanied by non-mitogenic responses such as modification in cell-cell adhesion, cell migration and invasion of matrix;[58,69] proteases important in these processes, which underlie breast tumorigenesis and metastasis, can be induced by IGF-I in breast cancer cells.[70] A less well understood though clearly very important function of the IGFs in malignant breast disease lies in their ability to rescue cells from apoptosis induced by growth factor deprivation or anticancer drugs.[71-73]

A number of agents used to treat breast cancer, including antiestrogens such as tamoxifen and ICI 182780,[62,74,75] reduce expression of IGFR1 and some of its downstream signaling components and this has been postulated to contribute to the antitumor activity of these drugs. Antibody blockade of the IGFR1 inhibits monolayer and anchorage-independent growth of the MCF-7 and MDA-MB-231 breast cancer cell lines in vitro.[2] Similarly, attenuation of IGFR1 expression using antisense RNA constructs reduced proliferation of the MCF-7 and MDA-MB-435 breast cancer cell lines in vitro,[76,77] and tumor growth and metastatic spread of MDA-MB-435s cells in vivo.[77] Collectively, these data support the argument that strategies designed to reduce IGFR1 expression or activation may prove beneficial in the treatment of breast cancer. However, two studies which employed dominant-negative mutants to functionally impair IGFR1 activity reported no reduction in the monolayer growth of MCF-7 or MDA-MB-435 cells in vitro,[58,78] indicating that different techniques employed to achieve the same end—attenuation of IGFR1 activity—may have markedly different outcomes in terms of modulating breast cell proliferation and tumor growth.

An important concept which has emerged over the last few years as a result of such studies is that there may be dissociation between the mitogenic, non-mitogenic and antiapoptotic actions of IGFs in breast cancer cells, indicating that factors other than expression of a functional IGFR1 govern the ability of IGF-I to promote tumor formation and growth, invasion and metastasis. This was exemplified in the groundbreaking study of Arteaga et al[1] which showed that although immunoneutralization of the IGFR1 blocked the growth of both MCF-7 and MDA-MB-231 cells in vitro, tumor formation of the MDA-MB-231 xenografts, but not MCF-7 xenografts, was inhibited in vivo. Similarly, expression of a dominant-negative receptor had no effect on monolayer growth of MCF-7 or MDA-MB-435s cells,[58,78] but anchorage-independent growth, cellular adhesion and invasiveness of mutant-transfected cells were reduced,[58,78] as was the number of metastases from the primary tumor when MDA-MB-435s transfectants were injected into nude mice.[78]

Differential responses to the proliferative and antiapoptotic actions of IGFs have also been demonstrated in breast cancer cells where cell death is induced by treatment with chemotherapeutic drugs[73] or serum-deprivation.[79] In MCF-7 cells, IGF-I promoted cell survival when apoptosis was induced by either of the cytotoxic drugs doxorubicin or paclitaxel, but the pathways activated by IGF-I differed depending on the apoptotic stimulus. In doxorubicin-treated cells, IGF-I activated only the PI-3 kinase pathway, while both the PI-3 kinase and p44/42 MAPK pathways were activated by IGF-I in paclitaxel-treated cells. This implied that IGF-I blocked apoptosis induced by both agents, but stimulated proliferation only in the paclitaxel-treated cells.[73] There may also be differential responses depending on the cell line, with IGF-I able to fully abrogate apoptosis triggered by serum-deprivation in MCF-7 cells, but achieving only partial rescue in serum-deprived T47D cells.[79]

The explanation for these data is not clear; however a study in ZR-75-1 suggests that the different cellular responses require different threshhold levels of IGFR1 expression.[80] Clones of the ZR-75-1 cell line expressing an inducible IGF-II expression vector did not proliferate in response to endogenous IGF-II unless the IGFR1 was constitutively co-expressed.[80] By contrast however, molecular responses such as expression of pS2 and CAT-activity of a transiently transfected AP1-CAT gene were not dependent upon overexpression of the IGFR1 for IGF-II-inducibility.[80]

The signaling pathways and intermediates downstream of the activated IGFR1 are involved to varying degrees in the different processes of breast cancer. The p44/42 MAPK/ERK pathway appears to be primarily involved in mediating a proliferative response to IGFs, while the PI-3 kinase/Akt pathway is more closely involved in IGF effects on apoptosis and migration.[71-73,81] A study showing that MCF-7 cells are dependent upon the PI-3 kinase pathway for IGF-stimulated proliferation rather than p44/42 MAPK indicates that there may be cell-specific differences in the signaling pathways which predominate to mediate a growth response to IGFs.[66] IRS-1 and SHC, substrates of IGFR1, have also been shown to have specific roles in the different actions of IGF-I in MCF-7 cells;[82] IRS-1 appears to be critical to cell survival, SHC is necessary for the formation and maintenance of cell-cell interactions, and both have been implicated in monolayer and anchorage-independent growth.[82] Thus overexpression or sustained activation of one of these intermediates relative to another could explain why there may be changes in one parameter in reponse to IGFs, with no apparent effect on another.

Other Proteins Involved in IGF Signaling in Breast Cancer

Not all IGF signaling in breast cancer occurs via the IGFR1, and there is good evidence that a number of other proteins are capable of mediating a growth response to IGFs. IGF-I can bind the insulin receptor, albeit with low affinity, and it, like the IGFR1, is overexpressed in many breast cancer cell lines and breast tumors.[83,84] Some of the growth stimulatory effects of IGF-II in breast cancer cells may also be mediated via the type A insulin receptor, an isoform which binds IGF-II with an affinity similar to that of IGFR1 for IGF-II, and which is highly expressed in breast cancer cell lines and tissues.[85] Mitogenic signal-transduction by the type 2 IGF/mannose-6-phosphate receptor (IGF2/M6PR) in MCF-7 breast cancer cells has also been proposed;[86] however this is controversial, and it is likely that the primary role of this receptor in the context of IGF action in breast cancer is in IGF-II degradation (see below). Insulin/IGFR1 hybrids which bind IGF-I are also abundant in breast cancer, with their expression often exceeding that of either the insulin receptor or IGFR1.[87] There is also evidence of heteromeric receptor formation between IGFR1 and ErbB-2, one of members of the EGF receptor family, in a mouse mammary cancer cell line.[88] In these cells, IGFR1 appears to direct ErbB-2 phosphorylation, but not vice versa, and the formation of the hybrid receptors is induced by heregulin and IGF-I in MCF-7 breast cancer cells.[88] Collectively, these findings underscore the importance of considering all potential signaling receptors when designing strategies which target the signaling pathways of the IGF system as a treatment for breast cancer.

The Role of the IGF2/M6PR in Breast Cancer

Studies in other cell systems indicate that the IGF2/M6P receptor is not involved in IGF signal transduction, but instead binds and internalizes IGF-II, leading to its degradation.[89] In doing so, the IGF2/M6PR removes IGF-II from the extracellular environment and reduces its growth-stimulatory potential by precluding its activation of other, signaling receptors. Equally important is its function in the binding and activation of mannose-6-phosphate-bearing moieties, including factors implicated in breast tumorigenesis such as transforming growth factor-β (TGF-β) and cathepsin D, and intracellular trafficking.[90,91] Thus changes in IGF2/M6PR expression or function will affect not only the IGF system, but also the activity of other factors involved in breast cancer cell proliferation, migration and invasion.

The potential significance of IGF2/M6PR in breast tumorigenesis was highlighted in a study using a rat model of mammary carcinoma, where regression of tumors in response to the therapeutic agent d-limonene, a monoterpene, was associated with increased expression of IGF2/M6PR; tumors which failed to respond to therapy lacked a rise in IGF2/M6PR.[92] A tumor-suppressor role for the IGF2/M6P receptor was formally proposed in 1996, when Hankins et al reported a high rate of loss of heterozygosity (LOH) of the IGF2/M6PR gene in breast tumors, coupled with missense mutations in the remaining allele in several instances.[93] These mutations may result in expression of a functionally altered receptor, incapable of IGF-II binding and degradation, or binding of its other ligands.[94] Subsequently, LOH was identified in poorly-differentiated ductal carcinoma in situ, but not well- to moderately-differentiated early invasive carcinoma.[95] The presence of missense mutations in the *Igf2r* gene to any significant degree in breast tumors has been disputed however,[96] and it has been proposed instead that the expressed receptor is readily saturated, and excess ligands are displaced to other sites where their action facilitates tumor progression. Underexpression of the IGF2/M6PR in breast tumor tissue compared with adjacent normal tissue has also been observed,[97] which would support this model of receptor saturation.

Interactions of the IGF and Steroid Hormone Systems in Breast Cancer

A relationship between the IGF and steroid hormone systems in breast cancer has been acknowledged for many years. The expression of IGFR1 in breast tumors correlates highly with expression of ER,[51,57,98,99] and early studies indicated that estrogen sensitizes breast cancer cells to the proliferative effects of insulin and IGF-I by increasing the expression of IGFR1.[98] The expression of many other components of the IGF system is also modulated by estrogen, and this may contribute to its growth-promoting effects in breast cancer. Upregulation of IGFR1, IRS-1, the p85 regulatory subunit of PI-3 kinase and IGF-II, and downregulation of IGF2/M6PR and IGFBP-3 in response to estrogen have been demonstrated in various breast cancer cell lines.[34,67,75,98-100] However, it is increasingly apparent that interaction of the two systems is highly complex, and exists on a number of levels.

The majority of breast cancer cell lines which show a proliferative response to IGF-I or –II are ER positive, and a synergistic response to the growth promoting effects of IGF-I and estrogen is well documented.[61,68,101] Overexpression of IGFR1 in MCF-7 cells, while having little effect on the actions of IGFs or estradiol alone, enhances the proliferative response to IGFs in the presence of estradiol;[61] conversely, reduced expression of the IGFR1 in the same cell line abrogates the synergistic effect of the two agents.[67] Sublines of MCF-7 cells which do not express ERα exhibit reduced IGFR1 and IRS-1 mRNA and protein expression, reduced IGF signaling, and no growth response to IGF-I or estrogen, compared with the parental line.[102] Re-expression of ERα in these cells restores their sensitivity to the proliferative effects of both IGF-I and estrogen, implying that a functional ER is a critical requirement for IGF signaling.

At the molecular level, multiple signaling components appear to be involved in the combined actions of estrogen and IGFs. The activity of Akt/protein kinase B downstream of PI-3 kinase is enhanced in the presence of estrogen and IGF,[72] and a synergistic effect of the two on expression and activation of various cell cycle components, including cyclin D1, cyclin E and pRb has also been identified.[67,68,103]

The mechanisms involved in the synergism between IGF-I and estradiol in breast cancer cells are now beginning to be identified, with complex interactions within the cell cycle machinery playing a key role. An MCF-7 subline which has a 50% reduction in the expression of IGFR1 exhibits an attenuated response to estrogen and IGF-I, attributable to decreased potentiation of IGF-I's effect on cyclin D1 and cyclin E expression, and phosphorylation of the retinoblastoma protein, Rb.[67] Expression of inhibitory p27^{Kip1} is also decreased by estrogen and IGF-I, with a consequent increase in the activity of G1 cyclin-cdk complexes to promote S-phase entry.[67] Another study also showed that estrogen downregulates expression of p21$^{Cip1/Waf1}$, thereby counteracting a slight increase in its expression in response to IGF-I and enabling the formation of active cyclin E-Cdk2 complexes lacking p21$^{Cip1/Waf1}$.[68]

The differential effect of IGFs on mitogenic and non-mitogenic actions in breast cancer cells has also been linked with the ER status of the cell, such that IGFR1 controls non-mitogenic processes regardless of estrogen receptor status, while a proliferative response to IGFs may depend on ER expression.[81,104] Fascinating data indicating more direct coupling of the IGFR1- and ER-activated signaling cascades is now emerging from studies in other cell types which may prove applicable in breast cancer. Phosphorylated p44/42 MAPK downstream of the activated IGFR1 and Ras can phosphorylate and activate ER in the absence of ligand;[105] furthermore, it has been shown that ERα can rapidly induce phosphorylation of the IGFR1.[106] These observations have given rise to a model evoking positive feedback activation of IGFR1/MAPK and ER signaling pathways; IGF activation of the IGFR1 signaling cascade causes phosphorylation of p44/42 MAPK which in turn phosphorylates ERα, thereby enabling it to bind and activate the IGFR1.[106] This has some experimental support in breast cancer cells. In MCF-7 cells ER-mediated activation of the MAPK pathway by estradiol occurs with kinetics similar to those of peptide mitogens,[107] and IGF-I stimulated p44/42 MAPK phosphorylation is enhanced by estradiol in wild type MCF-7 cells, but not in cells where IGFR1 expression is abrogated by antisense cDNA transfection.[67] Studies have also described IGF-I-enhancement of ER-mediated transactivation,[108] a phenomenon which has been variously attributed to the p44/42 MAPK,[105] Akt,[109] Src/JNK,[110] and pp90rsk1[111] signaling intermediates.

The IGF System in the Development of Hormone Resistance

The involvement of the IGF system in the development of hormone independence has been studied using cell models where hormone insensitivity is induced by estrogen deprivation over long periods in vitro. A clonal derivative of the MCF-7 ER+ve cell line which exhibits an estrogen responsive rather than dependent phenotype shows essentially complete loss of sensitivity to estrogen when IGF-II is overexpressed by way of an inducible vector.[112] Changes in basal and IGF-II induced proliferation could be blocked by anti-IGFR1 antibody, implying that both the loss of estrogen dependence and the development of hormone insensitivity are IGFR1-mediated processes.[112] In another study, MCF-7 cells overexpressing IRS-1 showed a reduction in estrogen requirement for growth in monolayer culture as well as in soft agar, and a complete loss of estrogen dependence for growth when IRS-1 is several-fold overexpressed.[113] High concentrations of IGF-I also abolish the estrogen requirement for DNA synthesis in MCF-7 cells overexpressing IGFR1.[114] While the mechanisms underlying these observations are still to be elucidated, they indicate that the IGF and steroid hormone systems are closely linked in breast cancer, and that targeting both pathways simultaneously may prove beneficial in breast cancer treatment.

The IGFBPs in Breast Cancer

The well-established role for the IGFBPs in controlling the bioactivity of IGF-I and -II in the circulation made them obvious candidates for regulating the growth-promoting actions of the IGFs in breast cancer cells. A correlation between increased IGFBP secretion by MCF-7 cells in response to retinoic acid and decreased IGF-stimulated cell proliferation[115,116] provided the first evidence supporting the postulate that IGFBPs have significant antiproliferative actions in breast cancer cells. This area has expanded enormously in recent years, and the development of strategies designed to exploit the growth inhibitory potential of the IGFBPs is now the focus of many researchers in the field.

Expression of IGFBPs in Breast Cancer Cells and Tumors

Each of the six IGFBPs has been identified in extracts of breast tumor tissue, although estimates of their relative levels of expression differ depending on the study, possibly because of different analytical techniques used.[117-120] It is clear, however, that IGFBP expression both in breast tumors and cancer cell lines in vitro correlates with the presence or absence of ER.[120,121] This relationship is strongest for IGFBP-3, which is highly expressed in tumors and cell lines lacking ER, and IGFBP-2, which is more highly expressed in ER-positive breast tumors and cells.[117,120,121] IGFBP-4 and –5 have also been reported to correlate positively with ER[120] although the relationship does not seem to be as strong as with IGFBP-2.[120] A very low expression level for IGFBP-1 has made its detection in tumor tissue difficult and it is still not clear that it is expressed at all;[117,120] however, IGFBP-1 mRNA and protein have been reported in three ER-negative breast cancer cell lines.[121,122] Unpublished data from our laboratory also indicates that IGFBP-6, although readily measured in tumor cytosols, shows no clear relationship with ER status. The significance of the relationship between ER and IGFBP expression remains unclear.

A number of well-recognized inhibitors of breast cancer cell proliferation have been shown to modulate expression of IGFBPs by cells in vitro (Table 2), and the significance of these changes has been revealed in studies utilizing antisense oligodeoxynucleotides to block protein expression of specific IGFBPs. IGFBP-3, and to a lesser extent IGFBP-5, appear to be particularly important in mediating the effects of a diverse group of antiproliferative agents, including antiestrogens,[123-125] TGF-β,[126] retinoic acid,[127] TNF-α,[128] and vitamin D analogs.[129,130]

Some studies have reported alterations in IGFBP expression associated with changes in hormone sensitivity and resistance to antiestrogens. An estrogen-independent subline of the ZR-75-1 breast cancer cell line showed reduced expression of IGFBP-3 compared with the estrogen-sensitive parental line,[131] while tamoxifen-resistant sublines had decreased expression of IGFBP-4 compared with the parent cell line, and increased secretion of IGFBP-3 in response to serum.[131] Decreased expression of IGFBP-2 in both tamoxifen-resistant and estrogen-insensitive ZR-75-1 sublines has also been documented.[132] It has also been suggested that tamoxifen-resistant breast carcinoma cells obtained from fine-needle biopsy of tumor tissue have inactivating mutations in IGFBP-2, and that restoration of wild-type IGFBP-2 expression via transfection with IGFBP-2 cDNA restores sensitivity to the growth-inhibitory effects of tamoxifen.[133] Two sublines of the MCF-7 breast cancer cell line which differ in their invasive ability were also found to have different IGFBP profiles, with the non-invasive subline secreting much lower amounts of IGFBPs that its invasive counterpart.[134] This study also showed that whereas MDA-MB-231 cells secrete most of the IGFBPs they synthesize, the majority of the IGFBPs produced by MCF-7 cells remain intracellular.[134] The cause of such differences and the functional significance of changes in IGFBP expression associated with the development of hormone insensitivity are not known.

Regulation of IGF Action by IGFBPs in Breast Cancer Cells

In breast cancer cells, as in many other cell types, specific IGFBPs have been shown to both enhance and inhibit the growth-promoting effects of the IGFs. In MCF-7 cells IGFBP-1 inhibited the mitogenic effect of IGF-I,[135] IGFBP-4 and IGFBP-5 had no effect, and IGFBP-2 and IGFBP-3 potentiated IGF-I stimulated DNA synthesis.[136] In the latter study, IGF-I activity in IGFBP-3 cDNA-transfected cells was also enhanced compared with control cells, and this was interpreted as IGFBP-3 protecting cells from IGF-mediated downregulation of its receptor.[136] In contrast with these findings, however, exogenous IGFBP-3 has been shown to inhibit serum-, estrogen- or IGF-stimulated DNA synthesis[137,138] and the antiapoptotic activity of IGF-I in MCF-7 cells.[125]

Although IGFBP-3 was first shown to have both IGF-inhibitory and –potentiating effects a number of years ago[139] the mechanism by which it can do so remains unknown. Different preparations of IGFBP-3 have been used in the MCF-7 studies refered to above, with the recombinant

Table 2. Factors regulating expression of IGFBPs in breast cancer cells

	Increased Expression	Decreased Expression
IGFBP-2	IGF-I[116] 17-β estradiol[118,170]	17-β estradiol[134]
IGFBP-3	retinoic acid[100,116,127,171,172] TGF-beta [126] IGF-I[116] cyclic AMP[100] vitamin D[171] tamoxifen (antiestrogen)[170,173] ICI 182780 (antiestrogen) [23,173] sodium butyrate[174] TNF-alpha[128]	17-β estradiol[100,123,170]
IGFBP-4	retinoic acid[116] 17-β estradiol[118,175,176]	progesterone[177] tamoxifen (antiestrogen)[173] RU 486 (antiestrogen)[177] ICI 164384 (antiestrogen)[177] ICI 182780 (antiestrogen)[173]
IGFBP-5	17-β estradiol[118,175] ICI 182780 (antiestrogen)[124,178] 1-25(OH)2 vitamin D[129] EB1089 (vitamin D analog) [129] KH1060 vitamin D analog [130]	progesterone[177] RU 486 (antiestrogen)[177] ICI 164384 (antiestrogen)[177] 17-β estradiol[124,134]
IGFBP-6	retinoic acid[100] cyclic AMP[100] 17-β estradiol[100]	

IGFBP-3 which enhanced IGF action expressed in CHO cells,[136] and inhibitory effects of IGFBP-3 observed with *E. coli*-expressed recombinant protein[137] or plasma-derived IGFBP-3.[138] These proteins differ in their degree and type of glycosylation, and probably also phosphorylation, which may have a bearing on their action. Alternatively, the MCF-7 lines maintained in different laboratories could undergo changes over time due to clonal expansion of a few cells within the population, leading to sublines with altered sensitivity to growth regulators such as IGFs and IGFBPs. It has also been suggested that IGFBP-3 might attenuate IGF activity without binding the ligand.[140] Preincubation of MCF-7 cells with IGFBP-3 reduced the ability of [Q(3)A(4)Y(15)L(16)]-IGF-I, which binds IGFR1 but not IGFBPs, to phosphorylate the IGFR1.[140] This effect was specific for IGFBP-3 and the IGFR1, because phosphorylation of the insulin receptor in response to insulin was not affected by IGFBP-3. The mechanism by which this occurs remains to be elucidated, but direct interaction of IGFBP-3 and IGFR1 did not appear to be involved.[140]

IGF-Independent Actions of IGFBPs in Breast Cancer

While many of the effects of IGFBPs on breast cancer cells clearly involve abrogation of IGF activity, a number of studies have also shown IGFBP effects on proliferation and apoptosis in the absence of IGFs. To a large extent, focus in recent years has shifted from the IGF-regulatory role of IGFBPs in breast cancer, to studying their intrinsic bioactivity.

One of the first reports of IGF-independent effects of IGFBP-3 in breast cancer cells was published by Oh et al who showed that exogenous, unglycosylated IGFBP-3 inhibited DNA synthesis and cell proliferation in Hs578T cells.[141] The IGF-independence of this effect was inferred from the inability of IGFBP-1 to mimic the effect, the failure of IGFs to elicit a growth response in Hs578T cells, and abrogation of IGFBP-3's effect by wild type IGFs or IGF analogs which have high affinity for IGFBPs but not by those with reduced affinity for the binding proteins. Subsequently it was reported in the same cell line that the growth inhibitory effects of endogenous IGFBP-3 induced in response to TGF-β are also IGF-independent.[126] In another study, T47D cells overexpressing IGFBP-3 as the result of transfection with cDNA proliferated more slowly in fetal calf serum than their vector-transfected counterparts.[142] This was also thought to reflect an IGF-independent effect of the endogenous IGFBP-3 because inhibition occurred in the presence of a theoretical molar excess of IGFs.[142]

Although other studies have been unable to confirm antiproliferative effects of IGFBP-3 in Hs578T cells,[143] a clear pro-apoptotic role for IGFBP-3 has been identified in this breast cancer cell line and a number of others. Transfection of IGFBP-3 cDNA into MCF-7 or T47D cells results in induction of apoptosis per se,[144] and increases the sensitivity of T47D cells to apoptosis induced by ionising radiation.[144] This appears to involve an increase in the ratio of pro-apoptotic to anti-apoptotic members of the Bcl-2 family,[144] in particular the Bax:Bcl-2 ratio. The change in Bax protein concentration is not accompanied by a corresponding increase in expression of its mRNA[144] suggesting that post-translational mechanisms are involved in its induction. Furthermore T47D cells lack wild-type p53, indicating that in these cells IGFBP-3's apoptotic activity is p53-independent.

IGFBP-3 also enhances the effects of other apoptotic agents in breast cancer cells. In Hs578T cells, preincubation with non-glycosylated IGFBP-3 accentuates apoptosis induced by ceramide,[143] antimycin A,[145] or paclitaxel;[146] however, by contrast with the apoptotic effect of endogenous IGFBP-3 in the MCF-7 and T47D transfectants,[144] exogenous IGFBP-3 alone did not induce apoptosis in Hs578T cells.[143,146] This may reflect a difference in the ability of endogenous compared with exogenous IGFBP-3 to induce apoptosis; alternatively, there may be cell-specific differences, perhaps at the level of expression of components of the apoptotic machinery, which result in differential sensitivity to IGFBP-3. The apoptotic trigger also appears to be an important determinant in the ability of IGFBP-3 to enhance apoptosis induced by other agents, because cell death stimulated by an integrin-binding peptide, Arg-Gly-Asp (RGD), is not accentuated by IGFBP-3.[147]

Mechanisms Involved in the IGF-Independent Actions of IGFBPs in Breast Cancer Cells

Very little is known of the mechanisms by which IGFBPs exert their intrinsic bioactivity. In Hs578T cells, growth inhibition induced by IGFBP-3 correlates with increased binding of IGFBP-3 to the cell monolayer, suggesting the presence of binding sites which might be mediating IGFBP-3's effect.[141] Affinity-labeling studies indicated the presence of IGFBP-3-binding species of 20-, 26- and 50-kDa on Hs578T cells,[148] and characterization of IGFBP-3 binding to Hs578T and MCF-7 cells using radiolabeled non-glycosylated IGFBP-3 as tracer revealed an affinity constant of ~8 x 10^{-9}M for both cell lines, and 3-fold more binding sites on Hs578T cells than MCF-7 cells.[149] Binding was weakly competed by IGFBP-5 and synthetic fragments of the mid-region of IGFBP-3, but not by other IGFBPs.[149] However, there is still no evidence that these binding sites represent receptors involved in IGFBP-3 signal transduction. More complete characterization of a protein which may be involved in mediating or modulating the effects of IGFBP-3 has also been reported.[150] This protein, identified from a yeast two-hybrid screen of Hs578T cells and designated 4-33, is expressed intracellularly and on the surface of Hs578T cells. Cell binding of IGFBP-3 is increased when 4-33 is overexpressed, and induction of apoptosis and decreased DNA synthesis is apparent in the presence of IGFBP-3 and 4-33.[150] While these correlative observations are consistent with 4-33 being involved in IGFBP-3's cellular actions, the mechanisms involved have not been elucidated.

In other cell types, the type V TGF-β receptor has been reported to be the putative IGFBP-3 receptor.[151] While this cannot be confirmed in breast cancer cells, a different link between IGFBP-3 and TGF-β signaling has been identified which indicates a requirement for both TGF-β and the type II TGF-b receptor (TGF-βRII) in IGFBP-3 signaling.[152] T47D cells transfected with TGF-βRII cDNA to restore TGF-β signaling capability became sensitive to growth inhibition by IGFBP-3 in the presence of TGF-β. In these transfectants, TGF-β and IGFBP-3 independently and synergistically stimulated phosphorylation of two intermediates of TGF-β signaling, Smad 2 and Smad 3, suggesting that IGFBP-3 inhibitory signaling in these cells requires an intact TGF-β signaling pathway.[152]

The potential for direct nuclear actions of IGFBP-3 and -5 in breast cancer cells is suggested by studies showing nuclear import of both proteins via their C-terminal domain nuclear localization sequence.[153] This is mediated principally by the importin beta nuclear transport factor,[154] and subsequent accumulation of the IGFBP in the nucleus suggests binding to nuclear components. The function of nuclear IGFBP-3 and IGFBP-5 in breast cancer cells remains an important question, although interaction between IGFBP-3 and the nuclear retinoid X receptor has been identified in other cells.[155]

The differential effects of the six IGFBPs on MCF-7 cell proliferation was noted earlier, and there is clearly IGFBP-specificity with respect to modulating apoptosis in breast cancer cells as well. In Hs578T cells IGFBP-3 enhances the apoptotic effect of ceramide, IGFBP-4 and IGFBP-5 inhibit it, and IGFBP-1, IGFBP-2, and IGFBP-6 have no effect. Similarly IGFBP-5, but not IGFBP-3, inhibits apoptosis induced by RGD peptides.[147] In view of the fact that IGFBP-3 and -5 share a number of structural features and are functionally similar in many other respects, it is surprising and fascinating that they appear to have diverse activities in breast cancer cells. Although the basis for IGFBP specificity remains largely undefined, structural elements inherent to the different proteins are undoubtedly key factors in determining their action. Thus by virtue of an RGD motif within its C-terminal domain, IGFBP-1 is able to induce dephosphorylation of focal adhesion kinase and apoptosis in a number of breast cancer cell lines in the absence of any other apoptotic stimulus.[156] Post-translational modification of an IGFBP may also affect its bioactivity. IGFBP-3[1-97], an amino-terminal fragment of IGFBP-3 generated by an MCF-7-derived protease, inhibited DNA synthesis in MCF-7 cells under conditions where intact IGFBP-3 was without effect.[138] Delineation of the structural determinants involved in IGFBP function and specificity will provide important clues as to their roles in the regulation of breast cancer cell proliferation.

The Role of IGFBPs in Breast Tumor Growth

Surprisingly few studies have extended the in vitro studies described above to investigate the effects of IGFBPs on breast tumor growth using in vivo models of breast cancer. MCF-7 breast cancer cells transfected with IGFBP-4 cDNA showed reduced tumor growth in vivo compared with control cells,[157] consistent with an observation in vitro in the same cell line of reduced responsiveness to IGFs and slower growth in low serum-containing medium.[157]

However, a study which examined the effect of administered IGFBP-1 on tumor growth in xenograft models of breast cancer revealed differences in response relating to the cell line from which the tumor was derived, and also differential effects of the IGFBP in vitro compared with in vivo. Polyethylene glycol-conjugated IGFBP-1 (PEG-IGFBP-1), which has increased half-life in the circulation compared with unconjugated IGFBP-1, inhibited malignant ascites formation from MDA-MB-435A-derived tumors, delayed but didn't inhibit growth of MDA-MB-231 tumors, and had no effect on the growth of MCF-7 tumors.[158] This contrasted with observations in vitro, where anchorage independent growth of the MCF-7 and MDA-MB-435A cells was inhibited by PEG-IGFBP-1, but no effect on MDA-MB-231 colony formation was seen. Thus, even from the limited number of studies investigating the in vivo actions of IGFBPs, it is apparent that their effects on tumor growth are IGFBP-specific, and are tissue-specific in the sense that not all breast tumors will respond to an IGFBP in the same way. Futhermore, observations regarding their actions in vitro will not necessarily translate to tumor growth in vivo, re-affirming the need to assess affects of IGFBPs on growth parameters in both models. This is becoming increasingly important in view of many observations which implicate some IGFBPs in the stimulation, rather than inhibition, of breast tumor growth.

As described earlier, circulating IGFBP-3 is not an independent prognostic indicator for breast cancer development, but a high concentration of IGFBP-3 in the circulation reduces the risk associated with elevated IGF-I in the pre-menopausal women.[25] This implies that a high concentration of IGFBP-3 is favorable in breast cancer. However, an early study analyzing IGFBPs in breast tumor extracts indicated an increase in total IGFBP content compared with adjacent histologically-normal tissue, suggesting that malignant transformation of the breast is associated with increased expression of IGFBPs.[117] Subsequently it was shown that high tumor levels of IGFBP-3 are associated with other indicators of poor prognosis, having an inverse correlation with ER, and a positive correlation with high S-phase, aneuploidy and tumor size.[44,159-161] This unexpected finding suggests that a high level of expression of IGFBP-3 is associated with increased tumor growth, and poor prognosis. IGFBP-4 is also associated with unfavorable outcome, because although there is a weak inverse correlation between its expression and tumor size and S-phase fraction, patients with large tumors expressing low levels of IGFBP-4 have improved disease-free survival compared with patients with large tumors and high IGFBP-4 content.[120] While it is difficult to reconcile these findings with the abundant experimental data which show that IGFBPs are growth inhibitory to breast cancer cells in vitro, recent studies from our laboratory showing that breast cells can become refractory to growth inhibition by IGFBP-3 may provide some explanation for such observations.

IGFBP Resistance in Breast Cancer

T47D breast cancer cells transfected with IGFBP-3 cDNA are initially growth inhibited by the expressed protein, but with increasing passage number in vitro become resistant to its antiproliferative effects and show instead enhanced growth compared with vector-transfectants.[142] This suggests that under some circumstances, IGFBP-3 may switch from functioning as an inhibitor of breast cancer cell proliferation, to a stimulator. While the mechanism involved in this change is not known, our studies have also indentified a link between activation of the p44/42 MAP kinase pathway downstream of Ras, and development of IGFBP-3 insensitivity in breast epithelial cells.

In the phenotypically normal breast cell line MCF-10A DNA synthesis is inhibited by IGFBP-3 in an IGF-independent manner;[162] however, if these cells are transfected with oncogenic H-*ras* cDNA, resulting in constitutive activation of Ras signaling pathways and malignant transformation, IGFBP-3 is no longer able to inhibit DNA synthesis. Sensitivity to IGFBP-3's inhibitory effects can be restored by blockade of p44/42 MAP kinase activation,[162] implying that the the Ras-MAPK pathway is involved in the development of IGFBP-3 insensitivity by a mechanism as yet unknown. This has obvious implications, given that signaling through Ras is upregulated or chronically activated in many breast cancer cell lines.[163] Our understanding of the causes of IGFBP-3 insensitivity will no doubt develop with a greater appreciation of the mechanisms by which it exerts its growth inhibitory effects.

Concluding Remarks

It is clear that our knowledge of the IGF axis in breast cancer has advanced considerably within the last decade or so, but there are still many facets which remain poorly understood, and many questions which remain unanswered. Why do different breast cancer cell lines and tumors respond disparately to anti-IGF strategies? What enables some IGFBPs to both inhibit and enhance IGF action? How do the IGFBPs exert their intrinsic bioactivity, what is the structural basis of their specificity and functional differences, and what regulates sensitivity to their effects? The fact that data derived in vitro do not always allow us to predict an in vivo response is evidence that there is still much to learn regarding the relationships between the IGF system and other growth factor and hormone systems in the breast, and interactions between the malignant epithelial cells and the stromal cells with which they are intimately involved in vivo. It is clear that defining the role of this important growth regulatory system in breast cancer, comprising as it does multiple ligands, receptors and regulatory proteins, will present a considerable challenge over the next few years.

References

1. Arteaga CL, Kitten LJ, Coronado EB et al. Blockade of the type I somatomedin receptor inhibits growth of human breast cancer cells in athymic mice. J Clin Invest 1989; 84:1418-1423.
2. Arteaga CL, Osborne CK. Growth inhibition of human breast cancer cells in vitro with an antibody against the type I somatomedin receptor. Cancer Res 1989; 49:6237-6241.
3. Peyrat JP, Bonneterre J, Hecquet B et al. Plasma insulin-like growth factor-1 (IGF-1) concentrations in human breast cancer. Eur J Cancer 1993; 29A:492-497.
4. Barni S, Lissoni P, Brivio F et al. Serum levels of insulin-like growth factor-I in operable breast cancer in relation to the main prognostic variables and their perioperative changes in relation to those of prolactin. Tumori 1994; 80:212-215.
5. Bruning PF, Van Doorn J, Bonfrer JM et al. Insulin-like growth-factor-binding protein 3 is decreased in early-stage operable pre-menopausal breast cancer. Int J Cancer 1995; 62:266-270.
6. Bohlke K, Cramer DW, Trichopoulos D et al. Insulin-like growth factor-I in relation to premenopausal ductal carcinoma in situ of the breast. Epidemiology 1998; 9:570-573.
7. Colletti RB, Roberts JD, Devlin JT et al. Effect of tamoxifen on plasma insulin-like growth factor I in patients with breast cancer. Cancer Res 1989; 49:1882-1884.
8. Lien EA, Johannessen DC, Aakvaag A et al. Influence of tamoxifen, aminoglutethimide and goserelin on human plasma IGF-I levels in breast cancer patients. J Steroid Biochem Mol Biol 1992; 41:541-543.

9. Lonning PE, Hall K, Aakvaag A et al. Influence of tamoxifen on plasma levels of insulin-like growth factor I and insulin-like growth factor binding protein I in breast cancer patients. Cancer Res 1992; 52:4719-4723.
10. Pollak M, Costantino J, Polychronakos C et al. Effect of tamoxifen on serum insulinlike growth factor I levels in stage I breast cancer patients. J Natl Cancer Inst 1990; 82:1693-1697.
11. Helle SI, Anker GB, Tally M et al. Influence of droloxifene on plasma levels of insulin-like growth factor (IGF)-I, Pro-IGF-IIE, insulin-like growth factor binding protein (IGFBP)-1 and IGFBP-3 in breast cancer patients. J Steroid Biochem Mol Biol 1996; 57:167-171.
12. Helle SI, Holly JM, Tally M et al. Influence of treatment with tamoxifen and change in tumor burden on the IGF-system in breast cancer patients. Int J Cancer 1996; 69:335-339.
13. Cobleigh MA. Breast cancer and fenretinide, an analogue of vitamin A. Leukemia 1994; 8:S59-63.
14. Decensi A, Formelli F, Torrisi R et al. Breast cancer chemoprevention:studies with 4-HPR alone and in combination with tamoxifen using circulating growth factors as potential surrogate endpoints. J Cell Biochem 1993; 17G:226-233.
15. Torrisi R, Parodi S, Fontana V et al. Effect of fenretinide on plasma IGF-I and IGFBP-3 in early breast cancer patients. Int J Cancer 1998; 76:787-790.
16. Kaulsay KK, Ng EH, Ji CY et al. Serum IGF-binding protein-6 and prostate specific antigen in breast cancer. Eur J Endocrinol 1999; 140:164-168.
17. Helle SI, Geisler S, Aas T et al. Plasma insulin-like growth factor binding protein-3 proteolysis is increased in primary breast cancer. Br J Cancer 2001; 85:74-77.
18. Lawrence JB, Conover CA, Haddad TC et al. Evaluation of continuous infusion suramin in metastatic breast cancer:impact on plasma levels of insulin-like growth factors (IGFs) and IGF-binding proteins. Clin Cancer Res 1997; 3:1713-1720.
19. Ferrari L, Zilembo N, Bajetta E et al. Effect of two-4-hydroxyandrostenedione doses on serum insulin-like growth factor I levels in advanced breast cancer. Breast cancer Res Treat 1994; 30:127-132.
20. Frost VJ, Helle SI, Lonning PE et al. Effects of treatment with megestrol acetate, aminoglutethimide, or formestane on insulin-like growth factor (IGF) I and II, IGF-binding proteins (IGFBPs), and IGFBP-3 protease status in patients with advanced breast cancer. J Clin Endocrinol Metab 1996; 81:2216-2221.
21. Bajetta E, Ferrari L, Celio L et al. The aromatase inhibitor letrozole in advanced breast cancer:effects on serum insulin-like growth factor (IGF)-I and IGF-binding protein-3 levels. J Steroid Biochem Mol Biol 1997; 63:261-267.
22. Toniolo P, Bruning PF, Akhmedkhanov A et al. Serum insulin-like growth factor-I and breast cancer. Int J Cancer 2000; 88:828-832.
23. Li BD, Khosravi MJ, Berkel HJ et al. Free insulin-like growth factor-I and breast cancer risk. Int J Cancer 2001; 91:736-739.
24. Byrne C, Colditz GA, Willett WC et al. Plasma insulin-like growth factor (IGF) I, IGF-binding protein 3, and mammographic density. Cancer Res 2000; 60:3744-3748.
25. Hankinson SE, Willett WC, Colditz GA et al. Circulating concentrations of insulin-like growth factor-I and risk of breast cancer. Lancet 1998; 351:1393-1396.
26. Dickson RB, Huff KK, Spencer EM et al. Induction of epidermal growth factor-related polypeptides by 17 beta-estradiol in MCF-7 human breast cancer cells. Endocrinology 1986; 118:138-142.
27. Huff KK, Kaufman D, Gabbay KH et al. Secretion of an insulin-like growth factor-I-related protein by human breast cancer cells. Cancer Res 1986; 46:4613-4619.
28. Kasid A, Lippman ME. Estrogen and oncogene mediated growth regulation of human breast cancer cells. J Steroid Biochem 1987; 27:465-470.
29. Minuto F, Del Monte P, Barreca A et al. Partial characterization of somatomedin C-like immunoreactivity secreted by breast cancer cells in vitro. Mol Cell Endocrinol 1987; 54:179-184.
30. Owens PC, Gill PG, De Young NJ et al. Estrogen and progesterone regulate secretion of insulin-like growth factor binding proteins by human breast cancer cells. Biochem Biophys Res Commun 1993; 193:467-473.
31. Myal Y, Shiu RP, Bhaumick B et al. Receptor binding and growth-promoting activity of insulin-like growth factors in human breast cancer cells (T-47D) in culture. Cancer Res 1984; 44:5486-5490.
32. Karey KP, Sirbasku DA. Differential responsiveness of human breast cancer cell lines MCF-7 and T47D to growth factors and 17 beta-estradiol. Cancer Res 1988; 48:4083-4092.
33. Ogasawara M, Sirbasku DA. A new serum-free method of measuring growth factor activities for human breast cancer cells in culture. In Vitro Cell Develop Biol 1988; 24:911-920.
34. Yee D, Cullen KJ, Paik S et al. Insulin-like growth factor II mRNA expression in human breast cancer. Cancer Res 1988; 48:6691-6696.
35. Cullen KJ, Yee D, Sly WS et al. Insulin-like growth factor receptor expression and function in human breast cancer. Cancer Res 1990; 50:48-53.
36. Huff KK, Knabbe C, Lindsey R et al. Multihormonal regulation of insulin-like growth factor-I-related protein in MCF-7 human breast cancer cells. Mol Endocrinol 1988; 2:200-208.
37. Dickson RB, Kasid A, Huff KK et al. Activation of growth factor secretion in tumorigenic states of breast cancer induced by 17 beta-estradiol or v-Ha-ras oncogene. Proc Natl Acad Sci USA 1987; 84:837-841.
38. Dickson RB,Lippman ME. Estrogenic regulation of growth and polypeptide growth factor secretion in human breast carcinoma. Endocrine Rev 1987; 8:29-43.
39. Baxter RC, Maitland JE, Raison RL et al. High molecular weight somatomedin-C/IGF-I from T47D human mammary carcinoma cells: Immunoreactivity and bioactivity. In: Spencer EM, ed. Insulin-like Growth Factors/Somatomedins. Berlin: Walter de Gruyter, 1983:615-621

40. Lee AV, Darbre P, King RJ. Processing of insulin-like growth factor-II (IGF-II) by human breast cancer cells. Mol Cell Endocrinol 1994; 99:211-220.
41. Quinn KA, Treston AM, Unsworth EJ et al. Insulin-like growth factor expression in human cancer cell lines. J Biol Chem 1996; 271:11477-11483.
42. Yee D, Paik S, Lebovic GS et al. Analysis of insulin-like growth factor I gene expression in malignancy: evidence for a paracrine role in human breast cancer. Mol Endocrinol 1989; 3:509-517.
43. Paik S. Expression of IGF-I and IGF-II mRNA in breast tissue. Breast cancer Res Treat 1992; 22:31-38.
44. Yu H, Levesque MA, Khosravi MJ et al. Associations between insulin-like growth factors and their binding proteins and other prognostic indicators in breast cancer. Br J Cancer 1996; 74:1242-1247.
45. Cullen KJ, Allison A, Martire I et al. Insulin-like growth factor expression in breast cancer epithelium and stroma. Breast cancer Res Treat 1992; 22:21-29.
46. Singer C, Rasmussen A, Smith HS et al. Malignant breast epithelium selects for insulin-like growth factor II expression in breast stroma:evidence for paracrine function. Cancer Res 1995; 55:2448-2454.
47. Giani C, Cullen KJ, Campani D et al. IGF-II mRNA and protein are expressed in the stroma of invasive breast cancers:an in situ hybridization and immunohistochemistry study. Breast cancer Res Treat 1996; 41:43-50.
48. Huynh H, Alpert L, Pollak M. Pregnancy-dependent growth of mammary tumors is associated with overexpression of insulin-like growth factor II. Cancer Res 1996; 56:3651-3654.
49. Peyrat JP, Bonneterre J, Beuscart R et al. Insulin-like growth factor 1 receptors in human breast cancer and their relation to estradiol and progesterone receptors. Cancer Res 1988; 48:6429-6433.
50. Foekens JA, Portengen H, van Putten WL et al. Prognostic value of receptors for insulin-like growth factor 1, somatostatin, and epidermal growth factor in human breast cancer. Cancer Res 1989; 49:7002-7009.
51. Papa V, Gliozzo B, Clark GM et al. Insulin-like growth factor-I receptors are overexpressed and predict a low risk in human breast cancer. Cancer Res 1993; 53:3736-3740.
52. Almeida A, Muleris M, Dutrillaux B et al. The insulin-like growth factor I receptor gene is the target for the 15q26 amplicon in breast cancer. Genes, Chromosomes & Cancer 1994; 11:63-65.
53. Resnik JL, Reichart DB, Huey K et al. Elevated insulin-like growth factor I receptor autophosphorylation and kinase activity in human breast cancer. Cancer Res 1998; 58:1159-1164.
54. Maor SB, Abramovitch S, Erdos MR et al. BRCA1 suppresses insulin-like growth factor-I receptor promoter activity:potential interaction between BRCA1 and Sp1. Molec Genetics Metab 2000; 69:130-136.
55. Webster NJ, Resnik JL, Reichart DB et al. Repression of the insulin receptor promoter by the tumor suppressor gene product p53:a possible mechanism for receptor overexpression in breast cancer. Cancer Res 1996; 56:2781-2788.
56. Turner BC, Haffty BG, Narayanan L et al. Insulin-like growth factor-I receptor overexpression mediates cellular radioresistance and local breast cancer recurrence after lumpectomy and radiation. Cancer Res 1997; 57:3079-3083.
57. Railo MJ, von Smitten K, Pekonen F. The prognostic value of insulin-like growth factor-I in breast cancer patients. Results of a follow-up study on 126 patients. Eur J Cancer 1994; 30A:307-311.
58. Surmacz E, Guvakova MA, Nolan MK et al. Type I insulin-like growth factor receptor function in breast cancer. Breast cancer Res Treat 1998; 47:255-267.
59. Schnarr B, Strunz K, Ohsam J et al. Down-regulation of insulin-like growth factor-I receptor and insulin receptor substrate-1 expression in advanced human breast cancer. Int J Cancer 2000; 89:506-513.
60. Lee AV, Hilsenbeck SG, Yee D. IGF system components as prognostic markers in breast cancer. Breast cancer Res Treat 1998; 47:295-302.
61. Daws MR, Westley BR, May FE. Paradoxical effects of overexpression of the type I insulin-like growth factor (IGF) receptor on the responsiveness of human breast cancer cells to IGFs and estradiol. Endocrinology 1996; 137:1177-1186.
62. Guvakova MA, Surmacz E. Tamoxifen interferes with the insulin-like growth factor I receptor (IGF-IR) signaling pathway in breast cancer cells. Cancer Res 1997; 57:2606-2610.
63. Jackson JG, Yee D. IRS-1 expression and activation are not sufficient to activate downstream pathways and enable IGF-I growth response in estrogen receptor negative breast cancer cells. Growth Horm IGF Res 1999; 9:280-289.
64. Musgrove EA, Sutherland RL. Acute effects of growth factors on T-47D breast cancer cell cycle progression. Eur J Cancer 1993; 29A:2273-2279.
65. Takahashi K, Suzuki K. Association of insulin-like growth-factor-I-induced DNA synthesis with phosphorylation and nuclear exclusion of p53 in human breast cancer MCF-7 cells. Int J Cancer 1993; 55:453-458.
66. Dufourny B, Alblas J, van Teeffelen HA et al. Mitogenic signaling of insulin-like growth factor I in MCF-7 human breast cancer cells requires phosphatidylinositol 3-kinase and is independent of mitogen-activated protein kinase. J Biol Chem 1997; 272:31163-31171.
67. Dupont J, Karas M, LeRoith D. The potentiation of estrogen on insulin-like growth factor I action in MCF-7 human breast cancer cells includes cell cycle components. J Biol Chem 2000; 275:35893-35901.
68. Lai A, Sarcevic B, Prall OW et al. Insulin/Insulin-like growth factor-I and estrogen cooperate to stimulate cyclin E-Cdk2 Activation and cell cycle progression in MCF-7 breast cancer cells through differential regulation of cyclin E and p21WAF1/Cip1. J Biol Chem 2001; 276:25823-25833.
69. Doerr ME, Jones JI. The roles of integrins and extracellular matrix proteins in the insulin-like growth factor I-stimulated chemotaxis of human breast cancer cells. J Biol Chem 1996; 271:2443-2447.
70. Dunn SE, Torres JV, Nihei N et al. The insulin-like growth factor-1 elevates urokinase-type plasminogen activator-1 in human breast cancer cells:a new avenue for breast cancer therapy. Mol Carcinog 2000; 27:10-17.

71. Dunn SE, Hardman RA, Kari FW et al. Insulin-like growth factor 1 (IGF-1) alters drug sensitivity of HBL100 human breast cancer cells by inhibition of apoptosis induced by diverse anticancer drugs. Cancer Res 1997; 57:2687-2693.
72. Ahmad S, Singh N, Glazer RI. Role of AKT1 in 17beta-estradiol- and insulin-like growth factor I (IGF-I)-dependent proliferation and prevention of apoptosis in MCF-7 breast carcinoma cells. Biochem Pharmacol 1999; 58:425-430.
73. Gooch JL, Van Den Berg CL, Yee D. Insulin-like growth factor (IGF)-I rescues breast cancer cells from chemotherapy-induced cell death--proliferative and anti-apoptotic effects. Breast cancer Res Treat 1999; 56:1-10.
74. Salerno M, Sisci D, Mauro L et al. Insulin receptor substrate 1 is a target for the pure antiestrogen ICI 182,780 in breast cancer cells. Int J Cancer 1999; 81:299-304.
75. Molloy CA, May FE, Westley BR. Insulin receptor substrate-1 expression is regulated by estrogen in the MCF-7 human breast cancer cell line. J Biol Chem 2000; 275:12565-12571.
76. Neuenschwander S, Roberts CT Jr, LeRoith D. Growth inhibition of MCF-7 breast cancer cells by stable expression of an insulin-like growth factor I receptor antisense ribonucleic acid. Endocrinology 1995; 136:4298-4303.
77. Chernicky CL, Yi L, Tan H et al. Treatment of human breast cancer cells with antisense RNA to the type I insulin-like growth factor receptor inhibits cell growth, suppresses tumorigenesis, alters the metastatic potential, and prolongs survival in vivo. Cancer Gene Therapy 2000; 7:384-395.
78. Dunn SE, Ehrlich M, Sharp NJ et al. A dominant negative mutant of the insulin-like growth factor-I receptor inhibits the adhesion, invasion, and metastasis of breast cancer. Cancer Res 1998; 58:3353-3361.
79. Xie SP, Pirianov G, Colston KW. Vitamin D analogues suppress IGF-I signalling and promote apoptosis in breast cancer cells. Eur J Cancer 1999; 35:1717-1723.
80. Abdul-Wahab K, Corcoran D, Perachiotti A et al. Overexpression of insulin-like growth factor II (IGFII) in ZR-75-1 human breast cancer cells:higher threshold levels of receptor (IGFIR) are required for a proliferative response than for effects on specific gene expression. Cell proliferation 1999; 32:271-287.
81. Bartucci M, Morelli C, Mauro L et al. Differential insulin-like growth factor I receptor signaling and function in estrogen receptor (ER)-positive MCF-7 and ER-negative MDA-MB-231 breast cancer cells. Cancer Res 2001; 61:6747-6754.
82. Nolan MK, Jankowska L, Prisco M et al. Differential roles of IRS-1 and SHC signaling pathways in breast cancer cells. Int J Cancer 1997; 72:828-834.
83. Milazzo G, Giorgino F, Damante G et al. Insulin receptor expression and function in human breast cancer cell lines. Cancer Res 1992; 52:3924-3930.
84. Papa V, Pezzino V, Costantino A et al. Elevated insulin receptor content in human breast cancer. J Clin Invest 1990; 86:1503-1510.
85. Sciacca L, Costantino A, Pandini G et al. Insulin receptor activation by IGF-II in breast cancers:evidence for a new autocrine/paracrine mechanism. Oncogene 1999; 18:2471-2479.
86. De Leon DD, Wilson DM, Powers M et al. Effects of insulin-like growth factors (IGFs) and IGF receptor antibodies on the proliferation of human breast cancer cells. Growth Factors 1992; 6:327-336.
87. Pandini G, Vigneri R, Costantino A et al. Insulin and insulin-like growth factor-I (IGF-I) receptor overexpression in breast cancers leads to insulin/IGF-I hybrid receptor overexpression:evidence for a second mechanism of IGF-I signaling. Clin Cancer Res 1999; 5:1935-1944.
88. Balana ME, Labriola L, Salatino M et al. Activation of ErbB-2 via a hierarchical interaction between ErbB-2 and type I insulin-like growth factor receptor in mammary tumor cells. Oncogene 2001; 20:34-47.
89. Oka Y, Rozek L, Czech M. Direct demonstration of rapid insulin-like growth factor II receptor internalization and recycling in rat adipocytes. Insulin stimulates 125I-insulin-like growth factor II degradation by modulating the IGF-II receptor recycling process. J Biol Chem 1985; 260:9435-9442.
90. Braulke T. Type-2 IGF receptor:A multi-ligand binding protein. Horm Metab Res 1999; 31:242-246.
91. Oates AJ, Schumaker LM, Jenkins SB et al. The Mannose 6-Phosphate/Insulin-Like Growth Factor 2 Receptor (M6P/IGF2R), a Putative Breast Tumor suppressor Gene. Breast cancer Res Treat 1998; 47:269-281.
92. Jirtle RL, Haag JD, Ariazi EA et al. Increased mannose 6-phosphate/insulin-like growth factor II receptor and transforming growth factor beta 1 levels during monoterpene-induced regression of mammary tumors. Cancer Res 1993; 53:3849-3852.
93. Hankins GR, De Souza AT, Bentley RC et al. M6P/IGF2 receptor:a candidate breast tumor suppressor gene. Oncogene 1996; 12:2003-2009.
94. Ellis MJ, Leav BA, Yang Z et al. Affinity for the insulin-like growth factor-II (IGF-II) receptor inhibits autocrine IGF-II activity in MCF-7 breast cancer cells. Mol Endocrinol 1996; 10:286-297.
95. Chappell SA, Walsh T, Walker RA et al. Loss of heterozygosity at the mannose 6-phosphate insulin-like growth factor 2 receptor gene correlates with poor differentiation in early breast carcinomas. Br J Cancer 1997; 76:1558-1561.
96. Rey JM, Theillet C, Brouillet JP et al. Stable amino-acid sequence of the mannose-6-phosphate/insulin-like growth-factor-II receptor in ovarian carcinomas with loss of heterozygosity and in breast-cancer cell lines. Int J Cancer 2000; 85:466-473.
97. Lemamy GJ, Roger P, Mani JC et al. High-affinity antibodies from hen's-egg yolks against human mannose-6-phosphate/insulin-like growth-factor-II receptor (M6P/IGFII-R): characterization and potential use in clinical cancer studies. Int J Cancer 1999; 80:896-902.
98. Stewart AJ, Johnson MD, May FE et al. Role of insulin-like growth factors and the type I insulin-like growth factor receptor in the estrogen-stimulated proliferation of human breast cancer cells. J Biol Chem 1990; 265:21172-21178.

99. Lee AV, Jackson JG, Gooch JL et al. Enhancement of insulin-like growth factor signaling in human breast cancer:estrogen regulation of insulin receptor substrate-1 expression in vitro and in vivo. Mol Endocrinol 1999; 13:787-796.
100. Martin JL, Coverley JA, Pattison ST et al. Insulin-like growth factor-binding protein-3 production by MCF-7 breast cancer cells:stimulation by retinoic acid and cyclic adenosine monophosphate and differential effects of estradiol. Endocrinology 1995; 136:1219-1226.
101. el-Tanani MK,Green CD. Insulin/IGF-1 modulation of the expression of two estrogen-induced genes in MCF-7 cells. Mol Cell Endocrinol 1996; 121:29-35.
102. Oesterreich S, Zhang P, Guler RL et al. Re-expression of estrogen receptor alpha in estrogen receptor alpha-negative MCF-7 cells restores both estrogen and insulin-like growth factor-mediated signaling and growth. Cancer Res 2001; 61:5771-5777.
103. Dufourny B, van Teeffelen HA, Hamelers IH et al. Stabilization of cyclin D1 mRNA via the phosphatidylinositol 3-kinase pathway in MCF-7 human breast cancer cells. J Endocrinol 2000; 166:329-338.
104. Lee AV, Guler BL, Sun X et al. Oestrogen receptor is a critical component required for insulin-like growth factor (IGF)-mediated signalling and growth in MCF-7 cells. Eur J Cancer 2000; 36:109-110.
105. Kato S, Endoh H, Masuhiro Y et al. Activation of the estrogen receptor through phosphorylation by mitogen-activated protein kinase. Science 1995; 270:1491-1494.
106. Kahlert S, Nuedling S, van Eickels M et al. Estrogen receptor alpha rapidly activates the IGF-1 receptor pathway. J Biol Chem 2000; 275:18447-18453.
107. Migliaccio A, Di Domenico M, Castoria G et al. Tyrosine kinase/p21ras/MAP-kinase pathway activation by estradiol-receptor complex in MCF-7 cells. EMBO J 1996; 15:1292-1300.
108. Lee AV, Weng CN, Jackson JG et al. Activation of estrogen receptor-mediated gene transcription by IGF-I in human breast cancer cells. J Endocrinol 1997; 152:39-47.
109. Martin MB, Franke TF, Stoica GE et al. A role for Akt in mediating the estrogenic functions of epidermal growth factor and insulin-like growth factor I. Endocrinology 2000; 141:4503-4511.
110. Feng W, Webb P, Nguyen P et al. Potentiation of estrogen receptor activation function 1 (AF-1) by Src/JNK through a serine 118-independent pathway. Mol Endocrinol 2001; 15:32-45.
111. Joel PB, Smith J, Sturgill TW et al. pp90rsk1 regulates estrogen receptor-mediated transcription through phosphorylation of Ser-167. Mol Cell Biol 1998; 18:1978-1984.
112. Daly RJ, Harris WH, Wang DY et al. Autocrine production of insulin-like growth factor II using an inducible expression system results in reduced estrogen sensitivity of MCF-7 human breast cancer cells. Cell Growth Differen 1991; 2:457-464.
113. Surmacz E, Burgaud JL. Overexpression of insulin receptor substrate 1 (IRS-1) in the human breast cancer cell line MCF-7 induces loss of estrogen requirements for growth and transformation. Clin Cancer Res 1995; 1:1429-1436.
114. Guvakova MA, Surmacz E. Overexpressed IGF-I receptors reduce estrogen growth requirements, enhance survival, and promote E-cadherin-mediated cell-cell adhesion in human breast cancer cells. Exptl Cell Res 1997; 231:149-162.
115. Fontana JA, Burrows MA, Clemmons DR et al. Retinoid modulation of insulin-like growth factor-binding proteins and inhibition of breast carcinoma proliferation. Endocrinology 1991; 128:1115-1122.
116. Adamo ML, Shao ZM, Lanau F et al. Insulin-like growth factor-I (IGF-I) and retinoic acid modulation of IGF-binding proteins (IGFBPs): IGFBP-2, -3, and -4 gene expression and protein secretion in a breast cancer cell line. Endocrinology 1992; 131:1858-1866.
117. Pekonen F, Nyman T, Ilvesmaki V et al. Insulin-like growth factor binding proteins in human breast cancer tissue. Cancer Res 1992; 52:5204-5207.
118. Figueroa JA, Jackson JG, McGuire WL et al. Expression of insulin-like growth factor binding proteins in human breast cancer correlates with estrogen receptor status. J Cell Biochem 1993; 52:196-205.
119. McGuire SE, Hilsenbeck SG, Figueroa JA et al. Detection of insulin-like growth factor binding proteins (IGFBPs) by ligand blotting in breast cancer tissues. Cancer Lett 1994; 77:25-32.
120. Yee D, Sharma J, Hilsenbeck SG. Prognostic significance of insulin-like growth factor-binding protein expression in axillary lymph node-negative breast cancer. J Natl Cancer Inst 1994; 86:1785-1789.
121. Clemmons DR, Camacho HC, Coronado E et al. Insulin-like growth factor binding protein secretion by breast carcinoma cell lines:correlation with estrogen receptor status. Endocrinology 1990; 127:2679-2686.
122. Yee D, Favoni RE, Lupu R et al. The insulin-like growth factor binding protein BP-25 is expressed by human breast cancer cells. Biochem Biophys Res Commun 1989; 158:38-44.
123. Huynh H, Yang X, Pollak M. Estradiol and antiestrogens regulate a growth inhibitory insulin-like growth factor binding protein 3 autocrine loop in human breast cancer cells. J Biol Chem 1996; 271:1016-1021.
124. Huynh H, Yang XF, Pollak M. A role for insulin-like growth factor binding protein 5 in the antiproliferative action of the antiestrogen ICI 182780. Cell Growth Differen 1996; 7:1501-1506.
125. Nickerson T, Huynh H, Pollak M. Insulin-like growth factor binding protein-3 induces apoptosis in MCF7 breast cancer cells. Biochem Biophys Res Commun 1997; 237:690-693.
126. Oh Y, Muller HL, Ng L et al. Transforming growth factor-beta-induced cell growth inhibition in human breast cancer cells is mediated through insulin-like growth factor-binding protein-3 action. J Biol Chem 1995; 270:13589-13592.
127. Gucev ZS, Oh Y, Kelley KM et al. Insulin-like growth factor binding protein 3 mediates retinoic acid- and transforming growth factor beta2-induced growth inhibition in human breast cancer cells. Cancer Res 1996; 56:1545-1550.

128. Rozen F, Zhang J, Pollak M. Antiproliferative action of tumor necrosis factor-alpha on MCF-7 breastcancer cells is associated with increased insulin-like growth factor binding protein-3 accumulation. Int J Oncol 1998; 13:865-869.
129. Rozen F, Pollak M. Inhibition of insulin-like growth factor I receptor signaling by the vitamin D analogue EB1089 in MCF-7 breast cancer cells: A role for insulin-like growth factor binding proteins. Int J Oncol 1999; 15:589-594.
130. Rozen F, Yang XF, Huynh H et al. Antiproliferative action of vitamin D-related compounds and insulin-like growth factor-binding protein 5 accumulation. J Natl Cancer Inst 1997; 89:652-656.
131. McCotter D, van den Berg HW, Boylan M et al. Changes in insulin-like growth factor-I receptor expression and binding protein secretion associated with tamoxifen resistance and estrogen independence in human breast cancer cells in vitro. Cancer Lett 1996; 99:239-245.
132. Maxwell P,van den Berg HW. Changes in the secretion of insulin-like growth factor binding proteins -2 and -4 associated with the development of tamoxifen resistance and estrogen independence in human breast cancer cell lines. Cancer Lett 1999; 139:121-127.
133. Giannios J, Alexandropoulos N, Koratzis A et al. Downregulation of oestrogen receptor in advanced breast cancer after lipofection with wild-type (w-t) insulin growth factor binding protein IGFBP-2 cDNA plasmid. Eur J Cancer 2000; 36:
134. Dubois V, Couissi D, Schonne E et al. Intracellular levels and secretion of insulin-like-growth-factor-binding proteins in MCF-7/6, MCF-7/AZ and MDA-MB-231 breast cancer cells. Differential modulation by estrogens in serum-free medium. Eur J Biochem 1995; 232:47-53.
135. McGuire WJ, Jackson JG, Figueroa JA et al. Regulation of insulin-like growth factor-binding protein (IGFBP) expression by breast cancer cells:use of IGFBP-1 as an inhibitor of insulin-like growth factor action. J Natl Cancer Inst 1992; 84:1336-1341.
136. Chen JC, Shao ZM, Sheikh MS et al. Insulin-like growth factor-binding protein enhancement of insulin-like growth factor-I (IGF-I)-mediated DNA synthesis and IGF-I binding in a human breast carcinoma cell line. J Cell Physiol 1994; 158:69-78.
137. Pratt SE, Pollak MN. Insulin-like growth factor binding protein 3 (IGF-BP3) inhibits estrogen-stimulated breast cancer cell proliferation. Biochem Biophys Res Commun 1994; 198:292-297.
138. Salahifar H, Firth SM, Baxter RC et al. Characterization of an amino-terminal fragment of insulin-like growth factor binding protein-3 and its effects in MCF-7 breast cancer cells. Growth Horm IGF Res 2000; 10:367-377.
139. De Mellow JS, Baxter RC. Growth hormone-dependent insulin-like growth factor (IGF) binding protein both inhibits and potentiates IGF-I-stimulated DNA synthesis in human skin fibroblasts. Biochem Biophys Res Commun 1988; 156:199-204.
140. Ricort JM, Binoux M. Insulin-like growth factor (IGF) binding protein-3 inhibits type 1 IGF receptor activation independently of its IGF binding affinity. Endocrinology 2001; 142:108-113.
141. Oh Y, Muller HL, Lamson G et al. Insulin-like growth factor (IGF)-independent action of IGF-binding protein-3 in Hs578T human breast cancer cells. Cell surface binding and growth inhibition. J Biol Chem 1993; 268:14964-14971.
142. Firth SM, Fanayan S, Benn D et al. Development of resistance to insulin-like growth factor binding protein-3 in transfected T47D breast cancer cells. Biochem Biophys Res Commun 1998; 246:325-329.
143. Gill ZP, Perks CM, Newcomb PV et al. Insulin-like growth factor-binding protein (IGFBP-3) predisposes breast cancer cells to programmed cell death in a non-IGF-dependent manner. J Biol Chem 1997; 272:25602-25607.
144. Butt AJ, Firth SM, King MA et al. Insulin-like growth factor-binding protein-3 modulates expression of Bax and Bcl-2 and potentiates p53-independent radiation-induced apoptosis in human breast cancer cells. J Biol Chem 2000; 275:39174-39181.
145. Perks CM, McCaig C, Holly JM. Differential insulin-like growth factor (IGF)-independent interactions of IGF binding protein-3 and IGF binding protein-5 on apoptosis in human breast cancer cells. Involvement of the mitochondria. J Cell Biochem 2000; 80:248-258.
146. Fowler CA, Perks CM, Newcomb PV et al. Insulin-like growth factor binding protein-3 (IGFBP-3) potentiates paclitaxel-induced apoptosis in human breast cancer cells. Int J Cancer 2000; 88:448-453.
147. Perks CM, Bowen S, Gill ZP et al. Differential IGF-independent effects of insulin-like growth factor binding proteins (1-6) on apoptosis of breast epithelial cells. J Cell Biochem 1999; 75:652-664.
148. Oh Y, Muller HL, Pham H et al. Demonstration of receptors for insulin-like growth factor binding protein-3 on Hs578T human breast cancer cells. J Biol Chem 1993; 268:26045-26048.
149. Yamanaka Y, Fowlkes JL, Wilson EM et al. Characterization of insulin-like growth factor binding protein-3 (IGFBP-3) binding to human breast cancer cells:kinetics of IGFBP-3 binding and identification of receptor binding domain on the IGFBP-3 molecule. Endocrinology 1999; 140:1319-1328.
150. Ingermann AR, Kim HS, Oh Y. Characterization of a functional receptor for insulin-like growth factor binding protein-3. Growth Horm IGF Res 2000; 10:A27.
151. Leal SM, Liu Q, Huang SS et al. The type V transforming growth factor beta receptor is the putative insulin-like growth factor-binding protein 3 receptor. J Biol Chem 1997; 272:20572-20576.
152. Fanayan S, Firth SM, Butt AJ et al. Growth inhibition by insulin-like growth factor-binding protein-3 in T47D breast cancer cells requires transforming growth factor-beta (TGF-beta) and the type II TGF-beta receptor. J Biol Chem 2000; 275:39146-39151.
153. Schedlich LJ, Young TF, Firth SM et al. Insulin-like growth factor-binding protein (IGFBP)-3 and IGFBP-5 share a common nuclear transport pathway in T47D human breast carcinoma cells. J Biol Chem 1998; 273:18347-18352.

154. Schedlich LJ, Le Page SL, Firth SM et al. Nuclear import of insulin-like growth factor-binding protein-3 and -5 is mediated by the importin beta subunit. J Biol Chem 2000; 275:23462-23470.
155. Liu BR, Lee HY, Weinzimer SA et al. Direct functional interactions between insulin-like growth factor-binding protein-3 and retinoid X receptor-alpha regulate transcriptional signaling and apoptosis. J Biol Chem 2000; 275:33607-33613.
156. Perks CM, Newcomb PV, Norman MR et al. Effect of insulin-like growth factor binding protein-1 on integrin signalling and the induction of apoptosis in human breast cancer cells. J Mol Endocrinol 1999; 22:141-150.
157. Huynh H, Beamer WD, Pollak M. Overexpression of insulin-like growth factor binding protein 4 (IGFBP-4) in MCF-7 breast cancer cells is associated with reduced responsiveness to insulin-like growth factors in vitro and reduced tumor growth in vivo. Int J Oncol 1997; 11:193-197.
158. Van den Berg CL, Cox GN, Stroh CA et al. Polyethylene glycol conjugated insulin-like growth factor binding protein-1 (IGFBP-1) inhibits growth of breast cancer in athymic mice. Eur J Cancer 1997; 33:1108-1113.
159. Rocha RL, Hilsenbeck SG, Jackson JG et al. Correlation of insulin-like growth factor-binding protein-3 messenger RNA with protein expression in primary breast cancer tissues:detection of higher levels in tumors with poor prognostic features. J Natl Cancer Inst 1996; 88:601-606.
160. Rocha RL, Hilsenbeck SG, Jackson JG et al. Insulin-like growth factor binding protein-3 and insulin receptor substrate-1 in breast cancer:correlation with clinical parameters and disease-free survival. Clin Cancer Res 1997; 3:103-109.
161. Yu H, Levesque MA, Khosravi MJ et al. Insulin-like growth factor-binding protein-3 and breast cancer survival. Int J Cancer 1998; 79:624-628.
162. Martin JL, Baxter RC. Oncogenic ras causes resistance to the growth inhibitor insulin-like growth factor binding protein-3 (IGFBP-3) in breast cancer cells. J Biol Chem 1999; 274:16407-16411.
163. Clark GJ, Der CJ. Aberrant function of the Ras signal transduction pathway in human breast cancer. Breast cancer Res Treat 1995; 35:133-144.
164. Reed MJ, Christodoulides A, Koistinen R et al. The effect of endocrine therapy with medroxyprogesterone acetate, 4-hydroxyandrostenedione or tamoxifen on plasma concentrations of insulin-like growth factor (IGF)-I, IGF-II and IGFBP-1 in women with advanced breast cancer. Int J Cancer 1992; 52:208-212.
165. Baldini E, Giannessi PG, Gardin G et al. In vivo cytokinetic effects of recombinant human growth hormone (rhGH) in patients with advanced breast carcinoma. J Biol Regulators Homeostatic Agents 1994; 8:113-116.
166. Helle SI, Geisler J, Anker GB et al. Alterations in the insulin-like growth factor system during treatment with diethylstilboestrol in patients with metastatic breast cancer. Br J Cancer 2001; 85:147-151.
167. Anderson E, Ferguson JE, Morten H et al. Serum immunoreactive and bioactive lactogenic hormones in advanced breast cancer patients treated with bromocriptine and octreotide. Eur J Cancer 1993; 29A:209-217.
168. Lissoni P, Barni S, Meregalli S et al. Modulation of cancer endocrine therapy by melatonin:a phase II study of tamoxifen plus melatonin in metastatic breast cancer patients progressing under tamoxifen alone. Br J Cancer 1995; 71:854-856.
169. Lahti EI, Knip M, Laatikainen TJ. Plasma insulin-like growth factor I and its binding proteins 1 and 3 in postmenopausal patients with breast cancer receiving long term tamoxifen. Cancer 1994; 74:618-624.
170. Yee D, Favoni RE, Lippman ME et al. Identification of insulin-like growth factor binding proteins in breast cancer cells. Breast cancer Res Treat 1991; 18:3-10.
171. Colston KW, Perks CM, Xie SP et al. Growth inhibition of both MCF-7 and Hs578T human breast cancer cell lines by vitamin D analogues is associated with increased expression of insulin-like growth factor binding protein-3. J Mol Endocrinol 1998; 20:157-162.
172. Shang Y, Baumrucker CR, Green MH. Signal relay by retinoic acid receptors alpha and beta in the retinoic acid-induced expression of insulin-like growth factor-binding protein-3 in breast cancer cells. J Biol Chem 1999; 274:18005-18010.
173. Pratt SE, Pollak MN. Estrogen and antiestrogen modulation of MCF7 human breast cancer cell proliferation is associated with specific alterations in accumulation of insulin-like growth factor-binding proteins in conditioned media. Cancer Res 1993; 53:5193-5198.
174. Tsubaki J, Choi WK, Ingermann AR et al. Effects of sodium butyrate on expression of members of the IGF-binding protein superfamily in human mammary epithelial cells. J Endocrinol 2001; 169:97-110.
175. Sheikh MS, Shao ZM, Clemmons DR et al. Identification of the insulin-like growth factor binding proteins 5 and 6 (IGFBP-5 and 6) in human breast cancer cells. Biochem Biophys Res Commun 1992; 183:1003-1010.
176. Qin C, Singh P, Safe S. Transcriptional activation of insulin-like growth factor-binding protein-4 by 17beta-estradiol in MCF-7 cells:role of estrogen receptor-Sp1 complexes. Endocrinology 1999; 140:2501-2508.
177. Coutts A, Murphy LJ, Murphy LC. Expression of insulin-like growth factor binding proteins by T-47D human breast cancer cells:regulation by progestins and antiestrogens. Breast cancer Res Treat 1994; 32:153-164.
178. Parisot JP, Leeding KS, Hu XF et al. Induction of insulin-like growth factor binding protein expression by ICI 182,780 in a tamoxifen-resistant human breast cancer cell line. Breast cancer Res Treat 1999; 55:231-242.

CHAPTER 24

The Role of the IGF System in Prostate Cancer

Charles T. Roberts, Jr.

Introduction

An important aspect of the IGF system in postnatal human physiology is its involvement in tumorigenesis. This was hardly unexpected, based upon the demonstrated role of IGF-I, in particular, as a potent growth regulator. Reports over the last several years of a strong association of circulating IGF-I levels with the risk of developing a number of important human cancers, as well as the ongoing studies of the molecular mechanisms of IGF action in cancer cells, have renewed interest in the therapeutic and diagnostic possibilities of the IGF system in cancer therapy. The sections below summarize these molecular and epidemiological data, particularly with respect to prostate cancer.

IGF Action in Prostate Growth and Development

A significant amount of data has been accumulated that suggests that the IGF system plays an important role in the normal growth and development of the prostate. Prostatic stromal cells and epithelial cells in primary culture secrete IGFBPs, and stromal cells produce IGF-II, and both stromal and epithelial cells express the IGF-I receptor and are responsive to IGF-I with respect to proliferation.[1-4] In vivo, the prostate epithelial cells that are considered to be the likely precursors to prostatic intraepithelial neoplasia and prostatic adenocarcinoma respond to both locally produced IGF-II and circulating IGF-I through paracrine and endocrine mechanisms, respectively. Additional support for an important role for IGF action in prostate growth has come from recent studies showing that systemic administration of IGF-I increases rat prostate growth,[5] that modulation of rat ventral prostate weight by the 5-α reductase inhibitor finasteride is associated with altered levels of IGF-I receptor and IGFBP-3 gene expression,[6] and that IGF-I-deficient mice exhibit decreased prostate size and complexity of prostate architecture.[7]

The IGF-IR and IGF-II in Tumorigenesis-Molecular Studies

Most primary tumors and transformed cell lines express easily detectable levels of IGF-II mRNA and protein, while some tumors over-express the IGF-I gene.[8] The IGF-II gene is imprinted in mouse and human, which restricts expression to the paternal allele.[9] IGF-II gene imprinting is relaxed in several cancers, including Wilms tumor, rhabdomyosarcoma, and lung cancer, as well as in Beckwith-Wiedemann syndrome, which predisposes to Wilms tumor.[10-12] Altered imprinting leads to overexpression of IGF-II, which could contribute to the development of the tumors as a consequence of increased proliferative signaling.

Numerous studies performed over the last 20 years have suggested that transformed cells express higher levels of the IGF-IR, although the specific molecular mechanisms responsible for the increased expression of the IGF-IR gene in tumors remain poorly characterized. Amplification of the IGF-IR locus at band 15q26 has been reported in a small number of breast and melanoma cases.[13] The presumed role of the IGF-IR in tumorigenesis involves increased IGF-IR expression and concomitant increased responsiveness to IGF in terms of proliferation and inhibition of apoptosis. Although this picture is probably accurate with respect to the pediatric tumors that are often associated with chromosomal translocations, such as Wilms tumor and rhabdomyosarcoma, the situation in the epithelial tumors that are more prevalent in adults is more complex.[14]

Insulin-Like Growth Factors, edited by Derek LeRoith, Walter Zumkeller and Robert Baxter. ©2003 Eurekah.com and Kluwer Academic / Plenum Publishers.

The original suggestion that the IGF-IR itself functions as an oncogene was based upon the behavior of fibroblasts in which the IGF-IR had been over-expressed,[15] a system a system with limited relevance to human cancer. Other studies suggesting that increased IGF-IR expression modulates radiosensitivity have also used IGF-IR over-expression in fibroblasts.[16] It should be noted, however, that a recent report has demonstrated that inhibition of IGF-IR activity by a selective kinase inhibitor in MCF-7 breast cancer cells resulted in increased radiosensitivity.[17]

The multitude of earlier studies describing over-expression of the IGF-IR in breast, prostate, and other tumor types have been, for the most part, based on analyses of tissue homogenates or established cancer cell lines for which appropriate normal controls do not exist. The apparent IGF-IR content of homogenates, in particular, can be affected by contamination with stroma, which would dilute IGF-IR content in normal epithelium or small tumors. Studies of IGF-IR expression in breast and prostate clinical samples that employed immunohistochemistry, or of matched cell lines corresponding to normal and tumor tissue, revealed that normal epithelium and early-stage tumors both express abundant IGF-IR, and that IGF-IR expression is significantly reduced in advanced, metastatic prostate and breast cancer.[18-22] Decreased IGF-IR expression is also seen in metastases in the TRAMP mouse model of prostate cancer.[23] Additionally, Plymate et al[24,25] reported that re-expression of the IGF-IR in metastatic prostate epithelial cells reversed their malignant phenotype and increased sensitivity to apoptosis. These studies suggested the rather novel hypothesis that reduced IGF-IR action was necessary for cancer progression. This view has been challenged by a recent report by Hellawell et al,[26] who reported that IGF-IR expression was decreased in some metastatic prostate cancer samples as compared to benign or carcinoma tissue, but was increased in a majority of samples studied (8 out of 12). In this study, however, the immunostaining for IGF-IR using a single β-subunit antibody was diffusely cytoplasmic in most samples, in contrast to the expected membrane localization reported by Chott et al,[20] using two different α-subunit antibodies. In a separate study, Nickerson, et al[27] found an increase in IGF-IR (and IGF-I) gene expression during progression to androgen dependence in LNCaP and LAPC-9 xenografts, but a decrease in IGF-IR expression in the LAPC-4 xenograft model. The increases seen in the LNCaP and LAPC-9 models were 3 to 4-fold, whereas the decrease in the LAPC-4 model was greater than 10-fold (androgen-dependent versus androgen-independent). Thus, the issue of the extent of IGF-IR over-or under-expression in prostate cancer progression is far from being settled.

Another confusing issue is the effect of modulating IGF-IR expression on prostate cancer cell behavior. In addition to the reports cited above on the anti-tumorigenic effects of IGF-IR re-expression,[24,25] there are other reports that antisense inhibition of IGF-IR gene expression[28] or expression of a dominant-negative IGF-IR in prostate cancer cells also inhibits tumorigenecity. These apparently contradictory findings can be reconciled, if one considers the possibility that there are different levels or "set points" of IGF-IR activity (determined either by IGF-IR gene expression or intrinsic receptor activity), and that deviation from this set point in either direction modulates cellular phenotype. This concept is supported by the earlier studies of Rubini et al[29] demonstrating that small changes in IGF-IR levels, in the physiological range, can modulate the mitogenic and transforming phenotype of mouse embryo cells.

The activation of the IGF-IR present in normal epithelium by elevated circulating levels of IGF-I may underlie the epidemiological data described in more detail below, whereas the subsequent decrease in IGF-IR (if substantiated by additional studies) may represent an attempt by established cancer cells to avoid the potential differentiating effects of IGF-I at sites of metastasis. Alternatively, decreased expression of the IGF-IR may protect tumor cells from a novel, non-apoptotic form of programmed cell death that has been recently described as being triggered by the unliganded IGF-IR.[30] It is clear, however, that the prevailing notion that the IGF-IR is routinely over-expressed in transformed cells is somewhat of an over-generalization.

IGF-I and Prostate Cancer-Epidemiological Studies

The potent mitogenic activity of IGF-I in cell culture made it an obvious candidate risk factor in cancer development, but it was not until 1998 that several prospective studies suggested that high circulating levels of IGF-I were associated with an increased risk of developing prostate cancer.[31-34] The strength of the association between IGF-I levels and prostate cancer risk was questioned in subsequent cross-sectional studies,[35,36] while Djavan et al,[37] in a prospective study, found that the IGF-I/PSA ratio was superior to IGF-I or PSA measurements alone for predicting prostate cancer risk. A meta-analysis of the data available to date in 2000[38] concluded that high circulating IGF-I levels (although still within normal limits) posed a risk equivalent to that associated with high testosterone. In a screening trial, Finne et al[39] did not find an association between serum IGF-I levels and prostate cancer risk, while Baffa et al[40] actually found that circulating IGF-I levels were lower in a group of patients undergoing radical prostatectomy as compared to age-matched controls. In additional prospective studies, however, Harman et al[41] and Stattin et al[42] found that IGF-I levels were associated with prostate cancer risk, and that this association was especially evident in younger men. A second meta-analysis of 14 case-controlled studies[43] concluded that high normal circulating IGF-I levels were indeed associated with prostate cancer risk. In a subsequent cross-sectional study, Latif et al[44] did not find a correlation between levels of IGF-I or IGFBP-3 and prostate cancer stage, while, in the latest prospective study,[32] circulating IGF-I levels appeared to be most predictive of advanced prostate cancer.

While the conclusions of this extensive series of studies appear contradictory, there is, in fact, some consistency. Prospective studies have consistently demonstrated an association between high circulating IGF-I levels and prostate cancer risk, while cross-sectional studies have generated variable results. These data are consistent with the hypothesis that high serum IGF-I levels in younger men predict the occurrence of advanced prostate cancer years later, while IGF-I levels at the time of diagnosis may not be informative. This hypothesis suggests that long-term exposure of prostate epithelial cells to high levels of serum-derived IGF-I increases the probability of initiating hyperplasia in the cellular precursors of prostatic intraepithelial neoplasia and subsequent prostate adenocarcinoma.

Corroboration of the relationship between IGF-I levels and prostate carcinogenesis at the molecular level has now come from analysis of transgenic mice with targeted expression of IGF-I in the basal prostatic epithelium. This dysregulated IGF-I biosynthesis resulted in the appearance of hyperplastic lesions resembling PIN by 6 months of age,[45] and prostatic adenocarcinomas or small cell carcinomas were eventually seen in 50% of the transgenics. Specifically, deregulated expression of IGF-I and constitutive activation of IGF-I receptors in basal epithelial cells resulted in tumor progression similar to that seen in human disease. These studies also provide additional evidence for the prostate basal epithelial cell as a precursor to prostate cancer.

IGF-II and Prostate Cancer-Epidemiological Studies

A potential contribution of IGF-II levels to prostate cancer risk is of potential interest, given that levels of circulating IGF-II in humans are consistently several-fold higher than those of IGF-I throughout life. The only study to date to directly address the relationship of IGF-II levels to prostate cancer risk[41] found that serum levels of IGF-II were inversely related to risk. These authors suggest that IGF-II may actually inhibit prostate cancer development. It is intriguing that Gnanapragasam et al[46] have reported that IGF-II increases androgen receptor expression in prostate stromal cells and LNCaP prostate carcinoma cells, since loss of wild-type androgen receptor function and androgen insensitivity is a hallmark of advanced disease.

Summary

As demonstrated by the studies described above, IGF action is an important component of normal prostate growth and development, as well as prostate cancer initiation and progression. The behavior of the IGF system in prostate tumorigenesis reflects both expected and unanticipated effects of this complicated signaling mechanism. In particular, the effects of IGF-I on prostate growth and tumor development in mice and the association of high circulating IGF-I concentrations with increased prostate cancer risk in humans are consistent with the well-characterized mitogenic properties of the IGFs, while the possible negative association of circulating IGF-II levels with prostate cancer risk and the counter-intuitive decrease in IGF-IR expression seen in advanced human and murine prostate cancer and cell lines derived from prostate metastases may require a re-evaluation of a number of the assumptions that underlie our current understanding of IGF system function in growth and development in general.

References

1. Cohen P, Peehl DM, Lamson G et al. Insulin-like growth factors (IGFs), IGF receptors, and IGF-binding proteins in primary cultures of prostate epithelial cells. J Clin Endocrinol Metab 1991; 73(2):401-407.
2. Peehl DM, Cohen P, Rosenfeld RG. The insulin-like growth factor system in the prostate. World J Urol 1994; 13:306-311.
3. Cohen P, Peehl DM, Rosenfeld RG. The IGF axis in the prostate. Horm Metab Res 1994; 26:81-84.
4. Boudon C, Rodier G, Lechevallier E et al. Secretion of insulin-like growth factors and their binding proteins by human normal and hyperplastic prostatic cells in primary culture. J Clin Endocrinol Metab 1996; 81(2):612-617.
5. Torring N, Vinter-Jensen L, Pederson SB et al. Systemic Administration of insulin-like growth factor I (IGF-I) causes growth of the rat prostate. J Urol 1997; 158(1):222-227.
6. Huynh H, Seyam RM, Brock GB. Reduction of ventral prostate weight by finasteride is associated with suppression of insulin-like growth factor I (IGF-I) and IGF-I receptor genes and with an increase in IGF binding protein 3. Cancer Res 1998; 58(2):215-218.
7. Ruan W, Powell-Braxton L, Kopchick JJ et al. Evidence that insulin-like growth factor I and growth hormone are required for prostate gland development. Endocrinology 1999; 140(5):1984-1989.
8. Quinn KA, Treston AM, Unsworth EJ et al. Insulin-like growth factor expression in human cancer cell lines. J Biol Chem 1996; 271(19):11477-83.
9. DeChiara TM, Robertson EJ, Efstratiadis A. Parental imprinting of the mouse insulin-like growth factor II gene. Cell 1991; 64:849-859.
10. Ogawa O, Becroft DM, Morison IM et al. Constitutional relaxation of insulin-like growth factor II gene imprinting associated with Wilms' tumor and gigantism. Nat Genet 1993; 5(4):408-412.
11. Suzuki H, Veda R, Takahashi T. Altered imprinting in lung cancer. Nat Genet 1994; 6:332-333.
12. Zhan S, Shapiro DN, Helman LJ. Activation of an imprinted allele of the insulin-like growth factor II gene implicated in rhabbdomyosarcoma. J Clin Invest 1994; 94:445-448.
13. Almeida A, Muleris M, Dutrillaux B et al. The insulin-like growth factor I receptor gene is the target for the 15q26 amplicon in breast cancer. Genes. Chromosomes Cancer 1994; 11(1):63-65.
14. DePhino RA. The age of cancer. Nature 2000; 408:248-254.
15. Kaleko M, Rutter WJ, Miller AD. Overexpression of the human insulin-like growth factor I receptor promotes ligand-dependent neoplastic transformation. Mol Cell Biol 1990; 10:464-473.
16. Turner BC, Haffty BG, Narayanan L et al. Insulin-like growth factor-I receptor overexpression mediates cellular radioresistance and local breast cancer recurrence after lumpectomy and radiation. Cancer Res 1997; 57(3079-3083).
17. Wen B, Deutsch E, Marangoni E et al. Tyrphostin AG1024 modulates radiosensitivity in human breast cancer cells. Br J Cancer 2001; 85:2017-2021.
18. Tennant MK, Thrasher JB, Twomey PA et al. Protein and mRNA for the type 1 insulin-like growth factor (IGF) receptor is decreased and IGF-II mRNA is increased in human prostate carcinoma compared to benign prostate epithelium. J Clin Endocrinol Metab 1996; 81:3774-3782.
19. Happerfield LC, Miles DW, Barnes DM et al. The localization of the insulin-like growth factor receptor 1 (IGFR-1) in benign and malignant breast tissue. J Pathol 1997; 183:412-417.
20. Chott A, Sun Z, Morganstern D et al. Tyrosine kinases expressed in vivo by human prostate cancer bone marrow metastases and loss of type 1 insulin-like growth factor receptor. Am J Pathol 1999; 155:1271-1279.
21. Schnarr B, Strunz K, Ohsam J et al. Down-regulation of insulin-like growth factor-I receptor and insulin receptor substrate-1 expression in advanced human breast cancer. Int J Cancer 2000; 89:506-513.
22. Damon SE, Plymate SR, Carroll JM et al. Transcriptional regulation of insulin-like growth factor-I receptor gene expression in prostate cancer cells. Endocrinology 2001; 142(1):21-27.
23. Kaplan PJ, Mohan S, Cohen P et al. The insulin-like growth factor axis and prostate cancer: lessons from the transgenic adenocarcinoma of mouse prostate (TRAMP) model. Cancer Res 1999; 59(9):2203-2209.
24. Plymate SR, Bae VL, Maddison L et al. Reexpression of the type 1 insulin-like growth factor receptor inhibits the malignant phenotype of simian virus 40 T antigen immortalized human prostate epithelial cells. Endocrinology 1997; 138(4):1728-1735.
25. Plymate SR, Bea VL, Maddison L et al. Type-1 insulin-like growth factor receptor reexpression in the malignant phenotype of SV40-T-immortalized human prostate epithelial cells enhances apoptosis. Endocr J 1997; 7(1):119-124.
26. Hellawell GO, Turner GD, Davies DR et al. Expression of the type 1 insulin-like growth factor receptor is up-regulated in primary prostate cancer and commonly persists in metastatic disease. Cancer Res 2002; 62(10):2942-2950.
27. Nickerson T, Chang F, Lorimer D et al. In vivo progression of LAPC-9 and LNCaP prostate cancer models to androgen independence is associated with increased expression of insulin-like growth factor I (IGF-I) and IGF-I receptor (IGF-IR). Cancer Res 2001; 61(16):6276-80.
28. Burfeind P, Chernicky CL, Rininsland F et al. Antisense RNA to the type 1 insulin-like growth factor receptor supresses tumor growth and prevents invasion by rat prostate cancer cells in vivo. Proc Natl Acad Sci USA 1996; 93(14):7263-7268.
29. Rubini M, Hongo A, D'Ambrosio C et al. The IGF-I receptor in mitogenesis and transformation of mouse embryo cells: Role of receptor number. Exp Cell Res 1997; 230:284-292.
30. Sperandio S, de Belle I, Bredesen DE. An alternative, nonapoptotic form of programmed cell death. Proc Natl Acad Sci USA 2000; 97:14376-14381.

31. Mantzoros CS, Tzonou A, Signorello LB et al. Insulin-like growth factor 1 in relation to prostate cancer and benign prostatic hyperplasia. Br J Cancer 1997; 76(9):1115-1118.
32. Chan JM, Stampfer MJ, Ma J et al. Insulin-like growth factor-I (IGF-I) and IGF binding protein-3 as predictors of advanced-stage prostate cancer. J Natl Cancer Inst 2002; 94(14):1099-1106.
33. Wolk A, Mantzoros CS, Andersson SO et al. Insulin-like growth factor 1 and prostate cancer risk: A population-based, case-control study. J Natl Cancer Inst 1998; 90(12):911-915.
34. Cohen P. Serum insulin-like growth factor I levels and prostate cancer risk—interpreting the evidence. J Natl Cancer Inst 1998; 90:876-876.
35. Cutting CW, Hunt C, Nisbet JA et al. Serum insulin-like growth factor-1 is not a useful marker of prostate cancer. BJU Int 1999; 83(9):996-999.
36. Kurek R, Tunn UW, Eckart O et al. The significance of serum levels of insulin-like growth factor-1 in patients with prostate cancer. BJU Int 2000; 85(1):125-129.
37. Djavan G, Bursa B, Seitz C et al. Insulin-like growth factor 1 (IGF-1), IGF-1 density, and IGF-1/PSA ratio for prostate cancer detection. Urology 1999; 54(4):603-606.
38. Shaneyfelt T, Husein R, Bubley G et al. Hormonal predictors of prostate cancer: A meta-analysis. J Clin Oncol 2000; 18(4):847-853.
39. Finne P, Auvinen A, Koistinen H et al. Insulin-like growth factor I is not a useful marker of prostate cancer in men with elevated levels of prostate-specific antigen. J Clin Endocrinol Metab 2000; 85(8):2744-2747.
40. Baffa R, Reiss K, El-Gabry EA et al. Low serum insulin-like growth factor 1 (IGF-1): A significant association with prostate cancer. Tech Urol 2000; 6(3):236-239.
41. Harman SM, Metter EJ, Blackman MR et al. Serum levels of insulin-like growth factor I (IGF-I), IGF-II, IGF-binding protein-3, and prostate-specific antigen as predictor of clinical prostate cancer. J Clin Endocrinol Metab 2000; 85(11):4258-4265.
42. Stattin P, Bylund A, Rinaldi S et al. Plasma insulin-like growth factor-I, insulin-like growth factor-binding proteins, and prostate cancer risk: a prospective study. J Natl Cancer Inst 2000; 92(23):1910-1917.
43. Shi R, Berkel HJ, Yu H. Insulin-like growth factor-I and prostate cancer: a meta-analysis. Br J Cancer 2001; 85(7):991-996.
44. Latif Z, McMillan DC, Wallace AM et al. The relationship of circulating insulin-like growth factor 1, its binding protein-3, prostate-specific antigen and C-reactive protein with disease stage in prostate cancer. BJU Int 2002; 89(4):396-399.
45. DiGiovanni J, Kiguchi K, Frijhoff A et al. Deregulated expression of insulin-like growth factor 1 in prostate epithelium leads to neoplasia in transgenic mice. Proc Natl Acad Sci USA 2000; 97(7):3455-3460.
46. Gnanapragasam VJ, McCahy PJ, Neal DE et al. Insulin-like growth factor II and androgen receptor expression in the prostate. BJU Int 2000; 86(6):731-735.

CHAPTER 25

IGFs and Sarcomas

Fariba Navid and Lee J. Helman

Abstract

Insulin-like growth factors (IGFs) and their receptors exert their effects on a variety of sarcomas through autocrine, paracrine and endocrine mechanisms. In addition to regulating the proliferation of normal as well as malignant cells of mesenchymal origin, IGFs and their receptors play a central role in transformation, survival, metastasis and differentiation of sarcomas. Most of these effects are mediated through the IGFI receptor (IGFIR). Three pediatric sarcomas, namely rhabdomyosarcoma, osteosarcoma and Ewing's family of tumors illustrate the pleiotropic effects of IGFs. In this chapter, we review our current understanding of the role of IGFs in the pathogenesis of these tumors. Mechanisms of overexpression of IGFII and the IGFIR receptor in these tumors are discussed including epigenetic modifications and effects of mutant p53 on the promoters of these genes. The potential of the IGF signaling pathway as a therapeutic target of sarcomas will also be discussed.

Introduction

Insulin-like growth factors (IGFs) and their receptors play a significant role in the proliferation, survival, transformation and metastatic process of a number of sarcomas. As in epithelial derived tumors, most of the actions of IGFs are mediated through the insulin-like growth factor I receptor (IGFIR) which is expressed on the surface of most, if not all, sarcomas that have been studied. The functional importance of this receptor in mesenchymal tumors was first suggested by experiments in which overexpression of the IGFIR in NIH3T3 fibroblasts resulted in a transformed phenotype.[1] Subsequently it has been shown that impairment of IGFIR function by antisense strategies, antibodies, or dominant negative constructs results in amelioration of its effects on proliferation and transformation.[2-4]

The bioavailability and bioactivity of IGFs is determined by IGF-binding proteins (IGFBP). Six classic IGFBPs have been identified in humans.[5] The role of IGFBPs in the pathogenesis of tumors is multifactorial. IGFBPs are regulated by a variety of different hormones and growth factors.[6] IGF independent effects of IGFBPs have been described.[7,8] Post-translational modification of IGFBP by proteolysis adds another degree of complexity to this system.[9-11] And, in some instances the same IGFBP can both inhibit or enhance IGF action depending on its concentration or phosphorylation status.[12] As we learn more about these binding proteins, it becomes clear that our understanding of their role in the pathogenesis of sarcomas is limited. IGFBPs are extensively reviewed in another chapter and will not be discussed here.

The detection of endogenous IGF ligands in sarcoma cells has been variable. Some tumor cells, like rhabdomyosarcoma, express an abundance of IGFII that can easily be detected by Northern analysis. In other sarcomas, like osteosarcomas, the presence of IGFI mRNA can only be detected by more sensitive methods like RT-PCR or RNase Protection Assays (RPA). The significance of this is unclear. However, because of the significant production of IGFs in the extracellular milieu of these tumors, as in the case of osteosarcoma and the surrounding bone, and the circulating IGFs in serum, IGFs are believed to mediate their effects via paracrine and endocrine as well as autocrine loops in a variety of sarcomas.

Insulin-Like Growth Factors, edited by Derek LeRoith, Walter Zumkeller and Robert Baxter. ©2003 Eurekah.com and Kluwer Academic / Plenum Publishers.

Our understanding of the role of IGFs in sarcomas comes largely from the study of these growth factors in three pediatric sarcomas, namely rhabdomyosarcoma, osteosarcoma and Ewing's sarcoma. In this chapter, the role of IGFs in each of these tumors will be discussed.

Rhabdomyosarcoma

Rhabdomyosarcoma (RMS) is the most common soft tissue sarcoma in childhood and is believed to arise from skeletal muscle progenitor cells. Two main histological subtypes of rhabdomyosarcoma have been described, embryonal and alveolar rhabdomyosarcoma. These two histological subtypes have distinct light microscopic appearance, vary in their clinical presentations and have distinct genetic alterations that have been shown to play a role in the pathogenesis of these tumors. Alveolar rhabdomyosarcoma has a characteristic light microscopic appearance of small, round cells lined up along spaces resembling pulmonary alveoli. Characteristically, alveolar rhabdomyosarcoma is associated with a translocation involving the DNA binding domain of PAX3 on the long arm of chromosome 2, or less commonly the DNA binding domain of PAX7 on the short arm of chromosome 1, and the transactivation domain of the FKHR gene on chromosome 13. These tumors generally affect older children in mid-late adolescence. In contrast, embryonal rhabdomyosarcoma is usually stroma-rich, less dense and has a spindle cell appearance. Embryonal rhabdomyosarcoma usually has a loss of heterozygosity at the 11p15.5 locus, affects younger children and is associated with a better prognosis.[13]

Regardless of histology, essentially all rhabdomyosarcomas overexpress insulin-like growth factor II (IGFII).[14-16] Several different mechanisms have been described that can lead to the deregulation/overexpression of IGFII in RMS. These mechanisms include increase in IGFII gene dosage either by loss of imprinting (LOI) or loss of heterozygosity (LOH) with paternal duplication, and by alterations in transcription. Each one of these mechanisms will be discussed in turn.

Genomic imprinting is defined as the differential expression of alleles depending on the parent of origin. Loss of imprinting, reactivation of the normally transcriptionally silent allele, has been recognized as one of the contributing factors to human disease and cancer.[17] A number of imprinted genes, including IGFII, are found in the chromosomal region 11p15.5. In most tissues, including adult and fetal skeletal muscle, the paternal IGFII allele is transcriptionally active and the maternally derived allele is silent. Activation of the maternal IGFII allele, i.e., loss of imprinting, resulting in double gene dosage of IGFII has been demonstrated in RMS patient tissue samples.[18,19] In embryonal RMS, double gene dosage of IGFII analogous to LOI of IGFII can result from loss of heterozygosity at the 11p15.5 locus where the maternal allele is lost and the paternal allele is duplicated, known as uniparental isodisomy.[20-22] Furthermore, unipaternal isodisomy at 11p15.5[23] as well as LOI of IGF2[24] has been described in Beckwith-Wiedemann syndrome (BWS), a rare fetal overgrowth syndrome associated with a higher incidence of embryonal tumors including RMS.[25] The increased incidence of embryonal tumors in patients with BWS supports a role for IGFII in the pathogenesis of these tumors.

The other proposed method of overexpression of IGFII in RMS is through the regulation of its promoters. The IGFII gene has a complex 5' upstream region composed of 4 promoters (P1-P4) which are active in a developmental and tissue specific manner. The P1 promoter is active in adult liver and chondrocytes and its expression is biallelic.[26] The P2, P3, and P4 promoters of IGFII are active in embryonic tissues and are imprinted.[27] The overexpression of IGFII in various tumors is derived mainly from fetally active P3 and P4 promoters.[28,29] Two transcription factors, AP-2 and p53, have been shown to alter the regulation of IGFII expression. AP-2 levels are increased in RMS and can increase IGFII levels by binding to the P3 promoter of IGFII.[30] Wildtype p53 inhibits IGFII gene expression by binding to the P4 promoter and by interfering with TATA-binding protein to the TATA motif of the P3 promoter.[31,32] Furthermore, restoration of wildtype p53 in an alveolar RMS cell line with a p53 mutation results in a five-fold decrease in IGFII expression.[32] Wildtype p53 can also inhibit the expression of IGFIR by interfering with the TATA-binding protein to the IGFIR promoter.[33] Thus, tumors with p53 mutations, the most common gene altered in human cancer, have the potential loss of suppression of both IGFII and its receptor, IGFIR, resulting in increased signaling through the IGF pathway.

IGFII overexpression in RMS appears to have a significant consequence since it functions as an autocrine growth factor in RMS. RMS cell lines grow in serum free media, secrete IGFII, display the

IGFIR on their surface and are growth inhibited when treated with a blocking antibody for the IGFIR, α IR3.[14,34] In addition, the actions of IGFII in rhabdomyosarcoma extend beyond stimulating cell growth. Exogenous IGFII can stimulate motility in RMS cells suggesting that this growth factor may play a role in the invasion and metastasis of RMS.[14,35]

In vivo, blocking the IGF signaling pathway in RMS cells also results in significant growth inhibition of tumors in immunodeficient mice. A number of different strategies have been employed to block this pathway in vivo. Immunodeficient mice subcutaneously injected with RMS cells transfected with a kinase-deficient IGFIR (acting as a dominant negative) showed decreased tumor growth compared to mice injected with mock transfected RMS clones.[3] Twice weekly subcutaneous injections of α IR3 near the site of tumor formation results in a dose-dependent suppression of tumor growth in nude mice.[4] Shapiro et al[2] demonstrated that clones of RH30, an alveolar RMS cell line, stably transfected with antisense IGFIR not only had a markedly decreased growth rate, but also had impaired colony formation in soft agar, and an impaired ability to form tumors in immunodeficient mice. In these three studies, the effect of blocking the IGF signaling pathway was cytostatic, not cytotoxic.

Recently, Hahn et al showed that IGFII is required for the formation of rhabdomyosarcoma in a mouse knock-out model of the gene patched (Ptch). In humans, mutations in the patched gene result in Gorlin Syndrome which is characterized by generalized overgrowth of the body, cysts, developmental abnormalities of the skeleton and a predisposition to benign and malignant tumors of the skin and central nervous system.[36] Mice heterozygous for Ptch (Ptch$^{neo67/+}$) have a 30% incidence of RMS. However, female Ptch$^{neo67/+}$ mice crossed with male mice heterozygous for Igf2 (Igf2$^{+/-}$) did not develop tumors suggesting that IGFII is required for the development of RMS in these genetically manipulated mice.[37,38]

IGFII is known to be an important fetal growth factor that plays a role in normal skeletal muscle growth and differentiation.[39] Human fetal skeletal muscle expresses high levels of IGFII whereas the expression of IGFII is markedly decreased in adult skeletal muscle. In order to elucidate the role of IGFII in the pathogenesis of RMS, the consequences of unregulated overexpression of IGFII and IGFIR has been investigated in a mouse myoblast cell line, C2C12, by several laboratories. Both the absence as well as the acceleration of differentiation in C2C12 cells transfected with IGFII or IGFIR have been reported.[40-43] The reasons for these inconsistencies remains unclear.

Wang et al[43] showed that cellular transformation of C2C12 cells in vitro and tumor formation in vivo could be achieved with overexpression of IGFII, IGFII and Pax-3 and IGFII and PAX3-FKHR. Benign tumors were formed in all groups except cells co-transfected with IGFII and PAX3-FKHR. These tumors showed a higher microvessel density and lower apoptotic rate, were poorly differentiated and invaded the normal muscle. This study and those described demonstrating a cytostatic, not cytotoxic effect when disrupting the IGF signaling pathway would suggest that increased levels of IGFII in cells of muscle lineage may be permissive to the transformation of these cells.

Osteosarcoma

Osteosarcoma is the most common bone tumor in childhood and usually occurs at the site of rapid bone growth during the adolescent growth spurt. The observations that growth hormone (GH) and IGFI are at their highest levels during the adolescent growth spurt[44] and that IGFs are known to play an important role in normal bone growth and development[45-47] have led to the hypothesis that altered IGF signaling may play a role in the unregulated growth and malignant transformation of bone cells.

IGFs play a role in proliferation and survival of osteosarcoma cell lines. In vitro data from a number of groups have shown that osteosarcoma cell lines and tissues[48] express IGFIR, are growth stimulated by IGFI and IGFII and depend on the presence of IGFI in serum free media to proliferate and survive. In addition, the proliferative effects of IGFs can be blocked by antisense and monoclonal antibodies to the IGFIR indicating that the observed mitogenic response is mediated through IGFIR.[45,49-53] Pollak et al[54] showed that suramin, a compound that non-selectively interferes with the interaction between growth factors and their receptors, could attenuate the binding of IGFI to its receptor resulting in decreased mitogenic response to IGFI in osteosarcoma cells. The effect was reversible suggesting a cytostatic effect on the cells. A comparison between the growth inhibitory effect of α IR3 (IGFIR monoclonal antibody) and suramin, in six osteosarcoma cell lines showed

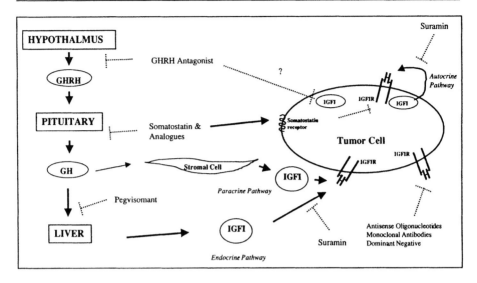

Figure 1. Strategies to inhibit IGF signaling in tumor cells. The dashed lines indicate the known targets of the various compounds developed to interfere with the IGF pathway.

greater growth inhibition with suramin suggesting the presence of other growth factor pathways may play a role in the proliferation of osteosarcomas.[50]

The most compelling evidence that IGFs play a role in the pathogenesis of osteosarcoma comes from in vivo experiments conducted in hypophysectomized animals. Pollak et al[55] implanted murine MGH-OGH osteosarcoma (determined in vitro to only be growth stimulated by IGFI and not other pituitary factors) into the gastrocnemius muscle of 12 hypophysectomized mice and 14 control mice. The tumors implanted in the hypophysectomized mice grew significantly slower than the implants in the control mice. In addition, hypophysectomy virtually abolished the metastatic behavior of MGH-OGS tumors (1 of 12 mice developed metastasis compared to 12 of 14 in the control group). Decreased serum IGFI levels were documented in the hypophysectomized mice. Similar observations were made in xenograft models using hypophysectomized animals for two closely related tumor types which occur later in life, namely chondrosarcoma and fibrosarcoma.[56,57]

Since the majority of osteosarcomas occur during the adolescent growth spurt when GH and IGFI levels are at their lifetime high, the potential to block the GH/IGF axis in this tumor could be of therapeutic benefit. Somatostatin analogues, growth hormone releasing hormone (GHRH) antagonists, and growth hormone receptor antagonists have all been developed and can potentially target tumors that are dependent on the hypothalamic/pituitary axis for growth and survival (Fig. 1).

Analogues of hypothalamic hormones, such as GHRH and LHRH, have been effective in decreasing pituitary hormone levels. GHRH antagonists suppress the growth of various tumors in vivo.[58-60] The decrease in tumor growth in these systems has been attributed to a decrease in GH release from the pituitary with subsequent decrease in IGFI production by the liver as well as a direct effect on IGFI and IGFII in tumor tissues.[61] A potent GHRH antagonist (MZ-4-71) has been used to treat athymic nude mice bearing xenografts of a human osteosarcoma cell line. The mice treated with MZ-4-71 had a significant decrease in tumor volume and a significant decrease in serum and tumor IGFI levels was compared to control mice.[62]

In addition to GHRH antagonists, somatostatin analogues have also been shown to have a growth inhibitory effect on osteosarcoma. Somatostatin is a tetradecapeptide first characterized as a growth hormone releasing factor inhibitor.[63,64] It is now known that somatostatin is normally present in a variety of human tissues and has a broad spectrum of biological actions which include inhibiting the release of various growth hormones as well as direct antiproliferative effects mediated by specific receptors located on tumor cells.[65] Several potent analogues of somatostatin, with longer half-lives than the native hormone, have been developed for therapeutic use. A decrease in tumor growth as well as a decrease in serum levels of IGFI and GH has been observed in animals harboring osteosarcoma or chondrosarcoma treated with several of these somatostatin analogues.[56,65-69] A phase I trial in 21 patients with metastatic and recurrent osteosarcoma using a long-acting somatostatin analogue was recently completed at the National Cancer Institute. No responses were observed in this trial; however, a 46% and 53% decrease in IGFI levels was observed at two weeks and eight weeks of therapy, respectively.[70]

Pegvisomant, a genetically engineered analogue of human growth hormone that functions as a growth hormone receptor antagonist, appears to decrease serum IGFI levels more than the somatostatin analogues and GHRH antagonists.[71,72] However, based on its mechanism of action this compound does not effect the IGFI production that is GH-independent. Therefore, it may be necessary to use the analogues that target the hypothalamic/pituitary axis in combination with each other and/or in combination with chemotherapeutic agents or with new agents targeting the downstream targets of cell signaling pathways in order to provide clinical benefit.

Ewing's Family of Tumors

Ewing's sarcoma family of tumors (ESFT) generally arise from bone and are the second most common bone tumor in childhood. A small percentage of these tumors may also arise in the soft tissue (extraosseous Ewing's sarcoma). ESFTs constitute a single pathologic entity that includes a spectrum of differentiation patterns from poorly differentiated Ewing's sarcoma to the most neuronally differentiated peripheral primitive neuroectodermal tumor (PNET).[73] The cell type of origin for this tumor is still uncertain, although a variable expression of neuronal immunohistochemical markers and ultrastructural features point to a neuroectodermal origin.

IGF signaling has been proposed to play a major role in the proliferation as well as the malignant behavior of ESFTs. The expression of seven growth factors (EGF, TGFα, IGFI, IGFII, NGF, TGFβ,and bFGF) was assessed in RNA samples from 6 ESFT cell lines and 8 ESFT patient samples by RT-PCR. All samples contained transcripts for TGFβ and IGFI. In addition, IGFIR was the only corresponding growth factor receptor consistently present in these samples.[74] Treatment with exogenous IGFI and IGFII results in an increased growth rate and motility of ESFT cells. Both effects induced by IGFs are abrogated in the presence of IGFIR antibody (α IR3).[74-76]

The in vitro observations demonstrating the role of IGFs in growth and malignant behavior of ESFT have been further substantiated by in vivo studies using α IR3 to block the IGFIR pathway in a Ewing's mouse xenograft model. Ewing's sarcoma cells form tumors when injected subcutaneously into athymic mice. In contrast, only four of nine mice injected with Ewing's sarcoma cells and treated α IR3 developed tumors. The four tumors from the α IR3 treated mice showed decreased mean tumor volume as well as increased number of apoptotic bodies on histological sections as compared to the untreated mice.[77]

Approximately 90-95% of ESFT express a characteristic reciprocal translocation which joins the NH2-terminal domain of EWS located on chromosome 22 to the DNA binding domain of an *Ets* family gene. The most common *Ets* family member to give rise to this translocation is FLI1, located on chromosome 11.[78,79] The genomic breakpoint resulting in the EWS-FLI1 translocation can be variable.[80] The resulting fusion protein from t(11;22), EWS-FLI1, plays a major role in the pathogenesis of ESFT.[81-83] Using fibroblast cell lines derived from IGFIR knock-out mice and wildtype littermates, Toretsky et al[84] showed that wildtype cells transfected to express the fusion protein were able to form soft agar colonies whereas the IGFIR deficient cells expressing the fusion protein did not form soft agar colonies; thus, implicating the IGF pathway as a necessary component in the transforming ability of EWS-FLI1.

IGF can also act as a survival factor and can protect ESFT cells from apoptosis in response to chemotherapeutic agents (Fig. 2). Toretsky et al[85] showed that two ESFT cell lines, TC71 and TC32, could be rescued from doxorubicin induced apoptosis by the addition of IGFI. Blocking the IGFIR pathway with suramin or α IR3 in vitro can enhance the cytotoxic effects of doxorubicin and vincristine in ESFT cells.[86] These observations would suggest that IGFIR is a survival factor for ESFT cells and as such can act as a mechanism of drug resistance. Thus one might postulate that patients whose tumors express lower IGFIR levels could be more sensitive to chemotherapy compared to patients with tumors expressing higher IGFIR levels. Interestingly, patients with EWS-FLI type I translocation, fusion between exon 7 of EWS and exon 6 of FLI1, seem to have a survival

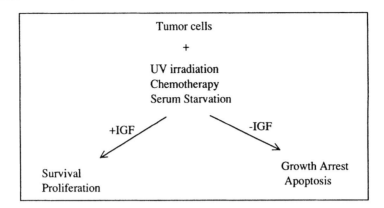

Figure 2. Increased IGF signaling in tumor cells can protect cells from apoptosis and growth arrest.

advantage[87,88], and this subset of patients also have lower IGFIR expression in their tumors as assessed by immunohistochemistry.[89]

Other Sarcomas

In addition to the three pediatric sarcomas discussed above, the expression of IGFI, IGFII and IGFIR has also been documented in a number of other sarcomas, including liposarcoma, synovial sarcoma, angiosarcoma, malignant fibrous histiocytoma, and leiomyosarcoma.[90-92] Sekyi-Otu et al[90] evaluated a panel of primary tumor samples representing a number of different sarcomas for IGFIR, IGFI and IGFII RNA levels using RT-PCR. All 29 mesenchymal tumors surveyed had some identifiable IGFIR level. 12 of these 29 samples had IGFIR levels comparable to or higher than MCF-7 cells (breast cancer cell line known to overexpress IGFIR). In addition, 22 of 28 samples and 17 of 27 samples had increased levels of IGFI and IGFII expression, respectively. Although the biological significance of these findings was not explored in this particular paper, these results would suggest that an autocrine/paracrine IGF signaling pathway may exist in these tumors.

Xie et al[92] studied a panel of 35 synovial sarcomas and correlated increased protein expression of IGFIR with increased lung metastases and increased proliferative rate as assessed by Ki-67 immunostaining. In contrast to Ewing's family of tumors, elevated expression of IGFIR did not correlate with the presence of one of three types of fusion protein generated from a characteristic translocation t(X;18)(p11.2;q11.2) found in synovial sarcoma.[93-97]

The changes in the expression pattern of IGFs from normal smooth muscle in the myometrium to benign uterine tumor (leiomyoma) to malignant leiomyosarcoma has been explored by several groups.[98-100] Increased IGFII expression in a subset of leiomyosarcomas is a consistent finding in these reports. In leiomyosarcomas with increased IGFII expression, the overall CpG methylation of the IGFII gene is low, whereas in normal uterus myometrium, leiomyomas and leiomyosarcomas with low IGFII expression the gene is methylated to partly demethylated.[101] These findings would suggest that the elevation of IGFII may be due to an epigenetic phenomenon. Similar to alveolar rhabdomyosarcoma, loss of imprinting at the IGFII allele has also been implicated as a mechanism of IGFII overexpression in leiomyosarcomas.[102] The role of IGFII in the pathogenesis of leiomyosarcoma has yet to be elucidated.

Conclusion

In this chapter, we have summarized the important role of the IGF signaling pathway in the pathogenesis of many sarcomas. A number of strategies have been devised to interfere with this pathway (Fig. 1). Preclinical xenograft models for these strategies have shown promise; however, the utility of the IGF pathway as a therapeutic target in humans with sarcomas remains to be determined. As discussed in the section on osteosarcomas, it is likely that many of the agents developed to target the IGF pathway will be most effective when used in combination other agents that enhance apoptosis and interfere with downstream signaling pathways.

References

1. Kaleko M, Rutter WJ, Miller AD. Overexpression of the human insulinlike growth factor I receptor promotes ligand-dependent neoplastic transformation. Mol Cell Biol 1990; 10(2):464-473.
2. Shapiro DN, Jones BG, Shapiro LH et al. Antisense-mediated reduction in insulin-like growth factor-I receptor expression suppresses the malignant phenotype of a human alveolar rhabdomyosarcoma. J Clin Invest 1994; 94(3):1235-1242.
3. Kalebic T, Blakesley V, Slade C et al. Expression of a kinase-deficient IGF-I-R suppresses tumorigenicity of rhabdomyosarcoma cells constitutively expressing a wild type IGF-I-R. Int J Cancer 1998; 76(2):223-227.
4. Kalebic T, Tsokos M, Helman LJ. In vivo treatment with antibody against IGF-1 receptor suppresses growth of human rhabdomyosarcoma and down-regulates p34cdc2. Cancer Res 1994; 54(21):5531-5534.
5. Shimasaki S, Ling N. Identification and molecular characterization of insulin-like growth factor binding proteins (IGFBP-1, -2, -3, -4, -5 and -6). Prog Growth Factor Res 1991; 3(4):243-266.
6. Mohan S. Insulin-like growth factor binding proteins in bone cell regulation. Growth Regul 1993; 3(1):67-70.
7. Kelley KM, Oh Y, Gargosky SE et al. Insulin-like growth factor-binding proteins (IGFBPs) and their regulatory dynamics. Int J Biochem Cell Biol 1996; 28(6):619-637.
8. Oh Y. IGF-independent regulation of breast cancer growth by IGF binding proteins. Breast cancer Res Treat 1998; 47(3):283-293.
9. Campbell PG, Novak JF, Yanosick TB et al. Involvement of the plasmin system in dissociation of the insulin-like growth factor-binding protein complex. Endocrinology 1992; 130(3):1401-1412.
10. Lalou C, Silve C, Rosato R et al. Interactions between insulin-like growth factor-I (IGF-I) and the system of plasminogen activators and their inhibitors in the control of IGF-binding protein-3 production and proteolysis in human osteosarcoma cells. Endocrinology 1994; 135(6):2318-2326.
11. Collett-Solberg PF, Cohen P. The role of the insulin-like growth factor binding proteins and the IGFBP proteases in modulating IGF action. Endocrinol Metab Clin North Am 1996; 25(3):591-614.
12. Flint DJ, Tonner E, Allan GJ. Insulin-like growth factor binding proteins: IGF-dependent and - independent effects in the mammary gland. J Mammary gland Biol Neoplasia 2000; 5(1):65-73.
13. Wexler LH, Helman LJ. Rhabdomyosarcoma and the Undifferentiated Sarcomas. In: Pizzo P, Poplack D, eds. Principles and Practice of Pediatric Oncology. 3rd ed. Philadelphia: Lippincott-Raven Publishers, 1997:799-830.
14. El-Badry OM, Minniti C, Kohn EC et al. Insulin-like growth factor II acts as an autocrine growth and motility factor in human rhabdomyosarcoma tumors. Cell Growth Differ 1990; 1(7):325-331.
15. Minniti CP, Tsokos M, Newton WA, Jr. et al. Specific expression of insulin-like growth factor-II in rhabdomyosarcoma tumor cells. Am J Clin Pathol 1994; 101(2):198-203.
16. Yun K. A new marker for rhabdomyosarcoma. Insulin-like growth factor II. Lab Invest 1992; 67(5):653-664.
17. Lalande M. Parental imprinting and human disease. Annu Rev Genet 1996; 30:173-195.
18. Zhan S, Shapiro DN, Helman LJ. Activation of an imprinted allele of the insulin-like growth factor II gene implicated in rhabdomyosarcoma. J Clin Invest 1994; 94(1):445-448.
19. Pedone PV, Tirabosco R, Cavazzana AO et al. Mono- and bi-allelic expression of insulin-like growth factor II gene in human muscle tumors. Hum Mol Genet 1994; 3(7):1117-1121.
20. Scrable H, Cavenee W, Ghavimi F et al. A model for embryonal rhabdomyosarcoma tumorigenesis that involves genome imprinting. Proc Natl Acad Sci USA 1989; 86(19):7480-7484.
21. Anderson J, Gordon A, McManus A et al. Disruption of imprinted genes at chromosome region 11p15.5 in paediatric rhabdomyosarcoma. Neoplasia 1999; 1(4):340-348.
22. Visser M, Sijmons C, Bras J et al. Allelotype of pediatric rhabdomyosarcoma. Oncogene 1997; 15(11):1309-1314.
23. Koufos A, Hansen MF, Copeland NG et al. Loss of heterozygosity in three embryonal tumors suggests a common pathogenetic mechanism. Nature 1985; 316(6026):330-334.
24. Weksberg R, Shen DR, Fei YL et al. Disruption of insulin-like growth factor 2 imprinting in Beckwith-Wiedemann syndrome. Nat Genet 1993; 5(2):143-150.
25. Wiedemann H. Tumors and hemihypertrophy associated with Wiedemann-Beckwith syndrome. Eur J Pediatr 1983; 141:129.
26. Ekstrom TJ, Cui H, Li X et al. Promoter-specific IGF2 imprinting status and its plasticity during human liver development. Development 1995; 121(2):309-316.
27. de Pagter-Holthuizen P, Jansen M, van der Kammen RA et al. Differential expression of the human insulin-like growth factor II gene. Characterization of the IGF-II mRNAs and an mRNA encoding a putative IGF-II-associated protein. Biochim Biophys Acta 1988; 950(3):282-295.
28. Scott J, Cowell J, Robertson ME et al. Insulin-like growth factor-II gene expression in Wilms' tumor and embryonic tissues. Nature 1985; 317(6034):260-262.
29. van Dijk MA, van Schaik FM, Bootsma HJ et al. Initial characterization of the four promoters of the human insulin- like growth factor II gene. Mol Cell Endocrinol 1991; 81(1-3):81-94.
30. Zhang L, Zhan S, Navid F et al. AP-2 may contribute to IGF-II overexpression in rhabdomyosarcoma. Oncogene 1998; 17(10):1261-1270.
31. Zhang L, Kashanchi F, Zhan Q et al. Regulation of insulin-like growth factor II P3 promotor by p53: a potential mechanism for tumorigenesis. Cancer Res 1996; 56(6):1367-1373.
32. Zhang L, Zhan Q, Zhan S et al. p53 regulates human insulin-like growth factor II gene expression through active P4 promoter in rhabdomyosarcoma cells. DNA Cell Biol 1998; 17(2):125-131.
33. Werner H, Karnieli E, Rauscher FJ et al. Wild-type and mutant p53 differentially regulate transcription of the insulin-like growth factor I receptor gene. Proc Natl Acad Sci USA 1996; 93(16):8318-8323.
34. De Giovanni C, Melani C, Nanni P et al. Redundancy of autocrine loops in human rhabdomyosarcoma cells: induction of differentiation by suramin. Br J Cancer 1995; 72(5):1224-1229.
35. Minniti CP, Kohn EC, Grubb JH et al. The insulin-like growth factor II (IGF-II)/mannose 6-phosphate receptor mediates IGF-II-induced motility in human rhabdomyosarcoma cells. J Biol Chem 1992; 267(13):9000-9004.
36. Gorlin RJ. Nevoid basal-cell carcinoma syndrome. Medicine (Baltimore) 1987; 66(2):98-113.

37. Hahn H, Wojnowski L, Specht K et al. Patched target Igf2 is indispensable for the formation of medulloblastoma and rhabdomyosarcoma. J Biol Chem 2000; 275(37):28341-28344.
38. Hahn H, Wojnowski L, Zimmer AM et al. Rhabdomyosarcomas and radiation hypersensitivity in a mouse model of Gorlin syndrome. Nat Med 1998; 4(5):619-622.
39. Tollefsen SE, Sadow JL, Rotwein P. Coordinate expression of insulin-like growth factor II and its receptor during muscle differentiation. Proc Natl Acad Sci USA 1989; 86(5):1543-1547.
40. Minniti CP, Maggi M, Helman LJ. Suramin inhibits the growth of human rhabdomyosarcoma by interrupting the insulin-like growth factor II autocrine growth loop. Cancer Res 1992; 52(7):1830-1835.
41. Quinn LS, Steinmetz B, Maas A et al. Type-1 insulin-like growth factor receptor overexpression produces dual effects on myoblast proliferation and differentiation. J Cell Physiol 1994; 159(3):387-398.
42. Stewart CE, James PL, Fant ME et al. Overexpression of insulin-like growth factor-II induces accelerated myoblast differentiation. J Cell Physiol 1996; 169(1):23-32.
43. Wang W, Kumar P, Epstein J et al. Insulin-like growth factor II and PAX3-FKHR cooperate in the oncogenesis of rhabdomyosarcoma. Cancer Res 1998; 58(19):4426-4433.
44. Henderson BE, Ross RK, Pike MC et al. Endogenous hormones as a major factor in human cancer. Cancer Res 1982; 42(8):3232-3239.
45. Scheven BA, Hamilton NJ, Fakkeldij TM et al. Effects of recombinant human insulin-like growth factor I and II (IGF- I/-II) and growth hormone (GH) on the growth of normal adult human osteoblast-like cells and human osteogenic sarcoma cells. Growth Regul 1991; 1(4):160-167.
46. Canalis E, McCarthy T, Centrella M. Growth factors and the regulation of bone remodeling. J Clin Invest 1988; 81(2):277-281.
47. Ernst M, Rodan GA. Increased activity of insulin-like growth factor (IGF) in osteoblastic cells in the presence of growth hormone (GH): positive correlation with the presence of the GH-induced IGF-binding protein BP-3. Endocrinology 1990; 127(2):807-814.
48. Burrow S, Andrulis IL, Pollak M et al. Expression of insulin-like growth factor receptor, IGF-1, and IGF-2 in primary and metastatic osteosarcoma. J Surg Oncol 1998; 69(1):21-27.
49. Kappel CC, Velez-Yanguas MC, Hirschfeld S et al. Human osteosarcoma cell lines are dependent on insulin-like growth factor I for in vitro growth. Cancer Res 1994; 54(10):2803-2807.
50. Benini S, Baldini N, Manara MC et al. Redundancy of autocrine loops in human osteosarcoma cells. Int J Cancer 1999; 80(4):581-588.
51. Raile K, Hoflich A, Kessler U et al. Human osteosarcoma (U-2 OS) cells express both insulin-like growth factor-I (IGF-I) receptors and insulin-like growth factor-II/mannose-6- phosphate (IGF-II/M6P) receptors and synthesize IGF-II: autocrine growth stimulation by IGF-II via the IGF-I receptor. J Cell Physiol 1994; 159(3):531-541.
52. Okazaki R, Conover CA, Harris SA et al. Normal human osteoblast-like cells consistently express genes for insulin-like growth factors I and II but transformed human osteoblast cell lines do not. J Bone Miner Res 1995; 10(5):788-795.
53. Pollak MN, Polychronakos C, Richard M. Insulinlike growth factor I: a potent mitogen for human osteogenic sarcoma. J Natl Cancer Inst 1990; 82(4):301-305.
54. Pollak M, Richard M. Suramin blockade of insulinlike growth factor I-stimulated proliferation of human osteosarcoma cells. J Natl Cancer Inst 1990; 82(16):1349-1352.
55. Pollak M, Sem AW, Richard M et al. Inhibition of metastatic behavior of murine osteosarcoma by hypophysectomy. J Natl Cancer Inst 1992; 84(12):966-971.
56. Redding TW, Schally AV. Inhibition of growth of the transplantable rat chondrosarcoma by analogs of hypothalamic hormones. Proc Natl Acad Sci USA 1983; 80(4):1078-1082.
57. Sekyi-Otu A, Bell R, Andrulis I et al. Metastatic behavior of the RIF-1 murine fibrosarcoma: inhibited by hypophysectomy and partially restored by growth hormone replacement. J Natl Cancer Inst 1994; 86(8):628-632.
58. Jungwirth A, Schally AV, Pinski J et al. Growth hormone-releasing hormone antagonist MZ-4-71 inhibits in vivo proliferation of Caki-I renal adenocarcinoma. Proc Natl Acad Sci USA 1997; 94(11):5810-5813.
59. Jungwirth A, Schally AV, Pinski J et al. Inhibition of in vivo proliferation of androgen-independent prostate cancers by an antagonist of growth hormone-releasing hormone. Br J Cancer 1997; 75(11):1585-1592.
60. Lamharzi N, Schally AV, Koppan M et al. Growth hormone-releasing hormone antagonist MZ-5-156 inhibits growth of DU-145 human androgen-independent prostate carcinoma in nude mice and suppresses the levels and mRNA expression of insulin-like growth factor II in tumors. Proc Natl Acad Sci U S A 1998; 95(15):8864-8868.
61. Csernus VJ, Schally AV, Kiaris H et al. Inhibition of growth, production of insulin-like growth factor-II (IGF- II), and expression of IGF-II mRNA of human cancer cell lines by antagonistic analogs of growth hormone-releasing hormone in vitro. Proc Natl Acad Sci USA 1999; 96(6):3098-3103.
62. Pinski J, Schally AV, Groot K et al. Inhibition of growth of human osteosarcomas by antagonists of growth hormone-releasing hormone. J Natl Cancer Inst 1995; 87(23):1787-1794.
63. Siler TM, VandenBerg G, Yen SS et al. Inhibition of growth hormone release in humans by somatostatin. J Clin Endocrinol Metab 1973; 37(4):632-634.
64. Brazeau P, Vale W, Burgus R et al. Hypothalamic polypeptide that inhibits the secretion of immunoreactive pituitary growth hormone. Science 1973; 179(68):77-79.
65. Schally AV. Oncological applications of somatostatin analogues. Cancer Res 1988; 48(24 Pt 1):6977-6985.
66. Reubi JC. A somatostatin analogue inhibits chondrosarcoma and insulinoma tumor growth. Acta Endocrinol (Copenh) 1985; 109(1):108-114.
67. Keri G, Erchegyi J, Horvath A et al. A tumor-selective somatostatin analog (TT-232) with strong in vitro and in vivo antitumor activity. Proc Natl Acad Sci USA 1996; 93(22):12513-12518.
68. Pinski J, Schally AV, Halmos G et al. Somatostatin analog RC-160 inhibits the growth of human osteosarcomas in nude mice. Int J Cancer 1996; 65(6):870-874.

69. Schally AV, Redding TW, Comaru-Schally AM. Potential use of analogs of luteinizing hormone-releasing hormones in the treatment of hormone-sensitive neoplasms. Cancer Treat Rep 1984; 68(1):281-289.
70. Mansky PJ, Liewehr DJ, Steinberg SM et al. Treatment of Metastatic Osteosarcoma with the Somatostatin Analog OncoLar: Significant Reduction of IGF-I Serum Levels. J Pediatr Hematol Oncol 2002; 24(6):440-446.
71. Trainer PJ, Drake WM, Katznelson L et al. Treatment of acromegaly with the growth hormone-receptor antagonist pegvisomant. N Engl J Med 2000; 342(16):1171-1177.
72. Parkinson C, Trainer PJ. Growth hormone receptor antagonists therapy for acromegaly. Baillieres Best Pract Res Clin Endocrinol Metab 1999; 13(3):419-430.
73. Dehner LP. Primitive neuroectodermal tumor and Ewing's sarcoma. Am J Surg Pathol 1993; 17(1):1-13.
74. Scotlandi K, Benini S, Sarti M et al. Insulin-like growth factor I receptor-mediated circuit in Ewing's sarcoma/peripheral neuroectodermal tumor: a possible therapeutic target. Cancer Res 1996; 56(20):4570-4574.
75. van Valen F, Winkelmann W, Jurgens H. Type I and type II insulin-like growth factor receptors and their function in human Ewing's sarcoma cells. J Cancer Res Clin Oncol 1992; 118(4):269-275.
76. Yee D, Favoni RE, Lebovic GS et al. Insulin-like growth factor I expression by tumors of neuroectodermal origin with the t(11; 22) chromosomal translocation. A potential autocrine growth factor. J Clin Invest 1990; 86(6):1806-1814.
77. Scotlandi K, Benini S, Nanni P et al. Blockage of insulin-like growth factor-I receptor inhibits the growth of Ewing's sarcoma in athymic mice. Cancer Res 1998; 58(18):4127-4131.
78. Delattre O, Zucman J, Melot T et al. The Ewing family of tumors—a subgroup of small-round-cell tumors defined by specific chimeric transcripts. N Engl J Med 1994; 331(5):294-299.
79. Turc-Carel C, Aurias A, Mugneret F et al. Chromosomes in Ewing's sarcoma. I. An evaluation of 85 cases of remarkable consistency of t(11;22)(q24;q12). Cancer Genet Cytogenet 1988; 32(2):229-238.
80. Zucman J, Melot T, Desmaze C et al. Combinatorial generation of variable fusion proteins in the Ewing family of tumors. Embo J 1993; 12(12):4481-4487.
81. Kovar H, Aryee DN, Jug G et al. EWS/FLI-1 antagonists induce growth inhibition of Ewing tumor cells in vitro. Cell Growth Differ 1996; 7(4):429-437.
82. May WA, Gishizky ML, Lessnick SL et al. Ewing sarcoma 11;22 translocation produces a chimeric transcription factor that requires the DNA-binding domain encoded by FLI1 for transformation. Proc Natl Acad Sci USA 1993; 90(12):5752-5756.
83. Toretsky JA, Connell Y, Neckers L et al. Inhibition of EWS-FLI-1 fusion protein with antisense oligodeoxynucleotides. J Neurooncol 1997; 31(1-2):9-16.
84. Toretsky JA, Kalebic T, Blakesley V et al. The insulin-like growth factor-I receptor is required for EWS/FLI-1 transformation of fibroblasts. J Biol Chem 1997; 272(49):30822-30827.
85. Toretsky JA, Thakar M, Eskenazi AE et al. Phosphoinositide 3-hydroxide kinase blockade enhances apoptosis in the Ewing's sarcoma family of tumors. Cancer Res 1999; 59(22):5745-5750.
86. Benini S, Manara MC, Baldini N et al. Inhibition of insulin-like growth factor I receptor increases the antitumor activity of doxorubicin and vincristine against Ewing's sarcoma cells. Clin Cancer Res 2001; 7(6):1790-1797.
87. de Alava E, Kawai A, Healey JH et al. EWS-FLI1 fusion transcript structure is an independent determinant of prognosis in Ewing's sarcoma. J Clin Oncol 1998; 16(4):1248-1255.
88. Zoubek A, Dockhorn-Dworniczak B, Delattre O et al. Does expression of different EWS chimeric transcripts define clinically distinct risk groups of Ewing tumor patients? J Clin Oncol 1996; 14(4):1245-1251.
89. de Alava E, Panizo A, Antonescu CR et al. Association of EWS-FLI1 type 1 fusion with lower proliferative rate in Ewing's sarcoma. Am J Pathol 2000; 156(3):849-855.
90. Sekyi-Otu A, Bell RS, Ohashi C et al. Insulin-like growth factor 1 (IGF-1) receptors, IGF-1, and IGF-2 are expressed in primary human sarcomas. Cancer Res 1995; 55(1):129-134.
91. Tricoli JV, Rall LB, Karakousis CP et al. Enhanced levels of insulin-like growth factor messenger RNA in human colon carcinomas and liposarcomas. Cancer Res 1986; 46(12 Pt1):6169-6173.
92. Xie Y, Skytting B, Nilsson G et al. Expression of insulin-like growth factor-1 receptor in synovial sarcoma: association with an aggressive phenotype. Cancer Res 1999; 59(15):3588-3591.
93. Clark J, Rocques PJ, Crew AJ et al. Identification of novel genes, SYT and SSX, involved in the t(X;18)(p11.2;q11.2) translocation found in human synovial sarcoma. Nat Genet 1994; 7(4):502-508.
94. Crew AJ, Clark J, Fisher C et al. Fusion of SYT to two genes, SSX1 and SSX2, encoding proteins with homology to the Kruppel-associated box in human synovial sarcoma. Embo J 1995; 14(10):2333-2340.
95. Ladanyi M, Antonescu CR, Leung DH et al. Impact of SYT-SSX Fusion Type on the Clinical Behavior of Synovial sarcoma: A Multi-Institutional Retrospective Study of 243 Patients. Cancer Res 2002; 62(1):135-140.
96. Skytting B, Nilsson G, Brodin B et al. A novel fusion gene, SYT-SSX4, in synovial sarcoma. J Natl Cancer Inst 1999; 91(11):974-975.
97. Kawai A, Woodruff J, Healey JH et al. SYT-SSX gene fusion as a determinant of morphology and prognosis in synovial sarcoma. N Engl J Med 1998; 338(3):153-160.
98. Hoppener JW, Mosselman S, Roholl PJ et al. Expression of insulin-like growth factor-I and -II genes in human smooth muscle tumors. Embo J 1988; 7(5):1379-1385.
99. Gloudemans T, Prinsen I, Van Unnik JA et al. Insulin-like growth factor gene expression in human smooth muscle tumors. Cancer Res 1990; 50(20):6689-6695.
100. Van der Ven LT, Roholl PJ, Gloudemans T et al. Expression of insulin-like growth factors (IGFs), their receptors and IGF binding protein-3 in normal, benign and malignant smooth muscle tissues. Br J Cancer 1997; 75(11):1631-1640.
101. Gloudemans T, Pospiech I, Van Der Ven LT et al. Expression and CpG methylation of the insulin-like growth factor II gene in human smooth muscle tumors. Cancer Res 1992; 52(23):6516-6521.
102. Vu TH, Yballe C, Boonyanit S et al. Insulin-like growth factor II in uterine smooth-muscle tumors: maintenance of genomic imprinting in leiomyomata and loss of imprinting in leiomyosarcomata. J Clin Endocrinol Metab 1995; 80(5):1670-1676.

CHAPTER 26

IGFs and Epithelial Cancer

Walter Zumkeller

Abstract

Insulin-like growth factors (IGFs), IGF receptors and IGF-binding proteins (IGFBPs) are expressed in epithelial cancer. The IGF system is involved in the regulation of the proliferation of both normal and malignant epithelial cells. Proliferative effects of IGFs are mediated predominantly through the type I IGF receptor (IGF-IR) and are modulated by specific IGF-binding proteins (IGFBPs). The role of the IGF system in the pathogenesis of prostate cancer, Wilms' tumors, colorectal adenoma as well as breast, ovarian and cervical cancer is discussed.

Introduction

Insulin-like growth factors (IGFs), IGF receptors and IGF-binding proteins (IGFBPs) play a pivotal role in the growth and development of epithelial cancer. The biological effects are mediated predominantly through the type I IGF receptor (IGF-IR). The IGF-IR blockade suppresses proliferation of epithelial cancer and thus might represent a strategy for cancer treatment. IGFBPs modify IGF-mediated effects by determining the bioavailability of IGFs. In addition, tumors secrete IGFBP specific proteases which increase the amount of free and thus bioactive IGFs.[1] This chapter focuses on the role of the IGF system in epithelial cancer, namely in of prostate cancer, Wilms' tumors, colorectal adenoma as well as breast, ovarian and cervical cancer.

Prostate Cancer

Prostate cancer represents a major public health care issue. Epidemiological and biological data implicate the IGF system in the pathophysiology of prostate cancer. Serum levels of IGF-I, IGFBP-1 and IGFBP-3 are associated with prostate cancer risk.[2] A meta-analysis of 14 case-control studies indicated that IGF-I and IGFBP-3 serum levels are higher in prostate cancer patients than in the controls.[3] In particular, high IGF-I serum levels appear to be a risk factor for prostate cancer,[4,5] but also IGFBP-2 serum levels are elevated in patients with prostate carcinoma and appear to correlate with prostate-specific antigen levels and the tumor stage.[6] In patients with clinically localized prostate cancer, preoperative plasma IGFBP-2 levels are inversely associated with biologically aggressive disease and disease progression whereas preoperative IGFBP-3 levels were decreased in patients with prostate cancer metastases.[7] Interestingly, IGFBP-2 and IGFBP-3 levels were lower in patients treated with prostate cancer that developed recurrent disease in comparison with those who were in remission.[8]

Sex-hormone induced prostatic carcinogenesis in a Noble rat model appears to be critically regulated by IGF-I.[9] Transgenic mice expressing human IGF-I in basal epithelial cells of prostate show IGF-I receptor activation and prostate tumorigenesis.[10] IGF-II mRNA expression is increased in human prostate carcinoma.[11] In benign prostatic hyperplasia, IGFBP-2 mRNA is decreased whereas IGFBP-5 mRNA is increased as compared to normal cells[12] and IGFBP-2 mRNA expression was increased in prostate intraepithelial neoplasia (PIN) as well as adenocarcinoma of the prostate whereas IGFBP-3 mRNA is overexpressed only in PIN.[13] In malignant human prostate, IGFBP-4 and –5 protein are increased as compared to benign prostate epithelium.[14] Prostate cells showed a decrease in IGFBP-2,-3, -5 and –6 mRNA expression as they became progressively more tumorigenic when cultured in medium supplemented with IGF-I.[15] IGFBP-3 is an antiapoptotic agent in prostate carcinoma cells, and 1,25(OH)$_2$D$_3$ and its analog EB1089 stimulated IGFBP-3 mRNA expression

Insulin-Like Growth Factors, edited by Derek LeRoith, Walter Zumkeller
and Robert Baxter. ©2003 Eurekah.com and Kluwer Academic / Plenum Publishers.

as well as inhibition of prostate cancer cell proliferation suggesting that the compounds are of potential therapeutic use in the treatment of prostate cancer.[16] IGFBP-3 mutants that do not bind IGFs stimulate apoptosis in human prostate carcinoma cells indicating that IGF-independent effects of IGFBP-3 may play a major role in apoptosis of prostate tumor cells.[17] Overexpression of IGFBP-4 in the malignant M12 prostate epithelial cell line inhibits colony formation in soft agar as compared with M12 controls.[18] Androgens stimulate IGFBP-5 mRNA expression in human prostate cancer cells.[19] Prostate carcinoma cells secrete cathepsin D which hydrolyzes endogenous IGFBPs thus modifying IGF-I in prostate cancer.[20]

During transformation of prostate epithelial cells from the benign to the metastatic state, a marked decrease in type I IGF receptors was observed. The decreased expression of the IGF-IR results from transcriptional expression of the IGF-IR, in part due to the increased expression of the WT1 tumor suppressor in metastatic prostate cancer.[21] Malignant phenotype in prostate cancer is associated with a significant decrease in IGF-IR expression and reexpression of the IGF-IR was found to result in decreased tumor growth.[22] Progression of LAPC-9 and LNCaP prostate cancer models to androgen independence is associated with increased expression of IGF-I and IGF-IR mRNA expression.[23]

The concept of antisurvival factor therapy as a component in the treatment of advanced prostate cancer with the aim of neutralizing the protective effect of these survival factors involves the combination of dexamethasone and somatostatin analogues which suppress GH-independent and GH-dependent IGF-I production respectively.[24] As a GHRH autocrine pathway was identified in prostate cell lines, GHRH antagonists could be used in order to block autocrine IGF production and thus prostate tumor growth.[25] Anti-estrogens were found to decrease IGF-IR levels in rat prostate and may thus be of potential use for the prevention or treatment of prostate cancer.[26] IGF-I-induced signalling through Akt both in normal prostate epithelial cells and Du145 prostate carcinoma cells is inhibited by black tea polyphenols.[27] This novel paradigm in prostate cancer treatment may have a significant impact on long-term prognosis, particularly in cases where other treatment regimens have had a limited effect.[28]

Wilms' Tumor

In Wilms' tumors, IGF-II mRNA expression is very high as compared with normal kidney[29-32] and IGF-II mRNA expression appears to be inversely correlated with nephroblastic differentiation.[33] IGF-II mRNA expression in fetal nephrogenesis is inversely related to normal epithelial differentiation and the persistence of IGF-II mRNA overexpression may thus lead to an aberrant differentiation in Wilms' tumors.[34]

Glypican 3 (GPC3) is coexpressed with IGF-II in Wilms' tumors and it was suggested that GPC3 may be involved in the development of embryonal tumors through a signalling pathway involving IGF-II.[35] However, IGF-II mRNA expression is not correlated to Wilms' tumor growth[36] and IGF-II peptide levels are not always increased[32] indicating posttranscriptional modification of expression. Cultured Wilms' tumor cells release various forms of 'big IGF-II' which may represent incorrectly processed IGF-II.[37] In serum of Wilms' tumor patients, the high molecular weight form of IGF-II was significantly reduced and IGFBP-2 substantially elevated as compared to normal controls.[38] The level of IGF-II peptides in Wilms' tumor can be low despite increased expression of IGF-II mRNAs which may be explained by the presence of antisense transcripts of the IGF-II gene in these tumors.[34] In contrast, anaplastic variants of Wilms' tumors express only marginal levels of IGF-II mRNA.[40]

Wilms' tumor-bearing mice show elevated IGF-II serum levels as compared to normal mice.[41] Forced expression of IGF-II from a retroviral construct suppressed tumor formation in nude mice grafts or tumorigenic fibroblasts.[42] Here, paracrine and endocrine effects of IGF-II appear to have opposite effects on cell biology. The Wilms' tumor suppressor gene WT1 functions as a potent repressor of IGF-II transcription in vivo indicating that WT1 negatively regulates blastemal cell proliferation by repressing IGF-II in the developing kidney.[43] WT1 binds to specific IGF-II RNA sequences and may thus regulate gene expression by transcriptional and posttranscriptional fashions.[44] Treatment of W13 blastemal Wilms' tumor cells with retinoic acid led to a suppression of growth associated with a downregulation of N-*myc* mRNA as well as an up-regulation of IGF-II mRNA.[45] The IGFBPs are important in this context as they determine the amount of free IGF-II at a cellular level. IGFBP-2 is detectable in conditioned culture media from Wilms' tumors[46] and

IGFBP-3 inhibits Wilms' tumor cell growth in culture.[47] Therefore, IGFBPs could be used in order to modulate the cell growth of Wilms' tumors in vitro and in vivo.

Type I IGF receptors are present in Wilms' tumors.[48] Using an antibody directed against the type I IGF receptor, inhibition of Wilms' tumor growth both in culture and in athymic mice was achieved.[49] Suramin inhibits WT growth by disruption of the interaction of IGF-II with the type I and type II IGF receptor.[50] WT1 suppresses the IGF-IR promoter and deletion of the WT1 gene in WT may lead to an overexpression of the receptor facilitating IGF-mediated tumor growth.[51] The inhibition of G401 Wilms' tumor cell proliferation by stable transfection with a WT1 expression vector is associated with a reduction in IGF-I mRNA levels.[52] Deletion or mutation of WT1 may result in increased expression of the IGF-IR.[53] Moreover, the type II IGF receptor is imprinted in a high number of Wilms' tumors indicating that a decreased IGF-IIR dosage limits IGF-II inactivation which would facilitate tumorigenesis.[54]

Colorectal Adenoma

Colorectal cancer risk shows a significant positive association with IGF-I levels in serum.[55,56] Sigmoidal epithelial cell proliferation correlated with serum IGF-I levels in patients with acromegaly indicating that IGF-I may constitute an increased risk for colonic neoplasms in acromegaly.[57] Elevated IGF-II serum levels were found in patients in colorectal adenomas and a significant fall in mean IGF-II levels was observed after adenoma removal.[58] However, a one-year GH replacement therapy leading to plasma IGF-I levels within the physiological range does not adversely affect colonic epithelial cell proliferation indicating that there is no risk of development of colorectal cancer.[59]

IGF-I plays an antiapoptotic role in HT29-D4 colon carcinoma cells which is based on the enhancement of the survival pathway initiated by TNF-α.[60] IGF-I induces the expression of vascular endothelial growth factor (VEGF) in COLO205 and LSLiM6 colon carcinoma cells thus promoting tumor progression through stimulation of angiogenesis in addition to its direct growth stimulatory effect.[61] IGF-I induces transient growth stimulation of colon carcinoma cells through Akt but subsequently transcriptional upregulation of the cdk inhibitor p27(kip1) results in growth arrest.[62] IGF-I treatment of liver-specific IGF-I-deficient mice bearing orthotopically transplanted colon adenocarcinoma showed increased rates of tumor development and liver metastasis as compared with saline-injected mice.[63] Thus, a novel approach for cancer immunogene therapy in colon carcinoma may consist in the cotransfection with genes encoding antisense IGF-I and mouse B7.1 molecules.[64]

Butyrate promotes apoptosis of colon carcinoma cells in vitro but IGF-II inhibits this pro-apoptotic effect downstream of histone deacetylase inhibition.[65] CaCo-2 colon carcinoma cells which constitutively express high levels of IGF-II show an increased proliferation rate, high IGF-I receptor number and an increased capability of anchorage-independent growth.[66] IGF-II increases the number of colon adenomas and the progression to carcinoma in the APC (Min/+) murine model of human familial adenomatous polyposis.[67]

The differentiation of HT29-D4 cells is characterized by a downregulation of IGF-I receptors and higher numbers of IGF-IR in undifferentiated may allow IGF-II to trigger a distinct signalling pathway involving the IGF-IR.[68] It was suggested that the induction of proliferation and tumor progression of colon cancer cells by the IGF-II/IGF-I receptor pathway is dependent on the activation of the type 2 prostaglandin endoperoxidase synthase/cyclooxygenase (COX2).[69] Colon carcinomas show also mutations of the IGF-II receptor.[70,71] Therefore, IGF-II that does not bind to the type II IGF receptor activates IGF-IR inducing cell proliferation.

Circulating IGFBP-2 levels were elevated in colorectal adenoma patients and fell after polypectomy.[58] IGFBP-2 was shown to inhibit anchorage-independent growth of HAT-29 colon carcinoma cells.[72] Short-chain fatty acids stimulate IGFBP-2 synthesis in Caco-2 cells thus modulating cell proliferation and differentiation.[73] Colon cancer extracts were able to degrade IGFBP-2 whereas normal tissue extracts had no effect indicating that the strong proteolysis of secreted IGFBP-2 in colon cancer tissue confers a growth advantage.[74] Furthermore, high risk colorectal adenoma patients showed decreased IGFBP-3 serum levels.[56] IGFBP-3 enhances p53-dependent apoptosis in human colonic adenoma-derived cells.[75] TGFβ_1 increases IGFBP-3 abundance while antisense oligonucleotides to IGFBP-3 block the growth-promoting effect of TGFβ_1 in colon carcinoma cell lines.[76] The proteolysis of IGFBP-4 by cell-bound plasmin increases IGF-II bioavailability in colon cancer cells leading to a modulation of cancer cell behavior.[77] Coincubation of IGFBP-6 decreased IGF-II-induced proliferation and colony formation of LIM 1215 colon cancer cells in agar.[65]

Breast Cancer

Among premenopausal women, circulating IGF-I concentration in serum correlates positively with the risk to develop breast cancer.[78,79] Fasting insulin level is associated with outcome in women with early breast cancer[80] and a high ratio of IGF-I to IGFBP-3 was associated with the risk of breast cancer indicating that measuring free IGF-I is more useful than total IGF-I with respect to the assessment of breast cancer risk.[81] Mammographic density which is one of the strongest predictors of breast cancer is positively associated with plasma IGF-I levels and is inversely correlated with plasma IGFBP-3 levels among premenopausal women.[82] Raloxifene administration in postmenopausal women with breast cancer significantly decreases IGF-I serum levels and may therefore be of potential value for breast cancer prevention.[83]

Transgenic mice overexpressing GH had a much reduced TGFβ$_1$ mRNA expression in the mammary gland similar to animals treated with somatostatin.[84] Both GH and IGF-I treatment of aging female rhesus monkeys led to a clear increase in mammary glandular size and epithelial proliferation index indicating that these factors are pivotal for mammary gland hyperplasia.[85] Transgenic mice expressing a GH antagonist have lower IGF-I levels and a decreased tumor incidence relative to controls indicating that perturbation of the GH-IGF-I axis may be responsible for an increased breast cancer risk.[86] IGF-I mRNA was expressed in the majority of fibroblasts derived from benign lesions whereas IGF-II mRNA expression was found in the majority of breast tumor-derived fibroblasts.[87] IGF-I-mediated induction of urokinase plasminogen activator-1 involves phosphatidylinositol 3-kinase and mitogen-activated protein kinase kinase-dependent pathways.[88] Transgenic mice expressing IGF-II develop mammary cancer, predominantly adenocarcinoma.[89] IGF-II mRNA expression is detectable at high levels in stroma of invasive breast cancer but not in normal breast.[90] A sharp upregulation of IGF-II was found in pregnancy accelerated breast cancer cell proliferation in a rat model.[91] MCF-7 cells overexpressing IGF-II showed marked morphological changes in anchorage-dependent culture and cloned also in soft agar in the absence of estrogen.[92] TGFβ$_1$, IGF-I and IGF-II act synergistically stimulate estradiol-17β hydroxysteroid dehydrogenase which converts oestrone to estradiol.[93] IGF-II protein showed a 3- to 4-fold increase in hormone-dependent (HD) lines growing in medroxyprogesterone acetate (MPA)-treated mice, as compared with HD-tumors growing in untreated mice.[94] IGF-II stimulates breast cancer cell growth via the insulin receptor isoform A indicating a novel autocrine/paracrine loop.[95]

Proteolysis of IGFBP-3 in plasma is increased in patients with breast cancers, particularly in cases with large tumors and metastatic disease.[96] Estrogens and TGFβ$_1$ show stimulatory and inhibitory activity respectively for MCF-7 breast cancer cells. In addition, estrogens and TGFβ$_1$ also stimulate and inhibit IGFBP-3 proteases respectively.[97] TNFα which inhibits proliferation and induces apoptosis of MCF-7 breast cancer cells also increased IGFBP-3 mRNA expression and decreased the number of IGF-I receptors in these cells.[98] Insulin induced a significant increase of IGFBP-5 mRNA expression in R3230AC mammary tumors in rats.[99] Treatment of breast cancer patients with tamoxifen increases IGFBP-1 and IGFBP-3 but decreases the amount of circulating free and total IGF-I.[100,101] Already at a low dose of 20 mg tamoxifen per day reduced IGF-I levels and increased IGFBP-1 levels in serum were found.[102] IGFBP-2, -3, -4 and -5 mRNA expression in rat mammary tumors increased significantly following estrogen ablation.[103] IGFBP-5 and IGFBP-6 mRNA were localized in the stromal component of breast tumors whereas IGFBP-2 mRNA was detectable in epithelial cells.[104] Both estradiol/testosterone and tamoxifen treatment in ovariectomized rhesus monkeys led to an increase of IGFBP-5 mRNA expression in the mammary epithelium.[105] Retinoic acid enhanced IGFBP-5 binding to the extracellular matrix in the canine mammary tumor cell line CMT-U335.[106] IGFBP-3 significantly accentuated ceramide- and paclitaxel-induced apoptosis in Hs578T human breast cancer cells.[107-109] Butyrate, a histone deacetylase inhibitor, activates the human IGFBP-3 promoter in breast cancer cells via an Sp1/Sp3 multiprotein complex.[110] Malignant transformation of MCF-10A human mammary epithelial cells by transfection with the *ras* oncogene abolished the inhibitory effect of IGFBP-3 possibly via activation of the MAP kinase.[111]

A significant higher ^{125}I-IGF-I binding capacity was found in malignant breast tumors as compared with benign breast tumors or normal breast tissue.[112] Estradiol upregulates IGF-IR mRNA in normal human breast tissue xenografts implanted into athymic nude mice.[113] IGF-IR immunoreactivity is found in most breast carcinomas and correlates significantly with ER status.[114] IGF-II mRNA expression and IGF-IR levels were higher whereas IGF-IIR levels were lower in murine mammary adenocarcinoma cells with high metastatic capacity in comparison with those cells showing low incidence of metastasis.[115] In MCF-7 human breast cancer cells, estrogen potentiates the IGF-I effect on IGF-IR signalling and modulates the activity of both p21 and p27.[116] Synergistic effects of IGF-I and 17β-estradiol potentiates the activity of the cyclin D1/CDK4 complex, which is involved in the triggering of progression through the cell cycle.[117] IGF-I reversed interleukin-4-induced apoptosis in breast cancer cells.[118] Estrogen receptor alpha (ERα) expression has been found to confer estrogen-mediated growth in ER-negative breast cancer cells and ERα appears to be pivotal for IGF signalling.[119] IGF-I partially restored growth in estrogen receptor (ER)-positive MCF-7 breast cancer cells treated with PI-3K and ERK1/ERK2 inhibitors indicating that IGF-IR growth-related functions may depend on ER expression.[120] IGF-IR overexpression stimulates aggregation of E-cadherin-positive MCF-7 breast cancer cells and thus would have anti-metastatic effects.[121] The activation of the IGF-IR in MCF-7 cells overexpressing IGF-IR induced a disruption of the polarized cell monolayer characterized by actin filament disassembly and tyrosine dephosphorylation of FAK, Cas and paxillin[122] and induces fascin spikes.[123] Progression of breast cancer is associated with a reduction of IGF-IR and insulin receptor substrate-1 (IRS-1) expression.[124] The inhibitory effect of apomorphine in breast cancer cells was associated with a reduction of phosphorylated IRS-1, a major intracellular substrate of the IGF-IR.[125] IGF-I enhances cell adhesion and motility in metastatic variants of human breast cancer cells which involves the insulin receptor substrate-2 (IRS-2).[126] The mitogenic function of IGF-I in human breast cancer cells is also linked to the induction of the prolyl isomerase Pin1 which stimulates cyclin D1 expression and RB phosphorylation.[127] The tumor suppressor gene PTEN blocks MAPK phosphorylation in response to insulin and IGF-I stimulation by inhibiting the phosphorylation of IRS-1 and IRS-1/Grb2/Sos complex formation in MCF-7 epithelial breast cancer cells which lead to cyclin D1 downregulation and cell growth inhibition.[128] Polymorphic CA repeats in the IGF-I gene appear to be associated with a higher risk for breast cancer.[129] IGF-I and heregulin induce the association of IGF-IR and ErbB2 in human breast cancer cells leading to tyrosine phosphorylation of both ErbB2 and ErbB3.[130] In SKBR3 human breast cancer cells overexpressing HER2/neu but few type I IGF receptors, trastuzumab (herceptin) significantly reduces proliferation but in cells overexpressing IGF-IR, trastuzumab had no effect.[131] Thus, modulating IGF-IR signalling may prevent resistance to trastuzumab. Tyrphostatin AG1024, a selective inhibitor of IGF-IR, induces apoptosis in MCF-7 cells which was associated with a downregulation of phospho-Akt1 expression, and increased expression of Bax, p53 and p21.[132] Akt1 is a downstream effector in IGF-I induced proliferation in MCF-7 breast carcinoma cells.[133] IGF-IIR levels were significantly lower in breast cancer cells than in normal cells indicating that this receptor could play a role as a tumor-suppressor gene.[134] However, no mutations of the IGF-IIR were found in breast cancer cell lines.[135]

Ovarian Cancer

IGFs, IGF receptors and IGFBPs are expressed in ovarian cancer.[136-138] IGFBP-1 levels are elevated in benign but not in the malignant ovarian cysts.[139] IGF-I and IGFBP-2 levels were significantly higher in cyst fluid from invasive malignant ovarian neoplasms as compared with benign neoplasms.[140] IGFBP-2 mRNA expression was up to 30-fold higher in malignant than in benign ovarian tumors. In addition, within malignant tissues IGFBP-2 mRNA levels correlated with the aggressiveness of the tumor.[141] IGFBP-2 serum levels were elevated in patients with epithelial ovarian cancer and correlated positively with the tumor marker, cancer antigen 125 (CA 125) whereas IGFBP-3 serum levels were significantly reduced.[142] In contrast, high IGFBP-3 expression was associated with a significantly reduced risk for disease progression in epithelial ovarian cancer.[143]

Human chorionic gonadotropin (hCG) suppresses cisplatin-induced apoptosis of ovarian cancer cells which was associated with a significant increase of IGF-I mRNA expression.[144] Estradiol stimulates IGFBP-3 and IGFBP-5 mRNA in PEO4 ovarian cancer cells.[145] IGFs increased the expression of 3β-hydroxysteroid dehydrogenase expression in a human ovarian thecal-like cell model.[146] GHRH antagonists potently inhibit growth and reduce tumorigenicity of UCI-107 ovarian cancer cells.[147] These GHRH antagonists appear to induce IGF-II production in OV-1063 human epithelial ovarian cancer cells.[148] GnRH agonist tryptorelin induces IGF-II mRNA expression in human ovarian adenocarcinoma IGROV-1 cells during a 12 h culture but not under prolonged GnRH expression.[149] Thus, the effect of GnRH agonists and GHRH antagonists may depend on culture conditions. The conflicting data have to be analyzed on the background of different experimental settings. In addition, GnRH agonists and GHRH antagonists have different short-term and long-term effects. IGF-II stimulates the synthesis of extracellular matrix glycoprotein tenascin-C which is overexpressed in the stroma of malignant ovarian tumors and thus may facilitate the migration of ovarian carcinoma cells.[150] Loss of imprinting (LOI) for IGF-II and H19 was found in epithelial ovarian cancer. In addition, loss of heterozygosity (LOH) of both IGF-II and H19 genes was associated with advanced ovarian cancer.[151,152]

Human ovarian adenocarcinoma show an increased expression of IGF-IR as compared with benign ovarian tumors and normal ovarian tissues.[153] Overexpression of the IGF-I receptor was seen in patients with recurrent or persistent epithelial ovarian cancer following chemotherapy as compared to the tumors at initial presentation.[154] The overexpression of the IGF-IR induces transformation and morphogenesis of rabbit ovarian mesothelial cells and the IGF-IR may downregulate the Fas expression inducing resistance to apoptosis of transformed ovarian mesothelial cells.[155] BRCA1 is a tumor suppressor gene frequently mutated in ovarian cancer and is capable of suppressing the IGF-IR promoter. Thus, activation of the overexpressed receptor by IGFs may then lead to ovarian cancer progression.[156] IGF-IR antisense oligodeoxynucleotides inhibited cell proliferation of human ovarian carcinoma cell lines.[157] Phosphorothioate antisense oligodeoxynucleotides against IGF-IR are effective in downregulating the IGF-IR in ovarian cancer cells[158] and may thus provide a new strategy for the therapy of this malignancy. LOH of IGF-IIR has been found in some ovarian cancers.[155] The myristylated COOH terminus of the IGF-IR can inhibit apoptosis of human ovarian carcinoma cells and inhibit tumorigenesis in nude mice.[159]

Cervical Cancer

Women with cervical cancer or advanced cervical intraepithelial neoplasia (CIN) have elevated levels of serum IGF-II as compared to normal controls indicating that IGF-II may be a useful marker for diagnosis and prognosis in cervical cancer.[160] Circulating IGF-I levels were higher and IGFBP-1 levels lower in postmenopausal endometrial cancer patients than in postmenopausal control subjects.[161] Both in human papilloma positive and negative cervical cancer cell lines downregulation of the type I IGF receptor led to a reversal of the transformed phenotype.[162] Autocrine secretion of IGF-II in HAT-3 cervical cancer cells stimulates mitogenesis of these cells whereas both antisense oligonucleotides to IGF-II and IGFBP-5 abrogated this effect.[163] It was suggested that autocrine production of IGF-II and overexpression of the type I IGF receptor are important factors controlling the proliferation of cervical carcinoma cells.[164] EGF stimulation of ECE16-1 ectocervical epithelial cells increases cell growth which is associated with a decrease in IGFBP-3 production.[165]

References

1. Zumkeller W. IGFs and IGFBPs: surrogate markers for diagnosis and surveillance of tumor growth? Mol Pathol 2001; 54:285-8.
2. Chokkalingam AP, Pollak M, Fillmore CM et al. Insulin-like growth factors and prostate cancer: a population-based case-control study in China. Cancer Epidemiol Biomarkers Prev 2001; 10:421-7.
3. Shi R, Berkel HJ, Yu H. Insulin-like growth factor-I and prostate cancer: a meta-analysis. Br J Cancer 2001; 85:991-6.
4. Pollak M, Beamer W, Zhang JC. Insulin-like growth factors and prostate cancer. Cancer Metastasis Rev 1998-99; 17:383-90.
5. Chan JM, Stampfer MJ, Giovannucci E et al. Plasma insulin-like growth factor-I and prostate cancer risk: a prospective study. Science 1998; 279:563-6.
6. Cohen P, Peehl DM, Stamey TA et al. Elevated levels of insulin-like growth factor-binding protein-2 in the serum of prostate cancer patients. J Clin Endocrinol Metab 1993; 76:1031-5.
7. Shariat SF, Lamb DJ, Kattan MW et al. Association of preoperative plasma levels of insulin-like growth factor I and insulin-like growth factor binding proteins-2 and -3 with prostate cancer invasion, progression, and metastasis. J Clin Oncol 2002; 20:833-41.
8. Yu H, Nicar MR, Shi R et al. Levels of insulin-like growth factor I (IGF-I) and IGF binding proteins 2 and 3 in serial postoperative serum samples and risk of prostate cancer recurrence. Urology 2001; 57:471-5.
9. Wang YZ, Wong YC. Sex hormone-induced prostatic carcinogenesis in the noble rat: the role of insulin-like growth factor-I (IGF-I) and vascular endothelial growth factor (VEGF) in the development of prostate cancer. Prostate 1998; 35:165-77.
10. DiGiovanni J, Kiguchi K, Frijhoff A et al.. Deregulated expression of insulin-like growth factor 1 in prostate epithelium leads to neoplasia in transgenic mice. Proc Natl Acad Sci USA 2000; 97:3455-60.

11. Tennant MK, Thrasher JB, Twomey PA et al. Protein and messenger ribonucleic acid (mRNA) for the type 1 insulin-like growth factor (IGF) receptor is decreased and IGF-II mRNA is increased in human prostate carcinoma compared to benign prostate epithelium. J Clin Endocrinol Metab 1996; 81:3774-82.
12. Cohen P, Peehl DM, Baker B et al. Insulin-like growth factor axis abnormalities in prostatic stromal cells from patients with benign prostatic hyperplasia. J Clin Endocrinol Metab 1994; 79:1410-5.
13. Tennant MK, Thrasher JB, Twomey PA et al. Insulin-like growth factor-binding protein-2 and -3 expression in benign human prostate epithelium, prostate intraepithelial neoplasia, and adenocarcinoma of the prostate. J Clin Endocrinol Metab 1996; 81:411-20.
14. Tennant MK, Thrasher JB, Twomey PA et al. Insulin-like growth factor-binding proteins (IGFBP)-4, -5, and -6 in the benign and malignant human prostate: IGFBP-5 messenger ribonucleic acid localization differs from IGFBP-5 protein localization. J Clin Endocrinol Metab 1996; 81:3783-92.
15. Plymate SR, Tennant M, Birnbaum RS et al. The effect on the insulin-like growth factor system in human prostate epithelial cells of immortalization and transformation by simian virus-40 T antigen. J Clin Endocrinol Metab 1996; 81:3709-16.
16. Huynh H, Pollak M, Zhang JC. Regulation of insulin-like growth factor (IGF) II and IGF binding protein 3 autocrine loop in human PC-3 prostate cancer cells by vitamin D metabolite 1,25(OH)2D3 and its analog EB1089. Int J Oncol 1998; 13:137-43.
17. Hong J, Zhang G, Dong F et al. Insulin-like growth factor (IGF)-binding protein-3 mutants that do not bind IGF-I and IGF-II stimulate apoptosis in human prostate cancer cells. J Biol Chem 2002; 277:10489-97.
18. Damon SE, Maddison L, Ware JL et al. Overexpression of an inhibitory insulin-like growth factor binding protein (IGFBP), IGFBP-4, delays onset of prostate tumor formation. Endocrinology 1998; 139:3456-64.
19. Gregory CW, Kim D, Ye P et al. Androgen receptor up-regulates insulin-like growth factor binding protein-5 (IGFBP-5) expression in a human prostate cancer xenograft. Endocrinology 1999; 140:2372-81.
20. Conover CA, Perry JE, Tindall DJ. Endogenous cathepsin D-mediated hydrolysis of insulin-like growth factor-binding proteins in cultured human prostatic carcinoma cells. J Clin Endocrinol Metab 1995; 80:987-93.
21. Damon SE, Plymate SR, Carroll JM et al. Transcriptional regulation of insulin-like growth factor-I receptor gene expression in prostate cancer cells. Endocrinology 2001; 142:21-7.
22. Plymate SR, Bae VL, Maddison L et al. Reexpression of the type 1 insulin-like growth factor receptor inhibits the malignant phenotype of simian virus 40 T antigen immortalized human prostate epithelial cells. Endocrinology 1997; 138:1728-35.
23. Nickerson T, Chang F, Lorimer D et al. In vivo progression of LAPC-9 and LNCaP prostate cancer models to androgen independence is associated with increased expression of insulin-like growth factor I (IGF-I) and IGF-I receptor (IGF-IR). Cancer Res 2001; 61:6276-80.
24. Koutsilieris M, Mitsiades C, Dimopoulos T et al. A combination therapy of dexamethasone and somatostatin analog reintroduces objective clinical responses to LHRH analog in androgen ablation-refractory prostate cancer patients. J Clin Endocrinol Metab 2001; 86:5729-36.
25. Chopin LK, Herington AC. A potential autocrine pathway for growth hormone releasing hormone (GHRH) and its receptor in human prostate cancer cell lines. Prostate 2001; 49:116-21.
26. Huynh H, Alpert L, Alaoui-Jamali MA et al. Co-administration of finasteride and the pure anti-oestrogen ICI 182,780 act synergistically in modulating the IGF system in rat prostate. J Endocrinol 2001; 171:109-18.
27. Klein RD, Fischer SM. Black tea polyphenols inhibit IGF-I-induced signaling through Akt in normal prostate epithelial cells and DU145 prostate carcinoma cells. Carcinogenesis 2002; 23:217-21.
28. Koutsilieris M, Mitsiades C, Dimopoulos T et al. Combination of dexamethasone and a somatostatin analogue in the treatment of advanced prostate cancer. Expert Opin Investig Drugs 2002; 11:283-93.
29. Reeve AE, Eccles MR, Wilkins RJ et al. Expression of insulin-like growth factor-II transcripts in Wilms' tumor. Nature 1985; 317:258-60.
30. Scott J, Cowell ME, Robertson ME et al. Insulin-like growth factor II in Wilms' tumors and embryonic tissues. Nature 1985; 317:260-2.
31. Schofield PN, Tate VE. Regulation of human IGF-II transcription in human fetal and adult tissues. Development 1987; 101:793-803.
32. Haselbacher GK, Irminger JC, Zapf J et al. Insulin-like growth factor II in human adrenal pheochromocytomas and Wilms' tumors: expression at the mRNA and protein level. Proc Natl Acad Sci USA 1987; 84:1104-6.
33. Yun K, Molenaar AJ, Fiedler AM et al. Insulin-like growth factor II messenger ribonucleic acid expression in Wilms tumor, nephrogenic rest, and kidney. Lab Invest 1993; 69:603-15.
34. Paik S, Rosen N, Jung W et al. Expression of insulin-like growth factor-II mRNA in fetal kidney and Wilms' tumor. An in situ hybridization study. Lab Invest 1989; 61:522-6.
35. Saikali Z, Sinnett D. Expression of glypican 3 (GPC3) in embryonal tumors. Int J Cancer 2000; 89:418-22.
36. Little MH, Ablett G, Smith PJ. Enhanced expression of insulin-like growth factor II is not a necessary event in Wilms' tumour progression. Carcinogenesis 1987; 8:865-8.
37. Schmitt S, Ren-Qiu Q, Torresani T et al. High molecular weight forms of IGF-II ('big-IGF-II') released by Wilms' tumor cells. Eur J Endocrinol 1997; 137:396-401.
38. Zumkeller W, Schwander J, Mitchell CD et al. Insulin-like growth factor (IGF)-I, -II and IGF binding protein-2 (IGFBP-2) in the plasma of children with Wilms' tumor. Eur J Cancer 1993; 29A:1973-7.
39. Baccarini P, Fiorentino M, D'Errico A et al. Detection of anti-sense transcripts of the insulin-like growth factor-2 gene in Wilms' tumor. Am J Pathol 1993; 143:1535-42.
40. Hazen-Martin DJ, Re GG, Garvin AJ et al. Distinctive properties of an anaplastic Wilms' tumor and its associated epithelial cell line. Am J Pathol 1994; 144:1023-34.
41. Ren-Qiu Q, Ruelicke T, Hassam S et al. Systemic effects of insulin-like growth factor-II produced and released from Wilms tumor tissue. Eur J Pediatr 1993; 152:102-6.
42. Schofield PN, Lee A, Cheetham JE et al. Tumor suppression associated with expression of human insulin-like growth factor II. Br J Cancer 1991; 63:687-92.

43. Drummond IA, Madden SL, Rohwer-Nutter P et al. Repression of the insulin-like growth factor II gene by the Wilms tumor suppressor WT1. Science 1992; 257:674-8.
44. Caricasole A, Duarte A, Larsson SH et al. RNA binding by the Wilms tumor suppressor zinc finger proteins. Proc Natl Acad Sci USA 1996; 93:7562-6.
45. Vincent TS, Re GG, Hazen-Martin DJ et al. All-trans-retinoic acid-induced growth suppression of blastemal Wilms' tumor. Pediatr Pathol Lab Med 1996; 16:777-89.
46. Vincent TS, Garvin AJ, Gramling TS et al. Expression of insulin-like growth factor binding protein 2 (IGFBP-2) in Wilms' tumors. Pediatr Pathol 1994; 14:723-30.
47. Qing RQ, Schmitt S, Ruelicke T et al. Autocrine regulation of growth by insulin-like growth factor (IGF)-II mediated by type I IGF-receptor in Wilms tumor cells. Pediatr Res 1996; 39:160-5.
48. Gansler T, Allen KD, Burant CF et al. Detection of type 1 insulinlike growth factor (IGF) receptors in Wilms' tumors. Am J Pathol 1988; 130:431-5.
49. Gansler T, Furlanetto R, Gramling TS et al. Antibody to type I insulinlike growth factor receptor inhibits growth of Wilms' tumor in culture and in athymic mice. Am J Pathol 1989; 135:961-6.
50. Vincent TS, Hazen-Martin DJ, Garvin AJ. Inhibition of insulin like growth factor II autocrine growth of Wilms' tumor by suramin in vitro and in vivo. Cancer Lett 1996; 103:49-56.
51. Werner H, Roberts CT Jr, Rauscher FJ 3rd et al. Regulation of insulin-like growth factor I receptor gene expression by the Wilms' tumor suppressor WT1. J Mol Neurosci 1996; 7:111-23.
52. Werner H, Shen-Orr Z, Rauscher FJ 3rd et al. Inhibition of cellular proliferation by the Wilms' tumor suppressor WT1 is associated with suppression of insulin-like growth factor I receptor expression. Mol Cell Biol 1995; 15:3516-22.
53. Werner H, Re GG, Drummond IA et al. Increased expression of the insulin-like growth factor I receptor gene, IGF1R, in Wilms tumor is correlated with modulation of IGF1R promoter activity by the WT1 Wilms tumor gene product. Proc Natl Acad Sci USA 1993; 90:5828-32.
54. Xu YQ, Grundy P, Polychronakos C. Aberrant imprinting of the insulin-like growth factor II receptor gene in Wilms' tumor. Oncogene 1997; 14:1041-6.
55. Kaaks R, Toniolo P, Akhmedkhanov A et al. Serum C-peptide, insulin-like growth factor (IGF)-I, IGF-binding proteins, and colorectal cancer risk in women. J Natl Cancer Inst 2000; 92:1592-600.
56. Renehan AG, Painter JE, Atkin WS et al. High-risk colorectal adenomas and serum insulin-like growth factors. Br J Surg 2001; 88:107-13.
57. Cats A, Dullaart RP, Kleibeuker JH et al. Increased epithelial cell proliferation in the colon of patients with acromegaly. Cancer Res 1996; 56:523-6.
58. Renehan AG, Painter JE, O'Halloran D et al. Circulating insulin-like growth factor II and colorectal adenomas. J Clin Endocrinol Metab 2000; 85:3402-8.
59. Beentjes JA, van Gorkom BA, Sluiter WJ et al. One year growth hormone replacement therapy does not alter colonic epithelial cell proliferation in growth hormone deficient adults. Clin Endocrinol (Oxf) 2000; 52:457-62.
60. Remacle-Bonnet MM, Garrouste FL, Heller S et al. Insulin-like growth factor-I protects colon cancer cells from death factor-induced apoptosis by potentiating tumor necrosis factor alpha-induced mitogen-activated protein kinase and nuclear factor kappaB signalling pathways. Cancer Res 2000; 60:2007-17.
61. Warren RS, Yuan H, Matli MR et al. Induction of vascular endothelial growth factor by insulin-like growth factor 1 in colorectal carcinoma. J Biol Chem 1996; 271:29483-8.
62. Ewton DZ, Kansra S, Lim S et al. Insulin-like growth factor-I has a biphasic effect on colon carcinoma cells through transient inactivation of forkhead1, initially mitogenic, then mediating growth arrest and differentiation. Int J Cancer 2002; 98:665-73.
63. Wu Y, Yakar S, Zhao L et al. Circulating insulin-like growth factor-I levels regulate colon cancer growth and metastasis. Cancer Res 2002; 62:1030-5.
64. Liu Y, Wang H, Zhao J et al. Enhancement of immunogenicity of tumor cells by cotransfection with genes encoding antisense insulin-like growth factor-1 and B7.1 molecules. Cancer Gene Ther 2000; 7:456-65.
65. Leng SL, Leeding KS, Whitehead RH et al. Insulin-like growth factor (IGF)-binding protein-6 inhibits IGF-II-induced but not basal proliferation and adhesion of LIM 1215 colon cancer cells. Mol Cell Endocrinol 2001; 174:121-7.
66. Zarrilli R, Romano M, Pignata S et al. Constitutive insulin-like growth factor-II expression interferes with the enterocyte-like differentiation of CaCo-2 cells. J Biol Chem 1996; 271:8108-14.
67. Hassan AB, Howell JA. Insulin-like growth factor II supply modifies growth of intestinal adenoma in Apc(Min/+) mice. Cancer Res 2000; 60:1070-6.
68. Garrouste FL, Remacle-Bonnet MM, Lehmann MM et al. Up-regulation of insulin/insulin-like growth factor-I hybrid receptors during differentiation of HT29-D4 human colonic carcinoma cells. Endocrinology 1997; 138:2021-32.
69. Di Popolo A, Memoli A, Apicella A et al. IGF-II/IGF-I receptor pathway up-regulates COX-2 mRNA expression and PGE2 synthesis in Caco-2 human colon carcinoma cells. Oncogene 2000; 19:5517-24.
70. Ouyang H, Shiwaku HO, Hagiwara H et al. The insulin-like growth factor II receptor gene is mutated in genetically unstable cancers of the endometrium, stomach, and colorectum. Cancer Res 1997; 57:1851-4.
71. Calin GA, Gafa R, Tibiletti MG et al. Genetic progression in microsatellite instability high (MSI-H) colon cancers correlates with clinico-pathological parameters: A study of the TGRbetaRII, BAX, hMSH3, hMSH6, IGFIIR and BLM genes. Int J Cancer 2000; 89:230-5.
72. Höflich A, Lahm H, Blum W et al. Insulin-like growth factor-binding protein-2 inhibits proliferation of human embryonic kidney fibroblasts and of IGF-responsive colon carcinoma cell lines. FEBS Lett 1998; 434:329-34.
73. Nishimura A, Fujimoto M, Oguchi S et al. Short-chain fatty acids regulate IGF-binding protein secretion by intestinal epithelial cells. Am J Physiol 1998; 275:E55-63.
74. Michell NP, Langman MJ, Eggo MC. Insulin-like growth factors and their binding proteins in human colonocytes: preferential degradation of insulin-like growth factor binding protein 2 in colonic cancers. Br J Cancer 1997; 76:60-6.

75. Williams AC, Collard TJ, Perks CM et al. Increased p53-dependent apoptosis by the insulin-like growth factor binding protein IGFBP-3 in human colonic adenoma-derived cells. Cancer Res 2000; 60:22-7.
76. Kansra S, Ewton DZ, Wang J et al. IGFBP-3 mediates TGF beta 1 proliferative response in colon cancer cells. Int J Cancer 2000; 87:373-8.
77. Remacle-Bonnet MM, Garrouste FL, Pommier GJ. Surface-bound plasmin induces selective proteolysis of insulin-like growth factor (IGF)-binding protein-4 (IGFBP-4) and promotes autocrine IGF-II bioavailability in human colon carcinoma cells. Int J Cancer 1997; 72:835-43.
78. Pollak MN. Endocrine effects of IGF-I on normal and transformed breast epithelial cells: potential relevance to strategies for breast cancer treatment and prevention. Breast cancer Res Treat 1998; 47:209-17.
79. Hankinson SE, Willett WC, Colditz GA et al. Circulating concentrations of insulin-like growth factor-I and risk of breast cancer. Lancet 1998; 351:1393-6.
80. Goodwin PJ, Ennis M, Pritchard KI et al. Fasting insulin and outcome in early-stage breast cancer: results of a prospective cohort study. J Clin Oncol 2002; 20:42-51.
81. Li BD, Khosravi MJ, Berkel HJ et al. Free insulin-like growth factor-I and breast cancer risk. Int J Cancer 2001; 91:736-9.
82. Byrne C, Colditz GA, Willett WC et al. Plasma insulin-like growth factor (IGF) I, IGF-binding protein 3, and mammographic density. Cancer Res 2000; 60:3744-8.
83. Torrisi R, Baglietto L, Johansson H et al. Effect of raloxifene on IGF-I and IGFBP-3 in postmenopausal women with breast cancer. Br J Cancer 2001; 85:1838-41.
84. Huynh H, Beamer W, Pollak M et al. Modulation of transforming growth factor beta1 gene expression in the mammary gland by insulin-like growth factor I and octreotide. Int J Oncol 2000; 16:277-81.
85. Ng ST, Zhou J, Adesanya OO et al. Growth hormone treatment induces mammary gland hyperplasia in aging primates. Nature Med 1997; 3:1141-4.
86. Pollak M, Blouin MJ, Zhang JC et al. Reduced mammary gland carcinogenesis in transgenic mice expressing a growth hormone antagonist. Br J Cancer 2001; 85:428-30.
87. Cullen KJ, Smith HS, Hill S et al. Growth factor messenger RNA expression by human breast fibroblasts from benign and malignant lesions. Cancer Res 1991; 51:4978-85.
88. Dunn SE, Torres JV, Oh JS et al. Up-regulation of urokinase-type plasminogen activator by insulin-like growth factor-I depends upon phosphatidylinositol-3 kinase and mitogen-activated protein kinase kinase. Cancer Res 2001; 61:1367-74.
89. Bates P, Fisher R, Ward A et al. Mammary cancer in transgenic mice expressing insulin-like growth factor II (IGF-II). Br J Cancer 1995; 72:1189-93.
90. Singer C, Rasmussen A, Smith HS et al. Malignant breast epithelium selects for insulin-like growth factor II expression in breast stroma: evidence for paracrine function. Cancer Res 1995: 55:2448-54.
91. Huynh H, Alpert L, Pollak M. Pregnancy-dependent growth of mammary tumors is associated with overexpression of insulin-like growth factor II. Cancer Res 1996; 56:3651-4.
92. Cullen KJ, Lippman ME, Chow D et al. Insulin-like growth factor-II overexpression in MCF-7 cells induces phenotypic changes associated with malignant progression. Mol Endocrinol 1992; 6:91-100.
93. Wong SF, Reimann K, Lai LC. Effect of transforming growth factor-beta1, insulin-like growth factor-I and insulin-like growth factor-II on cell growth and oestrogen metabolism in human breast cancer cell lines. Pathology 2001; 33:454-9.
94. Elizalde PV, Lanari C, Molinolo AA et al. Involvement of insulin-like growth factors-I and -II and their receptors in medroxyprogesterone acetate-induced growth of mouse mammary adenocarcinomas. J Steroid Biochem Mol Biol 1998; 67:305-17.
95. Sciacca L, Costantino A, Pandini G et al. Insulin receptor activation by IGF-II in breast cancers: evidence for a new autocrine/paracrine mechanism. Oncogene 1999; 18:2471-9.
96. Helle SI, Geisler S, Aas T et al. Plasma insulin-like growth factor binding protein-3 proteolysis is increased in primary breast cancer. Br J Cancer 2001; 85:74-7.
97. Salahifar H, Baxter RC, Martin JL. Differential regulation of insulin-like growth factor-binding protein-3 protease activity in MCF-7 breast cancer cells by estrogen and transforming growth factor-beta1. Endocrinology 2000; 141:3104-10.
98. Rozen F, Zhang J, Pollak M. Antiproliferative action of tumor necrosis factor-alpha on MCF-7 breast cancer cells is associated with increased insulin-like growth factor binding protein-3 accumulation. Int J Oncol 1998; 13:865-9.
99. Korc-Grodzicki B, Ren N, Hilf R. Insulin-like growth factor-binding proteins in R3230AC mammary tumors of intact and diabetic rats. Endocrinology 1993; 133:2362-70.
100. Campbell MJ, Woodside JV, Secker-Walker J et al. IGF status is altered by tamoxifen in patients with breast cancer. Mol Pathol 2001; 54:307-10.
101. Harper-Wynne CL, Sacks NP, Shenton K et al. Comparison of the systemic and intratumoral effects of tamoxifen and the aromatase inhibitor vorozole in postmenopausal patients with primary breast cancer. J Clin Oncol 2002; 20:1026-1035.
102. Bonanni B, Johansson H, Gandini S et al. Effect of low dose tamoxifen on the insulin-like growth factor system in healthy women. Breast cancer Res Treat 2001; 69:21-7.
103. Nickerson T, Zhang J, Pollak M. Regression of DMBA-induced breast carcinoma following ovariectomy is associated with increased expression of genes encoding insulin-like growth factor binding proteins. Int J Oncol 1999; 14:987-90.
104. Manni A, Badger B, Wei L et al. Hormonal regulation of insulin-like growth factor II and insulin-like growth factor binding protein expression by breast cancer cells in vivo: evidence for stromal epithelial interactions. Cancer Res 1994; 54:2934-42.
105. Zhou J, Ng S, Adesanya-Famuiya O et al. Testosterone inhibits estrogen-induced mammary epithelial proliferation and suppresses estrogen receptor expression. FASEB J 2000; 14:1725-30.

106. Oosterlaken-Dijksterhuis MA, Kwant MM, Slob A et al. IGF-I and retinoic acid regulate the distribution pattern of IGFBPs synthesized by the canine mammary tumor cell line CMT-U335. Breast cancer Res Treat 1999; 54:11-23.
107. Gill ZP, Perks CM, Newcomb PV et al. Insulin-like growth factor-binding protein (IGFBP-3) predisposes breast cancer cells to programmed cell death in a non-IGF-dependent manner. J Biol Chem 1997; 272:25602-7.
108. Perks CM, Bowen S, Gill ZP et al. Differential IGF-independent effects of insulin-like growth factor binding proteins (1-6) on apoptosis of breast epithelial cells. J Cell Biochem 1999; 75:652-64.
109. Fowler CA, Perks CM, Newcomb PV et al. Insulin-like growth factor binding protein-3 (IGFBP-3) potentiates paclitaxel-induced apoptosis in human breast cancer cells. Int J Cancer 2000; 88:448-53.
110. Walker GE, Wilson EM, Powell D et al. Butyrate, a histone deacetylase inhibitor, activates the human IGF binding protein-3 promoter in breast cancer cells: molecular mechanism involves an Sp1/Sp3 multiprotein complex. Endocrinology 2001; 142:3817-27.
111. Martin JL, Baxter RC. Oncogenic ras causes resistance to the growth inhibitor insulin-like growth factor binding protein-3 (IGFBP-3) in breast cancer cells. J Biol Chem 1999; 274:16407-11.
112. Jammes H, Peyrat JP, Ban E et al. Insulin-like growth factor 1 receptors in human breast tumor: localisation and quantification by histoautoradiographic analysis. Br J Cancer 1992; 66:248-53.
113. Clarke RB, Howell A, Anderson E. Type I insulin-like growth factor receptor gene expression in normal human breast tissue treated with oestrogen and progesterone. Br J Cancer 1997; 75:251-7.
114. Happerfield LC, Miles DW, Barnes DM et al. The localization of the insulin-like growth factor receptor 1 (IGFR-1) in benign and malignant breast tissue. J Pathol 1997; 183:412-7.
115. Guerra FK, Eijan AM, Puricelli L et al. Varying patterns of expression of insulin-like growth factors I and II and their receptors in murine mammary adenocarcinomas of different metastasizing ability. Int J Cancer 1996; 65:812-20.
116. Dupont J, Karas M, LeRoith D. The potentiation of estrogen on insulin-like growth factor I action in MCF-7 human breast cancer cells includes cell cycle components. J Biol Chem 2000; 275:35893-901.
117. Hamelers IH, van Schaik RF, van Teeffelen HA et al. Synergistic proliferative action of insulin-like growth factor I and 17 beta-estradiol in MCF-7S breast tumor cells. Exp Cell Res 2002; 273:107-17.
118. Gooch JL, Lee AV, Yee D. Interleukin 4 inhibits growth and induces apoptosis in human breast cancer cells. Cancer Res 1998; 58:4199-205.
119. Oesterreich S, Zhang P, Guler RL et al. Re-expression of estrogen receptor alpha in estrogen receptor alpha-negative MCF-7 cells restores both estrogen and insulin-like growth factor-mediated signaling and growth. Cancer Res 2001; 61:5771-7.
120. Bartucci M, Morelli C, Mauro L et al. Differential insulin-like growth factor I receptor signaling and function in estrogen receptor (ER)-positive MCF-7 and ER-negative MDA-MB-231 breast cancer cells. Cancer Res 2001; 61:6747-54.
121. Mauro L, Bartucci M, Morelli C et al. IGF-I receptor-induced cell-cell adhesion of MCF-7 breast cancer cells requires the expression of junction protein ZO-1. J Biol Chem 2001; 276:39892-7.
122. Guvakova MA, Surmacz E. The activated insulin-like growth factor I receptor induces depolarization in breast epithelial cells characterized by actin filament disassembly and tyrosine dephosphorylation of FAK, Cas, and paxillin. Exp Cell Res 1999; 251:244-55.
123. Guvakova MA, Boettiger D, Adams JC. Induction of fascin spikes in breast cancer cells by activation of the insulin-like growth factor-I receptor. Int J Biochem Cell Biol 2002; 34:685-98.
124. Schnarr B, Strunz K, Ohsam J et al. Down-regulation of insulin-like growth factor-I receptor and insulin receptor substrate-1 expression in advanced human breast cancer. Int J Cancer 2000; 89:506-13.
125. Chiarenza A, Scarselli M, Novi F et al. Apomorphine, dopamine and phenylethylamine reduce the proportion of phosphorylated insulin receptor substrate 1. Eur J Pharmacol 2001; 433:47-54.
126. Jackson JG, Zhang X, Yoneda T et al. Regulation of breast cancer cell motility by insulin receptor substrate-2 (IRS-2) in metastatic variants of human breast cancer cell lines. Oncogene 2001; 20:7318-25.
127. You H, Zheng H, Murray SA et al. IGF-1 induces Pin1 expression in promoting cell cycle S-phase entry. J Cell Biochem 2002; 84:211-6.
128. Weng LP, Smith WM, Brown JL et al. PTEN inhibits insulin-stimulated MEK/MAPK activation and cell growth by blocking IRS-1 phosphorylation and IRS-1/Grb-2/Sos complex formation in a breast cancer model. Hum Mol Genet 2001; 10:605-16.
129. Yu H, Li BD, Smith M et al. Polymorphic CA repeats in the IGF-I gene and breast cancer. Breast cancer Res Treat 2001; 70:117-22.
130. Balana ME, Labriola L, Salatino M et al. Activation of ErbB-2 via a hierarchical interaction between ErbB-2 and type I insulin-like growth factor receptor in mammary tumor cells. Oncogene 2001; 19:34-47.
131. Lu Y, Zi X, Zhao Y et al. Insulin-like growth factor-I receptor signaling and resistance to trastuzumab (Herceptin). J Natl Cancer Inst 2001; 93:1852-7.
132. Wen B, Deutsch E, Marangoni E et al. Tyrphostin AG 1024 modulates radiosensitivity in human breast cancer cells. Br J Cancer 2001; 85:2017-21.
133. Ahmad S, Singh N, Glazer RI. Role of AKT1 in 17beta-estradiol-and insulin-like growth factor I (IGF-I)-dependent proliferation and prevention of apoptosis in MCF-7 breast carcinoma cells. Biochem Pharmacol 1999; 58:425-30.
134. Lemamy GJ, Roger P, Mani JC et al. High-affinity antibodies from hen's-egg yolks against human mannose-6-phosphate/insulin-like growth-factor-II receptor (M6P/IGFII-R): characterization and potential use in clinical cancer studies. Int J Cancer 1999; 80:896-902.
135. Rey JM, Theillet C, Brouillet JP et al. Stable amino-acid sequence of the mannose-6-phosphate/insulin-like growth-factor-II receptor in ovarian carcinomas with loss of heterozygosity and in breast-cancer cell lines. Int J Cancer 2000; 85:466-73.
136. Yee D, Morales FR, Hamilton TC et al. Expression of insulin-like growth factor I, its binding proteins, and its receptor in ovarian cancer. Cancer Res 1991; 51:5107-12.

137. Beck EP, Russo P, Gliozzo B et al. Identification of insulin and insulin-like growth factor I (IGF I) receptors in ovarian cancer tissue. Gynecol Oncol 1994; 53:196-201.
138. Conover CA, Hartmann LC, Bradley S et al. Biological characterization of human epithelial ovarian carcinoma cells in primary culture: the insulin-like growth factor system. Exp Cell Res 1998; 238:439-49.
139. Seppälä M, Than G. Insulin-like growth factor-binding protein PP12 in ovarian cyst fluid. Arch Gynecol Obstet 1987; 241:33-5.
140. Karasik A, Menczer J, Pariente C et al. Insulin-like growth factor-I (IGF-I) and IGF-binding protein-2 are increased in cyst fluids of epithelial ovarian cancer. J Clin Endocrinol Metab 1994; 78:271-6.
141. Kanety H, Kattan M, Goldberg I et al. Increased insulin-like growth factor binding protein-2 (IGFBP-2) gene expression and protein production lead to high IGFBP-2 content in malignant ovarian cyst fluid. Br J Cancer 1996; 73:1069-73.
142. Flyvbjerg A, Mogensen O, Mogensen B et al. Elevated serum insulin-like growth factor-binding protein 2 (IGFBP-2) and decreased IGFBP-3 in epithelial ovarian cancer: correlation with cancer antigen 125 and tumor-associated trypsin inhibitor. J Clin Endocrinol Metab 1997; 82:2308-13.
143. Katsaros D, Yu H, Piccinno R et al. IGFBP-3. Expression and prognostic significance in epithelial ovarian cancer. Minerva Ginecol 2002; 54:15-24.
144. Kuroda H, Mandai M, Konishi I et al. Human chorionic gonadotropin (hCG) inhibits cisplatin-induced apoptosis in ovarian cancer cells: possible role of up-regulation of insulin-like growth factor-1 by hCG. Int J Cancer 1998; 76:571-8.
145. Krywicki RF, Figueroa JA, Jackson JG et al. Regulation of insulin-like growth factor binding proteins in ovarian cancer cells by oestrogen. Eur J Cancer 1993; 29A:2015-9.
146. McGee EA, Sawetawan C, Bird I et al. The effect of insulin and insulin-like growth factors on the expression of steroidogenic enzymes in a human ovarian thecal-like tumor cell model. Fertil Steril 1996; 65:87-93.
147. Chatzistamou I, Schally AV, Varga JL et al. Inhibition of growth and reduction in tumorigenicity of UCI-107 ovarian cancer by antagonists of growth hormone-releasing hormone and vasoactive intestinal peptide. J Cancer Res Clin Oncol 2001; 127:645-52.
148. Chatzistamou I, Schally AV, Varga JL et al. Antagonists of growth hormone-releasing hormone and somatostatin analog RC-160 inhibit the growth of the OV-1063 human epithelial ovarian cancer cell line xenografted into nude mice. J Clin Endocrinol Metab 2001; 86:2144-52.
149. Ho MN, Delgado CH, Owens GA et al. Insulin-like growth factor-II participates in the biphasic effect of a gonadotropin-releasing hormone agonist on ovarian cancer cell growth. Fertil Steril 1997; 67:870-6.
150. Wilson KE, Bartlett JM, Miller EP et al. Regulation and function of the extracellular matrix protein tenascin-C in ovarian cancer cell lines. Br J Cancer 1999; 80:685-92.
151. Chen CL, Ip SM, Cheng D et al. Loss of imprinting of the IGF-II and H19 genes in epithelial ovarian cancer. Clin Cancer Res 2000; 6:474-9.
152. Kim HT, Choi BH, Niikawa N et al. Frequent loss of imprinting of the H19 and IGF-II genes in ovarian tumors. Am J Med Genet 1998; 80:391-5.
153. Berns EM, Klijn JG, Henzen-Logmans SC et al. Receptors for hormones and growth factors and (onco)-gene amplification in human ovarian cancer. Int J Cancer 1992; 52:218-24.
154. van Dam PA, Vergote IB, Lowe DG et al. Expression of c-erbB-2, c-myc, and c-ras oncoproteins, insulin-like growth factor receptor I, and epidermal growth factor receptor in ovarian carcinoma. J Clin Pathol 1994; 47:914-9.
155. Coppola D, Saunders B, Fu L et al. The insulin-like growth factor 1 receptor induces transformation and tumorigenicity of ovarian mesothelial cells and down-regulates their Fas-receptor expression. Cancer Res 1999; 59:3264-70.
156. Maor SB, Abramovitch S, Erdos MR et al. BRCA1 suppresses insulin-like growth factor-I receptor promoter activity: potential interaction between BRCA1 and SP1. Mol Genet Metab 2000; 69:130-6.
157. Resnicoff M, Ambrose D, Coppola D et al. Insulin-like growth factor-1 and its receptor mediate the autocrine proliferation of human ovarian carcinoma cell lines. Lab Invest 1993; 69:756-60.
158. Muller M, Dietel M, Turzynski A et al. Antisense phosphorothioate oligodeoxynucleotide down-regulation of the insulin-like growth factor I receptor in ovarian cancer cells. Int J Cancer 1998; 77:567-71.
159. Hongo A, Yumet G, Resnicoff M et al. Inhibition of tumorigenesis and induction of apoptosis in human tumor cells by the stable expression of a myristylated COOH terminus of the insulin-like growth factor I receptor. Cancer Res 1998; 58:2477-84.
160. Mathur SP, Mathur RS, Young RC. Cervical epidermal growth factor-receptor (EGF-R) and serum insulin-like growth factor II (IGF-II) levels are potential markers for cervical cancer. Am J Reprod Immunol 2000; 44:222-30.
161. Ayabe T, Tsutsumi O, Sakai H et al. Increased circulating level of insulin-like growth factor-I and decreased circulating levels of insulin-like growth factor binding protein-1 in postmenopausal women with endometrial cancer. Endocr J 1997; 44:419-24.
162. Nakamura K, Hongo A, Kodama J et al. Down-regulation of the insulin-like growth factor I receptor by antisense RNA can reverse the transformed phenotype of human cervical cancer cell lines. Cancer Res 2000; 60:760-5.
163. Steller MA, Delgado CH, Zou Z. Insulin-like growth factor II mediates epidermal growth factor-induced mitogenesis in cervical cancer cells. Proc Natl Acad Sci USA 1995; 92:11970-4.
164. Steller MA, Delgado CH, Bartels CJ et al. Overexpression of the insulin-like growth factor-1 receptor and autocrine stimulation in human cervical cancer cells. Cancer Res 1996; 56:1761-5.
165. Hembree JR, Agarwal C, Eckert RL. Epidermal growth factor suppresses insulin-like growth factor binding protein 3 levels in human papillomavirus type 16-immortalized cervical epithelial cells and thereby potentiates the effects of insulin-like growth factor 1. Cancer Res 1994; 54:3160-6.

CHAPTER 27

Insulin-Like Growth Factors and Hematological Malignancies

Anne J. Novak and Diane F. Jelinek

Abstract

Immune cell development and homeostasis is a highly coordinated process influenced by a network of cells and soluble factors. The endocrine system, in particular the insulin-like growth factors (IGFs), has been shown to play a key role in development and maintenance of normal immune function. Additionally, there is a growing literature that indicates a role for IGFs in the biology of various hematopoietic malignancies, particularly multiple myeloma. In this article, we will review the role of IGFs and the type I IGF receptor in development of the immune system and lymphocyte function. Furthermore, we will review the evidence that links IGFs with the pathogenesis of hematological malignancies, with a particular emphasis on B malignancies.

Introduction

Orchestration of hematopoiesis and maintenance of normal immune function is a highly coordinated process regulated in part by a variety of cytokines and growth factors. Of interest, some of these same soluble factors often remain important following malignant transformation of hematopoietic lineage cells, most frequently as a result of inappropriate expression of these molecules and their specific receptors by the malignant cells. In this regard, insulin-like growth factors (IGF) are one of many soluble factors that immune cells encounter and there is increasing evidence that IGFs and the type I IGF receptor, IGF-IR, may also play a key cellular role in some hematopoietic malignancies. IGF-I and IGF-II are 70 and 67 amino acid peptides, respectively, that are structurally related to insulin. The majority of circulating IGFs found in serum are associated with specific binding proteins referred to as IGF-binding proteins (IGFBPs). Of interest, there are at least 6 different types of IGFBPs and these proteins may either inhibit or potentiate IGF-I-mediated signal transduction. There are two known IGF receptors designated types I and II. However, the majority of the cellular effects mediated by IGFs result from binding of ligand to the type I IGF receptor (IGF-IR), a type II receptor protein tyrosine kinase that triggers a downstream cascade of signal transduction. Signals transduced through the IGF-IR are known to result in 1) cell proliferation, 2) enhanced survival and suppression of apoptosis, and 3) establishment and maintenance of tumor cells.[1-3] The growth and survival-promoting properties of IGFs suggest therefore that inappropriate expression of IGF-I or IGF-II and the IGF-IR may contribute to loss of normal cell growth control. Although IGFs are typically not viewed as classical growth factors for hematopoietic cells, or as playing a significant role in the growth control of transformed hematopoietic cells, there is a growing literature that suggests otherwise. In this review, we will discuss the role of IGFs and the IGF-IR in the pathogenesis and maintenance of hematopoietic malignancies with a particular emphasis on IGF-I, the lymphocyic lineage of hematopoietic cells, and B lymphocyte malignancies.

IGFs and IGFBPs in the Immune System

IGF Expression

The development and regulation of the immune system is modulated by a variety of molecules, including IGFs. Although IGF-I and IGF-II expression is relatively widespread during fetal

Insulin-Like Growth Factors, edited by Derek LeRoith, Walter Zumkeller and Robert Baxter. ©2003 Eurekah.com and Kluwer Academic / Plenum Publishers.

development, expression of these factors in adults is moderately restricted.[4] The primary source of IGF-I is the liver where expression levels are controlled by growth hormone. However, IGF-I can also be expressed by a number of other cell types and regulation of expression is often achieved by a variety of extracellular soluble factors such as cytokines. With respect to the immune system, IGF-I is expressed by activated T and B cells, leukocytes, and macrophages.[5-8] Of these cell types, activated macrophages were shown to express the highest levels of IGF-I approaching levels seen in hepatic cells.[9] Of note, resting lymphocytes and monocytes/macrophages express very little IGF-I mRNA.

IGFs and Hematopoiesis

The actions of IGFs on the immune system are varied. IGF-I has been shown to be expressed in human bone marrow and to be one of the factors that may participate in hematopoiesis. For example, IGF-I and –II have been shown to stimulate erythroid and myeloid progenitors and to induce granulopoiesis in human bone marrow.[10-14] With respect to the humoral immune system, IGF-I stimulates primary B lymphopoiesis and protects murine-bone-marrow derived B lymphocytes from cytokine withdrawal.[15,16] Additionally, IGF-I was shown to potentiate murine IL-7 stimulated proliferation of early B cells.[17] Further evidence of a role for IGF-I and B cell development is suggested by observations that transgenic expression of IGF-I in mice results in increased spleen weight and mice treated with IGF-I in vivo have increased numbers of CD4+ T cells and splenic B cells.[15,18-20] A role for IGF-I in T cell development has also been suggested. For example, when IGF-I was given to diabetic rats, thymocyte DNA synthesis was increased as was the overall size of the thymus.[21] IGF-I was also shown to enhance thymus recovery following treatment of rats with cyclosporin or dexamethasone.[22,23] Finally, a role for IGF-II in T cell development has also been suggested by the selective thymic growth observed in transgenic mice overexpressing IGF-II.[7,24,25]

IGFs and Lymphocyte Function

IGF-I and -II have both been shown to have several effects on mature lymphocytes. Both peptides have been shown to enhance T cell proliferation by as much as three-fold, but it is important to note that in this study, T cells cultured with IGFs alone were not stimulated; rather, anti-CD3 antibodies and IL-2 were needed to demonstrate an effect.[26] Similar results were also observed by Kooijman et al.[7] It has also been demonstrated that IGF-I can protect T cells from undergoing Fas-induced apoptosis[27] and it may induce T cell migration.[28] There are few reports of the effects of IGFs on mature B cells, however, there are several studies that suggest a role for IGF-I, but not IGF-II, in B cell differentiation. Specifically, IGF-I was shown to augment IgE and IgG4 production by plasma cells and tonsillar B cells.[29,30] Enhancement of immunoglobulin synthesis by in vivo administration of IGF-I in mice has also been found.[31] IGF-I has been shown by one group to enhance human B cell proliferation,[32] but not by our group, therefore calling into question whether or not IGF-I can function as a growth factor for normal B cells.[33] In contrast to the few reports demonstrating an effect of IGF-I on normal B cells, there are a greater number of reports demonstrating an effect of IGFs on malignant B cells (discussed below).

IGFBPs and the Immune System

Serum IGFs largely exist in a complex with IGFBPs. This observation reflects the ability of IGFs to bind to IGFBPs with 5-50 times greater affinity as compared with the IGF-IR. There are six IGFBPs and IGFBP-3 is the primary serum carrier followed by IGFBP-2. IGFBPs are believed to have a number of biological functions, including their ability to modulate interaction of the IGFs with their receptors. Although IGFBPs were originally considered to counteract the activity of the IGFs, it is now realized that IGFBPs may also play a stimulatory role, perhaps as a result of their ability to increase the local concentration of IGFs in the vicinity of the IGF-IR, reviewed in.[34] There is an extensive literature regarding cellular expression of IGFBPs, however, a much smaller number of reports have demonstrated hematopoietic cell expression of IGFBPs. First, macrophages have been shown to express IGFBP-4.[35] IGFBP-4 is primarily an inhibitor of IGF action and it plays little role as a serum carrier. Because of its largely inhibitory role, the main function of this IGFBP may be to protect cells from overstimulation by IGFs.[36] Second, in normal human lymphocytes, IGFBP-2 and -3 mRNA was expressed in resting cells, and following activation, cells gained expression of IGFBP-4 and -5.[37] IGFBP-2, -3, and -5 have all been shown to be capable of IGF potentiation and IGFBP-2 and -3 can also inhibit IGF activity. Foll et al [38] demonstrated that human T cells expressed IGFBP-2 mRNA at higher levels than did B cells, and that IGFBP-2 expression appeared to precede T cell blast formation. This pattern was also observed when a panel of leukemic B and T cell lines were examined, i.e., IGFBP-2 was expressed at higher levels in T cell lines than B cell lines.[39] Third, using a murine pre-B cell line (Ba/F3), Hosokawa and colleagues demonstrated that ectopic expression of the Bcl-6 oncogene in these cells led to downregulation of a number of regulatory genes, including IGFBP-4.[40] Although it is clear from these studies that hematopoietic cells can express IGFBPs, the biological consequences of IGFBP expression or loss of expression by these cells has not been elucidated.

The Insulin-Like Growth Factor Receptor-I in the Immune System

IGF-IR

The effects of IGFs are primarily transmitted through the IGF-IR. The IGF-IR is synthesized as a single chain precursor that is subsequently cleaved into two separate α and β chains. The cell surface receptor complex is organized as a functional dimer of two α/β complexes. The α-subunit is entirely extracellular and responsible for ligand binding, whereas the β-subunit is largely intracellular and responsible for triggering downstream signaling events. For a review see.[41] Because of the important role that IGFs play in development and growth regulation, IGF-I receptor gene expression is tightly regulated. In general, high levels of IGF-IR mRNA are observed embryonically; however, in adult tissues, the levels of IGF-IR mRNA are very low. Of great interest, it has been shown that as little as a two-fold difference in IGF-IR number per cell can qualitatively change the nature of cellular responsiveness from enhanced survival to proliferation.[42] In corollary studies, antisense strategies resulting in reduction of IGF-IR mRNA levels and cell surface receptor numbers by as little as 30% severely reduced IGF-stimulated growth.[43] These results predict that overexpression of this receptor has great potential to contribute to inappropriate cellular proliferation.

IGF-IR Expression by Hematopoietic Cells

In human blood mononuclear cells, the IGF-IR is highly expressed on monocytes, natural killer cells, and CD4$^+$ T-helper cells, at an intermediate level on CD8$^+$ cytotoxic T cells, and at very low levels on B cells (Table 1).[44] With respect to the low level of B cell IGF-IR expression, one group failed to detect this receptor on B cells.[45] It has also been shown that IGF-IR expression levels on T cells increased following activation and another group suggested that IGF-IR levels might correlate with T cell differentiation stage.[46,8] As will be discussed in greater detail below, IGF-IR expression levels have often been observed to be elevated in human malignancies, including hematopoietic malignancies.

Upon binding of IGFs to the IGF-IR, the intrinsic tyrosine kinase activity of the receptor becomes activated resulting in receptor autophosphorylation.[4,47] The key features of the IGF-IR responsible for signal transduction include Y950 on the β-chain, which is the binding site for the two major substrates of the IGF-IR, an ATP-binding site at lysine 1003 without which activation does not occur, and a tyrosine kinase domain centered around tyrosines 1131, 1135, and 1136.[47] Tyrosine phosphorylation of the IGF-IR provides docking sites for src-homology 2 (SH2) or phosphotyrosine binding domain (PTB) containing substrates. The major substrate of the activated receptor is IRS-1 (insulin-receptor substrate 1) and it functions as a "docking" protein that can bring a variety of SH2 domain-containing proteins together, thereby triggering activation of several downstream signaling pathways.[41] Figure 1 provides a general schematic of how the IGF-IR is linked to activation of pathways important for cell proliferation (ras, raf, MAPK) and cell survival (Akt, Bad).

Regulation of IGF-IR Expression

The human IGF-I receptor is the product of a single-copy gene located at the distal end of human chromosome 15. The IGF-IR promoter lacks TATA and CAAT boxes, is very GC-rich, and can be distinguished from other "housekeeping" genes by the presence of a single start site of transcription contained within an initiator motif typical of genes highly regulated during differentiation and development.[48] The initiator motif results in an unusually long 5' untranslated region (UTR) of approximately 1 kb in the human gene. Although the basal promoter activity of the IGF-IR has

Table I. Relative IGFI-R expression on normal and malignant hematopoietic cells

Peripheral Blood Mononuclear Cells			
Monocytes	CD14$^+$	high	44
NK Cells	CD16$^+$	high	44
T lymphocytes	CD4$^+$	high	44
T lymphocytes	CD8$^+$	intermediate	44
B lymphoctyes	CD19$^+$	low/negative	44
Plasma Cells	CD38$^+$/CD45$^-$	low/negative	Unpublished observations
Malignant B Cells			
Multiple myeloma	CD38$^+$/CD45$^-$	low/intermediate	Unpublished observations
B-CLL	CD19$^+$/CD5$^+$	low/intermediate	Unpublished observations
Cell Lines			
KAS-6/1	Myeloma	high	33
ANBL-6	Myeloma	intermediate	33
KP-6	Myeloma	intermediate	33
DP-6	Myeloma	intermediate	33
RPMI 8226	Myeloma	intermediate	92, 94
U266	Myeloma	intermediate	92, 94
HL60	Myelomonocytic	intermediate	76-78
K562	Erythroleukemia	intermediate	79-80
Jurkat	T lymphoblastic leukemia	intermediate	81-82

been best studied in rat fibroblasts and maps to the proximal 450 bp of the 5'flanking region, the human promoter is very similar to that of rat.[49] There are numerous sites for the Sp1 zinc finger within this region in both the rat and human genes and the rat promoter has been shown to be functional in vivo as well as in vitro.[50, 51] Although Sp1 is ubiquitously expressed, numerous studies suggest that the levels of Sp1 expression are regulated during development in a tissue-specific manner and in response to a variety of external signals, including cytokines such as IL-2.[52,53,54] Because the expression of this gene is extremely low in adult tissues, it has also been suggested that in the postmitotic, fully differentiated cell, the IGF-I-R promoter is under constitutive inhibitory control.[41,47] For example, there is evidence that the p53 tumor suppressor gene product may play a role in the constitutive inhibition of the IGF-I-R in adult tissues.[55] Because p53 is frequently mutated in a variety of malignancies, these studies suggest a mechanism whereby IGF-I-R may become overexpressed during tumor progression.

IGFs and IGF-IR: A Pathway to Malignancy

Overexpression of IGFs and IGF-IR

The growth-promoting properties of IGFs as reviewed above suggest that inappropriate expression of IGFs may contribute to loss of normal cell growth control. In fact, IGFs have long been recognized as important mitogens in many types of malignancies.[1-3,56] Ample evidence suggests that many types of tumor cells may actually acquire the ability to express autocrine IGF-I or IGF-II, and in some cases this has been shown to interfere with cellular differentiation.[57] Moreover, it also has been shown that focal activation of the IGF-II gene in transgenic mice expressing the SV40 large T antigen delivers a crucial, second signal in this model of oncogene-induced tumorigenesis.[58] Lastly,

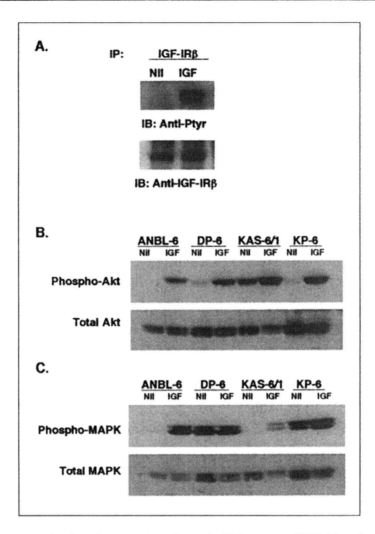

Figure 1. IGF-IR mediated signaling events in myeloma cells. (A) Serum starved KAS-6/1 myeloma cells were left unstimulated (Nil) or were stimulated with 200 ng/ml des-IGF-I for 10 min at 37 C. Cell lysates were precipitated with 1 mg of anti-IGF-IR, separated on a 7.5% SDS-polyacrylamide gel, and immunoblotted with an anti-phosphotyrosine antibody (top panel). The membrane was subsequently stripped and re-probed with an anti-IGF-IR antibody (lower panel). (B) ANBL-6, DP-6, KAS-6/1 and KP-6 myeloma cells were left unstimulated (Nil) or were stimulated with 200 ng/ml des-IGF-I for 10 min at 37C. Equivalent amounts of cell lysates were separated on a 7.5% SDS-polyacrylamide gel and immunoblotted with anti-phospho-Akt (top panel). The membrane was subsequently stripped and re-probed with anti-Akt (lower panel). (C) The membrane from (B) was subsequently stripped and re-probed with anti-phospho-MAPK (lower panel) and anti-MAPK (lower panel).

several reports have shown that IGF-I may be important in protecting cells from undergoing programmed cell death in response to either growth factor withdrawal or chemotherapeutic drugs.[16,59,60] In contrast to numerous reports that solid tumors may acquire expression of IGFs, or express these peptides at a higher level, there are very few reports suggesting that hematopoietic malignancies constitutively express more IGFs than do normal counterpart cells. However, there is some evidence in multiple myeloma that autocrine IGF expression may be induced by cytokines that stimulate tumor cell growth.[33]

As noted above, IGF-IR expression levels are tightly regulated in adult tissues. However, overexpression or constitutive activation of the IGF-IR has been found in several human cancers.[61,62] A role for the IGF-IR in neoplastic development has been strongly suggested by studies showing that transformed phenotypes can be reversed to non-transformed phenotypes by decreasing the absolute number of IGF-IRs. Different approaches have been used, including antisense expression plasmids or antisense oligonucleotides against either IGF-II [58] or IGF-I,[63,64] antisense against the IGF-IR,[43,65-68] antibodies to the IGF-IR,[69,70] and dominant negative mutants of the IGF-IR.[71-73] However, much of this work has been done in non-hematopoietic cells.

IGFs and the IGF-IR in Hematopoietic Malignancies

A functional role for the IGF-IR in hematopoietic malignancies has largely been demonstrated by the ability of IGF-IR monoclonal antibodies to block transformed cell growth. These results suggest that IGF-I may act in an autocrine or paracrine fashion to modulate cell proliferation.[74] The IGF-IR is expressed by a number of human leukemic cell lines including HL-60, Jurkat, and K-562 (Table 1).[75-82] Interestingly, an altered IGF-I receptor is expressed on the human promyelocytic leukemia cell line, HL-60, and there is evidence that this receptor contributes to cell growth.[83] In addition to the work done in cell lines, IGF-I has been found to enhance growth of freshly isolated human acute myelogenous leukemia and acute lymphoblastic leukemia blasts in vitro.[76,84] With respect to T cell malignancies, IGF-I has been shown to influence the growth of murine lymphoma cells bearing a pre-T cell phenotype.[85] Lastly, immature and mature primary human T lineage acute lymphoblastic leukemia cells were found to express high numbers of IGF-IR.[86] These authors also reported that human pre-B acute lymphoblastic leukemic cells expressed IGF-IR as well. Somewhat curiously, there are far more reports of a role for IGFs and the IGF-IR in B cell malignancies, particularly in multiple myeloma. This evidence will be discussed in greater detail below.

IGFs and the IGF-IR in B Cell Malignancies

B Cell Chronic Lymphocytic Leukemia (B-CLL)

B-CLL is the most common leukemia in the United States and a significant cause of morbidity and mortality in the older adult population.[87,88] The hallmark feature of this disease is the presence of elevated numbers of circulating clonal leukemic B cells that typically express CD19, CD23, CD5, and low levels of surface Ig.[89] This disease is not characterized by highly proliferative cells but rather by the presence of leukemic cells with significant resistance to apoptosis.[90,91] Because of our interest in this disease and the possible role of IGFs and the IGF-IR, we have begun to analyze B-CLL cells for expression levels of IGF-IR. Of interest, we have recently found that the IGF-IR is expressed at a higher level on B-CLL cells relative to normal peripheral blood B cells (unpublished observations). Because B-CLL cells are highly resistant to apoptosis, it is possible that the malignant B cells produce IGFs in an auto- or paracrine fashion to maintain survival. Work is currently underway to determine the significance of IGF-IR expression in B-CLL. Finally, using gene expression profiling, we have recently discovered that IGFBP-4 is overexpressed in leukemic B cells obtained from patients with B cell chronic lymphocytic leukemia relative to normal B cells obtained from normal individuals over age 60 (unpublished observations). This is an intriguing observation, however, the significance of this finding remains to be elucidated. Because IGFBP-4 is a potent inhibitor of IGF-I activity, and because IGF-I has been shown to positively regulate the growth of human pre-B cells, it is possible that overexpression of IGFBP4 underlies the dramatic suppression of normal B cells observed in B-CLL patients.

Multiple Myeloma

Multiple myeloma is a progressive disease characterized by the expansion of malignant plasma cells in the bone marrow. Moreover, this disease is often accompanied by the presence of osteolytic lesions. Although the most thoroughly investigated cytokine in this disease has been interleukin 6 (IL-6) because of its documented ability to function as a growth factor for myeloma cells, there is also a growing literature linking various aspects of IGF-I and its receptor with the biology of this disease. It has been known for a number of years that myeloma cell lines express IGF-IR and at higher levels than B lymphoblastoid cell lines.[92] Since this initial report, a number of multiple

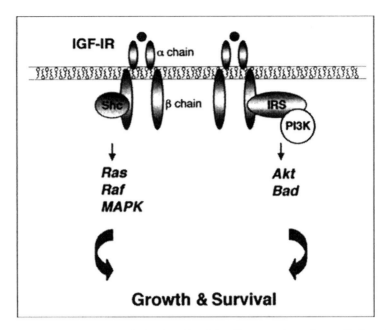

Figure 2. Schematic representation of IGF-IR mediated signaling events in myeloma cells. Experimental evidence exists demonstrating the ability of IGF-I to activate the Ras/MAPK and PI-3K signaling pathways in myeloma cells.

myeloma cell lines have been shown to express the IGF-IR and some have been demonstrated to be growth responsive to IGFs as well (Table 1).[29,33,93,94] In an effort to better understand how IGF-I stimulates myeloma cell growth, we have studied intracellular signaling events following IGF-I stimulation. As can be seen in Figure 1, stimulation of myeloma cell lines with des-IGF-I results in robust autophosphorylation of the IGF-IR and subsequent activation of Akt and MAPK. IGF-I-mediated activation of Akt in myeloma cell lines has also been reported by other investigators[95] and there is some evidence that loss of the PTEN tumor suppressor gene may result in high levels of Akt activation in myeloma cells.[96] In addition to Akt and MAPK, PI3-kinase and Bad have been shown to be downstream targets of IGF-IR mediated signaling in myeloma cells.[97] Phosphorylation of Bad induces its dissociation from Bcl-2 and Bcl-X_L resulting in cell survival. IGF-I has also been shown to protect myeloma cells from undergoing dexamethasone-induced apoptosis.[60] The growth enhancing effects of MAPK combined with the anti-apoptotic properties of Akt and Bad may therefore play a role in the survival and expansion of malignant myeloma cells (Fig. 2). In further support of a role for IGFs in multiple myeloma cell growth, it has been found that activation of IGF-IR in mouse plasma cells tumors is critical for malignant growth.[98] IGF-I has also been found to act as a chemoattractant factor for mouse 5T2 multiple myeloma cells.[99] The presence of IGF-I in the bone marrow microenvironment and expression of IGF-IR on myeloma cells may therefore contribute to the accumulation of these cells in the bone marrow. Of interest, this same group of investigators has also recently demonstrated that the bone marrow microenvironment itself may be capable of increasing expression of the IGF-IR, thereby further facilitating homing of myeloma cells to the bone marrow.[100] In summary, it is clear from the literature that IGFs play a role in the proliferation and survival of malignant myeloma cells and may also contribute to the selective recruitment of these cells to the bone marrow. It should be noted that the bulk of these data have been obtained using cell lines rather than primary tumor cells. We have shown in a preliminary report, however, that primary myeloma cells do indeed express elevated levels of IGF-IR.[101] In consideration of these data as well as our ongoing work on B-CLL, we have hypothesized that one key event that may occur during B cell transformation is the re-expression or elevation of the IGF-IR, thereby permitting altered tumor cell growth. This hypothesis is graphically depicted in Figure 3.

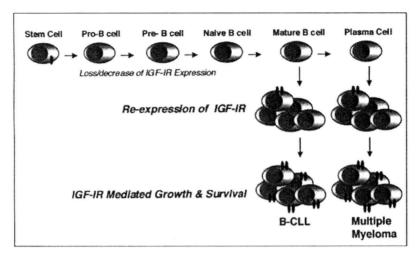

Figure 3. A possible role for the IGF-IR in the pathogenesis of B cell malignancies. This model suggests that IGF-IR levels decrease during normal B cell development and maturation into plasma cells and that re-expression or increased levels of IGF-IR may play an important role in B cell transformation.

Receptor Cross Talk

As previously mentioned, altered expression levels of IGFs or the IGF-IR are frequently observed in many human cancers. The mechanisms by which elevated levels of these molecules impact the growth and survival of tumor cells, however, is less clear. A complicating factor in this analysis is the likelihood that IGF-I signal transduction may not always occur in a linear fashion, i.e., in the absence of interaction with other receptors. Indeed, the IGF-IR has been shown to act in concert with other receptor systems and these "non-linear" patterns of signal transduction may underlie at least some of the effects of IGF on tumor cell growth and survival.

Thus, the IGF-IR communicates with and influences downstream signals transmitted through various growth factor receptors. For example, cross talk between the IGF-IR and the estrogen receptor has been well documented in human breast tumors.[102] The IGF-IR is expressed in a large number of human breast malignancies and the expression is positively correlated with the level of estrogen receptor.[103,104] Stimulation of breast cancer cells with IGF-I results in tyrosine phosphorylation and transcriptional activation of the estrogen receptor.[105,106] Moreover, physical association between IGF-IR and ERB-2 has been demonstrated in both murine and human breast cancer cells and results in ERB-2 phosphorylation.[107] The communication between IGF-IR and the estrogen or ERB-2 receptor in malignant cells indicates that IGF-IR expression may play an important role in the acquisition of the malignant phenotype.

In addition to growth factor receptor cross-talk, the IGF-IR has the ability to communicate with adhesion molecules, mainly integrins.[36,108] The convergence of signals between growth factor receptors and integrins has been well studied and is thought to play a key role in regulation of cell adhesion, migration, and survival.[108,109] Importantly, signals transmitted through integrins and growth factor receptors contribute to the metastatic cascade of numerous tumors.[110] The IGF-IR has been found to cooperate with the $\alpha v \beta 5$ integrin and to promote dissemination of malignant tumor cells.[111,112] In breast cancer cells, IGF-I mediated cell migration is dependent on integrin mediated adhesion events.[113] Additionally, full activation of IGF-IR mediated tyrosine kinase activity has been found to be dependent on integrin occupancy.[113] Convergence of the IGF-IR and integrin signaling cascades has also been demonstrated. p125^{FAK} focal adhesion kinase (FAK) as well as IRS-1 are activated upon both integrin ligation and IGF-IR stimulation.[114] Additionally, cell adhesion and subsequent activation of FAK results in the association of FAK and IRS-1 and modulation of IRS-1 expression.[115,116] Because integrins play an important role in survival and migration it is possible that IGF-IR and integrins coordinate their signals to influence the invasive and metastatic nature of tumor cells.

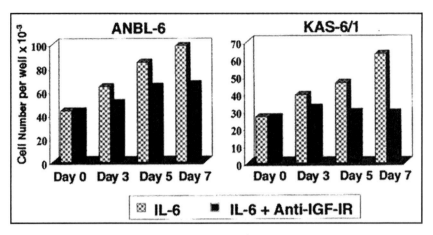

Figure 4. Anti-IGF-IR antibodies inhibit IL-6 stimulated myeloma cell proliferation. ANBL-6 and KAS-6/1 myeloma cells were cultured with 1 ng/ml IL-6 alone or in the presence of 1 μg/ml anti-IGF-IR (αIR3). The number of cells per well was quantitated on days 0, 3, 5, and 7. Values represent the mean of triplicate values.

While IGF-IR cross-talk has been elegantly demonstrated in various cell systems, the receptors that communicate with the IGF-IR in hematopoietic cells are poorly understood. In Jurkat T cells it has been shown that CD28 stimulation increases IGF-IR expression providing the cells with essential survival signals.[27] Signals transmitted through the IL-4 receptor and IGF-IR synergize in 32D myeloid progenitor cells to induce cell proliferation.[33,117] Recently our laboratory has reported that IGF-I directly stimulates myeloma cell growth and that it also enhances IL-6 stimulated multiple myeloma cell proliferation.[33] Furthermore, we demonstrate that blocking of the IGF-IR with antibodies results in inhibition of IL-6 mediated myeloma cell proliferation (Fig. 4). These results suggest that IL-6 stimulated growth of multiple myeloma cells may result in part from induction of autocrine IGF-I expression. While it is currently unclear how the IGF-IR and IL-6 elicit their effects in multiple myeloma cells we propose three possible scenarios: 1) the IGF-IR and the IL-6R transmit signals to the cells independently to induce additive cell proliferation, 2) the IGF-IR and the IL-6R synergize at some point in their signaling cascade to induce maximal cell proliferation, or 3) IL-6 stimulation may invoke subsequent autocrine IGF expression, resulting in further growth stimulation (Fig. 5).

In addition to potential IL-6 and IGF-I receptor cross-talk, a role for integrins in the pathogenesis of multiple myeloma has also been implicated.[118] Localization of multiple myeloma cells to the bone marrow microenvironment is thought to be regulated by interactions between the α4β1 integrin and vascular cell adhesion molecule-1 (VCAM), which is expressed on bone marrow stroma.[119,120] The interaction between α4β1 and VCAM was recently found to result in increased osteoclastogenic activity by myeloma cells, a possible mechanism for bone destruction.[121] Because IGF-IR can cross-talk with the integrin system, one can envision a scenario where stimulation of multiple myeloma cells with IGFs results in integrin activation thereby enhancing osteoclast activity and the pathogenesis of multiple myeloma.

Concluding Remarks

In this review, we have summarized what is currently known about the roles for IGFs and the type I IGF receptor in hematopoietic malignancies. Although the literature suggests that these molecules may be far more important in solid tumors than in the hematopoietic malignancies, there is compelling evidence that these may be important molecules to study in some diseases, such as multiple myeloma. Because of the multipotent effects of IGFs, e.g., enhanced cell proliferation, cell migration, and protection from apoptosis, alteration in expression levels of IGFs and the IGF-IR in human cancers is a phenomenon that clearly deserves further study. Given the dramatic cellular consequences of subtle changes in expression levels of the IGF-IR, there is an obvious need to elucidate the mechanisms that may underlie malignant cell overexpression of this receptor.

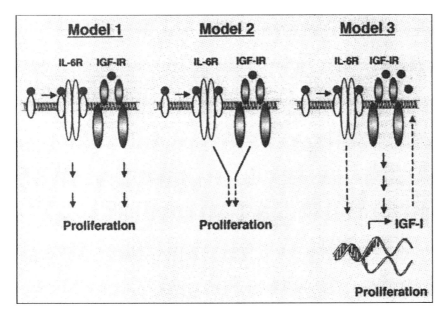

Figure 5. Speculative models of mechanisms by which IL-6 and/or IGF-I1 stimulate myeloma cell growth. IL-6 and IGF-I may activate two separate pathways (Model 1), IL-6 and IGF-I receptors may cross communicate (Model 2), and IL-6 stimulated myeloma cell growth may be facilitated by induction of an autocrine IGF-I pathway (Model 3).

References

1. Baserga R. The insulin-like growth factor I receptor: a key to tumor growth? Cancer Res 1995; 55(2):249-252.
2. Cullen KJ, Yee D, Rosen N. Insulin-like growth factors in human malignancy. Cancer Invest 1991; 9(4):443-454.
3. Macaulay VM. Insulin-like growth factors and cancer. Br J Cancer 1992; 65(3):311-320.
4. Cohick WS, Clemmons DR. The insulin-like growth factors. Annu Rev Physiol 1993; 55:131-153.
5. Sabharwal P, Varma S. Growth hormone synthesized and secreted by human thymocytes acts via insulin-like growth factor I as an autocrine and paracrine growth factor. J Clin Endocrinol Metab 1996; 81(7):2663-2669.
6. Brocardo MG, Schillaci R, Galeano A et al. Early effects of insulin-like growth factor-1 in activated human T lymphocytes. J Leukoc Biol 2001; 70(2):297-305.
7. Kooijman R, Willems M, Rijkers GT et al. Effects of insulin-like growth factors and growth hormone on the in vitro proliferation of T lymphocytes. J Neuroimmunol 1992; 38(1-2):95-104.
8. Kooijman RK, Scholtens LE, Rijkers GT et al. Differential expression of type I insulin-like growth factor receptors in different stages of human T cells. Eur J Immunol 1995; 25(4):931-935.
9. Rom WN, Basset P, Fells GA et al. Alveolar macrophages release an insulin-like growth factor I-type molecule. J Clin Invest 1988; 82(5):1685-1693.
10. Akahane K, Tojo A, Urabe A et al. Pure erythropoietic colony and burst formations in serum-free culture and their enhancement by insulin-like growth factor I. Exp Hema 1987; 15(7):797-802.
11. Kurtz A, Zapf J, Eckardt KU et al. Insulin-like growth factor I stimulates erythropoiesis in hypophysectomized rats. Proc Natl Acad Sci USA 1988; 85(20):7825-7829.
12. Merchav S, Tatarsky I, Hochberg Z. Enhancement of erythropoiesis in vitro by human growth hormone is mediated by insulin-like growth factor I. Br J Haematol 1988; 70(3):267-271.
13. Merchav S, Tatarsky I, Hochberg Z. Enhancement of human granulopoiesis in vitro by biosynthetic insulin-like growth factor I/somatomedin C and human growth hormone. J Clin Invest 1988; 81(3):791-797.
14. Muta K, Krantz SB, Bondurant MC et al. Distinct roles of erythropoietin, insulin-like growth factor I, and stem cell factor in the development of erythroid progenitor cells. J Clin Invest 1994; 94(1):34-43.
15. Jardieu P, Clark R, Mortensen D et al. In vivo administration of insulin-like growth factor-I stimulates primary B lymphopoiesis and enhances lymphocyte recovery after bone marrow transplantation. J Immunol 1994; 152(9):4320-4327.
16. Rodriguez-Tarduchy G, Collins MK, Garcia I et al. Insulin-like growth factor-I inhibits apoptosis in IL-3-dependent hemopoietic cells. J Immunol 1992; 149(2):535-540.
17. Landreth KS, Narayanan R, Dorshkind K. Insulin-like growth factor-I regulates pro-B cell differentiation. Blood 1992; 80(5):1207-1212.
18. Quaife CJ, Mathews LS, Pinkert CA et al. Histopathology associated with elevated levels of growth hormone and insulin-like growth factor I in transgenic mice. Endocrinology 1989; 124(1):40-48.

19. Mathews LS, Hammer RE, Behringer RR et al. Growth enhancement of transgenic mice expressing human insulin-like growth factor I. Endocrinology 1988; 123(6):2827-2833.
20. Clark R, Strasser J, McCabe S et al. Insulin-like growth factor-1 stimulation of lymphopoiesis. J Clin Invest 1993; 92(2):540-548.
21. Binz K, Joller P, Froesch P et al. Repopulation of the atrophied thymus in diabetic rats by insulin-like growth factor I. Proc Natl Acad Sci USA 1990; 87(10):3690-3694.
22. Beschorner WE, Divic J, Pulido H et al. Enhancement of thymic recovery after cyclosporine by recombinant human growth hormone and insulin-like growth factor I. Transplantation 1991; 52(5):879-884.
23. Hinton PS, Peterson CA, Dahly EM et al. IGF-I alters lymphocyte survival and regeneration in thymus and spleen after dexamethasone treatment. Am J Physiol 1998; 274(4 Pt 2):R912-920.
24. van Buul-Offers SC, de Haan K, Reijnen-Gresnigt MG et al. Overexpression of human insulin-like growth factor-II in transgenic mice causes increased growth of the thymus. J Endocrinol 1995; 144(3):491-502.
25. Kooijman R, van Buul-Offers SC, Scholtens LE et al. T cell development in insulin-like growth factor-II transgenic mice. J Immunol 1995; 154(11):5736-5745.
26. Johnson EW, Jones LA, Kozak RW. Expression and function of insulin-like growth factor receptors on anti-CD3-activated human T lymphocytes. J Immunol 1992; 148(1):63-71.
27. Walsh PT, O'Connor R. The insulin-like growth factor-I receptor is regulated by CD28 and protects activated T cells from apoptosis. Eur J Immunol 2000; 30(4):1010-1018.
28. Tapson VF, Boni-Schnetzler M, Pilch PF et al. Structural and functional characterization of the human T lymphocyte receptor for insulin-like growth factor I in vitro. J Clin Invest 1988; 82(3):950-957.
29. Kimata H, Yoshida A. Differential effect of growth hormone and insulin-like growth factor-I, insulin-like growth factor-II, and insulin on Ig production and growth in human plasma cells. Blood 1994; 83(6):1569-1574.
30. Kimata H, Fujimoto M. Growth hormone and insulin-like growth factor I induce immunoglobulin (Ig)E and IgG4 production by human B cells. J Exp Med 1994; 180(2):727-732.
31. Robbins K, McCabe S, Scheiner T et al. Immunological effects of insulin-like growth factor-I—Enhancement of immunoglobulin synthesis. Clin Exp Immunol 1994; 95(2):337-342.
32. Yoshida A, Ishioka C, Kimata H et al. Recombinant human growth hormone stimulates B cell immunoglobulin synthesis and proliferation in serum-free medium. Acta Endocrinol (Copenh) 1992; 126(6):524-529.
33. Jelinek DF, Witzig TE, Arendt BK. A role for insulin-like growth factor in the regulation of IL-6-responsive human myeloma cell line growth. J Immunol 1997; 159(1):487-496.
34. Jones JI, Clemmons C. Insulin-like growth factors and their binding proteins: Biological actions. Endocrine Rev 1995; 16(1):3-34.
35. Li YM, Arkins S, McCusker RH Jr et al. Macrophages synthesize and secrete a 25-kilodalton protein that binds insulin-like growth factor-I. J Immunol 1996; 156(1):64-72.
36. Jones JI, Doerr ME, Clemmons DR. Cell migration: Interactions among integrins, IGFs and IGFBPs. Prog Growth Factor Res 1995; 6(2-4):319-327.
37. Nyman T, Pekonen F. The expression of insulin-like growth factors and their binding proteins in normal human lymphocytes. Acta Endocrinologica 1993; 128(2):168-172.
38. Foll JL, Dannecker L, Zehrer C et al. Activation-dependent expression of the insulin-like growth factor binding protein-2 in human lymphocytes. Immunology 1998; 94(2):173-180.
39. Elmlinger MW, Wimmer K, Biemer E et al. Insulin-like growth factor binding protein 2 is differentially expressed in leukaemic B- and T-cell lines. Growth Reg 1996; 6(3):152-157.
40. Hosokawa Y, Maeda Y, Seto M. Target genes downregulated by the BCL-6/LAZ3 oncoprotein in mouse Ba/F3 cells. Biochem Biophys Res Comm 2001; 283(3):563-568.
41. LeRoith D, Werner H, Beitner-Johnson D et al. Molecular and cellular aspects of the insulin-like growth factor I receptor. Endocr Rev 1995; 16(2):143-163.
42. Rubini M, Hongo A, D'Ambrosio C et al. The IGF-I receptor in mitogenesis and transformation of mouse embryo cells: Role of receptor number. Exp Cell Res 1997; 230(2):284-292.
43. Neuenschwander S, Roberts CT, Jr., LeRoith D. Growth inhibition of MCF-7 breast cancer cells by stable expression of an insulin-like growth factor I receptor antisense ribonucleic acid. Endocrinology 1995; 136(10):4298-4303.
44. Kooijman R, Willems M, De Haas CJ et al. Expression of type I insulin-like growth factor receptors on human peripheral blood mononuclear cells. Endocrinology 1992; 131(5):2244-2250.
45. Lee PD, Rosenfeld RG, Hintz RL et al. Characterization of insulin, insulin-like growth factors I and II, and growth hormone receptors on human leukemic lymphoblasts. J Clin Endocrinol Metab 1986; 62(1):28-35.
46. Kozak RW, Haskell JF, Greenstein LA et al. Type I and II insulin-like growth factor receptors on human phytohemagglutinin-activated T lymphocytes. Cell Immunol 1987; 109(2):318-331.
47. Baserga R, Hongo A, Rubini M et al. The IGF-I receptor in cell growth, transformation and apoptosis. Biochimica et Biophysica Acta 1997; 1332(3):F105-126.
48. Sussenbach JS, Steenbergh PH, Holthuizen P. Structure and expression of the human insulin-like growth factor genes. Growth Regul 1992; 2(1):1-9.
49. Cooke DW, Bankert LA, Roberts CT, Jr. et al. Analysis of the human type I insulin-like growth factor receptor promoter region. Biochem Biophys Res Commun 1991; 177(3):1113-1120.
50. Beitner-Johnson D, Werner H, Roberts CT, Jr. et al. Regulation of insulin-like growth factor I receptor gene expression by Sp1: physical and functional interactions of Sp1 at GC boxes and at a CT element. Mol Endocrinol 1995; 9(9):1147-1156.
51. Werner H, Roberts CT, Jr., LeRoith D. The regulation of IGF-I receptor gene expression by positive and negative zinc-finger transcription factors. Adv Exp Med Biol 1993; 343:91-103.
52. Saffer JD, Jackson SP, Annarella MB. Developmental expression of Sp1 in the mouse. Mol Cell Biol 1991; 11(4):2189-2199.

53. Dharmavaram RM, Liu G, Mowers SD et al. Detection and characterization of Sp1 binding activity in human chondrocytes and its alterations during chondrocyte dedifferentiation. J Biol Chem 1997; 272(43):26918-26925.
54. Too CK. Induction of Sp1 activity by prolactin and interleukin-2 in Nb-2 T-cells: Differential association of Sp1-DNA complexes with Stats. Molec Cell Endo 1997; 129:7-16.
55. Werner H, Karnieli E, Rauscher FJ et al. Wild-type and mutant p53 differentially regulate transcription of the insulin-like growth factor I receptor gene. Proc Natl Acad Sci USA 1996; 93(16):8318-8323.
56. Yu H, Rohan T. Role of the insulin-like growth factor family in cancer development and progression. J Natl Cancer Inst 2000; 92(18):1472-1489.
57. Zarrilli R, Romano M, Pignata S et al. Constitutive insulin-like growth factor-II expression interferes with the enterocyte-like differentiation of CaCo-2 cells. J Biol Chem 1996; 271(14):8108-8114.
58. Christofori G, Naik P, Hanahan D. A second signal supplied by insulin-like growth factor II in oncogene-induced tumorigenesis. Nature 1994; 369(6479):414-418.
59. Sell C, Baserga R, Rubin R. Insulin-like growth factor I (IGF-I) and the IGF-I receptor prevent etoposide-induced apoptosis. Cancer Res 1995; 55(2):303-306.
60. Xu F, Gardner A, Tu Y et al. Multiple myeloma cells are protected against dexamethasone-induced apoptosis by insulin-like growth factors. Br J Haematol 1997; 97(2):429-440.
61. Baserga R, Resnicoff M, Dews M. The IGF-I receptor and cancer. Endocrine 1997; 7(1):99-102.
62. Werner H, LeRoith D. The role of the insulin-like growth factor system in human cancer. Adv Cancer Res 1996; 68:183-223.
63. Upegui-Gonzalez LC, Francois JC, Ly A et al. The approach of triple helix formation in control of gene expression and the treatment of tumors expressing IGF-I. Adv Exp Med Biol 2000; 465:319-332.
64. Ellouk-Achard S, Djenabi S, De Oliveira GA et al. Induction of apoptosis in rat hepatocarcinoma cells by expression of IGF-I antisense c-DNA. J Hepatol 1998; 29(5):807-818.
65. Resnicoff M, Sell C, Rubini M et al. Rat glioblastoma cells expressing an antisense RNA to the insulin-like growth factor-1 (IGF-1) receptor are nontumorigenic and induce regression of wild-type tumors. Cancer Res 1994; 54(8):2218-2222.
66. Resnicoff M, Coppola D, Sell C et al. Growth inhibition of human melanoma cells in nude mice by antisense strategies to the type 1 insulin-like growth factor receptor. Cancer Res 1994; 54(18):4848-4850.
67. Pass HI, Mew DJ, Carbone M et al. The effect of an antisense expression plasmid to the IGF-1 receptor on hamster mesothelioma proliferation. Dev Biol Stand 1998; 94:321-328.
68. Pass HI, Mew DJ, Carbone M et al. Inhibition of hamster mesothelioma tumorigenesis by an antisense expression plasmid to the insulin-like growth factor-1 receptor. Cancer Res 1996; 56(17):4044-4048.
69. Arteaga CL, Kitten LJ, Coronado EB et al. Blockade of the type I somatomedin receptor inhibits growth of human breast cancer cells in athymic mice. J Clin Invest 1989; 84(5):1418-1423.
70. Kalebic T, Tsokos M, Helman LJ. In vivo treatment with antibody against IGF-1 receptor suppresses growth of human rhabdomyosarcoma and down-regulates p34cdc2. Cancer Res 1994; 54(21):5531-5534.
71. Prager D, Li HL, Asa S et al. Dominant negative inhibition of tumorigenesis in vivo by human insulin-like growth factor I receptor mutant. Proc Natl Acad Sci USA 1994; 91(6):2181-2185.
72. Kalebic T, Blakesley V, Slade C et al. Expression of a kinase-deficient IGF-I-R suppresses tumorigenicity of rhabdomyosarcoma cells constitutively expressing a wild type IGF-I-R. Int J Cancer 1998; 76(2):223-227.
73. D'Ambrosio C, Ferber A, Resnicoff M et al. A soluble insulin-like growth factor I receptor that induces apoptosis of tumor cells in vivo and inhibits tumorigenesis. Cancer Res 1996; 56(17):4013-4020.
74. Baier TG, Jenne EW, Blum W et al. Influence of antibodies against IGF-I, insulin or their receptors on proliferation of human acute lymphoblastic leukemia cell lines. Leuk Res 1992; 16(8):807-814.
75. Blanchard MM, Barenton B, Sullivan A et al. Characterization of the insulin-like growth factor (IGF) receptor in K562 erythroleukemia cells; evidence for a biological function for the type II IGF receptor. Mol Cell Endocrinol 1988; 56(3):235-244.
76. Estrov Z, Meir R, Barak Y et al. Human growth hormone and insulin-like growth factor-1 enhance the proliferation of human leukemic blasts. J Clin Oncol 1991; 9(3):394-399.
77. Pepe MG, Ginzton NH, Lee PD et al. Receptor binding and mitogenic effects of insulin and insulin like growth factors I and II for human myeloid leukemic cells. J Cell Physiol 1987; 133(2):219-227.
78. Sinclair J, McClain D, Taetle R. Effects of insulin and insulin-like growth factor I on growth of human leukemia cells in serum-free and protein-free medium. Blood 1988; 72(1):66-72.
79. Tally M, Tang XZ, Enberg G et al. The binding of insulin-like growth factor II to the erythroleukemia cell line K562. Biosci Rep 1984; 4(12):1071-1077.
80. Spadoni GL, Tally M, Florell K et al. Determination of IGF-II levels in human serum using the erythroleukemia cell line K562. J Endocrinol Invest 1990; 13(2):97-102.
81. Lal RB, Rudolph DL, Folks TM et al. Over expression of insulin-like growth factor receptor type-I in T-cell lines infected with human T-lymphotropic virus types-I and -II. Leuk Res 1993; 17(1):31-35.
82. Cross RJ, Elliott LH, Morford LA et al. Functional characterization of the insulin-like growth factor I receptor on Jurkat T cells. Cell Immunol 1995; 160(2):205-210.
83. Kellerer M, Obermaier-Kusser B, Ermel B et al. An altered IGF-I receptor is present in human leukemic cells. J Biol Chem 1990; 265(16):9340-9345.
84. Oksenberg D, Dieckmann BS, Greenberg PL. Functional interactions between colony-stimulating factors and the insulin family hormones for human myeloid leukemic cells. Cancer Res 1990; 50(20):6471-6477.
85. Gjerset RA, Yeargin J, Volkman SK et al. Insulin-like growth factor-I supports proliferation of autocrine thymic lymphoma cells with a pre-T cell phenotype. J Immunol 1990; 145(10):3497-3501.
86. Baier TG, Ludwig WD, Schonberg D et al. Characterisation of insulin-like growth factor I receptors of human acute lymphoblastic leukaemia (ALL) cell lines and primary ALL cells. Eur J Cancer 1992; 28A(6-7):1105-1110.

87. Foon KA, Rai KR, Gale RP. Chronic lymphocytic leukemia: New insights into biology and therapy. Annal Int Med 1990; 113(7):525-539.
88. Rozman C, Montserrat E. Chronic lymphocytic leukemia. N Engl J Med 1995; 333(16):1052-1057.
89. Garand R, Robillard N. Immunophenotypic characterization of acute leukemias and chronic lymphoproliferative disorders: Practical recommendations and classifications. Hematol Cell Ther 1996; 38(6):471-486.
90. Reed JC. Molecular biology of chronic lymphocytic leukemia. Semin Oncol 1998; 25(1):11-18.
91. Osorio LM, Aguilar-Santelises M. Apoptosis in B-chronic lymphocytic leukaemia. Med Oncol 1998; 15(4):234-240.
92. Freund GG, Kulas DT, Way BA et al. Functional insulin and insulin-like growth factor-1 receptors are preferentially expressed in multiple myeloma cell lines as compared to B-lymphoblastoid cell lines. Cancer Res 1994; 54(12):3179-3185.
93. Georgii-Hemming P, Wiklund HJ, Ljunggren O et al. Insulin-like growth factor I is a growth and survival factor in human multiple myeloma cell lines. Blood 1996; 88(6):2250-2258.
94. Freund GG, Kulas DT, Mooney RA. Insulin and IGF-1 increase mitogenesis and glucose metabolism in the multiple myeloma cell line, RPMI 8226. J Immunol 1993; 151(4):1811-1820.
95. Tu Y, Gardner A, Lichtenstein A. The phosphatidylinositol 3-kinase/AKT kinase pathway in multiple myeloma plasma cells: Roles in cytokine-dependent survival and proliferative responses. Cancer Res 2000; 60(23):6763-6770.
96. Hyun T, Yam A, Pece S et al. Loss of PTEN expression leading to high Akt activation in human multiple myelomas. Blood 2000; 96(10):3560-3568.
97. Ge NL, Rudikoff S. Insulin-like growth factor I is a dual effector of multiple myeloma cell growth. Blood 2000; 96(8):2856-2861.
98. Li W, Hyun T, Heller M et al. Activation of insulin-like growth factor I receptor signaling pathway is critical for mouse plasma cell tumor growth. Cancer Res 2000; 60(14):3909-3915.
99. Vanderkerken K, Asosingh K, Braet F et al. Insulin-like growth factor-1 acts as a chemoattractant factor for 5T2 multiple myeloma cells. Blood 1999; 93(1):235-241.
100. Asosingh K, Gunthert U, Bakkus MH et al. In vivo induction of insulin-like growth factor-I receptor and CD44v6 confers homing and adhesion to murine multiple myeloma cells. Cancer Res 2000; 60(11):3096-3104.
101. Arendt BK, Witzig TE, Ansell SM et al. Overexpression of the type I insulin-like growth factor receptor in multiple myeloma. Blood 1999; 94 (Suppl.1):634a.
102. Zhang X, Yee D. Tyrosine kinase signalling in breast cancer: Insulin-like growth factors and their receptors in breast cancer. Breast cancer Res 2000; 2(3):170-175.
103. Smith CL. Cross-talk between peptide growth factor and estrogen receptor signaling pathways. Biology of Reproduction 1998; 58(3):627-632.
104. Peyrat JP, Bonneterre J. Type 1 IGF receptor in human breast diseases. Breast cancer Res Treat 1992; 22(1):59-67.
105. Lee AV, Weng CN, Jackson JG et al. Activation of estrogen receptor-mediated gene transcription by IGF-I in human breast cancer cells. J Endocrinol 1997; 152(1):39-47.
106. Kato S, Endoh H, Masuhiro Y et al. Activation of the estrogen receptor through phosphorylation by mitogen-activated protein kinase. Science 1995; 270(5241):1491-1494.
107. Balana ME, Labriola L, Salatino M et al. Activation of ErbB-2 via a hierarchical interaction between ErbB-2 and type I insulin-like growth factor receptor in mammary tumor cells. Oncogene 2001; 20(1):34-47.
108. Plopper GE, McNamee HP, Dike LE et al. Convergence of integrin and growth factor receptor signaling pathways within the focal adhesion complex. Mol Biol Cell 1995; 6(10):1349-1365.
109. Schwartz MA, Schaller MD, Ginsberg MH. Integrins: Emerging paradigms of signal transduction. Annu Rev Cell Dev Biol 1995; 11:549-599.
110. Parise LV, Lee J, Juliano RL. New aspects of integrin signaling in cancer. Semin Cancer Biol 2000; 10(6):407-414.
111. Jones JI, Prevette T, Gockerman A et al. Ligand occupancy of the alpha-V-beta3 integrin is necessary for smooth muscle cells to migrate in response to insulin-like growth factor. Proc Natl Acad Sci USA 1996; 93(6):2482-2487.
112. Brooks PC, Klemke RL, Schon S et al. Insulin-like growth factor receptor cooperates with integrin alpha v beta 5 to promote tumor cell dissemination in vivo. J Clin Invest 1997; 99(6):1390-1398.
113. Doerr ME, Jones JI. The roles of integrins and extracellular matrix proteins in the insulin-like growth factor I-stimulated chemotaxis of human breast cancer cells. J Biol Chem 1996; 271(5):2443-2447.
114. Baron V, Calleja V, Ferrari P et al. p125Fak focal adhesion kinase is a substrate for the insulin and insulin-like growth factor-I tyrosine kinase receptors. J Biol Chem 1998; 273(12):7162-7168.
115. Lebrun P, Mothe-Satney I, Delahaye L et al. Insulin receptor substrate-1 as a signaling molecule for focal adhesion kinase pp125(FAK) and pp60(src). J Biol Chem 1998; 273(48):32244-32253.
116. Lebrun P, Baron V, Hauck CR et al. Cell adhesion and focal adhesion kinase regulate insulin receptor substrate-1 expression. J Biol Chem 2000; 275(49):38371-38377.
117. Soon L, Flechner L, Gutkind JS et al. Insulin-like growth factor I synergizes with interleukin 4 for hematopoietic cell proliferation independent of insulin receptor substrate expression. Mol Cell Biol 1999; 19(5):3816-3828.
118. Cook G, Dumbar M, Franklin IM. The role of adhesion molecules in multiple myeloma. Acta Haematol 1997; 97(1-2):81-89.
119. Faid L, Van Riet I, De Waele M et al. Adhesive interactions between tumor cells and bone marrow stromal elements in human multiple myeloma. Eur J Haematol 1996; 57(5):349-358.
120. Sanz-Rodriguez F, Teixido J. VLA-4-dependent myeloma cell adhesion. Leuk Lymphoma 2001; 41(3-4):239-245.
121. Michigami T, Shimizu N, Williams PJ et al. Cell-cell contact between marrow stromal cells and myeloma cells via VCAM-1 and alpha(4)beta(1)-integrin enhances production of osteoclast-stimulating activity. Blood 2000; 96(5):1953-1960.

CHAPTER 28

Metabolic Effects of Insulin-Like Growth Factor I and Growth Hormone in Vivo: A Comparison

Nelly Mauras

Abstract

The metabolic effects of IGF-I are varied and remarkably similar in many respects to those of GH. IGF-I mediates some, but not all of the metabolic actions of GH in man. Both GH and IGF-I potently stimulate whole body protein synthesis rates in healthy subjects, with minimal effects in proteolysis, however in GH deficient individuals, GH and IGF-I stimulate both protein synthesis and degradation with a net anabolic effect. Both hormones, when administered to patients with GH deficiency states result in measurable changes in body composition with increased lean body mass and decreased adiposity. This is also observed when IGF-I is given to patients with GH-receptor mutations. These compounds have very different effects on carbohydrate metabolism, however. There is a potent glucose lowering effect observed after IGF-I administration, with improved insulin sensitivity despite marked lowering of circulating insulin concentrations, whereas GH therapy is associated with mild compensatory hyperinsulinemia, a reflection of relative insulin resistance. The latter observation makes IGF-I a potentially more convenient anabolic agent to use in conditions where carbohydrate metabolism is more likely to be impaired. GH increases lipolysis as well as lipid oxidation probably as a direct effect of GH in the adipocyte, however IGF-I increases lipid oxidation only when given chronically, most likely as a result of chronic insulinopenia. Both hormones have been shown in vitro to be anabolic in bone, however to date only GH has been shown to improve bone mineralization in humans. Because of the potent anabolic actions of these compounds, they have been tried in a variety of catabolic conditions in man. GH and IGF-I are both effective in reducing the protein wasting effects of glucocorticosteroids and mitigate some of the catabolic effects of severe hypogonadism in males. Similar to the documented positive effects of GH as replacement treatment in adults, data in patients with GH receptor mutations treated with IGF-I and studied by us recently suggest that IGF-I is beneficial in normalizing protein synthesis rates, bone calcium accretion and as well as decreasing adiposity and increasing lean body mass. A comparison of these and other effects of these hormones is provided in this review. Many more studies are still needed to fully elucidate the safety and efficacy of IGF-I for use in humans.

Introduction

Growth hormone (GH) is the most potent growth promoting agent available and its primacy controlling post natal somatic growth is unquestionable. GH has, however, a multiplicity of important metabolic functions in vivo besides promoting linear growth, some of which are mediated through insulin-like growth factor I (IGF-I), whereas some are not. GH interacts with 2 GH receptors (GHR) at 2 different binding sites. The GH/GHR interaction activates a complex signaling transduction cascade of events resulting in the generation of IGF-I. Most of the circulating IGF-I in plasma is of hepatic origin, however, ample experimental data support the concept that the locally generated IGF-I at the target sites might be more important in terms of the control of growth for e.g., than that generated systemically. Experimental animals with selective knock out of the hepatic

Insulin-Like Growth Factors, edited by Derek LeRoith, Walter Zumkeller
and Robert Baxter. ©2003 Eurekah.com and Kluwer Academic / Plenum Publishers.

IGF-I gene show a 70% decrease in circulating IGF-I concentrations yet normal post natal growth, indicating that the IGF-I generated at the site of the growing cartilage is critical for promotion of linear growth.[1]

IGF-I is a 70 amino acid polypeptide with significant (50%) structural homology with proinsulin.[2] It has both GH-like and insulin-like effects depending on the doses used, mode of delivery and length of treatment. In physiologic concentrations IGF-I acts via its type-I IGF-I receptor which itself has significant homology with the insulin receptor. Even though insulin is 100 times more potent activating its own receptor than IGF-I, in supraphysiologic or pharmacological concentrations IGF-I can also activate the insulin receptor, the latter responsible in part for the known hypoglycemic effects of IGF-I (see below).[3] Overall, IGF-I has a 7.5% equipotency with insulin.[3]

There are substantial similarities and as many differences in the effects of GH and IGF-I in man, indicating that the mechanisms of the effects of GH are clearly not all mediated via IGF-I. I will review some of the metabolic effects of IGF-I and specifically compare them with those of GH in the areas of protein, carbohydrate, lipid metabolism and body composition.

GH/IGF-I: Effects on Protein Metabolism

GH causes a significant increase in nitrogen retention as measured by nitrogen balance studies. Using different isotope dilution methods we can better compartmentalize if the changes in nitrogen retention are due to an increase in protein synthesis, decrease in protein breakdown or both. Administration of recombinant human (rh)GH to healthy volunteers results in a selective increase in whole body protein synthesis rates, with no effects in proteolysis.[4] When given to GH deficient subjects, however, both protein synthesis and protein degradation increase with a net anabolic effect.[5] GH has been tried in a variety of other catabolic conditions in man in an attempt to improve anabolism, improve nutrition and expedite recovery from severe injury. GH has been shown to improve nitrogen balance in debilitating catabolic conditions such as burned patients, subjects on parenteral nutrition, trauma victims, and after major surgery.[6-8] Because of these well documented and apparently potent anabolic effects, GH in high doses was used in a placebo controlled randomized trial of patients in the intensive care unit (ICU) in an attempt to ascertain whether, by improving nutritional status, it could improve the morbidity and decrease mortality in this setting. Surprisingly, results of the trial showed that those patients administered GH during a severe catabolic illness in ICU had a higher mortality than those receiving placebo, causing substantial concern on the use of this hormone in that specific setting.[9] The mechanisms for these observations are unknown, but whether GH treatment in an overwhelmingly ill patient induces the release or potentiates the action of specific cytokines is currently under investigation. The serious nature of the results of that trial should prompt caution, and GH should be discontinued in acutely ill patients in the hospital who are taking GH chronically.

GH, however, is clearly a potent protein anabolic agent which is routinely and safely used in a variety of other more chronic illnesses such as inflammatory bowel disease, cystic fibrosis or chronic renal failure for e.g.,[10] GH is also an FDA- approved indication for the treatment of AIDS wasting and to promote linear growth in children with chronic renal insufficiency, conditions which are severe and chronic.[11-12]

IGF-I, on the other hand, has very comparable effects as GH enhancing protein anabolism. When given to healthy volunteers in low doses, it selectively stimulates protein synthesis, with no effects on proteolysis.[13] However, in high doses, IGF-I suppresses proteolytic rates, effects indistinguishable to those of insulin.[14] In normally fed individuals, IGF-I, when used in combination with GH, does not show more potent protein anabolic effects than when each compound is given separately,[15] whereas in relative caloric deprivation, the co-administration of GH and IGF-I appears to be synergistic in enhancing a more positive protein balance.[16] Both GH and IGF-I have been used in severe GH deficient adults and found to have comparable effects enhancing protein synthesis;[5,17] this was also observed in post menopausal women treated with either GH or IGF-I for 1 month.[18] Both of these hormones have been tried as protein anabolic agents in a variety of experimental situations in man with very comparable results (see below under *GH/IGF-I: Comparison of Anabolic Effects*). Taken in aggregate, the available data suggest that IGF-I mediates the protein-anabolic actions of GH in man.

GH/IGF-I: Effects on Carbohydrate Metabolism

In infants, GH appears to be pivotal for the maintenance of normal glucose homeostasis and hepatic glucose production. Profound and persistent hypoglycemia is a common occurrence in neonatal GH deficiency and a major cause of morbidity. However, this critical role of GH in glucose homeostasis is markedly diminished in older children and adults, a transition which is poorly understood. Chronic administration of GH results in compensatory hyperinsulinemia and GH treatment is associated with the development of insulin resistance. The latter, however, is typically not associated with clinically significant disease and the majority of patients on GH do not have carbohydrate intolerance.[19] GH deficient adults given GH demonstrate a decrease in carbohydrate oxidation rates as measured by indirect calorimetry and an increase in glucose production rates as measured by stable isotope methods, both secondary to increased insulin resistance[5]. Again, neither of these findings typically results in any changes in glucose tolerance. In none of the pediatric series that monitor side effects of GH therapy is GH associated with any increase in the incidence of diabetes. Caution, however, should be exercised when using GH chronically in subjects in whom other risk factors for carbohydrate intolerance or diabetes are present such as the elderly, the obese, or those on glucocorticosteroids for example.

The effects of GH on carbohydrate metabolism are not mediated via IGF-I. IGF-I has insulin-like effects and its administration typically results in hypoglycemia despite the concomitant and potent suppression of circulating insulin concentrations.[13-14] Even though this effect may well be mediated through the insulin receptor, experimental mice knockouts of the insulin receptor gene show a potent glucose lowering effect of IGF-I, indicative that the hypoglycemic effect of IGF-I is mediated in part also through its own type I IGF-I receptor.[20] Studies on experimental mice knockouts for the liver IGF-I gene show a marked compensatory increase in insulin concentrations and a substantial decrease in the insulin-induced autophosphorylation of the insulin receptor and insulin receptor substrate in skeletal muscle, effects corrected by the administration of exogenous IGF-I.[21] These data indicate that hepatic-derived IGF-I is critical for the overall action of insulin in skeletal muscle.[21]

When IGF-I is administered to GH deficient subjects there is no change in carbohydrate oxidation rates, contrary to the decrease observed after GH treatment.[5] When given for 8w to patients with Laron's syndrome we observed a decrease in carbohydrate oxidation rates and an increase in hepatic glucose production rates, both of which probably reflect the insulinopenia caused at the portal level[22]. However, this is despite maintenance of normoglycemia, indicative of overall improvement in insulin sensitivity (see below under *GH/IGF-I: Comparison of Anabolic Effects in Man*). In general, IGF-I administration results in improved insulin sensitivity, the latter observation leading to the study of IGF-I as a potential adjunct in the treatment of diabetic patients.[23] Long term trials, most presently on hold because of the paucity of available IGF-I for human use, will be needed to better define the benefit of this hormone improving carbohydrate metabolism in selected patients. IGF-I offers clear theoretical advantages over GH treatment because of its insulin sensitizing effects.

GH/IGF-I Effects on Lipid Metabolism and Body Composition

GH receptors are expressed in human adipocytes and GH has been shown to have significant effects on fat metabolism.[24] The positive effects on lipids and deceased adiposity observed after GH treatment are some of the major benefits associated with GH replacement therapy in the adults.[25] GH therapy in GH deficient adults is associated with an improved lipolytic response of isolated adipocytes to epinephrine via an increased efficiency of the β adrenergic pathway and greater receptor number.[26-27] This increased sensitivity to the lipolytic effect of catecholamines is also observed in adipocytes of healthy adults treated with GH in culture.[28] The beneficial lipid profile observed after GH therapy, i.e., decreased total cholesterol, LDL cholesterol and Apo B100 concentrations (the main protein contents of LDL), is probably related in part to the induction of hepatic LDL receptors.[29] GH has been shown to have species and tissue specific effects on lipoprotein lipase (LPL) activity, the key enzyme in the regulation of the flux of free fatty acids (FFA), expressed both in the adipocyte and skeletal muscle. In hypophysectomized (hypox) rats, GH had no significant effect on LPL activity in adipose tissue, whereas in human adipocytes, GH has been shown to *decrease* LPL activity in GH deficient adults and in obese women, hence diminishing the flow of FFA to the

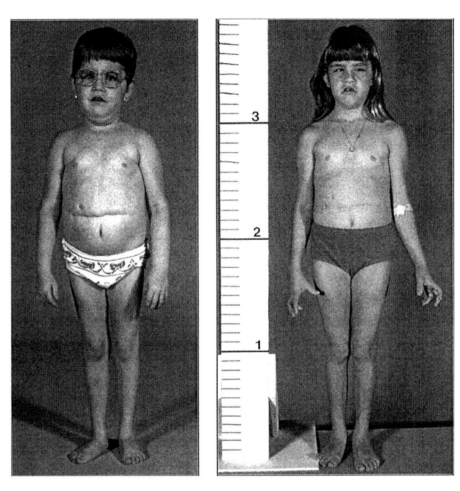

Figure 1. Characteristic features of a child with severe GH deficiency secondary to septo-optic dysplasia, left panel shows the patient at 6.9 years, right panel shows her 24 months after initiation of GH therapy.

adipocyte (antilipogenic).[26-27] This effect seems to be mostly post transcriptional as mRNA expression for the enzyme is unaffected by GH. In skeletal muscle of hypox rats however, GH *increases* LPL activity, indicating that the increased use of fatty acids in skeletal muscle tissue after GH therapy is not a result of increased availability of circulating FFAs.[30]

In children, GH therapy is well known to be associated with substantial changes in redistribution of body fat, from abdominal (android) to more peripheral (gynoid) pattern. When given to children with GH deficiency, the classical pudgy appearance on the hypopituitary child changes with a decrease in adiposity, and at times a remarkable "thinning out" of their physique (Fig.1). Utilizing gluteal and abdominal SC tissue from children treated with GH for three months, Rosenbaum et al found a marked decrease in abdominal adipocyte size, a reduced rate of lipogenesis and variable desensitization of the abdominal depots to the antilipolytic effects of insulin.[31] In obese children with Prader Willi syndrome treated with GH the increased growth velocity observed during treatment was accompanied by measurable reductions in percent fat mass, hence decreased adiposity.[32] GH administration, however, typically does not cause weight loss but rather a change in body composition, hence careful discussion with patients needs to include a realistic expectation of the anticipated results of treatment, particularly in adults.

The effects of GH on lipids/lipolysis are not IGF-I mediated as there are no functional type one IGF-I receptors in adipocytes.[24] IGF-I, both in vitro and in vivo, has been shown to have either no effect on lipolysis nor LPL activity and to have dual effects on lipid oxidation, decreasing oxidation rates after seven days of treatment,[17] and increasing lipid oxidation rates after eight weeks of similar treatment in GHD adults.[5] We, however observed substantial changes in body composition, with decreased adiposity and increased lean body mass, as well as increased lipolysis and lipid oxidation in a unique group of patients with GH receptor deficiency treated with IGF-I for eight weeks.[22] On the other hand, one year of treatment with IGF-I was not associated with any measurable improvement in lean body mass or decrease in adiposity in a group of healthy postmenopausal women reported recently.[33] Whether chronic, long term IGF-I therapy will result in sustained changes in adiposity will require further study, however, it is possible that the potent effects in decreasing adiposity and increasing lean body mass observed in patients with Laron's syndrome are a reflection of the severity and congenital nature of their deficiency and may not be replicated in normal, albeit elderly, subjects. This is similar to what is observed in GH deficient patients, whom are the most likely to show benefit from long term treatment as compared to normal elderly subjects.[25]

GH/IGF-I Effects on Bone

GH is clearly the principal regulator of epiphyseal growth and primarily responsible for normal linear growth. It stimulates cartilage uptake of proteoglycans and elongates the skeleton. Previous studies on the effects of GH on bone mineralization have yielded conflicting results, some showing no effects of GH on bone mass after six months of GH therapy, and others have shown an actual decrease in bone mineral density after treatment. Histomorphometric analysis of transiliac bone biopsies in GH deficient patients treated with GH for 1 year showed that only cortical thickness, not trabecular bone, increases after GH therapy. GH increases bone formation in GH-deficient states only after prolonged treatment (>18 months).[34] GH treatment is clearly important in maintaining the long term mineralization of the skeleton and profound deficiency is associated with osteopenia. IGF-I, particularly that locally generated, appears to mediate many of the anabolic actions of GH in bone.

IGF-I has been shown to have potent bone anabolic actions both in vitro and in vivo in the short term, stimulating collagen synthesis as well as increasing the proliferation of osteoblast precursors. In short term (6-7 days) experiments the administration of IGF-I to normal women fasted for 10 days, to women with anorexia nervosa, or to healthy young adult volunteers was associated with increased serum markers of bone formation.[35] However, recent data in postmenopausal women treated with IGF-I showed no changes in bone density after one year of treatment.[33] In addition, patients with GH receptor mutations and congenital IGF-I deficiency do not show any evidence of osteopenia or osteoporosis once their bone mineral density is corrected for vertebral size, as these patients are profoundly short.[36] Evidence thus far does not support a role for IGF-I in osteoporosis treatment but, similar to the use of GH chronically, much longer term studies maybe needed to demonstrate any potential beneficial effect.

GH/IGF-I: Use in Adult Replacement

Many of the salient features of the GH deficiency syndrome in adults can be improved with GH therapy.[25, 37] GH deficient adults have increased central adiposity, decreased lean body mass, decreased bone mineral density and an adverse plasma lipid profile. Even within six months of treatment GH increases lean body mass and decreases adiposity. Prolonged (>18 months) therapy also increases bone mineral density. Exercise capacity and skeletal muscle strength have also been shown to improve in GH deficient adults treated with GH. In addition, quality of life measures, including energy level, mood, sensitivity to pain and emotional lability can improve on GH replacement therapy. The data on the effects of GH on plasma lipids is rather mixed with some studies showing a lowering of LDL cholesterol concentrations and overall improvement of the lipid profile, while many others show no effect. GH has also been shown to improve the body composition of patients with AIDS wasting syndrome. There are, to date, no large studies examining the role of GH on muscle strength in GH sufficient states and data thus far do not support a role for GH as an ergogenic agent. Even though data available to date are reassuring, long term follow up of patients

Figure 2. A group of 10 young adult subjects with Laron's syndrome. All had an alternative splice mutation of codon 180, exon 6 of the GH receptor gene whom participated in the study in reference #22.

treated for many years are still needed in other to assess the long term safety of this hormone replacement in adults.

We have recently studied the specific protein anabolic effects of IGF-I in a human IGF-I gene knock out model, i.e., patients with GH receptor mutations who are incapable of generating IGF-I despite adequate GH output (Laron's syndrome) (Fig. 2). These were 10 patients from Ecuador (mean age 29 ± 2 years), cared for by Dr. Guevara Aguirre whom flew twice to the Clinical Research Center in Jacksonville for a series of complex metabolic studies using stable isotope tracers of leucine, glucose, glycerol, as well as indirect calorimetry and body composition analysis. They were studied before and after 8w of twice daily subcutaneous (SC) IGF-I.[22] Their mean IGF-I concentrations before treatment were 9 ± 2 ng/ml. In these patients IGF-I potently decreased the oxidation of protein, stimulated rates of protein synthesis and degradation, similar to the effects of GH in GH deficient adults.[5] In addition, using isotopic infusions of glycerol as an estimate of rates of whole body lipolysis and indirect calorimetry to measure substrate oxidation rates, we observed a significant increase in lipolytic rates as well as in fat oxidation after IGF-I therapy in these subjects. Body composition changed with a decrease in the percent fat mass and an increase in lean body mass even after only eight weeks of administration.[22] These results suggest that, similar to GH replacement of the GH deficient adult, IGF-I may be beneficial as a long term replacement of the rare GH receptor deficient individual.

GH/IGF-I: Comparison of Anabolic Effects in Man

A series of selected studies are discussed below which offer a direct comparison of the effects of both GH and IGF-I used in similar experimental situations in man.

GH/IGF-I in Glucocorticosteroid-Treated Subjects

Chronic glucocorticosteroid therapy can have a myriad of undesirable side effects which can greatly decrease the use of these potent anti inflammatory agents. Protein wasting, poor tissue healing, increased incidence of infections and accelerated bone loss for e.g., may all be observed after chronic steroid use.[10] When administered to growing children these effects of the steroids can be compounded further by a potent and significant suppression of linear growth. Since GH has been shown to have significant protein anabolic, bone protective and linear growth effects, studies have been performed to see if GH can diminish the protein-wasting effects of steroids in man. To eliminate the confounding variables of other medications and the underlying illness, Horber & Haymond, over a decade ago performed sophisticated metabolic studies in healthy volunteers given high doses of steroids for seven days in order to mimic the catabolic state of chronic steroid use.[4] Using stable and radioactive tracers of the essential amino acid leucine they showed that GH can essentially abolish the protein wasting effects of steroids in man. However, the combination of steroids and GH

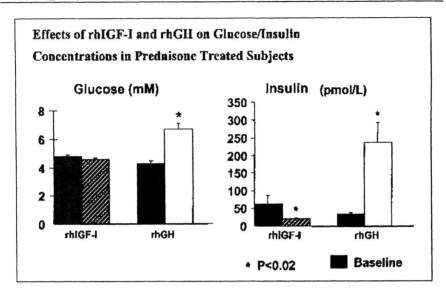

Figure 3. Changes in insulin and glucose concentrations in healthy young adult volunteers whom were treated with glucocorticosteroids and with either recombinant human (rh)GH or rhIGF-I for one week. Their data are reported in references #13 & 38.

substantially increased circulating insulin concentrations and plasma glucose concentrations were also increased, even though still within the normal range, indicating insulin resistance.[38]

Since IGF-I has similar protein-anabolic effects as GH, yet it improves insulin sensitivity, we repeated these experiments using IGF-I and found similar protein-anabolic effects in the glucocorticosteroid-treated subjects.[13] However, IGF-I treatment in these subjects was associated with normal glucose concentrations despite marked reductions in circulating insulin, indicative of the potent insulin-like activity of the peptide.[13] This offers at least a theoretical advantage over GH when long term treatment of a steroid-treated subject is considered (Fig. 3). Recently, we published the results of a 4 month trial in 10 children with profound growth retardation secondary to inflammatory bowel disease treated with GH.[39] We observed a significant improvement in their linear growth, decreased adiposity and improved lean body mass as well as increased measures of bone calcium accretion (using stable isotopes of calcium) and in markers of bone formation. Collectively, these and other data support the concept that both GH and IGF-I may play a role ameliorating the catabolic effects of steroids in man and that in children, at least GH offers hope overcoming the profound linear growth deficits that steroids can cause. Longer term studies using GH and IGF-I are still needed to determine the efficacy and safety of this approach.

GH/IGF-I in GH Deficiency

A small number of studies have directly compared the effects of GH vs. IGF-I in a variety of metabolic pathways in man. Recently, we studied a group of young, GH deficient adults using stable tracers of leucine, glucose and calcium as well as indirect calorimetry and body composition analysis before and after four weeks of daily SC GH, followed by daily IGF-I for four weeks, each subject as his/her own control, treated in random order. We observed that GH and IGF-I share common effects on protein, muscle and calcium metabolism yet divergent effects on lipid and carbohydrate metabolism in the GH deficient state.[5] These findings are similar to those reported by Hussain, et al after much shorter treatment of similar subjects for seven days.[17] Interestingly, a sub study of the pharmacokinetics of IGF-I in GH deficient adults showed that IGF-I is normally absorbed and distributed in a state of GH deficiency but it has much faster elimination kinetics than in normal subjects.[40] This is mostly secondary to the effect of the IGF-I binding proteins modulating the bioavailability of IGF-I in vivo.

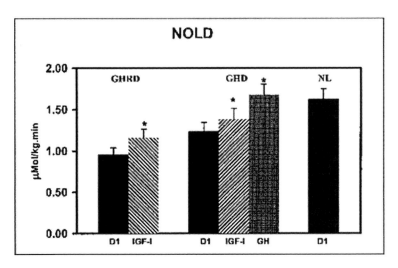

Figure 4. Changes in whole body protein synthesis as measured by non oxidative leucine disposal (NOLD) and in a group of young adult subjects with GH receptor deficiency (GHRD) treated with IGF-I for 8w, a group of GH deficient (GHD) young adults treated with both IGF-I and GH, and a similar group of normal controls. D1 represents baseline data. Reproduced with permission from J Clin Endocrinol Metab 85:3036, 2000.

We also compared the effect of GH and IGF-I treatment in GH deficient subjects with that of IGF-I treatment in GH receptor deficient individuals and observed that, as compared to controls, GH has the most potent effects normalizing rates of whole body protein synthesis.[22] IGF-I is nonetheless effective in both the GH- and GH receptor-deficient individuals enhancing whole body protein synthesis, strongly supportive of the potent protein-anabolic role of both of these hormones (Fig. 4).

GH/IGF-I in Hypogonadal Men

Another model of a severe catabolic state is that caused by profound hypogonadism in men. We have previously shown that when the hypogonadal state is induced by the administration of a GnRH analogue (GnRHa) to healthy young men, there is a significant decrease in protein synthesis rates, a decrease in muscle strength, a decrease in lean body mass an increase in adiposity, despite the maintenance of normal GH pulsatility and IGF-I production after 10 weeks of treatment.[41] We hence designed subsequent studies to ascertain if GH or IGF-I could diminish or abolish the catabolic effects of the hypogonadal state. Using similar metabolic tools as described above we studied a group of healthy young men at baseline, administered the GnRHa twice over six weeks, then for another four weeks added either daily SC GH or twice daily IGF-I and the studies repeated, each subject as his own control 10 weeks from baseline.[42] We observed similar effects of both hormones on whole body metabolism, with maintenance of protein synthesis and fat oxidation rates as well as preservation of fat free mass compared with the eugonadal state, preventing the decline observed during hypogonadism alone (Fig. 5).[42] This was amplified further by molecular assessment of important genes involved in normal muscle function using percutaneous skeletal muscle biopsies. After IGF-I treatment there was a decrease in IGF-I gene expression in skeletal muscle, similar to that observed during GnRHa alone, yet after GH treatment there was a marked increase in intramuscular IGF-I expression as well as in androgen receptor expression. Hence it appears that both GH and IGF-I may be beneficial maintaining lean body mass and protein balance during the hypogonadal state, even though systemic GH may be more potent than IGF-I affecting the skeletal muscle IGF-I system. These hormones may potentially play a role in the treatment of the catabolic effects of hypogonadism in man whom are unable to be treated with androgen therapy. The data also suggest that androgens are necessary for the full anabolic effect of these hormones to be observed in man.[42]

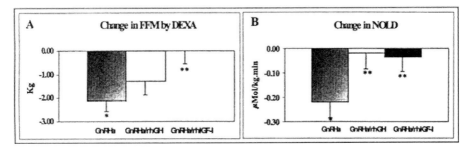

Figure 5. Comparison of the absolute changes in fat free mass (FFM), as measured by dual emission X ray absorptiometry, and whole body protein synthesis rates, as measured by non oxidative leucine disposal (NOLD) in healthy young men rendered hypogonadal with a GnRHa for 10w, as compared to those treated with GnRHa and rhIGF-I or GnRHa and rhGH.
* Represents significant differences within each group compared to baseline (p<0.01), ** represents the difference between a group and the GnRHa alone group (p<0.04). Reproduced with permission from J Clin Endocrinol Metab 86:2211, 2001

In Summary

GH and IGF-I have common metabolic effects and many of the metabolic actions of GH are indeed mediated by IGF-I, yet many are not. Both hormones are potent anabolic hormones in humans. Understanding their mechanism of action gives the clinician a logical framework to replace their patients and design adequate clinical studies to answer many of the questions regarding their use.

Acknowledgements

The author is grateful to Annie Rini and Susan Welch for the conduct of many of these studies to Brenda Sager and the Biomedical Analysis Laboratory for technical assistance, and to Morey Haymond, Randall J. Urban and Jaime Guevara-Aguirre for their collaboration. Supported by RO1 DK 51360 (NM), the Nemours Research Programs and GCRC RR 0585 (Mayo Clinic, Rochester, MN).

Table 1. Comparison of metabolic effects of GH and IFG-I in man

Effect	GH	IGF-I	Comments
Protein Synthesis	↑	↑	
Body Composition (FFM / % FM)	↑↓	↑↓	These effects are mostly evident in GH or GH receptor deficient states respectively.
Insulin Sensitivity	↓	↑	IGF-I affects glucose metabolism via both the insulin and the type I IGF-I receptor.
Lipolysis/Lipid Oxidation	↑	↓↑	IGF-I's effects are time dependent. Acutely, it suppresses lipid oxidation, chronically it increases it (mostly via insulin suppression).
Bone density	↑	No Effect after 1yr of Rx.	GH's effects are only observed with long term treatment.

References

1. Butler AA, LeRoith D. Mini review: Tissue-Specific Versus Generalized Gene Targeting of the IGF1 and IGF1r Genes and Their Roles in Insulin-Like Growth Factor Physiology. Endocrinology 2001; 142(5):1685-8.
2. Rinderknecht E, Humbel RE. The amino acid sequence of human IGF-I and its structural homology with proinsulin. J Biol Chem 1978; 253:2769-2776.
3. Werner H, Woloschak M, Stannard B et al. The insulin-like growth factor I receptor: molecular biology, heterogeneity, and regulation. In: LeRoith D, editor. Insulin-Like Growth Factors: Molecular & Cellular Aspects. Boca Raton: CRC Press, 1991:17-47.
4. Horber FF, Haymond MW. Human growth hormone prevents the protein catabolic side effects of prednisone in humans. J Clin Invest 1990; 86:265-272.
5. Mauras N, O'Brien KO, Welch S et al. IGF-I and GH treatment in GH deficient humans: differential effects on protein, glucose, lipid and calcium metabolism. J Clin Endocrinol Metab. 2000; 85:1686-1694
6. Lal SO, Wolf SE, Herndon DN. Growth hormone, burns and tissue healing. Growth Horm IGF Res 2000 Apr; 10 Suppl B:S39-43
7. Wilmore DW. The use of growth hormone in severely ill patients. Adv Surg 1999;33:261-74.
8. Hammarqvist F, Sandgren A, Andersson K et al. Growth hormone together with glutamine-containing total parenteral nutrition maintains muscle glutamine levels and results in a less negative nitrogen balance after surgical trauma. Surgery. 2001;129:576-586.
9. Takala J, Ruokonen E, Webster NR et al. Increased mortality associated with growth hormone treatment in critically ill adults. N Engl J Med. 1999; 9:341:785-792.
10. Mauras N. GH Therapy in the Glucocorticosteroid-dependent child: Metabolic and linear growth effects. Hormone Research 2001; 56 Suppl 1:13-18.
11. Mulligan K, Tai VW, Schambelan M. Use of growth hormone and other anabolic agents in AIDS wasting. J Parenter Enteral Nutr 1999; 23:S202-S209.
12. Haffner D, Schaefer F, Nissel R et al. Effect of growth hormone treatment on the adult height of children with chronic renal failure. German Study Group for Growth hormone Treatment in Chronic renal failure. N Engl J Med 2000; 343:923-930.
13. Mauras N, Beaufrere B. rhIGF-I enhances whole body protein anabolism and significantly diminishes the protein-catabolic effects of prednisone in humans, without a diabetogenic effect. J Clin Endocrinol Metab 1995; 80:869-874.
14. Turkalj I, Keller U, Ninnis R et al. Effect of increasing doses of recombinant human insulin-like growth factor-I on glucose, lipid, and leucine metabolism in man. J Clin Endocrinol Metab 1992; 75:1186-1191.
15. Mauras N. Combined recombinant human growth hormone and recombinant human insulin-like growth factor I: lack of synergy on whole body protein anabolism in normally fed subjects. J Clin Endocrinol Metab 1995; 80:2633-2637.
16. Kupfer SR, Underwood LE, Baxter RC et al. Enhancement of the anabolic effects of growth hormone and insulin-like growth factor I by use of both agents simultaneously. J Clin Invest 1993; 91:391-396.
17. Hussain MA, Schmitz O, Mengel A et al. Comparison of the effects of GH and IGF-I on substrate oxidation and on insulin sensitivity in GH-deficient humans. J Clin Invest 1994; 94:1126-1133.
18. Butterfield GE, Thompson J, Rennie MJ et al. Effect of rhGH and rhIGF-I treatment on protein utilization in elderly women. Am J Physiol 272 (Endocrinol Metab 35) 1997;E94-E99.
19. Saenger P, Attie KM, Dimartino-Nardi J et al. Metabolic consequences of 5-year growth hormone (GH) therapy in children treated with GH for idiopathic short stature. Genentech Collaborative Study Group . J Clin Endocrinol Metab 1998; 83:3115-3120.
20. Di Cola G, Cool MH, Accili D. Hypoglycemic effect of insulin-like growth factor-1 in mice lacking insulin receptors. J Clin Invest 1997; 99:2538-2544.
21. Yakar S, Liu JL, Fernandez AM et al. Liver specific igf-1 gene deletion leads to muscle insulin insensitivity. Diabetes 2001;50:1110-1118.
22. Mauras N, Martinez V, Rini A et al. Recombinant human IGF-I has significant anabolic effects in adults with GH receptor deficiency: studies on protein, glucose and lipid metabolism. J Clin Endocrinol & Metab 2000; 85:3036-3042.
23. Randomised placebo-controlled trial of human recombinant insulin-like growth factor I plus intensive insulin therapy in adolescents with insulin-dependent diabetes mellitus. Lancet 1997;35:1199-204.
24. DiGirolamo M, Eden S, Enberg G et al. Specific binding of human growth hormone but not insulin-like growth factors by human adipocytes. FEBS Lett 1986;205:15-19.
25. Vance ML, Mauras N. Growth hormone therapy in adults and children. N Eng J Med 1999; 341:1206-1216.
26. Beauville M, Harant I, Crampes F et al. Effect of long term rhGH administration in GH deficient adults on fat cell EPI response. Am J Physiol:Endocrinol Metab 1992;263:E467-E472.
27. Yang S, Bjorntorp P, Liu X et al. GH treatment of hypophysectomized rats increases catecholamine-induced lipolysis and the number of b adrenergic receptors in adipocytes: No differences in the effects of GH on different fat depots. Obes Res 1996;4:471-478.
28. Marcus C, Bolme P, Micha-Johansson G et al. GH increases the lipolytic sensitivity for catecholamines in adipocytes from healthy adults. Life Sci 1994;54:1335-41.
29. Rudling M, Norstedt G, Olivecrona H et al. Importance of GH for the induction of hepatic low density lipoprotein receptors. Proc Natl Acad Sci 1992;89:6983-6987.

30. Oscarsson J, Ottosson M, Vikman-Adolfsson K et al. GH but not IGF-I or insulin increases LPL activity in muscle tissues of hypophysectomized rats. J Endocrinol 1999;60:247-255.
31. Rosenbaum M, Gertner JM, Leibel RL. Effects of systemic growth hormone (GH) administration on regional adipose tissue distribution and metabolism in GH-deficient children. J Clin Endocrinol Metab 1989;69:1274-1281.
32. Myers SE, Carrel AL, Whitman BY et al. Sustained benefit after 2 years of growth hormone on body composition, fat utilization, physical strength and agility, and growth in Prader-Willi syndrome. J Pediatr. 2000 Jul;137:42-49.
33. Friedlander AL, Butterfield GE, Moynihan S et al. One year of insulin-like growth factor I treatment does not affect bone density, body composition, or psychological measures in postmenopausal women. J Clin Endocrinol Metab 2001;86:1496-503.
34. Ohlsson C, Bengtsson BA, Isaksson OG et al. Growth hormone and bone. Endocr Rev 1998; 19:55-79.
35. Grinspoon S, Baum H, Lee K et al. Effects of short-term recombinant human insulin-like growth factor I administration on bone turnover in osteopenic women with anorexia nervosa. J Clin Endocrinol Metab 1996;81:3864-3870.
36. Bachrach LK, Marcus R, Ott SM et al. Bone mineral, histomorphometry, and body composition in adults with GH receptor deficiency. J Bone Miner Res 2000;13:901.
37. Carroll PV, Christ ER, Bengtsson BA et al. Growth hormone deficiency in adulthood and the effects of growth hormone replacement: a review. Growth hormone Research Society Scientific Committee. J Clin Endocrinol Metab 1998;83:382-395.
38. Horber FF, Marsh HM, Haymond MW. Differential effects of prednisone and growth hormone on fuel metabolism and insulin antagonism in humans. Diabetes 1991;40:141-149.
39. Mauras N, George D, Evans J et al. Growth hormone has anabolic effects in glucocorticosteroid-dependent children with Inflammatory Bowel Disease: A pilot study. Metabolism 2002; 51:127-135.
40. Mauras N, Quarmby V, Bloedow DC. Pharmacokinetics of insulin-like growth factor-1 in hypopituitarism: Correlation with binding proteins. Am J Physiol:Endocrinol Metab 1999:E579-E584.
41. Mauras N, Hayes V, Welch S et al. Testosterone deficiency in young men: Marked alterations in whole body protein kinetics, strength and adiposity. J Clin Endocrinol Metab 1998;83:1886-1892.
42. Hayes VY, Urban RJ, Jiang J et al. Recombinant human growth hormone and recombinant human insulin-like growth factor I diminish the catabolic effects of hypogonadism in man: metabolic and molecular effects. J Clin Endocrinol Metab 2001;86:2211-2219.

CHAPTER 29

Insulin-Like Growth Factor II (IGF-II) and Non-Islet Cell Tumor Hypoglycemia (NICTH)

Naomi Hizuka, Izumi Fukuda, Yukiko Ishikawa and Kazue Takano

Summary

Extrapancreatic tumors associated with hypoglycemia (non-islet-cell tumor hypoglycemia, NICTH) is one of major causes of fasting hypoglycemia. In some patients with NICTH, insulin-like growth factor II (IGF-II) produced by and secreted from the tumor is thought to be a hypoglycemic agent. However, the mechanisms of the hypoglycemia is still unknown, because serum IGF-II (total IGF-II) levels are not always elevated in these patients. In this syndrome, the major form of IGF-II is big IGF-II that is not intact pro-IGF-II, but O-glycosylated pro-IGF-II-(E1-21) generated from abnormal processing. Increased bioavailability of IGF-II could be related to hypoglycemia. The big IGF-II circulates with IGFBP-3 as a binary complex but not as the ternary complex (150 kDa) of big IGF-II-IGFBP-3-ALS in this syndrome. After successful treatments, the hypoglycemia disappears, the circulating big IGF-II significantly decreases and the 150 kDa complex increases. As binary complexes of IGF-IGFBP pass through capillary, binary complex of the big IGF-II-IGFBP-3 in patients with NICTH might readily pass from the circulating to the extracellular space and access to the target tissue, resulting in acceleration of IGF bioavailability in vivo. Therefore, it has been suggested that the impaired formation of 150 kDa complex of big IGF-II could be one of the major causes of hypoglycemia. Although first line of the therapy of NICTH is surgical removal of the tumor, to control hypoglycemia, glucose infusion and administration of GH, glucocorticoid, and glucagon have been applied.

Introduction

Extrapancreatic tumors associated with hypoglycemia (non-islet-cell tumor hypoglycemia, NICTH) is one of major causes of fasting hypoglycemia.[1] In some patients with NICTH, insulin-like growth factor II (IGF-II) produced by and secreted from the tumor is thought to be a hypoglycemic agent.[2,3] However, the mechanisms of the hypoglycemia is still unknown, because serum IGF-II (total IGF-II) levels are not always elevated in these patients.[4-6] It has been reported that high molecular weight form of IGF-II (big IGF-II) increase in sera and tumors from the patients with IGF-II producing NICTH.[3,6-8] Until now, we have been able to investigate serum IGF-II in 51 patients with IGF-II producing NICTH, therefore, we report here clinical features and characterization of IGF-II and IGF binding proteins (IGFBPs) in this syndrome.

Clinical Features of IGF-II Producing NICTH

We characterized serum IGF-II from 51 patients (28 men and 23 women; age: 9 - 83 yrs) with IGF-II producing NICTH (Table 1). Serum IGF-II heterogeneity was analyzed by Western immunoblot (WIB)[6,8] (Fig. 1). In normal subjects, the majority of serum IGF-II was detected at 7.5 kDa, the expected size for IGF-II and a minor amount at 11 kDa. In contrast, in the sera from the 51 patients with NICTH, most of the circulating IGF-II migrated between 11 and 18 kDa, the fragment size predicted for big IGF-II, and a lesser amount at 7.5 kDa, expected for IGF-II. After successful

---

Insulin-Like Growth Factors, edited by Derek LeRoith, Walter Zumkeller and Robert Baxter. ©2003 Eurekah.com and Kluwer Academic / Plenum Publishers.

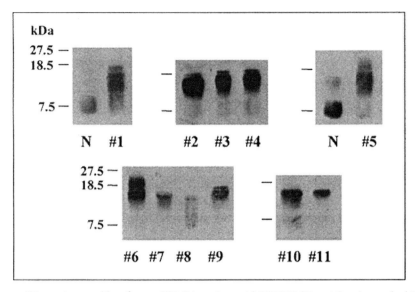

Figure 1. Western immunoblot of serum IGF-II in patients with NICTH (#1 to 11) and normal subjects (N). Redrawn from Hizuka N et al.[6]

Table 1. Tumors causing NICTH in 51 patients

Tumor	N	Tumor	N
Mesothelioma	6	Hepatocellular carcinoma	19
Leiomyosarcoma	3	Gastric cancer	7
Fibrosarcoma (liver, breast)	2	Colon cancer	2
Histiocytoma	1	Renal cell carcinoma	1
Liposarcoma	1	Prostate cancer	1
Hemangiopericytoma	1	Pancreatic cancer	1
Sarcoma (jejunum)	1	Adrenal cancer	1

Tumors from 4 patients were histologically unclassified.

removal of tumor, the hypoglycemia disappeared, and the circulating big IGF-II significantly decreased.[6] Theses data indicate that the presence of big IGF-II could be related to hypoglycemia in the NICTH syndrome.

It has been reported that the tumors causing NICTH were certain large, mostly mesenchymal origin tumors, such as mesothelioma, leiomyosarcoma, histiocytoma, fibrosarcoma, etc. However, in our series, there were also various carcinomas, such as hepatocellular carcinoma and gastric cancer, as well as mesenchymal origin tumors. In our study, there were no tumors less than 5 cm (diameter).

Serum IGF-II and IGF-I Levels

Serum IGF-II levels were elevated in only one third patients (Fig. 2), but serum IGF-I levels were low in all patients with NICTH. IGF-II/IGF-I ratios ranged from 16.4 to 64.2 with a mean of 35.0 ± 2.2.[6] The value was significantly higher than those for normal subjects. These data show that in the NICTH, serum IGF-II levels based on IGF-I levels are inappropriately elevated.

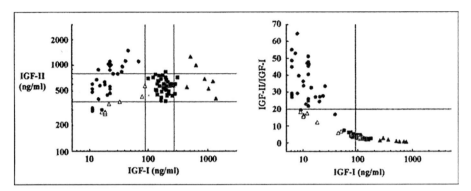

Figure 2. Serum IGF-I, IGF-II and IGF-II/IGF-I ratios in patients with NICTH (●), GH deficiency (△), acromegaly (▲), and normal subjects (■). Redrawn from Hizuka N et al.[6]

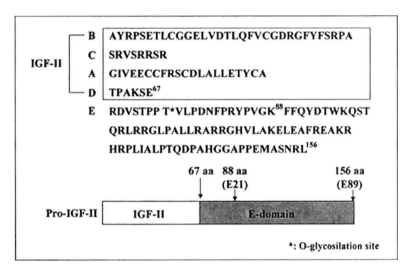

Figure 3. Amino acid sequence of human pro-IGF-II.

Characterization of Big IGF-II

IGF-II is synthesized as pro-IGF-II (Fig. 3) that consists of 67 amino acids of IGF-II with a carboxyl 89 amino acid extension (E domain), and the mature form is produced by cleavage of the E domain.[9] A relatively small amount of big IGF-II is yielded in the processing of pro-IGF-II. Big IGF-II has been isolated from normal sera. The big form, designated as pro-IGF-II-(E1-21), is produced by cleavage after the single lysine at position 21 of the E domain in the processing of pro-IGF-II.[10] The pro-IGF-II-(E1-21) is O-linked glycosylated on threonine at position 8 of the E domain.[11] It has been reported that big IGF-II in NICTH lacks normal E-domain O-linked glycosylation, suggesting that big IGF-II in NICTH might be generated by an abnormal processing of pro-IGF-II.[12] However, we have found that the apparent size of big IGF-II varies in sera from the patients with NICTH and there is a possibility that slower migration pattern of IGF-II might be due to a different size of sugar moiety attached to pro-IGF-II. Therefore, we investigated the effect of O-glycosidase digestion on migration of IGF-II and analyzed the results by WIB. By WIB analysis

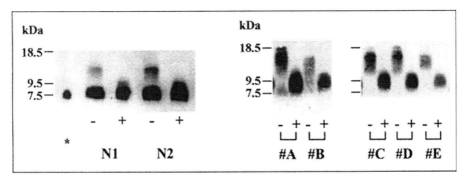

Figure 4. Effect of neuramidase and O-glycosidase treatment on size of big IGF-II in sera from patients with NICTH (#A to E) and normal subjects (N1, N2). The extracted IGF-II from sera were treated either without (-) or with (+) neuramidase and O-glycosidase, and then were analyzed by Western immunoblot. Molecular weight markers are indicated on the left. The asterisk refers to the recombinant hIGF-II. Redrawn from Hizuka N et al.[13]

the big IGF-II was reduced in size to 9.5 kDa in the enzyme treated sera of the ten patients with NICTH (Fig. 4). The migration pattern is similar to that observed in sera of normal subjects after O-glycosidase digestion. Our data suggest that the big IGF-II from patients with NICTH is O-glycosylated pro-IGF-II-(E1-21), but, the sizes of sugar moiety are larger than those from normal subjects suggesting abnormal glycosylation in NICTH.[13]

To investigate the characterization of big IGF-II further, pro-IGF-II in patients with NICTH by WIB using specific antibody against E10-21 (Ab E10-21) and E71-89 (Ab E71-89) of pro-IGF-II were investigated.[14,15] The 14-18 kDa protein was detected in sera from NICTH by WIB using Ab E10-21 and IGF-II antibody, but not Ab E71-89. These data suggest that big IGF-II in NICTH is not intact pro-IGF-II, but pro-IGF-II- (E1-21) generated from abnormal processing.

Circulating Form of IGF-II and Serum IGFBPs

In sera, most of IGF-II binds to IGFBPs, and it has been speculated that the IGFBPs might modulate IGF action. The binary complex of IGF-IGFBP readily passes from the circulating to the extracellular space and access to the target tissue as same as free form of IGF, but, the ternary complex of IGF-IGFBP-3-ALS does not.[16] Therefore, it has been suggested that the alternation of IGFBPs might be related to hypoglycemia by increased bioavailability of IGF-II. Serum IGFBPs and ALS were investigated in patients with NICTH. Serum IGFBP-2 levels increased, and serum levels of IGFBP-3 and ALS decreased. Serum IGFBP-6 levels increased in some patients. These alteration of IGFBPs and ALS were restored after successful removal of the tumors.

When normal serum was gel filtered at pH 7.4, 150kDa and 40kDa complexes of IGF-II with IGFBPs were found (Fig. 5). 150kDa complex consists of IGF, IGFBP-3 and ALS, and 40kDa complex consists of binary complex of IGF and IGFBPs. In patients with NICTH, the 150kDa and 40kDa complexes were not found but a peak of IGF-II with IGFBPs eluted as 60-80kDa was found (Fig. 5). In this fraction, four IGFBPs (IGFBP-1, -2, -3, -4) were found by Western ligand blot (WLB). In both normal subjects and the patients with NICTH, free form of 7.5kDa IGF-II was not increased. However, in serum from some patients with NICTH, a small peak of IGF-II eluted in approximately 17kDa was found. In the peak fraction (~17kDa), IGFBPs were not found by WLB and big form of IGF-II (11-18kDa) was observed by WIB. These data indicated that free form of big IGF-II was increased in sera from these patients. Furthermore, IGF-II eluted as 60-80kDa complex in patients with NICTH were mainly big form of IGF-II. These data suggest that big IGF-II binds IGFBPs, and the big IGF-II with IGFBP-3 does not bind ALS, resulting in no 150kDa complex formation. The decreased serum IGFBP-3 and ALS levels could be one of the causes of impaired formation of 150kDa complex in the NICTH.

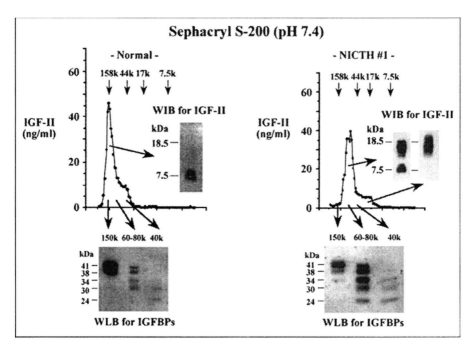

Figure 5. Representative elution profiles of serum IGF-II on Sephacryl S-200 (pH 7.4) in a normal adult (left) and a patient with NICTH (right). IGF-II and IGFBPs in in gel-filtered fractions were also analyzed by Western immunoblot and by Western ligand blot, respectively. Redrawn from Fukuda I et al.[13]

Mechanism of Hypoglycemia

In 1988, Daughaday et al reported that a tumor obtained from one patient with NICTH produced and secreted IGF-II.[3] Since IGF-II has a hypoglycemic effect in vivo, IGF-II produced by tumor is thought to be one of causes of hypoglycemia in patients with NICTH. However, serum IGF-II levels are not always elevated.[4-6] Therefore, it is hard to explain that the quantity of IGF-II is merely responsible for the development of hypoglycemia in patients with NICTH. We consider the possible mechanisms of hypoglycemia as follows:
 1. increased bioactivity of big IGF-II,
 2. increase of free form of big IGF-II,
 3. increased bioavailability of IGF-II, and
 4. suppression of GH secretion.

Increased bioactivity of big IGF-II could be related to hypoglycemia. However, it has been reported that insulin-like activity of big IGF-II did not increase in bioassay using adipocytes.[7] In our preliminary data, we did not find increased bioactivity of big IGF-II in adipocytes. Insulin-like action of IGF-II is acted through insulin and IGF-I receptor. Adipocytes have abundant insulin receptors, and muscle cells have abundant IGF-I receptors. We have not tested bioassay using muscle cell. Therefore, possible mechanism of increased bioactivity of big IGF-II has not been ruled out.

One possible mechanism of hypoglycemia is an increase of free form IGF-II. As described above, we found that free form of 7.5kDa IGF-II was not increased, but a free form of big IGF-II increased in sera from some patients with NICTH.[17] Frystyk et al also reported increased levels of circulating free insulin-like growth factors in patients with NICTH.[18] Thus the increase of free form of big IGF-II in serum might be one of the causes of hypoglycemia.

Increased bioavailability of IGF-II could be related to hypoglycemia. As mentioned above, the alternation of IGFBPs might be related to hypoglycemia by increased bioavailability of IGF-II. In the patients with NICTH, the big IGF-II circulates with IGFBP-3 as a binary complex but not as the ternary complex (150 kDa) of big IGF-II-IGFBP-3-ALS. After successful treatments, the

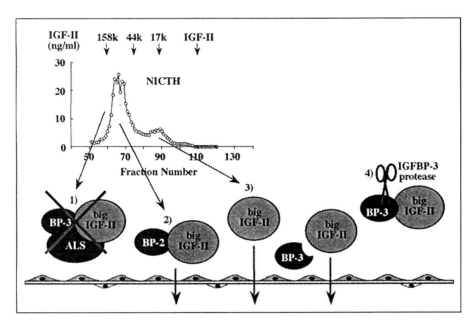

Figure 6. Possible mechanism of hypoglycemia in NICTH. 1) impaired formation of 150kDa complex, 2) increased big IGF-II-IGFBP-2 complex, 3) increase of free form of big IGF-II, and 4) increased IGFBP-3 protease activity.

hypoglycemia disappears, the circulating big IGF-II significantly decreases and the 150 kDa complex increases.[19,20] As binary complexes of IGF-IGFBP pass through capillary, binary complex of the big IGF-II-IGFBP-3 in patients with NICTH might readily pass from the circulating to the extracellular space and access to the target tissue, resulting in acceleration of IGF bioavailability in vivo. Therefore, it has been suggested that the impaired formation of 150 kDa complex of big IGF-II could be one of the major causes of hypoglycemia. Furthermore, increased binary complex of big IGF-II-IGFBP-2/IGFBP-6 could be related to hypoglycemia.

By Western immunoblot, IGFBP-3 was detected as 41/38 and 30kDa protein. It is considered that 41/38kDa protein is intact IGFBP-3, and 30kDa protein is degraded form. IGFBP-3 by WIB was only 30kDa in pregnant women.[16] We found that a patient with NICTH had only 30kDa IGFBP-3 same as pregnant serum. As the degraded IGFBP-3 by IGFBP-3 protease can not bind IGF, bioavailable IGF-II might increase. Therefore, in some patients with NICTH, increased IGFBP-3 protease activity might be related to hypoglycemia.

Furthermore, the all patients with NICTH have low serum IGF-I levels. The increased big IGF-II could suppress GH secretion, leading to low levels of IGF-I, IGFBP-3, and ALS. Thus, the suppression of GH secretion might be related to hypoglycemia. However, other unknown factors could be related to hypoglycemia.

Treatment of NICTH

First line of the therapy of NICTH is surgical removal of the tumor or specific chemotherapy for the tumors. However, successful treatments are often not possible, especially in malignant tumors that were vastly extended. To control hypoglycemia, glucose infusion and administration of GH, glucocorticoid, and glucagon have been applied.[20-23]

Acknowledgments

We are greatly indebted to the doctors who sent us valuable samples of NICTH. This work was supported in part by Grants-in-Aid for Scientific Research (No. 10671043, 14571074) from The Ministry of Education, Science and Culture, Japan.

References

1. Takayama-Hasumi S, Eguchi Y, Satoh A et al. Insulin autoimmune syndrome is the third leading cause of spontaneous hypoglycemic attacks in Japan. Diabetes Res Clin Practice 1990; 10:211-214.
2. Daughaday WH. Autocrine, paracrine and endocrine manifestation of insulin-like growth factor secretion by tumors. In: Spencer EM, ed. Modern Concepts of Insulin-Like Growth Factors. New York: Elsevier Science Publishing Co., Inc., 1991:557-565.
3. Daughaday WH, Emanuele MA, Brooks MH et al. Synthesis and secretion of insulin-like growth factor II by a leiomyosarcoma with associated hypoglycemia. N Engl J Med 1988; 319:1434-1440.
4. Zapf J, Walter H, Froesch ER. Radioimmunological determination of insulin-like growth factors I and II in normal subjects and in patients with growth disorders and extrapancreatic tumor hypoglycemia. J Clin Invest 1981; 68:1321-1330.
5. Widmer U, Zapf J, Froesch ER. Is extrapancreatic tumor hypoglycemia associated with elevated levels of insulin-like growth factor II? J Clin Endocrinol Metab 1982; 55:833-839.
6. Hizuka N, Fukuda I, Takano K et al. Serum Insulin-like growth factor II in 44 patients with non-islet cell tumor hypoglycemia. Endocr J 1998; 45 Suppl:S61-5.45.
7. Zapf J, Futo E, Peter M et al. Can "big" insulin-like growth factor II in serum of tumor patients account for the development of extrapancreatic tumor hypoglycemia? J Clin Invest 1992; 90:2574-84.
8. Enjoh T, Hizuka N, Perdue JF et al. Characterization of new monoclonal antibodies to human insulin-like growth factor-II and their application in Western immunoblot analysis. J Clin Endocrinol Metab 1993; 77:510-517.
9. Daughaday WH, Rotwein P. Insulin-like growth factors I and II. Peptide, messenger ribonucleic acid and gene structures, serum, and tissue concentrations. Endocr Rev 1989; 10:68-91.
10. Zumstein PP, Luthi C, Humbel RE. Amino acid sequence of a variant pro-form of insulin-like growth factor II. Proc Natl Acad Sci USA 1985; 82:3169-3172.
11. Hudgins WR, Hampton B, Burgess WH et al. The identification of O-glycosylated precursors of insulin-like growth factor II. J Biol Chem 1992; 25:267:8153-8160.
12. Daughaday WH, Trivedi B, Baxter RC. Serum "big insulin-like growth factor II" from patients with tumor hypoglycemia lacks normal E-domain O-linked glycosylation, a possible determinant of normal propeptide processing. Proc Natl Acad Sci USA 1993; 90:5823-5827.
13. Hizuka N, Fukuda I, Takano K et al. Serum high molecular weight form of insulin-like growth factor II from patients with non-islet cell tumor hypoglycemia is O-glycosylated. J Clin Endocrinol Metab 1998; 83:2875-7.
14. Khosla S, Ballard FJ, Conover CA. Use of site-specific antibodies to characterize the circulating form of big insulin-like growth factor II in patients with hepatitis C-associated osteosclerosis. J Clin Endocrinol Metab 2002; 87:3867-70.
15. Fukuda I, Hizuka N, Ishikawa Y et al. Characterization of big IGF-II in patients with non-islet cell tumor hypoglycemia [Abstract]. Program & Abstracts of The Endocrine Society's 83rd Annual Meeting, The Endocrine Society, 2001:185.
16. Rajaram S, Baylink DJ, Mohan S. Insulin-like growth factor-binding proteins in serum and other biological fluids: Regulation and functions. Endocr Rev 1997; 18:801-831.
17. Fukuda I, Hizuka N, Takano K et al. Circulating forms of insulin-like growth factor II (IGF-II) in patients with non-islet cell tumor hypoglycemia. Endocrinol Metab 1994; 1:89-95.
18. Frystyk J, Skjaerbaek C, Zapf J et al. Increased levels of circulating free insulin-like growth factors in patients with non-islet cell tumor hypoglycaemia. Diabetologia 1998; 41:589-94.
19. Baxter RC, Daughaday WH. Impaired formation of the ternary insulin-like growth factor-binding protein complex in patients with hypoglycemia due to nonislet cell tumors. J Clin Endocrinol Metab 1991; 73:696-702
20. Baxter RC, Holman SR, Corbould A et al. Regulation of the insulin-like growth factors and their binding proteins by glucocorticoid and growth hormone in nonislet cell tumor hypoglycemia. 1995 J Clin Endocrinol Metab 80:2700-2708.
21. Agus MS, Katz LE, Satin-Smith M et al. Non-islet-cell tumor associated with hypoglycemia in a child: successful long-term therapy with growth hormone. J Pediatr 1995; 127:403-407.
22. Teale JD, Marks V. Glucocorticoid therapy suppresses abnormal secretion of big IGF-II by non-islet cell tumors inducing hypoglycaemia (NICTH). Clin Endocrinol (Oxf) 1998; 49:491-498.
23. Hoff AO, Vassilopoulou-Sellin R. The role of glucagon administration in the diagnosis and treatment of patients with tumor hypoglycemia. Cancer 1998; 82:1585-1592.

CHAPTER 30

Diabetes

Tero Saukkonen and David B. Dunger

Summary

Insulin-like growth factor-I (IGF-I) has effects on insulin-stimulated glucose metabolism, increasing peripheral glucose disposal and decreasing hepatic glucose output, thus enhancing insulin sensitivity, however, the mechanisms are poorly understood. It is thought that IGF-I may act directly through its own type 1 IGF receptor, or indirectly by negative feedback depression of pituitary growth hormone (GH) secretion. The effects of IGF-I on glucose metabolism involve interactions with insulin and IGF binding proteins. This is most clearly evident in subjects with type 1 (insulin-dependent) diabetes mellitus (T1DM), a condition which is characterized by complete absence of endogenous insulin secretion. In T1DM, portal insulin levels are subnormal, leading to hepatic GH resistance and chronically low circulating levels of IGF-I and IGF-BP3, increased levels of IGFBP-1 and thus decreased IGF-I bioactivity. Recombinant human IGF-I (rhIGF-I) has convincingly been shown to improve insulin sensitivity and glycaemic control in T1DM as well as in type 2 diabetes (T2DM) and insulin resistance syndromes. However, the recent lack of supply of rhIGF-I has impeded further evaluation of its therapeutic potential. The GH/ IGF-I axis has also been intensively studied in relation to the pathogenesis of diabetic microvascular complications, yet differences between animal models and humans have made interpretation of some of the findings difficult. Future research will have to more precisely elucidate the complex regulation of circulating and local IGF-I bioactivity and bioavailability in order to define the role of IGF-I based therapies in management of diabetes and its complications.

Introduction

The isolation of IGF-I stemmed in part from the work of Froesch and colleagues, who noted a discrepancy between the amount of insulin-like activity detectable in serum before and after the addition of anti-insulin antibodies which they attributed to "non-suppressible insulin-like activity (NSILA)".[1] It was subsequently shown that this NSILA could be accounted for by circulating IGF-I and IGF-2. The subsequent discovery that IGF-I had considerable structural homology with human pro-insulin and that the IGF-I receptor was similar in structure and function to the insulin receptor gave weight to the early supposition that the non-suppressible insulin-like activity represented by circulating IGF-I and IGF-2 could have an important role in glucose homeostasis and in the pathogenesis and treatment of diabetes.

IGF-I and Glucose Metabolism

When recombinant IGF-I became available for study, Guler et al[2] were the first to show its acute hypoglycaemic effect in healthy human volunteers. It became apparent that IGF-I could improve rates of glucose disposal in skeletal muscle independent of insulin, presumably acting through its own type 1 receptor.[3] Subtle differences were observed however, in that IGF-I, in contrast to insulin, did not increase rates of glucose oxidation but did affect glucose transport and had a preferential effect on the rate of lactate formation.[3] The subtle interactions between IGF-I and insulin action were further highlighted by studies of healthy volunteers which demonstrated that IGF-I could increase glucose disposal, yet it directly suppressed insulin secretion.[4] The doses of IGF-I used in these studies were unlikely to have significant cross-reaction with the insulin receptors, and the

Insulin-Like Growth Factors, edited by Derek LeRoith, Walter Zumkeller and Robert Baxter. ©2003 Eurekah.com and Kluwer Academic / Plenum Publishers.

Table 1. Glucose turnover and substrate oxidation rates during hyperinsulinemic, euglycemic clamp on day 5 of continuous s.c. infusion of saline or IGF-I (10 μg/kg/h) treatment in eight healthy subjects.

	Saline	IGF-I
Glucose infusion rate (M-value) (mg/kg.min)	4.29 ± 1.24	5.49 ± 1.05*
Hepatic glucose output (mg/kg.min)	-0.46 ± 0.63	-1.62 ± 1.05*
Glucose appearance rate (Ra) (mg/kg.min)	3.39 ± 0.76	3.91 ± 0.85
Glucose disposal rate (Rd) (mg/kg.min)		
Oxidative	1.96 ± 0.48	2.38 ± 0.39*
Nonoxidative	2.33 ± 1.21	3.12 ± 0.98*
Lipid oxidation rate (mg/kg.min)	0.61 ± 0.14	0.69 ± 0.08
Protein oxidation rate (mg/kg.min)	0.85 ± 0.20	0.62 ± 0.13*

* $p < 0.05$
Reproduced with permission from Hussain MA, Schmitz O, Mengel A et al. Insulin-like growth factor I stimulates lipid oxidation, reduces protein oxidation, and enhances insulin sensitivity in humans. J Clin Invest 1993; 92(5):2249-56. © American Society for Clinical Investigation.

authors concluded that IGF-I could increase insulin sensitivity. Consequently, Hussein et al[5] showed that insulin stimulated oxidative and non-oxidative glucose disposal, representing the major components of insulin sensitivity, were enhanced during 5 days of subcutaneous IGF-I treatment (Table 1). Direct effects of IGF-I on insulin stimulated glucose metabolism have been harder to show in these normal volunteer studies; however, Boulware et al[6] reported an inhibitory effect of IGF-I on hepatic glucose production.

The mechanism whereby IGF-I can stimulate insulin-like actions through its own receptor is still poorly understood. The type 1 IGF-I receptor has a high affinity for IGF-I, while insulin only binds it to a very minor degree.[7,8] In contrast the insulin receptor has the greatest affinity for the insulin molecule with progressively weaker interactions with IGF-2 and least of all with IGF-I. Intracellular signalling pathways appear to be common to both peptides and their receptors,[9] however receptors with an IGF-I type intracellular domain seem to be more effective at stimulating DNA synthesis.[10] Hybrid receptors have been detected in muscle tissues which appear to have a greater affinity for IGF-I[11] but again, whether this has a role in glucose disposal in muscle is poorly understood.[12] The presence and function of the IGF binding proteins may also be important as these may effect the presentation of peptide to the receptor and subsequent signalling.

Interaction between Insulin and IGF-I and Its Binding Proteins

Unlike insulin, IGF-I is produced by a wide variety of tissues and there are no identifiable tissue stores. The liver contributes the greatest amount to circulating concentration, IGF-I being produced at a relative concentrate of 3 – 40 μg/kg/day.[13] IGF-I circulates at much greater concentrations in plasma than those of insulin and it is largely bound to IGFBP-3 and the acid labile sub-unit to form a ternary complex of approximately 150 kda. This large complex provides a circulating reservoir of IGF-I[14] with a half life of around 15 hours, whereas the half life of unbound IGF-I is 10 – 12 minutes.[15] Given that IGF-I circulates in the high affinity complex there has been much debate as to how IGF-I might be delivered to the tissues. It appears however that the affinity of IGFBPs to IGF-I is lost following cleavage by specific proteases, allowing IGF-I to bind to its receptor.[16] Furthermore, low-molecular weight binding proteins (IGFBP-1, -2 and -4) may be capable of crossing endothelial barrier, facilitating transcapillary transport of IGF-I.[17] These mechanisms may explain how IGFBPs regulate tissue bioactivity and bioavailability of IGF-I.[17]

Generation of hepatic IGF-I following activation of the GH receptor is dependent on insulin. Insulin enhances IGF-I production by direct regulation of the hepatic GH receptor[18] or by a permissive effect on post receptor events.[19] Circulating GH binding protein levels, representing the extracellular domain of the GH receptor,[20] are regulated by both insulin and nutritional status.[21] Insulin also has an important role in controlling IGF bioactivity by regulating circulating concentrations of the small molecular weight binding protein IGFBP-1.[22] Hepatocyte production of IGFBP-1 is inversely regulated by insulin,[23] and values in the plasma undergo a circadian variation with peak values observed overnight when insulin levels are at their lowest.[24] The acute suppressive effect of insulin on IGFBP-1 levels is thought to represent not only suppression of hepatic IGFBP-1 production but also possibly transport out of the circulation into the tissues.[25] Thus in the short term, insulin like effects of IGF-I may be regulated by BP-1. In the longer term, the insulin dependent protease activity which has been demonstrated in newly diagnosed subjects with type 1 diabetes[26] may indicate that insulin has a role in unlocking IGF-I from the major IGFBP-3/ALS complex.

The dependence of hepatic IGF-I generation and BP-1 levels on portal insulin levels[27] provides some explanation as to the nutritional regulation of growth. Prolonged fasting will result in reductions in IGF-I concentrations with elevations in IGFBP-1. As well as reducing the stimulus for growth these will also lead to diminished negative feedback for GH hypersecretion which, through its effects on lipolysis, may have a protective effect during fasting. In the longer term, the consequence of nutritional deprivation is insulin resistance, partly driven by GH hypersecretion and enhanced lipolysis, which may have the effect of diverting glucose as a fuel away from muscle, thus preserving brain glucose homeostasis.

The recently described specific hepatic IGF-I knockouts showed apparent normal growth[28,29] but they were insulin resistant with reductions in IGF binding proteins which might favour increased availability of IGF-I from the diminished circulating store.[30] In further studies, where these animals were crossed with ALS knockout animals, nearly complete loss of the circulating IGF-I reservoir led to growth failure.[31] Perhaps in insulin resistant states, such as starvation, suppression of IGFBP-1 and increased bioavailability of IGF-I may have a role in the continued provision of IGF-I for anabolism. In experimental models, IGF-I is known to have an anti-catabolic effect during starvation.[32]

These various strands of evidence suggest that through complex interaction with the binding proteins and insulin, IGF-I may have a role in glucose homeostasis and the determination of insulin sensitivity.

IGF-I and Diabetes

Type 1 Diabetes

The first evidence that abnormalities of the GH / IGF-I axis may be critical to the pathogenesis of metabolic abnormalities and complications of diabetes came from studies of type 1 diabetes (T1DM). Predictably, the insulinopenia associated with untreated T1DM leads to reduced levels of hepatic GH receptor numbers as reflected in decreased GHBP levels, reduced IGF-I expression and elevated IGFBP-1 levels.[33] Less predictably, however, these abnormalities are not reversed with standard or intensive insulin therapy. Introduction of insulin therapy causes a rise in GHBP levels, and concentrations of IGF-I and GHBP remain closely related to total insulin dose during childhood and puberty (Fig. 1).[34] However, subcutaneous administration of insulin dictates that considerable increases in insulin dose are required to achieve the appropriate portal insulin levels which, in normal subjects, are much greater than those seen in the periphery. Thus with standard insulin treatment, IGF-I levels tend to be low and IGFBP-1 levels high resulting in reduced IGF bioactivity (Fig. 2).[35] Complete correction of the GH/IGF-I axis only seems possible with portal administration of insulin[36] but at the moment this is not a clinically available option.

This reduced IGF bioactivity leads to compensatory high GH hypersecretion, and GH levels in T1DM show increased pulsed amplitude and baseline concentrations.[37] These GH pulse characteristics lead to increases in free fatty acid mobilisation, peripheral insulin resistance and enhanced ketogenesis.[38-41] Combined, these abnormalities have an important impact on metabolic control in treated patients with T1DM, particularly during puberty.[42]

The reduced IGF-I levels and elevated IGFBP-1 levels may also have a direct effect on insulin sensitivity which is not mediated through GH hypersecretion. In physiological studies using rhIGF-I and subjects with T1DM, Acerini et al[43] showed reductions in GH secretion associated with reduced insulin requirements for euglycaemia and some effects on hepatic glucose production rates. In

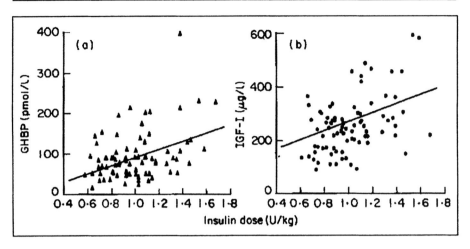

Figure 1. The relation between growth hormone binding protein (GHBP) (a) and IGF-I (b) concentrations and the total daily insulin dose in 104 pubertal subjects with diabetes. Reproduced with permission from Clayton KL, Holly JM, Carlsson LM et al. Loss of the normal relationships between growth hormone, growth hormone-binding protein and insulin-like growth factor-I in adolescents with insulin-dependent diabetes mellitus. Clin Endocrinol (Oxf) 1994;41(4):517-24. © Blackwell Science Ltd.

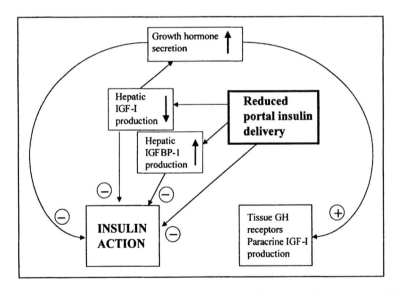

Figure 2. Schematic representation of the mechanisms for reduced insulin sensitivity in type 1 diabetes.

a further study, Crowne et al[44] showed that the effects of IGF-I on insulin requirements for euglycaemia could also be seen when GH levels were sustained using administration of exogenous hormone with somatostatin suppression of all endogenous GH production. Thus both GH hypersecretion in T1DM and relative IGF-I deficiency may affect insulin sensitivity in treated patients with T1DM.

Insulin Resistance Syndromes Including Type 2 Diabetes

Given the close structural homology between the insulin and IGF-I receptors it is not surprising that short stature is a feature of many genetic insulin resistant syndromes. In generalised insulin resistant syndromes such as Leprechaunism and Donohue Syndrome where subjects are either

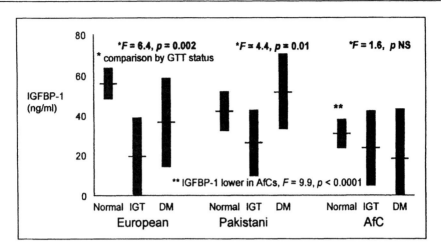

Figure 3. Fasting serum insulin-like growth factor binding protein 1 (IGFBP-1) plotted against glucose tolerance test (GTT) status, by ethnic group, Manchester, England, 1994–1998. IGT, impaired glucose tolerance; DM, newly detected diabetes mellitus; AfC, African Caribbeans. Reproduced with permission from Cruickshank JK, Heald AH, Anderson S et al. Epidemiology of the insulin-like growth factor system in three ethnic groups. Am J Epidemiol 2001;154(6):504-13. © Society for Epidemiologic Research.

homozygous for a single mutation or compound heterozygous for two distinct mutant alleles of the insulin receptor gene,[45] functional abnormalities of the IGF-I receptor have also been identified.[46] Insulin resistant syndromes are complex and in many patients the genetic defect has not been characterised. In these subjects, the involvement of post receptor signalling mechanisms common to the IGF-I receptor may determine whether the growth is also impaired and whether subjects may respond to rhIGF-I treatment.

Type 2 diabetes (T2DM) could also be considered as an insulin resistant condition although the pathophysiology is far more complex and abnormalities of the GH/IGF-I axis are not clearly defined. T2DM often develops as part of a symptom complex where patients have other features of metabolic syndrome including insulin resistance and hyperlipidaemia. Insulin resistance is often associated with reduced IGFBP-1 levels and normalisation of IGF-I[47] presumably because the defect is selective and, as in polycystic ovary syndrome, does not affect hepatic generation of these peptides. However, even within insulin resistant subjects, the levels to which IGFBP-1 is suppressed may vary between ethnic groups, suggesting that a degree of hepatic insulin resistance may also occur (Fig. 3).[48]

The transition from insulin resistance to impaired glucose tolerance or T2DM involves the loss of first phase insulin response and at that point one would predict that IGFBP-1 levels would increase and IGF-I generation would decrease but this has not ever been formally tested. Conversely, variation in IGF-I and IGFBP-1 levels may also have a role in the progression from insulin resistance to impaired glucose tolerance and T2DM. Epidemiological data from the Ely study indicate that subjects with lower IGF-I and increased BP-1 levels may be at higher risk of progression to T2DM despite similar degrees of insulin resistance at baseline and similar antecedent risk factors such as central obesity (N Wareham, personal communication). Whether variation in GH secretion plays a role in the pathogenesis of T2DM is less clear but it could occur when the disease has progressed to the point where portal insulin delivery is impaired with reduced IGF-I and elevated BP-1 levels and subsequent drive for GH secretion from reduced IGF bioactivity.

Links between small size at birth and subsequent risk for T2DM have revived interest in how early regulation of growth and insulin sensitivity might be linked. The hepatic IGF-I knockout mouse is severely insulin resistant,[30] supporting data from animal studies and more recent epidemiological studies that variation in IGF-I expression could be a risk factor for T2DM. However, the risk may be more related to IGF-I receptor expression in the muscle than IGF-I levels per se and LeRoith et al[49] recently replicated an IGF-I dominant negative receptor mutation in a transgenic mouse model. That animal went on to develop insulin resistance and impaired beta cell function, probably because insulin receptor expression in the muscle was reduced by the formation of IGF hybrids.[49] Increased IGF-I receptor hybrids have been reported in muscle extracts from subjects with T2DM but this area of study remains controversial.[12,50]

Finally association of IGF-I gene polymorphisms have been linked to the risk of T2DM. Subjects negative for the wild type 192 BP alleles present in 88% of the population had lower serum IGF-I levels and an increased odds ratio for the development of T2DM and myocardial infarction.[51]

The Use of rhIGF-I in Diabetes

Type 1 Diabetes

At an early stage it was predicted that the use of rhIGF-I could lead to the suppression of abnormally high GH levels and thus directly or indirectly to improvements in insulin sensitivity. Early physiological studies confirmed these predictions. At a dose of 40 μg/kg, chosen as it represented the estimated normal physiological production rate of this peptide, rh IGF-I given as a single subcutaneous injection led to an increase in IGF-I levels which were maximal after 2 hours and exhibited a half life of approximately 18 hours.[52] A modest rise in IGFBP-3 was also observed and levels of free IGF-I were generally less than 10% of the total.[52] The restoration of IGF-I levels to within the normal range led to suppression of GH hypersecretion with reductions in pulse amplitude and baseline concentrations. Concomitant with the reductions in GH levels, subcutaneous rhIGF-I administration resulted in improvements in insulin sensitivity and insulin requirements for euglycaemia overnight.[53] These effects were dose dependent but not necessarily directly related to reductions in GH levels as reductions in insulin requirements for euglycaemia were seen with doses as low as 20 μg/kg, where no change in GH secretion was observed.[43] These observations could not be explained by changes in glucagon secretion or other hormonal variables,[43] In further studies where physiological GH secretion was re-established overnight by somatostatin infusion and timed GH pulses,[54] decreases in insulin requirements and modest falls in IGFBP-1 were observed indicating that IGF-I might have direct effects on improving insulin sensitivity in T1DM.

As a result of these physiological studies, it was postulated that IGF-I replacement may have a role in improving glycaemic control in T1DM. Cheetham et al[55] studied 6 young adults with type 1 who were given rhIGF-I in a dose of 40 μg/kg/day in addition to their multiple injection therapy. Over a 28 day study period, sustained reductions in GH secretion, insulin requirements and a significant fall in Hba1C were observed.[55] This preliminary study was, however, unblinded and subsequently a randomised double blind placebo controlled study was completed by Quattrin and colleagues.[56] They reported similar reductions in insulin requirements and Hba1C in 22 adolescents with type 1 who were given rIGF1 in a dose of 80 μg/kg/day for 28 days in addition to conventional twice daily insulin therapy. A further study of adults with T1DM given rhIGF-I, 50 μg twice daily, for 19 days, reported no differences in HbA1C although significant reductions in both GH and insulin requirements were noted.[57]

Although the doses of rhIGF-I used in these preliminary studies were low, equivalent to 0.05 – 0.1 units/kg/day of insulin, it could have been argued that the same benefit could have been achieved by intensifying insulin therapy. This question was subsequently addressed by a large randomised double-blind placebo controlled trial in the UK and Sweden. In that study,[58] 53 subjects with T1DM were randomised to receive either placebo or rhGF-I in doses of 20 or 40 μg daily, given as a single subcutaneous injection, in addition to multiple injection therapy. Subjects were treated for a period of six months, during which time significant reductions in HbA1C were observed in subjects receiving the larger dose of 40 μg/kg/day (Fig. 4).[58] These improvements were achieved without the need to increase daily insulin dose and there was no associated change in body weight. There was however some loss in efficacy towards the end of the six month period which perhaps reflected poor compliance due to the need to administer five injections a day, although loss of bio-efficacy with prolonged rhIGF-I remains a possibility and will need to be excluded.

Severe Insulin Resistance

At an early stage it was proposed that rhIGF-I might improve glucose disposal by direct effects through the type 1 IGF-I receptor in patients with severe insulin resistance, and its effectiveness was

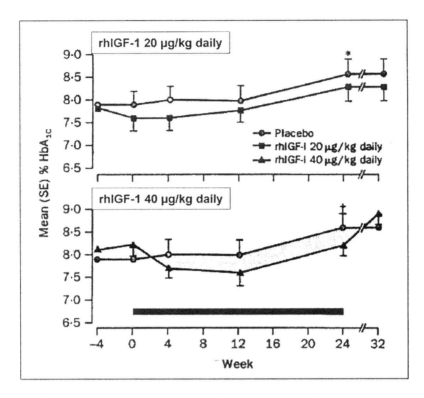

Figure 4. HbA1c (%) values during a 6-month trial of subcutaneous rhIGF-I therapy in adolescents with type 1 diabetes. Placebo vs. rhIGF-I 20 µg/kg per day, (analysis of covariance) *P=0.69; and placebo vs. rhIGF-I 40 µg/kg per day, †P=0.032. Solid bar, treatment period. Reproduced with permission from Acerini CL, Patton CM, Savage MO et al. Randomised placebo-controlled trial of human recombinant insulin-like growth factor I plus intensive insulin therapy in adolescents with insulin-dependent diabetes mellitus. Lancet 1997;350(9086):1199-204. © The Lancet.

first demonstrated by Schoenle and colleagues[59] in 3 patients with type A insulin resistance. They noted reductions in serum glucose and plasma insulin after two intravenous boluses of rhIGF-I (100ug/kg/dose with an interval of two hours). Subsequently, Morrow et al[60] reported that subcutaneous administration of rhIGF-I in a dose of 100 µg/kg twice daily for a month in 2 subjects with type A syndrome led to improved glycaemic control, reduced postprandial insulin levels and improved insulin sensitivity. Moses et al[61] subsequently extended these observations to a total of 6 subjects with various forms of severe insulin resistance, and Kuzuya et al[62] demonstrated similar findings in 8 patients, followed for up to 16 months, on treatment with IGF-I.

As discussed earlier in this chapter, the response to rhIGF-I may vary according to parallel involvement of the IGF-I receptor. In Rabson-Mendenhall Syndrome, which is defined by the presence of extreme insulin resistance, acanthosis nigricans and pineal hyperplasia, Quin et al initially reported acute glucose lowering effects of IGF-I but this did not lead to sustained clinical benefit.[63] In some cases of leprechaunism or Donohue Syndrome functional abnormalities of the type 1 IGF-I receptor have been identified and investigators have failed to show any glucose lowering effects of rhIGF-I.[64]

There seems little point in grouping all patients with severe insulin resistance together with respect to their response to IGF-I for as a group they differ not only in their underlying genetic abnormality but also in the severity of disease expression as it effects glucose, lipid and fat metabolism. Only detailed phenotypic and genotypic characterisation and treatment by standard protocol will resolve these issues.

Type 2 Diabetes

In patients with T2DM who failed therapy with diet or hypoglycaemic agents, insulin is often used as an alternative therapy. However, insulin may not necessarily correct the underlying insulin resistance and may lead to excessive weight gain. Thus it has been proposed that rhIGF-I could circumvent insulin resistance in these, generally obese patients, by reducing hyperglycaemia through effects at the type 1 IGF receptor. Furthermore, as discussed above, there may be other reasons to believe that increasing IGF-I levels may be beneficial in T2DM.

The initial studies of rhIGF-I in T2DM were carried out by Zenobi and colleagues[65] who investigated 8 patients during a five day subcutaneous infusion of rhIGF-I (120 µg/kg/day). They observed reductions in fasting and postprandial glucose levels, together with improvements in lipid metabolism. Longer term studies were subsequently carried out by Schalch and co-workers,[66] again using large subcutaneous doses of IGF-I up to 320 µg/kg/day which confirmed that IGF-I could improve glycaemic control, although treatment was generally poorly tolerated with many patients experiencing jaw tenderness, arthralgia, myalgia, hypotension and peripheral oedema.

Since those early studies, using very high doses, the trend has been towards exploring the effectiveness of lower doses of rhIGF-I in T2DM. Moses and co-workers[67] carried out an open label study of rhIGF-I of 6 weeks duration in 12 subjects. All of their subjects had poor glycaemic control at entry and 9 out of the 12 completed at least 4 weeks of treatment. In these patients there was a dramatic improvement in glycaemic control with reduction in HbA1C, fructosamine and fasting insulin levels, but perhaps most significantly the authors observed a substantial increase in insulin sensitivity index as measured during the intravenous glucose tolerance test with frequent sampling, after the method of Bergmann.[68] Even at these slightly reduced doses, 200 µg/kg/day, significant side effects were still observed. However, comparison of published clinical studies (Table 2) indicates that these side effects are dose related and indeed, they are rare or absent at physiological IGF-I doses (40 µg/kg/day or less). Studies reported by the Recombinant Human IGF-I in IDDM Study Group[70,71] support this view. In this large prospective study, 212 subjects with type 2 were randomised to receive either placebo or rhIGF-I in doses of 10, 20, 40 and 80 µg/kg twice daily for 12 weeks. Significant side effects were only observed in those subjects taking the largest doses of rhIGF-I, yet the same study indicated that even the lowest doses could produce significant improvements in glycaemic control.

GH/IGF-I Axis and Microvascular Complications of Diabetes

The use of rhIGF-I as a therapy in diabetes rekindled the debate as to the role of the GH/IGF-I axis in the pathogenesis of microvascular complications. Of particular concern was the study reported by Thraillkill et al[72] which indicated that large doses of rhIGF-I (120 and 140 µg/kg/day) in subjects with T1DM could be associated with progression of retinopathy scores during the course of 12 weeks of therapy. Furthermore, these high doses were associated with the development of papillitis and retinal haemorrhage. Although they speculated that this could have been a transient phenomenon related to improvements in glycaemic control (the so called normoglycaemic re-entry phenomenon which has been observed in previous studies), anxiety remained that IGF-I treatment could have effects on the progression of microangiopathic complications. However, these adverse effects in T1DM also seem to be dose related as the study of Acerini et al[58] using the lower dose of 40 µg once daily showed benefits in terms of glycaemic control without any deterioration in retinopathy scores or renal function as assessed by retinal photographs, glomerular filtration and microalbuminuria screening. The dose related risk of aggravation of diabetic complications in T1DM subjects treated with rhIGF-I may relate to the relative increases in bio-availability.[73]

The first suggestions that abnormalities of the GH/IGF-I axis and in particular exaggerated secretion of GH could be implicated in the pathogenesis of diabetic complications were made by Poulsen in 1953[74] and Lundbaek in 1970.[75] An early report from Merimee[76] indicated that increased serum IGF-I concentrations were associated with the development of progressive retinopathy, however, several subsequent large epidemiological studies failed to reveal any association between circulating IGF-I levels and risk of retinopathy.[77] In fact most studies indicated that individuals with diabetic microvascular complications had decreased circulating levels of IGF-I.[78,79] So much was the reversal in thought that a hypothesis of a protective effect of circulating IGF-I on target tissues susceptible to microvascular damage in diabetes such as the kidney, eye and neural tissue was

Table 2. Summary of recorded adverse events in clinical trials of rhIGF-I in patients with Type 1 and Type 2 diabetes

Study	Patients (n) Study Design	RhIGF-I Dose μg/kg/d	Treatment Duration	Adverse Effects	Retinopathy
Schalch et al[66]	Type 2 (12) open	180 - 320	5d	edema (facial/hand/feet) 50% jaw tenderness 67% postural hypotension 75% mild hypoglycaemia 17%	not reported
Jabri et al[69]	Obese type 2 (7) open	180-320	4-52d	facial edema 86% peripheral oedema 57% jaw tenderness 86% arthralgia 57% postural hypotension 43% hypoglycaemia 14%	not reported
Moses et al[67]	Type 2 (12) open	200	up to 6 weeks	facial and peripheral edema, jaw tenderness, arthralgias	not reported
Thrailkill et al[72]	Type 1 (169) randomized placebo-controlled	80-140	12 weeks	facial edema 9-11% peripheral oedema 14-23% jaw tenderness 16-26% arthralgia 18-25% headache 45-52% tachycardia 2-7%	early worsening of retinopathy in 12% of patients on rhIGF-I 120/140mg/kg/day
Carroll et al[57]	Type 1 (6) randomized placebo-controlled	100	19d	headache 67%	no changes
Quattrin et al[56]	Type 1 (22) placebo-controlled cross-over	80	28d	not different from placebo	not reported
Acerini et al[58]	Type 1 (36) randomized placebo-controlled	20-40	24 weeks	not different from placebo	no changes

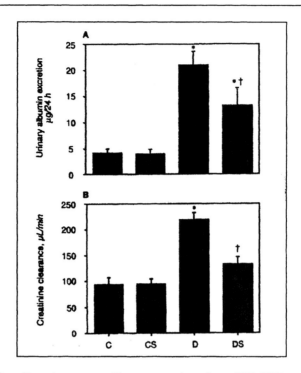

Figure 5. The effect of 3-week treatment with a somatostatin analogue, PTR-3173, on urinary albumin excretion (μg/24h) (A) and creatinine clearance (μL/min) (B) in non-obese diabetic mouse. C, untreated non-diabetic mice; CS, PTR-3173-treated non-diabetic mice; D, untreated diabetic mice; DS, PTR-3173-treated diabetic mice. Values are mean ± SEM. n = 7-10 in each group. * p<0.05 vs. C; † p<0.05 vs. D. Reproduced with permission from Landau D, Segev Y, Afargan M et al. A novel somatostatin analogue prevents early renal complications in the nonobese diabetic mouse. Kidney Int 2001;60(2):505-12. © Blackwell Science Ltd.

put forward by Jansen and Lamberts in 2000.[80] Certainly our own recent data on the risk of microalbuminuria in adolescent subjects with T1DM suggest that incipient nephropathy develops despite very low circulating levels of IGF-I.[81]

Low circulating levels of IGF-I and decreased IGF bioavailability may however be the stimulus for GH hypersecretion. Whereas there may be resistance at the level of the hepatic GH receptor, as this is dependent on portal insulin secretion, there is no evidence to suggest that other GH receptors are equally compromised. Early animal models of diabetic nephropathy and retinopathy were unhelpful in this respect as the rat with diabetes tends to have low GH levels rather than the elevated GH secretion observed in human subjects. Thus although streptozotocin induced diabetic rats showed rapid increases in renal IGF-I content[82] and sustained increases in renal cortical GHBP messenger RNA levels,[83] these could not be directly attributed to circulating GH levels. Nevertheless, paradoxically, somatostatin analogue therapy prevented kidney hypertrophy and increases in kidney IGF-I levels in STZ diabetic rats.[84]

Further insight into the pathogenesis of diabetic complications has come from the study of NOD mice where renal hypertrophy, associated with accumulation of renal IGF-I and IGFBP-1 when diabetes develops,[85] does seem to be related to GH hypersecretion. Transgenic mice overexpressing GH or IGF-I have enlarged glomeruli and the GH overexpressing mouse develops progressive glomerulosclerosis which has parallels to the pathogenesis of human diabetic nephropathy.[86,87] Recently studies using GH inhibitors have shown that some of the effects of diabetes on the kidney in mouse can be reversed (Fig. 5),[88,89] indicating that GH rather than circulating IGF-I may be the important abnormality. This idea was supported by studies of Cummings et al[90] showing that urinary IGF-I was associated with kidney volume and both urinary GH and IGF-I were associated

Table 3. Association between circulating levels of total IGF-I and retinopathy in cross-sectional studies

Study	Patients	Background	Proliferative-Exudative	Combined Background or Proliferative
Merimee et al[76]	Type 1 no retinopathy (8) retinopathy (25)	→	↑	
	Type 2 no retinopathy (18) retinopathy (29)	→	→	
Salardi et al[93]	Type 1 (children / adolescents) no retinopathy (8) retinopathy (25)			→
Hyer et al[94]	Type 1 and type 2 no retinopathy (84) retinopathy (287)	→	→	
Arner et al[78]	Type 1 no or little retinopathy (9) preproliferative/ proliferative retinopathy (9)		↓	
Dills et al[95]	Mainly type 1 (diagnosis <30yr, insulin treatment) no retinopathy (65) retinopathy (617)	→	→	
Dills et al[96]	Type 1 and type 2 (age at diagnosis >30yr) no retinopathy (822) retinopathy (93)		↑	
Janssen et al[97]	Type 1 no retinopathy (29) retinopathy (27)			→

↑ increased, ↓ decreased, → similar levels of IGF-I in patients with retinopathy compared to those without retinopathy, respectively

with microalbuminuria in children and adolescents with T1DM, while plasma IGF-I was not associated with either of kidney volume or microalbuminuria. Recently, a role for increased local IGF-I bioavailability was suggested by Shinada et al[91] based on the finding of decreased levels of intact IGFBP-3 in urine of diabetic patients with MA or overt nephropathy, a finding that was attributed to increased local serine protease activity. Nevertheless, even these studies may not completely resolve discussions of the pathogenesis of nephropathy in humans as some studies indicate that, in contrast to the mouse, IGF-I expression in the human kidney may be difficult to detect.[92]

The debate about the role of the GH/IGF-I axis in the pathogenesis of diabetic retinopathy has taken a similar course. Following on from those earlier studies of Merimee,[76] there has been little confirmation of the effects of circulating IGF-I on this complication (Table 3). Research has instead focused on local, i.e., vitreous concentrations of IGFs and IGFBPs. Retinal neovascularization and progression of diabetic retinopathy does seem to be associated with vitreous accumulation of IGF-I, IGF-II, IGFBP-1 and BP-3, although controversy exists upon whether this results from leakage of the blood-retinal barrier or from local synthesis of IGF-I and abnormal regulation of IGFBPs.[98,99] A direct role for IGF-I in retinopathy was suggested by Grant et al[100] who demonstrated that an intravitreal injection of IGF-I induced retinal neovascularization in rabbits. Although the changes in vitreal IGFs and IGFBPs have been associated with neovascularisation independently of the origin of underlying retinal ischemia[101] and may therefore not be specific for diabetic eye disease, this area has been suggested as a target for drug development to prevent retinopathy. Subcutaneous administration of a GH antagonist[102] or an IGF-I receptor antagonist[103] were shown to suppress retinal neovascularisation in a mouse model of proliferative retinopathy. Whether the dysregulation of vitreal IGFs and IGFBPs relates to circulating levels of GH is unclear, although it has to be remembered that very early studies showed that pituitary ablation suppressing GH levels actually led to a slowing in the progression of proliferative retinopathy.[74,75,104] Interestingly, in studies by Smith et al,[102] mice transgenic for GH did not have increased retinal neovascularisation despite greatly elevated serum IGF-I levels, and neither did mice transgenic for GH antagonist when exogenous IGF-I was given to increase serum IGF-I levels. This is consistent with the clinical findings that acromegalic patients with or without diabetes do not have an increased incidence of retinopathy.[105] Finally, insulin levels may also play a role as inferred by studies showing that rapid improvement in glycaemic control can actually lead to deterioration in retinopathy.[106]

The role of the GH/IGF-I axis in the development of diabetic neuropathy follows a similar pattern in that patients with sensor and autonomic neuropathy usually have decreased serum IGF-I levels.[107] Accordingly, both in vitro and in vivo animal studies suggest that replacement doses of IGF-I might prevent neuronal degeneration induced by hyperglycaemia.[108]

Conclusions and the Future Role of rhIGF-I Therapy

There is a complex interaction between the metabolic regulation and actions of IGF-I and those of insulin, which, together with sequence homology between the two peptides and their respective receptors, probably reflects their common evolutionary origins. Not only does insulin have an important role in modulating the growth promoting effects of IGF-I, but conversely IGF-I probably contributes to the normal regulation of glucose disposal and insulin sensitivity. However these effects of IGF-I are only evident in the fed state when insulin levels are replete. In insulin deficient subjects, such as those with T1DM, the effects of IGF-I depletion on metabolism are clearly evident. IGF-I deficiency may also play an important role in the development of β cell failure in type 2 subjects.[109] The full potential of rhIGF-I therapy in diabetes has not been evaluated because of the rather precipitate closure of drug development.[110] Initial reports of effects on diabetic complications could have been related to normoglycaemic re-entry or the high doses of peptide being used which might have exceeded the binding capacity of the plasma, exposing the tissues to relatively high levels of bioavailable and free IGF-I. Lower doses may have brought clinical efficacy without complications,[58] and given that only portal delivery of insulin will entirely normalise the GH IGF-I axis, particularly in T1DM, this may be the only satisfactory alternative therapy. Recently studies using a combination of IGF-I and IGFBP-3[111] (Fig. 6) have shown greater clinical tolerance with similar clinical efficacy in patients with T1DM and T2DM. Further clarification of the mode of action of this therapy in both type 1 and T2DM is needed and given the prior experience with rhIGF-I, long term safety data will be needed.

References

1. Froesch ER, Burgi H, Ramseier EB et al. Antibody-suppressible and non-suppressible insulin-like activities in human serum and their phusiological significance. An insulin assay with adipose tissue of increased precision and specificity. J Clin Invest 1963; 42:1816-1834.
2. Guler HP, Zapf J, Froesch ER. Short-term metabolic effects of recombinant human insulin-like growth factor I in healthy adults. N Engl J Med 1987; 317(3):137-40.
3. Dimitriadis G, Parry-Billings M, Bevan S et al. Effects of insulin-like growth factor I on the rates of glucose transport and utilization in rat skeletal muscle in vitro. Biochem J 1992; 285(Pt1):269-74.
4. Zenobi PD, Graf S, Ursprung H et al. Effects of insulin-like growth factor-I on glucose tolerance, insulin levels, and insulin secretion. J Clin Invest 1992; 89(6):1908-13.
5. Hussain MA, Schmitz O, Mengel A et al. Insulin-like growth factor I stimulates lipid oxidation, reduces protein oxidation, and enhances insulin sensitivity in humans. J Clin Invest 1993; 92(5):2249-56.
6. Boulware SD, Tamborlane WV, Matthews LS et al. Diverse effects of insulin-like growth factor I on glucose, lipid, and amino acid metabolism. Am J Physiol 1992; 262(1Pt1):E130-3.

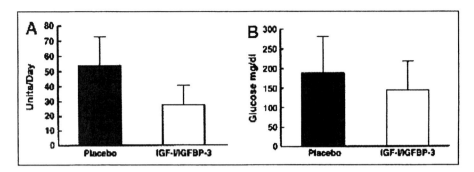

Figure 6. Daily total insulin dosage (panel A) and home-monitored serum glucose (panel B) during 2-week continuous subcutaneous infusion of placebo (solid bar) or rhIGF-I/IGFBP-3 (open bar) in 12 subjects with type 1 diabetes. Values are mean ± 1 SD. Reproduced with permission from Clemmons DR, Moses AC, McKay MJ et al. The combination of insulin-like growth factor I and insulin-like growth factor-binding protein-3 reduces insulin requirements in insulin-dependent type 1 diabetes: evidence for in vivo biological activity. J Clin Endocrinol Metab 2000;85(4):1518-24. © Endocrine Society.

7. Rechler MM, Nissley SP. Insulin-like growth factor (IGF)/somatomedin receptor subtypes: structure, function, and relationships to insulin receptors and IGF carrier proteins. Horm Res 1986; 24(2-3):152-9.
8. Steele-Perkins G, Turner J, Edman JC et al. Expression and characterization of a functional human insulin-like growth factor I receptor. J Biol Chem 1988; 263(23):11486-92.
9. Myers MG, Jr., Sun XJ, Cheatham B et al. IRS-1 is a common element in insulin and insulin-like growth factor-I signaling to the phosphatidylinositol 3'-kinase. Endocrinology 1993; 132(4):1421-30.
10. Lammers R, Gray A, Schlessinger J et al. Differential signalling potential of insulin- and IGF-1-receptor cytoplasmic domains. Embo J 1989; 8(5):1369-75.
11. Soos MA, Field CE, Siddle K. Purified hybrid insulin/insulin-like growth factor-I receptors bind insulin-like growth factor-I, but not insulin, with high affinity. Biochem J 1993; 290(Pt 2):419-26.
12. Spampinato D, Pandini G, Iuppa A et al. Insulin/insulin-like growth factor I hybrid receptors overexpression is not an early defect in insulin-resistant subjects. J Clin Endocrinol Metab 2000; 85(11):4219-23.
13. Wilton P. Treatment with recombinant human insulin-like growth factor I of children with growth hormone receptor deficiency (Laron syndrome). Kabi Pharmacia Study Group on Insulin-like Growth Factor I Treatment in Growth hormone Insensitivity Syndromes. Acta Paediatr 1992; 383(Suppl):137-42.
14. Binoux M, Hossenlopp P. Insulin-like growth factor (IGF) and IGF-binding proteins: comparison of human serum and lymph. J Clin Endocrinol Metab 1988; 67(3):509-14.
15. Guler HP, Zapf J, Schmid C et al. Insulin-like growth factors I and II in healthy man. Estimations of half-lives and production rates. Acta Endocrinol (Copenh) 1989; 121(6):753-8.
16. Clemmons DR. Role of insulin-like growth factor binding proteins in controlling IGF actions. Mol Cell Endocrinol 1998; 140(1-2):19-24.
17. Jones JI, Clemmons DR. Insulin-like growth factors and their binding proteins: biological actions. Endocr Rev 1995; 16(1):3-34.
18. Baxter RC, Turtle JR. Regulation of hepatic growth hormone receptors by insulin. Biochem Biophys Res Commun 1978; 84(2):350-7.
19. Maes M, Underwood LE, Ketelslegers JM. Low serum somatomedin-C in insulin-dependent diabetes: evidence for a postreceptor mechanism. Endocrinology 1986; 118(1):377-82.
20. Leung DW, Spencer SA, Cachianes G et al. Growth hormone receptor and serum binding protein: purification, cloning and expression. Nature 1987; 330(6148):537-43.
21. Hochberg Z, Hertz P, Colin V et al. The distal axis of growth hormone (GH) in nutritional disorders: GH-binding protein, insulin-like growth factor-I (IGF-I), and IGF-I receptors in obesity and anorexia nervosa. Metabolism 1992; 41(1):106-12.
22. Suikkari AM, Koivisto VA, Rutanen EM et al. Insulin regulates the serum levels of low molecular weight insulin-like growth factor-binding protein. J Clin Endocrinol Metab 1988; 66(2):266-72.
23. Cotterill AM, Cowell CT, Silink M. Insulin and variation in glucose levels modify the secretion rates of the growth hormone-independent insulin-like growth factor binding protein-1 in the human hepatoblastoma cell line Hep G2. J Endocrinol 1989; 123(3):R17-20.
24. Holly JM, Biddlecombe RA, Dunger DB et al. Circadian variation of GH-independent IGF-binding protein in diabetes mellitus and its relationship to insulin. A new role for insulin? Clin Endocrinol (Oxf) 1988; 29(6):667-75.
25. Bar RS, Boes M, Clemmons DR et al. Insulin differentially alters transcapillary movement of intravascular IGFBP-1, IGFBP-2 and endothelial cell IGF-binding proteins in the rat heart. Endocrinology 1990; 127(1):497-9.

26. Bereket A, Lang CH, Blethen SL et al. Insulin-like growth factor binding protein-3 proteolysis in children with insulin-dependent diabetes mellitus: a possible role for insulin in the regulation of IGFBP-3 protease activity. J Clin Endocrinol Metab 1995; 80(8):2282-8.
27. Brismar K, Fernqvist-Forbes E, Wahren J et al. Effect of insulin on the hepatic production of insulin-like growth factor-binding protein-1 (IGFBP-1), IGFBP-3, and IGF-I in insulin-dependent diabetes. J Clin Endocrinol Metab 1994; 79(3):872-8.
28. Yakar S, Liu JL, Stannard B et al. Normal growth and development in the absence of hepatic insulin-like growth factor I. Proc Natl Acad Sci USA 1999; 96(13):7324-9.
29. Sjogren K, Liu JL, Blad K et al. Liver-derived insulin-like growth factor I (IGF-I) is the principal source of IGF-I in blood but is not required for postnatal body growth in mice. Proc Natl Acad Sci USA 1999; 96(12):7088-92.
30. Yakar S, Liu JL, Fernandez AM et al. Liver-specific igf-1 gene deletion leads to muscle insulin insensitivity. Diabetes 2001; 50(5):1110-8.
31. Yakar S, Rosen CJ, Beamer WG et al. Circulating levels of insulin-like growth factor-I directly regulate bone growth and density. Studies utilizing the liver *Igf-1*-deficient and the *Als* gene-disrupted mice. Submitted for publication 2002.
32. O'Sullivan U, Gluckman PD, Breier BH et al. Insulin-like growth factor-1 (IGF-1) in mice reduces weight loss during starvation. Endocrinology 1989; 125(5):2793-4.
33. Arslanian SA, Menon RK, Gierl AP et al. Insulin therapy increases low plasma growth hormone binding protein in children with new-onset type 1 diabetes. Diabet Med 1993; 10(9):833-8.
34. Clayton KL, Holly JM, Carlsson LM et al. Loss of the normal relationships between growth hormone, growth hormone-binding protein and insulin-like growth factor-I in adolescents with insulin-dependent diabetes mellitus. Clin Endocrinol (Oxf) 1994; 41(4):517-24.
35. Dunger DB. Insulin and insulin-like growth factors in diabetes mellitus. Arch Dis Child 1995; 72(6):469-71.
36. Shishko PI, Kovalev PA, Goncharov VG et al. Comparison of peripheral and portal (via the umbilical vein) routes of insulin infusion in IDDM patients. Diabetes 1992; 41(9):1042-9.
37. Edge JA, Dunger DB, Matthews DR et al. Increased overnight growth hormone concentrations in diabetic compared with normal adolescents. J Clin Endocrinol Metab 1990; 71(5):1356-62.
38. Press M, Tamborlane WV, Sherwin RS. Importance of raised growth hormone levels in mediating the metabolic derangements of diabetes. N Engl J Med 1984; 310:810-5.
39. Edge JA, Matthews DR, Dunger DB. The dawn phenomenon is related to overnight growth hormone release in adolescent diabetics. Clin Endocrinol (Oxf) 1990; 33(6):729-37.
40. Perriello G, De Feo P, Torlone E et al. Nocturnal spikes of growth hormone secretion cause the dawn phenomenon in type 1 (insulin-dependent) diabetes mellitus by decreasing hepatic (and extrahepatic) sensitivity to insulin in the absence of insulin waning. Diabetologia 1990; 33(1):52-9.
41. Pal BR, Phillips PE, Matthews DR et al. Contrasting metabolic effects of continuous and pulsatile growth hormone administration in young adults with type 1 (insulin-dependent) diabetes mellitus. Diabetologia 1992; 35(6):542-9.
42. Amiel SA, Sherwin RS, Simonson DC et al. Impaired insulin action in puberty. A contributing factor to poor glycemic control in adolescents with diabetes. N Engl J Med 1986; 315(4):215-9.
43. Acerini CL, Harris DA, Matyka KA et al. Effects of low-dose recombinant human insulin-like growth factor-I on insulin sensitivity, growth hormone and glucagon levels in young adults with insulin-dependent diabetes mellitus. Metabolism 1998; 47(12):1481-9.
44. Crowne EC, Samra JS, Cheetham T et al. Recombinant human insulin-like growth factor-I abolishes changes in insulin requirements consequent upon growth hormone pulsatility in young adults with type I diabetes mellitus. Metabolism 1998; 47(1):31-8.
45. Taylor SI, Cama A, Accili D et al. Genetic basis of endocrine disease. 1. Molecular genetics of insulin resistant diabetes mellitus. J Clin Endocrinol Metab 1991; 73(6):1158-63.
46. Van Obberghen-Schilling EE, Rechler MM, Romanus JA et al. Receptors for insulinlike growth factor I are defective in fibroblasts cultured from a patient with leprechaunism. J Clin Invest 1981; 68(5):1356-65.
47. Frystyk J, Skjaerbaek C, Vestbo E et al. Circulating levels of free insulin-like growth factors in obese subjects: the impact of type 2 diabetes. Diabetes Metab Res Rev 1999; 15(5):314-22.
48. Cruickshank JK, Heald AH, Anderson S et al. Epidemiology of the insulin-like growth factor system in three ethnic groups. Am J Epidemiol 2001; 154(6):504-13.
49. Fernandez AM, Kim JK, Yakar S et al. Functional inactivation of the IGF-I and insulin receptors in skeletal muscle causes type 2 diabetes. Genes Dev 2001; 15(15):1926-34.
50. Federici M, Zucaro L, Porzio O et al. Increased expression of insulin/insulin-like growth factor-I hybrid receptors in skeletal muscle of noninsulin-dependent diabetes mellitus subjects. J Clin Invest 1996; 98(12):2887-93.
51. Vaessen N, Heutink P, Janssen JA et al. A polymorphism in the gene for IGF-I: functional properties and risk for type 2 diabetes and myocardial infarction. Diabetes 2001; 50(3):637-42.
52. Cheetham TD, Taylor A, Holly JM et al. The effects of recombinant human insulin-like growth factor-I (IGF-I) administration on the levels of IGF-I, IGF-II and IGF-binding proteins in adolescents with insulin-dependent diabetes mellitus. J Endocrinol 1994; 142(2):367-74.
53. Cheetham TD, Jones J, Taylor AM et al. The effects of recombinant insulin-like growth factor I administration on growth hormone levels and insulin requirements in adolescents with type 1 (insulin-dependent) diabetes mellitus. Diabetologia 1993; 36(7):678-81.

54. Crowne EC, Samra JS, Cheetham T et al. The role of IGF-binding proteins in mediating the effects of recombinant human IGF-I on insulin requirements in type 1 diabetes mellitus. J Clin Endocrinol Metab 2001; 86(8):3686-91.
55. Cheetham TD, Holly JM, Clayton K et al. The effects of repeated daily recombinant human insulin-like growth factor I administration in adolescents with type 1 diabetes. Diabet Med 1995; 12(10):885-92.
56. Quattrin T, Thrailkill K, Baker L et al. Dual hormonal replacement with insulin and recombinant human insulin-like growth factor I in IDDM. Effects on glycemic control, IGF-I levels, and safety profile. Diabetes Care 1997; 20(3):374-80.
57. Carroll PV, Umpleby M, Ward GS et al. rhIGF-I administration reduces insulin requirements, decreases growth hormone secretion, and improves the lipid profile in adults with IDDM. Diabetes 1997; 46(9):1453-8.
58. Acerini CL, Patton CM, Savage MO et al. Randomised placebo-controlled trial of human recombinant insulin-like growth factor I plus intensive insulin therapy in adolescents with insulin-dependent diabetes mellitus. Lancet 1997; 350(9086):1199-204.
59. Schoenle EJ, Zenobi PD, Torresani T et al. Recombinant human insulin-like growth factor I (rhIGF I) reduces hyperglycaemia in patients with extreme insulin resistance. Diabetologia 1991; 34(9):675-9.
60. Morrow LA, O'Brien MB, Moller DE et al. Recombinant human insulin-like growth factor-I therapy improves glycemic control and insulin action in the type A syndrome of severe insulin resistance. J Clin Endocrinol Metab 1994; 79(1):205-10.
61. Moses AC, Morrow LA, O'Brien M et al. Insulin-like growth factor I (rhIGF-I) as a therapeutic agent for hyperinsulinemic insulin-resistant diabetes mellitus. Diabetes Res Clin Pract 1995; 28(Suppl):S185-94.
62. Kuzuya H, Matsuura N, Sakamoto M et al. Trial of insulinlike growth factor I therapy for patients with extreme insulin resistance syndromes. Diabetes 1993; 42(5):696-705.
63. Quin JD, Fisher BM, Paterson KR et al. Acute response to recombinant insulin-like growth factor I in a patient with Mendenhall's syndrome. N Engl J Med 1990; 323(20):1425-6.
64. Backeljauw PF, Alves C, Eidson M et al. Effect of intravenous insulin-like growth factor I in two patients with leprechaunism. Pediatr Res 1994; 36(6):749-54.
65. Zenobi PD, Jaeggi-Groisman SE, Riesen WF et al. Insulin-like growth factor-I improves glucose and lipid metabolism in type 2 diabetes mellitus. J Clin Invest 1992; 90(6):2234-41.
66. Schalch DS, Turman NJ, Marcsisin VS et al. Short-term effects of recombinant human insulin-like growth factor I on metabolic control of patients with type II diabetes mellitus. J Clin Endocrinol Metab 1993; 77(6):1563-8.
67. Moses AC, Young SC, Morrow LA et al. Recombinant human insulin-like growth factor I increases insulin sensitivity and improves glycemic control in type II diabetes. Diabetes 1996; 45(1):91-100.
68. Bergman RN, Finegood DT, Ader M. Assessment of insulin sensitivity in vivo. Endocr Rev 1985; 6(1):45-86.
69. Jabri N, Schalch DS, Schwartz SL et al. Adverse effects of recombinant human insulin-like growth factor I in obese insulin-resistant type II diabetic patients. Diabetes 1994; 43(3):369-74.
70. RINDS. Safety profiling of rhIGF-I therapy in patients with NIDDM: a dose-ranging, placebo-controlled trial. Diabetes 1996; 45(Suppl 2):71A.
71. RINDS. Evidence from a dose-ranging study that recombinant insulin-like growth factor I (RhIGF-I) effectively and safely improves glycamic control in non-insulin dependent diabetes mellitus. Diabetes 1996; 45(Suppl 2):27A.
72. Thrailkill KM, Quattrin T, Baker L et al. Cotherapy with recombinant human insulin-like growth factor I and insulin improves glycemic control in type 1 diabetes. RhIGF-I in IDDM Study Group. Diabetes Care 1999; 22(4):585-92.
73. Moller N, Orskov H. Does IGF-I therapy in insulin-dependent diabetes mellitus limit complications? Lancet 1997; 350(9086):1188-9.
74. Poulsen JE. The Houssay phenomenon in man: recovery from retinopathy in a case of diabetes with Simmond's disease. Diabetes 1953; 2:7-12.
75. Lundbaek K, Christensen NJ, Jensen VA et al. Diabetes, diabetic angiopathy, and growth hormone. Lancet 1970; 2(7664):131-3.
76. Merimee TJ, Zapf J, Froesch ER. Insulin-like growth factors. Studies in diabetics with and without retinopathy. N Engl J Med 1983; 309:527-30.
77. Wang Q, Dills DG, Klein R et al. Does insulin-like growth factor I predict incidence and progression of diabetic retinopathy? Diabetes 1995; 44(2):161-4.
78. Arner P, Sjoberg S, Gjotterberg M et al. Circulating insulin-like growth factor I in type 1 (insulin-dependent) diabetic patients with retinopathy. Diabetologia 1989; 32(10):753-8.
79. Feldmann B, Jehle PM, Mohan S et al. Diabetic retinopathy is associated with decreased serum levels of free IGF-I and changes of IGF-binding proteins. Growth Horm IGF Res 2000; 10(1):53-9.
80. Janssen JA, Lamberts SW. Circulating IGF-I and its protective role in the pathogenesis of diabetic angiopathy. Clin Endocrinol (Oxf) 2000; 52(1):1-9.
81. Amin R, Ong K, Schultz C et al. Hormonal influences on the risk of microalbuminuria in adolescent girls with type 1 diabetes. A longitudinal study. In: ADA 61st Scientific Sessions. Philadelphia: 2001:698-P.
82. Flyvbjerg A, Thorlacius-Ussing O, Naeraa R et al. Kidney tissue somatomedin C and initial renal growth in diabetic and uninephrectomized rats. Diabetologia 1988; 31(5):310-4.
83. Landau D, Domene H, Flyvbjerg A et al. Differential expression of renal growth hormone receptor and its binding protein in experimental diabetes mellitus. Growth Horm IGF Res 1998; 8(1):39-45.

84. Flyvbjerg A, Frystyk J, Thorlacius-Ussing O et al. Somatostatin analogue administration prevents increase in kidney somatomedin C and initial renal growth in diabetic and uninephrectomized rats. Diabetologia 1989; 32(4):261-5.
85. Segev Y, Landau D, Marbach M et al. Renal hypertrophy in hyperglycemic non-obese diabetic mice is associated with persistent renal accumulation of insulin-like growth factor I. J Am Soc Nephrol 1997; 8(3):436-44.
86. Doi T, Striker LJ, Quaife C et al. Progressive glomerulosclerosis develops in transgenic mice chronically expressing growth hormone and growth hormone releasing factor but not in those expressing insulinlike growth factor-1. Am J Pathol 1988; 131(3):398-403.
87. Doi T, Striker LJ, Gibson CC et al. Glomerular lesions in mice transgenic for growth hormone and insulinlike growth factor-I. I. Relationship between increased glomerular size and mesangial sclerosis. Am J Pathol 1990; 137(3):541-52.
88. Segev Y, Landau D, Rasch R et al. Growth hormone receptor antagonism prevents early renal changes in nonobese diabetic mice. J Am Soc Nephrol 1999; 10(11):2374-81.
89. Landau D, Segev Y, Afargan M et al. A novel somatostatin analogue prevents early renal complications in the nonobese diabetic mouse. Kidney Int 2001; 60(2):505-12.
90. Cummings EA, Sochett EB, Dekker MG et al. Contribution of growth hormone and IGF-I to early diabetic nephropathy in type 1 diabetes. Diabetes 1998; 47(8):1341-6.
91. Shinada M, Akdeniz A, Panagiotopoulos S et al. Proteolysis of insulin-like growth factor-binding protein-3 is increased in urine from patients with diabetic nephropathy. J Clin Endocrinol Metab 2000; 85(3):1163-9.
92. Chin E, Bondy C. Insulin-like growth factor system gene expression in the human kidney. J Clin Endocrinol Metab 1992; 75(3):962-8.
93. Salardi S, Cacciari E, Ballardini D et al. Relationships between growth factors (somatomedin-C and growth hormone) and body development, metabolic control, and retinal changes in children and adolescents with IDDM. Diabetes 1986; 35(7):832-6.
94. Hyer SL, Sharp PS, Brooks RA et al. Serum IGF-1 concentration in diabetic retinopathy. Diabet Med 1988; 5(4):356-60.
95. Dills DG, Moss SE, Klein R et al. Is insulinlike growth factor I associated with diabetic retinopathy? Diabetes 1990; 39(2):191-5.
96. Dills DG, Moss SE, Klein R et al. Association of elevated IGF-I levels with increased retinopathy in late-onset diabetes. Diabetes 1991; 40(12):1725-30.
97. Janssen JA, Jacobs ML, Derkx FH et al. Free and total insulin-like growth factor I (IGF-I), IGF-binding protein-1 (IGFBP-1), and IGFBP-3 and their relationships to the presence of diabetic retinopathy and glomerular hyperfiltration in insulin-dependent diabetes mellitus. J Clin Endocrinol Metab 1997; 82(9):2809-15.
98. Burgos R, Mateo C, Canton A et al. Vitreous levels of IGF-I, IGF binding protein 1, and IGF binding protein 3 in proliferative diabetic retinopathy: a case-control study. Diabetes Care 2000; 23(1):80-3.
99. Spranger J, Buhnen J, Jansen V et al. Systemic levels contribute significantly to increased intraocular IGF-I, IGF-II and IGF-BP3 [correction of IFG-BP3] in proliferative diabetic retinopathy. Horm Metab Res 2000; 32(5):196-200.
100. Grant MB, Mames RN, Fitzgerald C et al. Insulin-like growth factor I acts as an angiogenic agent in rabbit cornea and retina: comparative studies with basic fibroblast growth factor. Diabetologia 1993; 36(4):282-91.
101. Meyer-Schwickerath R, Pfeiffer A, Blum WF et al. Vitreous levels of the insulin-like growth factors I and II, and the insulin-like growth factor binding proteins 2 and 3, increase in neovascular eye disease. Studies in nondiabetic and diabetic subjects. J Clin Invest 1993; 92(6):2620-5.
102. Smith LE, Kopchick JJ, Chen W et al. Essential role of growth hormone in ischemia-induced retinal neovascularization. Science 1997; 276(5319):1706-9.
103. Smith LE, Shen W, Perruzzi C et al. Regulation of vascular endothelial growth factor-dependent retinal neovascularization by insulin-like growth factor-1 receptor. Nat Med 1999; 5(12):1390-5.
104. Sharp PS, Fallon TJ, Brazier OJ et al. Long-term follow-up of patients who underwent yttrium-90 pituitary implantation for treatment of proliferative diabetic retinopathy. Diabetologia 1987; 30(4):199-207.
105. Flier JS, Moses AC. Diabetes in acromegaly and other endocrine disorders. In: DeGroot LJ, ed. Endocrinology. Philadelphia: W.B. Saunders, 1989:1389-1399.
106. Early worsening of diabetic retinopathy in the Diabetes Control and Complications Trial. Arch Ophthalmol 1998; 116(7):874-86.
107. Migdalis IN, Kalogeropoulou K, Kalantzis L et al. Insulin-like growth factor-I and IGF-I receptors in diabetic patients with neuropathy. Diabet Med 1995; 12(9):823-7.
108. Russell JW, Sullivan KA, Windebank AJ et al. Neurons undergo apoptosis in animal and cell culture models of diabetes. Neurobiol Dis 1999; 6(5):347-63.
109. Rhodes CJ. IGF-I and GH post-receptor signaling mechanisms for pancreatic beta-cell replication. J Mol Endocrinol 2000; 24(3):303-11.
110. Mitchell P. Cancelled IGF-1 trials bode ill for diabetic patients. Lancet 1997; 350(9091):1606.
111. Clemmons DR, Moses AC, McKay MJ et al. The combination of insulin-like growth factor I and insulin-like growth factor-binding protein-3 reduces insulin requirements in insulin-dependent type 1 diabetes: evidence for in vivo biological activity. J Clin Endocrinol Metab 2000; 85(4):1518-24.

CHAPTER 31

Insulin-Like Growth Factors in Critical Illness

Greet Van den Berghe

Abstract

Low circulating levels of insulin-like growth factor-I (IGF-I) and alterations in IGF-binding proteins (IGFBPs) mark the catabolic state of critical illness. The origin of these changes appears different during the first hours to days after onset and in the more chronic phase of critical illness.

The initial changes are low serum concentrations of IGF-I, IGFBP-3 and the acid-labile subunit (ALS) in the presence of activated growth hormone (GH) secretion, indicating peripheral GH-resistance. Reduced GH-receptor expression and/or impairment of GH signalling at the intracellular level may play a role. In addition, increased IGFBP-3 protease activity in serum as well as increased circulating levels of IGFBP-1 may alter IGF-I tissue availability. In prolonged critical illness, GH secretion is no longer elevated and often low, and its pattern is erratic and almost non-pulsatile. The reduced pulsatile component of GH secretion in this phase contributes to inadequate generation of GH-dependent IGF-I and binding proteins such as ALS, IGFBP-3 and IGFBP-5 and to impaired anabolism as indicated by biochemical markers. High IGFBP-1 and low insulin levels predict fatal outcome of prolonged critical illness and serum concentrations of IGFBP-2, IGFBP-4 and IGFBP-6 are uniformly elevated. Continuous infusion of GH-secretagogues reactivates pulsatile GH secretion in prolonged critical illness and evokes a proportionate rise in the GH-dependent, IGF-I and the ternary complex binding proteins IGFBP-3, ALS and GFBP-5, indicating recovery of GH responsiveness in this phase of illness. Only when GH-secretagogues are infused together with thyrotropin (TSH)—releasing hormone (TRH), whereby circulating levels of thyroid hormones are normalized, metabolic improvement ensues. This suggests a form of 'IGF-I resistance', which is resolved by also correcting the concomitant tertiary hypothyroidism.

These studies indicate a fundamental transition between acute and prolonged critical illness, from acute GH-resistance to a state of low (pulsatile) GH (and TSH) secretion but recovered GH responsiveness. Treatment with releasing factors (GH-secretagogues and TRH) takes advantage of active feedback inhibition loops and thus prevents overstimulation. Hence, this strategy may be a safer one than the administration of (high-doses) GH and/or IGF-I to counter the catabolic state in prolonged critical illness.

Introduction

By definition, critical illness is any condition requiring support of failing vital organ functions, either with mechanical aids (such as mechanical ventilation, hemodialysis or -filtration or cardiac assist devices) or pharmacological agents (such as inotropes or vasopressors), without which death would ensue.[1] According to this definition, critical illness can be caused by different types of insults such as surgical or traumatic injury or diseases often of an infectious origin. Patients suffering from such a life-threatening condition are generally treated in specialized intensive care units (ICUs), where permanent monitoring and high technological interventions are available and medical treatment is continuously adjusted in order to optimize chances for survival. In the predominantly surgical intensive care unit, two thirds of the admitted patients recover quickly and are discharged within 1-5 days. For the remaining one third of patients, onset of recovery does not follow within a few days of intensive medical care, hence critical illness becomes "prolonged" and intensive care is further

Insulin-Like Growth Factors, edited by Derek LeRoith, Walter Zumkeller and Robert Baxter. ©2003 Eurekah.com and Kluwer Academic / Plenum Publishers.

needed, often for weeks, sometimes months. Prolonged critical illness in this setting reflects a novel pathophysiological condition. Indeed, modern intensive care, providing mechanical ventilation, artifical feeding and vital organ support, has been available only for the last three decades and without intensive care, patients with such life-threatening conditions would not survive more than a few days. Prolonged critically ill patients suffer from a peculiar type of hypercatabolism which occurs despite adequate parenteral and / or enteral nutrition. Large amounts of protein continue to be lost from lean tissue, such as skeletal muscle, bone and solid organs, which causes impairment of vital functions, weakness and delayed or hampered recovery.[2,3] Furthermore, and in contrast to what occurs in the early phase of severe illnesses when lipolysis provides fatty acids for metabolism, prolonged critically ill patients no longer efficiently use fatty acids as metabolic substrates.[3] They store fat with feeding, both in adipose tissue and as fatty infiltrates in vital organs such as the pancreas and the liver.[1,2] The lean tissue hypercatabolism, consisting of accelerated breakdown of protein and impaired synthesis of protein in skeletal muscle and bone, [4,5,6] is a major, frustrating and resource-consuming clinical problem as it leads to persisting dependency on intensive medical care including mechanical ventilation, despite adequate and succesful treatment of the underlying disease that had initially warranted admission to the intensive care unit. In addition, patients become increasingly more susceptible to develop potentially lethal complications, mostly of an infectious origin. Indeed, mortality from prolonged critical illness is high: typically the risk of death is around 20% for adult patients with an ICU stay of >5 days and around 25% for those with an ICU stay of >21days.[4] Incidentally, male patients suffering from prolonged critical illness seem to have a higher risk of adverse outcome than female patients.[4] The reason for this gender difference in outcome remains obscure. In line with the foregoing is the inability of the classical scoring systems for severity of illness[7] to predict mortality in an individual long-stay intensive care patient. This enigma reflects the current absence of knowledge on the pathophysiological mechanisms underlying onset of recovery or, conversely, the lack of recovery in prolonged critically ill patients.

The complex system interrelating growth hormone (GH), the insulin-like growth factors (IGFs) and the IGF binding proteins (IGFBPs) is important for postnatal growth, differentiation, metabolic homeostasis and healing.[8,9] GH and the IGFs, either directly or indirectly, interfere with the function of almost every organ system in the body and the target tissue effects of IGFs are regulated by at least 6 binding proteins. Pronounced changes in the GH/IGF/IGFBP-axis occur immediately and uniformly in response to all types of severe illnesses, independent of its cause being a surgical or traumatic injury or a toxic/infectious challenge. It recently became clear that the nature of these endocrine changes differs when a lethal outcome of a severe illness is avoided by intensive care and critical illness becomes prolonged for weeks and months in the very unnatural setting of a modern ICU.[10]

This chapter gives an overview of the dynamic alterations within the GH/IGF/IGFBP-system in the human condition of intensive care-treated critical illness and the distinction between the acute and the chronic phase is specifically highlighted.

Changes within the IGF-System in the Acute Phase of Critical Illness

During the first hours or days after an acute, stressful insult, such as surgical or traumatic injury or severe infection, serum concentrations of IGF-I and the ternary complex, GH-dependent binding protein IGFBP-3 and its acid-labile subunit (ALS) decrease, all of which is preceded by a drop in serum levels of GH-binding protein (GHBP).[11,12] The latter was found to occur in parrallel with reduced GH-receptor expression in peripheral tissues in patients after elective abdominal surgery.[12] Animal models have shown that a rapid impairment of GH signalling at the intracellular level may also be involved.[13] As all models of acute illness, including the postoperative condition in patients, are accompanied by at least partial starvation, it is difficult to determine to what extent the alterations are due to the insult or to the concomitant malnutrition.[14] Furthermore, stress acutely stimulates GH secretion (Fig. 1).[15,16] In normal physiology, GH is released from the somatotropes in a pulsatile fashion, under the interactive control of the stimulatory hypothalamic GH-releasing hormone (GHRH) and the inhibitory somatostatin.[9] Since the 1980s, a series of synthetic GH-releasing peptides (GHRPs) and non-peptide analogues have been developed that potently release GH, through a specific G-protein coupled receptor located in the hypothalamus and the pituitary.[17,18] It now appears that there exists at least one highly conserved endogenous ligand for

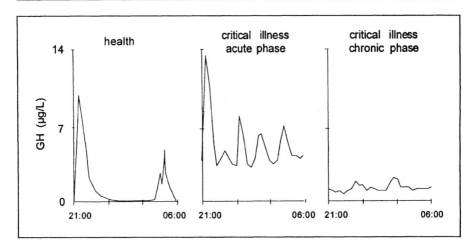

Figure 1. Nocturnal serum concentration profiles of GH illustrating the differences between the acute phase and the chronic phase of critical illness within an intensive care setting. Adapted, with permission, from.[10]

this receptor,[19] named "ghrelin", which originates in peripheral tissues such as the stomach as well as in the hypothalamic arcuate nucleus and which seems to be another key factor in the complex physiological regulation of pulsatile GH secretion. As orginally shown in rodents,[20] there is now also evidence that in the human,[21] the pulsatile nature of GH secretion is important for its metabolic effects.[4, 22] In the acute phase of stress, as after surgery, trauma or onset of sepsis, circulating GH levels become elevated and the normal GH profile, consisting of peaks alternating with virtually undetectable troughs, is altered: peak GH levels but even more so the interpulse concentrations are high and the GH pulse frequency is elevated (Fig. 1).[10,23,24] It is still unclear which factor ultimately controls the stimulation of GH release in response to stress. As in starvation,[25] more frequent withdrawal of the inhibitory somatostatin and/or an increased availability of stimulatory (hypothalamic and/or peripheral) GH-releasing factors could hypothetically be involved. The constellation of low circulating levels of IGF-I, IGFBP-3 and ALS and amplified GH secretion, as uniformly occurs in experimental human and animal models of acute stress and in acutely ill patients, is therefore a classical example of acquired peripheral GH resistance.[11,23] This is most likely brought about by the effects of cytokines such as TNF-α, IL-1 and IL-6. It has been hypothesised, but it remains unproved, that the primary events in acute illness are cytokine and/or starvation-induced reduced GH receptor expression and impairment of GH signalling at the intracellular level[13] and hence low circulating IGF-I levels which in turn, through reduced negative feedback inhibition, drive the abundant release of GH. The large amounts of GH may then exert direct (IGF-I independent) lipolytic, insulin-antagonising and immune stimulatory actions, while the indirect IGF-I-mediated somatotropic effects may be attenuated.[26, 27] This phenomenon would make sense in stressful conditions. Indeed, the set of alterations within the GH/IGF-I axis could contribute to the provision of metabolic substrates (glucose, free fatty acids, amino-acids such as glutamine) for vital organs such as the brain and the heart and for host defence and thus could be conceived as adaptive and beneficial for survival. However, the adapative nature of low IGF-availability is merely a theoretical concept as it has never been proven. An alternative interpretation could be that increased IGFBP-3 protease activity in plasma, also reported in acute illnesses,[11,28] results in facilitated dissociation of IGF-I from the ternary complex which could theoretically be an adaptive escape mechanism to secure IGF-I activity at the tissue level.[28] Once more, proof for such a phenomenon is hitherto lacking.

Circulating levels of the small, binary complex IGF-binding proteins, such as IGFBP-1 and IGFBP-2 have been reported to be elevated in the acute phase of critical illness.[11,29] It is unclear whether the changes in these IGF-binding proteins enhance or reduce the tissue effects of IGF-I. A high serum IGFBP-1 level on ICU admission has been found to be associated with a more negative nitrogen balance during the first two days of intensive care.[11] Part of this association can be explained by the effects of relative starvation, which is uniformly present in all models of acute stress, including acute critical illness in the ICU. Indeed, when patients are starved on admission to the ICU, they will lose more lean tissue over the next few days.

Changes within the IGF-System in the Chronic Phase of Critical Illness

In prolonged critical illness, the changes observed within the somatotropic axis are different. Firstly, the pattern of GH secretion becomes very chaotic and the amount of GH which is released in pulses is now much reduced compared to the acute phase (Fig. 1).[5,30,31] Moreover, although the non-pulsatile fraction is still somewhat elevated and the number of pulses is still high, mean nocturnal GH serum concentrations are hardly elevated when compared to the healthy, non-stressed condition, and substantially lower than in the acute phase of stress.[10] We observed that, when intensive care patients are studied from 7-10 days illness onward in the absence of drugs known to exert profound effects on GH secretion such as dopamine,[32,33] calcium entry blockers or glucocorticoids to name but a few, they present uniformly with mean nocturnal GH levels of about 1 µg/L, trough (interpulse) levels that are easily detectable and thus still elevated, and peak GH levels that hardly ever exceed 2 µg/L, and this, surprisingly, is independent of the patient's age, body composition and type of underlying disease.[4,10] Secondly, specifically the pulsatile component of GH secretion, which is substantially reduced, has been found to correlate positively with circulating levels of IGF-I, IGFBP-3 and ALS, which are all low.[5,30,31] In other words, the more the pulsatile GH secretion is suppressed, the lower the circulating levels of the GH-dependent IGF-I and ternary complex binding proteins become. This is not what one would expect if GH resistance were the primary cause of the low IGF-I levels, which would result in an inverse correlation or no correlation between GH secretion and circulating IGF-I. The recently documented elevated serum levels of GH-binding protein,[4] assumed to reflect GH receptor expression in peripheral tissues, in prolonged critically ill patients compared with those measured in a matched control group are in line with recovery of GH responsiveness with time during severe illness.[4,5] It seems that the lack of pulsatile GH secretion in the condition of prolonged stress is contributing to the low circulating levels of IGF-I and ternary complex binding proteins. Moreover, it was demonstrated that these low serum levels of GH-dependent IGF-I and binding proteins (IGFBP-3, ALS, IGFBP-5) are tightly related to biochemical markers of impaired anabolism, such as low serum osteocalcin and leptin concentrations during prolonged critical illness.[5] Together, these findings suggest that a relative hyposomatotropism, as demonstrated by a lack of pulsatile GH secretion, participates in the pathogenesis of the particular lean tissue wasting condition distinctively in the chronic phase of critical illness. In line with a higher risk for adverse outcome associated with male gender,[4] men appear to do worse than women in the sense that they lose more of the pulsatility and regularity within the GH-secretory pattern when critical illness progresses (despite indistinguishable total GH output) and concomitantly reveal even lower IGF-I and ALS levels than their female counterparts (Fig. 2).[4] It remains unclear if this sexual dimorphism within the GH/IGF-I axis is causally related to the gender difference in outcome of prolonged critical illness or merely reflects a casual association.

Serum concentrations of the small binding proteins IGFBP-2, IGFBP-4 and IGFBP-6 are clearly elevated in prolonged critical illness,[5,6] the cause of which still remains unclear. Also, the consequences of these changes for metabolism are not known. Serum IGFBP-1 concentrations in the chronic phase of critical illness are much lower than those observed in the acute phase, which is probably due to the effect of feeding. Indeed, parenteral and/or enteral feeding is built up over the first few days of intensive care to a maintenance of about 25 kCal per kilogram bodyweight per 24h, composed of normal and equilibrated amounts of glucose, protein and lipids. However, serum IGFBP-1 levels in prolonged critical illness still appear to correlate with lean tissue wasting, which occurs despite feeding.[4,5,10] Selectively in patients who subsequently won't survive, serum IGFBP-1 concentrations increase again when deterioration starts, a noticeable distinction between survivors and non-survivors present several weeks before death. Indeed, a high serum BP-1 concentration in the chronic phase of critical illness, in the fed state, seems to predict an adverse outcome of chronic critical illness (Fig. 3).[5 and Van den Berghe, unpublished observations] IGFBP-1 is distinct among the members of the IGFBP family in it being acutely regulated by metabolic stimuli.[34] Studies with cultured human liver explants suggest that the major regulatory influences on IGFBP-1 production are insulin, which is inhibitory, and hepatic substrate deprivation, which is stimulatory, acting through a cyclic AMP-dependent mechanism.[35,36] Moreover, an inverse correlation of IGFBP-1

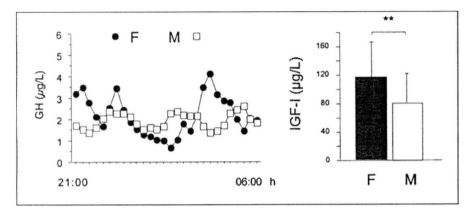

Figure 2. The more "feminized" pattern of GH secretion (more irregular and less pulsatile GH secretory pattern for an identical mean nocturnal GH level) in prolonged critically ill men compared to women is illustrated by the representative nocturnal (21:00h-06:00h) GH serum concentration series (sampling every 20 minutes) obtained in a male (squares) and a matched female (circles) patient. Concomitantly, protracted critically ill men have lower circulating levels of IGF-I than female patients. IGF-I results are presented as mean ± SD. ** P < 0.01 Adapted, with permission, from.[4]

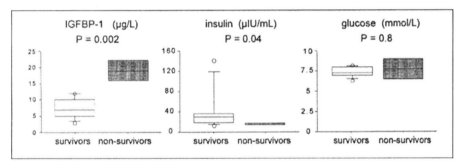

Figure 3. Serum IGFBP-1 concentration were found to be higher in non-survivors compared to survivors in prolonged critical illness. With permission, from [1]. Concomitantly, non-survivors revealed lower serum insulin levels for the same blood glucose level. Box plots represent medians, P25-P75 and P10-P90 and circles represent the absolute values for outliers.

with IGF-I and the GH-dependent proteins ALS and IGFBP-3 during critical illness is consistent with its inverse regulation by GH, as previously suggested.[37,38,39] The higher IGFBP-1 levels observed in prolonged critically ill patients who did not survive coincided with lower insulin concentrations compared to survivors, for the same range of blood glucose level; a surprising finding considering that these patients are thought to be insulin resistant (Fig. 3). Whether or not this indicates that also insulin secretion is becoming impaired in the long-stay intensive care patients remains unclear. It is clear however that in unfavourable metabolic conditions, the hepatocyte alters its production of IGF-regulatory proteins, for which the trigger might be reduced hepatocyte substrate availability (theoretically caused by either hepatic hypoperfusion or hypoxia, hypoglycemia, relative insulin deficiency or hepatic insulin resistance) leading to increased cyclic AMP production, which would both suppress IGF-I and ALS[40] and stimulate IGFBP-1.[35] It is unclear to what extent loss of GH pulsatility may contribute to this switch, but recent data[35] suggest that activation of hepatic IGF-I and ALS expression may require pulsatile GH, and animal studies similarly suggest that suppression of hepatic IGFBP-1 expression by insulin requires acute, rather than prolonged or non-pulsatile, GH action.[41] Further exploring the apparent link between serum IGFBP-1 levels, insulin and outcome of prolonged critical illness will shed new light on the pathophysiological processes crucial for recovery and survival. We recently showed that strict glycemic control below 110 mg/dL with intensive insulin therapy indeed substantially reduces morbidity and mortality of intensive care-dependent critical illness.[42]

Pathophysiology of Chronic Changes

As impaired pulsatile GH secretion in the chronic phase of critical illness seems to contribute to the low IGF-I and GH-dependent IGFBPs; the ensuing question is, what is its cause? Is the pituitary taking part in the "multiple organ failure syndrome" becoming unable to synthesise and secrete GH? Or, alternatively, is the lack of pulsatile GH secretion due to increased somatostatin tone and/or to a reduced stimulation by the endogenous releasing factors such as GHRH and/or ghrelin? Studying GH responses to administration of GH-secretagogues (GHRH and GHRP), in a dose which is known to evoke a maximal GH response in healthy volunteers, enables, to a certain extent, to differentiate between a primarily pituitary and a hypothalamic origin of the relatively impaired GH release in critically ill patients. Indeed, the combined administration of GHRH and GHRP appears to be the most powerful stimulus for pituitary GH release in humans.[43] A low GH response in critical illness would thus be compatible with a pituitary dysfunction and/or a high somatostatin tone whereas a high GH response would indicate reduced (hypothalamic) stimulation of the somatotropes.

We found that GH responses to a bolus injection of GHRP are high in long-stay intensive care patients and several-fold higher than the response to GHRH, the latter being normal or often subnormal.[44] GHRH+GHRP evokes a clear synergistic response in this condition, revealing the highest GH responses ever reported in a human study.[44] The exuberant GH responses to secretagogues exclude the possibility that the relatively impaired pulsatile GH secretion during protracted critical illness is due to a lack of pituitary capacity to synthesise GH or is due to accentuated somatostatin-induced suppression of GH release. Inferentially, one of the mechanisms which could be involved is reduced availability of ghrelin or another putative endogenous ligand for the GHRP receptor. Ultimately, the combination of low availability of somatostatin and of an endogenous GHRP-like ligand emerges as a plausible mechanism that clarifies:

  (i) the reduced GH burst amplitude,
  (ii) the increased frequency of spontaneous GH secretory bursts and
  (iii) the elevated interpulse levels as well as
  (iv) the striking responsiveness to GHRP alone or in combination with GHRH, and this without markedly increased responsiveness to GHRH alone. Female patients with prolonged critical illness have a markedly higher response to a bolus of GHRP compared to male patients, a difference which is lost when GHRH is injected together with GHRP (Fig. 4).[4] Less endogenous GHRH action in prolonged critically ill men, possibly due to the concomitant profound hypoandrogenism[4], accompanying loss of action of an endogenous GHRP-like ligand with prolonged stress in both genders, may explain this finding.

Effects of Releasing Factors in the Chronic Phase of Critical Illness

The hypothesis of reduced endogenous stimulation of GH secretion and recovery of peripheral GH sensitivity in patients who are critically ill for a prolonged duration was further explored by examining the effects of continuous infusion of GHRP with or without GHRH. Continuously infusing GHRP (1 µg/kg/h) alone, and even more so the combination of GHRH and GHRP (1+1 µg/kg/h), for up to two days was found to substantially amplify pulsatile GH secretion (> 6-fold and > 10-fold respectively) in this condition, without altering the relatively high burst frequency (Fig. 5).[30,31] Reactivated pulsatile GH secretion was accompanied by a proportionate rise in serum IGF-I (66% and 106%, respectively), IGFBP-3 (50% and 56%) and ALS (65% and 97%) indicating peripheral GH-responsiveness (Fig. 5).[5,31] The presence of considerable responsiveness to reactivated pulsatile GH secretion in these patients and the high serum levels of GHBP clearly delineates the distinct pathophysiological paradigm present in the chronic phase of critical illness as opposed to the acute phase, which is thought to be primarily a condition of GH-resistance. After two days treatment with GHRP, (near)normal levels of IGF-I, IGFBP-3, IGFBP-5 and ALS are reached and, as shown in a subsequent study, maintained for at least up to 5 days (Fig. 6).[5] GH secretion after 5

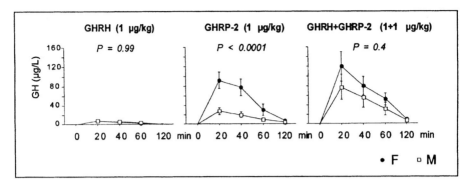

Figure 4. Responses (increments above baseline) of GH obtained 20, 40, 60 and 120 minutes after intravenous bolus administration of GHRH (1 μg/kg), GHRP-2 (1 μg/kg), and GHRH+GHRP-2 (1+1 μg/kg) in matched male and female protracted critically ill patients. Five men and 5 women were randomly allocated to each secretagogue group. Results are presented as mean ± SEM. Circles depict results from female and squares from male patients. P-values were obtained using repeated measures ANOVA. Adapted, with permission, from.[4]

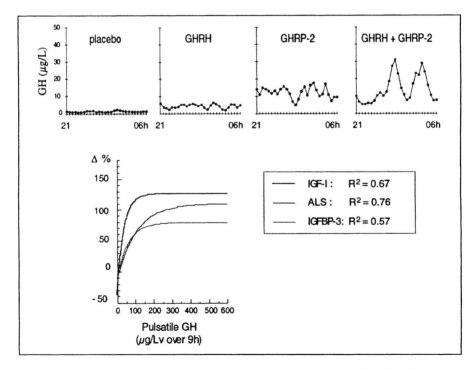

Figure 5. Nocturnal serum GH profiles in the prolonged phase of illness illustrating the effects of continuous infusion of placebo, GHRH (1 μg/kg/h), GHRP-2 (1 μg/kg/h), or GHRH+GHRP-2 (1+1 μg/kg/h). Exponential regression lines have been reported between pulsatile GH secretion and the changes in circulating IGF-I, ALS and IGFBP-3 obtained with 45h infusion of either placebo, GHRP-2 or GHRH+GHRP-2. They indicate that the parameters of GH responsiveness increase in proportion to GH secretion up to a certain point, beyond which further increase of GH secretion has apparently little or no additional effect. It is noteworthy that the latter point corresponds to a pulsatile GH secretion of approximately 200 μg/Lv over 9h, or less, a value that can usually be evoked by the infusion of GHRP-2 alone. In the chronic, non-thriving phase of critical illness, GH-sensitivity is clearly present, in contrast to the acute phase of illness, which is thought to be primarily a condition of GH resistance. From [10], with permission.

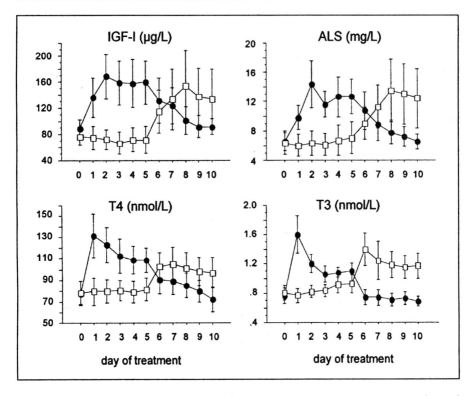

Figure 6. Serum concentrrations (mean ± SEM) of IGF-I, ALS, T4 and T3 in response to a randomized treatment with either five days GHRP-2+TRH infusion (1+1 µg/kg/h) followed by 5 days placebo (filled symbols) or 5 days placebo followed by five days GHRP-2+TRH infusion (1+1 µg/kg/h) (open symbols) in a group of 10 male and 4 female prolonged critically ill patients. All $P < 0.0001$ with ANOVA. The mean age of the patients was 68y. The mean intensive care stay at the time of study start was 40 days. Adapted with permission from.[1]

days treatment with GH-secretagogues was found to be lower than after two days, suggesting active feed-back inhibition loops which prevented overtreatment.[5,31] In this study, in which GHRP was infused together with TRH for 5 days, the self-limited endocrine responses induced a shift towards anabolism at the level of several peripheral tissues, as indicated by a rise in serum levels of osteocalcin, insulin and leptin and a decrease in urea production.[5] These peripheral responses seem to depend on the co-infusion of TRH with concomitant normalisation of the thyroid axis. Indeed, GHRP-2 infused alone evokes identical increments in serum concentrations of IGF-I, IGFBP-3 and ALS, but is devoid of the anabolic tissue responses that are present with the combined infusion of GHRP and TRH.[45] Usually, infusion of GHRP without GHRH suffices to reactivate pulsatile GH secretion sufficiently in the prolonged critically ill patient and to elicit the IGF-I and IGFBP responses in prolonged critical illness. However, in critically ill men, in particular those men who are being treated in the intensive care unit for a very long time (several weeks), it may be necessary to add a low dose of GHRH (0.1 µg/kg/h)[1, Van den Berghe G, unpublished observations] because of the simultaneous lack of endogenous GHRH activity accompanying the reduced availability of the GHRP-like ligand.[4]

Conclusion

The changes within the GH/IGF/IGFBP axis during the acute phase of critical illness are low serum concentrations of IGF-I, IGFBP-3 and ALS in the presence of enhanced GH secretion, indicating peripheral GH-resistance. Reduced GH-receptor expression and/or increased IGFBP-3 protease activity in serum as well as increased levels of IGFBP-1 may play a role. In the chronic phase of critical illness, GH secretion is no longer elevated and often low, and its pattern is erratic and almost non-pulsatile. The reduced pulsatile component of GH secretion in this phase contributes to inadequate generation of GH-dependent IGF-I and binding proteins such as ALS, IGFBP-3 and IGFBP-5 and to impaired anabolism as indicated by biochemical markers. High IGFBP-1 and low insulin levels predict fatal outcome of prolonged critical illness and serum concentrations of IGFBP-2, IGFBP-4 and IGFBP-6 are uniformly elevated. Continuous infusion of GH-secretagogues reactivates pulsatile GH secretion and evokes a proportionate rise in the GH-dependent, ternary complex IGF(BP)s whereas IGFBP-1 production is relatively suppressed. Only when circulating levels of thyroid hormones are concomitantly normalised, treatment with GH-secretagogues evokes metabolic improvement.

References

1. Van den Berghe G. Novel insights into the neuroendocrinology of critical illness. Eur J Endocrinol 2000; 143:1-13
2. Carroll PV. Protein metabolism and the use of growth hormone and insulin-like growth factor-I in the critically ill patient. GH & IGF Res 1999; 9:400-413
3. Streat SJ, Beddoe AH, Hill GL. Aggressive nutritional support does not prevent protein loss despite fat gain in septic intensive care patients. J Trauma 1987; 27:262-266
4. Van den Berghe G, Baxter RC, Weekers F et al. A paradoxical gender dissociation within the growth hormone / insulin-like growth factor I axis during protracted critical illness. J Clin Endocrinol Metab 2000; 85:183-192.
5. Van den Berghe G, Wouters P, Weekers F et al. Reactivation of pituitary hormone release and metabolic improvement by infusion of growth hormone releasing peptide and thyrotropin-releasing hormone in patients with protracted critical illness. J Clin Endocrinol Metab 1999; 84:1311-1323.
6. Van den Berghe G, Weekers F, Baxter RC et al. Five days pulsatile GnRH administration unveils combined hypothalamic-pituitary-gonadal defects underlying profound hypoandrogenism in men with prolonged critical illness. J Clin Endocrinol Metab 2001; 86:3217-3226
7. Knaus WA, Draper EA, Wagner DP et al. APACHE II: A severity of disease classification system. Crit Care Med 1985; 13:818-829.
8. LeRoith D. Insulin-like growth factors. N Engl J Med 1997; 336:633-640
9. Thorner MO, Vance ML, Laws ER et al. The anterior pituitary. In: JD Wilson, Foster DW, Kronenberg HM et al, eds. Williams Textbook of Endocrinology, edn 9, ch 9. Philadelphia: W.B. Saunders Company, 1998:249-340.
10. Van den Berghe G, de Zegher F, Bouillon R. Acute and prolonged critical illness as different neuroendocrine paradigms. J Clin Endocrinol Metab 1998; 83:1827-1834.
11. Baxter RC, Hawker FH, To C et al. Thirty day monitoring of insulin-like growth factors and their binding proteins in intensive care unit patients. GH & IGF Res 1998; 8:455-463.
12. Hermansson M, Wickelgren R, Hammerqvist F et al. Measurement of human growth hormone receptor messenger ribonucleic acid by a quantitative polymerase chain reaction-based assay: demonstration of reduced expression after elective surgery. J Clin Endocrinol Metab 1997; 82:421-428.
13. Mao Y, Ling PR, Fitzgibbons TP et al. Endotoxin-induced inhibition of growth hormone receptor signaling in rat liver in vivo. Endocrinology 1999; 140:5505-5515.
14. Isley WL, Underwood LE, Clemmons DR. Dietary components that regulate serum somatomedin-C concentrations in humans. J Clin Invest 1983; 71:175-182
15. Noel GL, Suh HK, Stone SJG et al. Human prolactin and growth hormone release during surgery and other conditions of stress. J Clin Endcorinol Metab 1972; 35:840-51.
16. Jeffries MK, Vance ML. Growth hormone and cortisol secretion in patients with burn injury. J Burn Care Rehab 1992; 13:391-395.
17. Bowers CY, Momany FA, Reynolds GA et al. On the in vitro and in vivo activity of a new synthetic hexapeptide that acts on the pituitary to specifically release growth hormone. Endocrinology 1984; 114:1537-1545.
18. Howard AD, Feighner SD, Cully DF et al. A receptor in pituitary and hypothalamus that functions in growth hormone release. Science 1996; 273:974-977.
19. Kojima M, Hosoda H, Date Y et al. Ghrelin is a growth-hormone-releasing acylated peptide from stomach. Nature 1999; 402:656-660.
20. Gevers EF, Wit JM, Robinson IC. Growth, growth hormone (GH) binding protein, and GH receptors are differentially regulated by peak and trough components of GH secretory pattern in the rat. Endocrinology 1996; 137:1013-1018.
21. Giustina A, Veldhuis JD. Pathophysiology of the neuroregulation of growth hormone secretion in experimental animals and the human. Endocr Rev 1998; 19:717-797.
22. Hindmarsh PC, Dennison E, Pincus SM et al. A sexually dimorphic pattern of growth hormone secretion in the elderly illness. J Clin Endocrinol Metab 1999; 84:2679-2685.
23. Ross R, Miell J, Freeman E et al. Critically ill patients have high basal growth hormone levels with attenuated oscillatory activity associated with low levels of insulin-like growth factor-1. Clin Endocrinol 1991; 35:47-54.

24. Voerman HJ, Strack van Schijndel RJM, de Boer H et al. Growth hormone: secretion and administration in catabolic adult patients, with emphasis on the critically ill patient. Neth J Med 1992; 41:229-244.
25. Hartman ML, Veldhuis JD, Johnson ML et al. Augmented growth hormone secretory burst frequency and amplitude mediate enhanced GH secretion during a two day fast in normal men. J Clin Endocrinol Metab 1992; 74:757-765.
26. Bentham J, Rodriguez-Arnao-J, Ross RJ. Acquired growth hormone resistance in patients with hypercatabolism. Horm Res 1993; 40:87-91.
27. Timmins AC, Cotterill AM, Cwyfan Hughes SC et al. Critical illness is associated with low circulating concentrations of insulin-like growth factors-I and -II, alterations in insulin-like growth factor binding proteins, and induction of an insulin-like growth factor binding protein-3 protease. Crit Care Med 1996; 24:1460-1466.
28. Gibson FA, Hinds CJ. Growth hormone and insulin-like growth factors in critical illness. Int Care Med 1997; 23:369-378.
29. Rodriguez-Arnao J, Yarwood Y, Ferguson C et al. Reduction in circulating IGF-I and hepatic IGF-I mRNA levels after ceacal ligation and puncture are associated with differential regulation of hepatic IGF-binding protien-1, -2 and -3 mRNA levels. J Endocrinol 1996; 151:287-292.
30. Van den Berghe G, de Zegher F, Veldhuis JD et al. The somatotropic axis in critical illness: effect of continuous GHRH and GHRP-2 infusion. J Clin Endocrinol Metab 1997; 82:590-599
31. Van den Berghe G, de Zegher F, Baxter RC et al. Neuroendocrinology of prolonged critical illness: effect of continuous thyrotropin-releasing hormone infusion and its combination with growth hormone-secretagogues. J Clin Endocrinol Metab 1998; 83:309-319.
32. Van den Berghe G, de Zegher F, Lauwers P et al. Growth hormone secretion in critical illness: effect of dopamine. J Clin Endocrinol Metab 1994; 79:1141-1146.
33. Van den Berghe G, de Zegher F. Anterior pituitary function during critical illness and dopamine treatment Crit Care Med 1996; 24:1580-1590.
34. Yeoh SI, Baxter RC. Metabolic regulation of the growth hormone independent insulin-like growth factor binding protein in human plasma. Acta Endocrinol 1988; 119:465-473.
35. Lewitt MS, Baxter RC. Regulation of growth hormone-independent insulin-like growth factor-binding protein (BP-28) in cultured human fetal liver explants. J Clin Endocrinol Metab 1989; 69:246-252.
36. Lewitt MS, Baxter RC. Inhibitors of glucose uptake stimulate the production of insulin-like growth factor binding protein (IGFBP-1) by human fetal liver. Endocrinology 1990; 126:1527-1533
37. Baxter RC. Circulating binding proteins for the insulin-like growth factors. TEM 1993; 4:91-96.
38. Norrelund H, Fisker S, Vahl N et al. Evidence supporting a direct suppressive effect of growth hormone on serum IGFBP-1 levels, experimental studies in normal, obese and GH-deficient adults. GH & IGF Res 1999; 9:52-60.
39. Olivecrona H, Hilding A, Ekström C et al. Acute and short-term effects of growth hormone on insulin-like growth factors and their binding proteins: serum levels and hepatic messenger ribonucleic acid responses in humans. J Clin Endocrinol Metab 1999; 84:553-560.
40. Delhanty PJD, Baxter RC. The regulation of acid-labile subunit gene expression and secretion by cyclic adenosine 3',5'-monophosphate. Endocrinology 1998; 139:260-265.
41. Hu M, Robertson DG, Murphy LJ. Growth hormone modulates insulin regulation of hepatic insulin-like growth factor binding protein-1 transcription. Endocrinology 1996; 137:3702-3709.
42. Van den Berghe G, Wouters P, Weekers F et al. Intensive Insulin Therapy in Critically Ill Patients. N Engl J Med 2001; 345:1359-1367
43. Micic D, Popovic V, Doknic M et al. Preserved growth hormone (GH) secretion in aged and very old subjects after testing with the combined stimulus GH-releasing hormone plus GH-releasing hexapeptide-6. J Clin Endocrinol Metab 1998; 83:2569-2572.
44. Van den Berghe G, de Zegher F, Bowers CY et al. Pituitary responsiveness to growth hormone (GH) releasing hormone, GH-releasing peptide-2 and thyrotropin releasing hormone in critical illness. Clin Endocrinol 1996; 45:341-351.
45. Van den Berghe G, Baxter RC, Weekers F et al. The combined administration of GH-releasing peptide-2 (GHRP-2), TRH and GnRH to men with prolonged critical illness evokes superior endocrine and metabolic effects than treatment with GHRP-2 alone. Clinical Endocrinology 2002; 56:656-669.

CHAPTER 32

Laron Syndrome:
Primary GH Insensitivity or Resistance

Zvi Laron

Abstract

An up to-date description of the etiology, clinical and laboratory pathology of the Laron Syndrome (Primary GH resistance or insensitivity). is presented. Laron syndrome is a unique model to study the physiological role of IGF-I(primary IGF-I deficiency) and the IGF-I /GH relationship. Investigations during follow-up of a large cohort of patients since birth into adult age have revealed the role of IGF-I on growth, development and metabolic mechanisms in all major systems of the body. The therapeutic experience with IGF-I throws light on the similarities and disimilarities between GH and IGF-I actions.

Laron syndrome (LS) is a unique model in man to study the intrauterine and long-term postnatal effects of primary IGF-I (insulin-like growth factor-I) deficiency, the physiological role of this hormone, and its pharmacological actions.[1]

History

In 1966 our group reported three siblings with clinical and biochemical features of isolated growth hormone deficiency (IGHD) but who had abnormally high serum GH levels.[2] Within two years we collected 22 patients belonging to 14 families of Oriental Jewish descent.[3] Growth hormone (GH) resistance was diagnosed by the inability of the high endogenous GH or administration of exogenous human GH (hGH) to raise the low somatomedin-C (insulin-like growth factor-I [IGF-I]) levels.[4,5]

The first hypothesis that the defect may be due to an abnormal GH molecule was disproved by showing that the circulating GH of these patients is normal by immunologic[6,7] and radio-receptor tests using either rabbit liver[8] or human GH receptors (GH-R).[9] Proof that the disease is due to a GH- receptor (GH-R) defect was shown by us in 1984 by demonstrating that[125]I-hGH does not bind to GH-R prepared from liver membranes obtained by biopsies performed on two patients.[10]

The description of serum growth hormone binding protein (GHBP), identical in structure with the extracellular domain of the GH-R was made in 1986.[11,12] The absence of GHBP in the serum of patients with Laron syndrome was reported in 1987.[13] Subsequent cloning[14] and characterization[14,15] of the GH-R and the introduction of new techniques in molecular biology enabled the identification of a series of molecular defects of the GH-R[15-18] or in its downstream pathways.[19] Clinically the GH resistance was shown by markedly reduced linear growth resulting in dwarfism and the inability of exogenously administered hGH to raise the low serum IGF-I and stimulate growth.[4,5] Clinical use and treatment with recombinant biosynthetic IGF-I was started in 1988.[20] Animal models were described in chicken[21] and a GH-R knock-out mouse was reported recently.[22,23] The main milestones are summarized in Table 1.

Nomenclature

The name of this syndrome underwent several changes. At first called pituitary dwarfism with high serum GH[2,3] it was named later Laron dwarfism.[24] Subsequently, by consensus of several groups

Insulin-Like Growth Factors, edited by Derek LeRoith, Walter Zumkeller and Robert Baxter. ©2003 Eurekah.com.

Table 1. Milestones in the description and elucidation of the pathology of Laron syndrome

Year	Observation	First Report	Ref.
1966	Description of dwarfism with high serum GH levels	Laron et al	2
1971	Low and unresponsive IGF-I (somatomedin) and growth	Laron et al	5
1984	Absence of GH binding to human liver membranes	Eshet et al	10
1986	Description of serum GHBP	Herington & Ymer	11
	Description of serum GHBP	Baumann et al	12
1987	Absence of GHBP in the serum of patients with Laron syndrome	Daughaday & Trivedi	13
1987	Cloning of GH receptor	Leung et al	14
1988	Initiation of IGF-I treatment	Laron et al	20
1989	Characterization of the GH-R gene and description of exon deletions	Godowski et al	15
1989	Missense mutations in the GH-R	Amselem et al	16
1993	Post GH-R defect	Laron et al	19
1996	Mutation in intracellular domain	Woods et al	18
1997	Mutation in the transmembrane domain of the GH-R	Silbergeld et al	17
1997	GH-R knock-out mouse	Zhou et al	21

GH = growth hormone; GH-R = growth hormone receptor; GHBP = growth hormone binding protein

GH resistance (insensitivity) was divided into primary called Laron syndrome (LS) and secondary resistance caused by antibodies to GH or associated with other diseases[25] (Table 2).

Geographical Distribution and Genetic Aspects

Since the first description of patients with LS in Israel of Oriental Jewish origin (Yemen, Iran, Middle East and North Africa)[2,3] our cohort has increased to 54 patients comprising Israeli Jews, Arabs, Druze and Palestinian Arabs.[26,27] Other patients have been diagnosed in many parts of the world, the majority originating in the Mediterranean and Middle East areas: Spain, Italy, North Africa and Moslem countries of Jewish, Arab, Turkish and Iranian origin or in descendants of subjects originating in these regions.[28,29] A large genetic isolate has been reported from Ecuador[30] assumed to be descendants of Jews converted to Christianity fleeing the Spanish Inquisition[31] and a smaller one in the Bahamas.[32] There are numerous patients in Turkey,[33] Iran,[34] Pakistan and India,[35,36] Saudia Arabia,[24] North Africa,[37] Spain[38] and Southern Italy.[39] In many of the above populations consanguinity was and is still in practice. (Fig. 1). Isolated patients not of Mediterranean or Oriental descent have been reported from Denmark,[40] The Netherlands,[41] Poland,[42] Russia,[43] Slovenia,[44] Japan,[45,46] Cambodia,[47] Vietnam,[48] Brazil,[49] Argentina,[50] and UK.[51] A more detailed but not complete listing appears in recent reviews of this syndrome.[28,29,52,53] Additional patients are being continuously diagnosed but most are still unreported or undiagnosed.

Analysis of the Israeli cohort led to the conclusion that Laron syndrome is caused by a fully penetrant autosomal recessive mechanism.[54]

Clinical Aspects

Several descriptions of the clinical aspects of Laron syndrome have been published in recent years.[27,55-57] The following descriptions are based mainly on 54 patients of the Israeli cohort (many of them having been followed closely from infancy into adult age; when appropriate comparison is made with the large Ecuadorian cohort[56] and other reports.

Table 2. Classification of growth hormone resistance (insensitivity)

---

1. **Primary GH resistance (insensitivity) Syndrome = Laron syndrome** (hereditary)
   a. GH receptor defects (quantitative and qualitative)
   b. Abnormalities of GH signal transduction (postreceptor defects)
   c. Primary defects of synthesis and action of IGF-I
2. **Secondary GH Resistance (insensitivity) Diseases** (acquired conditions; sometimes transitory).
   a. Circulating antibodies to GH that inhibit GH action
      (hGH gene deletion patients treated with hGH)
   b. Antibodies to the GH receptor
   c. GH insensitivity caused by malnutrition
   d. GH insensitivity caused by liver disease
   e. GH insensitivity caused by uncontrolled diabetes mellitus
   f. Other conditions

---

hGH = human growth hormone
Modified from Laron et al[25]

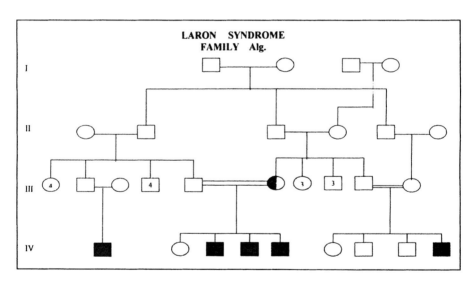

Figure 1. Pedigree with consanguinous marriages in families with Laron syndrome. Only the patients and one mother were investigated for molecular defects.

Gestation and Delivery

According to the mothers' accounts and available hospital records, the pregnancies with LS patients and deliveries were uneventful.[28,58] Some mentioned weaker movements of the fetus compared to normal siblings. Birth length (not available in all patients) revealed that the majority were short, measuring 42-46 cm, this was also found in patients from other countries. Birth weight was above 2500g in most newborns with LS, but some babies weighed below 2100g.

Congenital Malformations

One infant with LS is known to have had congenital dislocation of the hip joints. Other malformations reported in individual patients were slight aortic stenosis, cataract, and undescended testicles.[57]

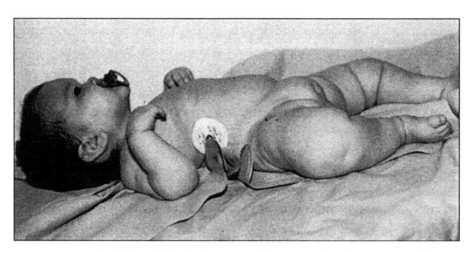

Figure 2. A one and a half year old girl with Laron syndrome. Note frontal recessions, saddle nose and marked obesity.

Features

Early Childhood

The young children resemble those with hereditary or congenital isolated GH deficiencies (Fig.2). They are short, obese and the boys have hypogenitalism and hypogonadism.[59] The head seems large but in effect, the head circumference is below the normal size or in the low normal ranges.[58,59] (Fig. 3). There is underdevelopment of the facial bones[60,61] (Fig. 4), this gives the typical appearance of protruding forehead, saddle nose and "sunset" look. The hair is sparse, thin, silky[27,62] and easy to pluck. If of Mediterranean or Mid-Eastern origin, the patients have blue sclerae.

The infants are obese, not by weight, but as evidenced by skinfold measurements or by soft tissue x-ray as the bones are thin and the muscles underdeveloped (Fig. 5). The motor development is slow. Onset of teething is delayed, and in most patients become defective[58,59] (Fig. 6). Subsequently they become crowded due to the small mandible (Fig. 7). Because of the bad quality of their teeth, neglect, and lack of financial resources for dental treatment, many of the young adult patients had their teeth extracted and wear complete prosthesis. The children and even some adult females have a very high-pitched voice[28] due to a narrow oropharynx.[63] Newborns and infants sweat profusely due to marked hypoglycemia (see Table 4),[3,55,59] crying at night until they receive a sweetened drink.

Body Proportions and Growth

The upper/lower segment ratio is above the norm for sex and age denoting short limbs for the trunk size.[64] The hands and feet are small (acromicia)[3,53,59] (Fig. 8). During infancy they wear dolls shoes as shoe shops do not carry their size, and most of their life they have difficulties in finding appropriate clothing.[65] From infancy on, linear growth is slow, the height deficit ranging between 4-10 height SDS below the median for normal height. If untreated, as most patients with LS are, their final height ranges between 116-142 cm in males and 108-136 cm in females.[66] The adult stature in the larger Ecuadorian cohort has been reported to be 95-124 cm for females and 106-141 cm in males.[56,67] Special growth charts of this syndrome have been derived from the longitudinal follow-up of untreated patients of the Israeli cohort.[68] They fit any primary GH or IGF-I deficiency. The growth of the hair and nails is also slow. Analysis of the height of parents and adult siblings revealed that male heterozygotes are of normal height but some of the mothers and sisters are below the 3rd percentile (Tanner chart).[59] It is noteworthy that none of the first degree relatives has a height above the 50th centile. The same was found in the Ecuadorian cohort of Laron syndrome patients.[67] Whether this is due to their being heterozygous for the disease or belonging to oriental ethnic groups with lower height standards, is as yet unsettled.

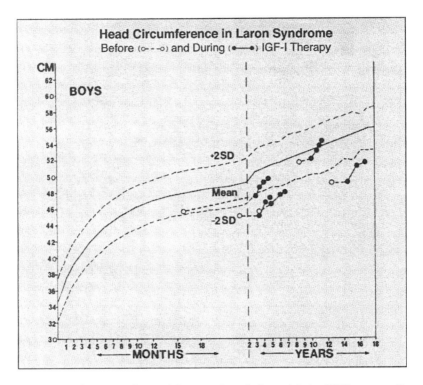

Figure 3. Head circumferences in 5 boys with Laron syndrome before and during IGF-I treatment. Reproduced with permission from Laron et al (2002).[27]

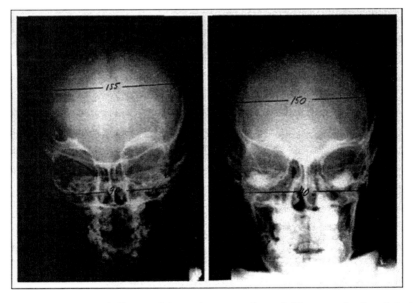

Figure 4. Anterior-posterior skull X-ray of a boy with Laron syndrome (left) compared to that of a healthy same aged child (right) Note underdevelopment of facial bones and absence of frontal sinus.

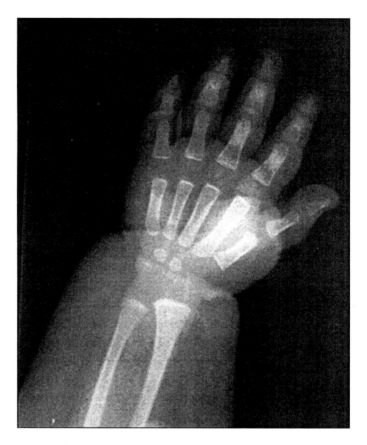

Figure 5. X-ray of the hand and forearm of a one year old girl with Laron syndrome. Note retarded skeletal maturation, marked obesity and thin muscles.

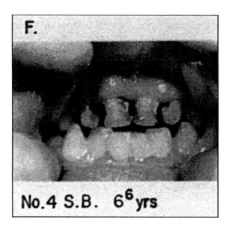

Figure 6. Defective teeth in a 6 1/2 year old girl with Laron syndrome.

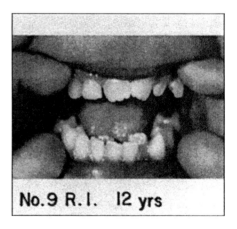

Figure 7. Crowding of permanent teeth in a 12 year old girl with Laron syndrome.

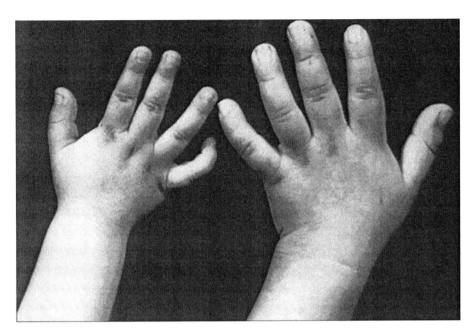

Figure 8. Acromicia in Laron syndrome. Size of hand of a 4 year old boy with Laron syndrome (left) compared with hand of a 4 year old healthy boy (right). Reprinted with permission from Laron et al (2002).[53]

Sexual Maturation

The genitalia and gonads are small since birth[28] which is easily evident in the males presenting a small penis[69] and testes (Fig. 9). In the girls the genitalia are also small[28] (Fig. 10) and the small size of the ovaries is evident by ultrasonography. Puberty is delayed, more so in boys than in girls[70,71] and patients with LS do not have a real pubertal growth spurt.[71] The sequence of appearance of pubertal signs in boys is as follows: testicular enlargement was found to begin between 13 and 16 years and axillary hair after age 16. The first conscious ejaculation, an important milestone in boys[72] which normally occurs at a mean age of 13 1/2 years was reported to take place between 17 and 21 years.[70] In girls, puberty is less retarded and menarche occurred between 13 to 15 years in most patients. Both sexes reach full sexual development. The final length of the penis in males is between 8-10 cm and the testicular volume 5-9 ml.[28,70] The breast of the females reach normal size and some appear large compared to body size[70] (Fig. 11). In early adulthood there are no difficulties in reproduction. Five of our female patients married to non LS husbands, have 1-3 normal children, so does one male patient. Another, married to a heterozygote female for this disease, has two children with LS. Another three couples have no children. Several of our unmarried patients have sexual relations, others never had.

Nutritional State

Since birth, children with LS are obese despite eating very little. Their obesity is progressive during childhood and adulthood, when it becomes excessive[73] (Figs. 2,11,12).

Skeletal System

Skeletal maturation is retarded starting in utero and slow in progressing. Closure of the epiphyseal cartilage in the long bones occurs after age 16-18 in girls and 20-22 years in boys. The fontanels and sutures of the skull close much later.[74] The diploe of the skull is very thin and the sinuses are underdeveloped (Fig. 13). The facial bones especially sphenoid and mandible are underdeveloped[60] (Fig. 4). The long bones are thin and already in young adult age, osteoporosis is evident[75] (Fig. 14).

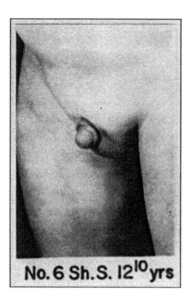
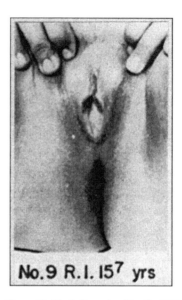

Figure 9. Pubis, penis and scrotum of a 12 10/12 year old boy with Laron syndrome. Note sexual retardation.

Figure 10. Genitalia and pubis of a 15 7/12 year old girl with Laron syndrome. Note marked sexual retardation.

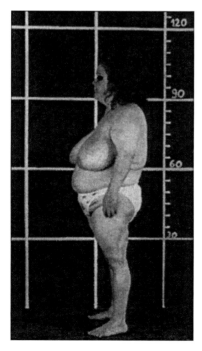
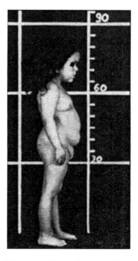

Figure 11. Lateral view of a 50 year old female with Laron syndrome Note: Dwarfism, obesity and large breasts.

Figure 12. Lateral view of a 12 year old girl with Laron syndrome Note Dwarfism, potruding forehead, saddle nose and obesity.

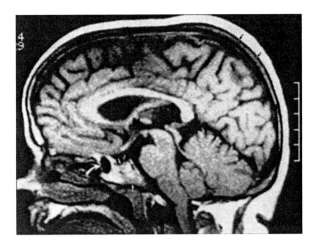

Figure 13. Lateral view of an MRI cut of the skull of an 68 year old patient with Laron syndrome. Note absence of frontal sinus, small sphenoid sinus (curved arrow) thin diploe (double arrows) and normal sized pituitary.

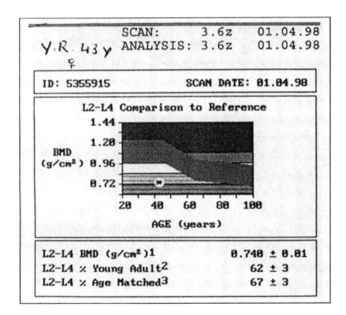

Figure 14. Severe osteoporosis of the lumbar spine of a 43 year old female with Laron syndrome

A recent finding in untreated patients was stenosis of the cervical spine (Fig. 15) and osteoarthritic changes of the atlanto-odontoid joint causing myelopathy in two patients.[63] Two untreated girls aged 1 1/2 and 12 years also showed cervical spine stenosis (Laron, unpublished). The diameter of the oro-pharynx was significantly smaller in the LS patients than in the same aged controls.[63]

Muscular System

Untreated patients with LS, both children and adults, have reduced muscle development and reduced muscle strength and endurance.[76]

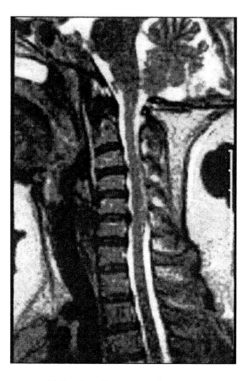

Figure 15. Spinal stenosis in a 51 old female with Laron syndrome as evidenced on MRI imaging.

Cardio-Pulmonary and Vascular Findings

The investigation of adult LS patients, revealed cardiomicria, reduced width of the cardiac muscle and a reduced left ventricular output.[77] Pulmonary function studies showed a reduced maximal aerobic capacity.[78]

Recent investigations found reduced branching in the retinal vascularization in patients with LS.[79]

Neuro-Psychological Development

Brain growth as evidenced by head circumference (Fig. 3) is below normal starting in utero[58,80] and motor development in infancy is slow and delayed.[2,3,58] Electroencephalograms performed during childhood in 14 patients showed a normal pattern in 12, an epileptic pattern in one and a paroxysmal pattern in another.[3] Pneumoencephalography (before the time of CT's) in 6 children revealed normal sized ventricles.[3] Skull CT performed in 2 patients was normal but MRI imaging of 9 untreated adult patients and 3 untreated children with LS, revealed diffuse parenchymal loss of various degrees in the adult patients only.[74] Three patients had localized atrophy in the occipital lobe and one had a lacunar infarct in the caudate nucleus. One patient had leucomalacia (see below).

Repeated psychological evaluations of untreated patients at various ages revealed an overall lower distribution pattern in intelligence tests than the same aged general population.[81] Of note is that greater deficits were recorded in the Performance IQ than the Verbal IQ, the latter showed some improvement with age.[82] In our cohort there is a great variability in the mental development of patients with LS; from normal intelligence (one Ph.D., two M.A.'s) to severe mental retardation. The patient with the most severe retardation had on MRI areas of periventricular leukomalacia.[74] A seemingly normal IQ reported for the patients with LS in Ecuador[83] is controversial.[84] MRI findings of the brain in one 11-year-old girl treated by IGF-I for 8 years were normal.[74] The physiological effects of IGF-I on the brain have been reviewed recently.[85]

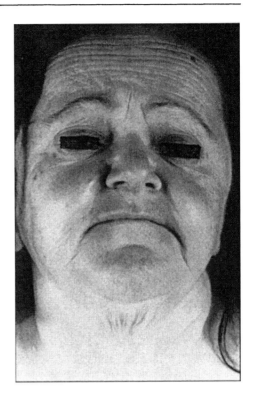

Figure 16. The face of a 50 year old female with Laron syndrome. Note increased wrinkles of the skin of the forehead, face and neck.

Sleep Disorders

Sleep disorders are a common feature of adult patients with LS (Laron unpublished). The narrow oro-pharynx leading to construction of the upper airways, and marked obesity predispose to sleep apnoea. One adult patient with severe breathing difficulties, due to obstructive sleep apnoea syndrome (OSAS) required the use of continuous positive airway pressure (CPAP).[86]

Ageing, Longevity and Mortality

Despite the appearance of early aging, such as thin and wrinkled skin (Fig. 16) patients with LS have a long life, over the seventies, both in our and in the Ecuadorian cohorts.[56,87] We registered one death, the sister of one of our patients. She died at age 3 of "meningoencephalitis" with convulsions, in 1950 before we recognized the disease. In the Ecuadorian cohort a series of deaths have been reported, attributed to untreated hypoglycemia infections in early age, and heart disease in adults.[56] Hypoglycemia cannot be excluded as the main or as a contributory factor in childhood death. Table 3 summarizes the main clinical characteristics of LS.

Laboratory Findings

Neonates and babies with LS suffer from severe hypoglycemia.[3,55,59] Low blood glucose having symptomatic or asymptomatic hypoglycemia (Table 4),[3,58] probably related to the state of fasting or feeding. Before age 6, children with LS have insulin-induced hypoglycemic non-responsiveness. The response to hypoglycemia challenge becomes normal during puberty,[28] by the development of counter-regulatory mechanisms.[58] In later life, some develop glucose intolerance.[88,89]

Serum alkaline phosphatase, inorganic phosphorus, and creatinine (the latter as an index of glomerular filtration) are low,[90] as are serum procollagens.[91] Serum total and LDL cholesterol are low or normal during early childhood but increase progressively with age and obesity to supra-normal levels.[73] Fasting free fatty acids are also high during severe hypoglycemia.[28]

Table 3. Main clinical characteristics of patients with primary IGF-I deficiency as exemplified by the classic form of Laron syndrome

- Dwarfism (progessing to -4 to -10 SDS)
- Obesity—general and visceral (progressive, starting in utero)*
- Small head circumference for age
- Protruding forehead
- Small facial bones (saddle nose)
- Sparse hair, big frontal recessions (alopecia in adult males)
- Defective teeth, crowding
- Delayed motor development
- Delayed skeletal maturation
- Delayed sexual development (small gonads)
- Reduced lean body mass
- Reduced bone density (osteoporosis in young adulthood)
- High pitched voice

* not so apparent in post-GH receptor defects

Table 4. Main biochemical and hormonal characteristics in patients with primary IGF-I deficiency as exemplified by the classic form of Laron syndrome

- Hypoglycemia (symptomatic and asymptomatic)
- High serum growth hormone
- Low serum IGF-I (non responsive to exogenous hGH)
- High serum IGFBP-1
- Low serum IGFBP-3
- Progressive insulin resistance
- Progressive rise in serum cholesterol

Hormones

Overnight fasting serum GH levels are high[2,3,58] and nocturnal pulses may reach peak levels of 200-300 ng/ml (μ/L).[92] The regulation of GH secretion is normal as evidenced by the normal number of 24 hour GH pulses, the response to stimulatory agents (insulin hypoglycemia, arginine, etc.)[3] and suppressive drugs such as somatostatin.[93] The administration of exogenous IGF-I suppresses serum hGH levels[94] proving a normal feedback mechanism. Despite the lifelong oversecretion of pituitary GH[92] the pituitary gland in patients with LS is not enlarged.[95] One possible explanation is that similar to other organs, pituitary gland growth is IGF-I dependent.

Serum IGF-I levels are very low, even undetectable,[4,5,58] and do not rise upon the administration of exogenous hGH,[5] evidence for the state of GH resistance in these patients. The number of unoccupied IGF-I binding sites (i.e., receptors) in the tissues is increased[96] being modulated by the circulating IGF-I levels.[97] Serum IGFBP-3, the GH stimulated IGF-BP is low,[98] but IGFBP-1 is elevated.[99] The levels of IGFBP-2 are normal.

Thyroid function is normal[100] as is adrenal hormone secretion.[3] Gonadotropin and sex hormone secretions are low but within normal limits correlating with the biological (skeletal) age and rise with puberty and adult age.[3] Prolactin may be elevated probably due to a drift phenomenon to the high GH secretion by the somatomammotropic cells.[101]

Despite the low glucose levels, serum insulin is relatively high denoting a state of insulin resistance.[89] This state, present already in infancy, increases with advancing age and with the progressive obesity. In few patients, glucose intolerance and hyperinsulinism was registered during an oral glucose tolerance test.[102] The insulin resistance can lead in adulthood to insulin exhaustion and even diabetes.[88,89]

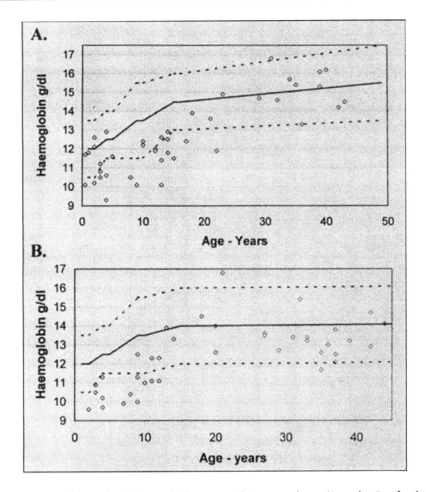

Figure 17. Hemoglobin values (along age) of patients with Laron syndrome (A= males, B = females). Reproduced from Sivan et al.[158]

Hematopoietic System

Untreated children and adolescents with LS, present with subnormal values for the erythropoietic indices, (red blood cell count, hemoglobin, hematocrit and mean corpuscular volume). During adult age the values become normal (Fig. 17).[158] These findings show, that during the active growth periods IGF-I is an essential component of the erythroid line formation. Its active role is further proven by the GH independent stimulating effect on erythropoiesis by IGF-I treatment of patients with LS.[103] Whether this effect is mediated by erythropoietin, is not known at present.

Untreated LS patients have an elevated monocyte count in the presence of a normal total neutrophil count and a tendency for a low lymphocyte count.[103]

Histological Examination of the Skin

Microscopic examination of skin biopsies of prepubertal patients with LS revealed moderately thickened elastin fibers with a tendency to cluster; in pubertal patients there was a reduction in the number of elastin fibers in the papillary layer and clustering of collagen fibrils in the reticular areas.[104] The skin in young adults is very thin and appears to have lost even more elastic fibers. These changes are part of the effects of IGF-I deficiency on connective tissue. Histological examinations of skin in the adult age group are missing.

Hair

Light and electromicroscopic studies of the hair of untreated children and adults with LS revealed four main structural defects in the hair shaft: pilli torti, grooving, pseudomonilethrix and trichorrhexis.[105,106] Whereas pili torti is a congenital abnormality, the other changes are usually acquired and likely to occur in genetically fragile hair.

The Molecular Pathology of Laron Syndrome—The GH Receptor (GH-R)

The structure of the human GH-R has been established by cloning.[14,15] The GH-R gene is located on the short arm of chromosome 5 (p13-p12).[107] It has 620 aminoacids preceded by an aminoacid signal sequence, and consists of 9 exons and spans over 87 Kb.[15] Exon 2 to 7 encodes the extracellular domain (EC) of 246 aminoacids, exon 8 corresponds to the transmembrane domain (TM = 24 aminoacids). The GH-R belongs to the superfamily of cytokine receptors. Unlike insulin receptors, cytokine receptors do not have intrinsic tyrosine kinase activity. Instead they are closely associated with protein kinases encoded by other genes. In the case of the GH-R, the kinase is Janus kinase 2 (JAK2). Receptor occupancy leads to autophosphorylation of JAK2, association of JAK2 with the GH-R and subsequent phosphorylation of the receptor itself.[108] The intracellular signaling cascade includes activation of mitogen-activated protein kinase (MAPK) and of latent transcription factors known as STATs (signal transducers and activators of transcription).[109] At the end of the signal transduction cascade is the transcription of specific genes such as IGF-I, IGFBP-3 (IGF binding protein-3), etc.[109]

Growth hormone receptors form homodimers in the course of binding a single GH molecule.[110] The initial event involves binding of site A of GH to a single receptor molecule and the second involves binding of site B of GH to a second receptor molecule. Sites A and B on the GH molecule differ, but the binding sites on the two receptor molecules are nearly identical.

Genetic Abnormalities of the GH-R

Deletions and mutations in the GH-R gene are responsible for the classical form of GH resistance, i.e., LS. A variety of gene defects have been observed: from exon deletions to nonsense, frameshift, splice and missense mutations of exons and introns (Table 5).[15,18,45] The majority are in the extracellular domain of the receptor: exons 2 to 7 and introns 2 to 7 resulting in the absence of circulating GHBP[13,125] (see below). Four reports describe mutations affecting the transmembrane domain (exon 8)[17,18,118] and another three mutations in the cytoplasmatic domain (exons 9 and 10).[45,46,51] It is of note that despite the great variability in the molecular defects of the gene, they all result in lack of GH signal transmission. Indeed, a single aminoacid substitution in the extracellular domain of the hGH receptor prevents ligand binding to the GH-R[121] as does defective membrane expression.[126] It is of note that in 37 patients investigated from the large Ecuadorian cohort the same mutation in exon 6 of the GH-R (E180 splice, A to G at 594) was found,[119] whereas in the smaller Israeli cohort a series of mutations were registered[15,111] (Shevah and Laron, unpublished).

In 1993 we described the first patients with a so-far unidentified post-receptor defect[19] resulting in the generation of IGFBP-3 but not of IGF-I denoting separate signaling pathways for the transcription of the two genes.

The GH Binding Protein (GHBP)

Herington et al[11] and Baumann et al[12] independently described a serum protein capable of binding GH with high affinity. This GH binding protein (GHBP) was shown to be identical in structure with the extracellular domain of the GH-R.[14] Between 30 and 50% of the circulating GH is bound to this protein. Its quantitative measurements revealed that its serum concentrations vary with age, being low in neonates and reaching maximal values in young adulthood,[127] as well as in pathological conditions.[128] Determination of serum GHBP can be used as a simple quantitative estimation of the extracellular domain of the GH-R,[129] its absence meaning a defect in this domain of the receptor.

Normal or elevated serum GHBP in typical LS patients denotes a defect in the transmembrane, intracellular or post GH-R areas.[17,18,51,127] A low serum GHBP concentration in relatives of patients with LS helps identify heterozygous carriers.[129]

Table 5. GH-R mutations reported in patients with Laron syndrome

Mutations	Molecular Defect	Nucleotide Change	Exon	Domain[a]	GHBP[b]	Reference (No.)
Deletion	Exons 3-5-6		3-5-6	EC	-	Godowski et al, 1989 (15)
Nonsense	W-15X	G→A at 83	2	EC		Shevah et al, (2003) (111)
	C38X	C→A at 168	4	EC	-	Amselem et al, 1991 (112)
	R43X	C→T at 181	4	EC	-	Amselem et al, 1991 (112)
	Q65X	C→T at 197	4	EC	-	Sobrier et al, 1997 (113)
	W80X	G→A at 293	5	EC	-	Sobrier et al, 1997 (113)
	L141X	T→A at 476	6	EC	-	Shevah et al, (2003) (111)
	W157X	G→A at 525	6	EC	-	Sobrier et al, 1997 (113)
	E183X	G→T at 601	6	EC	?	Berg et al, 1994 (114)
	R217X	C→T at 703	7	EC	-	Amselem et al, 1993 (115)
	Z224X	G→T at 724	7	EC	-	Kaji et al 1997 (45)
Frameshift	21delTT	del TT at 118	4	EC	-	Counts and Cutler, 1995 (116)
	36delC	del C at 162	4	EC	-	Sobrier et al, 1997 (113)
	46delTT	del TT at 192-193	4	EC	-	Berg et al, 1993 (117)
	230delT	del T at 744	7	EC	-	Sobrier et al, 1997 (113)
	230delAT	delAT at 743-744	7	EC	-	Berg et al, 1993 (117)
	309del C	del C at 981	10	IC	-	Kaji et al 1997 (45)
Splice	Intron 2	G→A at 70+1		EC	-	Sobrier et al, 1997 (113)
	Intron 4	G→A at 266+1		EC	-	Amselem et al, 1993 (115)
	Intron 5	G→A at 71+1		EC	-	Berg et al, 1993 (117)
	Intron 5	G→C at 130-1		EC	?	Berg et al, 1994 (114)
	Intron 5	G→C at 440-1		EC	-	Amselem et al, 1993 (115)
	Intron 6	G→ T at 189-1		EC	?	Berg et al, 1993 (117)
	Intron 6	G→T at 619-1		EC	-	Berg et al, 1993 (117)
	Intron 7	G→T at 785-1	7/8	EC/TM	+++	Silbergeld et al, 1997 (17)
			7/8	EC/TM	?	Shevah et al, 2002 (118)
	E180splice	A→G at 594	6	EC	+	Berg et al, 1992 (119)
	G223G	C→T at 723	7	EC	-	Sobrier et al, 1997 (113)
	G236splice	C→T at 766	7	EC	-	Baumbach et al, 1997 (32)
	R274T	G→C at 874	8	TM	++	Woods et al, 1996 (18)
	GHR(1-277)[c1]	G→A at 876	9	TM/IC	++	Iida et al, 1998 (46)
	GHR(1-277)[c2]	G→C at 876	9	TM/IC	+	Ayling et al, 1997 (51)
Missense	M-18L	A→T at 73	2	EC		Quinteiro et al, (2002) (120)
	C38S	T→A at 166	4	EC	?	Sobrier et al, 1997 (113)
	S40L	C→T at 173	4	EC	-	Sobrier et al, 1997 (113)
	W50R	T→C at 202	4	EC	-	Sobrier et al, 1997 (113)
	R71K	G→A at 266	4	EC	-	Amselem et al, 1993 (115)
	F96S	T→C at 341	5	EC	-	Amselem et al, 1989 (16)
	V125A	T→C at 428	5	EC	-	Amselem et al, 1993 (115)
	P131Q	C→A at 446	6	EC	?	Walker et al, 1998 (48)
	V144D	T→A at 485	6	EC	-	Amselem et al, 1993 (115)
	D152H	G→C at 508	6	EC	+	Duquesnoy et al, 1994 (121)
	I153T	T→C at 512	6	EC	-	Wojeik et al, 1998 (124)
	Q154P	A→C at 515	6	EC	?	Wojeik et al, 1998 (124)
	V155G	T→G at 518	6	EC	?	Wojeik et al, 1998 (124)
	R161C	C→T at 535	6	EC	-	Amselem et al, 1993 (115)
	Y178S	A→C at 587	6	EC		Oh et al, 1999 (122)
	Y208C	A→G at 677	7	EC		Enberg et al, 2000 (123)
	R211G	C→G at 685	7	EC	-	Amselem et al, 1993 (115)
	D244N	G→A at 784	7	EC		Enberg et al, 2000 (123)

[a] EC=extracellular; TM=transmembrane; IC=intracellular. [b] (?) not available; (-) undetectable; (+) detectable; (++) high levels; (+++) very high levels. [c] These two mutations were identified on the same GHR allele.

Table 6. Hormonal changes during IGF-I therapy of patients with Laron syndrome

Somatostatin	↑
GHRH	↓
hGH	↓
GHBP	↓
Insulin	↓
Glucagon	↓
TSH	↓± Slight and temporary
PRL	↓ (If elevated)
Gonadotropins	↑± More evident
Androgens	↑± With high doses

Treatment

The only treatment for LS is replacement therapy with recombinant biosynthetic IGF-I available since 1986.[130] Unfortunately, the amounts for clinical use were restricted and permitted treatment of only a small number of patients. Nevertheless, in addition to helping restore many of the auxological and metabolic defects, the replacement therapy enabled to learn the physiology and pharmacology of IGF-I in man.

Untreated patients with LS have an increased number of IGF-I receptors when compared to healthy subjects.[96] One week IGF-I administration reduced the number of specific binding sites using red blood cells (RBCS) to normal values.[97] An intravenous bolus injection of IGF-I (75 mg/kg) in the fasting state induced marked hypoglycemia in both children and adult patients with LS[20] as well as in healthy controls. The concomitant decrease of serum insulin proved that the hypoglycemia was IGF-I induced.[131] The injected IGF-I was more rapidly eliminated in the patients with LS than in control subjects (t 1/2 = 2.57 ± 0.67 vs. 4.43 ± 0.52 minutes)[132] due to the low concentration of serum IGFBP-3 in LS.[98] Intravenous IGF-I administration suppresses circulating GHRH, GH, TSH[20,131] and glucagon.[133] This is explained by stimulation of somatostatin secretion by IGF-I.[134] Short- and long-term clinical trials in patients with LS, have been performed first by our group and subsequently by investigators in the USA, Europe and Japan, each group using IGF-I from a different source, (exception Israel and Japan) but with an identical structure. The only difference is that whereas our group uses one subcutaneous injection per day administered before breakfast, the other groups administered two injections of IGF-I a day: morning and evening.

Short term and transitory effects are water and electrolyte retention and calciuria.[75,90] Administration of IGF-I for months or years persistently suppressed GH, and serum insulin, preventing hypoglycemia, and stabilizing blood glucose levels—provided that meals were regular.[135] This was also confirmed by Walker et al.[136] Sensitive markers of IGF-I activity are serum alkaline-phosphatase procollagen I (PICP), PIIINP,[91] serum phosphate and GFR.[90] IGF-I administration raised sex hormone binding protein[137] and decreased lipoprotein(a).[138]

IGF-I administration also affects its specific binding proteins participating in its own regulation of available free hormone. During the first weeks of administration IGF-I suppresses the IGFBP's[99] including IGFBP-3[98] but longer administration leads to the increased levels of IGFBP-3 as well as of its acid labile fraction[139,140] as well as of the other BP's. This finding has practical importance as it prolongs the biological half life of the administered IGF-I during long term treatment. and requires in most patients a progressive reduction of the IGF-I dose in order to prevent overdosage and undesirable effects.[141] During long-term treatment there is also a rise in IGFBP-1 and 2.

One of the major effects of IGF-I is acceleration of linear growth. Due to limitations in the availability of the drug despite ten years experience there are only few reports on long-term treatment of children with LS. Table 7 summarizes the only comparable data of published results.[142-145] Two 7-year treatments [123,146] and one 5-year report[44] have been published. It seems that once daily IGF-I administration[144] is as effective in promoting growth as is the twice daily IGF-I administration.[142,143,145] On the other hand, twice daily treatment or not decreasing the IGF-I dose with time[147] caused more adverse effects (Table 8). In the first year of treatment the growth velocity was

Table 7. Comparative linear growth response of children with Laron syndrome treated by IGF-I

Author	Year	Ref.	n	At Start			IGF-I dose µg/kg/d	Growth Velocity (cm/yr)			
				Age (yr) Range	BA (yr)	Ht SDS m		0 Before	1st Year	2nd	3rd
Ranke et al	1995	142	31	3.7 - 19	1.8 - 13.3	-6.5	40 - 120 b.i.d.	3.9 ± 1.8	(n = 26) 8.5± 2.1	(n=18) 6.4±2.2	
Backeljauw et al	1996	143	5	2-11	0.3 - 6.8	-5.6	80 - 120 b.i.d.	4.0	(n = 5) 9.3	(n = 5) 6.2	(n = 1) 6.2
Klinger & Laron	1995	144	9	0.5 - 14	0.2 - 11	-5.6	150-200 once	4.7±1.3	(n = 9) 8.2±0.8	(n = 6) 6±1.3	(n = 5) 4.8±1.3*
Guevarra-Aguirre	1997	145	(n = 15) (n = 15) (n = 6) 15	3.1 - 17	4.5 - 9.3		120 b.i.d.	3.4±1.4	8.8±11	6.4±1.1	5.7±1.4
			8				80 b.i.d.	3.0±1.8	(n = 8) 9.1±2.2	(n = 8) 5.6±2.1	

* The younger children had a growth velocity of 5.5 and 6.5 cm/yr.
BA = bone age; SDS = standard deviation score below the median for sex and age

Table 8. Adverse effects reported during IGF-I therapy of children with Laron syndrome

	Ranke et al[142]	Backeljauw et al[143]	Klinger and Laron[144]	Guevarra-Aguirre et al[145]
	n = 31	n = 5	N = 9	n = 22
Headaches (early)	21	2	-	?
Nausea, Vomiting	?	-	-	3
Hypoglycemia	13	+ early	1 (transitory)	6
Papilledema (transitory)	1	2	-	-
Bell's palsy	1	-	-	-
Lipohypertrophy	7	-	-	-
Enlargement of lymphoid tissue	3	4	1	?
Thickening of nose	5	2	2	?
Local reaction (transitory)	?	?	2	3
Transient tachycardia	?	?	2	16

higher than in subsequent years (Fig. 18). Concomitantly there was also a progressive growth of the extremities (hands, feet, chin and nose). Despite effectively stimulating linear growth, the growth response is not as intense as that of GH in GH deficiency.[146,148] During IGF-I treatment we also registered a fast catch-up of the head circumference even at ages 10-14[80,149] (Fig. 3) denoting brain growth, and a reduction in adipose tissue as measured by skinfold thickness. The latter was more accentuated in a group of adult patients with LS, treated for 9 months[75] and lasted only for one year at the most (Fig. 19). Recently we found the IGF-I treatment of patients with LS to have a significant stimulatory effect on erythroporesis[158] (Fig. 20).

As these patients suffer from osteoporosis[75] and muscle weakness[76] it is not certain that they can undergo limb lengthening[150] unless they are treated by IGF-I several years before and during the operation.

The dwarfism of untreated patients with LS, the underdevelopment of the skeletal and muscular system and the variable deficits in the neuro-psychomotor systems[66] limit their occupational opportunities and make relationships with the other sex difficult. These facts are a source of emotional suffering, even depression, necessitating continuous psychosocial counselling.[151]

All or most of the above consequences could be avoided if IGF-I treatment would be initiated once the diagnosis is made. It is to be regretted, to say the least, that in our advanced society, a treatable disease remains untreated with the exception of a small number of children on time limited clinical trials.[152,153]

The Pygmies

An interesting and controversial issue is whether Pygmies belong to LS. Merimee and Laron[154] summarizing the incomplete findings performed in this population in Africa conclude that the short stature of the African pygmies as in LS is due primarily, if not solely, to a deficiency of IGF-I resulting from a defect in the GH-R. As LS patients seem shorter than the pygmies and their IGF-I levels lower, the pgymies may have a different less complete genetic defect. Overall metabolic aberrations reflect a decrease in GH-R including a decrease in serum GHBP which points to a defect in the extracellular domain of the receptor. Eight out of 10 pygmies administered 5 mg hGH/b.i.d. for 5 days did not raise their serum IGF-I.[155] These findings support early conclusions that the pygmies are resistant to GH.[156] On the other hand, Geffner et al[157] reported IGF-I resistance in a HTLV-II transformed cell line from one Efe Pygmy, suggesting that the defect is with the IGF-IR program and not the GHR.

Figure 18. Growth of an untreated girl with Laron syndrome (A) compared to the growth of a girl treated with IGF-I.(B) Growth drawn on a special growth chart for this syndrome.[68] Note declining growth rate without treatment and growth stimulating effect of IGF-I, but inability to normalize height during 8 years of treatment.

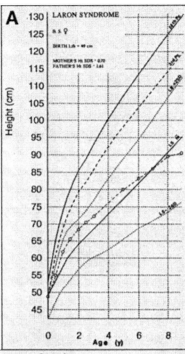

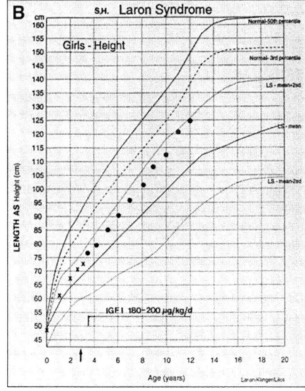

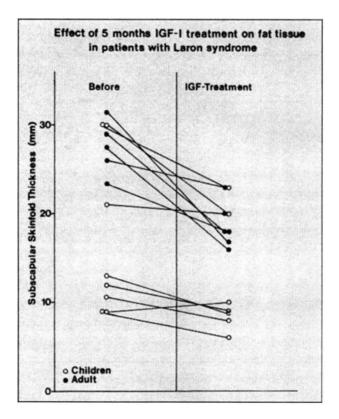

Figure 19. Reduction of subscapular skinfold thickness upon institution of IGF-I treatment in patients with Laron syndrome

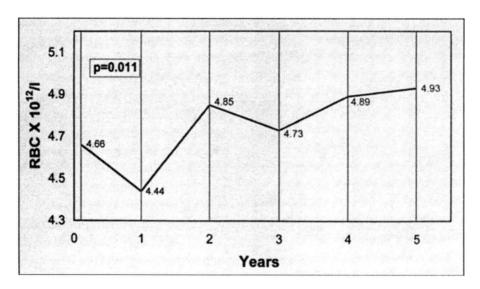

Figure 20. Effect of long term IGF-I treatment on red blood cell count in 7 children with Laron syndrome. Reproduced with permission from Sivan et al.[158]

Acknowledgements

We acknowledge the generous supply of IGF-I for more than 10 years from Fujisawa Pharmaceutical Co., Osaka, Japan. The author thanks Mrs. Gila Waichman in the preparation of the manuscript. and Miss Orit Shevah, MSc student, in assembling Table 5.

References

1. Laron Z, Parks JS, eds. Lessons from Laron syndrome (LS) 1966-1992. Pediatric and Adolescent Endocrinology, vol. 24. Basel-New York: Karger, 1993:367.
2. Laron Z, Pertzelan A, Mannheimer S. Genetic pituitary dwarfism with high serum concentration of growth hormone. A new inborn error of metabolism? Isr J Med Sci 1996; 2:152-155.
3. Laron Z, Pertzelan A, Karp M. Pituitary dwarfism with high serum levels of growth hormone. Isr J Med Sci 1968; 4:883-894.
4. Daughaday WH, Laron Z, Pertzelan, A et al. Defective sulfation factor generation: a possible etiological link in dwarfism. Trans Assoc Am Phys 1969; 82:129-138.
5. Laron Z, Pertzelan A, Karp M et al. Administration of growth hormone to patients with familial dwarfism with high plasma immunoreactive growth hormone. Measurement of sulfation factor, metabolic, and linear growth responses J Clin Endocrinol Metab 1971; 33:332-342.
6. Eshet R, Laron Z, Brown M et al. Immunoreactive properties of the plasma hGH from patients with the syndrome of familial dwarfism and high plasma IR-hGH. J Clin Endocrinol Metab 1973; 37:819-821.
7. Eshet R, Laron Z, Brown M et al. Immunological behavior of hGH from plasma of patients with familial dwarfism and high IR-hGH in a radioimmunoassay system using the cross-reaction between hGH and HCS. Horm Metab Res 1974; 6:79-81.
8. Jacobs LS, Sneid DS, Garland JT et al. Receptor-active growth hormone in Laron dwarfism. J Clin Endocrinol Metab 1976; 43:403-407.
9. Eshet R, Peleg S, Josefsberg Z et al. Some properties of the plasma hGH activity in patients with Laron-type dwarfism determined by a radioreceptor assay using human liver tissue. Horm Res 1985; 22:276-283.
10. Eshet R, Laron Z, Pertzelan A et al. Defect of human growth hormone in the liver of two patients with Laron type dwarfism. Isr J Med Sci 1984; 20:8-11.
11. Herington AC, Ymer S, Stevenson J. Identification and characterization of specific binding proteins for growth hormone in normal human sera. J Clin Invest 1986; 77:1817-1823.
12. Baumann G, Stolar MN, Amburn K et al. A specific GH-binding protein in human plasma: Initial characterization. J Clin Endocrinol Metab 1986; 62:134-141.
13. Daughaday WH, Trivedi B. Absence of serum growth hormone binding protein in patients with growth hormone receptor deficiency (Laron dwarfism). Proc Natl Acad Sci USA 1987; 84:4636-4640.
14. Leung DW, Spencer SA, Cachianes G et al. Growth hormone receptor and serum binding protein: purification, cloning and expression. Nature 1987; 330:537-543.
15. Godowski PJ, Leung DW, Meacham LR et al. Characterization of the human growth hormone receptor gene and demonstration of a partial gene deletion in two patients with Laron type dwarfism. Proc Natl Acad Sci USA 1989; 86:8083-8087.
16. Amselem S, Duquesnoy P, Attree O et al. Laron dwarfism and mutations of the growth hormone-receptor gene. N Engl J Med 1989; 321:989-995.
17. Silbergeld A, Dastot F, Klinger B et al. Intronic mutation in the growth hormone (GH) receptor gene from a girl with Laron syndrome and extremely high serum GH binding protein: extended phenotypic study in a very large pedigree. J Pediatr Endocrinol Metab 1997; 10:265-274.
18. Woods KA, Fraser NC, Postel-Vinay MC et al. A homozygous splice site mutation affecting the intracellular domain of the growth hormone (GH) receptor resulting in Laron syndrome with elevated GH-binding protein. J Clin Endocrinol Metab 1996; 81:1686-1690.
19. Laron Z, Klinger B, Eshet R et al. Laron syndrome due to a post-receptor defect: response to IGF-I treatment. Isr J Med Sci 1993; 29:757-763.
20. Laron Z, Klinger B, Erster B et al. Effect of acute administration of insulin-like growth factor I in patients with Laron type dwarfism. Lancet 1988; ii:1170-1172.
21. Scanes CG, Boone F, McCann-Levorse L et al. The sex-linked dwarf (dw) chicken: A model for Laron dwarfism? In: Laron Z, Parks JS, eds. Lessons from Laron syndrome (LS) 1966-1992. Pediatric and Adolescent Endocrinology, vol. 24. Basel-New York: Karger, 1993:313-318.
22. Zhou Y, Xu BC, Maheshwari HG et al. A mammalian model for Laron syndrome produced by targeted disruption of the mouse growth hormone receptor/binding protein gene (the Laron mouse). Proc Natl Acad Sci USA 1997; 94:13215-13220.
23. Kopchick JJ, Laron Z. Is the Laron mouse an accurate model of Laron syndrome? Mol Genet Metab 1999; 68:232-236.
24. Elders MJ, Garland JT, Daughaday WH et al. Laron's dwarfism: Studies on the nature of the defect. J Pediatr 1973; 83:253-263.
25. Laron Z, Blum W, Chatelain P et al. Classification of growth hormone insensitivity syndrome. J Pediatr 1993; 122:241.
26. Laron Z. Laron syndrome: From description to therapy. The Endocrinologist 1993; 3:21-28.
27. Laron Z. Growth hormone resistance (Laron syndrome). In: Chrousos G, ed. Hormone Resistance and Hypersensitivity States. New York: Lippincott Williams & Wilkins, 2002:251-267.

28. Laron Z. Laron type dwarfism (hereditary somatomedin deficiency): a review. In: Frick P, Von Harnack GA, Kochsiek GA, Prader A, eds. Advances in Internal Medicine and Pediatrics. Berlin-Heidelberg: Springer-Verlag, 1984:117-150.
29. Rosenfeld RG, Rosenbloom AL, Guevara-Aguirre J. Growth hormone (GH) insensitivity due to primary GH receptor deficiency. Endocr Rev 1994; 15:369-390.
30. Rosenbloom AL, Guevara-Aguirre J, Rosenfeld RG et al. The little women of Loja: growth hormone receptor deficiency in an inbred population of Southern Ecuador. N Engl J Med 1990; 323:1367-1374.
31. Rosenbloom AL, Guevara-Aguirre J. Lessons from the genetics of Laron syndrome. Trends Endocrinol Metab 1998; 9:276-282.
32. Baumbach L, Schiavi A, Bartlett R et al. Clinical, biochemical and molecular investigations of a genetic isolate of growth hormone insensitivity (Laron's syndrome). J Clin Endocrinol Metab 1997; 82:444-451.
33. Yordam N, Kandemir N, Erkul I et al. A review of the Turkish patients with growth hormone insensitivity (Laron type). Eur J Endocrinol 1995; 133:539-542.
34. Razzaghy-Azar M. Laron syndrome—a review of 8 Iranian patients with Laron syndrome. In: Laron Z, Parks JS, eds. Lessons from Laron syndrome (LS) 1966-1992. Pediatric and Adolescent Endocrinology, vol. 24. Basel-New York: Karger, 1993:93-100.
35. Maheshwari HG, Clayton PE, Mughal Z et al. In: Laron Z, Parks JS, eds. Lessons from Laron syndrome (LS) 1966-1992. Pediatric and Adolescent Endocrinology, vol. 24. Basel-New York: Karger, 1993:160-166.
36. Desai MP, Colaco P, Choksi CS et al. Endogenous growth hormone nonresponsive dwarfism in Indian children. In: Laron Z, Parks JS, eds. Lessons from Laron syndrome (LS) 1966-1992. Pediatric and Adolescent Endocrinology, vol. 24. Basel-New York: Karger, 1993:81-92
37. Rappaport R, Czernichow P, Prevost C et al Retard de croissance avec somathormone circulant elevee et incapacite de generer la somatomedine(nanisme de type Laron). Ann Pediatr (Paris) 1977; 24:63-67.
38. Alcaniz JJ, Salto L, Barcelo B. GH secretion in two siblings with Laron's dwarfism: The effects of glucose, arginine, somatostatin and bromocryptine. J Clin Endocrinol Metab 1978; 47:453-456.
39. Cohen A, Cordone G, Fasce G et al. An additional patient of Laron-type dwarfism in Italy. In: Laron Z, Parks JS, eds. Lessons from Laron syndrome (LS) 1966-1992. Pediatric and Adolescent Endocrinology, vol. 24. Basel-New York: Karger, 1993:244-253 and 313-318.
40. Kastrup KW, Andersen H, Hanssen KG. Increased immunoreactive plasma and urinary growth hormone in growth retardation with defective generation of somatomedin A (Laron's syndrome). Acta Pediatr Scand 1975; 64:613-618
41. Van den Brande JL, DU Caju MVL, Visser HKA et al. Primary somatomedin deficiency: case report. Arch Dis Child 1974; 49:297-305.
42. Walker JL, Ginalska-Malinowska M, Romer TE et al. Effects of the infusion of insulin-like growth factor 1 in a child with growth hormone insensitivity syndrome (Laron dwarfism). N Engl J Med 1991; 324:1483-1488.
43. Tiulpakov AN, Orlovsky IV, Natalya U et al. Growth hormone insensitivity (Laron syndrome) in a Russian girl of Slavic origin, caused by a common mutation of the GH receptor gene. J Endocrine Genet 1999; 1:95- 100.
44. Krzisnik C, Battelino T. Five year treatment with IGF-I of a patient with Laron syndrome in Slovenia (a follow-up report). J Pediatr Endocrinol Metab 1997; 10:443-447.
45. Kaji H, Nose O, Tajiri H et al. Novel compound heterozygous mutations of growth hormone (GH) receptor gene in a patient with GH insensitivity syndrome. J Clin Endocrinol Metab 1997; 82:3705-3709.
46. Iida K, Takahashi Y, Kaji H et al Growth hormone (GH) insensitivity syndrome with high serum GH-binding protein levels caused by a heterozygous splice site mutation of the GH receptor gene producing a lack of intracellular domain. J Clin Endocrinol Metab 1998; 83:531-537.
47. Vesterhus P. Growth hormone resistance. Genetic defects and therapy in Laron syndrome and similar conditions. Tidaakr Nor Laegeforen 1997; 117:948-951.
48. Walker JL, Crock PA, Behncken SN et al. A novel mutation affecting the interdomain link region of the growth hormone receptor in a Vietnamese girl and response to long-term treatment with recombinant human insulin-like growth factor-I and luteinizing hormone-releasing hormone analog. J Clin Endocrinol Metab 1998; 83:2554-2561.
49. Saldanha PH, Toledo PA. Familial dwarfism with high IRGH: Report of two affected sibs with genetic and epidemiologic considerations. Hum Genet 1981; 59:367-372.
50. Arriazu MC, Gonzalez-Aguilar P. The first reported case of Laron-type dwarfism in Argentine. In: Laron Z, Parks JS, eds. Lessons from Laron syndrome (LS) 1966-1992. Pediatric and Adolescent Endocrinology, vol. 24. Basel-New York: Karger, 1993:101-103.
51. Ayling RM, Ross R, Towner P et al A dominant-negative mutation of the growth hormone receptor causes familiar short stature. Nature Genet 1997; 16:13-14.
52. Woods KA, Savage MO. Laron syndrome: typical and atypical forms. Growth hormone resistance. Baillere's Clin Endocrinol Metab 1996; 10:371-387.
53. Laron Z. Growth hormone insensitivity (Laron syndrome). In: LeRoith D, ed. Reviews In Endocrine And Metabolic Disorders. Vol. 3. Norwell: Kluwer Academic Publishers, 2002:347-355.
54. Pertzelan A, Adam A, Laron Z. Genetic aspects of pituitary dwarfism due to the absence of biological activity of growth hormone. Isr J Med Sci 1968; 4:895-900.
55. Laron Z. Prismatic cases: Laron syndrome (primary growth hormone resistance). From patient to laboratory to patient. J Clin Endocrinol Metab 1995; 80:1526-1573.
56. Rosenbloom AL, Guevara-Aguirre J, Rosenfeld RG et al. Growth hormone receptor deficiency) in Ecuador. J Clin Endocrinol Metab 1999; 84:4436-4443.

57. Laron Z. Laron syndrome—Primary growth hormone resistance. In: Jameson JL, ed. Hormone Resistance Syndromes. Contemporary Endocrinology, Vol. 2. Totowa: Humana Press, 1999:17-37.
58. Laron Z, Pertzelan A, Karp M et al. Laron syndrome—A unique model of IGF-I deficiency. In: Laron Z, Parks JS, eds. Lessons from Laron syndrome (LS) 1966-1992. Pediatric and Adolescent Endocrinology, vol. 24. Basel: Karger, 1993:3-23.
59. Laron Z. Natural history of the classical form of primary growth hormone (GH) resistance (Laron syndrome). J Pediatr Endocrinol Metab 1999; 12:231-249.
60. Scharf A, Laron, Z. Skull changes in pituitary dwarfism and the syndrome of familial dwarfism with high plasma immunoreactive growth hormone. A roentgenologic study. Horm Metab Res 1972; 4:93-97.
61. Konfino R, Pertzelan A, Laron Z. Cephalometric measurements of familial dwarfism and high plasma immunoreactive growth hormone. Am J Orthod 1975; 68:196-201.
62. Laron Z, Klinger B, Grunebaum M. Laron type dwarfism. Special feature—Picture of the month. Am J Dis Child 1991; 145:473-474.
63. Kornreich L, Horev G, Schwarz M et al. Laron syndrome abnormalities: Spinal stenosis, os odontoideum, degenerative changes of the atlanto-odontoid joint, and small oropharynx. Am J Neuroradiology 2002; 23:625-631.
64. Arad I, Laron Z. Standards for upper/lower body segment/sitting height—subischial leg length, from birth to 18 years in girls and boys. In: Proceedings 1st International Congr. Auxology, Rome, 1977 Milan, Centro Auxologia Italiano di Piancavallo, 1979:159-164.
65. Shurka E, Laron Z. Adjustment and rehabilitation problems of children and adolescents with growth retardation. I. Familial dwarfism with high plasma immunoreactive human growth hormone. Isr J Med Sci 1975; 11:352-357.
66. Laron Z. Consequences of not treating children with Laron syndrome (primary growth hormone insensitivity). J Pediatr Endocrinol Metab 2001; 14(Suppl. 5):1243-1248.
67. Rosenbloom AL, Guevara-Aguirre J, Berg MA et al. Stature in Ecuadorians heterozygous for growth hormone gene E180 splice mutation does not differ from that of homozygous normal relatives. J Clin Endocrinol Metab 1998; 83:2373-2375.
68. Laron Z, Lilos P, Klinger B. Growth curves for Laron syndrome. Arch Dis Child 1993; 68:768-770.
69. Laron Z, Sarel R. Penis and testicular size in patients with growth hormone insufficiency. Acta Endocrinol 1970; 63:625-633.
70. Laron Z, Sarel R, Pertzelan A. Puberty in Laron-type dwarfism. Eur J Pediatr 1980; 134:79-83.
71. Pertzelan A, Lazar L, Klinger B et al. Puberty in 15 patients with Laron syndrome: a longitudinal study. In: Laron Z, Parks JS, eds. Lessons from Laron syndrome (LS) 1966-1992. Pediatric and Adolescent Endocrinology, vol. 24. Basel-New York: Karger, 1993:27-33.
74. Laron Z, Arad J, Gurewitch R et al. Age at first conscious ejaculation: a milestone in male puberty. Helv Paediatr Acta 1980; 35:13-30.
73. Laron Z, Klinger B. Body fat in Laron syndrome patients: effect of insulin-like growth factor I treatment. Horm Res 1993; 40:16-22.
74. Kornreich L, Horev G, Schwarz M et al. Craniofacial and brain abnormalities in Laron syndrome (primary growth hormone insensitivity). Eur J Endocrinol 2002; 146:499-503.
75. Laron Z, Klinger B. IGF-I treatment of adult patients with Laron syndrome: Preliminary results. Clin Endocrinol (Oxf) 1994; 41:631-638.
76. Brat O, Ziv I, Klinger B et al. Muscle force and endurance in untreated and human growth hormone or insulin-like growth factor-I-treated patients with growth hormone deficiency or Laron syndrome. Horm Res 1997; 47:45-48.
77. Feinberg MS, Scheinowitz M, Laron Z. Echocardiographic dimensions and function in adults with primary growth hormone resistance (Laron syndrome) Am J Cardiol 2000; 85:209-213.
78. Laron Z, Gaides M, Scheinowitz M et al. Untreated patients with primary GH insensitivity (Laron syndrome) have reduced maximal aerobic capacity. (Abst P1-123). Horm Res 2000; 53 (Suppl. 2):37.
79. Hellstrom A, Carlsson B, Niklasson A et al. Reduced retinal vascularisation in patients with defects in the GH or IGF-I-receptor genes. J Clin Endocrinol Metab 2002; in press.
80. Laron Z, Lazar L, Klinger B. Growth hormone, insulin-like growth factor I and brain growth and function. In: Castells S, Wisniewski KE, eds. Growth hormone Treatment in Down's syndrome. Chichester: Wiley and Sons, 1993:151-161.
81. Frankel JJ, Laron Z. Psychological aspects of pituitary insufficiency in children and adolescents with special reference to growth hormone. Isr J Med Sci 1968; 4:953-961.
82. Galatzer A, Aran O, Nagelberg N et al. Cognitive and psychosocial functioning of young adults with Laron syndrome. In: Laron Z, Parks JS, eds. Lessons from Laron syndrome (LS) 1966-1992. Pediatric and Adolescent Endocrinology, vol. 24. Basel-New York: Karger, 1993:53-60.
83. Kranzler JH, Rosenbloom AL, Martinez V et al. Normal intelligence with severe insulin-like growth factor I deficiency due to growth hormone receptor deficiency: A controlled study in a genetically homogeneous population. J Clin Endocrinol Metab 1998; 83:1953-1958.
84. Laron Z, Galatzer A. A comment on normal intelligence in growth hormone receptor deficiency. J Clin Endocrinol Metab 1999; 83:4528.
85. Laron Z. Growth hormone (GH) and insulin-like growth factor-I (IGF-I) effects on the brain. Chapter 85. In: Pfaff D, Arnold A, Etgen A, Fahrbach S, Rubin RT, eds. Hormones, Brain and Behavior. Vol. 5. San Diego: Academic Press, 2002:75-96.
86. Dagan Y, Abadi J, Lifschitz A et al. Severe obstructive sleep apnoea syndrome in an adult patient with Laron syndrome. GH & IGF Res 2001; 11:247-249.

87. Laron Z. Effects of growth hormone and insulin-like growth factor-I. Deficiency on ageing and longevity. In: Goode J, ed. Endocrine Facets of Ageing. Novartis Foundation Symposium No. 242. Chichester: John Wiley & Sons, 2002:125-142.
88. Laron Z, Avitzur Y, Klinger B. Carbohydrate metabolism in primary growth hormone resistance (Laron syndrome) before and during insulin-like growth factor-I treatment. Metabolism 1995; 44 (Suppl. 4):113-118.
89. Laron Z, Avitzur Y, Klinger B. Insulin resistance in Laron syndrome (primary insulin-like growth factor-I [IGF-I] deficiency) and effect of IGF-I replacement therapy. J Pediatr Endocrinol Metab 1997; 10(Suppl.1):105-115.
90. Klinger B, Laron Z. Renal function in Laron syndrome patients treated by insulin-like growth factor-I. Pediatr Nephrology 1994; 8:684-688.
91. Klinger B, Jensen LT, Silbergeld A et al. Insulin-like growth factor-I raises serum procollagen levels in children and adults with Laron syndrome. Clin Endocrinol 1996; 45:423-429.
92. Keret R, Pertzelan A, Zeharia A et al. Growth hormone (hGH) secretion and turnover in three patients with Laron-type dwarfism. Isr J Med Sci 1988; 34:75-79.
93. Laron Z, Pertzelan A, Doron M et al. The effect of dihydrosomatostatin in dwarfism with high plasma immunoreactive growth hormone. Horm Metab Res 1977; 9:338-339.
94. Laron Z, Klinger B, Jensen LT et al. Biochemical and hormonal changes induced by one week of administration of rIGF-I to patients with Laron type dwarfism. Clin Endocrinol 1991; 35:145-150.
95. Kornreich L, Horev G, Schwarz M et al. Pituitary size in patients with Laron syndrome. Eur J Endocrinol 2003; in press.
96. Eshet R, Dux Z, Silbergeld A et al. Erythrocytes from patients with low serum concentrations of IGF-I have an increase in receptor sites for IGF-I. Acta Endocrinol (Copenh) 1991; 125:354-358.
97. Eshet R, Klinger B, Silbergeld A et al. Modulation of insulin like growth factor I (IGF-I) binding sites on erythrocytes by IGF-I treatment in patients with Laron syndrome (LS). Regulatory Peptides 1993; 48:233-239.
98. Laron Z, Klinger B, Blum WF et al. IGF binding protein 3 in patients with LTD: effect of exogenous rIGF-I. Clin Endocrinol 1992; 36:301-304.
99. Laron Z, Suikkari AM, Klinger B et al. Growth hormone and insulin-like growth factor regulate insulin-like growth factor binding protein in Laron type dwarfism, growth hormone deficiency and constitutional growth retardation. Acta Endocrinol 1992; 127:351-358.
100. Klinger B, Ionesco A, Anin S et al. Effect of insulin-like growth factor I on the thyroid axis in patients with Laron-type dwarfism and healthy subjects. Acta Endocrinol (Copen) 1992; 127:515-519.
101. Silbergeld A, Klinger B, Schwartz H et al. Serum prolactin in patients with Laron type dwarfism: effect of insulin-like growth factor I. Horm Res 1992; 37:160-164.
102. Laron Z, Karp M. Carbohydrate metabolism in the syndrome of familial dwarfism and high plasma immunoreactive growth hormone (Laron type dwarfism). In: Podolsky S, Wiswanathan M, eds. Secondary diabetes: The spectrum of the diabetic syndrome. New York: Raven Press, 1980:363-371.
103. Laron Z, Lilos P, Sivan B. The effect of IGF-I deficiency and replacement on the hematopoietic system in patients with primary GH insensitivity (Laron syndrome). (Abst. OR1-26) Horm Res 2002; 58(Suppl 2):9.
104. Abramovici A, Josefsberg Z, Mimouni M et al. Histopathological features of the skin in hypopituitarism in Laron type dwarfism. Isr J Med Sci 1983; 19:515-519.
105. Laron Z, Ben Amitai D, Lurie R. Impaired hair growth and structural defects in patients with Laron syndrome (primary IGF-I deficiency). (Abst. 1-449). Pediatr Res 2001; 49:76A.
106. Lurie R, Ben-Amitai D, Laron Z. Laron syndrome/growth hormone (GH) insensitivity/ a unique model to explore the effect of insulin like growth factor I (IGF-I) deficiency on human hair. J Am Acad Derma 2003; in press.
107. Barton DE, Foellmer BE, Woods WI et al. Chromosome mapping of the growth hormone receptor gene in man and mouse. Cytogenet Cell Genet 1989; 50:137-141.
108. Argetsinger LS, Campbell GS, Yang X et al. Identification of JAK2 as a growth hormone receptor-associated tyrosine kinase. Cell 1993; 74:237-244.
109. Waxman DJ, Frank SJ. Growth hormone action: signalling via JAK/STAT-coupled receptor. In: Conn PM, Means AR, eds. Principles of molecular regulation. Totowa: Humana Press, 2000:55-83.
110. de Vos AM, Ultsch, M, Kossiakioff AA. Human growth hormone and extracellular domain of its receptor: crystal structure of the complex. Science 1992; 255:306-312.
111. Shevah O, Borrelli P, Rubinstein M et al. Identification of 2 novel mutations in the human growth hormone receptor gene. J Investic Endocrinol 2003: 16:in press.
112. Amselem S, Sobrier ML, Duquesnoy P et al. Recurrent nonsense mutations in the growth hormone receptor from patients with Laron dwarfism. J Clin Invest 1991; 87:1098-1102.
113. Sobrier ML, Dastot F, Duquesnoy P et al. Nine novel growth hormone receptor gene mutations in patients with Laron syndrome. J Clin Endocrinol Metab 1997; 82:435-437.
114. Berg MA, Peoples R, Perez-Jurado L et al. Receptor mutations and haplotypes in growth hormone receptor deficiency:a global survey and identification of the Ecuadorean E180splice mutation in an oriental Jewish patient. Acta Paediatr 1994; Suppl. 399:112-114.
115. Amselem S, Duquesnoy P, Duriez B et al. Spectrum of growth hormone receptor mutations and associated haplotypes in Laron syndrome. Hum Molec Genet 1993; 2:355-359.
116. Counts DR, Cutler GB. Growth hormone insensitivity syndrome due to point deletion and frame shift in the growth hormone receptor. J Clin Endocrinol Metab 1995; 80:1978-1981.

117. Berg MA, Argente J, Chernausek S et al. Diverse growth hormone receptor gene mutations in Laron syndrome. Am J Hum Genet 1993; 52:998-1005.
118. Shevah O, Nunez O, Rubinstein M, Laron Z. Intronic mutation in the growth hormone (GH) receptor gene in a Peruvian girl with Laron syndrome. J Pediatr Endocrinol Metab 2002; 15:1039-1040.
119. Berg MA, Guevara-Aguirre J, Rosenbloom AL et al. Mutation creating a new splice site in the growth hormone receptor genes of 37 Ecuadorean patients with Laron syndrome. Hum Mutation 1992; 1:124-134.
120. Quinteiro C, Castro-Feijoo L, Loidi L et al. Novel mutation involving the translation initiation codon of the growth hormone receptor gene (GHR) in a patient with Laron syndrome. J Pediatr Endocrinol Metab 2002; 15:1041-1045.
121. Duquesnoy P, Sobrier ML, Duriez B et al. A single amino acid substitution in the exoplasmic domain of the human growth hormone (GH) receptor confers familial GH resistance (Laron syndrome) with positive GH-binding activity by abolishing receptor homodimerization. Embo J 1994; 13:1386-1395.
122. Oh PS, Kim IS, Moon YH et al. A molecular genetic study on the two Korean patients with Laron syndrome. 81st Annual Meeting of the Endocrine Society, San Diego, CA, 1999. Abstract P1-26:140.
123. Enberg B, Luthman H, Segnestam K et al. Characterization of novel missense mutations in the GH receptor gene causing severe growth retardation. Eur J Endocrinol 2000; 143:71-76.
124. Wojcik J, Berg MA, Esposito N et al. Four contiguous amino acid substitutions, identified in patients with Laron syndrome, differently affect the binding affinity and intracellular trafficking of the growth hormone receptor. JCEM 1998; 83:4481-4489.
125. Baumann G, Shaw MA, Winter RJ. Absence of plasma growth hormone-binding protein in Laron-type dwarfism. J Clin Endocrinol Metab 1987; 65:814-816
126. Duquesnoy P, Sobrier ML, Amselem S et al. Defective membrane expression of human growth hormone (GH) receptor causes Laron-type GH insensitivity syndrome. Proc Natl Acad Sci USA 1991; 88:10272-10276.
127. Silbergeld A, Lazar L, Erster B et al. Serum growth hormone binding protein activity in healthy neonates, children and young adults—correlation with age, height and weight. Clin Endocrinol 1989; 31:295-303.
128. Baumann G, Shaw MA, Amburn K. Circulating growth hormone binding proteins. J Endocrinol Invest 1994; 17:67-81.
129. Laron Z, Klinger B, Erster B et al. Serum GH binding protein activity identifies the heterozygous carriers for Laron type dwarfism. Acta Endocrinol 1989; 121:603-608.
130. Niwa M, Sato Y, Saito Y et al. Chemical synthesis, cloning and expression of genes for human somatomedin C (insulin like growth factor I) and 59Val somatomedin C. Ann NY Acad Sci 1986; 469:31-52.
131. Laron Z, Klinger B, Silbergeld A et al. Intravenous administration of recombinant IGF-I lowers serum GHRH and TSH. Acta Endocrinol 1990; 123:378-382.
132. Klinger B, Garty M, Silbergeld A et al. Elimination characteristics of intravenously administered rIGF-I in Laron type dwarfs (LTD). Develop Pharmacol Therap 1990; 15:196-199.
133. Takano H, Hizuka N, Asakawa K et al. Effects of s.c. administration of recombinant human insulin like growth factor-I (IGF-I) on normal human subjects. Endocrinol Japon 1990; 37:309-317.
134. Gil-Ad I, Koch Y, Silbergeld A et al. Differential effect of insulin-like growth factor I (IGF-I) and growth hormone (GH) on hypothalamic regulation of GH secretion in the rat. J Endocrinol Invest 1996; 19:542-547.
135. Laron Z, Anin S, Klinger B. Long-term IGF-I treatment of children with Laron syndrome. In Laron Z, Parks JS. eds. Lessons from Laron syndrome (LS) 1966-1992. Pediatric and Adolescent Endocrinology, vol. 24. Basel-New York: Karger, 1993:226-236.
136. Walker JL, Van Wyk JJ, Underwood LE. Stimulation of statural growth by recombinant insulin-like growth factor I in a child with growth hormone insensitivity syndrome (Laron type). J Pediatr 1992; 121:641-646.
137. Gafny M, Silbergeld A, Klinger B et al. Comparative effects of GH, IGF-I and insulin on serum sex hormone binding globulin. Clin Endocrinol 1994; 41:169-175.
138. Laron Z, Wang XL, Klinger B et al. Growth hormone increases and insulin-like growth factor-I decreases circulating lipoprotein(a). Eur J Endocrinol 1997; 136:377-381.
139. Kaneti H, Karasik A, Klinger B et al. Long-term treatment of Laron type dwarfs with insulin-like growth factor I increases serum insulin-like growth factor—binding protein 3 in the absence of growth hormone activity. Acta Endocrinol 1993; 128:144-149.
140. Kaneti H, Silbergeld A, Klinger B et al. Long-term effects of insulin-like growth factor (IGF)-I on serum IGF-I, IGF-binding protein-3 and acid labile subunit in Laron syndrome patients with normal growth hormone binding protein. Eur J Endocrinol 1997; 137:626-630.
141. Laron Z, Klinger B, Silbergeld A. Serum insulin-like growth factor-I (IGF-I) levels during long-term IGF-I treatment of children and adults with primary GH resistance (Laron syndrome). J Pediatr Endocrinol Metab 1999; 12:145-152.
142. Ranke MB, Savage MO, Chatelain PG et al. Insulin-like growth factor I improves height in growth hormone insensitivity: Two years' results. Horm Res 1995; 44:253-264.
143. Backeljauw PF, Underwood LE, the GHIS Collaborative Group. Prolonged treatment with recombinant insulin-like growth factor-I in children with growth hormone insensitivity syndrome—a clinical Research Center study. J Clin Endocrinol Metab 1996; 81:3312-3317.
144. Klinger B, Laron Z. Three year IGF-I treatment of children with Laron syndrome. J Pediatr Endocrinol Metab 1995; 8:149-158.

145. Guevara-Aguirre J, Rosenbloom AL, Vasconez O et al. Two year treatment of growth hormone (GH) receptor deficiency with recombinant insulin-like growth factor-I in 22 children:comparison of two dosage levels and to GH treated GH deficiency. J Clin Endocrinol Metab1997; 82:629-633.
146. Backeljauw PF, Underwood LE, and The GHIS Collaborative Group. Therapy for 6.5-7.5 years with recombinant insulin-like growth factor I in children with growth hormone insensitivity syndrome: A clinical research center study. J Clin Endocrinol Metab 2001; 86:1504-1510.
147. Klinger B, Anin S, Silbergeld A et al. Development of hyperandrogenism during treatment with insulin-like growth factor-I (IGF-I) in female patients with Laron syndrome. Clin Endocrinol 1998; 48:81-87
148. Laron Z, Klinger B. Comparison of the growth-promoting effects of insulin-like growth factor I and growth hormone in the early years of life. Acta Paediatr 2000; 88:38-41.
149. Laron Z, Anin S, Klipper-Aurbach Y et al. Effects of insulin-like growth factor on linear growth, head circumference and body fat in patients with Laron-type dwarfism. Lancet 1992; 339:1258-1261.
150. Laron Z, Klinger B. Are patients with Laron syndrome candidates for limb lengthening? In: Laron Z, Mastragostino S, Romano C et al, eds. Limb lengthening: For whom, when and how? London-Tel Aviv: Freund Publishing House, 1995:79-91.
151. Laron Z. Final height and psychosocial outcome of patients with Laron syndrome (primary GH resistance): a model of genetic IGF-I deficiency. In: Gilli G, Benso L, Schell L, eds.. Human growth from conception to maturity. London: Smith-Gordon Publ., 2002:215-225.
152. Laron Z. Somatomedin-1 (recombinant insulin-like growth factor-I). Clinical pharmacology and potential treatment of endocrine and metabolic disorders BioDrugs 1999; 11:55-70.
153. Laron Z. Clinical use of insulin-like growth factor-I—Yes or no? Paediatr Drugs 1999; 1:155-159.
154. Merimee TJ, Laron Z. The pgymy. In: Merimee TJ, Laron Z, eds. Growth hormone, IGF-I and growth: New views of old concepts. London-Tel Aviv: Freund Publishing House Ltd., 1996:217-240.
155. Merimee TJ, Baumann G, Daughaday W. Growth hormone binding protein II. Studies in pygmies and normal statured subjects. J Clin Endocrinol Metab 1990; 71:1183-1188.
156. Rimoin DL, Merimee TJ, Rabinowitz D et al. Peripheral subresponsiveness to human growth hormone in the African Pygmies. N Engl J Med 1969; 281:1383-1388.
157. Geffner ME, Bailey RC, Bersch N et al. Insulin-like growth factor-I unresponsiveness in an Efe Pygmy. Biochem Biophys Res Comm 1993; 193:1216-1223.
158. Sivan B, Lilos P, Laron Z. Effects of insulin-like growth factor-I (IGF-I) deficiency and replacement therapy on the haematopoietic system in patients with Laron syndrome, primary growth hormone (GH) insensetivity. JPEM 2003; 16: in press.

Index

A

α2β1 integrin 307, 311
α5β1 integrin 59, 287, 310
Acid labile subunit (ALS) 48, 51, 57, 59, 65, 121-123, 127, 129-131, 178, 212, 220, 224, 226, 230, 263, 264, 434, 437-439, 443, 457-465
Acromegaly 124, 131, 250, 318, 319, 329, 333, 338, 401, 437
Acute renal failure 254-256
Adipose tissue 76, 425, 458, 484
Akt/PKB 110, 152
Alzheimer's disease 153, 162, 178
Antisense 18, 111, 138, 160, 197, 208, 266, 284-286, 305, 348, 357-364, 370, 373, 374, 386, 390, 392, 400, 401, 404, 412, 415
Apoptosis 1, 22, 29-31, 35, 37, 45, 47, 87, 91, 92, 104, 107, 108, 110-112, 116, 127, 137, 147, 148, 159, 162-165, 167, 168, 177, 181-187, 190, 197, 199, 205, 216, 240, 248, 254, 255, 265-269, 271, 272, 278, 279, 281, 284, 286, 287, 294-300, 306, 308, 310, 311, 318, 340-343, 346-348, 350, 351, 353, 362-364, 367, 370, 371, 375-377, 381-386, 394, 395, 400-404, 410, 411, 415, 416, 418
Astrocytes 79, 140, 147, 152, 159-161, 175
Autocrine action 121, 124
Axonal regeneration 168

B

B cell 411, 412, 415-417
Bcl family 148, 162, 164
Bcl-2 162, 164, 165, 168, 237, 266, 295, 296, 376, 416
Big IGF-II 93, 434-439
Binding affinities 12, 13, 32-34, 55, 58, 59, 97, 263, 266
Binding proteins 29, 33, 48, 50, 55, 64, 73, 75, 83, 91, 95, 96, 121, 138, 140, 159, 188, 192, 194, 206, 208, 209, 219, 220, 231, 236, 244, 246, 247, 252, 253, 255, 256, 262-265, 270, 281, 284, 287, 297, 304, 306-308, 318, 333, 338, 340, 350-352, 359, 367, 376, 390, 391, 399, 410, 429, 434, 441-445, 453, 457-460, 465, 467, 469, 480, 482

Blood brain barrier (BBB) 140
Bone acquisition 206, 212, 214, 215
Bone formation 65, 78, 206, 209, 211-213, 215, 219-224, 226-228, 230, 231, 270, 427, 429
Brain development 83, 125, 137, 138, 140, 151, 152, 154
BRCA1 23, 24, 342, 343, 350, 370, 404
Breast 34, 37, 114, 127, 173, 188, 189, 192, 194, 196-199, 215, 266-268, 270-272, 281, 282, 284-287, 289, 297-299, 306, 307, 310, 312, 313, 317-325, 333, 339, 340, 346, 347, 350, 353, 362, 367-378, 385, 386, 395, 399, 402, 403, 417, 434, 473, 475
Breast cancer 34, 37, 114, 188, 192, 194, 196-199, 266-268, 270-272, 281, 282, 284-287, 289, 297-310, 317, 319-324, 333, 339, 340, 347, 350, 353, 362, 367-378, 386, 395, 402, 403, 417

C

C. elegans 104-107, 114, 283
Caspase 111, 148, 162, 164-167, 237, 267, 295, 296
Catabolism 253, 458
Cell proliferation 29, 92, 116, 149, 168, 178, 188, 190, 192, 199, 220, 223, 226, 227, 228, 237, 255, 266, 267, 268, 281, 282, 284, 294, 296, 308, 325, 342, 360, 361, 367, 369, 371, 372, 374, 375, 377, 378, 400, 401, 402, 404, 410, 411, 412, 415, 418
Ceramide 295, 296, 297, 300, 376, 377, 402
Chondroitin sulfate 122, 305
Chronic lymphocytic leukemia 415
Chronic myeloid leukemia (CML) 108, 360
Chronic renal failure 69, 226, 244, 247, 251, 252, 253, 254, 255, 256, 424
Collagen 110, 122, 206, 209, 211, 283, 305, 306, 307, 308, 311, 312, 427, 479
Colorectal cancer 329, 401
Cys-rich domain 6, 14, 18, 26

D

Dentate gyrus 137, 144, 147, 148, 149, 153, 154
Depression 441, 484
Diabetes 67, 72, 73, 80, 177, 179, 244, 248, 250, 251, 253, 425, 441, 443, 444, 445, 446, 447, 448, 449, 450, 452, 453, 468, 478
Disulfide link 5, 48
Drosophila 4, 5, 105, 107, 108, 152, 283
Dwarfism 106, 152, 249, 467, 469, 475, 479, 484

E

Elastin 305, 479
Endonucleolytic cleavage 91, 93, 98, 99
Epidermal growth factor (EGF) 6, 32, 104, 108, 124, 162, 172, 188, 189, 192-194, 196, 246, 266, 306, 361, 362, 372, 394, 404
Epidermal growth factor receptor (EGFR) 5, 6, 361
Epithelial cancer 399
Estrogen 50, 75, 77, 188, 189, 190, 192, 197, 198, 208, 210, 212, 266, 267, 286, 306, 317, 319, 320, 326, 339, 352, 353, 368, 369, 370, 372, 373, 374, 400, 402, 403, 417
Estrogen receptor (ER) 24, 192, 196-199, 208, 267, 286, 317, 368, 370, 372-374, 377, 403, 417
Ewing's family of tumors 390, 394, 395
Extracellular matrix 48, 58, 59, 169, 195, 209, 210, 220, 222, 228, 231, 265, 282, 300, 304-309, 311, 312, 402, 404

F

Fibroblast 2, 30-32, 34, 35, 37, 69, 80, 83, 104, 106, 108-110, 114, 115, 124, 160, 189, 197, 209, 210, 227, 235, 246, 265, 266, 269, 272, 273, 282, 284, 287, 304-311, 347-349, 359, 369, 386, 390, 394, 400, 402, 413
Fibronectin 1, 2, 7, 9, 11, 13, 26, 27, 33, 34, 59, 173, 300, 306-308, 310, 311
Fibronectin type III (FnIII) domain 5, 7, 8, 11, 26, 27
Focal adhesion kinase (FAK) 28, 169, 172, 173, 270, 299, 306, 310, 377, 403, 417
FSH 79, 92, 131

G

G protein 32, 37, 458
Gα_i 22, 31, 32, 37
GH binding protein (GHBP) 105, 252, 443, 444, 450, 458, 462, 467, 469, 480, 481, 483, 484
GH receptor defect 248, 468, 479
Ghrelin 459, 462
GHRH antagonist 343, 393, 394, 400, 403, 404
GHRP 458, 462-465
Glucocorticosteroid 423, 425, 428, 429
Glucose metabolism 431, 441, 442
Glucose transport 143, 147, 151-153, 441
Glycosylation 2, 24, 33, 48, 55, 56, 58, 69, 209, 220, 222, 264, 273, 340, 348, 375, 436, 437
Granzyme B 35
Grb10 22, 31
Growth hormone (GH) 1, 8, 32, 64, 65, 69, 71-76, 78-80, 83, 91, 106, 121-124, 127, 129, 130, 131, 133, 189, 190, 206, 208, 212, 215, 222-227, 237, 244-256, 263, 305, 312, 318, 319, 320, 324, 338, 339, 342, 343, 369, 392-394, 400-402, 411, 423-431, 434, 437-439, 441-446, 448, 450-452, 457-465, 467-470, 478-482, 484
GSK3 β 152

H

Heparan sulfate 308, 309
Heparin binding 56-59, 220, 264, 265, 270, 309
Heparin binding domain 56, 57, 220, 264
HER2/neu 361, 362, 403
HNF-1α 71-73, 82, 83
HNF-3β 50, 71, 72, 83, 95, 97
Human umbilical vein endothelial cells (HUVEC) 273
Huntington's disease 178
Hypoglycemia 93, 425, 434, 435, 437-439, 461, 470, 477-479, 482, 485

I

IL-3 108, 109, 111, 112, 114, 116
IL-6 208, 415, 418, 419, 459
Imprinting 92, 340, 385, 404

Insulin 1, 2, 6, 8, 11, 13-18, 22-32, 34, 37, 48, 50, 64, 65, 69, 72-75, 80, 82, 84-92, 95, 104-106, 108, 121, 124, 127, 130, 137, 140, 145, 147, 152, 153, 158, 164, 171, 177, 188, 192, 198, 206, 235, 236, 244, 245, 253, 256, 262, 263, 266, 269, 281, 304, 305, 312, 317-320, 325, 326, 329, 331, 338, 340, 343, 346, 347, 357, 359, 363, 370-372, 375, 390, 391, 399, 402, 403, 410, 412, 423-426, 429, 431, 434, 438, 441-448, 450, 452, 453, 457-462, 464, 465, 467, 477-480, 482, 483
Insulin-like growth factor 1 (IGF-1) 6, 13, 14, 16, 17, 22-27, 29-32, 34, 36, 37, 79, 111, 114, 137-140, 142-154, 244-256, 321-323, 326-328, 330, 332, 363
Insulin-like growth factor binding protein (IGFBP) 16, 33, 48, 49-60, 80, 91, 121-125, 127, 129-131, 138, 140, 158-161, 168, 173, 178, 188, 194-196, 198, 199, 206-213, 215, 216, 219-231, 236, 244-248, 250-255, 262-274, 281, 284-289, 294, 297-300, 304, 306-311, 313, 318-333, 338, 340, 342, 359, 367-369, 372-378, 385, 387, 390, 399, 400-404, 410-412, 415, 434, 437-439, 441-443, 445, 446, 450-453, 457-465, 478-480, 482
  IGFBP-1 48-53, 55-59, 80, 125, 127, 159, 173, 194, 195, 198, 209, 210, 246-248, 250-252, 262-265, 269, 270, 284, 287, 299, 310, 368, 369, 374, 375, 377, 399, 402-404, 437, 441-443, 445, 446, 450, 451, 457, 459-461, 465, 478, 479, 482
  IGFBP-2 48-56, 58, 59, 139, 140, 158-160, 168, 194-196, 198, 209, 216, 246, 247, 252, 254, 269-271, 284, 308, 309, 340, 368, 369, 374, 377, 399-403, 411, 412, 437-439, 457, 459, 460, 465, 478
  IGFBP-3 48-51, 53-59, 123, 127, 129-131, 159, 178, 194, 195, 198, 199, 208-212, 215, 223, 224, 226, 227, 230, 247, 248, 252, 263-272, 274, 281, 284-287, 289, 297-300, 307-309, 318-332, 338, 340, 342, 350, 368, 369, 372, 374-378, 385, 387, 399, 400-404, 411, 434, 437-439, 442, 443, 446, 451, 453, 457-465, 478-480, 482
  IGFBP-4 48-56, 59, 125, 159, 161, 194, 196, 198, 209, 210, 212, 213, 220, 226, 264, 271, 284, 287, 298, 307, 308, 359, 369, 374, 377, 378, 399-401, 411, 412, 415, 457, 460, 465

  IGFBP-5 48-51, 53-59, 139, 140, 148, 159, 160, 194, 195, 198, 207, 209-211, 219-231, 246, 248, 250, 263-265, 270, 271, 284, 287, 297-300, 307-311, 374, 376, 377, 399, 400, 402-404, 457, 460, 462, 465
  IGFBP-6 48, 49, 51-53, 55, 56, 58, 59, 158, 161, 194, 209, 264, 271, 284, 287, 297, 308, 368, 374, 377, 402, 437, 439, 457, 460, 465
Insulin-like growth factor (IGF) receptor 1, 2, 91, 121, 124, 152, 158, 159, 196, 199, 206, 211, 212, 216, 228, 229, 263, 269, 281, 284, 287, 297, 304, 367, 370, 399-401, 403, 404, 441, 448
  IGF-1R 1, 8, 11, 13, 14, 22-32, 34, 36, 37, 93, 105, 121, 122, 124, 125, 127, 130, 131, 137, 138, 140, 142, 151, 152, 244, 246, 247, 250, 251, 255, 363, 412
  IGF-2R 22, 24, 32-37, 198
Insulin resistance 65, 130, 253, 340, 423, 425, 429, 441, 443-448, 461, 478, 479
Integrin αVβ3 304
Integrin binding 59, 269
Interferon-γ (IFN-γ) 353
IRS-1 22, 28-32, 104, 108-112, 114-116, 130, 164, 171, 197, 198, 212, 306, 311, 349, 350, 370-373, 403, 412, 417

J

JAK 22, 28, 32, 64, 78, 79, 252
JNK 29, 31, 162, 164-168, 178, 296, 373

K

Kidney 34, 37, 80, 92, 95, 125, 130, 244-248, 250, 251, 254-256, 268, 305, 346, 351, 352, 400, 448, 450, 451

L

L domains 1, 5, 6, 26, 48, 49, 51-59
Laminin 4, 6, 8, 110, 251, 306, 311
Laron syndrome 467-481, 483, 485, 487
Leiomyosarcoma 395, 434, 435
Leukemia 35, 108, 331, 346, 360, 412, 415
Leukemia inhibitory factor (LIF) 35
Lipid metabolism 130, 424, 425, 448
Lung cancer 297, 319, 329, 332, 385
Lymphokine-activated killer cells (LAK) 361
Lysosomal enzymes 22, 32-36, 236

M

Macrophage 80, 83, 294, 353, 359, 411
Mammary gland 125, 127, 188-197, 294, 297, 320, 339, 363, 402
Mannose 6-phosphate receptor 282
MAP kinase (MAPK) 22, 27-29, 32, 82, 111, 113, 166, 170-173, 237, 267, 270, 296, 298, 306, 310, 371, 373, 378, 402, 403, 412, 414, 416, 480
MEF2 family 235
Metabolism 130, 131, 133, 143, 221, 223, 226, 235, 247, 281, 304, 359, 423-425, 429-431, 441, 442, 447, 448, 452, 458, 460
Metastasis 318, 362, 370, 371, 386, 390, 392, 393, 401, 403
MHC-I 361, 364
MHC-II 361
Mitochondria 31, 111, 164, 165, 237, 294-298
Molecular mechanism 83, 211, 219, 221, 223, 235-237, 240, 241, 266, 310-312, 346, 350, 351, 385
mRNA 2, 24, 34, 36, 48, 49, 51, 64-69, 71-83, 91-101, 124, 130, 131, 138-140, 142, 145, 150, 151, 158-161, 170, 172-175, 189-198, 208-212, 221-223, 227, 236, 237, 240, 245, 246, 250-252, 254, 267, 281, 282, 284, 287, 305, 331, 346-353, 357, 358, 360, 369, 372, 374, 385, 390, 399-404, 411, 412, 426
Muscle 24, 32, 59, 64, 65, 73, 78, 83, 92, 124, 125, 127, 130, 131, 152, 159, 169, 176, 177, 220, 224, 228, 235-237, 240, 241, 244, 247, 250, 253, 282, 284, 286, 287, 307-312, 391-393, 395, 425-427, 429, 430, 438, 441-443, 445, 446, 458, 475, 476, 484
Myelination 137, 145, 150-152, 158, 162, 173-176, 178
Myeloma 287, 410, 412, 414-416, 418, 419
Myf-5 235
Myoblasts 35, 92, 109, 235, 237-241, 272, 308, 392
MyoD 114, 235, 236, 240
Myofiber 236, 240, 241

N

N-methyl-D-aspartate (NMDA) 162
Natural killer (NK) cell 361, 412
Nephropathy 450, 451
Nerve growth factor (NGF) 162, 359
Neurite 160, 162, 169-173, 177, 179
Neurodegenerative disease 158, 161, 162, 166, 168, 174, 178, 300
Neurogenesis 137, 147-149, 153
NICTH 434-439
Nitric oxide (NO) 162
Null mutant mice 121, 124, 125, 127, 129-131, 133, 190

O

Obesity 106, 471, 473, 475, 477-479
Olfactory bulb 140, 142, 144, 148, 153, 270, 308, 309
Oligodendrocyte 79, 137, 140, 148-152, 162, 173-175, 178
Osteoprotogerin (OPG) 211, 215
Osteosarcoma 49, 51, 210, 211, 220-223, 227, 228, 230, 284, 339, 350, 390-395
Ovarian cancer 331, 350, 403, 404

P

14-3-3 proteins 22, 29, 182, 297, 348
p53 23, 24, 165, 196, 266, 284, 297, 298, 346, 349-351, 354, 360, 370, 376, 390, 391, 401, 403, 413
Paracrine action 121, 124, 125, 213, 229
Phospholipase C (PLC) 114
Phosphorylation 3, 9, 22, 24, 27-29, 31, 33, 37, 48, 55-57, 59, 72, 75, 76, 78, 110, 114, 116, 130, 145-147, 152-154, 164, 167, 168, 171-173, 178, 194, 197, 198, 209, 220, 222, 229, 252, 253, 264, 265, 268-270, 272, 281-283, 287, 289, 299, 300, 304, 306, 307, 311, 312, 340, 348-350, 354, 361, 370, 372, 373, 375-377, 390, 403, 412, 416, 417, 425, 480
PI3 kinase 72, 78, 82, 311
Post-transcriptional processing 91, 98
Primary GH resistance 467, 468
Primary IGF deficiency 224, 467, 479
Pro-IGF-II 93, 434, 436, 437
Pro-IGF-II-(E1-21) 434, 436, 437

Progesterone 50, 189, 192, 198, 223, 306, 369, 370, 374, 402
Projection neuron 137, 138, 142, 144, 145, 147-150, 152
Proliferin 34, 35
Promoter 23, 24, 49-51, 64-69, 71-73, 75-83, 93-99, 114, 115, 125, 127-130, 150, 176-178, 190, 196, 208, 210, 212, 227, 236, 265, 266, 272, 283, 346, 349-354, 358, 359, 362, 363, 370, 390, 391, 401, 402, 404, 412, 413
Prostate 110, 127, 215, 264, 266, 267, 269, 271-273, 284, 286, 297, 318-320, 325-329, 331, 333, 339-341, 347, 385-387, 399, 400, 434
Protein kinase C 75, 80, 81, 83, 284, 296, 361
Protein metabolism 424
Pygmies 484

R

R cells 106-108, 362
Receptor chimeras 5, 11-14
Reproductive system 79
Retinoic acid 35, 51, 95, 222, 223, 266, 269, 284, 305, 320, 374, 400, 402
Retinoid X receptor 268, 272, 377
Retinopathy 448-452
Rhabdomyosarcoma 37, 100, 346, 350, 352, 385, 390-392, 395

S

Schwann cell 159, 161, 162, 165, 166, 173, 175, 176
SH2-B 22, 31
Shc 22, 28, 105, 109, 116, 170-172
Signaling 22-24, 27-32, 36, 37, 71, 73, 75, 78, 100, 104-106, 108, 110-116, 130, 137, 145, 152-154, 162, 164-166, 168-173, 175, 178, 188, 190, 192, 194, 197, 198, 212, 219, 228-230, 236, 237, 240, 241, 252, 266-268, 270-272, 281-284, 286, 287, 289, 304, 306, 310-312, 338-341, 343, 347, 349, 350, 370-373, 376, 378, 385, 387, 390-392, 394, 395, 412, 414, 416-418, 423, 480
Skeletal muscle 65, 78, 92, 127, 177, 235-237, 240, 244, 247, 253, 308, 391, 392, 425, 426, 427, 430, 441, 458
Smad 376

Somatostatin 250, 400, 402, 444, 446, 450, 458, 459, 462, 478, 482, 483
Sphingomyelin 295
Steroid hormone 223, 328, 352, 372, 373
Stress 96, 162, 164, 263, 295, 298, 458-460, 462
Suramin 392-394, 401
SV40 T antigen 112
Synaptogenesis 137, 138, 145, 152
Synovial sarcoma 395

T

Tenascin 305, 308, 309
Thyroglobulin 35, 54, 127
Thyroid 35, 92, 95, 127, 237, 269, 331, 346, 457, 464, 465, 478
Transcription 23, 24, 32, 49-51, 64-69, 71-83, 91, 93-99, 112-114, 116, 127, 162, 164-166, 172, 173, 192, 208, 223, 230, 235-237, 240, 252, 268, 283, 287, 300, 346, 349-354, 357, 358, 362, 363, 391, 400, 412, 480
Transcription factor 23, 32, 50, 64, 71-73, 76, 81, 95-99, 113, 114, 164-166, 172, 173, 192, 208, 230, 235-237, 240, 283, 346, 349-354, 391, 480
Transforming growth factor-β (TGF-β) 34, 37, 189, 198, 208, 210, 221, 223, 266-268, 270-273, 281-289, 305, 372, 374, 376, 394, 401, 402
Transgenic mice 71, 79, 127, 137, 150, 170, 173, 175, 190, 213, 236, 248, 250, 251, 339, 359, 387, 399, 402, 411, 413, 450
Translation 48, 65, 67-69, 71, 75, 80, 83, 91, 97, 99, 100, 145, 147, 250, 357, 358
Triple-helix forming oligonucleotide (TFO) 357, 362, 363
Tumor 298, 300, 360, 377, 399-404
Tumor hypoglycemia 93, 434
Tumor necrosis factor (TNF) 4, 6, 8, 75, 78, 80, 211, 266, 273, 284, 295, 353, 362, 364, 374, 401, 459
Tumor suppressor 22-24, 37, 110, 196, 198, 266, 284, 340, 346, 349-354, 370, 413, 416
Type 1 IGF receptor 1, 121, 124, 281, 297, 367, 441, 448
Type I TGF-β receptor 268
Type II TGF-β receptor 268, 281, 287, 376
Tyrosine kinase domain 1, 3, 8, 11, 22, 26, 27, 31, 348, 412

V

Vascular cell adhesion molecule-1 (VCAM) 418
Vitronectin 306, 307, 310-312

W

Wilms' tumor 96, 273, 348, 351, 352, 399, 400, 401
WT1 23, 24, 96, 97, 346, 351-354, 358, 400, 401